MICROSOMES
AND
DRUG OXIDATIONS

DR OTTO ROSENTHAL

MICROSOMES AND DRUG OXIDATIONS

Proceedings of the Third International Symposium
Berlin, July 1976

Editors

VOLKER ULLRICH
Institute of Physiological Chemistry
University of the Saarland

IVAR ROOTS
Institute for Clinical Pharmacology
Free University of Berlin

ALFRED HILDEBRANDT
Institute for Clinical Pharmacology
Free University of Berlin

RONALD W. ESTABROOK
Southwestern Medical School
University of Texas

and

ALLAN H. CONNEY
Department of Biochemistry and Drug Metabolism
Roche Research Laboratories

Published as a Supplement to
BIOCHEMICAL PHARMACOLOGY

PERGAMON PRESS

OXFORD · NEW YORK · TORONTO · SYDNEY · PARIS · FRANKFURT

1977

U.K.	Pergamon Press Ltd., Headington Hill Hall, Oxford OX3 0BW, England
U.S.A.	Pergamon Press Inc., Maxwell House, Fairview Park, Elmsford, New York 10523, U.S.A.
CANADA	Pergamon of Canada Ltd., 75 The East Mall, Toronto, Ontario, Canada
AUSTRALIA	Pergamon Press (Aust.) Pty. Ltd., 19a Boundary Street, Rushcutters Bay, N.S.W. 2011, Australia
FRANCE	Pergamon Press SARL, 24 Rue des Ecoles, 75240 Paris, Cedex 05, France
WEST GERMANY	Pergamon Press GmbH, 6242 Kronberg-Taunus, Pferdstrasse 1, Frankfurt-am-Main, West Germany

Copyright © 1977 Pergamon Press Ltd.

All Rights Reserved. No part of this publication may be reproduced, stored in a retrieval system or transmitted in any form or by any means: electronic, electrostatic, magnetic tape, mechanical, photocopying, recording or otherwise, without permission in writing from the publishers

First edition 1977

Library of Congress Cataloging in Publication Data

Symposium on Microsomes and Drug Oxidations, 3d, Berlin, 1976.
Microsomes and drug oxidations.

Includes bibliographical references.
1. Drug metabolism--Congresses. 2. Microsomes--Congresses. 3. Oxidation, Physiological--Congresses. 4. Cytochrome P-450--Congresses. I. Ullrich, Volker. II. Title. [DNLM: 1. Cytochrome P-450--Metabolism--Congresses. 2. Microsomes--Metabolism--Congresses. 3. Oxidation-reduction--Congresses. 4. Drugs--Metabolism--Congresses. QH603.M4 M626 1976]
RM300.S86 1976   615.7   77-3358
ISBN 0-08-021523-8

In order to make this volume available as economically and rapidly as possible the authors' typescripts have been reproduced in their original form. This method unfortunately has its typographical limitations but it is hoped that they in no way distract the reader.

Printed and bound in Great Britain by Butler & Tanner Ltd, London and Frome

CONTENTS

LIST OF PARTICIPANTS	xv
PREFACE	xxix
Otto Rosenthal, the Man and his Work *HERBERT REMMER*	xxxi

I BIOCHEMISTRY

Lipid Structure of Liver Microsomal Membranes, Limitations of a Model *A. STIER, W. KÜHNLE and R. RÖSEN*	1
Interactions between Cytochrome P-450 and NADPH-Cytochrome P-450 Reductase in the Microsomal Membrane *CHUNG S. YANG*	9
A Study of Cytochrome P-450 Reduction by Sodium Dithionite *R. E. EBEL, D. H. O'KEEFFE and J. A. PETERSON*	17
Investigation of Microsomal Hydroxylating System at Sub-Zero Temperatures *P. DEBEY, M. LEON, C. BALNY and P. DOUZOU*	23
Isolation of Microsomal Electron Transport Components and Reconstitution of the Electron Transfer Activity on Liposomal Vesicles *W. DUPPEL, J. POENSGEN, V. ULLRICH and G. DAHL*	31
Differences in Substrate Affinity and Catalytic Specificity of Inducible and Non-Inducible Forms of Cytochrome P-450 in Rabbit Liver *T. FISCHBACH and W. LENK*	39
Effects of Buffer Composition on the Low-Spin ESR Spectrum of Cytochrome P-450 *D. H. O'KEEFFE, R. E. EBEL and J. A. PETERSON*	47
Kinetics of Tiamutin^R Interaction with Cytochrome P-450 from Rabbit Liver *I. SCHUSTER, C. FLESCHURZ and ILSE HELM*	51
Different Phases of Hydroxylase Induction in Liver Microsomes of Female Mice during Inhalation of Cyclohexane and D,L-Camphor *G. MOHN*	59

Protein-Lipid Interactions in the Liver Microsomal
   Hydroxylase System
   M. INGELMAN-SUNDBERG ... 67

Further Studies of the Fluroxene Mediated Destruction of
   Hepatic Microsomal Cytochromes P-450 *in Vitro*
   K. M. IVANETICH, J. A. MARSH, J. J. BRADSHAW and
   L. S. KAMINSKY ... 76

Purification of Membrane-Bound Oxygenases: Isolation of
   two Electrophoretically Homogeneous Forms of Liver
   Microsomal Cytochrome P-450
   M. J. COON, D. P. BALLOU, D. A. HAUGEN, S. O. KREZOSKI,
   G. D. NORDBLOM and R. E. WHITE ... 82

Microsomal Hemeproteins $P-450_{LM-2}$ and $P-450_{LM-4}$: Comparative
   Structural Studies
   K. DUS, D. CAREY, R. GOEWERT and R. A. SWANSON ... 95

Hepatic Microsomal Ethanol Oxidizing System:
   Isolation and Reconstitution
   R. TESCHKE, K. OHNISHI, Y. HASUMURA and C. S. LIEBER ... 103

Characterization of NADPH-Cytochrome c (P-450) Reductase
   Purified by Biospecific Affinity Chromatography
   Utilizing Spectrophotometric and Electron Paramagnetic
   Resonance Techniques
   Y. YASUKOCHI, J. A. PETERSON and B. S. S. MASTERS ... 111

Relationship between the Membrane Lipids and Substrate-
   Cytochrome P-450 Binding Reaction in Bovine Adreno-
   Cortical Microsomes
   S. NARASIMHULU ... 119

Kinetic and Spectral Evidence for Multiple Species of
   Cytochrome P-450 in Liver Microsomes
   G. POWIS, R. E. TALCOTT and J. B. SCHENKMAN ... 127

Differences in the Interactions of R and S Warfarin
   with Hepatic Microsomal Cytochrome P-450
   M. J. FASCO, L. J. PIPER and L. S. KAMINSKY ... 136

Multiple Forms of Housefly Cytochrome P-450
   J. CAPDEVILA and M. AGOSIN ... 144

Catalytic Hydroxylation Activities and Some Other
   Properties of Partially Purified Rabbit Pulmonary
   Microsomal Cytochrome P-450
   E. ARINÇ and R. M. PHILPOT ... 152

Hepatic Microsomal and Solubilized Mixed-Function Oxidase
   Systems from the Little Skate, *Raja erinacea*, A Marine
   Elasmobranch
   J. R. BEND, R. J. POHL, E. ARINÇ and R. M. PHILPOT ... 160

Aminimide Polymers as Supports for Affinity Chromatography, Under Low and High Pressure, of Microsomal Enzymes
J. C. M. TSIBRIS, J. E. EPPERT, C. A. KETCHUM, W. C. SHERRILL, R. D. SAFIAN and C. M. WEST — 170

Sex Differences in the Pattern of Cytochromes P-450 in Rat Liver Microsomes
R. KAHL, M. BUECKER and K. J. NETTER — 177

Characterization of Multiple Forms of Highly Purified Cytochrome P-450 from the Liver Microsomes of Rats, Mice and Rabbits
W. LEVIN, D. RYAN, M.-T. HUANG, J. KAWALEK, P. E. THOMAS, S. B. WEST and A. Y. H. LU — 185

The Mechanism of Cytochrome P-450 Action
V. ULLRICH — 192

Oxygen Reactions of the P-450 Heme Protein
S. G. SLIGAR, B. S. SHASTRY and I. C. GUNSALUS — 202

The Peroxidase Nature of Cytochrome P-450
A. D. RAHIMTULA and P. J. O'BRIEN — 210

Cytochrome P-450 Catalyzed Oxene Transfer from Various Donors
F. LICHTENBERGER and V. ULLRICH — 218

Association of Type I, Type II, and Reverse Type I Difference Spectra with Absolute Spin State of Cytochrome P-450 Iron
D. W. NEBERT, K. KUMAKI, M. SATO and H. KON — 224

NMR Relaxation Time Studies of Substrate Interactions with P-450 and other Hemoproteins
R. F. NOVAK, I. M. KAPETANOVIC and J. J. MIEYAL — 232

The Interaction of Polyhalogenated Methanes with Ferrous Cytochrome P-450
C. R. WOLF, D. MANSUY, W. NASTAINCZYK and V. ULLRICH — 240

Substrate-Elicited Dissociation of the Isosafrole Metabolite-Cytochrome P-450 Complex and the Consequential Reactivation of Monooxygenation
C. R. ELCOMBE, M. DICKINS, B. C. SWEATMAN and J. W. BRIDGES — 247

The Interaction of Halothane with Liver Microsomal Cytochrome P-450
W. NASTAINCZYK and V. ULLRICH — 254

The Formation of Hydrogen Peroxide During Hepatic Microsomal Electron Transport Reactions
J. WERRINGLOER — 261

Contents

Oxygen Transfer in Microsomal Oxidative Desulfuration H. KEXEL, E. SCHMELZ and H.-L. SCHMIDT	269
Redox and Ligand Dynamics in P-450$_{cam}$ – Putidaredoxin Complexes T. C. PEDERSON, R. H. AUSTIN and I. C. GUNSALUS	275
The Inhibition of Mixed-Function Oxidation Reactions by Amphetamines in Liver and Lung Microsomes M. R. FRANKLIN	284
The Physiological Implications of Drug Oxidations to the Cell S. ORRENIUS, P. MOLDÉUS, H. THOR and J. HÖGBERG	292
Nicotinamide Nucleotide Systems and Drug Oxidation in the Liver Cell H. SIES and K. WEIGL	307
The Role of Reducing Equivalents Generated in Mitochondria in Hepatic Mixed-Function Oxidation R. G. THURMAN, M. LURQUIN, R. EVANS and F. C. KAUFFMAN	315
Destruction of Cytochrome P-450 by Linoleic Acid Hydroperoxide E. H. JEFFERY, D. NERLAND, R. el-AZHARY and G. J. MANNERING	323
Stoichiometry of Oxygen Uptake, NADPH Oxidation and Ethylmorphine N-Demethylation by Hepatic Microsomes J. L. HOLTZMAN, R. P. MASON and R. R. ERICKSON	331
Structural and Functional Interaction of Rough Endoplasmic Reticulum (RER) and Mitochondria during Synthesis of Hepatic Cytochrome P-450 P. J. MEIER, M. A. SPYCHER and U. A. MEYER	339
Mechanism of N-Oxide Formation by Hepatic Mixed Function Oxidases P. HLAVICA and M. KEHL	346
A Preliminary Comparative Study on the Influence of Cysteamine and Metyrapone on Mixed-Function Oxygenase Activities in Variously Pretreated Liver Microsomes from Rats and Mice R. MULL, W. HINKELBEIN, J. GERTZ and K. FLEMMING	354
Inhibition of Rat Liver Microsomal DT Diaphorase by Benzo (a) Pyrene Metabolite(S) C. LIND and L. ERNSTER	362
Characteristics of the Hepatic Monooxygenase Catalyzing the Metabolism of Prostaglandin A$_1$ (PGA$_1$) in Guinea Pigs and Rats. I. Effects of Inducers D. KUPFER and J. NAVARRO	370

Studies on a Cytochrome P-450-dependent Hydroxylase System 377
   Active on Steroids in *Bacillus megaterium*
   A. BERG, K. CARLSTRÖM M. INGELMAN-SUNGBERG, J. RAFTER
   and J.-A. GUSTAFSSON

II PHARMACOLOGY

Impact of Drug Monoxygenases in Clinical Pharmacology 385
   A. BRECKENRIDGE, M. R. BENDING and G. BRUNNER

The Role of Liver Nuclei in the Formation of DNA Binding 395
   Products from Benzo (a) Pyrene
   B. JERNSTRÖM, H. VADI and S. ORRENIUS

Mechanism of Microsomal Activation of Benzo (a) Pyrene to 403
   Diol-Epoxides: The Separation and Characterization of
   Intermediates and Products
   S. K. YANG, P. P. ROLLER and H. V. GELBOIN

The Metabolism of Benzo (a) Pyrene in Cell Cultures 411
   and Homogenates from Different Human Fetal Tissues
   O. PELKONEN and P. KORHONEN

Aryl Hydrocarbon Hydroxylase in Cultured Human Lymphocytes 418
   B. PAIGEN, H. L. GURTOO, J. MINOWADA and K. PAIGEN

Assignment of a Human Gene for Aryl Hydrocarbon Hydroxylase 426
   Expression to Chromosome 2
   F. J. WIEBEL, S. BROWN, J. D. MINNA and H. V. GELBOIN

Oxidative Metabolism of Carcinogens by Trout Liver Resulting 435
   in Protein Binding and Mutagenicity
   J. AHOKAS, R. PÄÄKKÖNEN, K. RÖNNHOLM, V. RAUNIO,
   N. KÄRKI and O. PELKONEN

Drug-Drug Interactions via Inhibition of Microsomal Enzymes 442
   Involved in Metabolism of Epoxides Produced by Microsomal
   Monooxygenase
   F. OESCH, H. R. GLATT and P. BENTLEY

Levels of Glutathione and Glutathione-Metabolizing Enzymes 447
   in Rat Lung
   M. S. MORON, J. W. DePIERRE, K. JACOBSSON and B. MANNERVIK

The Disposition of Benzo (a) Pyrene in Isolated Perfused 453
   Rat Lung
   K. VÄHÄKANGAS, K. NEVASAARI, O. PELKONEN and N. T. KÄRKI

Epoxide Hydratase in Rat Lung 459
   J. SEIDEGÅRD, J. W. DePIERRE, M. S. MORON,
   K. A. M. JOHANNESEN and L. ERNSTER

Monooxygenase Catalyzed Activation of Thiono-Sulfur Containing Compounds to Reactive Intermediates
R. A. NEAL, T. KAMATAKI, A. L. HUNTER and G. CATIGNANI — 467

Stimulatory Effects of Polychlorinated Biphenyls (PCB) on Cytochromes P-450 and P-448 Mediated Microsomal Oxidations
A. P. ALVARES — 476

Depression of Hepatic Cytochrome P-450-dependent Monooxygenase Systems with Administered Interferon Inducing Agents
K. W. RENTON and G. J. MANNERING — 484

Alterations of Hepatic Microsomal Mixed Function Oxygenase Dependent H_2O_2 Formation due to "Uncoupling" and Induction as Demonstrated *in vivo* by Elimination Kinetics of Ethanol in Guinea Pig
A. G. HILDEBRANDT, L. LEHNE, I. ROOTS and M. TJOE — 492

The Microsomal Metabolism of Carcinogenic and/or Therapeutic Hydrazines
R. A. PROUGH, M. L. COOMES and D. L. DUNN — 500

Role of Hydroxylated Metabolites of Phenytoin in Dose-Dependency
A. J. GLAZKO, T. CHANG, E. MASCHEWSKE, A. HAYES and W. A. DILL — 508

Species Differences in Drug Metabolism: *in vivo* Parameters of Hepatic Drug-Metabolising Enzyme Activity in the Baboon
W. H. DOWN — 516

Evidence for Binding of Lidocaine to Two Catalytically Different Sites of Liver Microsomal Cytochrome P-450
C. V. BAHR, I. HEDLUND, B. KARLEN and H. GRASDALEN — 520

Differential Influence of Physiological, Pharmacological and Pathological Alterations on Hepatic and Entrahepatic Drug Metabolism
T. E. GRAM, B. I. SIKIC, C. L. LITTERST and E. G. MIMNAUGH — 527

Metabolism of Hexobarbital Enantiomers and Interaction with Cytochrome P-450 in Male and Female Mice and Rats
J. NOORDHOEK, A. P. VAN DEN BERG, E. M. SAVENIJE-CHAPEL and E. KOOPMAN-KOOL — 534

The Mechanism of Degradation of Endogenous Heme and Cytochrome P-450 by Heme Oxygenase
M. D. MAINES — 543

Induction of Cytochrome P-448 by 3-Methylcholanthrene in the Rat during Inhibition of Protein Synthesis *in vivo*
G. F. KAHL, B. ZIMMER, T. GALINSKY, H. G. JONEN and R. KAHL — 551

Stimulation of Liver Growth and Mixed-Function Oxidase by   559
α-Hexachlorocyclohexane: Separation of Inductive
Pathways
R. SCHULTE-HERMANN

Induction of Cytochrome P-450 by Long-Term Infusion of   568
Phenobarbital
O. ROSENTHAL, H. M. VARS, H. SCHLEYER, D. Y. COOPER,
S. S. LEVIN and J. TOUCHSTONE

Metabolism and Mutagenicity of 2-Acylaminofluorenes and   575
Related Compounds
S. S. THORGEIRSSON, P. J. WIRTH and W. L. NELSON

In Vivo Parameters of Drug Metabolism - Differences in   581
Specificity Towards Inducing Agents
I. ROOTS, B. LEY and A. G. HILDEBRANDT

Studies on the Mechanism by which Disulfiram and   589
Diethyldithiocarbamate Affect Drug Metabolism
M. MARSELOS, P. ALAKUIJALA, M. LANG and
R. TÖRRÖNEN

Alterations in the Activities of Drug Metabolizing Enzymes   597
in Rat Liver Microsomes by Dietary and in vitro
Incorporated Cholesterol
M. LANG and O. HÄNNINEN

Steroid - 16α-Hydroxylase in Rat Liver: Biochemical and   605
Biological Properties
P. KREMERS, A. AZHIR-AMIRSOLEYMANIE, J. DeGRAEVE
and J. E. GIELEN

Correlation between Drug, Lipid and Carbohydrate Metabolism   610
in the Rat after Chronic Beta-Adrenoreceptor Antagonist
Treatment
E. FELLENIUS, K. O. BORG, B. EKLUND, K.-J. HOFFMANN,
B.-M. MAGNUSSON, I. SKÅNBERG, M. WALLBORG and B. WALLIN

Conversion of Trichloroethylene to Carbon Monoxide by   615
Microsomal Cytochrome P-450
P. S. TRAYLOR, W. NASTAINCZYK and V. ULLRICH

Stereoselective in vitro Aromatic Oxygenation of Chiral   622
1,4-Benzodiazepin-2-Ones
S. RENDIĆ, V. ŠUNJIĆ, F. KAJFEŽ and N. BLAŽEVIĆ

Effects of Disease States on Drug Disposition in Man   628
E. S. VESELL

Isolation of Rat Liver Epoxide Hydratase: Properties   646
and Substrate Specificity of the Pure Enzyme
P. BENTLEY and F. OESCH

Reduction of a Tertiary Amine N-Oxide and an Arene Oxide in Relation to the Reduction Rate of Cytochrome P-450 in Rat Liver Microsomes
R. KATO, K. IWASAKI and H. NOGUCHI ... 654

Microsomal Oxidation of Isoproterenol and Irreversible Protein Binding of Metabolites
M. SCHEULEN, H. KAPPUS and H. M. BOLT ... 661

Benzo (α) Pyrene Metabolism by Microsomes and Isolated Epithelial Cells from Rat Small Intestine
R. GRAFSTRÖM, S. J. STOHS, M. D. BURKE, P. MOLDÉUS and S. ORRENIUS ... 667

Characterization of the Drug Monooxygenase System in Mouse Small Intestine
P. WOLLENBERG and V. ULLRICH ... 675

Formation of Alkylation and Carbamylation Intermediates and Cytochrome P-450 Catalyzed Monooxygenation of the 2-Chloroethylnitrosoureas CCNU and Methyl CCNU
D. J. REED and H. E. MAY ... 680

Proposed Mechanism for the Reductive Glycosidic Cleavage of Daunomycin (NSC 82151) and Adriamycin (NSC 123127)
D. W. YESAIR, S. McNITT and L. BITTMAN ... 688

Covalent Binding of Styrene Oxide to Rat Liver Macromolecules *in vivo* and *in vitro*
J. MARNIEMI, E. SUOLINNA, N. KAARTINEN and H. VAINIO ... 698

III OUTLOOK

The Regulation of Human Drug Metabolism by Nutritional Factors
A. KAPPAS, A. P. ALVARES, K. E. ANDERSON, W. A. GARLAND, E. J. PANTUCK and A. H. CONNEY ... 703

The Bay-Region Theory: A Quantum Mechanical Approach to Aromatic Hydrocarbon-Induced Carcinogenicity
D. M. JERINA and R. E. LEHR ... 709

The Activation and Induction of Biphenyl Hydroxylation and Chemical Carcinogenesis
D. V. PARKE ... 721

Integrated Electron Transfer: Iron States and Regulation
I. C. GUNSALUS ... 730

Purification of Liver Microsomal Cytochrome P-450: Hopes and Promises
W. LEVIN ... 735

Active Oxygen - Fact or Fancy 748
 R. W. ESTABROOK and J. WERRINGLOER

Index 759

LIST OF PARTICIPANTS

Achtert, Günther, Kali-Chemie Pharma, Abt. Pharmakokinetik, D-3000 Hannover, Hans-Böckler-Allee 20, Germany.

Agosin, Moises, Department of Zoology, University of Georgia, Riverbend Research Laboratories, Athens, Georgia 30602, USA.

Ahokas, Jorma, T., Department of Pharmacology, University of Oulu, SF-90220 Oulu 22, Finland.

Altmann, Hans-Jürgen, Abt. Toxikologie, Max-von-Pettenkofer-Institut des Bundesgesundheitsamtes, Postfach, D-1000 Berlin 45, Germany.

Alvares, Alvito P., The Rockefeller University of New York, N.Y. 10021, USA.

Arvela, Pentti I., Department of Pharmacology, Free University of Berlin, Thielallee 69/73, D-1000 Berlin 33, Germany.

Baars, A.J., Department of Pharmacology, Subfaculty Pharmacy, Sylvius Laboratories of the State University, Wassenaarseweg 72, Leiden, The Netherlands.

Von Bahr, Christer, Department of Clinical Pharmacology, Huddinge University Hospital, S-14186 Huddinge, Sweden.

Batt, Anne-Marie, Faculté des Sciences Pharmaceutiques et Biologiques, Laboratoire de Biochimie Pharmacologique, 5-7, rue A. Lebrun-B.P. 403, F-54001 Nancy Cedex, France.

Belvedere, Giorgio, Instituto di Ricerche Farmacologiche "Mario Negri", Via Eritrea, 62, I-20157 Milano, Italy.

Bend, John R., Pharmacology Branch, National Institute of Environmental Health Sciences, P.O. Box 12233, Research Triangle Park, North Carolina 27709, USA.

Bentley, Philip, Abt. für biochemische Pharmakologie, Pharmakologisches Institut der Universität Mainz, Obere Zahlbacher Straße 67, D-6500 Mainz, Germany.

Berg, Anders, KEMJ I, Karolinska Institutet, S-104 01 Stockholm 60, Sweden.

Berggren, Margareta, Department of Forensic Medicine, Karolinska Institutet, S-10401 Stockholm 60, Sweden.

Beyhl, Friedrich E., Hoechst Aktiengesellschaft, Pharma Biochemie Nord D528, D-6230 Frankfurt (M) 80, Postfach, Germany.

Bock, Karl Walter, Institut für Toxikologie, Wilhelmstrasse 56, D-7400 Tübingen, Germany.

Böttcher, Irmgard, Schering AG, Müllerstr. 170-178, D-1000 Berlin 65, Germany.

List of Participants

Boyd, G.S., Department of Biochem., University of Edinburgh, Medical School, Edinburgh, EH 8 9 AG, Scotland, U.K.

Breckenridge, Alasdair M., University of Liverpool, Department of Clinical Pharmacology, New Medical Building, Ashton Street, P.O.Box 147, Liverpool, L69 3BX, England, U.K.

Brunner, G., Med. Universitätsklinik, Humboldtallee 1, D-3400 Göttingen, Germany.

Budczies, Babette, Abt. Toxikologie, Max-von-Pettenkofer-Institut des Bundesgesundheitsamtes, Postfach, D-1000 Berlin 45, Germany.

Burchell, Brian, Department of Biochemistry, Medical Sciences Intitute, The University, Dundee, DD1 4HN, Scotland, U.K.

Busch, U., Firma Dr. Karl Thomae, D-7950 Biberach/Riß, Germany.

Capdevila, Jorge, Department of Zoology, University of Georgia, Riverbend Research Laboratories, Athens, Georgia 30602, USA.

Carlström, Kjell, Hormone Laboratory, Sabbatsberg Hospital, S-113 82 Stockholm, Sweden.

Charuel, Claude, Laboratoires Pfizer, Centre de Recherches, B.P. 42, F-37400-Amboise, France.

Cho, Arthur K., Department of Pharmacology, School of Medicine, The Center for the Health Sciences, University of California, Los Angeles, Cal. 90024, USA

Conney, Allan H., Hoffmann-LaRoche, Inc., Department of Biochemistry and Metabolism, Nutley, New Jersey 07110, USA.

Coomes, Marguerite L., 809 Denton, Denton, Tx. 76201, USA.

Coon, M.J., Department of Biological Chemistry, The University of Michigan Medical School, Ann Arbor, Michigan 48109, USA.

Cooper, David Y., 5013 Ravdin Ctyd.Bldg., Hospital of the Univ., 34th Spruce St. Philadelphia, Pa. 19104, USA

Cox, P.J., Institute of Cancer Research, Royal Cancer Hospital, Chester Beatty Res. Inst., Fulham Road, London SW 3 6JB, England, U.K.

Cremer-Schlede, Eva, Rüdesheimer Platz 1, D-1000 Berlin 33, Germany

Davies, Donald S., Department of Clinical Pharmacology, Royal Postgraduate Medical School, Ducane Road, London W12 OHS, England, U.K.

Debey, Marie-Pascale, Institute de Biologie Physico-Chimique, 13 Rue P.et M. Curie, F-75005 Paris, France.

Denk, Helmut, A-1230 Wien, Rudolf-Zellergasse 51, Austria.

List of Participants

Dickins, Maurice, Dept. of Drug Metabolism, Wellcome Research Laboratories, Langley Court, Beckenham, Kent BR3 3BS, England, U.K.

Deutschmann, G., Physiologisch-chemisches Institut der Universität des Saarlandes, D-6650 Homburg/Saar, Germany.

Diehl, Horst, Universität Bremen, D-2800 Bremen 33, Achtergasse, Germany.

Dietsch, Agnès, Laboratoire de Biochimie Vegétale, Institut de Botanique, 28, rue de Goethe, F-67 000 Strasbourg, France.

Down, William H., Huntingdon Research Centre, Huntingdon, Cambridgeshire PE18 6 ES, England, U.K.

Duppel, Wilfried, Universität des Saarlandes, Fachrichtung 3.3, Physiologische Chemie, D-6650 Homburg/Saar, Germany.

Dus, Karl M., Dept. of Biochemistry, St. Louis University School of Medicine, 1402 South Grand Blvd., St. Louis, Missouri 63104, USA.

Dybing, Erik, Department of Environmental Toxicology, National Institute of Public Health, Oslo 1, Norway.

Ebel, D., University of Texas, Health Science Center at Dallas, South Western Med. School, 5323 Harry Hines Blvd., Dallas, Texas 75235, USA.

Eben, Anneliese, Bayer AG, Institut für Toxikologie, Friedrich-Ebert-Straße 217, D-5600 Wuppertal, Germany.

Elcombe, C.R., Pharmakologisches Institut der Universität Mainz, Obere Zahlbacher Str. 67, D-6500 Mainz, Germany.

Ernster, Lars, Dept. of Biochem., University of Stockholm, Box 6409, S-113-82 Stockholm, Sweden.

Estabrook, Ronald W., University of Texas, Health Science Center at Dallas, South Western Med. School, 5323 Harry Hines Blvd., Dallas, Texas 75235, USA.

Fann, William E., Departments of Psychiatry and Pharmacology, Baylor College of Medicine, 1200 Moursund Ave., Houston, Texas 77025, USA.

Fauran, François, Laboratoires Sarget, B.P. 100- Ave. du Pr. Kennedy, F-33701 Merignac, France.

Fellenius, Erik, AB Hässle, Analytical Chemistry and Biochemistry, Fack, S-43120, Mölndal, Sweden.

Franklin, Michael, R., University of Utah, College of Pharmacy, Salt Lake City, Utah 84112, USA.

Garbe, Andreas, E. Merck, Institut Grafing, Am Feld 32, 8018 Grafing bei München, Germany.

Gelboin, Harry V., National Cancer Institute, National Institute of Health, Rm. 3E 24, Bethesda, Maryland 20014, USA.

List of Participants

Gielen, J., Laboratoire de Chimie Médicale, University of Liège, Bd. de la Constitution 151, B-4020 Liège, Belgium.

Gillette, James R., Laboratory of Chemical Pharmacology, National Heart and Lung Institute, Laboratory of Chemical Pharmacology, National Institutes of Health, Bethesda, Maryland 20014, USA.

Glatt, Hans R., Pharmakologisches Institut, Obere Zahlbacher Straße 67, D-5600 Mainz, Germany.

Glazko, A.J., Research Laboratories, Parke, Davis & Company, 2800 Plymouth Rd., Ann Arbor, Michigan 48106, USA.

Golan, Mario, Wiesbadenerstr. 28 B, D-6503 Mainz-Kastel, Germany.

Gräfström, Roland, Dept. of Forensic Medicine, Karolinska Institutet, S-10401 Stockholm 60, Sweden.

Gram, Theodore E., Laboratory of Toxicology, National Cancer Institute NIH, Bethesda, Md. 20014, USA.

Gunsalus, I.C., Department of Biochemistry, University of Illinois, Urbana, Ill. 61801, USA.

Gustafsson, Jan-Åke, Department of Chemistry, Karolinska Institutet, S-10401 Stockholm 60, Sweden.

Hänninen, Osmo, Dept. Physiology, Univ. Kuopio, P.O.Box 138, SF 70101 Kuopio 10, Finland.

Haroz, Richard K., Battelle, Research Center, 7, route de Drize, 1227 - Carouge/ Genève, Switzerland.

Heinemeyer, Gerhard, Institut für Klinische Pharmakologie, Klinikum Steglitz, Hindenburgdamm 30, D-1000 Berlin 45, Germany.

Herken, H., Institut für Pharmakologie der Freien Universität Berlin, Thielallee 69, D-1000 Berlin 33, Germany.

Herzberg, Mala, Ichilov Hospital, Department of Chemical Pathology, Tel Aviv, Israel.

Herzberg, Wolfgang, Wilhelmallee 4, D-2000 Hamburg 55, Germany.

Hesse, Sigrun, Gesellschaft f. Strahlen-u. Umweltforschung mbH. München, Abt. für Toxikologie, D-8042 Neuherberg, Ingolstädter Landstrasse 1, Germany.

Hildebrandt, Alfred, Institut für Klinische Pharmakologie, Klinikum Steglitz, Hindenburgdamm 30, D-1000 Berlin 45, Germany.

Hirata, Masaharo, Department of Forensic Medicine, Karolinska Institutet, S-104 01 Stockholm 60, Sweden.

Hlavica, P., Pharmakologisches Institut der Universität, D-8000 München 15, Nußbaumstr. 26, Germany.

List of Participants

Hoensch, Harald, Hausserstraße 140, D-7400 Tübingen, Germany.

Hoffbauer, R.W., Institut für Humangenetik, Martinistraße 52, D-2000 Hamburg 20, Germany.

Högberg, Johan, Rättsmedicinska Institutionen, Karolinska Institutet, S-104 01 Stockholm 60, Sweden.

Holtzman, J.L., Clinical Pharmacology Unit, Veterans Administration Hospital and Department of Pharmacology, University of Minnesota, Minneapolis, Minn. 55417, USA.

Ihlefeld, Heimbert, Universität Bremen, Studienbereich II, D-2800 Bremen, Achterstrasse, Germany.

Ingelman-Sundberg, Magnus, Department of Chemistry, Karolinska Institutet, Fack, S-104 01 Stockholm, Sweden.

Jakobsson, Sten, Dept. Forensic Medicine, Karolinska Institutet, S-10401 Stockholm 60, Sweden.

Jauch, Rolf, 795 Biberach, Dr. Karl Thomae GmbH, Germany.

Jeffery, Elizabeth, Dept. Pharmacology, 105 Millard Hall, University of Minnesota, Minneapolis, MN 55455, USA.

Jerina, D., Laboratory of Chemistry, National Institute of Health, Bethesda, Md. 20014, USA.

Jernström, Bengt, Dept. of Forensic Medicin, Karolinska Institutet, S-10401 Stockholm 60, Sweden.

Johannesen, Karin, University of Stockholm, Arrhenius Laboratory, Department of Biochemistry, Fack, S-104 05 Stockholm, Sweden.

Joly, Jean-Gil, Hospital St-Luc, Centre de Recherches Cliniques, 1058 St-Denis, Montreal, Quebec, Canada H2X 3J4.

Jonen, H.G., Dept. of Pharmacology, University of Mainz, Obere Zahlbacher Str. 67, Hochhaus, D-6500 Mainz, Germany.

Jonsson, John, University of Uppsala, Psychiatric Research Department, Ulleråker Hospital, 75017 Uppsala, Sweden.

Kärki, Niilo, Department of Pharmacology, University of Oulu, SF-90220 Oulu 22, Finland.

Kahl, G.F., Dept. of Pharmacology, University of Mainz, Obere Zahlbacher Str. 67, Hochhaus, D-6500 Mainz, Germany.

Kahl, Regine, Dept. of Pharmacology, University of Mainz, Obere Zahlbacher Str. 67, Hochhaus, D-6500 Mainz, Germany.

Kaminsky, Laurence, New York State Health Dept., Division of Labs. and Research, New Scotland Ave., Albany, N.Y. 12201, USA.

List of Participants

Kappas, A., Rockefeller University, Dept. of Biochemistry and Drug Metabolism, New York, N.Y. 10021, USA.

Kato, Ryuichi, 1-6, 2-Chome, Kashima, Yodogawa-ku, Osaka, 532, Japan.

Kaufmann, Rolf, Inst. für Toxikologie, Abt. f. Zelltoxikologie, Wilhelmstrasse 56, D-7400 Tübingen, Germany.

Kewitz, Helmut, Institut für Klinische Pharmakologie, Klinikum Steglitz, Hindenburgdamm 30, D-1000 Berlin 45, Germany.

Kexel, Hugo, Lehrstuhl für Allgem. Chemie und Biochemie der Techn. Universität München, D-8050 Freising-Weihenstephan, Germany.

Kiese, Manfred, Pharmakologisches Institut der Universität München, Nußbaumstr. 26, D-8000 München 2, Germany.

Kim, Sangduk, Hels. Res. Inst., 3420 N. Broad St., Temple Univ., Philadelphia, Pa., USA.

Koch-Weser, Jan, Centre de Recherche Merrell, International, 16 rue d'Ankara, 67000 Strasbourg, France.

Kremers, P., Laboratory of Chemistry, Laboratoire de Chimie Médicale, University of Liège, Bd. de la Constitution 151, B-4020 Liège, Belgium.

Kupfer, David, Worcester Foundation, Shrewsbury, MA. 01545, USA.

Larsen, John Chr., Institute of Toxicology, National Food Institute, 19 Mørkhøj Bygade, DK-2860 Søborg, Denmark.

Lang, Matti, Dept. of Physiol., Univ. Kuopio, P.O. Box 138, SF-70101 Kuopio 10, Finland.

Lech, John J., Dept. of Pharmacology, RM 112, Medical College of Wisconsin, 561 N 15 Th. St., Milwaukee, Wis. 53233, USA.

Lenk, Werner, Pharmakologisches Institut, Nußbaumstraße 26, D-8000 München 2, Germany.

Levin, Wayne, Dept. Biochemistry & Drug Metabolism, Hoffmann LaRoche Inc., Nutley, N.J. 07110, USA.

Lieber, Charles S., Bronx Veterans Administration Hospital, Section of Liver Disease & Nutrition, 130 W.Kingsbridge Rd., Bronx, New York 10468, USA.

Lind, Christina, University of Stockholm, Arrhenius Laboratory, Dept. of Biochemistry, Fack, S-10405 Stockholm, Sweden.

Lindeke, Björn, The Biomedical Center, University of Uppsala, Box 574, S-75123 Uppsala, Sweden.

Lissner, R., c/o Merck, Med. Res./Medical Biochemistry, Frankfurter Str. 250, D-6100 Darmstadt, Germany.

List of Participants

Lübke, Klaus, Schering AG, Abteilung Allgemeine Biochemie, Müllerstr. 170-178, D-1000 Berlin 65, Germany.

Maines, Mahin D., 283 Ashford Ave., Dobbsferry, New York 10522, USA.

Mannering, G.J., Department of Pharmacology, 105 Millard Hall, University of Minnesota Medical School, Minneapolis, Minnesota 55455, USA.

Marlowe, Carolyn G., Department of Pharmacology, University of Kentucky, College of Medicine, Lexington, Kentucky 40506, USA.

Marniemi, Jukka, Department of Physiology, University of Turku, Kiinamyllynk, 10, SF-20520 Turku 52, Finland.

Marselos, Marios, Dept. of Physiology, University of Kuopio, SF-70101 Kuopio, Finland.

Marsh, J., Physiology Department, Medical School, Observatory, Cape, Republic of S.A.

Masters, Bettie Sue S., Department of Biochemistry, The Univ. of Texas, Health Science Center at Dallas, 5323 Harry Hines Blvd., Dallas, Texas 75235,USA.

Mattern, Ina E., Medical-Biological Laboratory TNO, P.O. Box 45, Rijswijk 2100, The Netherlands.

Meyer, Urs A., Div. of Clinical Pharmacology, CL-25, Kantonsspital, CH-8091 Zürich, Switzerland.

Mercier, Michel, Faculté de Medecine - Ecole de Pharmacie, Tour van Helmont, U.C.L. 73 69 Av. Emmanuel Mounier, 73, 1200 Bruxelles, Belgium.

Mieyal, John J., Department of Pharmacology, North Western University Medical School, 303 Chicago Ave., Chicago, Illinois 60611, USA.

Mitchell, Jerry R., Laboratory of Chemical Pharmacology, National Heart and Lung Institute, National Institutes of Health, Bethesda, Maryland 20014, USA.

Mohn, Gertrud, Physiologisch-Chemisches Institut der Universität des Saarlandes, LKH Bau 44, D-6650 Homburg/Saar, Germany.

Moldéus, Peter, Dept. of Forensic Medicine, Karolinska Institutet, S-10401 Stockholm 60, Sweden.

Moron, Maria, University of Stockholm, Arrhenius Laboratory, Department of Biochemistry, Fack, S-104 05 Stockholm, Sweden.

Mosser, Jacqueline, Laboratories Sarget B.B. 100 - Ave du Pr. Kennedy, F-33701 Merignac, France.

Mull, Robert, Inst. für Biophysik und Strahlenbiologie, D-7800 Freiburg i.Br., Albertstr. 23, Germany.

Muller-Eberhard, Ursula, Scripps Clinic & Research Foundation, Dept. of Biochemistry , 476 Prospect St., La Jolla, Ca. 92037, USA.

Narasimhulu, Skakunthola, Harrison Dept. for Surgical Research, B-703, Richards Bldg., University of Pennsylvania, Philadelphia, Pa. 19174, USA.

Nastainczyk, Wolfgang, Physiologisch-Chemisches Institut der Universität des Saarlandes, D-6650 Homburg/Saar, Germany.

Neal, Robert, Department of Biochemistry, Vanderbilt University School of Medicine, Nashville, Tennessee 37232, USA.

Nebert, Daniel W., National Institute of Child Health & Human Development, National Institutes of Health, Bethesda, Md. 20014, USA.

Netter, K.J., Pharmakologisches Institut der Universität Mainz, Langenbeckstraße 1, D-6500 Mainz, Germany.

Noordhoek, J., Dept. Pharmacology, Medical Faculty, Erasmus University, P.O. Box 1738, Rotterdam, The Netherlands.

Novak, Raymond F., 400 Custer Ave., Chicago, Ill. 60202, USA.

Oesch, Franz, Pharmakologisches Institut, Universität Mainz, Obere Zahlbacher Str. 77, D-6500 Mainz, Germany.

Okuda, Kyuichio, Physiologisch-Chemisches Institut der Universität des Saarlandes, D-6650 Homburg/Saar, Germany.

Omura, Tsuneo, Dept. of Biology, Society of Science, Kyushu University, Fukuoka 812, Japan.

Orrenius, Sten, Department of Forensic Medicine, Karolinska Institutet, S-10401 Stockholm 60, Sweden.

Orton, Terence C., Safety of Medicines Department, ICI Pharmaceuticals Division, Mereside, Alderley Park, Macclessfield, Cheshire, England, U.K.

Paigen, Beverly, Department of Molecular Biology, Roswell Park Memorial Institute, Buffalo, N.Y., USA.

Paigen, Kenneth, Department of Molecular Biology, Roswell Park Memorial Institute, Buffalo, N.Y., USA.

Parke, D.W., Dept. of Biochem., Univ. of Surrey, Guildford, Surrey GU2 5XH England, U.K.

Pawlak, Andrzey, Dept. of Human Genetics, Medical School, Poznán 60 781, ul. Swiecickiego 6, Poland.

Pederson, Thomas C., Department of Biochemistry, University of Illinois, Urbana, Illinois 61801, USA.

List of Participants

Pelkonen, Olavi, Department of Pharmacology, University of Oulu, SF-90220 Oulu 22, Finland.

Peterson, Julian A., Department of Biochemistry, University of Texas Health Science Center, 5323 Harry Hines Blvd., Dallas, Tx. 75235, USA.

Philpot, Richard M., National Institute of Environmental Health Sciences, P.O.Box 12233, Research Triangle Park, North Carolina 27709, USA.

DePierre, Joseph W., University of Stockholm, Arrhenius Laboratory, Dept. of Biochemistry, Fack, S-104 05 Stockholm, Sweden.

Pitré, Davide, Bracco Indistria Chimica S. p.A., Via E. Folli, 50, I-20132-- Milano, Italy.

Poensgen, Jutta, Physiologisch-Chemisches Institut der Universität des Saarlandes, D-6650 Homburg/Saar, Germany.

Powis, G., Dept. Pharmacology, Yale Medical School, 333 Cedar Street, New Haven, Conn. 06510, USA.

Prough, Russell A., Dept. of Biochemistry, The University of Texas Health Science Center, 5323 Harry Hines Blvd., Dallas, Tx. 75235, USA.

Pyykkö, Kaija, Institute of Clinical Sciences, University of Tampere, Teiskontie 35, SF-33520 Tampere 52, Finland.

Rafter, Joseph James, Kemiska Institutionen, Karolinska Institutet, Solnavägen 1, S-104 01 Stockholm 60, Sweden.

Rahimtula, Anver, Department of Biochemistry, Memorial University of Newfoundland, St. Johns, Newfoundland, Canada.

Reed, Donald J., Dept. of Biochemistry and Biophysics, Oregon State University, Corvallis, Oregon 97331, USA.

Remmer, H., Institut für Toxikologie der Universität Tübingen, Wilhelmstraße 56, D-7400 Tübingen, Germany.

Rendic, Slobodan, Institut für Pharmazeutische Chemie der Westfälischen Wilhelms-Universität, D-4400 Münster/W., Hittorfstr. 58-62, Germany.

Roots, Ivar, Institut für Klin. Pharmakologie, Klinikum Steglitz, Hindenburgdamm 30, D-1000 Berlin 45, Germany.

Rosenthal, Otto, 570 B Dulles Bldg., Hospital of University of Pennsylvania, Philadelphia, Pa. 19104, USA.

Rubin, Alan, Lilly Laboratory for Clinical Research, Wishard Memorial Hospital, 1001 West 10th St., Indianapolis, Indiana 46202, USA.

Ruf, Hans-Heinrich, Physiologisch-Chemisches Institut der Universität des Saarlandes, D-6650 Homburg/Saar, Germany.

Sato, Ryo, Institute for Protein Research, Osaka University, 5311, Yamadakami, Suita, Osaka 565, Japan.

List of Participants

Schenkman, John B., Department of Pharmacology, Yale University School of Medicine, 333 Cedar Street, New Haven, Conn. 06517, USA.

Scheulen, M., Institut für Toxikologie, Wilhelmstr. 56, D-7400 Tübingen, Germany.

Schleyer, Heinz, 5014 Ravdin Courtyard Building, Hospital, Univ. of Pennsylvania, Philadelphia, Pa. 19104, USA.

Schmassmann, H.U., Pharmakologisches Institut, Obere Zahlbacher Str.67, Hochhaus, D-6500 Mainz, Germany.

Schmelz, Eike, Lehrstuhl für Allgem. Chemie und Biochemie der Technischen Universität München, D-8050 Freising-Weihenstephan, Germany.

Schmid, Jochen, c/o Dr. Karl Thomae GmbH, Birkendorfer Str. 65, D-7950 Biberach/Riß, Germany.

Schmidt, H.-L., Lehrstuhl für Allgem. Chemie und Biochemie der Technischen Universität München, D-8050 Freising-Weihenstephan, Germany.

Schmoldt, Achim, Pharmakologisches Institut der Universität Hamburg, Martinstr. 52, D-2000 Hamburg 20, Germany.

Schoene, Klaus, Institut für Aerobiologie, D-5948 Schmallenberg, Grafschaft, Germany.

Schoppler, J., Dept. für Expt. Toxikologie, Schering AG, Müllerstraße 170/178, D-1000 Berlin 65, Germany.

Schreiber, Eric, Chem. Fabrik v. Heyden, Donaustaufer Str. 378, D-8400 Regensburg, Germany.

Schulman, Martin, Department of Forensic Medicine, Karolinska Institutet, S-10401 Stockholm 60, Sweden.

Schulte-Hermann, Rolf, Institut für Toxikologie und Pharmakologie der Philipps-Universität, Pilgrimstein 2, D-3550 Marburg/L., Germany.

Schüppel, R., Institut für Pharmakologie der Technischen Universität, Büttenweg 17, D-3300 Braunschweig, Germany.

Schuster, Ingeborg, Sandoz Forschungsinstitut GmbH, Abt. Biochemie, Brunnerstraße 59, A-1235 Wien, Austria.

Schwartzkopff, Thomas, Institut für Klinische Pharmakologie, Klinikum Steglitz, Hindenburgdamm 30, D-1000 Berlin 45, Germany.

Seidegard, Janeric, University of Stockholm, Arrhenius Laboratory, Department of Biochemistry, Fack, S-10405 Stockholm, Sweden.

Sies, Helmut, Institut für Physiol. Chemie, Goethestr. 33, D-8000 München 2, Germany.

List of Participants

Siest, Gérard, Faculté des Sciences Pharmaceutiques et Biologiques, Laboratoire de Biochemie, 5-7 rue Albert Lebrun - B.P. 403, F-54001 Nancy Cedex, France.

Sipal, Z., Department of Biochemistry, Charles University, Albertov 2030, 128 40 Praha 2, Czechoslovakia.

Sligar, Stephen G., Department of Biochemistry, University of Illinois, Urbana, Illinois 61801, USA.

Snyder, Robert, Department of Pharmacology, Thomas Jefferson University, Philadelphia, Pennsylvania, USA.

Sotaniemi, Eero A., Clinical Research Unit, Dept. Intern. Med., University of Oulo, SF-90220, Oulo 22, Finland.

Spitzauer, Peter, Institut für Ökologische Chemie der GSF, D-8042 Neuherberg, Post Oberschleißheim, Ingolstädter Landstr. 1, Germany.

Staudinger, Hansjürgen, Holbeinstraße 3, D-7800 Freiburg i. Br., Germany.

Stier, A., Max-Planck-Institut für Biophysikalische Chemie, D-3400 Göttingen-Nikolausberg, Am Faßberg, Germany.

Stitzel, Robert, Department of Pharmacology, West Virginia Univerity, Morgantown, West Virginia 26506, USA.

Tanaka, Minoru, Central Research Laboratories, Sankyo Co., Ltd., No. 2-58, Hiromachi 1-chome, Shinagawa-ku, Tokyo, 140, Japan.

Temple, David J., Institut für Pharm. Chem. der Universität Frankfurt, Georg-Voigt-Str. 14, D-6000 Frankfurt am Main, Germany.

Temple, Cherrie, Institut für Pharm. Chem. der Universität Frankfurt, Georg-Voigt-Str. 14, D-6000 Frankfurt am Main, Germany.

Tennekes, H.A., Tunstall Laboratory, Shell Research Ltd., Broad Oak Road, Sittingbourne, Kent, ME9 8AG, U.K.

Terriere, Leon C., Department of Entomology, Oregon State University, Corvallis, Oregon 97331, USA.

Teschke, Rolf, II. Medizinische Universitätsklinik, Moorenstr. 5, D-4000 Düsseldorf 1, Germany.

Thenot, Jean-Paul, Institute for Lipid Research, Baylor College of Medicine, Houston, Texas 77030, USA.

Thor, Hjördis, Dept. Forensic Medicine, Karolinska Institutet, S-10401 Stockholm 60, Sweden.

Thorgeirsson, Snorri S., 9305 Kingsley Ave., Bethesda, Maryland 20014, USA.

Thurman, Ronald G., Department Biochemistry and Biophysics, Univ. of Pennsylvania, Philadelphia, Pa. 19174, USA.

List of Participants

Traylor, Teddy G., Physiologische Chemie der Universität des Saarlandes, D-6650 Homburg/Saar, Germany.

Traylor, Patricia S., Physiologische Chemie der Universität des Saarlandes, D-6650 Homburg/Saar, Germany.

Trifilieff, Elisabeth, Laboratoire du Pr Ourisson, Institut de Chimie, 1, rue Blaise Pascal, F-67008 Strasbourg, Cedex, France.

Tsibris, John C.M., Biochemistry Department, Box J-245, University of Florida, Gainesville, Fl. 32610, USA.

Tunek, Anders, Dept. of Enviromental Health, University of Lund, Sölvegatan 21, S-223 62 Lund, Sweden.

Tuong Chi Cuong, Parcor, Drug Metabolism Department, 195 Route D'Espagne, F-31023 Toulouse, France.

Uehleke, H., Bundesgesundheitsamt Berlin, Postfach, D-1000 Berlin 33, Germany.

Ullrich, Volker, Physiol.-chem. Institut der Universität des Saarlandes, D-6650 Homburg/Saar, Germany.

Vadi, Helena, Dept. of Forensic Medicine, Karolinska Institutet, S-104 01 Stockholm 60, Sweden.

Vähäkangas, Kirsi, Department of Pharmacology, University of Oulu, Kajaanintie 52, SF-90220 Oulu 22, Finland.

Vainio, Harri, Institute of Occupational Health, Department of Industrial Hygiene and Toxicology, Haartmaninkatu 1, SF-00290 Helsinki 29, Finland.

Vatsis, Kostas P., Department of Biological Chemistry, Medical Science Building I, Medical School, The University of Michigan, Ann Arbor, Michigan 48104, USA.

Vesell, Elliot S., Department of Pharmacology, The Milton S. Hershey Medical Center, Hershey, Pennsylvania 17033, USA.

Viljakainen, Eila, Clinical Research Unit, Dept. Intern Med. University of Oulu, SF-90220 Oulu 22, Finland.

Visser, R., Middelgeest 4, Leiderdorp, The Netherlands.

Volz, Manfred, Radiochem. Labor der Hoechst AG, 623 Ffm-Höchst, Germany.

Villeneuve, Jean-Pierre, c/o Dr. Jean-Gil Joly, Clinical Research Center, Hospital St. Luc, 1058 St.-Denis, Montreal, Quebec, Canada.

Waddell, William J., Department of Pharmacology, University of Kentucky, College of Medicine, Lexington, Kentucky 40506, USA.

Welch, Richard M., Wellcome Research Laboratories, RTP, Research Triangle Park, No. Carolina, USA.

List of Participants

Wende, Peter, Institut für Physiologische Chemie der Universität des Saarlandes, D-6650 Homburg/Saar, Germany.

Wendt, Gernot, c/o Hoffmann-LaRoche & Co, AG, CH-4002 Basel, Postfach, Switzerland.

Werringloer, Jürgen, Dept. of Biochemistry, The University of Texas, Southwestern Medical School at Dallas, Tx. 75235, USA.

Weymann, Jürgen, Knoll AG Chemische Fabriken, Postfach 210805, D-6700 Ludwigshafen, Germany.

Wiebel, F.J., National Institute of Health, National Cancer Institute, Bldg. 37 3D28, Bethesda, Md. 20014, USA.

Wolf, Roland C., Institut für Physiologische Chemie der Universität des Saarlandes, D-6650 Homburg/Saar, Germany.

Wolff, Thomas, Gesellschaft f. Strahlen- u. Umweltforschung mbH München, Abteilung für Toxikologie, D-8042 Neuherberg, Ingolstädter Landstraße 1, Germany.

Wollenberg, Peter, Institut für Physiologische Chemie der Universität des Saarlandes, D-6650 Homburg/Saar, Germany.

Yang, C.S., Department of Biochemistry, New Jersey Medical School, Newark, N.J. 07103, USA.

Yang, Shen K., Bldg. 37, Rm 2E24, National Cancer Institute, National Institutes of Health, Bethesda, Maryland 20014, USA.

Yasukochi, Yukio, Department of Biochemistry, The Univ. of Texas Health Science Center at Dallas, 5323 Harry Hines Blvd., Dallas, Texas 75235, USA.

Yesair, David W., Life Science Section, Arthur D. Little, Inc., Cambridge, Mass. 02140, USA.

Zehnder, K. c/o Biopharmazeutische Abtlg. Sandoz AG, CH-4002 Basel, Switzerland.

Ziegler, Daniel M., The University of Texas at Austin, Calyton Foundation Biochemical Institute, Austin, Texas 78712, USA.

Zimmer, Arno, c/o Dr. Karl Thomae GmbH, D-7950 Biberach, Riss, Germany.

PREFACE

Humans are chronically exposed to numerous non-nutrient dietary chemicals, to drugs used for the treatment or prevention of diseases, and to increasingly large numbers of man-made environmental pollutants. Because of these considerations, it is critical to examine carefully the biochemical properties of the enzymes that metabolize foreign chemicals and the many factors that influence the metabolism of chemicals to toxic and non-toxic products. These considerations led to earlier symposia in Bethesda, Maryland (1968), and Palo Alto, California (1972), and to the present Third International Symposium on Microsomes and Drug Oxidations in Berlin, West Germany, on July 21-24, 1976. The objective of organizing this Symposium was to provide an overview of our current knowledge of the oxidative metabolism of drugs, carcinogens, and various other environmental chemicals. The present volume attests to the success of this meeting.

The philosophy of the Symposium was to provide a common meeting ground for biochemists, pharmacologists, toxicologists, and clinicians. Considerable personal interactions occurred among scientists from the many countries who attended, and a spirit of friendship and scientific challenge was present throughout the meeting. The Symposium was especially distinguished by the presentation of an honorary doctor of medicine degree from the Free University to Professor Otto Rosenthal (University of Pennsylvania).

On behalf of all the participants, the organizers are indebted to the staff of the Klinikum Steglitz of the Free University of Berlin for excellent arrangements and kind hospitality. In particular, a debt of gratitude is due to Dr. P. Traylor for expert help in the final preparations of the manuscripts, and we thank the Stiftung Volkswagenwerk, Free University of Berlin, Paul Martini-Stiftung, the Schering Corp., the Sandoz Co., the Borroughs Wellcome Co., and Hoffmann-La Roche Inc. for their generous support. In addition, we are indebted to the publishers of *Hoppe-Seyler's Zeitschrift für Physiologische Chemie* for printing the abstracts and to Pergamon Press Ltd. for the present publication.

Volker Ullrich
Ivar Roots
Alfred G. Hildebrandt
Ronald W. Estabrook
Allan H. Conney

OTTO ROSENTHAL, THE MAN AND HIS WORK

Herbert Remmer

Institute of Toxicology, University of Tübingen, Wilhelmstraße 56, D-7400 Tübingen, W. Germany

Lieber Herr Rosenthal: But now I should use the American manner of addressing you: Dear Otto, dear colleagues and friends, members of this Faculty of the Free University of Berlin: I am deeply moved by having the privilege to honour Professor Rosenthal!

A Berliner addresses you as a Berliner! You lived in this beloved, marvellous and great city for 35 years, the best years of your life, before you were banished and went over to the United States.

As a student of the Berlin University and a former member of this Faculty I address you, who was also a student of this University in the 20th and a research assistant and professor until you had to leave.

It speaks now a friend to his fatherly friend, a colleague to his admired colleague, who remembers so many hours talking with you about the new situation in our country and discussing the latest news about cytochrome P-450, when I visited Philadelphia, and we were sitting in your small and modest office room which no German professor would accept! You have never been pretentious. Your reserved but warm appearance and, sometimes, shy behaviour might be the reason why many scientists, even friends, overlooked your unique contributions in biochemistry, particularly, in the field which is the topic of this symposium.

Thus, a scientist, having worked about several aspects of cytochrome P-450, addresses a senior scientist whom he admires because of his imagination to use procedures which he transfered from Berlin to the Harris Department of Biochemistry, belonging to the Clinic of Surgery in Philadelphia.

I refer to the Warburg Apparatus and the reaction vessels which came from Berlin, reminding all the students and younger people here in the audience, who are probably not so familiar with these fundamental work of Otto Rosenthal and his associates, Ron Estabrook and David Cooper, that it was his idea to use the method he learnt here in Dahlem and in the Charité, the old University Clinic, today in the eastern part, but formerly in the center of our now divided city. With this procedure Otto Rosenthal has been able to obtain such great scientific achievements. Figure 1 presents the vessel in which microsomes from bovine adrenal cortex were incubated with NADPH and 17-hydroxyprogesterone. The 21-hydroxylation of this steroid could be stopped by adding CO. Irradiation with light released the inhibiting effect of CO. The so called chemical action spectrum delivered a maximal energy for reversing the diminished enzyme activity produced by CO at a wavelength of 450 nm. The idea of cytochrome P-450 was born (1).

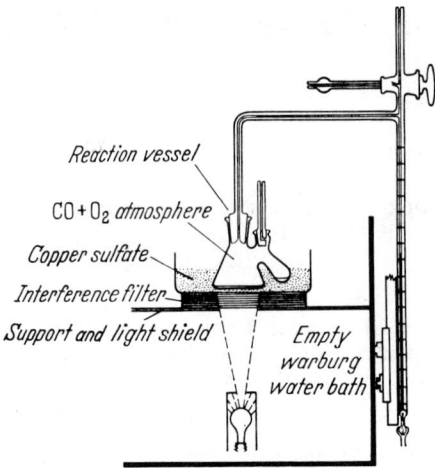

Fig. 1. The arrangement for light irradiation employed for the determination of the light reversibility of CO-inhibition of cortexolone formation.

This ingenious concept, dear Otto, would have never been created if scientific work has not been originated in this city!

I am too young to tell from own experience something about the stimulating scientific life in Berlin during your early years. It is no exaggeration if I say that here in Berlin, 50 years ago, in the 20th, lived and worked world's scientific elite! Let me quote Werner Heisenberg, one of the leading atom-physicists, Nobel-Price-Winner, who deceased just recently, several months ago, who wrote in his memoirs that he, as a young man of 26 years of age and as a Bavarian - my German friends know what a Berliner thinks about people from Bavaria -, was invited to give a seminar about his new atomic concept before the most distinguished society here in Berlin. He describes his feeling and says: "The University of Berlin was at that time viewed as a most prominent strong-hold of physics in Germany and in the world. Here worked at the same time Max Planck, Albert Einstein, von Laue and Nernst. Here Planck discovered the quanten-theory, and Rubens confirmed it with measurements of thermic radiation, and here Einstein in the year 1916 formulated the concept of relativity and the theory of gravitation." (2)

But I should continue now and mention that just in the neighbourhood, here nearby in Dahlem, stand even today well preserved the buildings of the former Kaiser-Wilhelm-Gesellschaft. Most of them belong now to the Free University and some to the Max-Planck-Society. In one, Herr Rosenthal, you should remember, was located the Institute of Biology. The director was the famous geneticist Goldschmidt. In this building worked during the 20th an enthusiastic crew of young scientists, all becoming Nobel-Price-Winners in later days. I should remind you of such names, as Otto Warburg, Hans Krebs, Fritz Lippmann, Meyerhoff, Ochoa. Excuse me, if I miss one or the other name!

Also famous pharmacologists, as Nachmansohn and Blaschko, worked in this outstanding department for awhile.

In the neighbouring building beyond the road was situated the Department of Chemistry; even today it is used as Institute of Chemistry of the Free University. In the basement Otto Hahn possessed a very small laboratory where he performed his outstanding experiments proving the scission of uranium, which initiated a new era.

And, Herr Rosenthal, if you walked from the Department of Chemistry just over the street, you would arrive at a building, formerly the Department of Biochemistry of the Kaiser-Wilhelm-Gesellschaft, and now the Department of Pharmacology of the Free University. The Director in those days was the famous Neuberg. In this department, you remember, you tried your first steps on the road to science. Even before you ended your medical studies, you worked in this institute. Not satisfied, I suppose, with the informations the students received, and wanting to know more, you probably went every day from the Charité to Dahlem and worked as a student in the laboratory of Neuberg. I should mention your first paper which appeared in 1923 in the "Biochemische Zeitschrift"; Neuberg and Rosenthal are the authors (3). This road on which you started passed the Charité where you stayed for nearly ten years, and ended in Philadelphia at the Harris Foundation.

In one dark year the road you went was blocked. The flourishing scientific life in Dahlem was almost deadly hurted, and science in Germany, even after nearly 50 years, never recovered again to this old and highly successful standard. This embarrassing situation in 1933 was masterly described by Hans Selye in his book "From Dream to Discovery" (4). Let me quote him:
"The great Galileo was exiled from his home and had to wander from city to city, until, thrown into prison, he was asked to disclaim his discoveries as despicable errors. Descartes, after also being exiled, was forced to lead a hazardous existence as soldier, physician, philosopher, and physicist until he died in a foreign land. Vesalius had to live a vagabond's life, being severely penalized for imaginary crimes and eventually dying of hunger. Copernicus did not dare publish his great discoveries, and it was only on the day of his death that he finally saw them in print. Kepler, though pensioned by his emperor, lived in misery, because he never received his pension. Let no one think that this kind of persecution could happen only in the dark Middle Ages; it can happen in a highly cultured, modern state right now. It happened a short time ago in Hitler's Germany. The science of Albert Einstein, Otto Loewi, Otto Warburg, and many others among the nations' and the world's greatest minds was labeled Jewish and, hence, unacceptable. All these men had to flee their country in shame . . ." Can this be ever forgotten?

Honouring you, Otto Rosenthal, we honour with you all the nameless scientists of Jewish origin who were exiled, and found, as you did, a new home and a place of peace for scientific work in the United States!

It is an old law in history that the 'genius loci', the spirit of a place and of an epoch, if it vanishes, can never be revived. Let us wish that all the young people here in this Free University, in the Klinikum, students, assistants and professors, which have the privilege to teach and to work as scientists in a university with an old tradition, be it here in Steglitz or nearby in Dahlem, in the old buildings of the Kaiser-Wilhelm-Gesellschaft, might sometimes breath and feel the air of this spirit, enabling them to experience the imagination and creativity of a generation who lived and worked here 50 years ago.

Celebrating Otto Rosenthal and his contribution, we all might feel some of the 'genius loci' of science which governed in the 20th here over Dahlem and in Berlin!

REFERENCES

(1) R. W. Estabrook, D. Y. Cooper, and O. Rosenthal, The light reversible carbon monoxide inhibition of the steroid C21-hydroxylase system of the adrenal cortex, Biochem. Z. 338, 741 (1963).

(2) Heisenberg, W., Der Teil und das Ganze, Piper Verlag, München, 1969, p. 90.

(3) C. Neuberg and O. Rosenthal, Über die Cellase der Taka-Diastase, Biochem. Z. 143, 399 (1923).

(4) Selye, H., From Dream to Discovery, McGraw-Hill, New York, 1964, p. 122.

I BIOCHEMISTRY

LIPID STRUCTURE OF LIVER MICROSOMAL MEMBRANES, LIMITATIONS OF A MODEL

A. Stier, W. Kühnle and R. Rösen

Max Planck Institute for Biophysical Chemistry, 34 Göttingen, Germany

SUMMARY

Rigid and fluid lipid areas are present in rabbit liver microsomes depending on temperature.
Rigid lipid areas (boundary lipid) form haloes around the cytochrome P-450 system.
The boundary lipid undergoes a transition from rigid to fluid state around 35°C.
The results suggest that the lateral organization of lipids in microsomal membranes exhibits the dynamics of a multiphase system. The lipid phase diagram is modulated by conformation of cytochrome P-450 and of other integral membrane proteins and by many physiological and experimental factors perturbing lipid structure. But the lipid states may determine assemblage of the cytochrome P-450 system as well as conformation of its enzyme components and therefore biotransformation of drugs qualitatively as well as quantitatively. This may impose limitations for microsomes as an in vitro model.

RIGID AND FLUID AREAS IN LIVER MICROSOMES

Part of the lipids of microsomal membranes form lamellae existing in a state of high fluidity as shown previously by spinlabel experiments (Ref 1). An order parameter (S = 0.71 at 23°C) which is a measure for the angular deviation of the amplitude of anisotropic rotation of a spin-labeled fatty acid chain from the normal to the membrane (Ref 2) and therefore a measure of lipid fluidity can be evaluated (see Fig. 4). It lies in the lower range of the order parameters measured so far by this method in biological membranes. The fluidity of this part of the membrane allows for a high rate of lateral diffusion (several μm/sec) of a stearic acid within it.
Other lipids are segregated into areas which are not accessible for the fatty acid spin probe due to their low fluidity. Proton magnetic resonance spectra show (see Fig. 1), underlying the sharp resonances of the cholinomethyl, methylene and methyl groups of lipids in a highly fluid state, broad bands representing immobilized lipids that are nearly absent in microsomal lipid extracts, redispersed by ultrasonication, or are strongly reduced in microsomal preparations to which cholate has been added. The sharp band of the cholinomethyl group is remarkable, it is absent

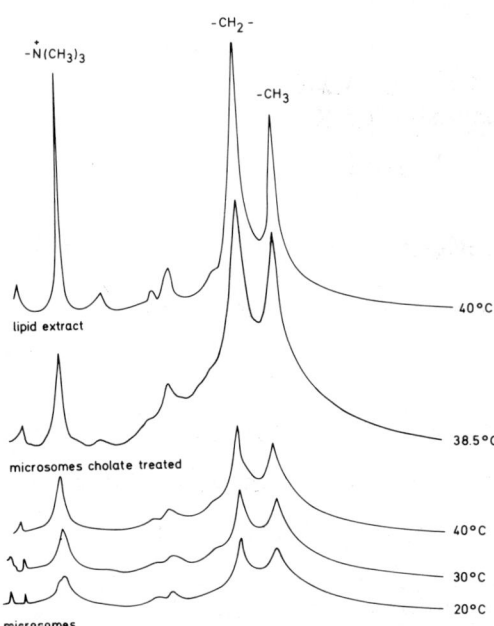

Fig. 1. Proton NMR spectra of liver microsomes (15 mg/ml protein) from phenobarbital stimulated rabbits, of cholate (final conc. 10^{-3} M) treated microsomes and of resuspended microsomal lipid extracts (0.6% W/W) in 0.01 M potassium phosphate D_2O buffer p_D 7.4 recorded by Fourier transform technique on a Bruker WH 270 spectrometer. Vertical scaling factors are different. Microsomes have been washed 3 times in D_2O and once in the buffer used for final suspension. Lipid extracts (Ref 4) have been resuspended by low power ultrasonication at 40°C under N_2 after exhaustive evaporation of solvents under high vacuum (from Ref 3).

in microsomes (Ref 5). This indicates a quite different protein lipid interaction of these two membrane types: The lipid layer in liver microsomes may extend over large areas devoid of surface proteins, allowing rapid lateral phase separation processes, rapid lipid exchange with other cell membranes, and with regard to biotransformation of drugs unhindered substrate access to and product release from the membrane. The strongest immobilization is suffered by the methylene groups, an effect which can also be seen in carbon-13 natural abundance NMR spectra (Ref 19). The amount of immobilized lipid is distinctly higher in microsomes from untreated animals compared to phenobarbital pretreated ones, and much higher for both microsomal preparations at 40°C than at 20°C. It should be noted that the values in Table 1 may not be taken as absolute, there being unresolved doubts about interpretation of spectra from differently sized vesicles (Ref 7). The higher amount of lipid immobilization in microsomal membranes enriched in cytochrome P-450 might give a clue to the topography of the rigid lipid areas (see below).

The sharp methylene and methyl resonances in proton NMR spectra increase with temperature in a nonlinear way as shown by the upbend curves in Fig. 2a in contrast to microsomal preparations to which cholate or hexobarbital had been added and also in contrast to the resuspended lipid extract of these microsomes (see Fig. 2b). This upbend shape of the curves which is less noticeable in microsomes from untreated animals might originate from a superposition of two different temperature gradients of fluidity belonging to two different lipid areas within the membrane, one being fairly fluid over the whole temperature range from 20-40°C, the other existing in a rigid crystalline state below

TABLE 1 Proportion of Fluid Lipid Appearing in Proton Resonances of Rabbit Liver Microsomes [1]

	Temp (°C)	$-CH_2$ [2] (%)	$-CH_3$ [2] (%)	$-N^+(CH_3)_3$ (%)
Phenobarbital	24	6	19	40
	40	20	53	60
Controls	24	10	39	60
	40	25	67	88

[1] Areas below sharp resonances of H^1 NMR spectra taken in the presence of an external standard from microsomal suspensions and from lipid extracted from these suspensions and resuspended by ultrasonication in the same volume of buffer have been compared (from Ref 4).

[2] determined by delayed Fourier transform technique (Ref 7).

about 35°C and undergoing a kind of lipid phase transition at that point. A similar temperature dependence of the signal height of bulk methylene resonances in the carbon-13 NMR spectra can be seen (Ref 19).

BOUNDARY LIPID AND THE CYTOCHROME P-450 COMPLEX

Is the rigid lipid associated with the cytochrome P-450 system? Is its thermotropic polymorphism correlated to a conformational change of cytochrome P-450? A cytochrome P-450 preparation (Ref 8) in which cytochrome P-450 accounts for about 50% of proteins can occur in 2 states: A metastable state shows a NMR-spectrum composed of sharp and broad methylene and methyl resonance bands (similar to microsomes). It can be irreversibly converted to a state associated with strong immobilization of lipids. The conversion sets in abruptly when the temperature reaches 36°C (Fig. 3). The strong lipid immobilization can also be seen from the high order parameter of a stearic acid spinlabel incorporated into this preparation (see Fig. 4), it is much higher than in resuspended lipid extracts of the same preparation and compared to the corresponding preparations of microsomes. The temperature curve of the cytochrome P-450 preparation shows a distinct inflection at 36°C. These experiments do not simulate the original situation in microsomes but indicate different thermotropic polymorphic conformational states of cytochrome P-450 which are associated with different states of the lipid. Other examples of functions of endoplasmic reticulum membrane of liver depending critically on temperature between 30 - 40°C are aniline hydroxylation (Ref 10), UDP-glucuronyltransfer (Ref 11), CCl_4 stimulated lipid peroxidation (Ref 12), NADPH dependent reduction of a stearic acid nitroxide (Ref 1), NADPH oxidation (Ref 13), formation of a stable nitroxide from 2-aminonaphthalene (Ref 14) and biliary excretion of lead (Ref 15).

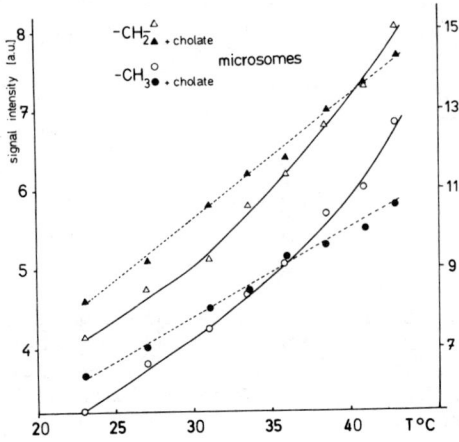

Fig. 2a. Temperature dependence of signal intensities of sharp resonances of methylene (triangles) and methyl resonances (circles) of proton NMR spectra of liver microsomes (15 mg/ml protein) from phenobarbital stimulated rabbits, in presence (closed signs, dashed curves, right scale) and absence (open signs, continuous curves, left scale) of cholate (final conc. 10^{-3} M) as determined by application of delayed Fourier transform technique (Ref 7) in presence of an external standard. For other experimental conditions see Fig. 1 (from Ref 3).

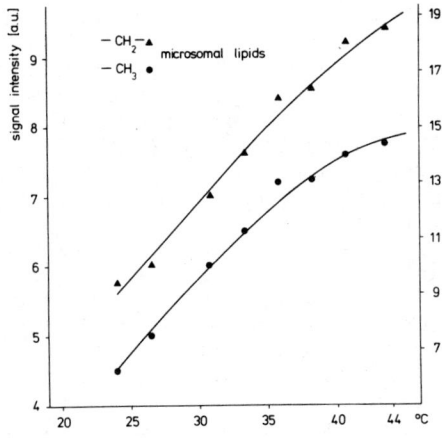

Fig. 2b. Temperature dependence of signal intensities of methylene resonances (triangles, right scale) and methyl resonances (discs, left scale) of proton NMR spectra of sonicated lipids (0.45% W/W) of rabbit liver microsomes (same preparation as in Fig. 2a).

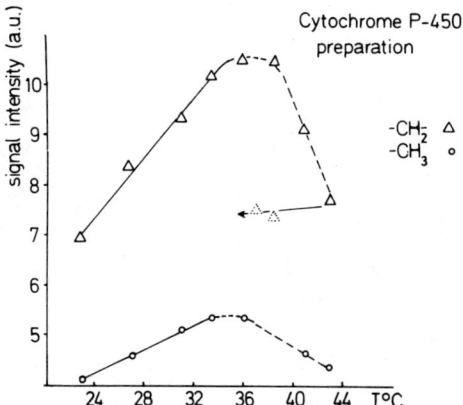

Fig. 3. Temperature dependence of the methylene (triangles) and methyl resonances (circles) of Proton NMR spectra of a cytochrome P-450 enriched membrane fraction (8nM cytochrome P-450/mg protein) of liver microsomes from phenobarbital stimulated rabbits. The fraction was separated from bulk membrane constituents by column chromatography on a AH Sepharose 4B of cholate treated microsomes (Ref 8) and dialyzed against 0.01 M potassium phosphate D_2O buffer p_D 7.4 (from Ref 3).

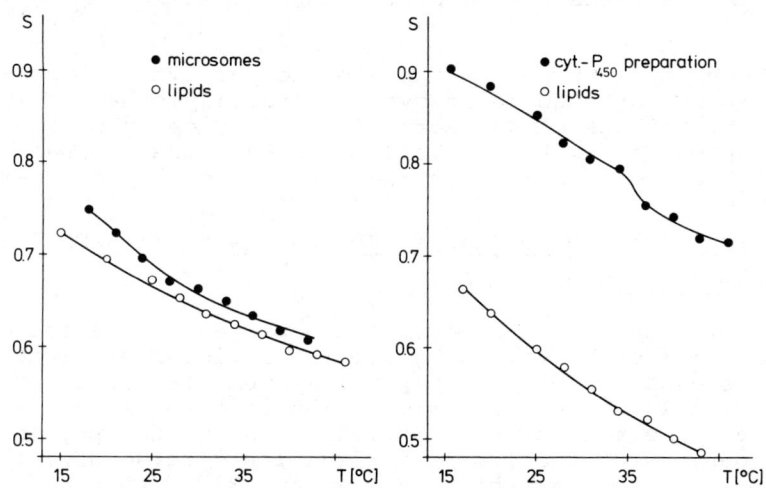

Fig. 4. Temperature dependence of the order parameter S of ESR spectra of a fatty acid spinlabel (N-oxyl-4,4'-dimethyloxazolidine derivative of stearic acid, the doxyl group positioned on C-5, Syva Inc. Palo Alto) incorporated in rabbit liver microsomes, in a cytochrome P-450 preparation described in Fig. 3 and in resuspended lipid extracts of the two preparations recorded on a Varian E9 spectrometer. The small step seen in the curve of the cytochrome P-450 preparation around 36°C was consistently observed in 5 experiments using preparations from different animals (from Ref 3).

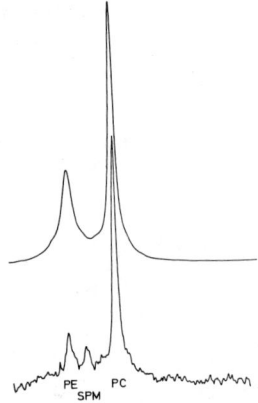

Fig. 5. Phosphorous NMR spectra of lipid extracts (Ref 4) of liver microsomes from phenobarbital stimulated rabbits (upper spectrum) and of a cytochrome P-450 enriched microsomal fraction prepared as described in Fig. 3 (lower spectrum) recorded on a Bruker WH 270 Fourier transform NMR spectrometer (from Ref 3). Note the reduced proportion of phosphatidylethanolamine (PE) relative to sphingomyelin (SPM) and phosphatidylcholine (PC) appearing in the P-resonances of the lower spectrum.

From the P-NMR spectra we see a preference of cytochrome P-450 for lecithin and sphingomyelin over phosphatidylethanolamine (see Fig. 5). Perhaps it reflects also an asymmetry of its transversal distribution with a preference to face the luminal site in view of the known asymmetric distribution of lipids (Ref 9).

PROSPECTS: MICROSOMAL MEMBRANES - A MULTIPHASE SYSTEM, LIMITATIONS OF A MODEL

In conclusion these results may be interpreted in the following way: The lateral organization of lipids in microsomal membranes exhibits the dynamics of a multiphase system. The lipid phase diagram is modulated by the conformation of cytochrome P-450 and of other integral membrane proteins and by many physiological and experimental factors perturbing lipid structure. The whole system is in metastable state around 37°C, to put it in another way, on a phase border in a phase diagram. But the lipid states may determine conformation and assemblage of the cytochrome P-450 system (as shown schematically in Fig. 6) and therefore influence biotransformation reactions quantitatively and qualitatively: cytochrome P-450 may act in the assembled complex as mixed function oxygenase, in the disassembled state as oxidase or peroxidase.

Moreover the lipid phase separation dynamics - due to the cooperativity of the supramolecular structure of lipid bilamellae (Ref 16) - may link the functions of the cytochrome P-450

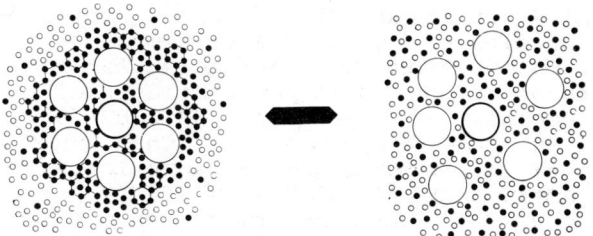

Fig. 6. Two interconvertible states of lateral organization of cytochrome P-450 (large light circles), of cytochrome P-450 reductase (large heavy circles) and of 2 types of membrane lipids (small open and closed circles) in a multiphase model of liver microsomes.

system to other functions localized in areas of the endoplasmic reticulum membrane distant from the latter (e.g. biotransformation phase II reactions), an aspect pertinent to entanglement of cytochrome P-450 in the physiology as well as pathology of the liver (Ref 17 and 18).

This concept of lipid dynamics imposes limitations for microsomes as an in vitro model for biotransformation of drugs which have been surveyed more comprehensively (Ref 19). Loss of control of some ER functions in isolated microsomes may perturb lipid structure and therefore influence the biotransformation system, especially in the cases where these functions are lipid mediated linked to the functions of the cytochrome P-450 system. Some of the experimental conditions may shift the membrane to a stable but nontheless irrelevant state with regard to the in vivo situation. Thereby the results may become even more reproducible but erroneous as is often found in biological assays involving membranes if these are "aged". The better model might be hepatocytes in which the nearly complete product pattern can be detected in situ by carbon-13 NMR spectroscopy as pilot studies using carbon-13 labeled fatty acids show.

REFERENCES

(1) A. Stier and E. Sackmann, Spin label as enzyme substrates heterogeneous lipid distribution in liver microsomal membranes, Biochim. Biophys. Acta 311, 400-408 (1973).
(2) W.L. Hubbel and H.M. McConnell, Molecular motion in spin-labeled phospholipids and membranes, J. Amer. Chem. Soc. 93, 314-326 (1971).
(3) A. Stier and R. Rösen, in preparation.
(4) E.G. Bligh and W.J. Dyer, Rapid method of total lipid extraction and purification, Can. J. Biochem. Physiol. 37, 911-917 (1957).
(5) J.D. Robinson, N.J.M. Birdsall, A.G. Lee and J.C. Metcalfe, ^{13}C and ^{1}H nuclear magnetic resonance relaxation measurements of the lipids of sarcoplasmic reticulum membranes, Biochemistry 11, 2903-2909 (1972).

(6) P.A. Kroon, M. Kainosho and S.I. Chan, Proton magnetic resonance studies of lipid bilayer membranes Experimental determination of inter- and intramolecular nuclear relaxation rates in sonicated phosphatidylcholine bilayer vesicles, Biochim. Biophys. Acta 433, 282-293 (1976).
(7) C.H.A. Seiter, G.W. Feigenson, S.I. Chan and M. Hsu, Delayed Fourier transform proton magnetic resonance spectroscopy, J. Amer. Soc. 94, 2535-2537 (1972).
(8) Y. Imai and R. Sato, An affinity column method of partial purification of cytochrome P-450 from phenobarbital-induced rabbit liver microsomes, J. Biochem. 75, 689-697 (1974).
(9) O. Nilson and G. Dallner, Distribution of constitutive enzymes and phospholipids in microsomal membranes of rat liver, FEBS Lett. 58, 190-193 (1975).
(10) J.B. Schenkman, The effects of temperature and substrates on component reactions of hepatic microsomal mixed-function oxidase, Mol. Pharmacol. 8, 178-188 (1972).
(11) S. Eletr, D. Zakim and D.A. Vessey, A spin-label study of the role of phospholipids in the regulation of membrane-bound microsomal enzymes, J. Mol. Biol. 78, 351-362 (1973).
(12) T.F. Slater, personal communication.
(13) J. Chayen, Histochemistry of drug metabolism systems, Fourth European workshop on drug metabolism, Mainz, 1974.
(14) A. Stier and R. Clauss, in preparation.
(15) C.D. Klaasen and D.W. Shoeman, Biliary excretion of lead in rats, rabbits, and dogs, Toxicol. appl. Pharmacol. 29, 434-446 (1974).
(16) E.J. Shimshick and H.M. McConnell, Lateral phase separation in phospholipid membranes, Biochemistry 12, 2351-2360 (1973).
(17) F.O. Schmitt, D.M. Schneider and D.M. Crothers (eds.), Functional linkage in biomolecular systems, Raven Press, New York, 1975.
(18) A. Stier, Membrane fluidity, in T.F. Slater (ed.), Biochemical mechanisms of liver injury, Academic Press, London, in press.
(19) A. Stier, Lipid structure and drug metabolizing enzymes, Biochem. Pharmacol. 25, 109-113 (1976).
(20) A. Stier and E. Sube, unpublished results.

INTERACTIONS BETWEEN CYTOCHROME P-450 AND NADPH-CYTOCHROME P-450 REDUCTASE IN THE MICROSOMAL MEMBRANE

Chung S. Yang

*Department of Biochemistry, New Jersey Medical School, CMDNJ
Newark, New Jersey 07103 U.S.A.*

The endoplasmic reticulum of liver cells contains a monooxygenase system which catalyzes the biotransformation of steroids, fatty acids, drugs, carcinogens, and various other xenobiotics (1-3). In this enzyme system, the reducing equivalents from NADPH are transferred through NADPH-cytochrome P-450 reductase to cytochrome P-450 which in turn catalyzes the oxygenation of various substrates. Cytochrome P-450 can account for as much as fifteen percent of the microsomal protein, and the number of this hemoprotein can be 10-25 times greater than that of the reductase molecule in the membrane (4). The existence of a lipophilic mobile electron carrier between these two enzymes has never been demonstrated. It appears that one reductase molecule has to interact with a great number of cytochrome P-450 molecules for efficient catalysis. The NADPH-cytochrome P-450 reductase is an amphipathic protein with a molecular weight of about 80,000 daltons and cytochrome P-450 is an integral membrane protein with a molecular weight around 50,000 daltons (5). The organization and mobility of these enzymes in the membrane are not very well understood (6-9). This paper reports the interactions of these two enzymes in the membrane and attempts to elucidate the nature of the organization of the microsomal monooxygenase enzymes.

Organizational Models of the Monooxygenase Enzymes

Two different concepts of organization of cytochrome P-450 and the reductase are illustrated by the rigid and nonrigid models in Fig. 1. With the rigid model, the reductase molecule is surrounded by many cytochrome P-450 molecules to form individual electron transfer complexes and translational diffusion of these proteins is not required for catalysis. With the nonrigid model, both the reductase and cytochrome P-450 are mainly surrounded by phospholipids; lateral mobility, at least short ranged diffusion, of these enzymes is required for their catalytic functions. The nonrigid model does not imply that the monooxygenase enzymes are randomly distributed in the membrane; rather certain domains of the membrane may have high local concentrations of these enzymes.

Chemical Modification Studies

As a first test of these two models, the chemical modification approach of Franklin and Estabrook (6) was used. The study is based on the rationale that, for example, when 50% of the reductase is inactivated, only 50% of the cytochrome P-450 can be reduced enzymically according to the rigid model. Based on the nonrigid model, however, it can be predicted that more than 50%

or almost all of the existing cytochrome P-450 should be reducible by NADPH, although it may proceed at a slow rate. Mersalyl (sodium O-[(3-hydroxy-mercuri-2-methoxypropyl)carbamyl]phenoxyacetate) was used to inactivate the NADPH-dependent reductase activity and its effect on the enzymic reduction of cytochrome P-450 was measured. The results of this study (7) favor the nonrigid over the rigid model. For example, when approximately 75% of the reductase was inactivated, about 70% of the cytochrome P-450 could still be reduced in 5 min. A series of experiments also showed that when the reductase activity was inhibited to different extents (up to 85% inhibition) by different amounts of mersalyl, almost all the cytochrome P-450 molecules in the system could still be reduced by NADPH within a period of 20-25 min. These results suggest that cytochrome P-450 and the reductase do have translational mobility, although the rate and extent of the movement may not be as great as have been suggested for the cytochrome b_5 reduction system (10-11).

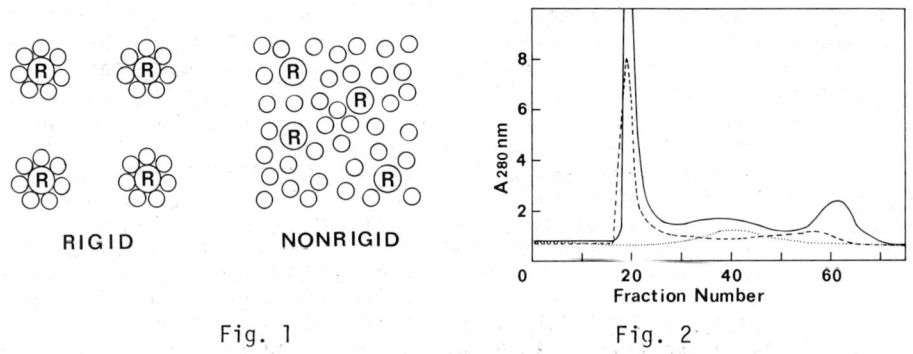

Fig. 1   Fig. 2

Fig. 1. The rigid and nonrigid models of the organization of NADPH-cytochrome P-450 reductase, ®, and cytochrome P-450, O.

Fig. 2. Elution profiles of microsomes, cytochrome P-450, and enriched microsomes. Sonicated control microsomes were incubated with solubilized cytochrome P-450 (from phenobarbital-pretreated rats) at 37° for 30 min. The sample was applied to a Sepharose 4B column (1.5 x 40 cm) and eluted with Buffer A (0.1 M potassium buffer, pH 7.4, containing 5 mM $MgCl_2$ and 0.1 mM EDTA). One ml fractions were collected. The traces are: ———, 0.7 ml of microsomes (4.2 mg protein) incubated with 0.4 ml of cytochrome P-450 (55 nmol); ----, 0.7 ml of microsomes incubated with 0.4 ml of a buffer containing 0.05 M potassium phosphate, pH 7.4, 0.1 mM EDTA, 0.1 mM DTT and 20% glycerol, in which the cytochrome P-450 sample had been dialyzed; ...., 0.2 ml of cytochrome P-450 incubated with 0.4 ml of Buffer A.

Incorporation of Purified Cytochrome P-450 into the Microsomal Membrane

According to the nonrigid model, purified cytochrome P-450, when added to microsomes, should be able to incorporate into the membrane, to receive electrons from the reductase, and to catalyze monooxygenase reactions. This has been demonstrated experimentally (8). The addition of partially purified cytochrome P-448 (isolated from 3-methylcholanthrene-pretreated rats) to control microsomes greatly enhanced the microsomal benzo[a]pyrene hydroxylase activity and a maximum of 5-fold enhancement was observed (8). This is due to the binding or incorporation of the cytochrome P-450 into the membrane. As shown in Fig. 2, the enriched microsomes can be separated from the unbound cytochrome P-450 by gel filtration with a Sepharose 4B column. Similar results were also obtained with binding studies using cytochrome P-448. The relationship between the bound or incorporated cytochrome P-448 and the enhanced benzo[a]pyrene hydroxylase activity was also studied. It was observed that the enhanced hydroxylase activity was proportional to the amount of cytochrome P-448 incorporated and the cytochrome P-448 was about 2.5 times as active as the endogenous cytochrome P-450 (8). The results suggest that the added cytochrome P-448 can incorporate into the membrane and become a functional part of the monooxygenase system.

Previously, we have reported that upon binding to microsomes, the added cytochrome P-448 can be reduced by NADPH, but were unable to determine whether the exogenous cytochrome P-448 can obtain electrons directly from the reductase molecules or indirectly from the endogenous cytochrome P-450 molecules (8). To examine these possibilities, the effects of added cytochrome P-448 on linoleic acid hydroperoxide pretreated microsomes were studied (Table 1). The hydroperoxide treatment inactivated more than 95% of the cytochrome P-450 in control microsomes but retained about 70% of the reductase activity. As a consequence, less than 2% of the benzo[a]pyrene hydroxylase activity was retained. The microsomes, however, can regain hydroxylase activity after incubating with cytochrome P-448. Similar results were also observed with microsomes from phenobarbital or 3-methylcholanthrene pretreated animals. It is also noted that after incubation with cytochrome

TABLE 1   Effect of Cytochrome P-448 on Benzo[a]pyrene Hydroxylase Activities

Microsomes	Microsomal protein (mg)	Cytochrome P-450 (nmol)	Reductase activity (unit)	Hydroxylase Activity	
				Without P-448	With P-448
Control	0.36	0.34	1.02	0.354	0.588
Treated control	0.38	0.01	0.73	0.005	0.726
PB	0.39	0.78	0.93	0.420	0.740
Treated PB	0.43	0.03	0.76	0.027	0.842
MC	0.29	0.62	0.91	0.659	0.714
Treated MC	0.42	0.07	0.96	0.042	0.604

Microsomes, with quantities indicated, were preincubated with 3.6 nmol of cytochrome P-448 in 0.25 ml of Buffer A at 37° for 30 min. The benzo[a]pyrene hydroxylase activity was assayed (8) at 37° with an incubation period of 5 min, and is expressed as nmol of product formed per min per unit of reductase activity. Control, PB, and MC microsomes were obtained from control, phenobarbital-pretreated, and 3-methylcholanthrene-pretreated rats, respectively. After treatment with linoleic acid hydroperoxide, they are referred to as "Treated Microsomes".

P-448, the hydroperoxide treated control and PB microsomes had higher activities than the corresponding untreated ones. This is probably because that in the treated microsomes the incorporated cytochrome P-448 does not have to compete with the less active endogenous cytochrome P-450 for the reductase molecules. These results suggest that the bound cytochrome P-448 molecules have a direct access to the NADPH-cytochrome P-450 reductase molecule and a heme to heme electron transfer from the endogenous cytochrome P-450 to the added cytochrome P-448 is not required for the enhanced catalysis.

Cytochrome P-450 can also bind to microsomes and enhance oxidative demethylase activity (Fig. 3). Upon the addition of cytochrome P-450 to freshly thawed microsomes, no enhancement in demethylase activity was observed.

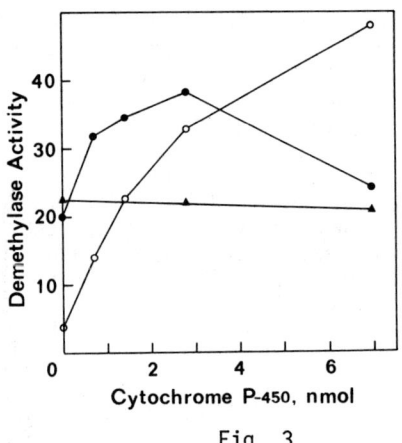 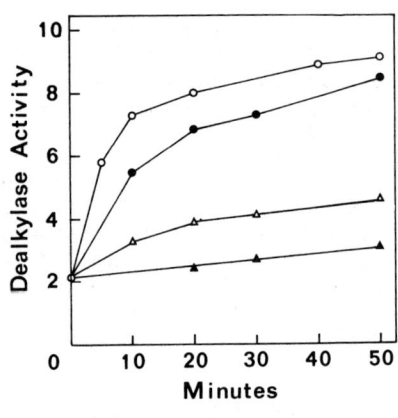

Fig. 3   Fig. 4

Fig. 3. Effects of solubilized cytochrome P-450 on microsomal benzphetamine demethylase activity. Microsomes were incubated with different amounts of cytochrome P-450 in 0.3 ml of Buffer A at 37° for 30 min; then more buffer and substrates were added for the assay of benzphetamine demethylase activity. The activity was assayed by the method of Thomas et al. (12), and is expressed as nmol of formaldehyde formed per 10 min at 37°. Each assay contained control microsomes (▲――▲), sonicated microsomes (●――●), or linoleic acid hydroperoxide treated microsomes (0――0) corresponding to 0.32, 0.32, or 1.26 mg of protein, respectively.

Fig. 4. Effect of temperature on the incorporation of cytochrome P-448 into microsomes. Sonicated control microsomes (0.48 mg protein) were incubated with 1.4 nmol of cytochrome P-448 in a volume of 0.22 ml in a cuvette at 37° (0――0), 30° (●――●), 20° (△――△), or 0° (▲――▲) for the periods indicated. Buffer A (at 20°) and substrates were added, and the cuvette was equilibrated at 20° for 2-3 min. The reaction was initiated with 80 nmol of NADPH and the rate was measured at 20°. One (fluorescence) unit of dealkylase activity corresponds to the production of 16 pmol of 7-hydroxycoumarin per min.

However, when the microsomes were converted into smaller vesicles by a brief sonication, significant enhancement in benzphetamine N-demethylase activity was seen. A more dramatic increase in the demethylase activity was demonstrated when the microsomes had been pretreated with linoleic acid hydroperoxide. Such treatment inactivated about 90% of the cytochrome P-450 molecules but retained most of the reductase activity.

The microsomal ethoxycoumarin dealkylase activity was also enhanced by the addition of partially purified cytochrome P-448. The effect of preincubation on the extent of enhancement is shown in Fig. 4. The enhancement is not instantaneous and the results suggest that a temperature-dependent incorporation of cytochrome P-448 into the microsomes is required. At 37°, it took 10 min to attain most of the enhancement and 30-40 min to reach a plateau. Such a preincubation requirement was also observed for the binding of cytochrome P-448 or P-450 to microsomes and for the enhancement of benzo[a]pyrene hydroxylase and benzphetamine demethylase activities.

All the results described so far are consistent with a nonrigid model based on structural considerations. The data, however, do not indicate whether translational diffusion is involved in the catalysis of the monooxygenase reactions. The rate and extent of the lateral movement of these enzymes in the membrane remain to be determined. If we assume that these enzymes have the diffusion rate of rhodopsin in outer rod segments (13), then a collision rate of 16 per msec is predicted if these two enzymes are separated by 100 Å (14). Such a rate is too fast to be rate-limiting in the enzymic reduction of cytochrome P-450 (9) and in monooxygenase reactions.

Effect of Temperature on Monooxygenase Reactions

The effect of temperature on the rate of the ethylmorphine demethylase reaction of PB microsomes is shown in Arrhenius plots (Fig. 5). A linear plot was not observed. Rather, the data can best be fitted by two straight lines of different slopes. This was observed when either NADPH or NADP was used to initiate the reaction. Similar breaks were also observed with the NADPH-dependent demethylation of benzphetamine, aminopyrine, and p-nitroanisole. The results of these experiments are summarized in Table 2. These reactions have activation energies of 10-12 and 19-21 Kcal per mol at temperature ranges above and below the break temperature (at about 24°), respectively. A similar break was also observed in the ethylmorphine demethylation reaction catalyzed by control microsomes, except that the activation energy at temperatures below the break is 16-18 Kcal per mol. The results are consistent with the results of Duppel and Ullrich (15), obtained with the dealkylation of 7-ethoxycoumarin and p-nitroanisole; but differ from those of Schenkman (16) and Holtzman and Carr (17). The effect of temperature on the NADPH-dependent reduction of cytochrome c was also studied and a break in the Arrhenius plot was not observed (data not shown), consistent with previous results (16, 17). The results suggest that the NADPH-dependent reductase alone is not responsible for the break in the Arrhenius plots of monooxygenase reactions. The observed break is probably due to a membrane effect, since treating microsomes with 30% glycerol abolished the break of the Arrhenius plot (data not shown). The crystal-liquid crystal phase transition of phospholipid has been used to explain the discontinuity of the Arrhenius plots of the reactions catalyzed by many membrane enzymes (18-20). However, it is not known whether the present results can be attributed to the phase transition of membrane lipids. Thus, although the presently observed break in Arrhenius plots is possibly due to the interactions of cytochrome P-450 and the reductase in the membrane, alternative interpretations are equally possible. For example, different rate-limiting steps may be involved under different experimental conditions and the change of temperature may cause a shift in the rate-limiting steps of the monooxygenase reactions.

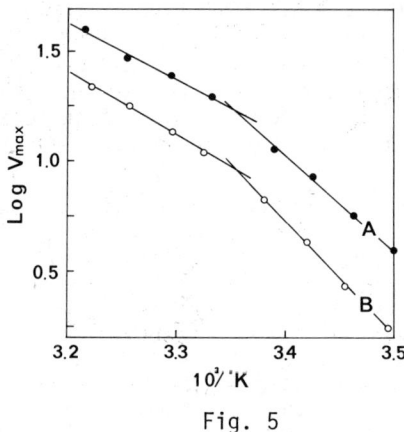

Fig. 5

Fig. 5. Arrhenius plots of the ethylmorphine demethylase reaction. The reaction mixture contained 1.5 mg microsomal protein, a NADPH-generating system, and 0.65, 1.00, 2.00, or 5.00 mM of ethylmorphine. In set A, the reactions were initiated by NADPH. The V_{max} values (in nmol HCHO/min/mg) were obtained from double-reciprocal plots. In set B, the assay mixture contained 2.5 mg of microsomal protein and the reaction was initiated by NADP.

TABLE 2 Energy of Activation of Microsomal Demethylase Reactions

Substrate	Number of Experiments	Break Temperature	E_a (38-25°) Kcal/mol	E_a (23-12°) Kcal/mol
Ethylmorphine	14	24.4 ± 1.4°	12.3 ± 2.1	21.5 ± 1.9
Benzphetamine	6	23.9 ± 1.2°	10.1 ± 0.7	18.9 ± 1.4
Aminopyrine	4	24.0 ± 1.8°	11.7 ± 1.2	20.6 ± 0.8
p-Nitroanisole	6	23.9 ± 1.6°	9.7 ± 0.3	19.8 ± 1.5

With ethylmorphine as a substrate, V_{max} values were used for the Arrhenius plot. With other substrates, the observed velocity at selected substrate concentrations, i.e., benzphetamine at 0.2 or 0.5 mM, aminopyrine at 1.0 or 2.0 mM, and p-nitroanisole at 0.5 or 1.0 mM, were used. Since the substrate concentration did not affect the activation energy of the reactions, results were summarized and expressed as mean ± standard deviation.

Requirements of a Nonrigid Model

Considering the approximate dimensions of cytochrome P-450, the reductase, and the microsomal membrane, it is difficult to envision how a reductase molecule can be rigidly associated and react efficiently with 20-25 cytochrome P-450 molecules. In a rigid model, are some of the cytochrome P-450 molecules located in the inner circle surrounding the reductase and others in the outer circle? In such an arrangement, where are the cytochrome b_5 and cytochrome b_5 reductase molecules which are known to interact with the cytochrome P-450 system? Do the outer circle cytochrome P-450 molecules obtain electrons from the reductase through the inner circle cytochrome P-450 molecules during active catalysis? Since the reductase probably has only one site that can transfer electrons to cytochrome P-450, even the electron transfer from the reductase to all the inner circle cytochrome P-450 molecules would not be favorable if these molecules were rigidly associated in a molecular complex in which rotational mobility is hindered.

Although translational diffusion of monooxygenase enzymes in the membrane is a feature of the nonrigid model, it is not known whether such a process is rate-limiting in the enzymic reduction of cytochrome P-450. Even if the diffusion is rate-limiting for cytochrome P-450 reduction, the Arrhenius plot of the cytochrome P-450 reduction rate may or may not show a break above 4°, depending on the properties of the phospholipids surrounding the system. The properties of the lipids may be altered due to the binding to proteins and a mixture of different phospholipid molecules may not show a clear thermally-induced phase transition. This point has been clearly demonstrated in the NADH-dependent reduction of cytochrome c, a reaction in which translational diffusion of cytochrome b_5 and cytochrome b_5 reductase is involved (10, 11). In this case, a break in the Arrhenius plot of the reduction of cytochrome c cannot be observed with microsomes, but can be demonstrated when the two enzymes were incorporated into dimyristoyl lecithin liposomes (14). Thus, the lack of a break in the Arrhenius plot of the "fast phase" of the reduction of cytochrome P-450 in microsomes (9, 21), may not be in conflict with the nonrigid model. In addition, the interesting observation that 48% of the cytochrome P-450 could still be reduced in the "fast phase" when 95% of the NADPH-cytochrome P-450 reductase was removed from the microsomes by trypsin digestion (21), is consistent with the nonrigid model. According to the nonrigid model, it can be predicted that the addition of purified NADPH-cytochrome P-450 reductase to microsomes should enhance some monooxygenase activities. This has recently been demonstrated (22). Thus, the above results and discussions are in favor of a nonrigid organization of cytochrome P-450 and NADPH-cytochrome P-450 reductase in the membrane. The nonrigid model also requires the translational diffusion of the monooxygenase enzymes in catalysis, a property which remains to be demonstrated.

Acknowledgments

I am grateful to Mr. F. S. Strickhart and Mr. L. P. Kicha for capable technical assistance, and to the American Cancer Society for the Faculty Research Award. This work was supported by Grant CA-16788 from the National Cancer Institute and Grant No. 472 from the Nutrition Foundation.

References

1. S. Orrenius and L. Ernster, in <u>Molecular Mechanisms of Oxygen Activation</u> (Ed. O. Hayaishi)p. 215, Academic Press, New York (1974).
2. J. R. Gillette, D. C. Davis and H. A. Sasame, <u>Ann. Rev. Pharmacol.</u> 12, 57 (1972).
3. E. C. Miller and J. A. Miller, in <u>The Molecular Biology of Cancer</u> (Ed. H. Busch) p. 377, Academic Press, New York (1974).
4. R. W. Estabrook, M. Franklin, J. Baron, A. Shigematsu and A. Hildebrandt, in <u>Drugs and Cell Regulation</u> (Ed. E. Mihich) p. 227, Academic Press, New York (1971).
5. M. J. Coon, D. A. Haugen, F. P. Guengerich, J. L. Vermilion and W. L. Dean, in <u>The Structural Basis of Membrane Function</u> (Eds., Y. Hatefi and L. Djavadi-Ohaniance) p. 409, Academic Press, New York (1976).
6. M. R. Franklin and R. W. Estabrook, <u>Arch. Biochem. Biophys.</u> 143, 318 (1971).
7. C. S. Yang, <u>FEBS. Lett.</u> 54, 61 (1975).
8. C. S. Yang and F. S. Strickhart, <u>J. Biol. Chem.</u> 250, 7968 (1975).
9. R. W. Estabrook, J. Werringloer, B. B. S. Masters, H. Jonen, T. Matsubara, R. Ebel, D. O'Keeffe, and J. A. Peterson, in <u>The Structural Basis of Membrane Function</u> (Eds. Y. Hatefi and L. Djavadi-Ohaniance) p. 429, Academic Press, New York (1976).
10. M. J. Rogers and P. Strittmatter, <u>J. Biol. Chem.</u> 249, 895 (1974).
11. M. J. Rogers and P. Strittmatter, <u>J. Biol. Chem.</u> 249, 5565 (1974).
12. P. E. Thomas, A. Y. H. Lu, D. Ryan, S. B. West, J. Kavalek, and W. Levin, <u>J. Biol. Chem.</u> 251, 1385 (1976).
13. R. A. Cone, <u>Nature</u> 247, 438 (1974).
14. P. Strittmatter and M. J. Rogers, <u>Proc. Nat. Acad. Sci. U.S.A.</u> 72, 2658 (1975).
15. W. Duppel and V. Ullrich, <u>Biochim. Biophys. Acta</u> 426, 399 (1976).
16. J. B. Schenkman, <u>Mol. Pharmacol.</u> 8, 178 (1972).
17. J. L. Holtzman and M. L. Carr, <u>Arch. Biochem. Biophys.</u> 150, 227 (1972).
18. J. K. Raison, <u>J. Bioenergetics</u> 4, 285 (1973).
19. S. Eletr, D. Zakin and D. A. Vessey, <u>J. Mol. Biol.</u> 78, 351 (1973).
20. A. Stier and E. Sackmann, <u>Biochim. Biophys. Acta</u> 311, 400 (1973).
21. J. A. Peterson, R. E. Ebel, D. H. O'Keeffe, T. Matsubara and R. W. Estabrook, <u>J. Biol. Chem.</u> 251, 4010 (1976).
22. J. T. Miwa and A. K. Cho, <u>Life Sci.</u> 18, 983 (1976).

A STUDY OF CYTOCHROME P-450 REDUCTION BY SODIUM DITHIONITE

Richard E. Ebel, David H. O'Keeffe, and Julian A. Peterson

Department of Biochemistry, The University of Texas Health Science Center, Dallas, Texas 75235

ABSTRACT

The reduction of both the membrane bound, rat liver microsomal and the soluble, bacterial cytochromes P-450 by sodium dithionite has been studied as a function of temperature (4-37°) and reductant concentration. The results indicate that in this temperature range (a) the reduction of the microsomal enzyme is a biphasic process; each phase is first order at a given dithionite concentration and the percentage of slower phase is relatively constant (70-80%) and independent of the presence or absence of exogenous substrate; (b) the reduction of cytochrome P-450$_{cam}$ is a first order process throughout the temperature range employed and independent of the presence or absence of d-camphor; (c) neither the fast or slow phase of reduction of microsomal cytochrome P-450 nor the single phase of cytochrome P-450$_{cam}$ reduction exhibits a break in the Arrhenius plot; (d) the presence of substrate results in a significant decrease in the rate constant for the dithionite dependent reduction; and (e) the reduction by dithionite does not appear to be a second order process. These findings are compared with previously obtained data from a study of the NADPH-dependent reduction of microsomal cytochrome P-450.

INTRODUCTION

One of the first steps in the microsomal cytochrome P-450 - catalyzed hydroxylation reaction is the donation of an electron to the hemin iron. This reaction, catalyzed by NADPH - cytochrome P-450 reductase, has received much attention in the literature and its kinetics are complex, exhibiting at least a biphasic time course. Utilizing computer-assisted dual-wavelength stopped-flow spectrophotometry (1), the temperature dependence of this reaction was examined in detail. The results of this investigation led us to propose a unique spatial arrangement for the membrane bound enzymes of this microsomal electron transport system (2). In an attempt to test this model, the kinetics of reduction of microsomal cytochrome P-450 catalyzed by sodium dithionite were investigated and compared to similar studies with the soluble cytochrome P-450$_{cam}$ isolated from <u>Pseudomonas putida</u>.

MATERIALS AND METHODS

Microsomes were isolated from phenobarbital (PB) pretreated rats as previously described (2). The reduction of cytochrome P-450 was monitored by following the appearance of the ferrous-carbon monoxide complex (450 minus 490 nm) in a buffer system kept totally anaerobic using an enzymatic oxygen scavenging system (3). The reaction was not a single first order process but the data could be analyzed using a nonlinear estimation program which fit the data to equation 1 which describes the sum of two concurrent first order processes.

$$A_t = A_\infty [(1-F_s)(1-e^{-k_f t}) + F_s(1-e^{-k_s t})] \quad \text{(Eq. 1)}$$

where
A_t = absorbance at time, t
A_∞ = final absorbance
F_s = fraction slow phase
k_f = rate constant for fast phase
k_s = rate constant for slow phase

Cytochrome P-450$_{cam}$ was isolated from Pseudomonas putida grown on d-camphor (4). Camphor-free enzyme was prepared from the camphor-bound form by chromatography on a Sephadex G-10 column equilibrated with 50 mM MOPS pH 7.0 which had been neutralized with Tris base. Sodium dithionite solutions were prepared by either adding the solid directly to or by dilution of a concentrated solution of sodium dithionite (0.5-1.0M) into the buffer system above. All manipulations of sodium dithionite solutions were carried out in a glove bag flushed with argon.

RESULTS AND DISCUSSION

The NADPH-dependent cytochrome P-450 reductase reaction has previously been shown to consist of two concurrent first order processes (2,3). The Arrhenius plots of the NADPH-dependent reduction reaction showed two distinct patterns which were a function of the lipophilicity of exogenous substrate present. In the first pattern the fast phase of the reaction shows no discontinuity in the Arrhenius plot while the slow phase shows a break at about 20°. This pattern (i.e., with a break in the slow phase but not in the fast phase) was observed with microsomes isolated from phenobarbital-induced rats in the presence of hexobarbital, ethylmorphine, or in the absence of exogenous Type I compounds, as well as, with microsomes from untreated rats in the presence of hexobarbital. Among these only slight differences in activation energies and the fractions of fast and slow phases were observed. In the case of PB microsomes with hexobarbital, the activation energy for the fast phase was approximately 18 kcal while that for the slow phase was about 47 and 0 kcal above and below the break point, respectively. The amount of cytochrome P-450 reduced in the fast phase was approximately 50% at 4° and this value increased to about 75% at the higher temperatures. The other pattern of Arrhenius plot was observed with the addition of more hydrophobic substrates, such as, androstanedione or d-camphor (2). In these cases there was no discontinutiy in either the fast or slow phase. The activation energies for the fast and slow phases were 15 and 17 kcal, respectively. It should be noted that at 25° the presence of Type I compounds increased the rate of reduction of cytochrome P-450 in the fast phase but at 37° there was no significant difference.

These data led us to propose a model for the spatial arrangement of the NADPH-cytochrome P-450 reductase and cytochrome P-450 in the microsomal membrane (2). The reductase was envisioned as lying mostly above the membrane (5) while the cytochrome P-450 molecules are embedded in the membrane. Approximately 50% of the cytochromes P-450 (approximately 10 per reductase) were pictured as clustered around the reductase molecule and are viewed as those reduced in the fast phase. The remainder of the cytochromes

P-450 are distributed away from the reductase and are those reduced in the slow phase. Apparently translational motion of the molecules or of the cluster is required for reduction in the slow phase, since there is a break in the Arrhenius plot. The observation of a break in this temperature region probably indicates, by analogy to other membrane enzyme systems, that a fluidity change of the membrane lipids occurs in the region of these proteins. When the added substrate was d-camphor or androstanedione, no break was observed. The assumption was made that the lack of a break was due to the presence of a significant amount of hydrophobic substrate molecules in the microsomal membrane, which results in the depression of the temperature of the membrane fluidity change (break in the Arrhenius plot) to a value lower than the range used in the study.

In an effort to further develop the above model, i.e. that the biphasic kinetics of NADPH-dependent reduction of cytochrome P-450 is due to a unique spatial arrangement of the enzymes involved and that the break in the Arrhenius plots observed in several instances is the result of this spatial arrangement and the interaction with the microsomal membrane lipids, the reduction of cytochrome P-450 by sodium dithionite was investigated. Initially, the assumption was made that the reduction of cytochrome P-450 by dithionite would be a bimolecular process; (direct interaction of cytochrome P-450 and dithionite) which would exhibit pseudo-first order kinetics and that the microsomal membrane would not influence this process i.e., no break in the Arrhenius plot would be observed. However, the reduction of microsomal cytochrome P-450 by sodium dithionite was not a simple pseudo-first order process. Instead the reaction was biphasic and, when analyzed using equation 1, was consistent with the reaction proceeding through two concurrent first order processes. Approximately 70-80% of the reaction was slow phase over the entire temperature range employed (4-37°) and was independent of the presence or absence of exogenous Type I substrate. The dithionite-dependent reduction of cytochrome P-450$_{cam}$, in contrast, displayed first order kinetics over the same temperature range in the presence or absence of its substrate, d-camphor. An Arrhenius plot of the reduction of microsomal cytochrome P-450 by 25 mM dithionite in the presence of 1 mM hexobarbital shows that neither phase displays a break (Fig. 1). Comparable data with NADPH are shown as dashed lines. The activation energy for the dithionite-dependent reduction of microsomal cytochrome P-450

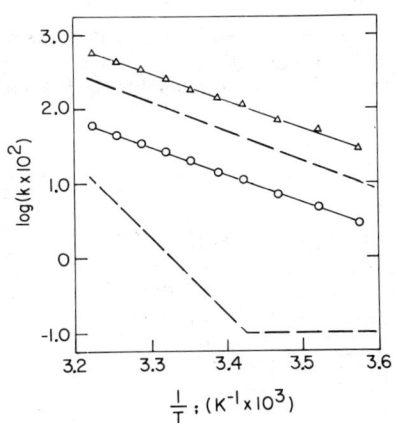

Fig. 1 Arrhenius plots of dithionite-dependent reduction of cytochrome P-450$_{LM}$ in the presence of 1 mM hexobarbital (△, fast phase; ○, slow phase); NADPH-dependent reduction shown with broken line (- - -)

in both the fast and slow phase was approximately 17 kcal. In spite of the fact that some of these results differed from our initial expectations for the dithionite-dependent reduction reaction, the observations that the percentages of fast and slow phases were constant over the temperature range employed and independent of substrate, as well as the lack of a break in the Arrhenius plot, support the assumptions used in developing the model for the NADPH-dependent process.

Neither phase of reduction of microsomal cytochrome P-450 by dithionite is a second order process in that doubling the dithionite concentration does not result in a doubling of the rate constant for reduction (Fig. 2). The shape of the Arrhenius plot, however, does not change as a function of dithionite concentration. Similarly, the dithionite dependent reduction of the soluble, bacterial cytochrome P-450$_{cam}$, while displaying first order kinetics at each dithionite concentration, did not show a second order kinetic pattern as a function of dithionite concentration (Fig. 3). The presence of exogenous Type I substrates in the case of microsomal cytochrome P-450 or camphor in the case of cytochrome P-450$_{cam}$ significantly diminished the rate of reduction (both fast and slow phase rates were slower for cytochrome P-450$_{LM}$) observed with dithionite. Interestingly, the rate of reduction of cytochrome P-450$_{cam}$ was comparable to that of the slow phase of reduction of microsomal cytochrome P-450.

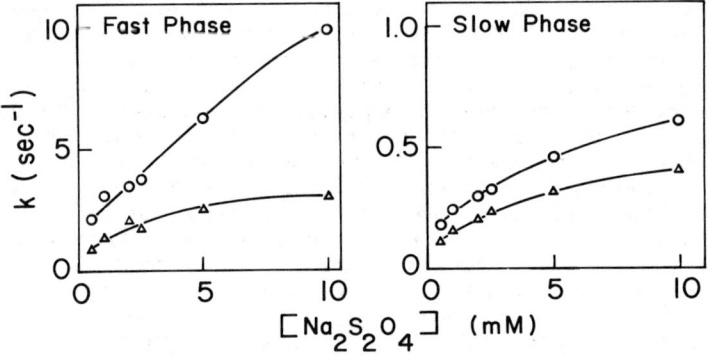

Fig. 2 Rate constants for reduction of microsomal cytochrome P-450 as a function of dithionite concentration at 37° (○, no exogenous substrate; △ , 1 mM hexobarbital)

Fig. 3 Rate constants for reduction of cytochrome P-450$_{cam}$ as a function of dithionite concentration at 37° (○, camphor-free; △ , 1 mM d-camphor)

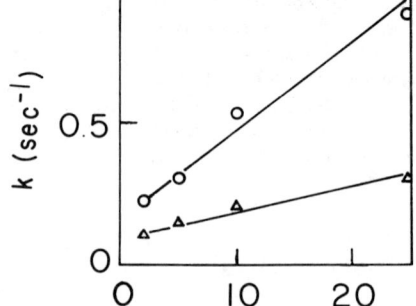

At least three possibilities exist which might provide an explanation for either or both phases of the dithionite dependent reduction of microsomal cytochrome P-450. The first would involve reduction of the NADPH-cytochrome P-450 reductase by dithionite and subsequent enzymatic reduction of cytochrome P-450. This is unlikely because of the different Arrhenius patterns observed for the slow phase of reduction. In the presence of hexobarbital the NADPH dependent reduction shows a break at about $20°$ with activation energies of 47 and 0 kcal above and below the break point, respectively, while no break and an activation energy of 17 kcal was observed with dithionite (Fig. 1). For this to be a reasonable mechanism for the fast phase of reduction, one would be required to postulate that the NADPH-dependent reduction of the reductase itself would be rate-limiting in the normal enzymatic reaction, since the rate of reduction of cytochrome P-450 is greater with dithionite, at relatively high concentrations, than with NADPH at a saturating level (0.3 mM). That is, assuming that the reductase reduced by NADPH or dithionite catalyzes the reduction of cytochrome P-450 in an enzymatically indistinguishable manner, then the reduction of the flavin moieties of the reductase (FAD, FMN) from the half-reduced to the fully-reduced state (6) would be the rate-limiting step, which seems at this time to be an unlikely possibility. A second alternative would involve the cytochrome b_5 reductase-cytochrome b_5 system. The NADH-dependent reduction of cytochrome P-450, presumably through this system, is a slow process relative to the NADPH-dependent reduction. For instance, at $37°$ in the presence of 1 mM hexobarbital, the NADH (0.3mM) -dependent reduction of cytochrome P-450 appears to be a first order process with a rate constant of 0.03 per second as compared to the NADPH (0.3mM) -dependent reduction which is biphasic with rate constants of 2.0 and 0.13 per second. Arguments similar to the above for the NADPH-dependent system would exclude this as a reasonable pathway for the dithionite dependent reduction of cytochrome P-450. The third and most probable explanation would involve direct reduction of cytochrome P-450 by dithionite. This is supported by the observation that the purified bacterial protein can be readily reduced by dithionite and exhibits several kinetic properties similar to the microsomal system. However, unlike the bacterial enzyme, the reduction of microsomal cytochrome P-450 is biphasic. The physical basis of the biphasic reduction process may argue for the existence of two types of cytochrome P-450, physically different proteins and/or membranous environment, which would be reduced by a common mechanism but at different rates. However, an equally plausible explanation would suggest that a single form of cytochrome P-450 could be reduced in two kinetically distinct ways, e.g. via both an inner and an outer sphere electron transfer process. The observation that the dithionite dependent reduction of both microsomal and bacterial cytochrome P-450 does not obey second order kinetics as a function of dithionite concentration excludes, by definition, the possibility that this reduction reaction proceeds via a simple bimolecular collision. It may be speculated, therefore, that the rate-limiting step involves the formation of a cytochrome P-450 - dithionite complex prior to reduction.

Acknowledgement - This work was supported in part by Research Grants GM 16488 and GM 19036 from the United States Public Health Service and Research Grant I-405 from The Robert A. Welch Research Foundation; J.A.P. is the recipient of Research Career Development Award GM 30962 from the United States Public Health Service; D.H.O. is the recipient of a Robert A. Welch Research Foundation Postdoctoral Fellowship.

REFERENCES

1. Peterson, J.A. and Mock, D.M., Dual wavelength/stopped flow spectrophotometry:Computer Acquisition and Analysis, Anal. Biochem., 68, 545-553, (1975)
2. Peterson, J.A., Ebel, R.E., O'Keeffe, D.H., Matsubara, T., and Estabrook, R.W., Temperature dependence of cytochrome P-450 reduction. A model for NADPH-cytochrome P-450 reductase:cytochrome P-450 interaction, J. Biol. Chem., 251, 4010-4016, (1976)
3. Matsubara, T., Baron, J., Peterson, L.L., and Peterson, J.A., NADPH-cytochrome P-450 reductase Arch. Biochem. Biophys., 172, 463-469, (1976)
4. Peterson, J.A., Camphor binding by Pseudomonas putida cytochrome P-450, Arch. Biochem. Biophys., 144, 678-693, (1971)
5. Masters, B.S.S., Baron, J., Taylor, W.E., Isaacson, E.L., and LoSpalluto, J.J., Immunological studies of electron transport chains involving cytochrome P-450. I. Effects of antibodies to pig liver microsomal reduced triphosphopyridine nucleotide-cytochrome c reductase and the non-heme iron protein from bovine adrenocortical mitochrondria, J. Biol. Chem., 246, 4143-4150, (1971)
6. Masters, B.S.S., Prough, R.A., and Kamin, H., Properties of the stable aerobic and anaerobic half-reduced states of NADPH-cytochrome c reductase, Biochemistry, 14, 607-613, (1975)

INVESTIGATION OF MICROSOMAL HYDROXYLATING SYSTEM AT SUB-ZERO TEMPERATURES

Pascale Debey, Monique Leon, Claude Balny, and Pierre Douzou

Institut de Biologie Physico-Chimique, 13 rue P. et M. Curie, 75005, Paris, France,
U-128 INSERM, B.P. 5051, 34033 Montpellier Cedex, France

INTRODUCTION

There are still many unsolved problems concerning the multi-enzyme hydroxylating system of rat liver microsomes. These include the function of each individual component and their interaction, the molecular mechanism of the oxygen activation and hydroxylation, and the nature of the membrane environment. Low temperature investigations in fluid mixed solvents are a useful possible approach and examples of this procedure (1, 2) using polyols as "antifreeze" will be described. Reactions of the terminal cytochrome P-450 hydroxylase were studied and when possible where compared to those of the soluble bacterial cytochrome P-450. Two aspects of this work will be developed in this communication.

1) Selective inhibition of reactions within a given sequence and definition of "out-of-equilibrium" and "uncoupling" conditions.

2) Sub-zero temperature chromatography as a method to isolate labile components such as native enzymes or transient intermediates.

RESULTS

<u>The Low Temperature Inhibition of Single Reactions Within a Chain :
Definition of "Out-of equilibrium" and "Uncoupling" Conditions</u>

A preliminary study of individual activities of the various enzymes of the assembly (NADH and NADPH cytochrome c reductase, NADPH cytochrome P-450 reductase, NADH and NADPH cytochrome b_5 reductase) in mixed solvent between room temperature and -30°C showed that the normal sequence of events and the hydroxylating membrane-bound function are preserved (3).

However, some kinetics may be slowed down to such an extend that certain reactions are practically stopped for hours, and this could through the multi-components system out of thermodynamic equilibrium for dynamic reasons. For instance the NADH and NADPH reductions of cytochrome P-450 are too slow to be recorded below -5°C, whereas the corresponding reductions of cytochrome b_5 are still rather rapid (half times of 50 sec and 3 min for respective reductions by NADH and NADPH at -10°C). This results in the accumulation of reduced cytochrome b_5 in the presence of fully oxidized cytochrome P-450 (3).

Similarly, the whole reaction sequence leading to hydroxylation can be "resolved" into several steps, each one being "uncoupled" from the others at selected temperatures. Accordingly a 3-steps procedure was developed in order to form, accumulate, and then decompose the reaction product of reduced P-450 and oxygen. The reaction was monitored indirectly using the luminescent oxidation of luminol by an oxidizing species (4).

The microsomal system was first fully reduced by NADPH in the presence of luminol at +10°C (first step). In this stable state, the suspension was brought to -30°C and O_2 rapidly added (second step). At that stage no emission was recorded. Upon progressive heating (third step) two luminescent signals were recorded (Fig. 1). The first one appears at -20°C, increases to a maximum at -16°C, and moves back down to zero intensity at about -5°C, whereas the second one starts at 0°C to reach a constant plateau at +20°C. Both emissions are sensitive to inhibitors of cytochrome P-450.

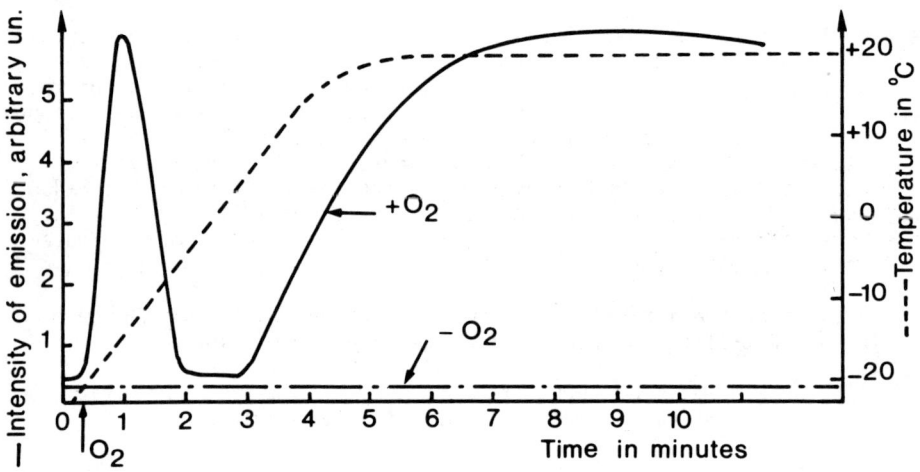

Fig. 1. Thermoluminescence of microsomes in presence of luminol. Microsomal suspension (3.5 mg/ml) reduced by NADPH ($5\ 10^{-4}$ M) at +10°C. Solvent, 1:1 (v/v) mixture of ethylene glycol and aqueous buffer.

The only component of the assembly eventually able to react with oxygen at -16°C is cytochrome P-450. Thus the luminescence recorded at that temperature probably results from the decomposition of an intermediate (presumably oxygenated) compound of cyt.P-450. At the temperature where the electron flow is resumed (0°C) (3), the steady state intensity of the second emission process is directly controlled by the steady-state concentration of cyt.P-450 reacting with oxygen.

Direct spectrophotometric evidence of such an oxy-ferrous compound was obtained with a similar experimental procedure. After full reduction in presence of the substrate hexobarbital, the microsomal suspension was allowed to rapidly react with oxygen at selected temperatures between -20°C and -45°C. Figure 2 shows differential optical spectra obtained at -45°C (Fig. 2.a) and -20°C (Fig. 2.b) and their evolution as a function of time.

At -45°C the spectrum recorded within 20 sec after oxygen addition ("dead time" of the method) is closely similar to that reported by Estabrook (5) on microsomes and to that of "Complex I" described more recently by Guengerich et al. (6) by addition of oxygen to purified cytochrome P-450 at room temperature. It is progressively converted into a "Complex II" - type spectrum (6) (compare spectra 1 and 2 Fig. 2.a), stable over 30 minutes.

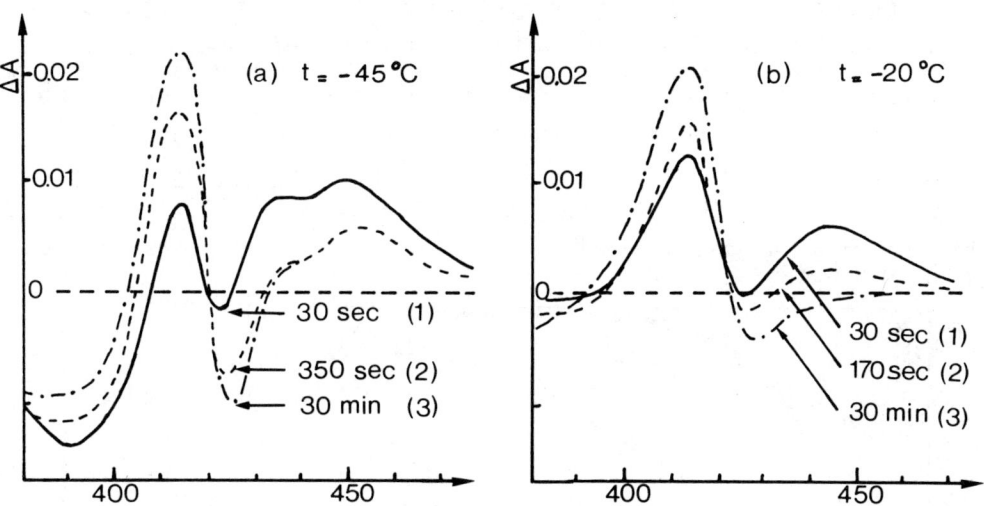

Fig. 2. Difference spectra obtained after addition of oxygen to a reduced microsomal suspension (∿ 3 mg proein/ml) at respectively (a) -45°C, (b) -20°C.

At -20°C the first spectrum is similar to a "Complex II" and further decays rather rapidly (Fig. 2.b). Similar but less intense spectra are obtained in absence of substrate.

Present results deserve the following comments. The rapid transformation of "Complex I"-type into "Complex II"-type spectrum occurs at temperatures as low as -45°C, a result suggesting the influence of other components of the electron transport chain (such as cytochrome b_5) on the stability of "Complex I". On the other hand the optical differential spectra show that cytochrome b_5 is reoxidized during the evolution of the two above compounds. Such an oxidation of cytochrome b_5 was not observed by Guengerich with purified preparations of cytochromes P-450 and b_5. Finally, according to the observed stabilities of "Complex I" and "Complex II"-type spectra, it is suggested that the thermoluminescence of luminol between -20°C and -10°C described above could be due to the decomposition of "Complex II".

These preliminary observations and comments deserve further investigations to elucidate the structural differencies between the two compounds, as well as their conversion and decomposition mechanisms.

A good model to provide the insight and experience necessary to understand the actual fate of microsomal P-450 oxy-compound is given by the oxy-ferro compound of soluble bacterial cytochrome P-450. Autoxidizable at room temperature, but easily stabilized at sub-zero temperature (7), it is known to be enzymatically decomposed by the reduced putidaredoxin (Pd^-), which plays the role of both an effector and an electron donor (8). It was studied by similar two-steps procedure involving direct spectrophotometric recordings. In the first step, camphor bound cyt.P-450 (Fe_S^{3+}) and Pd^0 were both fully reduced at +10°C in the absence of oxygen and in the presence of low camphor concentrations. In the second step, the solution was then brought to a selected low temperature (-10°C to -40°C) and O_2 added. Even at -40°C it is not possible to accumulate the rapidly formed oxy-ferrous compound ($Fe_S^{2+} - O_2$), which is immediately decomposed into Fe^{3+} by Pd^- by a strictly <u>monophasic, Pd^- independent</u> reaction (7) ($k = 10^{-2}$ sec^{-1} at -40°C, E_A = 11.5 Kcal/Mole) (Fig. 3). This indicates that the recorded process is rate-limited by the decomposition of one intermediate.

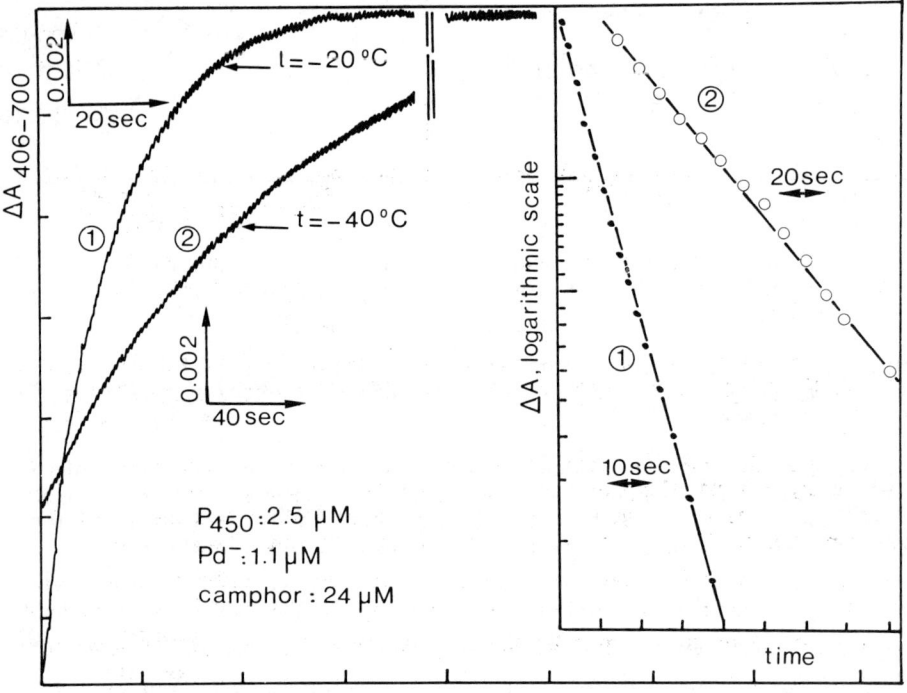

Fig. 3. Decomposition of the substrate bound oxy-ferro compound of bacterial cytochrome P-450 by Pd^-. Solvent as in Fig. 1. The insert shows the semi-logarithmic plot of the kinetics.

The present procedure allows one to "quench" any reaction other than the decomposition of the oxy-ferrous compound, particularly the binding of camphor to the free ferric cytochrome which would be the starting step of enzyme recycling. These conditions provide the "uncoupling" essential to a precise analysis of the process.

The enzymatic decomposition certainly involves several steps and transient species which cannot be seen in the present experimental conditions. The procedure should be now extended to lower temperatures or relayed by "trapping" procedures such as described by Chance et al. for mitochondrial enzymes (9).

Finally it is important to note that the oxy-ferrous compound, almost indefinitely stabilized at sub-zero temperatures, is quickly destroyed at -40°C by putidaredoxin. This suggests that microsomal P-450 could have a similar fate in presence of electron donors such as cytochrome b_5 or NADPH reductase at higher temperature.

Low Temperature Purification of Enzymes in their Native State or the Form of Unstable Transitory Compounds

The understanding of the functionning of a multi-enzyme assembly as a whole can often be achieved through a parallel study of the integrated system and of the reactions performed by each individual component after their isolation and purification. Low temperature could be a very useful parameter since it might stabilize labile enzymes either in their native form (i.e. the membrane solubilized cytochrome P-450) (10), or in the form of transient intermediates (i.e. oxygenated compound of bacterial cytochrome P-450) (11).

Purification of native enzymes.
The separation of the enzymes of the microsomal hydroxylating system by DEAE cellulose chromatography after solubilization by deoxycholate according to the method of Lu et al. (12) was carried out in a polyol-water mixture at selected temperatures (10) as shown on the Fig. 4.

From this figure it may be seen that the chromatographic resolution was perfect and similar to that obtained in a classical way at +4°C. Calculated on the basis of a similar cytochrome b_5 concentration in both aliquots, the yield of cytochrome P-450 (or P-420) was 20 times higher at -20°C than in normal conditions at +4°C, a result which in itself justifies the use of the low temperature procedure. Finally we must note the inversion of the elution positions of flavoproteins and cytochromes which all eluted at lower ionic strength. The origin of this modification is not yet thoroughly understood and its significance and its possible applications to chromatography are under close examination in this Laboratory.

Transient intermediates.
The low temperature stabilization of short-lived functional enzyme intermediates may be used to isolate them from other chemicals such as substrates or oxido-reducing agents used for their preparation and still present in excess in the reaction medium.

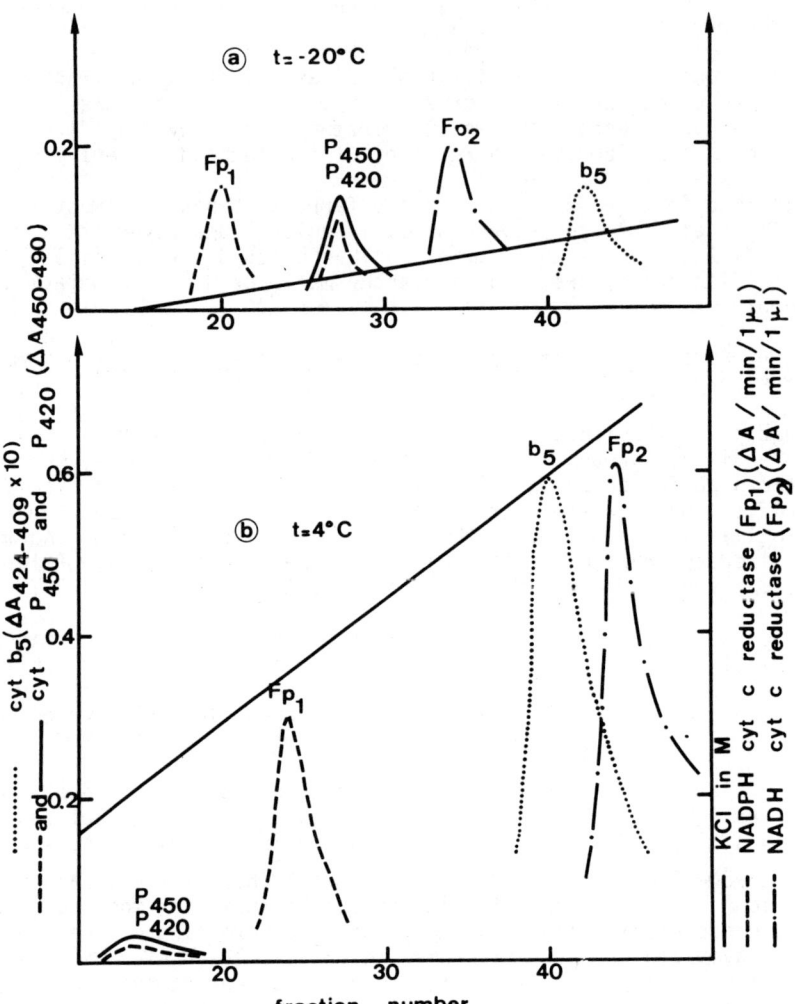

Fig. 4. Elution pattern after chromatography of the microsomal multi-enzyme system, (a) at -20°C in a 1:1 (v/v) mixture of ethylene glycol and buffer, (b) at +4°C in aqueous buffer.

In this case it was used to purify the camphor-bound and camphor-free oxyferrous compounds of bacterial cytochrome P-450. The $Fe^{2+}_S - O_2$ compound was prepared in the presence of excess camphor at -20°C, where it is almost indefinitely stable. It was next separated by chromatography on LH 20 gel at -20°C from the exogenous photochemical reducing system (proflavine + EDTA), and from bound camphor with an elution solvent free from this substrate. All the tubes contained 100 % of pure complex with neither ferric or ferrous cytochrome, nor any contaminants (proflavine or even H_2O_2 produced during its reoxidation). In the elution peak tube the cytochrome concentration was only 3 times lower than in the starting solution, thus providing stock solutions of pure and concentrated compound as reactants for further experiments.

CONCLUSION

It has been shown through several examples that use of sub-zero temperatures, inhibiting some reactions and stabilizing intermediates, provides suitable conditions for an analysis at the molecular level of individual enzyme-catalyzed reactions of multi-components systems.

Other interesting applications of the procedure may be set up ; for instance study of the membrane fluidity through comparative kinetic recordings of membrane-bound and solubilized enzymes, including cytochrome P-450 in its "native" conformation, now under investigation in this Laboratory.

ACKNOWLEDGEMENTS

The authors are greatly indebted to Mrs. Evelyne Begard for her expert technical assistance.

REFERENCES

(1) P. Douzou, Methods of Biochemical Analysis 22, Wiley, New York, 1974.

(2) P. Douzou, Low temperature biochemistry, Trends in Biochem. Sciences 1, 25, 1976.

(3) P. Debey, C. Balny, and P. Douzou, Enzyme assay in microsomes below zero degrees, Proc. Natl. Acad. Sci. USA 70, 2633, 1973.

(4) P. Debey, C. Balny, and P. Douzou, Low temperature studies of microsomal cytochrome P-450 : Release of oxidizing species, FEBS Letters 46, 75, 1974.

(5) R.W. Estabrook, A.G. Hildebrandt, J. Baron, K.J. Netter, and K. Liebman, A new spectral intermediate associated with cytochrome P-450 function in liver microsomes, Biochem. Biophys. Res. Commun. 42, 132, 1971.

(6) F.P. Guengerich, D.P. Ballou, and M.J. Coon, Spectral intermediates in the reaction of oxygen with purified liver microsomal cytochrome P-450, Biochem. Biophys Res. Commun. 70, 951, 1976

(7) P. Debey, E. Begard, and I.C. Gunsalus, Univalent electron transfers from putidaredoxin to bacterial cytochrome P-450 studied at sub-zero temperature : II) Decomposition of the oxy-ferrous cytochrome by reduced putidaredoxin, <u>Biochemistry</u> submitted, 1976.

(8) I.C. Gunsalus, J.R. Meeks, J.D. Lipscomb, P. Debrunner, and E. Münck, <u>Molecular Mechanisms of Oxygen Activation</u>, Academic Press, New York and London, 1974.

(9) B. Chance, N. Graham, and V. Legallais, Low temperature trapping method for cytochrome oxidase oxygen intermediates, <u>Anal. Biochem.</u> 67, 552, 1975.

(10) C. Balny, C. Le Peuch, and P. Debey, Low temperature column chromatography : Application to microsomal hydroxylating system, <u>Anal. Biochem.</u> 63, 321, 1975.

(11) P. Debey, C. Balny, and P. Douzou, The sub-zero temperature chromatographic isolation of transient intermediates of a multi-step cycle : purification of the substrate-bound oxy-ferrous cytochrome P-450, <u>FEBS Letters</u>, submitted, 1976.

(12) A.Y.H. Lu, and M.J. Coon, Role of hemoprotein P-450 in fatty acid co-hydroxylation in a soluble enzyme system from liver microsomes, <u>J. Biol. Chem.</u> 243, 1331, 1968.

ISOLATION OF MICROSOMAL ELECTRON TRANSPORT COMPONENTS AND RECONSTITUTION OF THE ELECTRON TRANSFER ACTIVITY ON LIPOSOMAL VESICLES

W. Duppel, J. Poensgen, V. Ullrich[*] and G. Dahl[†]

[*] Department of Physiological Chemistry, [†] Department of Physiology I, University of the Saarland, Hamburg/Saar, G.F.R.

ABSTRACT

The four components of the microsomal electron transport system, cytochrome P450, cytochrome b_5, NADPH cytochrome P450 reductase and NADH cytochrome b_5 reductase have been purified from a single microsomal preparation of phenobarbital-pretreated rats. After solubilization of the microsomes with the nonionic detergent Renex 690 the components were separated on DEAE-Sephadex and further purified individually.
Liposomal vesicles were prepared by sonication of lipids, extracted from microsomes by organic solvents. Liposomes, sonicated for 15 min, were incubated with cytochrome b_5 or NADPH cytochrome P450 reductase or both, and the protein-loaded vesicles were separated from the free components by gel-filtration on Ultrogel AcA 22. Free cytochrome b_5 was reduced only very slowly by NADPH in the presence of NADPH cytochrome P450 reductase, while the addition of NADPH to liposomes loaded with cytochrome b_5 and the reductase resulted in a very rapid reduction of the cytochrome. O-dealkylation activity of 7-ethoxycoumarin could be restored by incubation of cytochrome P450 with liposome-bound reductase, while free detergent - or trypsin-solubilized reductase only showed negligible activities under these conditions. These experiments show the importance of a proper membrane environment for the microsomal electron transport.

INTRODUCTION

The pioneering work in Coon's laboratory showed a specific phospholipid requirement for the reconstitution of monooxygenation activity with detergent-solubilized microsomal electron transport components (1,2). Autor et al. showed that no formation of aggregates or membrane-like structures occurred when phospholipids were added to the solubilized electron transport components in order to reconstitute hydroxylation activity (3).
It has been shown by Strittmatter and coworkers for cytochrome b_5 and NADH cytochrome b_5 reductase (4,5) that these proteins consist of a hydrophilic and a hydrophobic part, the latter anchoring the molecule in the membrane. A similar structure has been proposed for the NADPH cytochrome P450 reductase by Welton et al. (6). Therefore these proteins are easily incorporated into phospholipid membranes and it has been shown for the cytochrome b_5 system, that the electron transport is considerably faster if both components are anchored in the membrane (7). Since these proteins belong to the membraneous system of the endoplasmic reticulum, kinetic data from experiments with membrane-bound proteins should be more physiologically relevant than those from the solubilized components. In the present study we isolated the four microsomal electron transport components by means of a nonionic detergent and investigated the effect of the membrane on the electron transfer between various components.

METHODS

Male Sprague-Dawley rats weighing 100-150 g were pretreated with phenobarbital for three days and microsomes were prepared from the perfused livers. The microsomal pellets were solubilized by a slightly modified procedure of a purification method for NADPH cytochrome P450 reductase described by Dignam and Strobel (8). The subsequent chromatography of the 100 000 xg supernatant on DEAE-Sephadex A25 yielded three well separated protein peaks which were further purified individually. The first peak contained cytochrome P450 as well as NADH cytochrome b_5 reductase, which was separated from cytochrome P450 by rechromatography on a hydroxyl apatite column as described by Kamataki et al. (9). The cytochrome P450 fraction was further treated with 15 % polyethylene glycol, calcium phosphate gel, and - to remove excess of detergent- with Bio-Beads SM2 prior to the reconstitution with the other components. The final preparation contained up to 12.5 nmol P450 per mg of protein, was free of cytochrome b_5, and NADPH cytochrome P450 reductase, but still contained some detergent.

Cytochrome b_5 was further purified to homogeneity by procedures described by Spatz and Strittmatter (4). For further purification the NADPH cytochrome P450 reductase was treated according to the method of Dignam and Strobel (8) to yield a specific activity of more than 20 μmol of cytochrome c reduced per mg and min with an additional treatment of Bio-Beads SM2 prior to the incorporation into liposomes.

Lipids were prepared from freeze-dried microsomes by extraction with chloroform/methanol (2:1) and treated as described by Folch et al. (10). For the preparation of liposomes, aliquots of the final solution were dried under a stream of nitrogen and sonicated in 0.1 M Tris-buffer pH 7.7 for 15 min at $0^{\circ}C$.

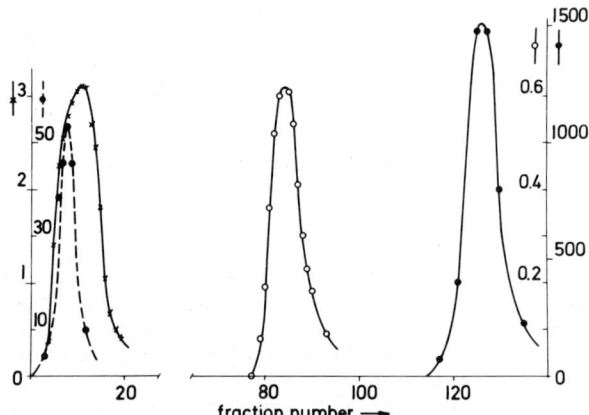

Fig. 1 Separation of microsomal components on DEAE-Sephadex A25

The 100 000 xg supernatant of Renex-solubilized and protamine sulfate-treated microsomes (1000 mg of protein) was loaded onto a 2.5 x 40 cm DEAE Sephadex A25 column, equilibrated with 0.1 M Tris buffer pH 7.7 containing 0.12 % Renex, 20 % glycerol, 10^{-4}M dithiothreitol and 10^{-3}M EDTA. The first peak was eluted by washing with the equilibration buffer. Starting at fraction 30 a linear gradient from 0 to 0.4 M KCl was used to elute the other proteins.
— • — nmol cytochrome c reduced per min and ml
-- • -- μmol ferricyanid reduced per min and ml
— o — $A_{413 \ nm}$ (cytochrome b_5)
— x — $A_{416 \ nm}$ (cytochrome P450)

Liposomes were incubated with NADPH cytochrome P450 reductase or with the reductase and cytochrome b_5 for 30 min at 30°C and subjected to gel chromatography on an Ultrogel AcA 22 column. The reduction of cytochrome b_5 was monitored by the increase of absorbance at 423 nm, and the cytochrome c reductase activity was determined as described by Omura and Takesue (11). O-dealkylation activity of 7-ethoxycoumarin was measured as described by Ullrich and Weber (12).

RESULTS AND DISCUSSION

When the solubilized microsomes were applied to a DEAE Sephadex A25 column, three distinct well separated protein peaks were obtained as shown in Fig. 1. The first one contained cytochrome P450 and the NADH-cytochrome b_5 reductase, which were eluted with the equilibration buffer. Cytochrome b_5 was eluted from the column at the beginning of a linear KCl gradient and at the end of this gradient a sharp peak of NADPH cytochrome P450 reductase emerged. The fractions obtained were further purified separately by procedures given under "Methods". Before incubation with liposomes the purified proteins were treated with Bio-Beads SM2 in order to further decrease their detergent content. The liposomes used for incubation were sonicated for 15 min and consisted mainly of single-shelled vesicles with a diameter of 200-500 Å, while liposomes only shaken with buffer for 5 min were multi-layered and had a diameter of 3000-6000 Å (Fig. 2 a,b).

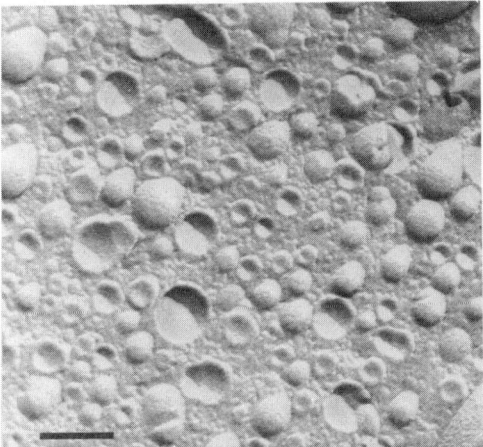 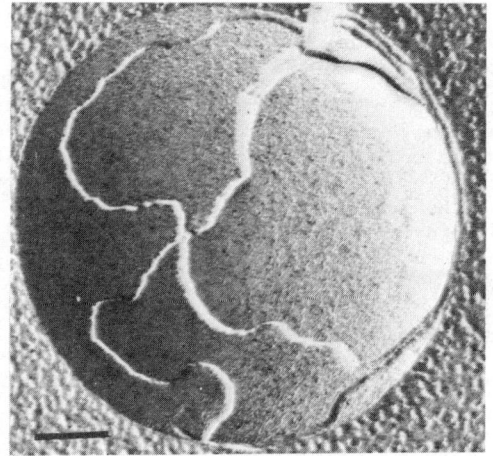

Fig. 2 Freeze-fractured liposomal vesicles

Liposomal vesicles were prepared from microsomes as described under "Methods"
left: Electron micrograph of freeze-fractured vesicles sonicated for 15 min in 0.1 M Tris buffer.
right: same, but shaken for 5 min. Magnification 120 000 x. Bar represents 1000 Å.

As known from experiments in Strittmatter's laboratory the detergent-solubilized amphipatic molecule of cytochrome b_5 is readily incorporated into liposomal vesicles whereas trypsin-solubilized cytochrome b_5 is not bound (13). Since a similar amphipatic structure has been proposed for the NADPH cytochrome P450 reductase by Welton et al. (6) and it is known from experiments in Coon's laboratory that only reductase solubilized by detergents from the microsomal membrane -but not by proteolytic digestion -is able to support hydroxylation activity (1), we investigated whether the binding of these two forms of reductase to liposomes is different.

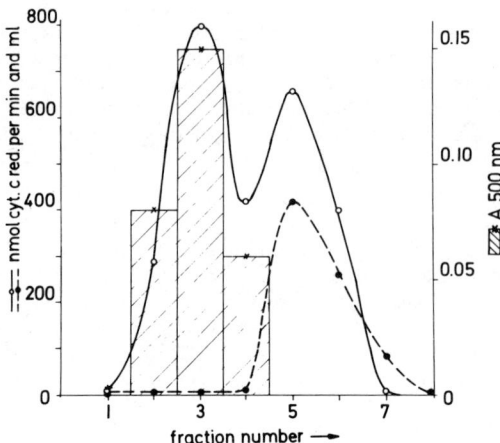

Fig. 3 Separation of liposome-bound and free NADPH cytochrome P450 reductase on Ultrogel AcA 22

Reductase (0.7 unit) was incubated with 200 µl of liposomes (30 mg/ml) for 30 min at 30°C and the free and liposome-bound proteins were separated on a 0.5 x 15 cm Ultrogel AcA 22 column (solid line). Trypsin-treated reductase (0.2 unit) was treated in the same manner (dashed line).

As seen in Fig. 3 the two reductase preparations incubated with liposomes behave differently when separated on Ultrogel AcA 22. Renex-solubilized reductase is bound to liposomes and eluted in the void volume of the column, whereas trypsin-treated reductase is not bound to liposomes and is eluted later, close to the fractions of unbound reductase.

Free detergent-solubilized reductase or trypsin-treated reductase, when incubated with cytochrome P450 supported only very low O-dealkylation activity, while liposome-bound reductase was able to support an even higher monoxygenation activity under the same conditions than microsomes from the same preparation (Table 1).

TABLE 1 Rate of 7-Ethoxycoumarin Dealkylation

Microsomes and Different Reconstituted Systems	nmol Umbelliferone formed per nmol Cyt. P450 and min.
Microsomes (PB-induced)	1.3
Lip. - Red. + P450	2.9
Deterg.-solub. Red. + P450	0.06
Trypsin-treated Red. + P450	0.03

Liposome-bound reductase (0.38 units) as well as free detergent-solubilized and trypsin-treated reductase (0.32 and 0.4 units, respectively) were incubated with 1.6 nmol cytochrome P450 for 30 min at 30°C and the rate of 7-ethoxycoumarin dealkylation was determined after the addition of a known amount of umbelliferone as fluorescence standard. The rate of microsomal dealkylation activity of the same preparation is included for comparison.

Since the electron transfer from NADH via NADH cytochrome b_5 reductase to cytochrome b_5 is very rapid under optimal conditions and has been examined in more detail in Strittmatter's laboratory already (7), we investigated the effect of membrane-bound NADPH cytochrome P450 reductase on the electron transport from NADPH to membrane-bound cytochrome b_5.

When detergent-solubilized NADPH cytochrome P450 reductase and detergent-solubilized cytochrome b_5 were incubated together with liposomes, liposome-bound fractions of the proteins could be separated from unbound fractions by gel-chromatography on Ultrogel AcA 22 as shown in Fig. 4.

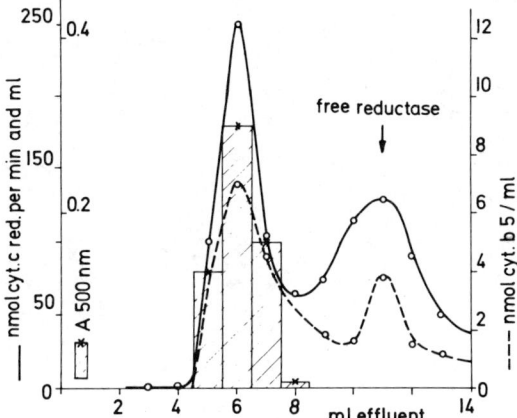

Fig. 4 Separation of free and liposome-bound enzymes on Ultrogel AcA 22

NADPH cytochrome P450 reductase (1.6 units), 50 nmol of cytochrome b_5 and 4 mg of liposomes were incubated for 30 min at 30°C. A 1-ml sample was passed through a 1 x 20 cm Ultrogel AcA 22 column. The fraction in which free reductase was eluted from the same column is indicated by an arrow.

The reduction of cytochrome b_5 by NADPH occurs rapidly if the cytochrome and the reductase are bound simultaneously to the membrane, whereas it is very slow if only one of the proteins is incorporated. If the membrane structure is destroyed by the addition of deoxycholate to such a preparation in a final concentration of 0.2 %, the rapid reduction of cytochrome b_5 is abolished. Fig. 5 shows that free detergent-solubilized cytochrome b_5 is reduced only very slowly by liposome-bound reductase, whereas part of the liposome-bound cytochrome b_5 is reduced extremely rapidly. Stopped-flow experiments which are currently in progress are needed to show the actual velocity of this reaction, which is too fast to be measured by conventional methods.

Further experiments with different reductase concentrations bound to liposomes indicated that only those cytochrome b_5 molecules are reduced rapidly, which are bound to a liposomal vesicle also containing a reductase molecule. With increasing reductase concentrations more vesicles carry both reductase and cytochrome b_5, and a higher percentage of cytochrome b_5 is reduced rapidly (Table 2). This explanation is supported by the fact that a mixture of two populations of liposomes - one with cytochrome b_5 bound and one with the reductase bound to them - only shows a slow reduction of cytochrome b_5.

More quantitative experiments are needed to show whether one reductase molecule on a liposome is able to reduce only one cytochrome b_5 molecule, or a certain number (defined complex), or all cytochrome b_5 molecules on one liposomal vesicle (free lateral diffusion). The latter model has been proposed for the NADH-mediated cytochrome b_5 reduction by Rogers and Strittmatter (14) and

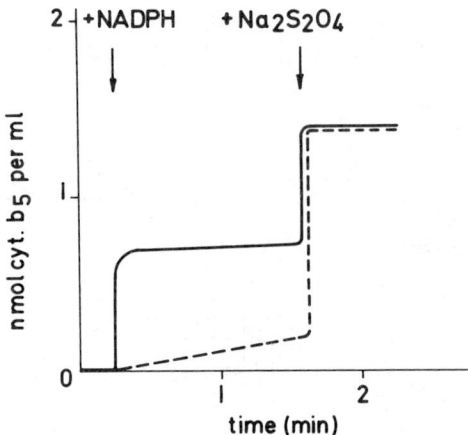

Fig. 5 Reduction of free and liposome-bound cytochrome b_5 by liposome-bound NADPH cytochrome P450 reductase

The reduction was started by the addition of NADPH and followed by the change of absorbance at 423 nm. Complete reduction was achieved by the addition of solid sodium dithionite. Solid line: reductase and cytochrome b_5 bound to liposomes; dashed line: liposome-bound reductase acting on free cytochrome b_5.

for the NADPH catalyzed cytochrome P450 dependent monooxygenation by Yang (15), while a rigid model with defined complexes of reductase and cytochrome P450 has been proposed by Franklin et al. (16).

TABLE 2 Correlation of Liposomal Reductase Concentration with Percentage of Rapidly Reduced Cytochrome b_5

Reductase Incubated per mg Lipid (mU)	Reductase Bound per mg Lipid (mU)	Cytochrome b_5 Reduced Rapidly (%)
58	15	10
192	59	26
350	75	30
625	220	53

Increasing amounts of NADPH cytochrome P450 reductase were incubated with fixed amounts of cytochrome b_5 and liposomes. Free and liposome-bound proteins were separated on Ultrogel AcA 22. The reduction of liposome-bound cytochrome b_5 was initiated by the addition of NADPH, and after complete reduction by dithionite the percentage of rapidly reduced cytochrome b_5 was calculated.

The arrangement of the proteins in the membrane needs more detailed investigations, but our experiments show that binding of the microsomal electron transfer components to the membrane occurs easily and that the electron transport is very much facilitated if the components are incorporated in a defined manner in the membrane bilayer.

ACKNOWLEDGEMENTS

This work was supported by the Sonderforschungsbereich 38 "Membranforschung" of the Deutsche Forschungsgemeinschaft.

REFERENCES

1) A.Y.H. Lu, K.W. Junk and M.J. Coon, Resolution of the cytochrome P450-containing w-hydroxylation system of liver microsomes into three components, J. Biol. Chem. 244, 3714 (1969).
2) H.W. Strobel, A.Y. H. Lu, J. Heidema and M.J. Coon, Phosphatidylcholine requirement in the enzymatic reduction of hemoprotein P450 and in fatty acid, hydrocarbon and drug hydroxylation, J. Biol. Chem. 245, 4851 (1970).
3) A.P. Autor, R.M. Kaschnitz, J.K. Heidema and M.J. Coon, Sedimentation and other properties of the reconstituted liver microsomal mixed-function oxidase system containing cytochrome P450, reduced triphosphopyridine nucleotide-cytochrome P450 reductase, and phosphatidylcholine, Mol. Pharmacol. 9, 93 (1973).
4) L. Spatz and P. Strittmatter, A form of cytochrome b_5 that contains an additional hydrophobic sequence of 40 amino acid residues, Proc. Nat. Acad. Sci. USA 68, 1042 (1971).
5) L. Spatz and P. Strittmatter, A form of reduced nicotinamide adenine dinucleotide-cytochrome b_5 reductase containing both the catalytic site and an additional hydrophobic membrane binding segment, J. Biol. Chem. 248, 793 (1973).
6) A.F. Welton, T.C. Pederson, J.A. Buege and S.D. Aust, The molecular weight of NADPH-cytochrome c reductase isolated by immunoprecipitation from detergent-solubilized rat liver microsomes, Biochem. Biophys. Res. Commun. 54, 161 (1973).
7) M.J. Rogers and P. Strittmatter, Lipid-protein interactions in the reconstitution of the microsomal reduced nicotinamide adenine dinucleotide-cytochrome b_5 reductase, J. Biol. Chem. 248, 800 (1973).
8) J.D. Dignam and H.W. Strobel, Preparation of homogeneous NADPH cytochrome P450 reductase from rat liver, Biochem. Biophys. Res. Commun. 63, 845 (1975).
9) T. Kamataki, M.C.M. Lee Lin, D.H. Belcher and R.A. Neal, Studies on the metabolism of parathion with an apparently homogeneous preparation of rabbit liver cytochrome P450, Drug Metab. Disposition 4, 180 (1976).
10) J. Folch, M. Lees and G.H. Sloan-Stanley, A simple method for the isolation and purification of total lipids from animal tissues, J. Biol. Chem. 226, 497 (1957).
11) T. Omura and S. Takesue, A new method for simultaneous purification of cytochrome b_5 and NADPH cytochrome c reductase from rat liver microsomes, J. Biochem. (Tokyo) 67, 249 (1970).
12) V. Ullrich and P. Weber, The O-dealkylation of 7-ethoxycoumarin by liver microsomes, Hoppe-Seyler's Z. Physiol. Chem. 353, 1171 (1972).
13) P. Strittmatter, M.J. Rogers and L. Spatz, The binding of cytochrome b_5 to liver microsomes, J. Biol. Chem. 247, 7188 (1972).
14) M.J. Rogers and P. Strittmatter, Evidence for random distribution and translational movement of cytochrome b_5 in endoplasmic reticulum, J. Biol. Chem. 249, 895 (1974).

15) C.S. Yang, The association between cytochrome P450 and NADPH-cytochrome P450 reductase in microsomal membrane,
FEBS Lett. 54, 61 (1975).
16) M.R. Franklin and R.W. Estabrook, On the inhibitory action of mersalyl on microsomal drug oxidation: A rigid organization of the electron transport chain,
Arch. Biochem. Biophys. 143, 319 (1971).

DIFFERENCES IN SUBSTRATE AFFINITY AND CATALYTIC SPECIFICITY OF INDUCIBLE AND NON-INDUCIBLE FORMS OF CYTOCHROME P-450 IN RABBIT LIVER

Tilman Fischbach and Werner Lenk

Pharmakologisches Institut der Universität München

Reports on substrate affinity and catalytic specificity of inducible and non-inducible monooxygenase activity in rabbit liver are scanty, and this may be the reason why recently published results on 3-MCh-inducible monooxygenase activity in rabbit liver microsomes (1, 2) have been interpreted as either poor catalytic activity of cytochrome P-448 to all substrates or as specific catalytic activity to some substrates which are not yet known (2).

In a study on the dependence of the specific ω-, (ω-1)-, (ω-2)-, and (ω-3)-hydroxylation activity of rabbit liver microsomes on inducers, like PB or 3-MCh, alterations in the rates of the specific oxygenations were observed due to the induction process. Some specific oxygenations, however, were not altered, indicative of the participation of non-inducible forms of cytochrome P-450 in the aforementioned specific oxygenation. As substrates served the 4-chloroanilides of the lower fatty acids (acetic (1a), propionic (1b), butyric (1c), and valeric acid (1d)), and phenacetin whose N-hydroxylation was of particular interest.

$$\text{I} \quad \begin{array}{l} \text{a: R} = CH_3- \\ \text{b: R} = CH_3-CH_2- \\ \text{c: R} = CH_3-CH_2-CH_2- \\ \text{d: R} = CH_3-CH_2-CH_2-CH_2- \end{array}$$

The results of these experiments are shown in Fig. 1.

PB-inducible form(s) of cytochrome P-450 did not participate in the ω-hydroxylation of the acetic moiety, probably it does not meet the steric requirements for this oxygenase, but it did hydroxylate the propionic residue in ω-position, leaving the (ω-1)-carbon atom unaffected. The butyric and valeric residues were hydroxylated in ω- as well as in (ω-1)-position, indicative of an alteration in the catalytic specificity of PB-inducible form(s) of cytochrome P-450 with increasing chain length.

3-MCh-inducible form(s) of cytochrome P-450, on the other hand, catalyzed

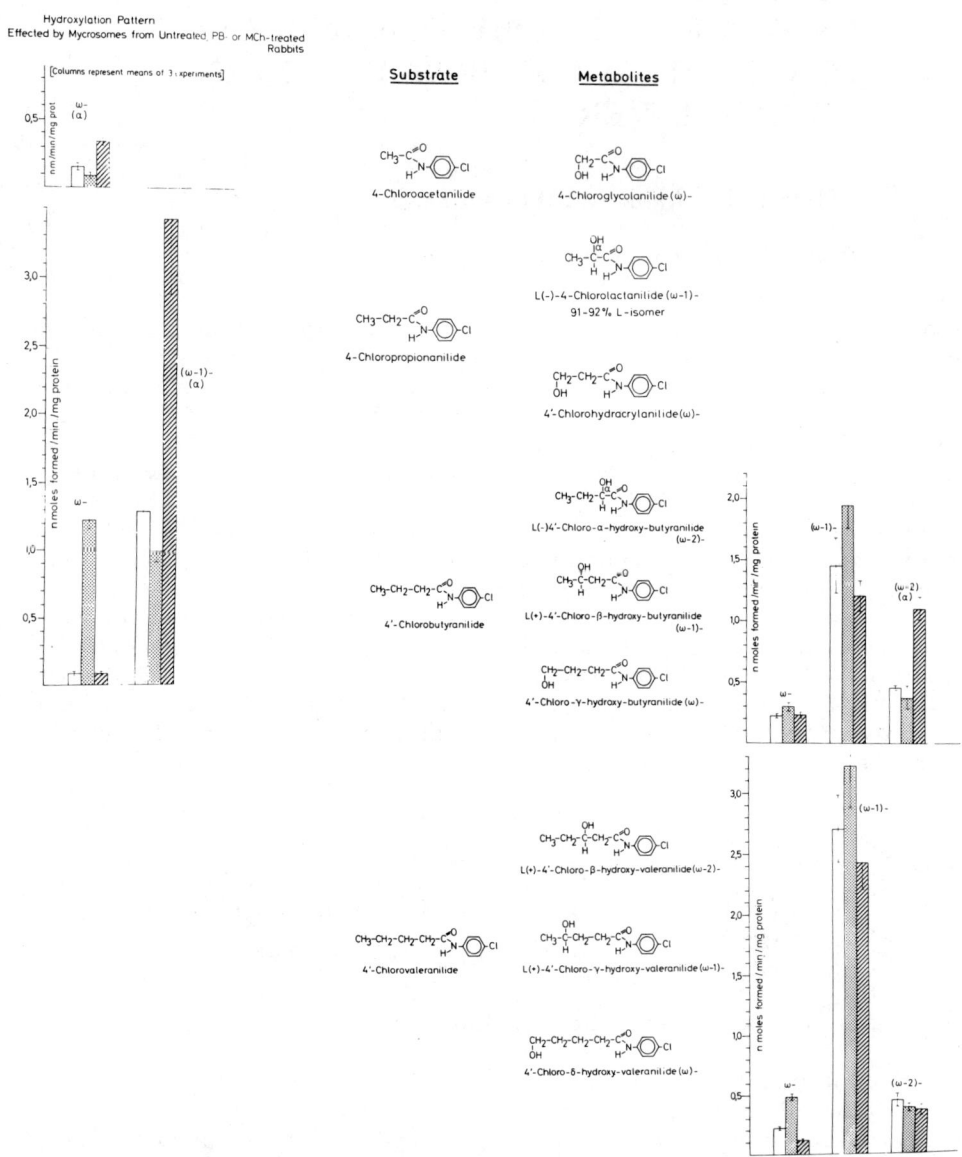

Fig. 1. Alterations in the hydroxylation pattern of the acetic, propionic, butyric, and valeric acid residues due to induction of rabbits with either phenobarbital or 3-methylcholanthrene. Columns represent means of 3 experiments (S. E. indicated by bars) and mean (from left to right: specific activity of untreated, PB-, and 3-MCh-stimulated rabbit liver microsomes.

the ω-hydroxylation of the acetic residue, a position α to the carboxamide moiety, also the (ω-1)-hydroxylation of the propionic residue, a position α to the carboxamide moiety, leaving the ω-carbon atom unaffected, and it catalyzed the (ω-2)-hydroxylation of the butyric residue, also a position α to the carboxamide group. 3-MCh-inducible cytochrome P-448 did not catalyze the (ω-2)- or (ω-3)-hydroxylation of the valeric residie, positions ß or α to the carboxamide group. The rate of hydroxylation of the carbon atom in (ω-3)-position by normal microsomes, PB- or 3-MCh-stimulated microsomes was too low to enable the determination of the amount of this specific hydroxylation product.

These results show that both the PB-inducible as well as the 3-MCh-inducible form(s) of cytochrome P-450 display well defined catalytic properties and that form(s) of cytochrome P-450 are involved in the oxygenation of aliphatic carbon atoms whose catalytic activities could not be enhanced by induction with either PB or 3-MCh. Differences in catalytic specificity of both forms are depicted in Fig. 2.

The slight decrease in the specific (ω-2)-hydroxylation activity of (Id) due to the induction with either PB or 3-MCh can be explained with an increase in the amount of the PB-inducible or the 3-MCh-inducible oxygenase at the expense of the non-inducible form(s) of cytochrome P-450 during the induction process, catalyzing the (ω-2)-hydroxylation of the valeric residue.

Figure 3 shows part of the results of experiments with phenacetin as substrate. The similarity of the hydroxylation pattern of the ethyl moiety in phenacetin with that integrated in 4-chloropropionanilide (Ib) suggest that the residues which substitute the ethyl group, obviously do not affect the mechanism of enzymic hydroxylation, but determine the affinity to the oxygenases. A similar, although not identical hydroxylation pattern was observed with 7-ethoxycoumarine and rat liver microsomes (3).

It is noteworthy that with (1b) and phenacetin as substrates, specific ω-hydroxylation activity of the ethyl moiety was low, but enhanced 10-14 times by PB-induction, whereas with the butyric and valeric residues whose ω- and (ω-1)-carbon atoms are attacked by the PB-inducible oxygenase(s), specific ω-hydroxylation was low and (ω-1)-hydroxylation was rather high, but both were increased only 1.2 - 2.1 times. Whereas differences in the specific ω- and (ω-1)-hydroxylation activity reflect differences in substrate affinity, differences in the proportional increase due to PB-induction cannot be explained at present.

The hydroxylation pattern of the three different carbon atoms in position α to the carboxamide group is nearly identical: on induction with PB, the specific hydroxylation activity decreased and increased on induction with 3-MCh. Obviously, during the induction with PB, the amount of that form of cytochrome P-450 which does not participate in these hydroxylations was increased at the expense of that which does participate.

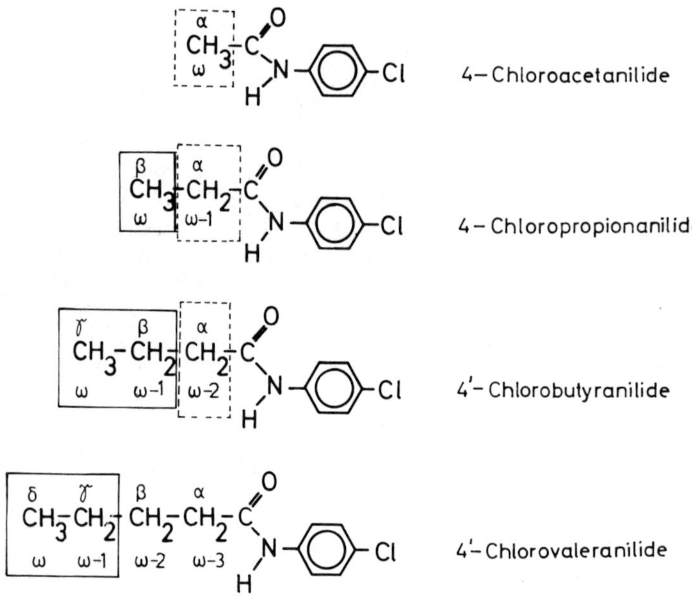

$CH_3-C(=O)-N(H)-C_6H_4-Cl$ 4−Chloroacetanilide
(α/ω position on CH₃)

$CH_3-CH_2-C(=O)-N(H)-C_6H_4-Cl$ 4−Chloropropionanilide
(β/ω on CH₃, α/ω-1 on CH₂)

$CH_3-CH_2-CH_2-C(=O)-N(H)-C_6H_4-Cl$ 4'−Chlorobutyranilide
(γ/ω, β/ω-1, α/ω-2)

$CH_3-CH_2-CH_2-CH_2-C(=O)-N(H)-C_6H_4-Cl$ 4'−Chlorovaleranilide
(δ/ω, γ/ω-1, β/ω-2, α/ω-3)

Fig. 2. Catalytic specificity of PB- or 3-MCh-inducible form(s) of cytochrome P-450 in rabbit liver microsomes in respect to ω-, (ω-1)-, and (ω-3)-hydroxylation of aliphatic C-atoms. Broken lines surround those C-atoms attacked by 3-MCh-inducible form(s), solid lines surround those C-atoms attacked by PB-inducible form(s) of cytochrome P-450.

It can also be seen from the graphs that in those cases where induction with PB caused an increase in ω- or (ω-1)-hydroxylation activity, treatment of the animals with 3-MCh caused a decrease in the same specific activities. Obviously, during the induction with 3-MCh, the amount of those form(s) of cytochrome P-450 not being involved in these specific oxygenations enhanced at the expense of those participating in these specific oxygenations. Reports by Kremers and Gielen (this volume) on alterations of the rate of 16α-hydroxylation of pregnenolone, testosterone, and progesterone due to induction of rats with either PB or 3-MCh reveal the same mechanism for the alteration of enzyme activities during induction. A report on the same subject concerning ω-hydroxylation of the acetic residue and the 9-hydroxylation of N-(2-fluorenyl)-acetamide by rabbit or guinea pig liver microsomes and its N-hydroxylation by rabbit liver microsomes has recently appeared (4). This is in accordance with the findings of Welton and Aust (5) who observed an increase in the amount of the hemoproteins 4 and 5 and a decrease in hemoproteins 2 and 3 on inducing rats with PB.

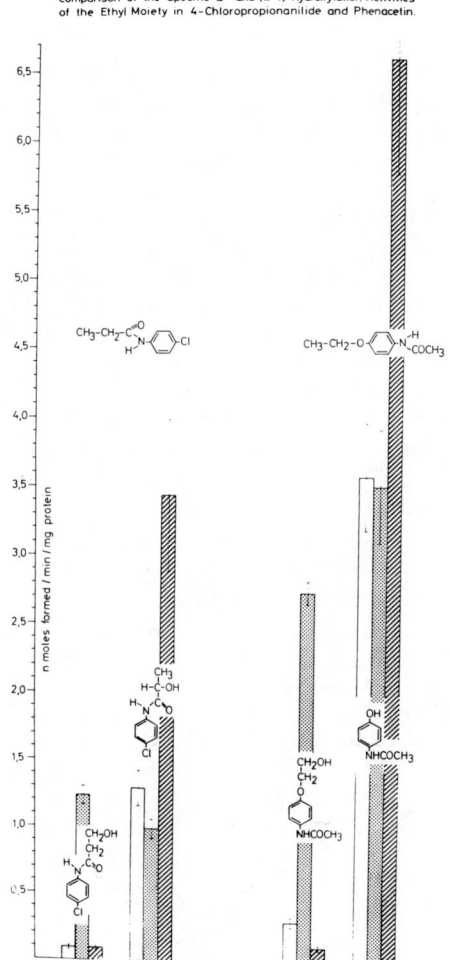

Fig. 3. Similarities in the hydroxylation pattern of the ethyl moiety in 4-chloropropionanilide (1b) and phenacetin.
For experimental details see caption of Fig. 1.

These results, although differing in details from those reported by earlier investigators (6-13) (vide infra), add to the current knowledge of the mechanism of ω- and (ω-1)-hydroxylation and help to understand the results obtained with substrates of medium and long chain length.

Different results were obtained when the effect of inducers (PB, B[a]P or 3-MCh) on the specific ω-, (ω-1)-, (ω-2)-, and (ω-3)-hydroxylation activity of carbon atoms in different positions of saturated chains was investigated. Substrates employed in these investigations were (in the order of decreasing chain length): n-octadecanoic, n-hexadecanoic, and n-dodecanoic acid, n-heptane, n-hexane, n-pentane, and n-propyl-p-nitrophenyl-ether (=PNPE). As the enzyme source served liver microsomes from untreated, PB-, B[a]P- or 3-MCh-stimulated rats, mice, guinea pigs or rabbits. Increased ω-hydroxylation following induction of either rats or rabbits with PB was observed only by Lu et al. (7) with n-dodecanoic acid as substrate and by Mitoma et al. (9) with PNPE as substrate and PB-stimulated guinea pig microsomes as the enzyme source. However, Björkhem and Danielson (8) found that PB-induction of rats did not affect the rate of ω-hydroxylation of n-dodecanoic acid.

Treatment of rats with 3-MCh or B[a]P caused a decrease in ω-hydroxylation activity with n-octadecanoic acid (6) or n-heptane (10) and did not affect ω-hydroxylation of n-hexane (11) or PNPE (9). Inducing mice with PB decreased ω-hydroxylation with n-hexane, induced guinea pigs ω-hydroxylated PNPE much slower.

Instead, PB-treatment of rats, mice or guinea pigs caused an increase in (ω-1)-hydroxylation observed with n-octadecanoic, n-hexadecanoic, and n-dodecanoic acid (8), as well with n-heptane (10), n-hexane (11), and PNPE (9); in contrast, (ω-1)-hydroxylation of n-pentane by PB-stimulated mouse liver microsomes was not significantly increased.

Treatment of rats with either 3-MCh or B[a]P did not affect (ω-1)-hydroxylation of n-heptane (10), n-hexane (11) or PNPE (9), whereas 3-MCh-stimulated guinea pig liver microsomes hydroxylated PNPE at a much higher rate in (ω-1)-position (9).

Hydroxylation of the (ω-2)-carbon atom in n-heptane (10), n-hexane (11) or PNPE (9) was found to increase after treatment of rats with either 3-MCh or B[a]P, whereas 3-MCh-stimulated guinea pig liver microsomes hydroxylated PNPE in (ω-2)-position with a much lower rate (9).

Hydroxylation rates of the (ω-3)-carbon atom of n-heptane effected by rat liver microsomes, although distinctly lower than those of the (ω-2)-carbon atom, increased much more on treatment with B[a]P than with PB (10).

REFERENCES

(1) S. A. Atlas, S. S. Thorgeirsson, A. R. Boobis, K. Kumaki, and D. W. Nebert, Differential induction of murine Ah locus-associated monooxygenase activities in rabbit liver and kidney, Biochem. Pharmac. 24, 2111 (1975).

(2) J. C. Kawalek and A. Y. H. Lu, Reconstituted liver microsomal enzyme system that hydroxylates drugs, other foreign compounds, and endogenous substrates, Mol. Pharmacol. 11, 201 (1975).

(3) V. Ullrich, U. Frommer, and P. Weber, Characterisation of cytochrome P-450 species in rat liver microsomes, I: Differences in the O-dealkylation of 7-ethoxycoumarin after pretreatment with phenobarbital and 3-methylcholanthrene, Hoppe-Seyler's Z. Physiol. Chem. 354, 514 (1973).

(4) K. Benkert, W. Fries, M. Kiese, and W. Lenk, N-(9-Hydroxy-9H-fluoren-2-yl)-acetamide and N-(9-oxo-9H-fluoren-2yl)-acetamide: Metabolites of N-(9H-fluoren-2yl)-acetamide, Biochem. Pharmac. 24, 1375 (1975).

(5) A. F. Welton and S. D. Aust, Multiplicity of cytochrome P_{450} hemoproteins in rat liver microsomes, Biochem. Biophys. Res. Commun. 56, 898 (1974).

(6) F. Wada, H. Shibata, M. Goto, and K. Sakamoto, Participation of the microsomal electron transport system involving cytochrome P-450 in ω-oxidation of fatty acids, Biochim. Biophys. Acta 162, 518 (1968).

(7) A. Y. H. Lu, K. W. Jung, and M. J. Coon, Resolution of the cytochrome P-450 containing ω-hydroxylation system of liver microsomes into three components, J. Biol. Chem. 244, 3714 (1969).

(8) I. Björkhem and H. Danielsson, ω- And (ω-1)-oxidation of fatty acids by rat liver microsomes, Eur. J. Biochem. 17, 450 (1970).

(9) C. Mitoma, R. L. Dehn, and M. Tanabe, In vitro metabolic studies on propyl p-nitrophenyl ether, Biochim. Biophys. Acta 237, 21 (1971).

(10) U. Frommer, V. Ullrich, Hj. Staudinger, and S. Orrenius, The monooxygenation of n-heptane by rat liver microsomes, Biochim. Biophys. Acta 280, 487 (1972).

(11) U. Frommer, V. Ullrich, and S. Orrenius, Influence of inducers and inhibitors on the hydroxylation pattern of n-hexane in rat liver microsomes, FEBS Lett. 41, 14 (1974).

(12) A. Krämer, Hj. Staudinger, and V. Ullrich, Effect of n-hexane inhalation on the monooxygenase system in mice liver microsomes, Chem.-Biol. Interactions 8, 11 (1974).

(13) U. Frommer, V. Ullrich, and Hj. Staudinger, Hydroxylation of aliphatic compounds by liver microsomes, I: The distribution pattern of isomeric alcohols, Hoppe-Seyler's Z. Physiol. Chem. 351, 903 (1970).

EFFECTS OF BUFFER COMPOSITION ON THE LOW-SPIN ESR SPECTRUM OF CYTOCHROME P-450

David H. O'Keeffe, Richard E. Ebel, and
Julian A. Peterson

Department of Biochemistry, The University of Texas Health Science Center at Dallas, Dallas, Texas 75235

ABSTRACT

Liver microsomal cytochrome P-450 (P-450$_{LM}$; EC 1.14.14.1) prepared from phenobarbital pretreated rats has been shown to be predominantly in the low spin state. In this study the effect of buffer composition on the ESR spectrum, i.e. asymmetry in the g_x and g_y signals, has been examined. In potassium phosphate (KPi) or morpholinopropane sulfonate (MOPS) buffer the spectrum of cytochrome P-450$_{LM}$ was symmetric with single g_x and g_y signals. The addition of Tris, sucrose, or Tris-sucrose resulted in varying degrees of asymmetry in these signals. Similar variations were observed with the soluble, bacterial cytochrome P-450$_{cam}$. In each buffer the double integrated intensity of the low-spin spectrum remains constant. Buffer composition seems important with respect to the hemin environment since different low-spin spectra are obtained in different buffers. These data indicate that extreme care must be exercised when interpreting ESR spectra with respect to possible cytochrome P-450 heterogeneity and that conditions for studying substrate interactions must be carefully chosen.

INTRODUCTION

Electron spin resonance (ESR) spectroscopy serves as an important probe of the electronic nature of the cytochrome P-450 hemin iron. For example, the results of a recent ESR spectral study of rat liver microsomal cytochrome P-450 were used to support the existence of at least two forms of the enzyme in the microsomal membrane (1). To further investigate the nature of the electronic environment of the cytochrome P-450 hemin iron, we examined the low-spin ESR spectrum as a function of the source of the enzyme and of the buffer in which it was suspended. The data obtained indicated that buffer composition significantly alters the spectrum; however, given a similar solvent milieu, the spectrum of this hemeprotein from several sources is nearly identical. These results suggest that the previously observed changes in the cytochrome P-450$_{LM}$ ESR spectrum may not be relevant to the question of the existence of multiple forms of cytochrome P-450 in the microsomal membrane.

MATERIALS AND METHODS

Microsomes were prepared as described previously (2) from livers of male Sprague-Dawley rats which received daily intraperitoneal injections of phenobarbital (PB) in saline (80 mg/kg) or 3-methylcholanthrene (3-MC) in corn oil (20 mg/kg) for four days. Microsomes were suspended in 0.1 $\underline{M}$ KPi pH 7.4, 0.1 m$\underline{M}$ EDTA at a protein concentration of approximately 50 mg per ml. The final protein concentration was 20 mg per ml (50-70 µ$\underline{M}$ cytochrome P-450) for cytochrome P-450$_{LM}$ ESR samples. Cytochrome P-450$_{cam}$ was isolated from Pseudomonas putida grown on d-camphor (3). Camphor-free enzyme was prepared from the camphor-bound form by chromatography on a Sephadex G-10 column equilibrated with 50 m$\underline{M}$ MOPS pH 7.0 which had been neutralized with Tris base. Rat hepatocytes were isolated from PB induced animals by the procedure of Zahlten and Stratman (4).

Electron spin resonance (ESR) spectra were recorded with a Varian Model E-4 ESR Spectrometer equipped with a variable temperature probe. A PDP-11 minicomputer (Digital Equipment Corp.) was used to collect and process the ESR spectral data.

The chemicals used were of the highest quality commercially available.

RESULTS AND DISCUSSION
=======================

The low-spin ESR spectrum of cytochrome P-450$_{LM}$ (PB) in buffers of various composition is shown in Fig. 1. The spectrum of the microsomal enzyme in MOPS, and 2-methyl-2-amino-1-propanol (MAP), as well as in KPi (see Fig. 3), is symmetric, i.e. the g_x and g_y signals are singlets, with only slight variability in g-values and half-band widths. The spectrum obtained for cytochrome P-450$_{LM}$ with the microsomal fraction suspended in Tris and Tris-sucrose shows varying amounts of asymmetry, readily apparent in the g_x and g_y signals. We have also observed splitting of these same cytochrome P-450$_{LM}$ signals when microsomes are frozen in sucrose alone or in high concentrations of MAP (1$\underline{M}$). With microsomal cytochrome P-450 the symmetry/asymmetry of the g_x and g_y ESR signals appears to be dependent upon the cation present except in the case of sucrose. Sucrose, in addition to causing asymmetry in its own right, seems to potentiate the effect of Tris.

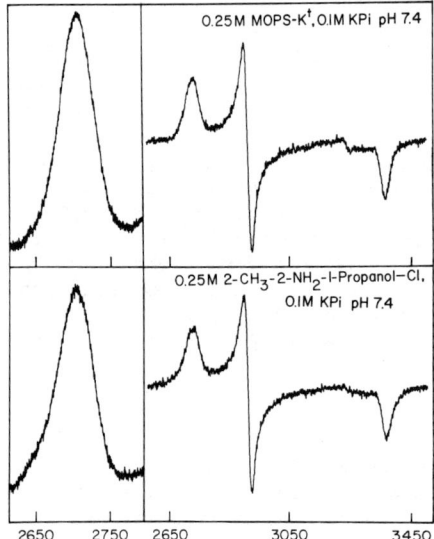

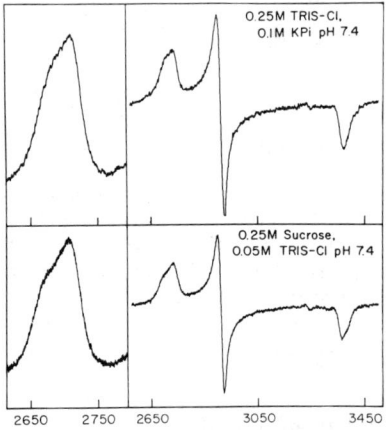

Fig. 1 Low-spin ESR spectra of cytochrome P-450$_{LM}$ (PB); -170°, power 50 mW, modulation amplitude 12.5, frequency 9.16, x-axis (Oe)

Variations in the low-spin ESR spectrum of camphor-free cytochrome P-450$_{cam}$ are shown in Fig. 2. The symmetry/asymmetry in the g_x and g_y signals is again dependent upon buffer composition. The broadening of the g_y and g_y ESR signals when the bacterial enzyme is frozen in sucrose or Tris-phosphate (Fig. 2) resembles that observed for cytochrome P-450$_{LM}$ (Fig. 1); however splitting of these signals is not as readily apparent. When cytochrome P-450$_{cam}$ is frozen in Tris-chloride (in the presence or absence of sucrose) these signals are dramatically narrowed and a concomitant increase in intensity is observed (Fig. 2), suggesting a more uniform environment about the hemin iron. Notice also that when the bacterial enzyme is frozen in Tris (also MAP, not shown) the symmetry characteristics are dependent upon the anion present. For instance, the g_x and g_y signals are sharp singlets in the presence of Tris or MAP neutralized with HCl but are broadened and in certain instances asymmetric with these buffers neutralized with H_3PO_4. The signals are also symmetric when the enzyme is frozen in MOPS or KPi, although in KPi broadening is apparent (Fig. 3).

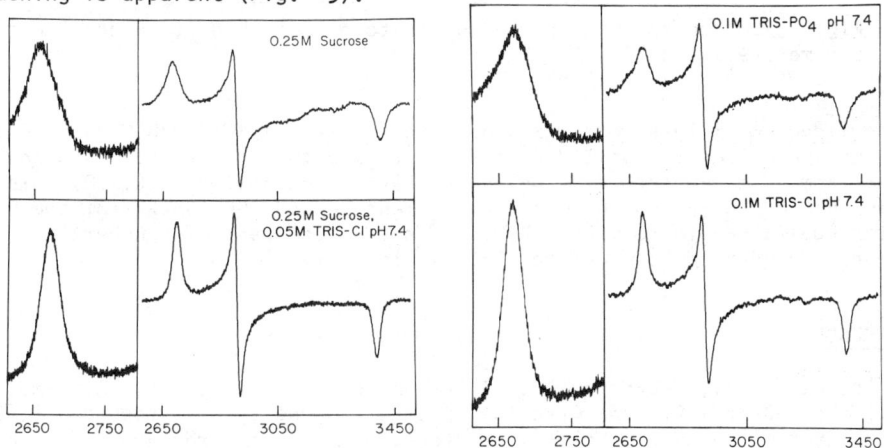

Fig. 2 Low-spin ESR spectra of camphor-free cytochrome P-450$_{cam}$; instrument parameters as in Fig. 1

In KPi the low-spin ESR spectrum of cytochromes P-450$_{cam}$ and P-450$_{LM}$ (PB and 3-MC) is essentially identical (Fig. 3). This spectrum, in turn, is very similar to that observed for cytochrome P-450 in isolated hepatocytes. Thus, given a similar milieu, the low-spin ESR signals of cytochrome P-450 are nearly identical. These results clearly demonstrate that the electronic environment of the hemin iron (active site) of cytochrome P-450 is very similar regardless of source.

Since the presence of various buffers has a significant effect on the line shape and g values for the microsomal and bacterial cytochromes P-450, experiments designed to investigate the electronic state of the hemin iron of cytochrome P-450 as a function of the presence or absence of exogenous substrate or inhibitor must be performed under precisely controlled conditions. These observations suggest that the appearance of asymmetric low-spin ESR signals for microsomal cytochrome P-450 cannot be used to support or negate arguments for a heterogeneous population of cytochromes P-450.

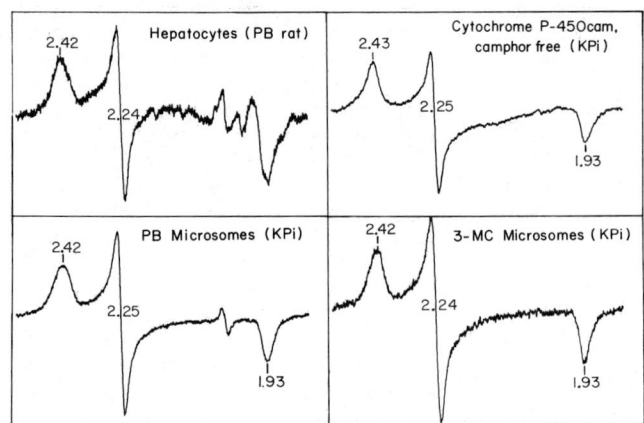

Fig. 3 Low-spin ESR specta of cytochromes P-450; instrument parameters as in Fig. 1

Acknowledgement - This work was supported in part by Research Grants GM 16488 and GM 19036 from the United States Public Health Service and Research Grant I-405 from The Robert A. Welch Research Foundation; J.A.P. is the recipient of Research Career Development Award GM 30962 from the United States Public Health Service; D.H.O. is the recipient of a Robert A. Welch Research Foundation Postdoctoral Fellowship.

REFERENCES

1. Grasdalen, H., Backstrom, D., Eriksson, L.E.G., Ehrenberg, A., Moldeus, P., Von Bahr, C., and Orrenius, S., Heterogeneity of cytochrome P-450 in rat liver microsomes: Selective interaction of metyrapone and SKF-525A with different fractions of microsomal cytochrome P-450, FEBS Lett., 60, 294-299, (1975)
2. Peterson, J.A., Ebel, R.E., O'Keeffe, D.H., Matsubara, T., and Estabrook, R.W. , Temperature dependence of cytochrome P-450 reduction--A model for NADPH-cytochrome P-450 reductase:cytochrome P-450 interaction, J. Biol. Chem., 251, 4010-4016, (1976)
3. Peterson, J.A. , Camphor binding by Pseudomonas putida cytochrome P-450, Arch. Biochem. Biophys., 144, 678-693, (1971)
4. Zahlten, R.N. and Stratman, F.W. , The isolation of hormone-sensitive rat hepatocytes by a modified enzymatic technique, Arch. Biochem. Biophys., 163, 600-608, (1974)

KINETICS OF TIAMUTIN[R] INTERACTION WITH CYTOCHROME P-450 FROM RABBIT LIVER

Ingeborg Schuster, Christine Fleschurz and Ilse Helm

Sandoz Forschungsinstitut, A-1235 Wien, Austria

ABSTRACT

The association reaction between Tiamutin[R] and partially purified Cytochrome P 450 from rabbit liver microsomes gives rise to a TYPE I spectral change and can be observed by stopped-flow kinetics. The second order rate constant at 25°C in a glycerol containing buffer is about 45 $\mu M^{-1} s^{-1}$, the dissociation rate constant about 1.1 s^{-1}. This association reaction takes also place at very high Tiamutin[R] concentrations where in equilibrium studies only modified TYPE II complex is observed. The formation of modified TYPE II change proceeds rather slowly and from its dependence on Tiamutin[R]-concentration can obviously not be interpreted in terms of a binding reaction. The parallelism between modified TYPE II emergence and unspecific binding of the drug to attached phospholipids gives an indication that accumulation of Tiamutin[R] in the lipid bulk might induce a conformational change at the TYPE I site of the protein.

INTRODUCTION

The diterpenoid antibiotikum Tiamutin[R] (14-deoxy-14[(2'-di-ethyl-amino-ethyl)-mercaptoacetoxy]-mutilin hydrogen fumarate) is highly enriched in liver owing to microsomal binding (1). The high capacity of microsomal lipids (about 1 m drug is bound per 5 m phospholipids) together with a medium affinity account nearly completely for binding of the drug (1). Only a small portion of total binding is contributed by the interaction with Cytochrome P 450, which is responsible for the degradation of the drug.

Depending on Tiamutin[R] concentration the specific interaction with Cytochrome P 450 leads to different changes in the absorption spectrum of the protein (1) which can be observed by means of difference spectrophotometry (2). At low Tiamutin[R] a TYPE I spectral change emerges, which diminishes at rising drug concentration due to formation of modified TYPE II complex. This change is accompanied by a variation in the kinetics of Tiamutin[R] hydroxylation (3). At low drug concentration high specificity of binding is coupled with a small $v_{max.}$, whereas at high drug concentration the binding constant is identical with the affinity of microsomal phospholipids for the drug and the maximal velocity is highly increased. The appearance of different spectral changes has been discussed on the basis of two different models (1). According to the assumption of Orrenius et al. (4) there might exist two independent sites, TYPE I and TYPE II which become superimposed. The second interpretation is based on the fact that modified TYPE II emergence is paralleled by the accumulation of the drug in the lipid surrounding of the protein. The interaction with the lipids might influence the TYPE I binding to Cytochrome P 450, so that a qualitatively different interaction occurs.

A distinction between these possibilities can be made by investigations on the kinetics of Tiamutin^R-Cytochrome P 450 interaction. Under appropriate experimental conditions the binding reaction should be in a time range observable by stopped-flow-kinetics. With this technique the kinetics of ligand-Cytochrome P 450 interactions have been examined only with soluble Cytochrome P 450 from Pseudomonas putida till now (5).

This report describes some equilibrium and kinetic properties of Cytochrome P 450-Tiamutin^R interactions and hence proposes a model for the nature of modified Type II change.

MATERIALS AND METHODS

Materials

Tiamutin^R a derivative of pleuromutilin was used throughout the experiments (in ref. (1) the hydrated form of the compound with very similar chemical and biochemical properties was described). A ^{3}H-labelled form of the compound was prepared by F. Battig (Sandoz Forschungsinstitut) with a specific activity of 22 μCi/mg. The chemical structure of the compound is given in Fig. 1.

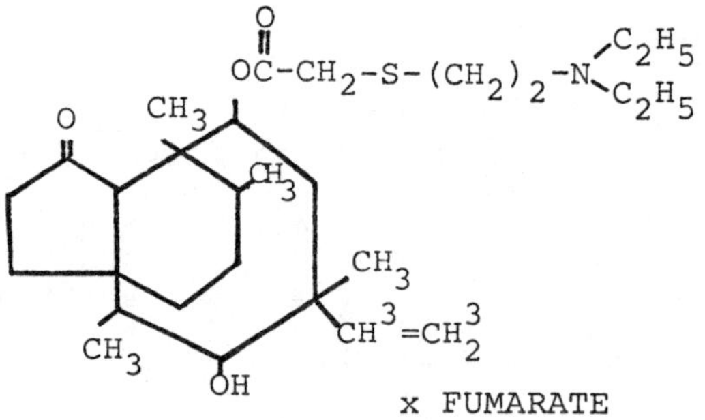

Fig. 1. Chemical structure of ^{3}H-labelled Tiamutin^R.
(14-deoxy-14-[(2'-diethylamino-ethyl)-mercapto-acetoxy]-mutilin hydrogen fumarate)

The purity of the drug was checked by thin layer chromatography in suitable solvent systems.

Chemicals and biochemicals were usually obtained from Merck (Darmstadt) and Boehringer (Mannheim). DEAE-Cellulose, Calcium phosphate gel and butylated hydroxytoluene were obtained from Sigma, Renex 690 from Atlas chemical company.

Experiments were usually performed in a mixture of 40% isotonic KCl-Tris buffer (50 mM Tris, 0.12 M KCl; pH 7.4) and 60% Tris acetate-glycerol buffer (Tris acetate 10 mM, glycerol 20%, EDTA 0.1 mM).

Methods

Male rabbits (New Zealand, 1.5 kg) were given a daily dose of 5 mg/kg Tiamutin^R i.p. over 8 days. After the treatment the animals were killed, the livers immediately excised and microsomes prepared after a standard procedure (1). Though these microsomes contained only about 1 nm Cytochrome P 450 per mg protein, the proportion of Tiamutin^R binding Cytochrome was increased compared with microsomes from untreated animals.

Cytochrome P 450 was partially purified by the procedure of van der Hoeven and Coon (6) to about 3 nm per mg protein. The content of Cytochrome P 450 was determined with the aid of carbon monoxide difference spectra of dithionite reduced samples (7).

Equilibrium measurements
Binding studies were performed by equilibrium dialysis technique as described in (1). ^{3}H-labelled Tiamutin^R in a concentration range between 0.1 μM and 2 mM was equilibrated against a Cytochrome P 450 solution (about 2 - 4 μM in Cytochrome P 450). In the mixed KCl-Tris-Tris acetate buffer equilibrium was established within 5 hours at 25°C.

For radioactivity measurements samples were put into 9 ml Instagel (Packard Instruments GmbH), counted in a Packard Tri-Carb 3375 liquid scintillation spectrometer and the resulting data corrected for losses from quenching.

Difference spectrophotometry was done on a Perkin-Elmer EPS 3T spectrophotometer supplied with an Ulbrich sphere to eliminate losses from stray light as far as possible. Titration of Cytochrome P 450 containing samples was performed by adding small volumes of Tiamutin^R solution to one cuvette, identical volumes of buffer to the other cuvette and by taking the difference spectra between 340 and 700 nm.

Stopped-flow measurements
A dual wavelength equipment with a kHz chopper, as described by Hess et al. (8) was used. The monochromators were set at 393 nm and 420 nm respectively. A total dead time of about 2 ms was determined. A mixing ratio of 5 volumes Cytochrome P 450-sample and 1 volume Tiamutin^R solution was chosen. The final concentrations of Cytochrome P 450 were between 1.7 μM and 7.5 μM, of Tiamutin^R between 1 μM and 2.7 mM. All experiments were performed at 25°C in a KCl-Tris/Tris acetate buffer mixture (pH 7.4).

RESULTS AND DISCUSSION

Equilibrium Measurements

Binding experiments with partially purified Cytochrome P 450 by equilibrium dialysis showed a far higher number of binding sites than Cytochrome P 450 molecules present in the solution. The Tiamutin^R binding characteristics (K_A about 900 M^{-1}, n about 60 moles per mol Cytochrome P 450; pH 7.4, 25°C) make it plausible that presumably phospholipids remained attached to Cytochrome P 450 during solubilization and partial purification procedure. As in intact microsomes (1) different spectral changes are observed also in partially purified Cytochrome P 450 (Fig. 2).

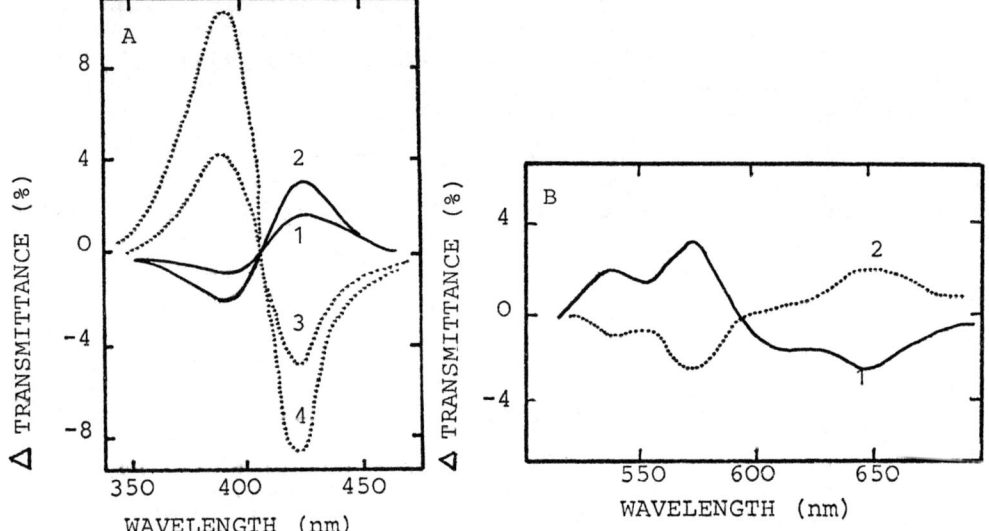

Fig. 2. Spectrophotometric titrations of partially purified Cytochrome P 450 with Tiamutin^R
KCl/Tris-Tris acetate buffer pH 7.4, 25°C.
A) Cyt P 450 13.4 µM, total concentration of Tiamutin^R: (1) 0.5 µM, (2) 60 µM, (3) 0.36 mM, (4) 0.61 mM.
B) Cyt P 450 30 µM, total concentration of Tiamutin^R: (1) 32 µM, (2) 1.1 mM.

At low Tiamutin^R concentrations a TYPE I spectral change emerges. The maximal amount of Cytochrome P 450 involved in this interaction was calculated after (9) to be around 9 - 10% of total Cytochrome P 450. Using the free concentration data for Tiamutin^R, as obtained by equilibrium dialysis, an affinity for TYPE I interaction of about 40 μM^{-1} was determined (KCl-Tris/Tris acetate buffer; pH 7.4, 25°C). The modified TYPE II spectral change visible at high Tiamutin^R concentrations is paralleled by a weak high capacity binding presumably to Cytochrome-attached phospholipids.

Kinetic Measurements

The kinetics of Tiamutin^R binding were investigated by measuring the rate of absorbance change at 393 nm minus the change at 420 nm in the dual wavelength stopped-flow apparatus. In order to obtain pseudo-first order kinetics a molar excess of Tiamutin^R of about

8 to 1000 was used in a series of experiments in which the final concentration of Cytochrome P 450 was changed from 1.7 to 7.5 µM.

Examples for the kinetic traces obtained at different Tiamutin^R concentrations are shown in Fig. 3.

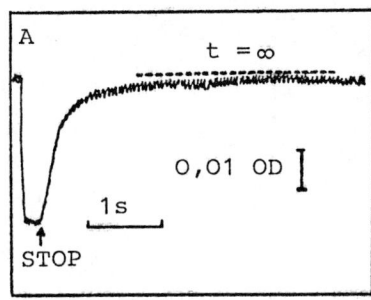 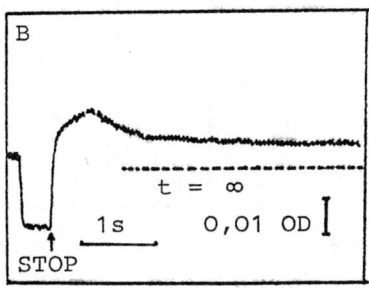

Fig. 3. Reaction of Tiamutin^R with partially purified Cyt P 450, observed by stopped flow kinetics. Oscilloscopic traces.
KCl-Tris/Tris acetate buffer pH 7.4, 25°C.
A) final concentration: Tiamutin^R 4.2 µM, Cytochrome P 450 7.2 µM
B) final concentration: Tiamutin^R 0.5 mM, Cyt P 450 6.4 µM.

At all Tiamutin concentrations investigated one rapid process can be observed. The sign and magnitude of its amplitude is in quantitative agreement with TYPE I spectral change (Fig. 4).

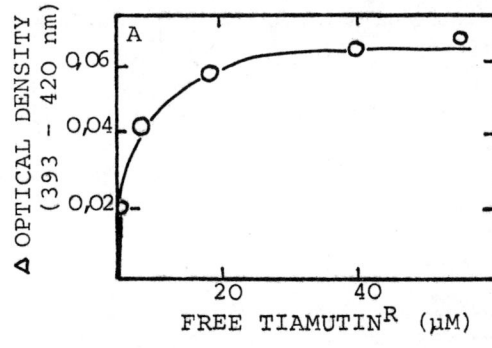 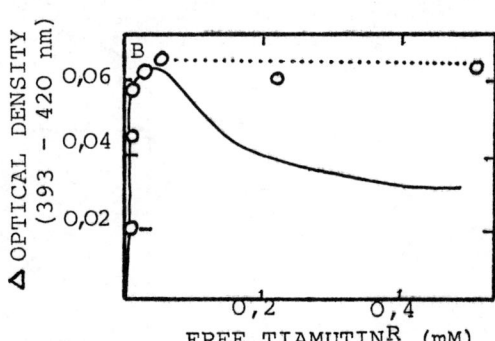

Fig. 4. Comparison of TYPE I spectral change from equilibrium titration and amplitude of fast reaction. Conditions as in Fig. 2 and 3; Circles - amplitudes of fast reaction from stopped flow kinetics; line - spectrophotometric titration in equilibrium; concentration of Tiamutin^R: A) low, B) high.

Up to free Tiamutin^R concentrations of 0.22 mM the data could be analyzed to be pseudo first order kinetics (at higher concentrations the reaction became too quick to be analyzed any more). The apparent rate constants show a linear relationship on free Tiamutin^R (Fig. 5).

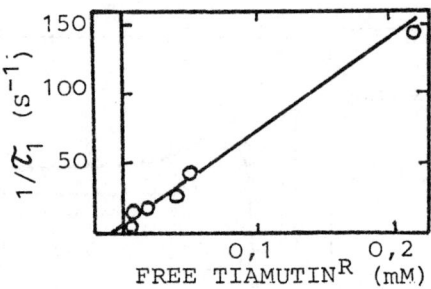

Fig. 5. Dependence of rapid reaction rate on free Tiamutin^R concentration. Conditions as in Fig. 3 and 4.

Obviously this process resembles the TYPE I binding process of Tiamutin^R to Cytochrome P 450. From the slope of the straight line in Fig. 5 an association rate constant can be calculated of about 45 $\mu M^{-1} s^{-1}$. With the association constant of 40 μM^{-1} the dissociation rate constant is determined to be 1.1 s^{-1}.

At higher Tiamutin^R concentrations the rapid binding process is followed by slower reactions inverse in the sign of their amplitudes Fig. 3. At least 3 different processes can be separated on the time axis. The two slowest processes contributed to a smaller extent to the total amplitude. The high amplitudes of the quickest of the three processes parallel well the formation of modified TYPE II complex as observed by optical titration in equilibrium (Fig. 6).

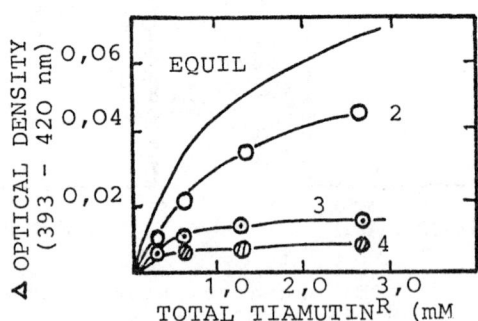

Fig. 6. Comparison of modified TYPE II spectral change from equilibrium titration (EQUIL) and amplitudes of the 3 slow reactions: 2, 3, 4. Conditions as in Fig. 3 and 4.

The slow reactions can be treated as first order processes and apparent rate constants can be evaluated. However the dependence of these constants on free Tiamutin^R concentration shows that modified TYPE II spectral change cannot be interpreted with a simple binding reaction of Tiamutin^R to Cytochrome P 450 (Fig. 7).

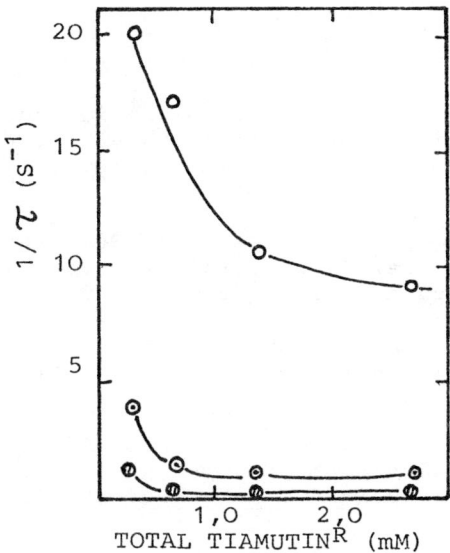

Fig. 7. Dependence of slow reaction rates on Tiamutin^R concentration. (Reaction 2, 3, 4 as in Fig. 6).

Contrary to binding processes these reactions slow down at increasing Tiamutin^R and finally become independent on ligand concentration.

Both types of reactions - the rapid binding process and the slow TYPE II changes - can be observed in intact microsomes too. However due to the less Cytochrome P 450 content of microsomes and the high noise due to the turbidity of the samples TYPE I reaction is only seen as a rapid burst which cannot be analyzed. TYPE II changes exhibit very similar kinetic traces as in samples with partially purified Cytochrome P 450.

The kinetic results give some insight into ligand-Cytochrome P 450 interaction: Binding of the lipophilic Tiamutin^R to Cytochrome P 450 gives rise to a TYPE I complex. This reaction proceeds also at high Tiamutin^R concentrations where only modified TYPE II change is seen in equilibrium. Cooperative effects in TYPE I binding were not detected, however cannot be excluded: Under our experimental conditions no reliable kinetic traces were obtained at degrees of Cytochrome P 450 saturation less than 30% (this means at very low Tiamutin^R concentration).

As it was proposed in (1) modified TYPE II change resembles surely not binding to an independent site but very likely a conformational change of Cytochrome P 450 which leads to the disappearance of TYPE I complex (see also the loss of 'high spin' absorption at 647 nm in the optical titration, Fig. 2). The fact that modified TYPE II emergence shows the same course as unspecific binding to microsomal phospholipids gives evidence that the accumulation of Tiamutin^R at these sites induces some change in the protein TYPE I site.

REFERENCES

1) I. Schuster, Ch. Fleschurz, I. Helm, On the interaction of a lipophilic durg with different sites of rat-liver microsomes. Equilibrium studies with a substituted Pleuromutilin, Eur. J. Biochem. 51, 511 (1975).

2) S. Narasimhulu, D.Y. Cooper, O. Rosenthal, Spectrophotometric properties of a triton-clarified steroid 21-hydroxylase system of adrenocortical microsomes, Life Sci. 4, 2101 (1965).

3) I. Schuster, Interactions of drugs with liver microsomes, in: Progress in Molecular and Subcellular Biology (Hahn, F., ed.) Springer, Berlin, in press.

4) S. Orrenius, B.J. Wilson, Ch. von Bahr, J.B. Schenkman, On the significance of drug induced spectral changes in liver microsomes, in: Biological Hydroxylation Mechanisms (Boyd, G.S., Smellie, R.M.S., eds.) 55 - 77. Acad. Press, London, 1972.

5) B.W. Griffin, J.A. Peterson, Camphor binding by pseudomonas putida cytochrome P 450. Kinetics and thermodynamics of the reaction, Biochemistry 11, 4740 (1972).

6) T.A. van der Hoeven, M.J. Coon, Preparation and properties of partially purified cytochrome P 450 and reduced nicotinamide adenine dinucleotide phosphate-cytochrome P 450-reductase from rabbit liver microsomes, J. Biol. Chem. 249, 6302 (1974).

7) T. Omura, R. Sato, The carbon monoxide-binding pigment of liver microsomes, I. Evidence for its hemoprotein nature, J. Biol. Chem. 239, 2370 (1964).

8) B. Hess, H. Kleinhans, H. Schlüter, A kHz dual wavelength spectrometer, design and experimental test, Hoppe Seyler's Z. Physiol. Chem. 351, 515 (1970).

9) H. Diehl, J. Schädelin, V. Ullrich, Studies on the kinetics of cytochrome P 450 reduction in rat liver microsomes, Z. Physiol. Chem. 351, 1359 (1970).

DIFFERENT PHASES OF HYDROXYLASE INDUCTION IN LIVER MICROSOMES OF FEMALE MICE DURING INHALATION OF CYCLOHEXANE AND D,L-CAMPHOR

Gertrud Mohn

Department of Physiological Chemistry, University of the Saarland, 6650 Homburg/Saar, German Federal Republic

ABSTRACT

In the liver microsomes of female mice we find two induction phases during inhalation of cyclohexane or D,L-camphor. During the 1st 24 h the apparent molar activity of the ethylumbelliferone dealkylase decreases very much with both of the inducers. With cyclohexane there also appears a large decrease of the affinity for ethylumbelliferone and cyclohexane in the microsomes, followed immediately by a similar increase. In the 2nd phase, during which the molar ethylumbelliferone dealkylase activity is constant, the affinity for ethylumbelliferone and cyclohexane also becomes constant.

INTRODUCTION

Recently we reported: The quotient Q may be taken as the apparent molar ethylumbelliferone dealkylase activity in the liver microsomes of female mice treated with cyclohexane or D,L-camphor (Ref. 1-3).

$$Q = \frac{mU \text{ ethylumbelliferone dealkylase}}{mU \text{ NADPH-cyt.P-450 reductase} \times \text{cyt.P-450 concentration}}$$

Theoretically the kinetics of microsomal hydroxylases should be very complex because these enzymes are bound to the microsomal membrane and because of the heterogenity of cyt.P-450 (Ref.4,5). Therefore it is necessary to present further arguments for the validity of our reference parameter [reductase activity x cyt.P-450 concentration], which was found empirically.

METHODS

Female white mice (23-25 g) of our own inbred strain inhaled 0.2 mM cyclohexane or 0.05 mM D,L-camphor in airtight cages for 0.25-4 days. They received water ad libitum and were fed in fresh

air with Altromin food each day at 8-10 o'clock. 1 or 2 h before death, the animals were also placed in fresh air. (Further details see Ref. 1 and 3).
Microsomal protein, the cytochromes P-450 and b_5, the ethylumbelliferone dealkylase and the NADPH-cyt.P-450 reductase activity are determined according to well known methods (Ref.1).

RESULTS

In contrast to results with liver microsomes of male rats (Ref. 6,7), in the liver microsomes of our female mice the activatability of the NADPH-cyt.P-450 reduction by substrates is very low (Table 1). After treating the animals with cyclohexane, D,L-camphor or phenobarbital, we can assume that our reductase values are closely related to the reductase concentration in the microsomes (Ref.8).

TABLE 1 Percentage Increase in the NADPH-cyt.P-450 Reductase Activity in the Presence of Substrates

Substrate	Inducers			
	Cyclo-hexane	D,L-camphor	Phenobar-bital	3-Methylcho-lanthrene⊕⊕
---	100	100	100	100
0.1 mM ethylum-belliferone	+ 5	+ 10⊕	+ 7⊕	+ 57⊕
1 mM aminopyrine	0	+ 9	+ 13⊕	+ 5
10 mM cyclohexane	− 10	− 9	+ 3	− 4

⊕ $p < 0.05$; ⊕⊕ the MC-experiments are done with female black mice C57BL/6N obtained from Nebert.
The reaction mixture contained 100 mM Tris (pH 7.6), 0.8 or 2 mg microsomal protein/ml, 1 mM $MgCl_2$, 8 mM Glucose, ca. 100 U catalase/ml, 8 U glucose oxidase/ml, 0.1 mM NADPH and a trace of silicon (No 0860 Roth/Karlsruhe). CO was bubbled through the suspension for 5 min, then a stop cock plunger containing NADPH and glucose oxidase was set upon the cuvette with CO flushing through the gas phase in the cuvette for another 5 min.

Dependent on the induction time with cyclohexane or camphor, the apparent molar ethylumbelliferone dealkylase activity Q decreases quickly during the 1st 24 h, but then remains constant (Ref. 1-3). Thus, these curves indicate two different phases of induction, the alteration of the hydroxylase population and a phase of augmentation, during which the composition of the hydroxylase system remains constant.

The reductase activity alone or the cyt.P-450 concentration alone or the product [cyt.P-450 concentration x cyt.b_5 concentration] is not a suitable reference parameter for the dealkylase (Ref.1, 3). It is indeed possible to test the suitability of such parameters by kinetic measurements with microsomes. For instance, in the binding spectra of cyclohexane with liver microsomes of cyclohexane-treated mice, we receive continuous time curves only by taking the P-450 concentration alone as reference parameter, because the substrate is bound to cyt.P-450 and not to a complex between P-450 and reductase (Ref.9-11).

Contrary to the values of ΔE_{max}/nmoles cyt.P-450 the K_S-values of the cyclohexane binding spectra are considerably altered during induction with cyclohexane. Figure 1 shows a large peak after 12 h of CH-inhalation which disappears thereafter. It is the same peak, which can be seen in the K_m-values of ethylumbelliferone and in the K_i-values of cyclohexane in the dealkylase reaction in Fig. 2 and 3. Especially the K_m-curve of ethylumbelliferone, but also the K-values of cyclohexane show two different induction phases. The duration of the 1st phase in the K_m-curve of ethylumbelliferone is exactly the same as in the curve of the molar ethylumbelliferone activity.

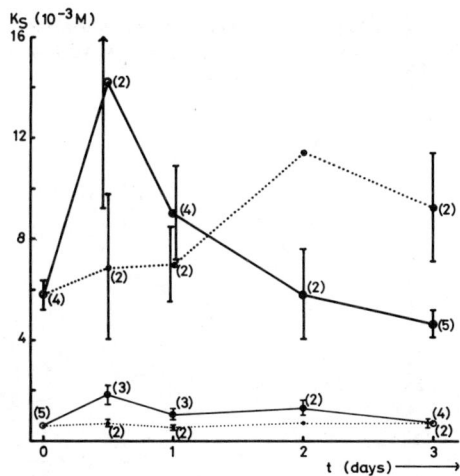

Fig. 1 The K_S-values of cyclohexane in the binding spectra of liver microsomes from cyclohexane-treated female mice as a function of the induction time

Numbers in brackets are the number of investigated groups. Each group consists of 4-8 mice. Bars represent ± SE. Lower and higher curves: titration with ethanolic solution of 0.2 or 2 M cyclohexane respectively. Dotted lines: controls.

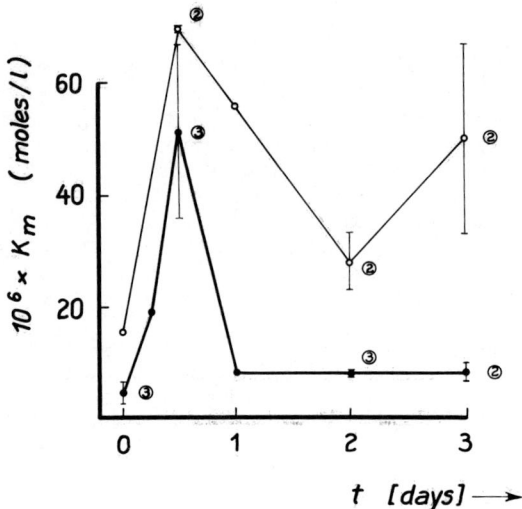

Fig. 2 The dependence of K_m-values of ethylumbelliferone on the induction time with cyclohexane.
Thin line: K_m at higher ethylumbelliferone concentrations.

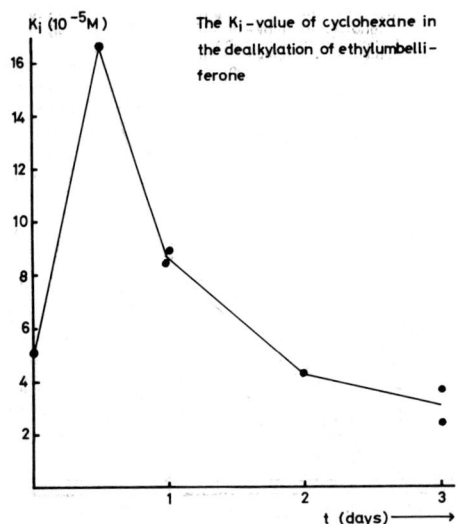

Fig. 3 The K_i-values of cyclohexane in the dealkylation of ethylumbelliferone as a function of the time of induction with cyclohexane
K_m of ethylumbelliferone is determined in the presence of 0.5 mM cyclohexane.

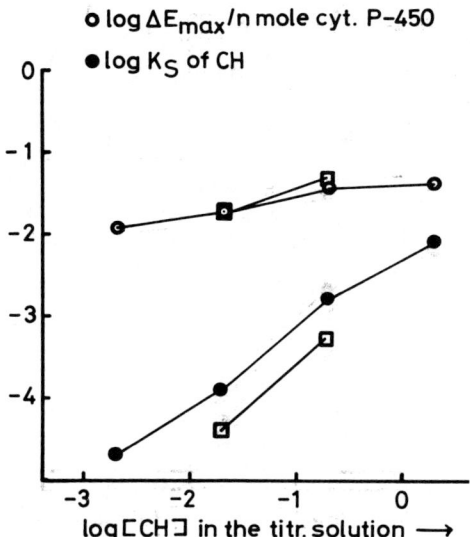

Fig. 4 Binding spectra of cyclohexane in the liver microsomes of mice treated with cyclohexane for 3 days: dependence of the kinetic parameters on the cyclohexane concentration in the titration solution.
○ Log. ΔE_{max}/nmole cyt.P-450 and ● log. K_S with cyclohexane dissolved in ethanol; □ corresponding values with cyclohexane dissolved in dimethylsulphoxide.

In Fig. 1 and especially in Fig. 4, we see the strong dependence of the K_S-values of cyclohexane on the range of concentrations studied. The higher the relation of solvent concentration to cyclohexane concentration in the microsomal suspension, the lower is the apparent K_S-value of cyclohexane. A similar behavior of the K_S-values of cyclohexane can be seen in the mouse liver microsomes after induction with camphor, phenobarbital or 3-methylcholanthrene. In contrast, the K_S-values of aminopyrine are independent of the concentration range used (Table 2).

The strong dependence of K_S- and K_m-values of lipophilic substrates (see also Ref. 12) especially indicates the importance of the solubility in the microsomal membrane for the apparent K-values. We can assume that the peaks of the K-curves of cyclohexane and ethylumbelliferone (Fig. 1-3) are caused by a change of their solubility in the microenvironment of the hydroxylase. Interestingly, this change seems to be a specific effect of cyclohexane. During induction with D,L-camphor, phenobarbital or 3-methylcholanthrene a similar peak in the K_m of ethylumbelliferone cannot be found.

DISCUSSION

Former experiments have shown two different phases of induction with cyclohexane and D,L-camphor (Ref. 1-3). The diagrams of the apparent molar dealkylase activity Q dependent on the induction time have a rapidly decreasing and a horizontal part. The 1st phase must be caused by an alteration of the substrate specificity of the hydroxylase system. This takes place mainly by an alteration of the cyt.P-450 population, but can also be influenced by a variation in the composition of the lipid areas in the membrane (Ref. 13) accompanied by an alteration of the partition coefficient of lipophilic substrates favourable or unfavourable for their enrichment in the microenvironment of the hydroxylases.

Not only the apparent affinity for ethylumbelliferone, but also that for cyclohexane decreases during CH-inhalation whereas cyt. P-450 concentration and reductase activity are already increasing (Ref. 1). On the other hand, the decrease in the molar ethylumbelliferone dealkylase activity cannot be caused only by lipid variation, because we have an analoguous Q-curve during camphor inhalation although there is no K_m-peak of ethylumbelliferone after 12 h.

TABLE 2 Kinetic Parameters in the Binding Spectra of Aminopyrine

Inducer	Conc. in the titr. solution	$K_S \pm SD$ mM	ΔE_{max}/nmole cyt. P-450 $\pm$ SD
Cyclohexane	0.05 and 0.4 M in DMSO or H_2O	0.25 $\pm$ 0.05	0.012 $\pm$ 0.001
D,L-Camphor	0.02 M, 0.05 M, 0.1 and 0.2 M in H_2O	0.24 $\pm$ 0.09	0.015 $\pm$ 0.004
Phenobarbital	0.05 and 0.4 M in DMSO or H_2O	0.22 $\pm$ 0.02	0.020 $\pm$ 0.002

4,5 mg microsomal protein were diluted with 0.1 M Tris (pH 7.6) to 3 ml. 5 µl substrate solution was placed into the sample cuvette and 5 µl solvent into the reference cuvette. This was repeated 7-9 times. Induction time: 3 days.
Liver microsomes from controls and MC-treated mice give no type I spectrum with aminopyrine.

Obviously, a reorganization of the lipid areas of the hydroxylase system or of the whole microsomal membrane takes place in the CH-microsomes. During the following 12-24 h of cyclohexane inhalation all the K-values decrease almost as quickly as they increased before. Thus, in the 2nd induction phase not only the apparent molar ethylumbelliferone dealkylase activity but also the K-values of ethylumbelliferone and cyclohexane become constant. K_m of ethylumbelliferone remains about 80% higher than in microsomes of untreated animals whereas K_S of cyclohexane returns to the starting point especially at lower substrate concentration.

ACKNOWLEDGMENT

I thank Mrs. Eve-Marie Philipp for her excellent technical assistance in these experiments.

REFERENCES

1. G.Mohn and E.-M.Philipp, Zeitlicher Verlauf der Induktion des Hydroxylasesystems der Mäuselebermikrosomen durch Cyclohexan, Hoppe-Seylers Z. Physiol.Chem. 355, 364 (1974)

2. G.Mohn, Differences caused by different inducing substances in the specificity of hydroxylases and in the type of hydroxylating cytochrome in liver microsomal fractions, Biochem.Soc.Transact. 3, 972 (1975)

3. G.Mohn, P.Niederlaender and C.Kessler, The relative ethylumbelliferone dealkylase activity characterizing the effect of different inducers on the hydroxylase system in the liver microsomes of female mice, in preparation

4. A.Y.H.Lu, W.Levin, S.B.West, M.Jacobsohn, D.Ryan, R.Kuntzman and A.H.Conney, Different substrate specificities of the cytochrome P-450 fractions from control and phenobarbital-treated rats, J.Biol.Chem. 248, 456 (1973)

5. A.P.Alvares and P.Siekevitz, Gel electrophoresis of partially purified cytochromes P-450 from liver microsomes of various-treated rats, Biochem.Biophys.Res.Comm. 54, 923 (1973)

6. P.L.Gigon, T.E.Gram and J.R.Gillette, Studies on the rate of reduction of hepatic microsomal cytochrome P-450 by NADPH: Effect of drug substrates, Mol.Pharmacol. 5, 109 (1969)

7. H.Diehl, J.Schädelin and V.Ullrich, Studies on the kinetics of cytochrome P-450 reduction in rat liver microsomes. Hoppe-Seylers Z.Physiol.Chem. 350, 1359 (1970)

8. G.Mohn, The effect of hydroxylase substrates on the reduction rate of cyt.P-450 by NADPH in the liver microsomes of female mice, Symp.Drug.Metabolism, Guilford, (1976)

9. H.Remmer, J.B.Schenkman, R.W.Estabrook, H.Sasame, J.R.Gillette, S.Narasimhulu, D.Y.Cooper and O.Rosenthal, Drug interaction with hepatic microsomal cytochrome, Mol.Pharmacol. 2, 187 (1966)

10. Estabrook, R.W., Cytochrome P-450, its function in the oxidative metabolism of drugs, in "Handbook of Experimental Pharmacology" Vol. 28/2 (Brodie,B.B. and Gillette,J.R., Eds.), pp.264-284, Springer, Berlin, 1971

11. J.Werringloer and R.W.Estabrook, Heterogenity of liver microsomal cytochrome P-450: The spectral characterization of reactants with reduced cytochrome P-450, Arch.Biochem. Biophys. 167, 270 (1975)

12. J.R.Hayes, M.U.K.Mgbodile and T.C.Cambell, Dependence of K_m and V_{max} on substrate concentration for rat hepatic microsomal ethylmorphine N-demethylase, Biochem.Pharmacol. 22, 1517 (1973)

13. S.C.Davison and E.D.Wills, Studies on the lipid composition of the rat liver endoplasmatic reticulum after induction with phenobarbitone and 20-methylcholanthrene, Biochem.J. 140, 461 (1974)

PROTEIN-LIPID INTERACTIONS IN THE LIVER MICROSOMAL HYDROXYLASE SYSTEM

Magnus Ingelman-Sundberg

Department of Chemistry, Karolinska Institutet, Stockholm, Sweden

ABSTRACT

Protein-lipid interactions in the liver microsomal hydroxylase system have been studied with delipidated purified cytochrome P-450 by reconstitution with phospholipid and with binding studies with the detergent, Triton X-100. It was found that upon recombination with phospholipid, delipidated P-450 was renatured, a process that was followed by the increase in absorption at 450 nm of the reduced, CO-bound complex. The renaturation, measured spectrophotometrically, was correlated to an increased $NaIO_4$-supported 6β-hydroxylation of androstenedione. In NADPH-supported 6β-hydroxylation about 100 times more phospholipid was necessary to get maximal lipid stimulation. It is suggested that the lipid has a dual function in the liver microsomal system, (A) binding to P-450 thereby inducing the active conformation of the enzyme and (B) forming a fluid environment that facilitates electron transport from NADPH to P-450.

In binding studies with Triton X-100 it was found that P-450 bound its own weight in detergent thereby dissociating from a tetrameric to a monomeric form which was catalytically active in 6β-hydroxylation of androstenedione in both $NaIO_4$ and NADPH-supported reactions.

INTRODUCTION

In 1968 Lu and Coon (1) resolved the liver microsomal hydroxylase system into three components - P-450, reductase and a heat stable factor. In 1970 (2) it was shown that the heat stable factor consisted of phosphatidylcholine (PC). However, it is still not known by what mechanism(s) the phospholipid acts in P-450-linked hydroxylation reactions.

The specificity for PC in the hydroxylation reaction is not complete; other amphipatic compounds such as other phospholipids and non-ionic detergents can replace PC (3). It thus seems as if the physical properties of an amphiphile rather than its specific chemical structure is the essential character of the lipid.

Duppel and Ullrich (4) found that the fluidity of the microsomal membrane affected the rate of electron transport from NADPH to P-450 via the reductase. This indicated a role for the phospholipid in the reduction of P-450. The same statement has also been made by Narasimhulu (5) who found that in the absence of lipid or detergent no reduction of P-450 occurred in delipidated adrenal cortex microsomes.

In order to evaluate the role(s) of the phospholipid in the liver microsomal hydroxylase system three types of P-450-lipid interactions have been studied,
(A) between PC and purified delipidated liver microsomal P-450
(B) between PC and neutral fats and liver microsomal P-450 and
(C) between the non-ionic detergent Triton X-100 and purified P-450.

INTERACTIONS OF PURIFIED DELIPIDATED P-450 WITH PC

Cytochrome P-450 and NADPH-cytochrome P-450 reductase were prepared according to the method of Coon and collaborators (6) from phenobarbital induced male rabbit livers. The specific content of P-450 in the preparations was 10-15 nmoles/mg protein. The fraction with the highest concentration of P-450 (the phenobarbital induced form, $P-450LM_2$ (6)) was according to SDS-gelelectrophoresis essentially homogeneous. The purified reductase had a specific activity of about 1,000 U/mg protein.

For delipidation, 50 nmoles of $P-450LM_2$ in 3 ml of 5 mM phosphate buffer containing 10^{-4} M EDTA was passed through a 100 ml column of Sephadex LH-20. This step removed all detergent that was not very tightly bound to the protein (7). The preparation was lyophilized and the residual powder extracted twice with 5 ml of 1-propanol and thereafter twice with 5 ml of acetone at -20 C, before it was dried under a stream of nitrogen. This extraction procedure reduced the phospholipid content from about 42 nmoles of phospholipid per mg of protein, i.e. 3.2 nmoles/nmole of P-450, to 17 nmoles/mg protein. For details of this procedure, see ref. (8).

Incorporation of delipidated P-450 into PC-vesicles

Egg yolk phosphatidylcholine (5-1,000 μg) was suspended in 50 mM phosphate buffer, pH 7.4, and sonicated under nitrogen at 28 C in order to form vesicles. Purified, delipidated and lyophilized P-450 (0.1-1 mg) was suspended in 1 ml of 100 mM phosphate buffer and added to the PC-suspension which was then sonicated for 1 min, incubated at 37 C for 5 min and subsequently sonicated for 30 sec. The specific content of P-450 decreased due to the delipidation procedure to about 2.1 nmoles/mg protein. Fig. 3, however, shows that incorporation of the protein into the PC vesicles increased the amount of active P-450, absorbing at 450 nm in the CO-bound, reduced state, by over 300%, i.e. to a specific content of about 7.4 nmoles/mg protein. This value corresponds to about 55% of the original concentration. Saturation of renaturable P-450 was obtained in the figure at 0.5 μg, i.e. 11 nmoles of PC per nmole of P-450. The renaturation of P-450 was, as shown in Fig. 1A, also well correlated to catalytic activity in 6β-hydroxylation of androstenedione in $NaIO_4$-supported hydroxylation reactions.

$NaIO_4$ catalyzes the formation of oxygenated P-450 in the absence of dioxygen (9) and therefore the present experiment measured the catalytic activity of P-450 alone; no reductase or NADPH were added. Thus, the results indicate a role for PC or related phospholipids for the formation of the catalytically active form of P-450. Since part of the renatured P-450 has its origin in P-420, the renaturation probably takes place by a conformational change in the delipidated P-450 by the addition of lipid. Conformational changes following reconstitution of delipidated enzymes with lipids have also been suggested for other types of enzymes (10-12).

INTERACTIONS OF THE P-450-DEPENDENT HYDROXYLASE SYSTEM WITH PC AND NEUTRAL FATS

Whereas very small amounts of lipid were necessary to stimulate $NaIO_4$-supported P-450 catalyzed hydroxylation of androstenedione about 100 times more PC per nmole of P-450 was necessary to saturate stimulation of NADPH-supported 6β-hydroxylation of androstenedione (see Fig. 1B). In this case, non-delipidated purified P-450 was used with some remaining detergent still attached to

P-450. The 6β-hydroxylation supported by NaIO$_4$ was not stimulated by lipid in a significant manner in any concentration. The results indicate that the phospholipid in this case acts in a more unspecific way and works at the level of electron transport from NADPH to P-450. These results are in agreement with those of Ullrich (4) and Narasimhulu (5) mentioned above.

Hydroxylations of androstenedione dependent on the composition of neutral fat

It has for a long time been known that feeding rats with different types of fats in their diet affects the composition of lipids in the liver microsomes and also affects the kinetics of different P-450-dependent hydroxylation reactions (13-15). In order to study these effects in a more simple and well-controlled system the purified liver microsomal hydroxylase system was reconstituted with egg yolk PC together with neutral fats containing saturated or unsaturated fatty acids, with or without cholesterol. The results concerning the 6β- and 16α-hydroxylations of androstenedione incubated with P-450LM$_2$, purified reductase and different types of lipids are summarized in Fig. 2.

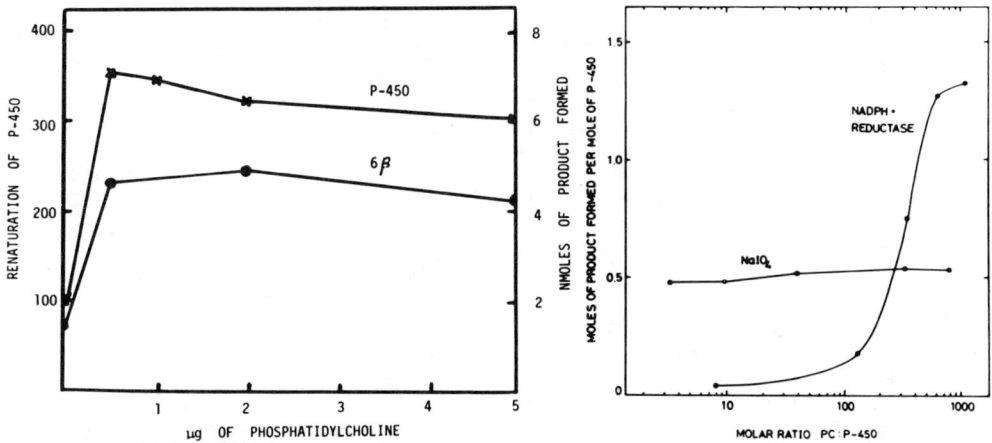

Fig. 1A (left) Renaturation of delipidated, purified rabbit liver cytochrome P-450 (P-450LM$_2$) by incorporation into PC-vesicles. Conditions as described in text. The activity of the 6β-hydroxylase active on androstenedione at different PC concentrations using the NaIO$_4$-supported system is shown (o——o). Incubations were performed as described previously (9).

Fig. 1B (right) Reactivation by PC of NADPH- and NaIO$_4$-supported 6β-hydroxylation of androstenedione catalyzed by purified non-delipidated P-450LM$_2$. One nmole of non-delipidated P-450 containing 3.2 nmoles of phospholipid and some residual detergent was added to a tube with 150 U of NADPH-cytochrome P-450 reductase containing 2 nmoles of phospholipid. A suspension of PC, previously sonicated for 5 min, in 50 mM phosphate buffer, pH 7.4, was then added and the total volume adjusted to 2 ml with phosphate buffer. The tube was sonicated for 1 min, incubated at 37 C for 3 min and sonicated again for 1 min. NADPH and androstenedione were added to give 2 mM and 71.4 μM concentrations respectively, and the tube subsequently incubated at 37 C for 1 h. In NaIO$_4$-supported hydroxylation reactions reductase and NADPH were omitted and NaIO$_4$ added to give a concentration of 10 mM. Incubations were performed for 5 min at 37 C.

Both hydroxylations were stimulated twice as much if saturated rather than unsaturated fatty acids were present in the triglycerides in the reconstitution lipid mixture. In both cases PC was present in 50% by weight and was essential for stimulating activity. When cholesterol (30% w/w) was added together with PC and saturated fats no stimulatory effect of the hydroxylation was obtained. The effect of cholesterol was thus inhibitory. One can conclude that changes of fluidity of the environmental lipid caused by altered ratio of unsaturated to saturated fatty acids affect the rate of hydroxylations. The mechanism behind this may be that the changes in the lipid environment affect the type of interaction between the reductase and P-450 resulting in decreased or increased electron flow. The effects seen on the drug metabolism (13,15) by feeding the rats with different types of lipids may therefore at least partially reflect direct membrane effects.

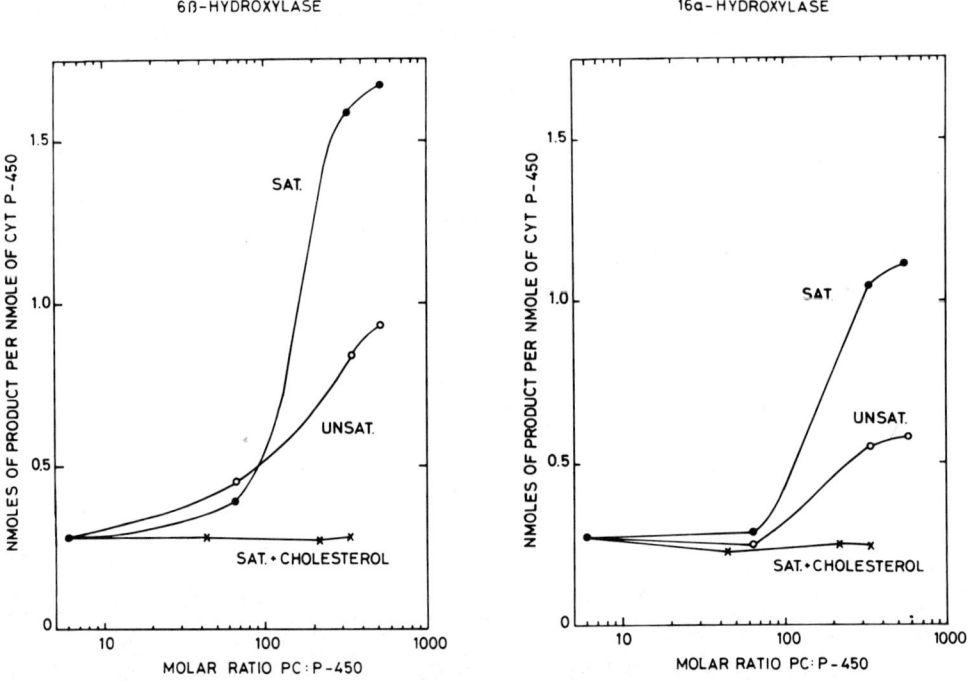

Fig. 2. Reactivation by PC and neutral fats of NADPH-supported 6β- (left) and 16α-hydroxylation of androstenedione catalyzed by purified non-delipidated P-450LM$_2$. Conditions as described in Fig. 1B except that reconstitution mixture consisted of PC:tripalmitin:stearic acid 5:4:1 (by weight) (SAT.), PC:trilinolein:oleic acid 5:4:1 (by weight) (UNSAT.) or PC:tripalmitin:stearic acid:cholesterol 5:4:1:5 (by weight) (SAT. + CHOLESTEROL).

The results presented above indicate that for the catalytic function of P-450 only about 10 times excess (in molar ratio) of PC is necessary, while for electron transport reconstitution nearly 1,000 times excess of PC is needed. When the reactivation curves in Figs. 1 and 2 in NADPH-supported hydroxylations were visualized in linear plots, sigmoidal kinetics were obtained for the reactivation. The same principal results have also been obtained for delipidated oligomycin sensitive ATPase (16). Here ATPase activity was stimulated at very small concentrations of phospholipid while stimulation of the $^{32}P_i$-ATP exchange activity (ion transport) was obtained at more than 10 times this concentration with sigmoidal kinetics. Thus, in these two cases it is probable that two types of interaction of the proteins with phospholipid exist: one specific with high affinity and one unspecific forming the fluid environment necessary for electron or ion transport.

BINDING OF TRITON X-100 TO CYTOCHROME P-450

Triton X-100 is an effective, nondenaturing detergent which has been widely used to solubilize membrane proteins. Triton X-100 does not cause gross changes in protein structure (17) and has been previously successfully used to determine subunit composition of membrane proteins (18,19). This has been possible since hydrophilic proteins like bovine serum albumin, hemoglobin, catalase, cytochrome c and ovalbumin (which have been used here) do not bind this detergent (18,20) and could be used as standard proteins in ultracentrifugation and gelfiltration for determination of sedimentation coefficient and Stokes radius.

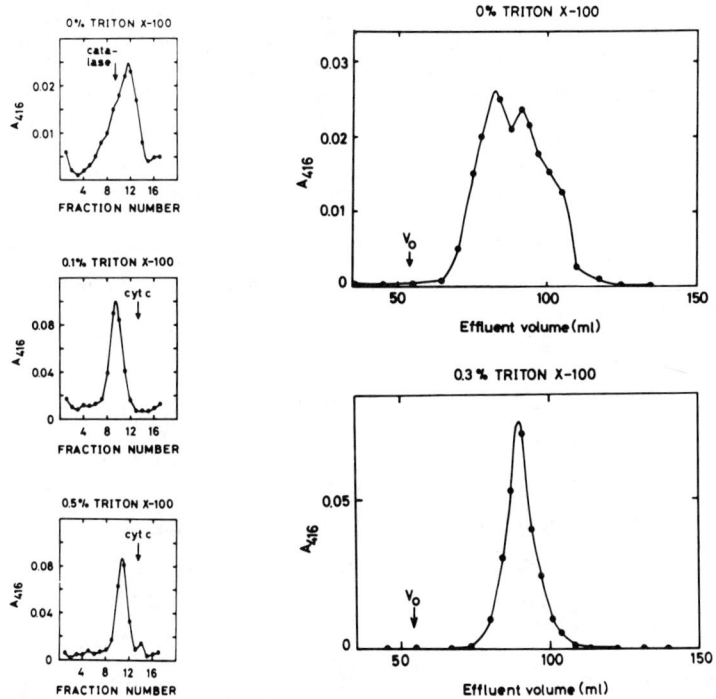

Fig. 3. <u>Sucrose gradient centrifugation (left) and gelfiltration on Ultrogel^R ACA-54 of P-450LM$_2$-Triton X-100 complexes.</u> Sucrose gradients (5-20%) were made in 50 mM phosphate buffer, pH 7.4, containing the indicated amount of Triton X-100. P-450LM$_2$ was equilibrated with the phosphate-Triton buffer by incubation at 37 C for 5 min and thereafter at 4 C for 2 h. 2 Nmoles of P-450LM$_2$ were then layered on the gradient in a volume of 0.2 ml and the tubes were centrifuged at 230,000 x g for 16 h or in the case without Triton for 5 h. Reference proteins used were catalase (11.3 S), bovine serum albumin (4.6 S), he-

moglobin (2.8 S) and cytochrome c (1.7 S). The tubes were punctured and fractions of 0.2 ml collected from the bottom. P-450LM$_2$ was assayed by its absorbance at 416 nm.

Ultrogel^R ACA-54 columns (2.2 x 80 cm) were equilibrated with 50 mM phosphate buffer, pH 7.4, containing the appropriate concentration of Triton X-100. The columns were calibrated with the above mentioned standard proteins and with ovalbumin and alcohol dehydrogenase (from yeast). P-450LM$_2$, 10 nmoles, was equilibrated in 3 ml of the Triton buffer as in the case of ultracentrifugation studies and applied on the columns. P-450 was assayed by absorbance at 416 nm. The void volume (V_0) of the columns is indicated in the figure.

Purified P-450LM$_2$ was equilibrated with Triton X-100 and the complexes formed were characterized by ultracentrifugation in sucrose gradients and by gelfiltration on Ultrogel^R ACA-34. The Ultrogel^R column gave a linear relationship between K_{av} and log Mw in the range of K_{av} 0.4 to 0.9. Outside this range it was not possible to perform determinations of Stokes radius. In Fig. 3 the patterns obtained by ultracentrifugation of the purified protein in 0, 0.1 and 0.5% Triton X-100 are shown. When no Triton had been added P-450 sedimented as a heterogeneous peak. The major component of the peak had a sedimentation coefficient of 8.1 S. In solutions of 50 mM phosphate buffer with more than 0.2% Triton, P-450 formed homogeneous complexes with the detergent with $s_{20,w}$ values of 3.4 (cf. Table 1). The same principal results were obtained by gelfiltration studies with and without added Triton X-100. Without detergent very heterogeneous forms of P-450 were eluted. With Triton X-100 homogeneous complexes were formed with a Stokes radius of about 46 Å (Fig. 3).

Table 1   Characterization of P-450LM$_2$-Triton X-100 complexes

NADPH and NaIO$_4$-supported 6β-hydroxylation of androstenedione was measured as described previously (11).

Amount of Triton X-100 %	$s_{20,w}$ S	$\bar{v}$ cm^3/g	a Å	6β-hydrox. of A^4-dione nmol prod./nmole P-450		M Daltons	No. of subunits
				NADPH	NaIO$_4$		
0	8.1	0.815	52	0.612	0.86	266,000	4
0.1	3.5	0.825	48	0.645	0.74	104,600	1
0.3	3.4	0.825	46	0.645	0.61	97,800	1
0.5	3.4	0.825	46	0.540	0.52	97,800	1
1.0	3.4	0.825	46	0.450	0.38	97,800	1

Attempts were made to determine the partial specific volume $\bar{v}$ both from a theoretical basis and in a practical way. P-450 was layered into 3 ml sucrose gradients (40-70% w/w) with 0, 0.1 and 0.3% TX-100 and the tubes were centrifuged for 4 days at 230,000 x g. Fractions of 0.2 ml were collected from the bottom, the sucrose concentrations were determined refractometrically and P-450 was measured by its absorbance at 416 nm. In all three cases P-450 concentrated as very narrow bands at a density of about 1.2 g/cm^3 (cf. Table 1) which corresponds to $\bar{v}$-values of about 0.82. It is now well established that molecular weights of proteins as well as of protein-detergent complexes (18-20) can be calculated from the formula:

$$M = \frac{6 \times \eta_{20,w} \times s_{20,w} \times a \times N}{1 - \bar{v} \times \rho_{20,w}}$$

in which M is the molecular weight; $\eta_{20,w}$ is viscosity of water at 20 C; $s_{20,w}$ sedimentation coefficient; $\rho_{20,w}$ density of water at 20 C; a Stokes radius and N Advogadros number. When introducing the values from Table 1 into the formula, a molecular weight of 266,000 for the major P-450 band without added detergent was obtained and 97,800 for complexes where maximum Triton X-100 is bound. P-450LM$_2$ has a minimum molecular weight of 50,000 according to SDS-gel-electrophoresis (6). Calculations from this value revealed that 0.9 mg Triton X-100 is bound per mg of P-450, i.e. about 74 nmoles per nmole P-450. These values are similar to those obtained with other membrane proteins (18,21,22). In all these calculations it has been assumed that under equilibration with Triton X-100 every remaining Renex 690 molecule (detergent used during purification) has been exchanged with the Triton detergent. A molecular weight of 266,000 for the major form of P-450LM$_2$ without Triton attached indicates that the protein exists as a tetramer with about 0.25 mg of detergent per mg of P-450 still bound. The correspondingly high value for the partial specific volume of 0.815 is also in agreement with a high detergent content. Upon equilibration with Triton X-100 the monomer is formed. This form is still catalytically active in both NaIO$_4$- and NADPH-supported 6β-hydroxylation of androstenedione (cf. Table 1).

The calculated molecular weights are very dependent on the $\bar{v}$-values obtained. In order to check the experimental value, the partial specific volume was calculated from the amino acid composition (23,24), assuming 51.12% of the P-450-Triton X-100 complex being protein, amount of Triton X-100 bound using $\bar{v} = 0.91$ for the detergent (20) and the carbohydrate content (25) with $\bar{v}$-values given in ref. (26). The results obtained gave a theoretical value for $\bar{v}$ of 0.823 for the P-450LM$_2$-Triton X-100 complexes which is in good agreement with the experimental value.

ACKNOWLEDGEMENTS

I am indebted to Professor J.-Å. Gustafsson for valuable discussions and to Miss Inger Johansson for excellent technical assistance.

This work was supported by grants from Magnus Bergvalls Stiftelse and the Swedish Medical Research Council (grant No. 13X-2819).

REFERENCES

1. Lu, A.Y.H., and Coon, M.J., Role of hemoprotein P-450 in fatty acid ω-hydroxylation in a soluble enzyme system from liver microsomes, J. Biol. Chem. 243, 1331 (1968).
2. Strobel, H.W., Lu, A.Y.H., Heidema, J. and Coon, M.J., Phosphatidylcholine requirement in the enzymatic reduction of hemoprotein P-450 and in fatty acid, hydrocarbon, and drug hydroxylation, J. Biol. Chem. 245, 4851 (1970).
3. Lu, A.Y.H., Levin, W., and Kuntzman, R., Reconstituted microsomal enzyme system that hydroxylates drugs, other foreign compounds and endogenous substrates. VII. Stimulation of Benzphetamine N-demethylation by lipid and detergent, Biochem. Biophys. Res. Commun. 60, 266 (1974).
4. Duppel, W., and Ullrich, U., Membrane effects on drug mono-oxygenation activity on hepatic microsomes, Biochim. Biophys. Acta 426, 399 (1976).

5. Narasimhulu, S., Role of phospholipids in adrenocortical microsomal hydroxylations: Activation of lipid depleted microsomal preparations by nonionic detergents, Adv. Exp. Med. Biol. 58, 271 (1975).
6. Haugen, D.A., van der Hoeven, T.A., and Coon, M.J., Purified liver microsomal cytochrome P-450, separation and characterization of multiple forms, J. Biol. Chem. 250, 3567 (1975).
7. Gaylor, J.L., and Delwiche, C.U., Removal of nonionic detergents from proteins by chromatography on Sephadex LH-20, Anal. Biochem. 28, 361 (1969).
8. Ingelman-Sundberg, M., Phospholipids and detergents as conformational effectors in the liver microsomal hydroxylase system, Biochem. Biophys. Res. Commun. In press (1976).
9. Hrycay, E.G., Gustafsson, J.-A., Ingelman-Sundberg, M., and Ernster, L., Sodium periodate, sodium chlorite, organic hydroperoxides and H_2O_2 as hydroxylating agents in steroid hydroxylations catalyzed by partially purified cytochrome P-450, Biochem. Biophys. Res. Commun. 66, 209 (1975).
10. Zakim, D., Regulation of microsomal enzymes byphospholipids. I. The effect of phospholipases and phospholipids on glucose 6-phosphatase, J. Biol. Chem. 245, 4953 (1970).
11. Vessey, D.A., and Zakim, D., Regulation of microsomal enzymes by phospholipids. II. Activation of hepatic Uridine Diphosphate-Glucoronyltransferase, J. Biol. Chem. 246, 4649 (1971).
12. Cunningham, C.C., and Hager, L.P., Crystalline pyruvat-oxidase from Escherichia coli.II. Activation by phospholipids, J. Biol. Chem. 246, 1575 (1971).
13. Norred, W.P., and Wade, A.G., Dietary fatty acid-induced alterations of hepatic microsomal drug metabolism, Biochem. Pharm. 21, 2887 (1972).
14. Laitinen, M., Enhancement of hepatic drug metabolism with dietary cholesterol in the rat, Acta pharmacol. et toxicol. 39, 241 (1976).
15. Hietanen, E., Laitinen, M., Vainio, H., and Hänninen, O., Dietary fats and properties of endoplasmatic reticulum: II. Dietary lipid induced changes in activities of drug metabolizing enzymes in liver and duodenum of rat, Lipids 10, 467 (1975).
16. Kagawa, Y., and Racker, E., Partial resolution of the enzymes catalyzing oxidative phosphorylation. XXV. Reconstitution of vesicles catalyzing $^{32}P_i$-Adenosine Triphosphate exchange, J. Biol. Chem. 246, 5477 (1971)
17. Makino, S., Reynolds, J.A., and Tanford, C., The binding of deoxycholate and Triton X-100 to proteins, J. Biol. Chem. 248, 4926 (1973).
18. Clarke, S., The size and detergent binding of membrane proteins, J. Biol. Chem. 250, 5459 (1975).
19. Simons, K., Helenius, A., and Garoff, H., Solubilization of the membrane proteins from Semliki Forest virus with Triton X-100, J. Mol. Biol. 80, 119 (1973).
20. Tanford, C., Nozaki, Y., Reynolds, I., and Makino, S., Molecular characterization of proteins in detergent solutions, Biochemistry 13, 2369 (1974).
21. Walter, H., and Hasselbach, W., Properties of the calcium-independent ATPase of the membranes of the sarcoplasmic reticulum delipidated by the nonionic detergent Triton X-100, Eur. J. Biochem. 36, 110 (1973).
22. Osborn, B.H., Sardet, C., and Helenius, A., Bovine Rhodopsin: Characterization of the complexes formed with Triton X-100, Eur. J. Biochem. 44, 383 (1974).
23. Dus, K., Litchfield, W., Miguel, A.G., van der Hoeven, T.A., Haugen, D.A., Dean, W.L., and Coon, M.J., Structural resemblance of cytochrome P-450 isolated from Pseudomonas putida and from rabbit liver microsomes, Biochem. Biophys. Res. Commun. 50, 15 (1974).
24. Cohn, E.J., and Edsall, J.T., Proteins, Amino Acids and Peptides as Dipolar Ions, Reinhold, New York, 1943, pp 370-375.

25. Coon, M.J., The purification of membrane-bound mono-oxygenases, presentation at the Third International Symposium on Microsomes and Drug Oxidations, Berlin, 1976.
26. Gibbons, R.A., Glycoproteins, Elsevier, Amsterdam, 1966, p 61.

FURTHER STUDIES OF THE FLUROXENE MEDIATED DESTRUCTION OF HEPATIC MICROSOMAL CYTOCHROMES P-450 IN VITRO

Kathryn M. Ivanetich, Julia A. Marsh, Jean J. Bradshaw, and Laurence S. Kaminsky

Department of Physiology & Medical Biochemistry, University of Cape Town Medical School, Cape Town, South Africa

ABSTRACT

Fluroxene (2,2,2-trifluoroethyl vinyl ether) is known to mediate the destruction of cytochromes P-450 *in vivo* and *in vitro* (1,2). Structural analogues of fluroxene do not mediate the destruction of cytochromes P-450 in phenobarbital induced microsomes, suggesting that the capability for destroying cytochromes P-450 in phenobarbital induced microsomes resides neither in the vinyl nor in the trifluoroethyl functional group alone. In contrast, known metabolites of fluroxene can apparently decrease the levels of cytochromes P-450 in phenobarbital induced microsomes and could possibly be responsible for the destruction of cytochromes P-450 observed in the presence of fluroxene, NADPH, and phenobarbital induced microsomes *in vitro*. Inhibition of the fluroxene mediated destruction of cytochromes P-450 by carbon monoxide, SKF 525A and anaerobic conditions indicates that the destruction of cytochromes P-450 by fluroxene requires the interaction of fluroxene with the cytochrome P-450 dependent drug metabolising pathway. A comparison of K_m and V_{max} values for the fluroxene mediated destruction of cytochromes P-450 in microsomes from variously induced rats indicates that both cytochrome P-450 and cytochrome P-448 may catalyze this reaction.

The disappearance of cytochromes P-450 in the presence of fluroxene, NADPH and induced microsomes is found to follow a first order rate equation. An observed first order rate constant of $5 \times 10^{-2} \text{min}^{-1}$ was calculated for this reaction. The maximum extent of destruction of cytochromes P-450 in differently induced microsomes indicates that both cytochrome P-448 and cytochrome P-450 appear to be destroyed *in vitro*.

INTRODUCTION

Relatively few compounds are known to destroy hepatic microsomal cytochromes P-450 *in vivo* and/or *in vitro* without affecting other microsomal proteins. Examples of this type of compound include allyl-*iso*-propylacetamide (AIA) (3,4), allyl containing barbiturates (5) and the volatile anaesthetic agent fluroxene (2,2,2-trifluoroethyl vinyl ether) (1,2). In the cases of fluroxene and AIA, the process has been clearly demonstrated to involve destruction of the heme moiety of existing type P-450 cytochromes (1,5). With regard to the fluroxene mediated destruction of cytochromes P-450, the questions of (i) the identity of reactive species destroying the heme moiety of cytochromes P-450, (ii) the mechanism of this process, and (iii) the product(s) of the altered heme moiety are currently being investigated in our laboratories. We report herein the results of our initial investigations of the mechanism of destruction of hepatic microsomal cytochromes P-450 by fluroxene *in vitro*. In this manuscript the terminology 'cytochrome P-448' or 'cytochrome P-450' refers to the specific enzyme indicated, whereas 'type P-450 cytochromes' or 'cytochromes P-450' refers to the heterogeneous group of enzymes including cytochrome P-450 and cytochrome P-448.

EXPERIMENTAL

Microsomes were isolated by gel filtration from control or induced male Wistar rats weighing 180 to 220 g as described earlier (1). Experimental procedures and assays are as described elsewhere (1,6,7).

RESULTS AND DISCUSSION

Destruction of Cytochromes P-450 by Analogues and Metabolites of Fluroxene in Phenobarbital Induced Microsomes *in vitro*

It has been reported earlier (1) and is confirmed in Table I that type P-450 cytochromes are destroyed following incubation of hepatic microsomes with fluroxene and NADPH *in vitro*. The destructive process appears to be specific for the heme moiety of cytochromes P-450 in that the levels of other microsomal proteins such as cytochrome b_5 or NADPH-cytochrome c reductase (1) or of added exogenous heme are unaffected (7). The fluroxene mediated destruction of cytochromes P-450 in microsomes from phenobarbital induced rats is not mimicked by compounds similar in structure to fluroxene such as ethyl vinyl ether, divinyl ether or 2,2,2-trifluoroethyl ethyl ether (Table 1), suggesting that no portion of the fluroxene molecule, either the vinyl or the trifluoroethyl portion, alone retains the ability to destroy cytochromes P-450 in phenobarbital induced microsomes *in vitro*. The metabolite 2,2,2-trifluoroethanol does not affect the levels of cytochromes P-450 *in vitro*, whereas the metabolites trifluoroacetic acid and bicarbonate and the proposed metabolite trifluoroacetaldehyde slightly lower the levels of cytochromes P-450 in phenobarbital induced microsomes (Table 1). The sum of the losses of cytochromes P-450 seen in the presence of trifluoroacetic acid, trifluoroacetaldehyde and bicarbonate approximately equal the loss of cytochromes P-450 seen in the presence of fluroxene alone. These results suggest that for phenobarbital induced microsomes, metabolites of fluroxene may be responsible for the fluroxene mediated destruction of cytochromes P-450 *in vitro*.

TABLE 1 The Interaction of Analogues and Metabolites of Fluroxene with Cytochromes P-450 in Phenobarbital Induced Microsomes *in vitro*

COMPOUND	BINDING[a] (TYPE I)	METABOLISM[b]	% DESTRUCTION OF[c] CYTS P-450
$CF_3CH_2OCH=CH_2$ (Fluroxene)	+	+	25%
$CH_3CH_2OCH=CH_2$	+	+	0%
$CH_2=CHOCH=CH_2$	+	+	0%
$CF_3CH_2OCH_2CH_3$	+	+	0%
CF_3CH_2OH	+	+	0%
CF_3CHO	+	±	8%
$CF_3COO^-Na^+$	−	−	8%[d]
$HCO_3^-Na^+$	−	−	16%[d]

[a] Assayed for by the appearance of a difference spectrum. Microsomes: 2 mg protein/ml 0.02 M Tris-HCl, pH 7.4.

[b] Measured by carbon monoxide inhibitable NADPH consumption.

[c] Assayed by the method of Omura and Sato (8) following incubation of the compound, microsomes, EDTA and NADPH or NADPH generating system at 30° for 30 min. Reference cuvettes for assays contained only microsomes which had been incubated as for the samples. Initial levels were approximately 2.5 nmol cytochromes P-450/mg microsomal protein.

[d] With or without NADPH present.

The Effects of Inhibitors on the Fluroxene Mediated Destruction of Cytochromes P-450 and on the Cytochrome P-450 Dependent Metabolism of Fluroxene *in vitro*

The effects of inhibitors on the fluroxene mediated destruction of cytochromes P-450 and the cytochrome P-450 dependent metabolism of fluroxene to 2,2,2-trifluoroethanol were investigated in phenobarbital induced microsomes *in vitro*. The metabolism of fluroxene and the fluroxene mediated destruction of cytochromes P-450 are fully inhibited under totally anaerobic conditions. Carbon monoxide ($CO:O_2$, 80:20; v/v) and SKF 525A inhibit the metabolism of fluroxene by approximately 60%, but fully inhibit the fluroxene mediated destruction of cytochromes P-450. The results of these studies indicate that the destruction of cytochromes P-450 by fluroxene may require the functioning of the cytochrome P-450 dependent enzyme system, and that the destruction reaction is more sensitive to inhibitors than is the reaction in which fluroxene is converted to 2,2,2-trifluoroethanol.

K_m and V_{max} Values for the Fluroxene Mediated Destruction of Cytochromes P-450 *in vitro*

The destruction of cytochromes P-450 *in vitro* as a function of fluroxene concentration was investigated in variously induced microsomes (Table 2). The K_m values for the destruction of cytochromes P-450 by fluroxene were calculated as 0.8 mM, 3.3 mM and 1.5 mM for phenobarbital induced, 3-methylcholanthrene induced and control microsomes, respectively. V_{max} values of approximately 0.5 nmol of cytochromes P-450 destroyed/mg microsomal protein/7 min were obtained for both types of induction compared to 0.2 nmol cytochromes P-450 destroyed/mg microsomal protein/10 min in control microsomes, indicating that the fluroxene mediated destruction of cytochromes P-450 is enhanced by elevated levels of cytochrome P-448 or cytochrome P-450.

In contrast, the cytochrome P-450 dependent conversion of fluroxene to its major metabolite, 2,2,2-trifluoroethanol, appears to be catalyzed only by cytochrome P-450. This conclusion is drawn from a comparison of K_m and V_{max} values for the conversion of fluroxene to 2,2,2-trifluoroethanol in microsomes from variously induced animals (7,9) and is in agreement with earlier studies using NADPH consumption as a measure of the metabolism of fluroxene *in vitro* (6). The observation that the ratios of trifluoroethanol produced to cytochromes P-450 destroyed vary widely in differently induced microsomes (Table 2) is consistent with the involvement of different type P-450 cytochromes in the two reactions.

TABLE 2 The Effect of Fluroxene Concentration on the Fluroxene Mediated Destruction of Cytochromes P-450 *in vitro*

INDUCTION[a]	[CYTS P-450] (nmol/mg mic. protein)	K_m (mM)	V_{max} (nmol/mg mic. protein/7 min)	TFE produced[b] / Cyts P-450 destroyed
None	1.2	1.5	0.2[c]	225
PB	2.7	0.8	0.5	295
MC	2.1	3.3	0.5	105

[a] Abbreviations used are PB, phenobarbital; MC, 3-methylcholanthrene; TFE, 2,2,2-trifluoroethanol; Cyts P-450, total type P-450 cytochromes; mic., microsomal.
[b] Calculated from (7).
[c] nmol/mg mic. protein/10 min.

The Fluroxene Mediated Destruction of Cytochromes P-450 *in vitro* as a Function of Time

The destruction of cytochromes P-450 in the presence of 30 mM fluroxene was monitored as a function of time in microsomes from variously induced rats (Table 3). For control microsomes the extent of destruction was not great enough to determine the order of the reaction. In the case of 3-methylcholanthrene and phenobarbital induced microsomes, the destruction of cytochromes P-450 followed first order kinetics with respect to the disappearance of cytochromes P-450. Pseudo first order rate constants of approximately $5 \times 10^{-2} \text{min}^{-1}$ were obtained for both types of microsomes. The maximum amount of destruction in these experiments was 1.0/2.0 and 1.0/2.5 nmol cytochromes P-450 destroyed in 3-methylcholanthrene and phenobarbital induced microsomes, respectively. The extent of destruction of cytochromes P-450 observed in these experiments is greater than could be accounted for by the destruction of either cytochrome P-448 or cytochrome P-450 alone. This is in contrast to AIA, which appears to decrease the levels of cytochrome P-450 preferentially while not greatly affecting cytochrome P-448 (3).

TABLE 3 The Fluroxene Mediated Destruction of Cytochromes P-450 *in vitro* as a Function of Time[a]

INDUCTION[b]	[CYTS P-450] (nmol/mg mic. protein)	k_1 (min^{-1})	MAXIMUM DESTRUCTION CYTS P-450 (nmol/mg mic. protein)
None	1.2	–	0.25
PB	2.5	4.3×10^{-2}	1.0
MC	2.0	4.9×10^{-2}	1.0

[a] Microsomes (2 mg protein/ml 0.02 M Tris-HCl buffer, pH 7.4) incubated at 30°. Samples contained microsomes, fluroxene (30 mM), NADPH generating system (7) and EDTA and were assayed against reference cuvettes containing only microsomes incubated as for samples.
[b] Abbreviations used are PB, phenobarbital; MC, 3-methylcholanthrene; Cyts P-450, total type P-450 cytochromes; mic., microsomal.

The observation that fluroxene, which was until recently a medically used anaesthetic agent, can destroy cytochromes P-450 *in vivo* and *in vitro* should stimulate investigations into the capabilities of other medically used drugs for affecting the levels of cytochromes P-450 in a similar fashion. The abilities of drugs to destroy specifically cytochromes P-450 would have important implications for the elucidation of drug-drug interactions *in vivo*.

REFERENCES

(1) K.M. Ivanetich, J.A. Marsh, J.J. Bradshaw and L.S. Kaminsky, Fluroxene (2,2,2-trifluoroethyl vinyl ether) mediated destruction of cytochrome P-450 *in vitro*, Biochem. Pharmacol. 24, 1933 (1975).

(2) K.M. Ivanetich, J.J. Bradshaw, J.A. Marsh, G.G. Harrison and L.S. Kaminsky, The role of cytochrome P-450 in the toxicity of fluroxene (2,2,2-trifluoroethyl vinyl ether) anaesthesia *in vivo*, Biochem. Pharmacol. 25, 773 (1976).

(3) W. Levin, M. Jacobson and R. Kuntzman, Incorporation of Radioactive-δ-aminolevulinic acid into microsomal cytochrome P-450: Selective breakdown of the hemoprotein by allylisopropylacetamide and carbon tetrachloride, Archs. Biochem. Biophys. 148, 262 (1972).

(4) F. De Matteis, Rapid loss of cytochrome P-450 and haem caused in the liver microsomes by the porphyrogenic agent 2-allyl-2-isopropylacetamide, FEBS Lett. 6, 343 (1970).

(5) W. Levin, M. Jacobson, E. Sernatinger and R. Kuntzman, Breakdown of cytochrome P-450 heme by secobarbital and other allyl containing barbiturates, Drug Metab. Dispos. 1, 275 (1973).

(6) K.M. Ivanetich, J.J. Bradshaw, J.A. Marsh and L.S. Kaminsky, The interaction of hepatic microsomal cytochrome P-450 with fluroxene (2,2,2-trifluoroethyl vinyl ether) *in vitro*, Biochem. Pharmacol. 25, 779 (1976).

(7) J.A. Marsh, J.J. Bradshaw, S. Lucas, G. Sapeika, L.S. Kaminsky and K.M. Ivanetich, An investigation of the mechanism of the fluroxene mediated destruction of hepatic microsomal cytochromes P-450 *in vitro*, submitted for publication.

(8) T. Omura and R. Sato, The carbon monoxide-binding pigment of liver microsomes. I. Evidence for its hemoprotein nature, J. Biol. Chem. 239, 2370 (1964).

(9) K.M. Ivanetich, J.A. Marsh and L.S. Kaminsky, The metabolism of fluroxene and 2,2,2-trifluoroethyl ethyl ether *in vitro*, in preparation.

ACKNOWLEDGEMENTS

This work was supported by grants from the Medical Research Council and UCT Staff Research Fund.

PURIFICATION OF MEMBRANE-BOUND OXYGENASES: ISOLATION OF TWO ELECTROPHORETICALLY HOMOGENEOUS FORMS OF LIVER MICROSOMAL CYTOCHROME P-450*

M. J. Coon, D. P. Ballou, D. A. Haugen,[†] S. O. Krezoski, G. D. Nordblom, and R. E. White[‡]

Department of Biological Chemistry, Medical School, The University of Michigan, Ann Arbor, Michigan 48109, U.S.A.

ABSTRACT

Evidence has been obtained from enzyme fractionation that liver microsomal cytochrome P-450 (P-450$_{LM}$) exists in multiple forms. Two of these have been isolated in a homogeneous state. They are designated according to their relative electrophoretic mobilities on polyacrylamide gel in the presence of sodium dodecyl sulfate as P-450$_{LM_2}$ and P-450$_{LM_4}$. P-450$_{LM_2}$, which was isolated from phenobarbital-induced rabbits, has a subunit molecular weight of 48,700, and P-450$_{LM_4}$ from β-naphthoflavone-induced rabbits has a subunit molecular weight of 55,300. These proteins also differ in their optical and electron paramagnetic resonance spectra, catalytic activities, and other properties. Studies on the mechanism of action of P-450$_{LM_2}$ have shown that it produces hydrogen peroxide during substrate hydroxylation in the presence of molecular oxygen in a reconstituted enzyme system containing NADPH, NADPH-cytochrome P-450 reductase, and phosphatidylcholine. Peroxides also serve as oxygen donors for substrate hydroxylation, catalyzed by P-450$_{LM_2}$ in the absence of molecular oxygen, NADPH, and the reductase. P-450$_{LM_2}$ is a two-electron acceptor under anaerobic conditions; the reduced protein reacts with O_2 to form two different oxyferro complexes, as shown by stopped flow methods. The rate of decomposition of the second complex appears to be altered by the presence of cytochrome b_5 or NADPH-cytochrome P-450 reductase. A scheme is presented to account for these and other related findings on the mechanism of P-450$_{LM}$-catalyzed hydroxylation reactions.

INTRODUCTION

A little over eight years ago, at the first International Symposium on Microsomes and Drug Oxidations held in Bethesda, we reported the solubilization and reconstitution of the P-450$_{LM}$-containing hydroxylation system of liver micro-

---

*This research was supported by Grant PCM76-14947 from the National Science Foundation and Grant AM-10339 from the United States Public Health Service.
†Postdoctoral Fellow, United States Public Health Service. Present address, Argonne National Laboratory, Argonne, Illinois.
‡Postdoctoral Fellow, United States Public Health Service.

somes[1] (1). Resolution of the system was accomplished by treatment of the microsomal membranes with a detergent in the presence of glycerol and other protective agents, followed by ion exchange column chromatography, and yielded three components: cytochrome P-450, NADPH-cytochrome P-450 reductase, and a heat-stable factor subsequently identified as phosphatidylcholine (2-4). More recently, we have devoted our efforts to the purification and characterization of these components with the goal of understanding (a) the basis for the remarkably broad substrate specificity of the liver microsomal hydroxylating system; (b) the interactions among the three main components, and also their interactions with other microsomal proteins; and (c) the mechanism of oxygen activation and insertion into substrates, catalyzed by cytochrome P-450.

The present paper will briefly summarize our present understanding of the properties of the components of the microsomal hydroxylating system. Due to limitations of space we will describe primarily the studies in our own laboratory, but major contributions made by others will also be indicated.

Evidence for the Occurrence of Multiple Forms of $P-450_{LM}$

The question of whether the numerous activities now attributed to $P-450_{LM}$ reside in one or more forms of this pigment has been the subject of much investigation. The variable enzyme activities toward different substrates observed in microsomes of animals treated with several inducing agents suggested that numerous catalysts might be involved (5-8). However, kinetic data obtained with microsomal suspensions (9-11) and the reconstituted system (12) indicated that a number of substrates act as mutually competitive inhibitors, thereby showing that they may be acted on by a single enzyme. Spectral evidence was presented suggesting the occurrence of two forms of cytochrome P-450 in liver microsomes from animals induced by PB or by polycyclic aromatic hydrocarbons (13-17). Studies on genetic regulation also supported the occurrence of a separate enzyme for the hydroxylation of aryl hydrocarbons (18). With the availability of methods for the solubilization and fractionation of liver microsomal proteins, such techniques were applied to microsomes obtained after the administration of different inducing agents to animals (19), and it was demonstrated that the substrate specificity resides in the different cytochrome fractions, as assayed in reconstituted systems (20-22).

Finally, some of the forms of $P-450_{LM}$ have recently been obtained in an apparently homogeneous state, thereby permitting an unequivocal assignment of different properties to different cytochromes. We have reported the isolation of the PB-inducible form of $P-450_{LM}$ in an electrophoretically homogeneous state from rabbits (23,24), as well as the separation and characterization of other forms with different catalytic and physical properties (25). More recently, we have characterized in detail electrophoretically homogeneous $P-450_{LM_2}$ and $P-450_{LM_4}$, isolated from PB- and BNF-induced microsomes, respectively (26), and shown that these proteins also appear to be homogeneous by the Ouchterlony immunochemical technique (27,28). Highly purified $P-450_{LM}$ has recently been ob-

---

[1]The abbreviations used are: $P-450_{LM}$, liver microsomal cytochrome P-450; PB, phenobarbital; BNF, β-naphthoflavone; and SDS, sodium dodecyl sulfate.

tained from induced animals by two other laboratories. The isolation of the cytochrome from 3-methylcholanthrene-treated rats and rabbits and from PB-treated rats has been reported by investigators at Hoffmann-La Roche (29,30). Immunochemical studies by the same group (31) showed that P-450$_{LM}$ from the carcinogen-treated rats was homogeneous but that the preparation from the drug-treated rats contained at least three hemeproteins. The cytochrome has also been isolated in electrophoretically homogeneous form from PB- and 3-methylcholanthrene-treated rabbits by investigators in Japan (32,33). The latter preparation appeared to contain bound carcinogen, thereby accounting for the observed optical spectrum.

Nomenclature of Multiple Forms of P-450$_{LM}$

As discussed at a separate session immediately preceding this Symposium, the naming of the multiple forms of P-450$_{LM}$ is difficult in view of present inadequate knowledge of their properties and functions. In our view, several methods which have been proposed have serious drawbacks, as follows: (a) Spectral method: the terms P-450 and P-448 are inadequate because there appear to be as many as six forms of the cytochrome in rabbit liver microsomes with rather similar spectra as the reduced CO complexes (25,26), and none of the proteins so far purified has such a complex with a maximum exactly at 450 nm. (b) Induction method: the terms P-450(PB) and P-450(BNF) have been suggested to identify the cytochromes by their inducers, but unfortunately the protein induced by BNF is present at about half as great a level when no inducer or PB is administered to the animals (26). Furthermore, some of the forms are not known to be inducible. (c) Substrate method: it would seem inadvisable to use this method until the substrate specificities are better delineated; the evidence so far available with the rabbit liver cytochromes suggests that they all tend to have a broad spectrum of activities, but with some quantitative differences (25). We have, instead, been using an electrophoretic method of designating our various forms of P-450$_{LM}$ which has proved useful and reproducible (25,26). The proteins are submitted to discontinuous polyacrylamide gel electrophoresis in the presence of SDS by the procedure of Laemmli (34), which we find gives about the best resolution and reproducibility of the various electrophoretic procedures available. Beginning with the major microsomal band of greatest electrophoretic mobility, the bands are numbered according to decreasing mobility and increasing molecular weight, as shown in Fig. 1. Thus, P-450$_{LM_2}$ is the homogeneous PB-inducible form, and P-450$_{LM_4}$ the homogeneous BNF-inducible form; four additional forms have been partially purified and shown to have typical spectra in the 450 nm region as their reduced CO complexes. As an example of the usefulness of the electrophoretic method, the purified cytochrome from 3-methylcholanthrene-induced rabbits, kindly furnished by Dr. A. Y. H. Lu, and the purified cytochrome from 2,3,7,8-tetrachlorodibenzo-p-dioxin-treated rabbits, kindly furnished by Dr. E. F. Johnson and Dr. U. Muller-Eberhard (35), are electrophoretically indistinguishable from our P-450$_{LM_4}$. Accordingly, one may assume for the present that our three laboratories are working with the same protein and avoid using multiple names based on other criteria, such as the inducer used. It is hoped, however, that a nomenclature based on the functions of these versatile catalysts will eventually be possible.

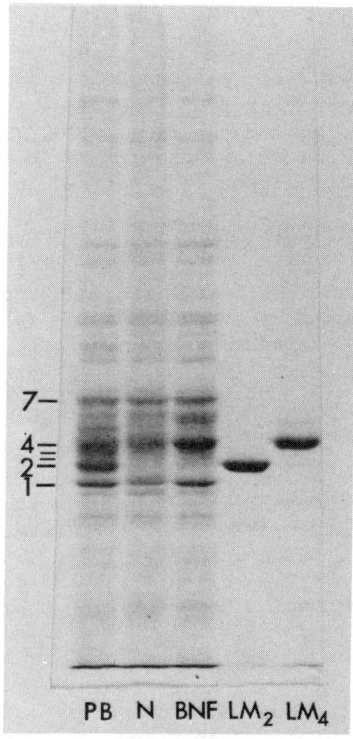

Fig. 1. Polyacrylamide slab gel electrophoresis of $P-450_{LM}$ fractions. The preparations were treated with SDS and mercaptoethanol at 100° and submitted to electrophoresis by the method of Laemmli (34) with a 7.5% separating gel. Migration was from top to bottom. The amounts of protein used were as follows: PB-induced microsomes, normal (uninduced) microsomes, or BNF-induced microsomes, 6 μg of protein each; purified $P-450_{LM_2}$ or LM_4, 1 μg of protein each.

RESULTS AND DISCUSSION

Purification and Characterization of $P-450_{LM_2}$ and $P-450_{LM_4}$

The two forms of cytochrome which have been isolated from rabbit liver microsomes and thoroughly characterized, PB-inducible $P-450_{LM_2}$ and BNF-inducible $P-450_{LM_4}$, have molecular weights of 48,700 and 55,300, respectively (26). These values, which were determined by amino acid analysis (kindly carried out by Dr. Karl M. Dus, St. Louis University Medical School) and also include heme and carbohydrate (28), agree well with those obtained by calibrated SDS-polyacrylamide gel electrophoresis. The purest preparations obtained have a cytochrome content of 20.1 nmol of $P-450_{LM_2}$ per mg of protein and 17.2 nmol of $P-450_{LM_4}$ per mg of protein, and both contain one molecule of heme per polypeptide chain. Cytochrome b_5, NADPH-cytochrome c reductase, and NADH-cytochrome c reductase were not detectable in these purified proteins. During purification, the phospholipid content was reduced to about 0.4 molecule per molecule of polypeptide, which is near the limit of detection.

Whereas $P-450_{LM_2}$ is present only in PB-induced microsomes, $P-450_{LM_4}$ occurs in significant amounts even without induction by BNF or polycyclic hydrocarbons. $P-450_{LM_4}$ isolated from control or PB-induced microsomes is indistinguishable from $P-450_{LM_4}$ isolated from BNF-induced microsomes in its chemical and physical properties, at least by the methods so far applied (26). Four other forms of $P-450_{LM}$ varying in subunit molecular weight from 46,000 to 60,000 have also been partially purified but will not be described in this report.

The optical spectra of purified $P-450_{LM_2}$ and $P-450_{LM_4}$ (the latter from BNF-induced animals) are shown in Fig. 2. Whereas oxidized $P-450_{LM_2}$ has a Soret band at 418 nm typical of the low spin state, the corresponding band of LM_4 is shifted to the blue and is indicative of the high spin state. This difference is seen most clearly when detergent has been largely removed from the latter cytochrome. In the course of such spectral studies it was found that the spectrum of oxidized $P-450_{LM_4}$ was highly dependent upon the concentration of protein and certain other components in the solution, whereas that of $P-450_{LM_2}$ was unaffected. For example, the addition to $P-450_{LM_4}$ of Renex 690 (a non-ionic detergent used in our enzyme purification procedures) at a concentration of 0.5% caused a fairly complete change from high to low spin spectrum, whereas 0.1% Renex gave an intermediate effect. Spectra were also determined in the presence of octylamine, since Jefcoate et al. (36) have used this compound for the quantitative determination of the high and low spin forms of cytochrome P-450 in rabbit liver microsomes. The resulting spectra with the purified cytochromes were as expected. In addition, EPR studies were carried out at liquid helium temperatures, and $P-450_{LM_2}$ was found to have a spectrum almost exclusively that of a low spin ferric hemeprotein, with g values in the high field region of 1.93, 2.25, and 2.43. In contrast, the EPR spectrum of purified $P-450_{LM_4}$ was at least 75% high spin and only partly low spin; the g values were 3.84 and 8.36 in the low field region and 1.93, 2.26, and 2.42 in the high field region. The change from high to low spin state we have observed upon the addition of Renex to purified $P-450_{LM_4}$ may possibly be due to the release of BNF or some other endogenous substrate bound to the isolated enzyme. It may be noted that Hashimoto and Imai (33) observed a similar change upon the addition of Emulgen, a nonionic detergent, to cytochrome P-450 isolated from microsomes of 3-methylcholanthrene-treated rabbits and thought to contain some of the carcinogen in a tightly bound form.

As shown in Table 1, $P-450_{LM_2}$, LM_4, and $LM_{1,7}$ (a mixture of LM_1 and LM_7) all have broad substrate specificity but differ in the turnover number with particular substrates (25). The position-specific oxygenation of benzo[a]pyrene by the different forms of purified $P-450_{LM}$ has been reported elsewhere (37).

Purification and Characterization of NADPH-Cytochrome P-450 Reductase

The detergent-solubilized liver microsomal NADPH-cytochrome c reductase retains the ability to reduce cytochrome P-450 (4), its natural substrate, and therefore to function in a reconstituted hydroxylation system (3). In contrast, cytochrome c reductase preparations purified after solubilization with a protease such as bromelain (38) or with a lipase (39) believed to contain a protease (40) do not retain activity toward cytochrome P-450 (41). Proteolytic solubilization apparently removes part of the polypeptide chain (42). As shown by gel filtration studies, the intact (detergent-solubilized) reductase is capable of binding to purified $P-450_{LM_2}$, whereas the protease-treated re-

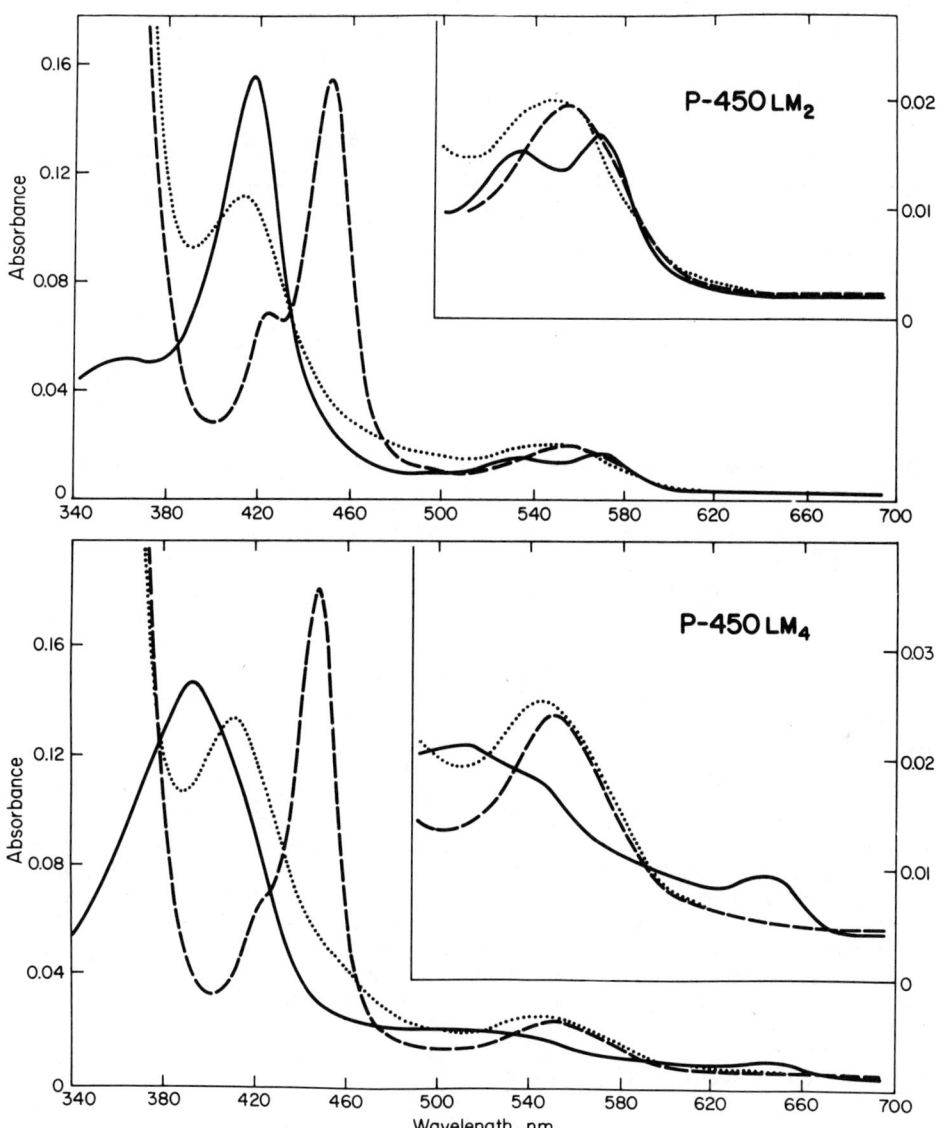

Fig. 2. Absolute spectra of purified P-450$_{LM_2}$ and P-450$_{LM_4}$. The latter was isolated from BNF-induced microsomes. The concentrations, based on heme analysis, were 1.40 nmol of P-450$_{LM_2}$ and 1.53 nmol of P-450$_{LM_4}$ per ml of 0.1 $\underline{M}$ phosphate buffer, pH 7.4, containing 20% glycerol. The spectra of the oxidized forms, reduced forms, and reduced CO complexes are shown by solid, dotted, and dashed curves, respectively.

TABLE 1 Substrate Specificity of Different Forms of Cytochrome P-450[a]

Substrate	Position of hydroxyl group in product	Activity of $P\text{-}450_{LM}$ (nmol/min/nmol P-450)		
		LM_2	LM_4	$LM_{1,7}$
Benzphetamine		66	3.0	7.5
p-Nitroanisole		6.2	0.6	0.6
Aniline		7.4	5.3	6.4
Biphenyl	2	0.7	0.3	0.4
"	4	5.4	0.4	0.6
Testosterone	6β	0.02	0.02	0.32
"	7α	0.02	0.04	0.02
"	16α	0.43	0.02	0.07
Benzpyrene		0.04	Trace	0.5

[a]The reaction mixtures were at 30° and contained $P\text{-}450_{LM}$ (0.1 nmol per ml), other components of the reconstituted system, and various substrates as described previously (25).

ductase is not (27). We have obtained the detergent-solubilized NADPH-cytochrome P-450 reductase from rat liver microsomes in highly purified form and shown that it contains both FMN and FAD (43,44). Both flavin nucleotides are also present in partially purified reductase from rabbit liver microsomes (23). Iyanagi and Mason (45) first reported the presence of both flavins in various reductase preparations. We now have evidence for the existence of two forms of NADPH-cytochrome P-450 reductase differing slightly in apparent minimum molecular weight as judged by electrophoresis; the two forms have been purified from both rat (43,46) and rabbit liver microsomes (26,47). Recently, Dignam and Strobel (48) have obtained an apparently homogeneous reductase preparation from rat liver and have made use of NADP-Sepharose affinity column chromatography to isolate the enzyme in high yield (49). Also, Yasukochi and Masters (50,51) have used 2',5'-ADP-Sepharose affinity column chromatography to obtain the apparently homogeneous enzyme from both rat and pig liver in very good yield; Dr. Masters has discussed this work at the present symposium.

Studies on the Mechanism of Action of $P\text{-}450_{LM}$

Several aspects of the mechanism of action of liver microsomal cytochrome P-450 have been clarified by studies with enzymes purified as described above.

Hydrogen peroxide formation and stoichiometry of the hydroxylation reaction.
According to the general equation for a mixed-function oxidation reaction (52), the expected stoichiometry of a $P\text{-}450_{LM}$-catalyzed hydroxylation reaction would correspond to the consumption of NADPH and O_2 and the formation of product in equimolar amounts, as follows:

$$RH + O_2 + NADPH + H^+ \longrightarrow H_2O + ROH + NADP^+ \tag{1}$$

In Equation 1, RH represents the substrate and ROH the product; with benzphetamine as substrate the reaction is believed to result in the conversion of the N-methyl group to an N-hydroxymethyl group, which decomposes spontaneously to yield formaldehyde and desmethylbenzphetamine (benzylamphetamine). Experimentally, it has proved quite difficult to establish the expected stoichiometry for hydroxylation reactions in liver microsomal suspensions. The values obtained have varied somewhat with the conditions employed, and various procedures have been used to correct for the endogenous rates of NADPH and O_2 utilization (53-58). In earlier experiments on the stoichiometry of benzphetamine demethylation in partially purified enzyme preparations, it appeared that added catalase reduced the oxygen uptake to the expected value (12). We have recently studied the stoichiometry of the hydroxylation of several substrates in a reconstituted system containing highly purified P-450$_{LM_2}$ and reductase and have found that hydrogen peroxide production in this catalase-free system varies with the substrate added. Peroxide formation represents endogenous NADPH oxidase activity (apparently due to autoxidation of the cytochrome), as follows:

$$O_2 + NADPH + H^+ \longrightarrow H_2O_2 + NADP^+ \tag{2}$$

The NADPH-dependent formation of H_2O_2 in liver microsomes was first reported by Gillette et al. (59) and has also been observed by others (60-63). Our results indicate that the expected 1:1:1 stoichiometry for O_2 consumption, NADPH oxidation, and product formation is obtained for various substrates only when a correction is made for the O_2 and NADPH contributing to the formation of hydrogen peroxide (Equation 2).

<u>Peroxide utilization for substrate hydroxylation</u>. The ability of liver microsomal suspensions to utilize organic hydroperoxides for the hydroxylation of various substrates and the possible role of cytochrome P-450 in these reactions have been reported by several laboratories (64-69). In a recent study we have shown that highly purified P-450$_{LM_2}$ free of other known electron carriers catalyzes the hydroperoxide-dependent hydroxylation of a variety of substrates in the absence of NADPH, NADPH-cytochrome P-450 reductase, and molecular oxygen (70). The addition of phosphatidylcholine is necessary for maximal activity. The ferrous form of P-450$_{LM_2}$ is not involved in peroxide-dependent hydroxylation reactions, as indicated by the lack of inhibition by carbon monoxide.

With cumene hydroperoxide present, a variety of substrates are attacked, including N-methylaniline, cyclohexane, benzphetamine, and aminopyrine. With benzphetamine as the substrate, cumene hydroperoxide may be replaced by other peroxides, including hydrogen peroxide, or by peracids or sodium chlorite. A study of the stoichiometry, as shown in Table 2, indicated that equimolar amounts of N-methylaniline, formaldehyde, and cumyl alcohol (α,α-dimethybenzyl alcohol) are formed in the reaction of N,N-dimethylaniline with cumene hydroperoxide. Accordingly, the reaction may be written as follows:

$$RH + XOOH \longrightarrow ROH + XOH \tag{3}$$

In Equation 3, RH is the substrate and XOOH the starting peroxide. In other studies it was shown that $H_2^{18}O$ is incorporated only slightly into cyclohexanol in the reaction of cyclohexane with cumene hydroperoxide. Thus, it appears that the oxygen atom in cyclohexanol is derived primarily from the peroxide, in accord with Equation 3.

TABLE 2 Stoichiometry of Peroxide-supported Demethylation Reaction[a]

Component determined	Amount of component (nmol)				
	Complete system	P-450$_{LM}$ omitted	Substrate omitted	Net change	Ratio relative to N-methylaniline
N-methylaniline	518	0	0	+518	1.0
Formaldehyde	438	0	0	+438	0.8
Cumyl alcohol	684	0	161	+523	1.0

[a]The reaction mixtures were incubated for 5 min at 30° and contained 2.1 nmol of P-450$_{LM_2}$, 7.0 μmol of N,N-dimethylaniline, 3.0 μmol of cumene hydroperoxide, and the usual buffer and phospholipid, in a final volume of 1.5 ml. The various components were determined as described elsewhere (70).

Electron uptake by P-450$_{LM_2}$ and formation of oxyferro intermediates. Evidence has been presented elsewhere that P-450$_{LM_2}$ is a two-electron acceptor under anaerobic conditions (71,72). Two electrons were taken up from dithionite, from NADPH in the presence of NADPH-cytochrome P-450 reductase, or from a photochemical system, and reoxidation of reduced P-450$_{LM_2}$ by molecular oxygen restored a state where two electrons from dithionite were required for re-reduction. Although these unexpected findings indicate the presence of an electron acceptor in addition to the heme iron atom, the second acceptor has not yet been identified. It is tentatively called factor C; A_1, A_2, and B were used to refer to the fractions obtained from resolution of the microsomal hydroxylation system and subsequently identified as cytochrome P-450, NADPH-cytochrome P-450 reductase, and phosphatidylcholine (1-4).

More recently, stopped flow spectrophotometry has shown the occurence of two distinct spectral intermediates in the reaction of oxygen with the reduced form of highly purified P-450$_{LM_2}$ (73). As indicated by difference spectra, Complex I (with maxima at 430 and 450 nm) is rapidly formed (within 10 msec) and then decays to form Complex II (with a broad maximum at 440 nm), which resembles the intermediate seen in steady state experiments. The latter complex appears to be similar to that observed by Estabrook et al. (74) with microsomal suspensions and by Gunsalus et al. (75) and Ishimura et al. (76) in the reaction of oxygen with the reduced form of highly purified P-450$_{cam}$ from Pseudomonas putida.

The rate constants of the individual reactions with several different substrates present are shown in Table 3. The rate constants for the formation of Complex I are very large, whereas those for the decay of Complex II and the decomposition of the latter, with regeneration of the oxidized cytochrome, are much smaller (73). The last step, which is rate-limiting in the overall process, involves the formation of hydrogen peroxide and also, when substrate is present, the formation of product. In other experiments, k_3 was found to be increased to 90 min^{-1} when cytochrome b_5 was added. Similar experiments with NADPH-cytochrome P-450 reductase added were more difficult to interpret, but provided indirect evidence that k_3 was increased to 23 min^{-1}. For comparison, the turnover number of the enzyme for benzphetamine hydroxylation at 12° is about 23 min^{-1}. Cytochrome b_5 was not oxidized during the time required

TABLE 3 Rate Constants for Individual Reactions:

$$P\text{-}450_{LM}^{red.} \xrightarrow{k_1} \text{Complex I} \xrightarrow{k_2} \text{Complex II} \xrightarrow{k_3} P\text{-}450_{LM}^{ox.}$$

Substrate	Rate constants (min^{-1})		
	k_1	k_2	k_3
None	≥ 60,000	210	12
Benzphetamine	≥ 60,000	270	3
Cyclohexane	≥ 60,000	215	7
p-Nitroanisole	≥ 60,000	205	27

The data were obtained at 12° under conditions described elsewhere (73).

for Complex II to form the oxidized cytochrome. Thus, it appears that cytochrome b_5 may act as an effector, in addition to its well-known function as a microsomal electron carrier. Precedent for such a dual function for an electron carrier associated with cytochrome P-450 is provided by the work of Lipscomb et al. (77) indicating an effector role for putidaredoxin in the bacterial hydroxylation system.

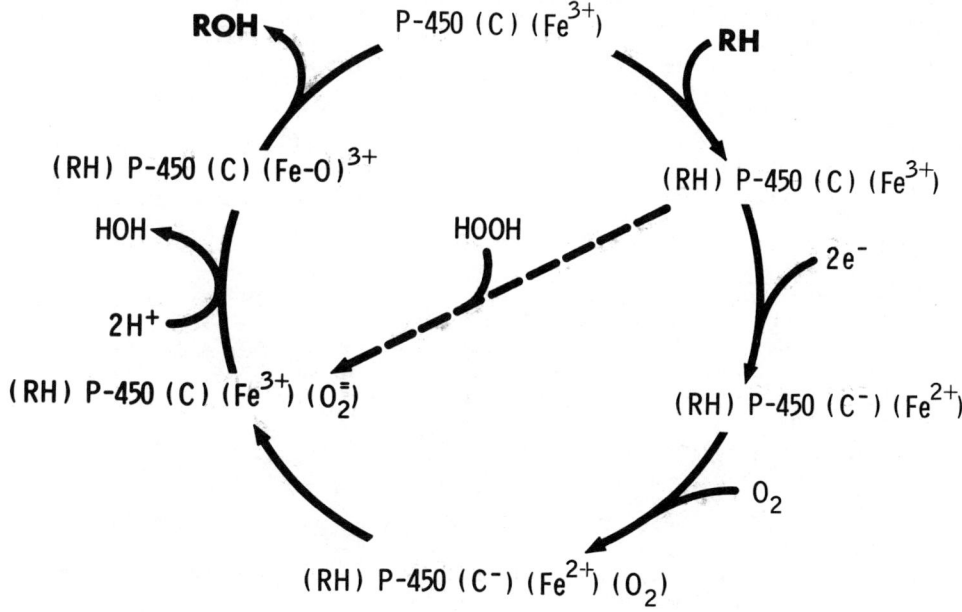

Fig. 3. Proposed scheme for mechanism of catalysis of P-450$_{LM}$, where RH represents the substrate, ROH the product, and C an unidentified electron acceptor.

The accompanying scheme (Fig. 3) showing the proposed mechanism of catalysis by P-450$_{LM}$ is in accord with the evidence presently available from this and other laboratories. It indicates the uptake of two electrons prior to oxygenation; this may possibly happen in two one-electron transfers, but the kinetic aspects are not yet known. Hydrogen peroxide utilization (dashed line) is indicated in a reaction which bypasses part of the cycle. Whether peroxide liberation by the enzyme is a reversal of this process is not clear. The possible role of factor C is indicated in order to balance the reactions according to the known electron uptake, but it should be emphasized that the factor may not necessarily be functional in electron transfer for substrate hydroxylation. These are important questions for further study.

REFERENCES

1. Coon, M. J., and Lu, A. Y. H. (1969) in Microsomes and Drug Oxidations (Gillette, J. R., et al., Eds.) pp. 151-166, Academic Press, New York.
2. Lu, A. Y. H., and Coon, M. J. (1968) J. Biol. Chem. 243, 1331-1332.
3. Lu, A. Y. H., Junk, K. W., and Coon, M. J. (1969) J. Biol. Chem. 244, 3714-3721.
4. Strobel, H. W., Lu, A. Y. H., Heidema, J., and Coon, M. J. (1970) J. Biol. Chem. 245, 4851-4854.
5. Gillette, J. R. (1966) Adv. Pharmacol. 4, 219-261.
6. Conney, A. H. (1967) Pharmacol. Rev. 19, 317-366.
7. Gelboin, H. V. (1967) Adv. Cancer Res. 10, 1-81.
8. Gillette, J. R. (1969) Fed. Eur. Biochem. Soc. Symp. 16, 109-124.
9. Rubin, A., Tephly, T. R., and Mannering, G. J. (1964) Biochem. Pharmacol. 13, 1007-1016.
10. Tephly, T. R., and Mannering, G. J. (1968) Mol. Pharmacol. 4, 10-14.
11. Wada, F., Shimakawa, H., Takasugi, M., Kotake, T., and Sakamoto, Y. (1968) J. Biochem. (Tokyo) 64, 109-113.
12. Lu, A. Y. H., Strobel, H. W., and Coon, M. J. (1970) Mol. Pharmacol. 6, 213-220.
13. Sladek, N. E., and Mannering, G. J. (1966) Biochem. Biophys. Res. Commun. 24, 668-674.
14. Alvares, A. P., Schilling, G., Levin, W., and Kuntzman, R. (1967) Biochem. Biophys. Res. Commun. 29, 521-526.
15. Hildebrandt, A., Remmer, H., and Estabrook, R. W. (1968) Biochem. Biophys. Res. Commun. 30, 607-612.
16. Nebert, D. W., and Gelboin, H. V. (1968) J. Biol. Chem. 243, 6250-6261.
17. Jefcoate, C. R. E., and Gaylor, J. L. (1969) Biochemistry 8, 3464-3472.
18. Gielen, J. E., Goujon, F. M., and Nebert, D. W. (1972) J. Biol. Chem. 247, 1125-1137.
19. Comai, K., and Gaylor, J. L. (1973) J. Biol. Chem. 248, 4947-4955.
20. Lu, A. Y. H., Kuntzman, R., West, S., Jacobson, M., and Conney, A. H. (1972) J. Biol. Chem. 247, 1727-1734.
21. Lu, A. Y. H., and Levin, W. (1972) Biochem. Biophys. Res. Commun. 46, 1334-1339.
22. Nebert, D. W., Heidema, J. K., Strobel, H. W., and Coon, M. J. (1973) J. Biol. Chem. 248, 7631-7636.
23. van der Hoeven, T. A., and Coon, M. J. (1974) J. Biol. Chem. 249, 6302-6310.
24. van der Hoeven, T. A., Haugen, D. A., and Coon, M. J. (1974) Biochem. Biophys. Res. Commun. 60, 569-575.
25. Haugen, D. A., van der Hoeven, T. A., and Coon, M. J. (1975) J. Biol. Chem. 250, 3567-3570.

26. Haugen, D. A., and Coon, M. J. (1976) J. Biol. Chem., in press.
27. Coon, M. J., Haugen, D. A., Guengerich, F. P., Vermilion, J. L, and Dean, W. L., (1976) in The Structural Basis of Membrane Function (Hatefi, Y., and Djavadi-Ohaniance, L., Eds.) pp. 409-427, Academic Press, New York.
28. Dean, W. L. (1976) Doctoral Thesis, The University of Michigan.
29. Ryan, D., Lu, A. Y. H., Kawalek, J., West, S. B., and Levin, W. (1975) Biochem. Biophys. Res. Commun. 64, 1134-1141.
30. Kawalek, J. C., Levin, W., Ryan, D., Thomas, P. E., and Lu, A. Y. H. (1975) Mol. Pharmacol. 11, 874-878.
31. Thomas, P. E., Lu, A. Y. H., Ryan, D., West, S. B., Kawalek, J., and Levin, W. (1975) J. Biol. Chem. 251, 1385-1391.
32. Imai, Y., and Sato, R. (1974) Biochem. Biophys. Res. Commun. 60, 8-14.
33. Hashimoto, C., and Imai, Y. (1976) Biochem. Biophys. Res. Commun. 68, 821-827.
34. Laemmli, U. K. (1970) Nature 227, 680-685.
35. Johnson, E. F., and Muller-Eberhard, U. (1976) Abstracts, Meeting of American Chemical Society, San Francisco.
36. Jefcoate, C. R. E., Calabrese, R. L., and Gaylor, J. L. (1970) Mol. Pharmacol. 6, 391-401.
37. Wiebel, F. J., Selkirk, J. K., Gelboin, H. V., Haugen, D. A., van der Hoeven, T. A., and Coon, M. J. (1975) Proc. Nat. Acad. Sci. USA 72, 3917-3920.
38. Aust, S. D., Rorrig, D. L., and Pederson, T. C. (1972) Biochem. Biophys. Res. Commun. 47, 1133-1137.
39. Masters, B. S. S., Williams, C. H., Jr., and Kamin, H. (1967) Methods Enzymol. 10, 565-573.
40. Buege, J. A., and Aust, S. D. (1972) Biochim. Biophys. Acta 286, 433-436.
41. Coon, M. J., Strobel, H. W., and Boyer, R. F. (1973) Drug Metab. Disp. 1, 92-97.
42. Satake, H., Imai, Y., and Sato, R. (1972) Abstracts, Ann. Meeting Jap. Biochem. Soc.
43. Vermilion, J. L., and Coon, M. J. (1974) Biochem. Biophys. Res. Commun. 60, 1315-1322.
44. Vermilion, J. L., and Coon, M. J. (1976) in Flavins and Flavoproteins (Singer, T. P., Ed.) pp. 674-678, Elsevier, Amsterdam.
45. Iyanagi, T., and Mason, H. S. (1973) Biochemistry 12, 2297-2308.
46. Vermilion, J. L. (1976) Doctoral Thesis, The University of Michigan.
47. Coon, M. J., Haugen, D. A., Dean, W. L., Vermilion, J. L., French, J. S., and Vatsis, K. P. (1976) Symposium on Recent Advances in Drug Metabolism, American Chemical Society Symposium Series, in press.
48. Dignam, J. D., and Strobel, H. W. (1975) Biochem. Biophys. Res. Commun. 63, 845-852.
49. Dignam, J. D., and Strobel, H. W. (1976) Biochemistry, in press.
50. Yasukochi, Y., and Masters, B. S. S. (1976) Fed. Proc. 35, 1654.
51. Yasukochi, Y., and Masters, B. S. S.(1976) J. Biol. Chem., in press.
52. Mason, H. S. (1957) Science 125, 1185-1188.
53. Orrenius, S. (1965) J. Cell. Biol. 26, 713-723.
54. Estabrook, R. W., and Cohen, B. (1969) in Microsomes and Drug Oxidations (Gillette, J. R., et al., Eds.) pp. 95-109, Academic Press, New York.
55. Stripp, B., Zampaglione, N., Hamrick, M., and Gillette, J. R. (1972) Mol. Pharmacol. 8, 189-196.
56. Sasame, H. A., Mitchell, J. R., Thorgeirsson, S., and Gillette, J. R. (1973) Drug Metab. Disp. 1, 150-155.
57. Shoemaker, D. D., and Hamrick, M. E. (1974) Biochem. Pharmacol. 23, 2325-2327.

58 Staudt, H., Lichtenberger, F., and Ullrich, V. (1974) Eur. J. Biochem. 46, 99-106.
59 Gillette, J. R., Brodie, B. B., and LaDu, B. N. (1957) J. Pharmacol. Exp. Ther. 119, 532-540.
60 Thurman, R. G., Ley, H. G., and Scholz, R. (1972) Eur. J. Biochem. 25, 420-430.
61 Boveris, A., Oshino, N., and Chance, B. (1972) Biochem. J. 128, 617-630.
62 Hildebrandt, A. G., Speck, M., and Roots, I. (1973) Biochem. Biophys. Res. Commun. 54, 968-975.
63 Hildebrandt, A. G., and Roots, I. (1975) Arch. Biochem. Biophys. 171, 385-392.
64 Kadlubar, F. F., Morton, K. C., and Ziegler, D.M. (1973) Biochem. Biophys. Res. Commun. 54, 1255-1261.
65 Rahimtula, A. D., and O'Brien, P. J. (1974) Biochem. Biophys. Res. Commun. 60, 440-447.
66 Ellin, Å., and Orrenius, S. (1975) FEBS Lett. 50, 378-381.
67 Rahimtula, A. D., and O'Brien, P. J. (1975) Biochem. Biophys. Res. Commun. 62, 268-275.
68 Hrycay, E. G., Gustafsson, J. A., Ingelman-Sundberg, M., and Ernster, L. (1975) FEBS Lett. 56, 161-165.
69 Hrycay, E. G., Gustafsson, J. A., Ingelman-Sundberg, M., and Ernster, L. (1975) Biochem. Biophys. Res. Commun. 66, 209-216.
70 Nordblom, G. D., White, R. E., and Coon, M. J. (1976) Arch. Biochem. Biophys. 175, 524-533.
71 Ballou, D. P., Veeger, C., van der Hoeven, T. A., and Coon, M. J. (1974) FEBS Lett. 38, 337-340.
72 Guengerich, F. P., Ballou, D. P., and Coon, M. J. (1975) J. Biol. Chem. 250, 7405-7414.
73 Guengerich, F. P., Ballou, D. P., and Coon, M. J. (1976) Biochem. Biophys. Res. Commun. 70, 951-956.
74 Estabrook, R. W., Hildebrandt, A. G., Baron, J., Netter, K. J., and Leibman, K. (1971) Biochem. Biophys. Res. Commun. 42, 132-139.
75 Gunsalus, I. C., Tyson, C. A., Tsai, R., and Lipscomb, J. D. (1971) Chem.-Biol. Interactions 4, 75-78.
76 Ishimura, Y., Ullrich, V., and Peterson, J. A. (1971) Biochem. Biophys. Res. Commun. 42, 140-146.
77 Lipscomb, J. D., Sligar, S. G., Namtvedt, M. J., and Gunsalus, I. C. (1976) J. Biol. Chem. 251, 1116-1124.

MICROSOMAL HEMEPROTEINS P-450$_{LM-2}$ AND P-450$_{LM-4}$: COMPARATIVE STRUCTURAL STUDIES*

K. Dus, D. Carey, R. Goewert, and R. A. Swanson

Biochemistry Department, St. Louis University Medical School, St. Louis, Missouri 63104

ABSTRACT

Although the two major hemeproteins of rabbit liver microsomes, termed P-450$_{LM-2}$ and P-450$_{LM-4}$, are readily distinguishable by their differences in size, substrate specificity, and immunochemical properties, there are nevertheless indications of extensive structural similarity between them. Except for higher values of cysteine and tryptophan in P-450$_{LM-4}$ the amino acid compositions correlate quite well. The BrCN digest of P-450$_{LM-4}$ shows much higher cross reactivity with anti P-450$_{CAM}$ antibodies than does the parent protein; similar small hemepeptides were isolated from the BrCN digests of P-450$_{LM-2}$ and P-450$_{LM-4}$. Photoaffinity labeling of both hemeproteins with 3'-azidophenobarbital at their substrate binding sites and subsequent digestion with trypsin yielded labeled peptides of similar composition; these labeled peptides were shown to be part of the respective hemepeptides. Comparative peptide mapping of the tryptic digests of the proteins revealed substantial similarity of overall peptide patterns.

INTRODUCTION

We reported (1) that cytochrome P-450$_{CAM}$ of the camphor methylene hydroxylase of P. putida (2) is able to cross react substantially with anti P-450$_{LM-2}$ antibodies and that P-450$_{LM-2}$, in turn, reacts equally well with antibodies against P-450$_{CAM}$. More recently, however, the immunochemical properties of cytochromes P-450$_{LM-2}$ and P-450$_{LM-4}$ were shown to differ significantly (3,4). Thus the question arose whether P-450$_{LM-2}$ is structurally more closely related to P-450$_{CAM}$, a protein of bacterial origin, than to P-450$_{LM-4}$ of liver microsomes.

On the other hand it is known that P-450$_{CAM}$ and P-450$_{LM-2}$ are of similar length (1) while P-450$_{LM-4}$ is at least 15% larger (3). Since the presence of the additional polypeptide segment could account for the altered immunochemical properties of the P-450$_{LM-4}$ molecule if it covered much of the surface, it seemed important to search for the possible existence of common structural features not directly measured by immunochemical tests with native proteins. We report here on a series of experiments aimed directly at the structure of the heme and substrate binding sites which can be made readily available by a combination of highly selective chemical cleavage and photoaffinity labeling of the sites.

---

*Supported by NIH grant GM21726 and NSF grant PCM75-23480 to K. Dus.

Experimental Procedures and Results

Table 1 compares the amino acid compositions of cytochromes P-450$_{LM-2}$ and P-450$_{LM-4}$ of rabbit liver microsomes to that of crystalline cytochrome P-450$_{CAM}$ of $\underline{P. putida}$. Highly purified protein preparations, P-450$_{LM-2}$ with specific content of 21 nmoles and P-450$_{LM-4}$ with 12.5 nmoles per mg protein, respectively, were obtained from Dr. M. J. Coon (5). The values represent

Table 1

AMINO ACIDS	Hemeproteins			Small BrCN-Derived Hemepeptides of		
	P. putida	Rabbit Liver, PB-induced				
	P-450$_{CAM}$	P-450$_{LM-2}$	P-450$_{LM-4}$	P-450$_{CAM}$	P-450$_{LM-2}$	P-450$_{LM-4}$
CySO$_3$H	6	6	7	1	1	1
Asx	36	38	44	3	4	4
MetSo$_2$	9	7	10	—	—	2
Thr	19	23	24	2	2	2
Ser	21	30	31	3	3	4
Glx	55	42	46	3	4	6
Pro	27	24	28	3	2	2
Gly	26	32	38	8	3	6
Ala	34	23	29	6	3	6
Val	24	26	32	4	2	2
Ile	24	19	22	2	2	2
Leu	40	54	58	3	5	4
Tyr	9	9	11	1	1	1
Phe	17	31	28	2	2	2
His	12	11	11	1	1	2
LyS	13	19	26	2	2	2
HSerL.	—	—	—	1	1	1
Trp	1	1	7	—	—	—
Arg	24	29	30	2	3	1
Total	397	424	482	47	41	50

the average of several determinations obtained from acid hydrolyzate under reducing conditions or after performate oxidation. Tryptophan was determined after hydrolysis with mercaptoethane sulfonic acid. The residue numbers are based on the occurrence of 11 histidines in both proteins and on the average recoveries near integers of most stable aliphatic residues. As expected, the amino acid compositions of the two microsomal hemeproteins show a preponderance of hydrophobic and acidic amino acids; for most types of amino acids there is close correlation between the two proteins except that P-450$_{LM-4}$ contains 6 additional tryptophans as well as one additional residue of cysteine and 3 of methionine. The polypeptide molecular weights calculated from these amino acid compositions are 47,600 daltons for P-450$_{LM-2}$ and 54,000 daltons for P-450$_{LM-4}$. These values are in good agreement with the molecular weights of 48,700 and 55,300 daltons, respectively, estimated for these polypeptides from SDS-acrylamide gel electrophoresis (5).

Electrofocusing experiments with several P-450$_{LM-2}$ preparations ranging in purity from 12 to 20 nmoles P-450 per mg protein were carried out either in 7% polyacrylamide gel or in 0.8% agarose gel if the resolved hemeproteins were to be recovered for subsequent amino acid analysis. As indicated in Fig. 1, only one hemeprotein band was observed in these preparations. After extensive electrodialysis to remove tightly bound ampholytes acid hydrolysis and amino acid analysis of this material closely confirmed the compositions previously determined with highly purified preparations (5). Thus the amino acid composition of P-450$_{LM-2}$ listed in Table 1 corresponds to a hemeprotein with a p_I of 5.4. The aggregation state of the preparation is unknown but it may be assumed that the hexameric form is present, as described by Coon et al. (3). In contrast, electrofocusing experiments with P-450$_{LM-4}$ could not be completed because the heme color faded out before the protein had reached its isoelectric point.

Figure 1

ELECTROFOCUSING: 0.8% AGAROSE / 2% AMPHOLYTES pH 5-8

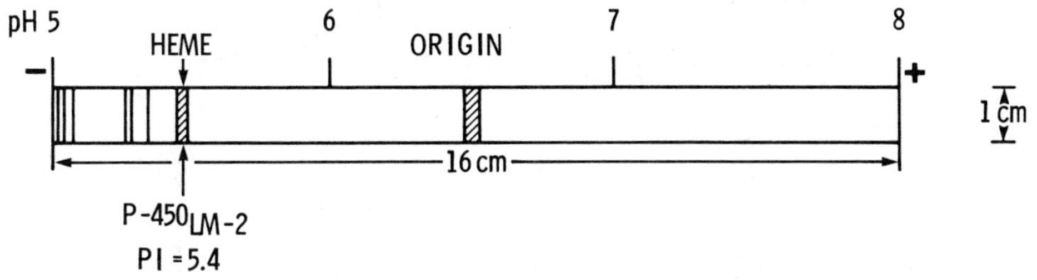

CONDITIONS: Applied 100µl of partially purified hemeprotein solution of 0.72mg/ml and a specific P-450 content of 13.1n moles/mg protein. Well depth 3mm, 5°C, 2h at 100V/.5mA.

Limited BrCN degradation of the P-450 substrate complex was previously found to produce, although in low yield, a small heme-containing peptide from P-450$_{CAM}$ and P-450$_{LM-2}$. The same experimental approach has now also yielded a corresponding peptide from P-450$_{LM-4}$. These small "hemepeptides" can be readily resolved from larger heme-containing peptides and all uncolored BrCN fragments of the digests by sieve chromatography on Sephadex G-100; they are further purified using thin layer chromatography or DEAE-Sephadex A-25 chromatography.

Figure 2 contains a comparison of spectra of the hemeproteins and the corresponding small BrCN-derived hemepeptides. Independent of the Soret maxima of the parent hemeprotein the Soret maximum of the resulting small hemepeptide is found to be at 390 nm in 20% acetic acid as the solvent. Although the hemepeptides of P-450$_{CAM}$ and P-450$_{LM-2}$ contain close to stoichiometric amounts of heme which can be extracted by the same acid-acetone procedure used for the hemeproteins, the hemepeptide of P-450$_{LM-4}$ loses its heme gradually during purification. So far no attempt to generate comparable small hemepeptides by degradation with proteolytic enzymes has been successful.

The formation of small hemepeptides during limited BrCN degradation seems to be based on the peculiar properties of P-450 hemeproteins. The tightly structured segment of polypeptide chain including both the heme and substrate binding sites could be considered a separate structural domain comparable perhaps to the domains found in antibody molecules or other specialized proteins. The fortuitous spacing of two methionine residues, one at each end of this domain and rather exposed, in each of the three P-450 hemeproteins

Figure 2   Comparison of Soret Maxima of Hemeproteins
and their BrCN-Derived Small Hemepeptides

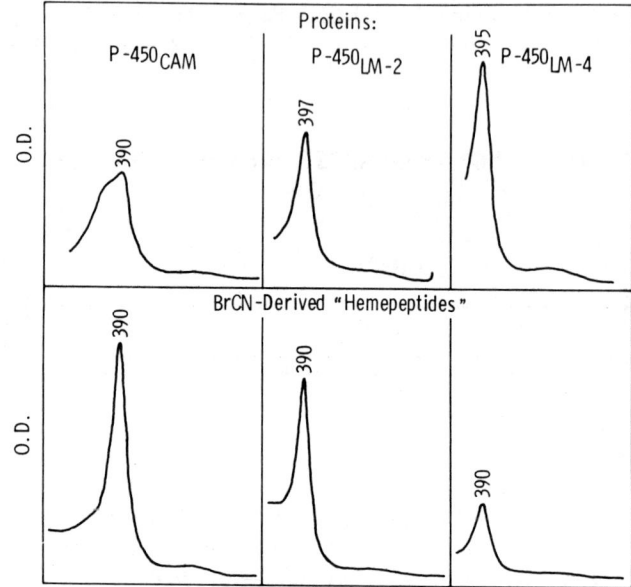

Wavelength (nm) dissolved in 20% Acetic acid

so far investigated, makes it possible to attain cleavage at these residues preferentially. Despite these advantages and the small size of the hemepeptides (below 50 residues) the low yield (10-15%) in which they are obtained has so far precluded their scrutiny by rigorous sequence analysis.

Table 1 also lists the amino acid compositions of the small BrCN-derived hemepeptides of $P-450_{CAM}$, $P-450_{LM-2}$ and $P-450_{LM-4}$. These peptides correspond to about one tenth of the size of their parent proteins and are similar in composition, including at least one residue each of cysteine and histidine, except that the peptide of $P-450_{LM-4}$ contains in addition two undegraded residues of methionine.

It is of great significance that the purified small hemepeptides are still able to cross react substantially with antibodies made against their parent proteins. This observation indicates that they have been derived from the surface of the hemeproteins and that they have retained most of their characteristic structural features throughout the purification procedure. In Fig. 3 we demonstrate that the hemepeptides of $P-450_{LM-2}$ and $P-450_{LM-4}$ are also able to cross react significantly with anti $P-450_{CAM}$ antibodies. This finding is most surprising in the case of the hemepeptide of $P-450_{LM-4}$ because the parent protein does not measurably cross react with antibodies made against either $P-450_{CAM}$ or $P-450_{LM-2}$. From this we conclude that $P-450_{LM-4}$

Figure 3

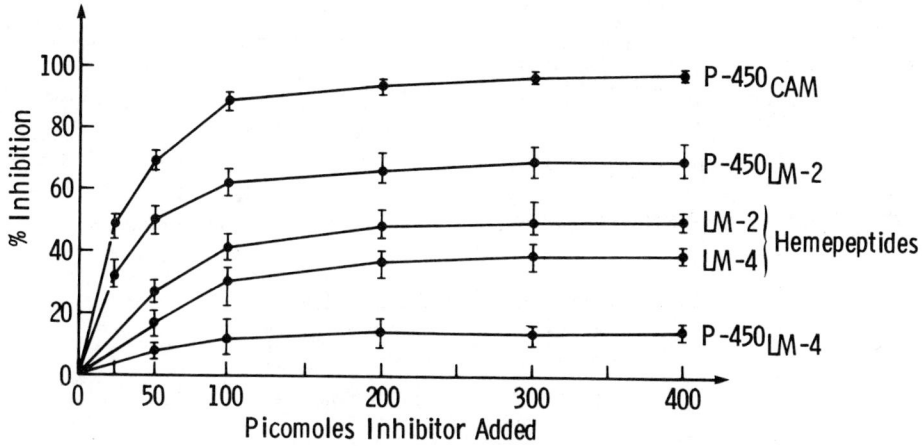

Inhibition of Binding Between ^{125}I-Labled Cytochrome P-450$_{CAM}$, Liver Microsomal P-450 Cytochromes LM-2 and LM-4 and their BrCN-Derived Small Hemepeptides

contains some portions, including the hemepeptide area, which are homologous to P-450$_{CAM}$ and P-450$_{LM-2}$; these portions of the polypeptide chain, however, are not exposed on the surface of the native hemeprotein to be recognized immunochemically. Since P-450$_{LM-4}$ contains a polypeptide chain which is about 60 residues longer than that of P-450$_{LM-2}$, and at least 85 amino acid residues longer than that of P-450$_{CAM}$ (Table 1) the surface of this protein may consist largely of the additional peptide portion. Thus recognition of antibodies specific for the smaller P-450 hemeproteins will be suppressed.

Another important aspect of the small BrCN-derived hemepeptides is the presence of the substrate binding site. From the study of the spectral shift associated with substrate binding (type I shift or blue shift) it has long been suggested that the substrate is bound close to the heme group since its presence profoundly influences the position of the maximum of the Soret band. At first we assumed that the maximum at 390 nm of the hemepeptide derived from P-450$_{CAM}$ implied the presence of the substrate but soon we learned that this maximum applied to all small hemepeptides (Fig. 2) whether the protein-substrate complex had a maximum at this wavelength or not. The significance of this feature is still not understood. For further investigation we designed tritiated photoaffinity labels (Fig. 4) on the structural basis of substrate or inhibitor analogues in the hope that they would still bind sufficiently tightly to permit specific binding at the site and give high efficiency of covalent linkage during subsequent photolysis. In agreement with numerous reports in the literature which followed Porter's original discovery (6) we found that aromatic azides were eminently suited for this purpose. One substrate and one inhibitor derivative were synthesized using a minimal number of reaction steps with good overall yields.

Figure 4 PHOTOAFFINITY LABELING OF P-450 HEMEPROTEINS

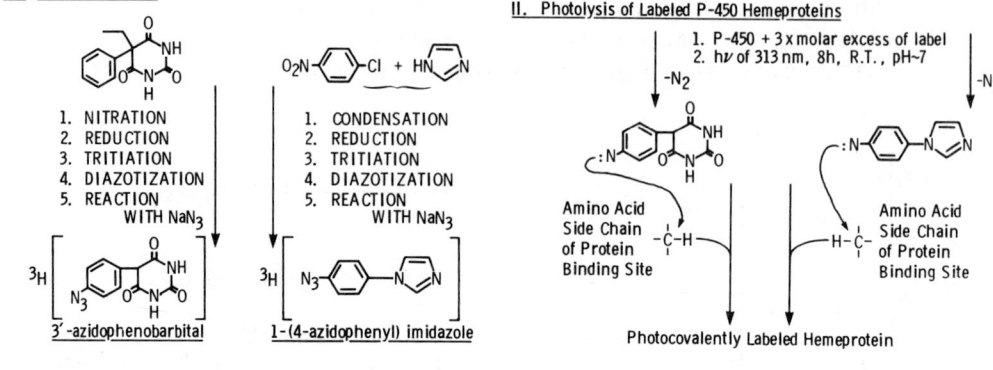

Both affinity labels were tritiated at the level of the amino compound by catalytic exchange reactions. Due to strong residual binding affinities the molar ratios of the azides could be limited to a 3x molar excess during photolysis. Exposure times of 8h at 313 nm corresponding to 4 life times of the labels did not cause measurable alterations of the spectra of the hemeproteins. The photolyzed reaction mixture was incubated with a large excess of cold substrate or inhibitor and extensively dialyzed. The extent of photocovalent incorporation of label was calculated from the ratio of actual counts to counts expected at saturation of one site. It is of interest that under these conditions $P-450_{CAM}$ was not susceptible to binding by APB.

The number and location of binding sites were determined as follows. One half of an APB-labeled $P-450_{LM}$ preparation was digested with BrCN and chromatographed. The other half was performate oxidized and then digested with trypsin. The resolved small hemepeptide was also performate oxidized and digested with trypsin. Both tryptic digests, that of the protein and that of the hemepeptide, were then chromatographed side by side on the same sheet of paper. Fig. 5 shows this chromatographic comparison. Except for a small amount of undigested protein remaining at the origin, most of the radioactive label (43%) migrated very fast, slightly behind the solvent front. A minor broad peak in the center of the chromatogram was also noted. In contrast, the hemepeptides were completely digested and only one radioactive peak (61%), coincident with the major peak of the protein digest, was found proving that the substrate binding site of the protein is an integral part of the hemepeptide.

It should be noted that photocovalent attachment of the substrate analogue significantly stabilized the conformation of the P-450-substrate complex thus leading to greatly increased yields (45%) of the small BrCN-derived hemepeptide.

Figure 5

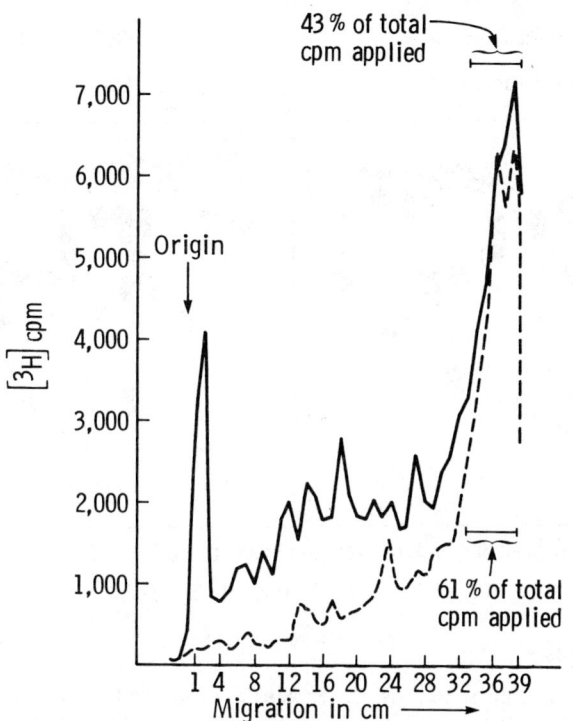

Chromatographic Comparison of Tryptic Peptides Obtained from APB-Cytochrome P-450$_{LM-4}$ (———) and Its BrCN-Derived Small Hemepeptides (- - -)

APB = 3'-Azidophenobarbital
Descending Chromatography on Whatman No. 3MM in n-BuOH/AcOH/H$_2$O //4/1/5

Encouraged by the similarity of chromatographic mobilities of the APB-labeled tryptic peptides of P-450$_{LM-4}$ we extended the comparison to the total tryptic digest of both liver microsomal hemeproteins by comparative peptide mapping (Fig. 6). To assure complete tryptic digestion to soluble peptides heme extraction and subsequent oxidation of the cysteines was found to be essential. All black colored spots in Fig. 6 indicate peptides found in roughly identical locations in the corresponding maps. Although the compositions of these peptides are still unknown, the preliminary comparison of mobilities in chromatography and electrophoresis indicates the existence of homologous portions of sequence in these liver microsomal cytochromes.

From a comparison of tryptic peptide maps of labeled and unlabeled P-450$_{LM-2}$ and P-450$_{LM-4}$ preparations we conclude that the spots marked by an arrow in Fig. 6 correspond to the peptides photocovalently labeled by APB. It is of interest to note that these spots are in roughly equivalent positions in both peptide maps again suggesting close similarity of APB binding sites in cytochromes P-450$_{LM-2}$ and P-450$_{LM-4}$.

Figure 6 Comparative Peptide Mapping of Rabbit Liver Microsomal P-450 Hemeproteins
(Phenobarbital-Induced; Tryptic Digest after Performate Oxidation)

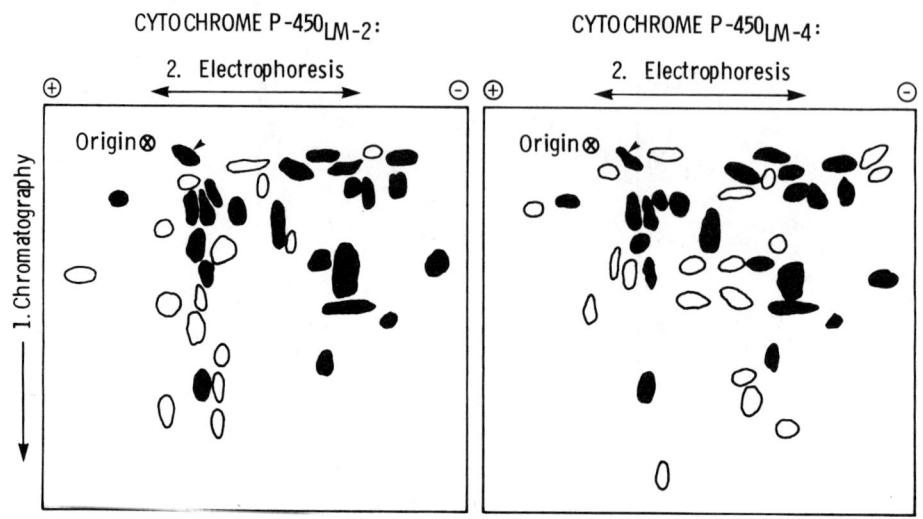

Chromatography: Descending on Whatman No. 3MM, n-BuOH/AcOH/H_2O //4/1/5, 16 Hrs.
Electrophoresis: 3.8 KV, 60'; $10°C$; Pyr/AcOH/H_2O //1/20/289, pH = 3.6

REFERENCES

(1) K. Dus, W. J. Litchfield, A. G. Miguel, T. A. van der Hoeven, D. A. Haugen, W. L. Dean, and M. J. Coon, Structural resemblance of cytochrome P-450 isolated from Pseudomonas putida and from rabbit liver microsomes, Biochem. Biophys. Res. Commun. 60, 15 (1974).

(2) Gunsalus, I. C., Meeks, J. R., Lipscomb, J. D., Debrunner, P., and Muenck, E. (1974) Molecular Mechanisms of Oxygen Activation, Academic Press, Inc., New York.

(3) Coon, M. J., Haugen, D. A., Guengerich, F. P., Vermilion, J. L., and Dean, W. L. (1976) The Structural Basis of Membrane Function, Academic Press, Inc., New York.

(4) P. E. Thomas, A. Y. H. Lu, D. Ryan, S. B. West, J. Kawalek, and W. Levin, Multiple forms of rat liver cytochrome P-450: immunochemical evidence with antibody against cytochrome P-448, J. Biol. Chem. 251, 1385 (1976).

(5) D. A. Haugen and M. J. Coon, Properties of electrophoretically homogeneous phenobarbital-inducible and β-naphthoflavone-inducible forms of liver microsomal cytochrome P-450, J. Biol. Chem. (1976), in press.

(6) G. W. J. Fleet, R. R. Porter, and J. R. Knowles, Affinity labelling of antibodies with aryl nitrene as reactive group, Nature 224, 54 (1969).

HEPATIC MICROSOMAL ETHANOL OXIDIZING SYSTEM: ISOLATION AND RECONSTITUTION

Rolf Teschke,* Kunihiko Ohnishi, Yasushi Hasumura, and Charles S. Lieber

Laboratory of Liver Disease, Nutrition and Alcoholism, Veterans Administration Hospital, Bronx, New York 10468, and Mount Sinai School of Medicine (CUNY), New York, New York 10029, USA

ABSTRACT

The NADPH-dependent microsomal alcohol oxidizing system which metabolizes methanol, ethanol, propanol and butanol was separated from both catalase and alcohol dehydrogenase activities by column chromatography on DEAE-cellulose. Activity of the microsomal system was recovered in fractions containing NADPH-cytochrome c reductase, phospholipids as well as three different forms of cytochrome P-450, as judged by cyanide titration. The microsomal ethanol oxidizing system was reconstituted with partially purified cytochrome P-450 and NADPH-cytochrome c reductase in the presence of a synthetic phospholipid.

Introduction

It was generally assumed that ethanol metabolism proceeds exclusively via ADH**, an enzyme of the cell sap of the hepatocyte. This concept is satisfactory at low ethanol concentrations since the oxidation of ethanol is almost completely abolished under these conditions by pyrazole, a potent inhibitor of ADH (1). At intermediate and higher concentrations, ethanol metabolism becomes less sensitive to pyrazole, a finding which suggested the operation of a non-ADH pathway for ethanol metabolism (1-7). A variety of studies have supported the concept that catalase, another enzyme capable of oxidizing ethanol in vitro, does not play a significant role in ethanol metabolism and that it cannot account quantitatively for the ADH-independent pathway (4-12). A NADPH-dependent microsomal ethanol oxidizing system (MEOS)** was described with characteristics similar to those of other microsomal drug metabolizing systems (2, 3, 13), and considerable evidence has now accumulated that MEOS is responsible for that fraction of ethanol metabolism not mediated by ADH (3-7).

The present study will focus on the biochemical nature of MEOS, its isolation from catalase and ADH and its reconstitution with partially purified cytochrome P-450, NADPH-cytochrome c reductase and synthetic phospholipids.

---

*Present address: Second Department of Medicine, University of Düsseldorf, 4000 Düsseldorf, West Germany
**Abbreviations: ADH - alcohol dehydrogenase
MEOS - microsomal ethanol oxidizing system

Isolation of the Microsomal Ethanol Oxidizing System (MEOS)

Following solubilization of the microsomal membranes by ultrasonication and treatment with deoxycholate, MEOS activity was recovered after column chromatography on DEAE-cellulose in fractions eluted with increasing KCl concentrations (14-19). MEOS activity was found only in fractions containing all three microsomal components cytochrome P-450, NADPH-cytochrome c reductase and phospholipids, whereas the combination of NADPH-cytochrome c reductase and phospholipids in the absence of cytochrome P-450 was completely inactive with regard to ethanol oxidation (14, 15). The microsomal column fraction exhibiting MEOS activity was then tested for the presence of various forms of cytochrome P-450 which were determined spectrophotometrically after cyanide titration (20). Three different types of cytochrome P-450 including form I were detected in our preparation (Fig. 1) which differs thereby from the reconstituted microsomal enzyme system of Ryan et al. (21) which lacks form I of cytochrome P-450. Supportive evidence that cytochrome P-450 is involved in MEOS activity was provided by the inhibitory effect of CO on the activity of the isolated MEOS (16) and by the fact that destruction of cytochrome P-450 (including form I) resulted in a concomitant loss of MEOS activity (22). Of particular interest was the observation that the isolated microsomal system metabolized not only ethanol (a modified type II binding substrate), but also type I binding ones such as aminopyrine and benzphetamine as well as compounds exhibiting a type II binding spectrum such as aniline (16).

Differentiation of MEOS from Alcohol Dehydrogenase

After column chromatography of solubilized microsomes on DEAE-cellulose, ADH activity was recovered prior to the fractions exhibiting NADPH-dependent MEOS activity (14-16). These findings therefore substantiate our previous thesis that hepatic microsomes contain an ethanol oxidizing system which operates independently of ADH activity (2). To rule out the possibility that trace amounts of ADH activity not detectable by enzymatic assay might still be present in the column fraction exhibiting MEOS activity, the potent ADH inhibitor pyrazole was used at final concentrations up to 1 mM. Under these experimental conditions, no decrease of the NADPH-dependent MEOS activity was observed. To rule out the possibility that ADH might have some facultative rather than obligatory role in microsomal ethanol oxidation, ADH was added to the fraction containing the isolated MEOS, and NADPH-dependent ethanol oxidation was determined. Under these experimental conditions, there was no increase of the NADPH-dependent MEOS activity; actually a slight decrease of MEOS activity was observed (16). Other characteristics of the isolated MEOS, including pH optimum (7.4 for MEOS vs. greater than 9 for ADH), optimum cofactor (NADPH for MEOS vs. NAD for ADH) and K_m values for ethanol (7-10 mM for MEOS vs. lower than 2 mM for ADH), provide further evidence against an obligatory or even facultative role of ADH in MEOS activity, in contradiction with other viewpoints (23-25).

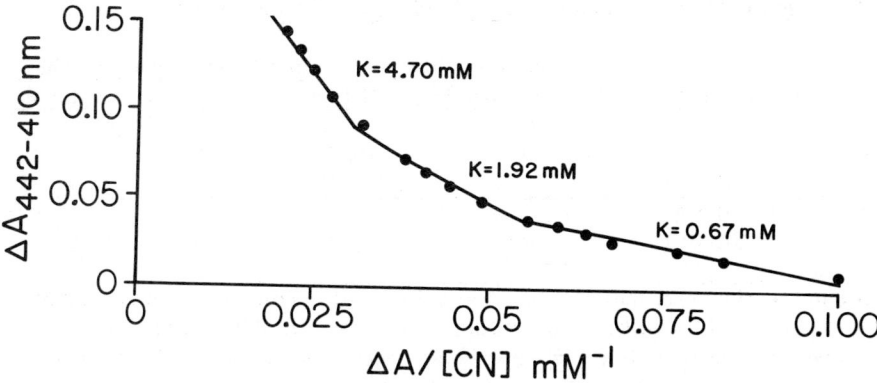

Fig. 1. Determination of various forms of cytochrome P-450 by titration with cyanide
The column fraction exhibiting MEOS activity was prepared as described previously (16). Following desalting by Sephadex G 25 column chromatography, the three different forms of cytochrome P-450 were determined spectrophotometrically after cyanide titration in a range of 0.1 to 16 mM using 3 mg protein/3 ml assay (20). Differences in absorbance between 442 nm and 410 nm were plotted against the absorbance over cyanide concentration (Hofstee plot), and the different affinities of cytochrome P-450 for cyanide were evaluated.

Differentiation of MEOS from Catalase Activity

Of major concern was the differentiation of MEOS from catalase, since both enzymes are capable of oxidizing ethanol in vitro and since during subcellular preparation of crude microsomes contamination with catalase is unavoidable. The observation that MEOS can clearly be separated from catalase activity by DEAE-cellulose column chromatography (14-19) indicates that catalase cannot be incriminated as an obligatory component in the activity of MEOS. That this is indeed the case was verified by the observation that replacement of the NADPH generating system by a H_2O_2 producing one was completely ineffective in sustaining ethanol oxidation in the isolated MEOS fraction (Table 1); the H_2O_2 dependent ethanol oxidation could only be promoted by the catalase fraction. These findings therefore substantiate that MEOS and catalase are different enzymes. Moreover, catalase inhibitors such as azide (16) and cyanide (Fig. 2) were ineffective in blocking the activity of the NADPH-dependent MEOS in its isolated form. On the other hand, addition of catalase to the isolated microsomal fraction resulted in an increase of the ethanol oxidation in the presence of a NADPH generating system (Fig. 2), suggesting that under certain conditions catalase could play some facultative role in connection with NADPH-dependent microsomal H_2O_2 generation. This additional pathway involving catalase, however, appears to be of limited importance in crude microsomes since the rate of microsomal H_2O_2 generation is extremely low, as discussed previously (16). The catalase-mediated enhancement was completely abolished by sodium cyanide at concentrations which left the rate of the NADPH-dependent microsomal ethanol oxidation unchanged in the

TABLE 1 Substrate specificity of catalase and the microsomal alcohol oxidizing system following column chromatography of solubilized microsomes.

Fractions	Generating system	nmoles aldehyde / min / flask produced from:			
		Methanol	Ethanol	Propanol	Butanol
Catalase	H_2O_2	16.8	19.6	0.3	0
	NADPH	0	0	0	0
MEOS	H_2O_2	0	0	0	0
	NADPH	11.5	26.6	16.6	10.6

The catalase fraction was eluted in the void volume after column chromatography on DEAE-cellulose of solubilized microsomes, whereas the MEOS fraction was recovered with 0.4 M KCl (16). Substrate specificity of catalase or the microsomal system was determined with a H_2O_2 generating system (glucose-glucose oxidase) or a NADPH generating system, respectively (16). Aliquots of the catalase fraction (1.2 mg protein/3 ml assay) or the MEOS fraction (4.5 mg protein/3 ml assay) were used for the assays.

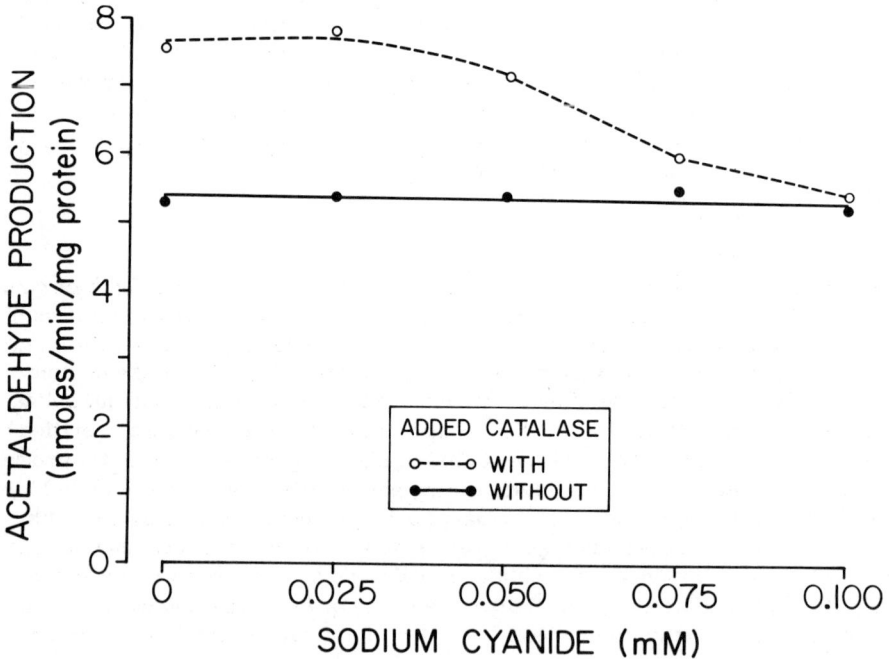

Fig. 2. Effect of sodium cyanide and catalase on the activity of the NADPH-dependent MEOS following isolation on DEAE-cellulose column chromatography.
The column fraction exhibiting MEOS activity was prepared and incubated with a NADPH generating system as described previously (16). When indicated, bovine liver catalase (900 Sigma units/3 ml assay) was included in the assay system. Sodium cyanide was employed at final concentrations up to 0.1 mM.

absence of catalase (Fig. 2). Therefore, the insensitivity of MEOS activity to cyanide when catalase is effectively blocked provides further evidence that catalase is not obligatory for MEOS.

Differentiation of MEOS from catalase was also achieved with studies concerning various alcohols as substrates. Whereas the MEOS fraction eluted by column chromatography oxidized not only ethanol but also methanol, propanol and butanol in a reaction requiring NADPH, catalase recovered in the void volume oxidized with a H_2O_2 generating system only lower aliphatic alcohols such as methanol and ethanol but not higher ones such as butanol (Table 1). Similar findings were reported in total hepatic microsomes (26). The microsomal alcohol oxidizing system can therefore be distinguished from catalase by a variety of characteristics including differences in substrate specificity. Therefore, the contention of some (27, 28) that catalase is an obligatory component of MEOS activity is unwarranted. Actually, this claim has now been retracted by one of the groups (25).

Reconstitution of MEOS with Microsomal Components

To define the role of cytochrome P-450, NADPH-cytochrome $\underline{c}$ reductase and phospholipid in MEOS activity, the constituents of the microsomal membranes were partially purified and tested for their capacity to oxidize ethanol to acetaldehyde when recombined. Cytochrome P-450 was partially purified by protease treatment and subsequent column chromatography on DEAE-cellulose using a stepwise elution of a KCl gradient (20). NADPH-cytochrome $\underline{c}$ reductase was partially purified according to a modification of the method of Levin et al. (29) with the omission of the Emulgen 911 step. The effect of each of the two microsomal components and synthetic phospholipid on ethanol oxidation is shown in Table 2. Whereas each of the components sustained no or negligible oxidation, the rate of acetaldehyde production was strikingly increased when the three components were combined. The rate of acetaldehyde production was dependent on the concentration of cytochrome P-450 and NADPH-cytochrome $\underline{c}$ reductase (30). Of particular interest was the finding that this reconstituted microsomal system was capable of metabolizing not only ethanol, but also propanol, butanol and benzphetamine. Moreover, the activity of the reconstituted MEOS was completely insensitive to the catalase inhibitor azide and was not sustained when NADPH was replaced by a H_2O_2 producing system (30). These findings therefore show that MEOS can be reconstituted with partially purified cytochrome P-450, NADPH-cytochrome $\underline{c}$ reductase and lecithin in a way similar to other microsomal drug metabolizing enzymes. Since various forms of cytochrome P-450 are present in the microsomal membranes (20, 31, 32), further studies are necessary to determine to what extent MEOS and other microsomal drug metabolizing enzymes differ by their cytochrome P-450 type.

TABLE 2 Reconstitution of MEOS with partially purified microsomal components.

	Acetaldehyde formed nmoles/min/ml
P-450 + Reductase + Lecithin	0.88
Reductase	0.22
P-450	0
Lecithin	0

The reaction mixture, in a final volume of 3.0 ml, contained (per ml) 100 µmoles of phosphate buffer (pH 7.4), 5µ moles of magnesium chloride, 1 µmole of disodium-EDTA, 1µ mole of NADPH, 50µ moles of ethanol, 0.30 nmole of cytochrome P-450, 200 units of reductase and 17 µg of L-α-dioleoyl lecithin. The reaction mixture was preincubated at 37°C for 5 min and the reaction was initiated by the addition of NADPH. Acetaldehyde formed in incubates during the incubation time of 0, 5, 10 and 15 min was directly measured by gas-liquid chromatography (30).

Acknowledgements:
The excellent technical assistance of Ms. Nancy Lowe is gratefully acknowledged. Parts of this study were supported by the Medical Research Service of the Veterans Administration and USPHS grants AA00224 and AM12511.

REFERENCES

(1) N. Grunnet, B. Quistorff, and H. I. D. Thieden, Rate-limiting factors in ethanol oxidation by isolated rat-liver parenchymal cells, Eur. J. Biochem. 40, 275 (1973).
(2) C. S. Lieber, and L. M. DeCarli, Hepatic microsomal ethanol oxidizing system: In vitro characteristics and adaptive properties in vivo, J. Biol. Chem. 245, 2505 (1970).
(3) C. S. Lieber, and L. M. DeCarli, The role of the hepatic microsomal ethanol oxidizing system (MEOS) for ethanol metabolism in vivo, J. Pharmacol. Exp. Ther. 181, 279 (1972).
(4) R. Rognstad, Isotopic probes into pathways of ethanol metabolism, Arch. Biochem. Biophys. 163, 544 (1974).
(5) J. Papenberg, J. P. von Wartburg, and H. Aebi, Metabolism of ethanol and fructose in the perfused rat liver, Enzym. Biol. Clin. 11, 237 (1970).
(6) R. Teschke, Y. Hasumura, and C. S. Lieber, Hepatic ethanol metabolism: Respective roles of alcohol dehydrogenase, the microsomal ethanol oxidizing system and catalase, Arch. Biochem. Biophys. 175, 635 (1976).
(7) S. Matsuzaki, and C. S. Lieber, Increase of ADH-dependent and -independent ethanol oxidation in isolated hepatocytes of rats fed ethanol chronically, Fed. Proc. 35, 707 (1976).
(8) G. R. Bartlett, Does catalase participate in the physiological oxidation of

alcohols?, Quart. J. Stud. Alc. 13, 583 (1952).
(9) F. W. Kinard, G. H. Nelson, and M. G. Hay, Catalase activity and ethanol metabolism in the rat, Proc. Soc. Exp. Biol. Med. 92, 772 (1956).
(10) M. E. Smith, Interrelations in ethanol and methanol metabolism. J. Pharmacol. 134, 233 (1961).
(11) D. Lester, and G. D. Benson, Alcohol oxidation in rats inhibited by pyrazole, oximes and amides, Science 169, 282 (1970).
(12) E. Feytmans, and F. Leighton, Effects of pyrazole and 3-amino-1,2,4-triazole on methanol and ethanol metabolism by the rat, Biochem. Pharmacol. 22, 349 (1973).
(13) C. S. Lieber, and L. M. DeCarli, Ethanol oxidation by hepatic microsomes: Adaptive increase after ethanol feeding, Science 162, 917 (1968).
(14) R. Teschke, Y. Hasumura, J.-G. Joly, H. Ishii, and C. S. Lieber, Microsomal ethanol-oxidizing system (MEOS): Purification and properties of a rat liver system free of catalase and alcohol dehydrogenase, Biochem. Biophys. Res. Commun. 49, 1187 (1972).
(15) E. Mezey, J. J. Potter, W. D. Reed, Ethanol oxidation by a component of liver microsomes rich in cytochrome P-450, J. Biol. Chem. 248, 1183 (1973).
(16) R. Teschke, Y. Hasumura, and C. S. Lieber, Hepatic microsomal ethanol oxidizing system: Solubilization, isolation and characterization, Arch. Biochem. Biophys. 163, 404 (1974).
(17) R. Teschke, Y. Hasumura, and C. S. Lieber, Hepatic microsomal alcohol oxidizing system: Affinity for methanol, ethanol, propanol and butanol, J. Biol. Chem. 250, 7397 (1975).
(18) R. Teschke, Y. Hasumura, and C. S. Lieber, Hepatic microsomal alcohol oxidizing system in normal and acatalasemic mice: Its dissociation from the peroxidatic activity of catalase-H_2O_2, Molec. Pharmacol. 11, 841 (1975).
(19) R. J. M. Corrall, L. C. Yu, B. A. Rosner, J. M. Margolis, H. M. Rodman, W. Kam, and B. R. Landau, Stereospecificity of the microsomal ethanol-oxidizing system, Biochem. Pharmacol. 24, 1825 (1975).
(20) K. Comai, and J. L. Gaylor, Existence and separation of three forms of cytochrome P-450 from rat liver microsomes, J. Biol. Chem. 248, 4947 (1973).
(21) D. Ryan, A. Y. H. Lu, S. West, and W. Levin, Multiple forms of cytochrome P-450 in phenobarbital and 3-methylcholanthrene-treated rats, J. Biol. Chem. 250, 2157 (1975).
(22) Y. Hasumura, R. Teschke, and C. S. Lieber, Hepatic microsomal ethanol oxidizing system (MEOS): Dissociation from reduced nicotinamide adenine dinucleotide phosphate-oxidase and possible role of form 1 of cytochrome P-450. J. Pharmacol. Exp. Ther. 194, 469 (1975).
(23) K. P. Vatsis, and M. P. Schulman, Evidence against the involvement of cytochrome P-450 in alcohol oxidation by mouse liver microsomes, Tenth International Congress of Biochemistry, July 25-31, 1976, Hamburg, Germany. Abstracts p. 667 (1976)
(24) K. J. Isselbacher, and E. A. Carter, Ethanol oxidation by liver microsomes. Evidence against a separate and distinct enzyme system. Biochem. Biophys. Res. Commun. 39, 530 (1970).
(25) R. G. Thurman, and H. J. Brentzel, The role of alcohol dehydrogenase in microsomal ethanol oxidation and the adaptive increase in ethanol metabolism due to chronic treatment with ethanol, Alcoholism: Clinical and Experimental Research, in press (1977).

(26) R. Teschke, Y. Hasumura, and C. S. Lieber, NADPH dependent oxidation of methanol, ethanol, propanol and butanol by hepatic microsomes, Biochem. Biophys. Res. Commun. 60, 851 (1974).

(27) R. G. Thurman, H. G. Ley, and R. Scholz, Hepatic microsomal ethanol oxidation. Hydrogen peroxide formation and the role of catalase, Eur. J. Biochem. 25, 420 (1972).

(28) J. M. Khanna, H. Kalant, and G. Lin, Metabolism of ethanol by rat liver microsomal enzymes, Biochem. Pharmacol. 19, 2493 (1970).

(29) W. Levin, D. Ryan, S. West, and A. Y. H. Lu, Preparation of partially purified lipid-depleted cytochrome P-450 and reduced nicotinamide adenine dinucleotide phosphate-cytochrome c reductase from rat liver microsomes, J. Biol. Chem. 249, 1747 (1974).

(30) K. Ohnishi, and C. S. Lieber, Reconstitution of the hepatic microsomal ethanol oxidizing system, Fed. Proc. 35, 706 (1976).

(31) J.-G. Joly, H. Ishii, and C. S. Lieber, Microsomal cyanide-binding cytochrome: Its role in hepatic ethanol oxidation, Gastroenterol. 62, 174 (1972).

(32) D. A. Haugen, T. A. Van Der Hoeven, and M. J. Coon, Purified liver microsomal cytochrome P-450, J. Biol. Chem. 250, 3567 (1975).

CHARACTERIZATION OF NADPH-CYTOCHROME c(P-450) REDUCTASE PURIFIED BY BIOSPECIFIC AFFINITY CHROMATOGRAPHY UTILIZING SPECTROPHOTOMETRIC AND ELECTRON PARAMAGNETIC RESONANCE TECHNIQUES

Yukio Yasukochi, Julian A. Peterson and Bettie Sue Siler Masters

Department of Biochemistry, The University of Texas Health Science Center at Dallas, 5323 Harry Hines Boulevard, Dallas, Texas 75235

ABSTRACT

Utilizing NADPH-cytochrome c (P-450) reductase (E.C.1.6.99.2) which was purified in high yields to homogeneity after biospecific affinity chromatography of detergent-solubilized preparations on Sepharose 4B-bound N6-(6-aminohexyl)-adenosine 2',5'-bisphosphate (2',5'-ADP-Sepharose 4B, Ref 1),[1] determinations of physical, spectral, and catalytic properties of the flavoprotein were performed. The resulting preparations from either pig or rat liver microsomes exhibited a molecular weight of 78,000 g. mole^{-1} as estimated by SDS polyacrylamide gel electrophoresis and contained 1 mole each of FAD and FMN. The catalytic turnover number for cytochrome c reduction, based on flavin content, was 1360 min^{-1} at 25°C. for the pig enzyme. These preparations possess identical spectral properties to the previously described proteolytically solubilized enzyme. In addition, absorbance spectral titrations of the detergent-solubilized, affinity-chromatographed reductase from either species demonstrated that the air-stable, reduced form of NADPH-cytochrome c (P-450) reductase contains two electron equivalents, confirming the previous results of Masters, et al. (2) obtained with proteolytically solubilized preparations from both pig and rat liver microsomes. Aerobic titration of the reductase with NADPH to the air-stable, reduced state produced a g=2 free radical signal in quantitative EPR studies which confirmed the existence of more than one electron in this oxidation-reduction state of the flavoprotein.

INTRODUCTION

The efforts of many laboratories in recent years have been directed toward the purification of hepatic microsomal constituents (3-9) with the ultimate goal of reconstituting a variety of monooxygenase activities, involving such substrates as steroids, carcinogens, drugs, and fatty acids, in order that a deeper understanding of such systems may be achieved.

A number of procedures have been employed for the purification of NADPH-cytochrome c (P-450) reductase from detergent-solubilized hepatic microsomes (3-6), but these procedures required multiple steps and most resulted in low

---

[1]Sepharose 4B-bound N6-(6-aminohexyl)-adenosine 2',5'-biphosphate will be referred to throughout the paper as 2',5'-ADP-Sepharose 4B.

percentage yields (3-5). The recent development by Yasukochi and Masters (1) of a biospecific affinity chromatography technique for the purification of this reductase from either pig or rat liver by a two- or three-step chromatography procedure facilitates the production of homogeneous enzyme in high percentage yields. Such a procedure is essential for the purification of large quantities of enzyme required for EPR spectroscopy and stopped-flow kinetic studies, in addition to other experiments involving chemical and physical characterization. The present paper will report some recent results obtained with these NADPH-cytochrome c (P-450) reductase preparations in experiments designed to further understand the mechanism of interaction of this flavoprotein with a variety of electron acceptors, ultimately to include its physiological acceptor, cytochrome P-450.

METHODS

Solubilization and Purification of Pig and Rat Liver Microsomal NADPH-Cytochrome c (P-450) Reductase

The procedures employed for the purification of both pig and rat liver microsomal reductases have been described in detail (1). Since the publication of these experimental methods, we have successfully employed the product, Agarose-Hexane-Adenosine 2',5'-Diphosphate (AG 2',5'-ADP, Type 2), recently marketed by P-L Biochemicals, Inc., in addition to the product, 2',5'-ADP-Sepharose 4B, produced by Pharmacia Fine Chemicals and kindly supplied to us by Dr. Robert Bywater of that Company.

Reagents for Assay and Titration

Peptostreptococcus elsdenii flavodoxin was a generous gift from Dr. Stephen G. Mayhew at Agricultural University in Wageningen, The Netherlands. NADPH was purchased from P-L Biochemicals, Inc. Potassium ferricyanide was obtained from Baker Chemical Co. and its concentration was determined at the beginning and end of each titration experiment using the $\varepsilon_M 420$ nm = 1.03×10^3 M^{-1} cm^{-1}. Other chemicals were reagent grade obtained from commercial sources.

Determination of Protein Concentration

Protein concentration was determined by the method of Lowry, et al. (10) as modified by Dulley and Grieve (11) for solutions which contain detergents.

Assay of NADPH-Cytochrome c Reductase Activity

All previous determinations of NADPH-cytochrome c reductase activity in this laboratory were performed in 0.05 M potassium phosphate buffer, pH 7.7, 0.1 mM EDTA at 25°C (12). In order to compare the activities obtained with our preparations to the data of Vermilion and Coon (13) and Dignam and Strobel (14), it has been essential to present activity data determined in 0.3 M potassium phosphate, pH 7.7, 0.1 mM EDTA at 30° C. Under these assay conditions, a doubling of specific activity was obtained (Table I).

The NADPH-cytochrome c (P-450) reductase preparations used in the present studies, purified from pig liver microsomes, were at least 85% reducible by excess NADPH and the turnover numbers were greater than 1300 min^{-1} in 0.05 M potassium phosphate buffer, pH 7.7, 0.1 mM EDTA at 25°C. with cytochrome c as electron acceptor. (See figure legends for details.)

Spectrophotometric Titrations and Kinetic Assays

All spectrophotometric titration studies were performed on a Cary Model 14R recording spectrophotometer thermostatically controlled at 25° C. under aerobic conditions. Kinetic assays were performed on a Beckman 25 ratio recording spectrophotometer at 25° C. or 30° C., as indicated.

Electron Paramagnetic Resonance (EPR) Spectroscopy

EPR experiments were performed with a Varian E-4 Spectrometer interfaced with a PDP-11 minicomputer. The cell used for the anaerobic photolytic reduction of flavodoxin in the EPR cavity was a flat cell obtained from Varian Associates. The measurements of NADPH-cytochrome c (P-450) reductase were performed in the same cell under aerobic conditions at room temperature, so that direct comparison could be made between the EPR signals obtained.

RESULTS

Comparison of NADPH-Cytochrome c (P-450) Reductase Purified by Affinity Chromatography on 2',5'-ADP-Sepharose 4B from Pig and Rat Liver Microsomes

The data of Table 1 report specific activities under both sets of assay conditions described in Methods for both pig and rat liver reductase preparations.

TABLE 1 Comparison of Specific Activities and Turnover Numbers of NADPH-Cytochrome c (P-450) Reductase[a]

	Specific Activity (μmoles min^{-1}mg^{-1})	Turnover Number (min^{-1})
Pig Liver Reductase	17.8[b]	1360[b]
	37.4[c]	2856[c]
Rat Liver Reductase	21.9[b]	1365[b]
	43.8[c]	2730[c]

[a] NADPH-cytochrome c reductase activities were assayed as described in Methods under two sets of experimental conditions.

[b] Assayed in 0.05 M. potassium phosphate buffer, pH 7.7, 0.1 mM EDTA, at 25°C.

[c] Assayed in 0.3 M. potassium phosphate buffer, pH 7.7, 0.1 mM EDTA, at 30°C.

Although the numbers presented do not represent the highest specific activities obtainable, the data illustrate that the affinity-chromatographed flavoprotein is spectrally pure, as indicated by the turnover numbers calculated on the basis of flavin. Comparison of these turnover numbers with those obtained with proteolytically solubilized reductase under identical assay conditions indicates that identical activities are obtained with the detergent-solubilized preparations when cytochrome c is the electron acceptor.

Aerobic Titration of the Air-Stable Half-Reduced Pig Liver Microsomal NADPH-Cytochrome c (P-450) Reductase

In order to determine the electron transfer properties of the detergent-solubilized, affinity-chromatographed reductase, the experiment shown in Fig. 1 was performed. The enzyme was reduced with excess NADPH (4 mole NADPH per

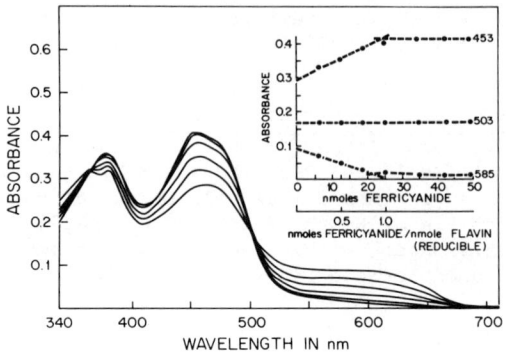

Fig. 1. Aerobic titration of affinity-chromatographed pig liver detergent-solubilized NADPH-cytochrome c reductase with $K_3Fe(CN)_6$. The reductase (31.2 nmoles total flavin) in 0.05 mM potassium phosphate buffer, pH 7.7, containing 0.1 mM EDTA, 20% glycerol, and 0.1% sodium deoxycholate, was titrated with 2.1 mM $K_3Fe(CN)_6$ in the above buffer at 25°C. To obtain the air-stable, half-reduced enzyme, 4 mol of NADPH/mol of flavin was added aerobically and the flavoprotein was allowed to reoxidize to the half-reduced state, which was confirmed by monitoring at 340 nm and 453 nm. Curve 1, half-reduced enzyme; Curves 2-6 show equilibrium absorption spectra after the addition of 6.3, 12.6, 18.9, 25.2 and 34.7 nmoles $K_3Fe(CN)_6$, respectively. The total volume change was less than 5.5%, and was not corrected for upon each addition. The inset shows the changes occurring at 453, 503 and 585 nm during the titration and the abscissa indicates total nmoles of $K_3Fe(CN)_6$ as well as the ratio of $K_3Fe(CN)_6$ to reducible flavin (in this case, 85% based on the addition of 50-fold excess NADPH under aerobic conditions). The specific activity of the flavoprotein used in this experiment was 36 μmole $min^{-1}mg^{-1}$ at 30°C in 0.3 M potassium phosphate, pH 7.7, and the turnover number was 1360 min^{-1} at 25°C.

mole total flavin, i.e., 126 nmoles of NADPH) aerobically and allowed to reoxidize in air until there was no further change at 340 nm or 453 nm, indicating complete oxidation of the NADPH and no further oxidation of flavin, respectively. Then, potassium ferricyanide was added in aliquots resulting in the various equilibrium absorption spectra shown in Fig. 1. The changes in absorbance at 453, 503, and 585 nm are indicated in the figure in the inset in which the abscissa indicates both total nmoles of potassium ferricyanide added and the ratio of oxidant to reducible flavin. The results show that one mole of potassium ferricyanide per mole of enzymatically reducible flavin is required to reoxidize the flavoprotein to its fully oxidized state, substantiating the conclusion that two electrons remain in the stable, half-reduced form of the reductase (containing one mole each of FAD and FMN) in agreement with the data of Masters, et al. (2) and Kamin (15) on proteolytic-

Electron Paramagnetic Resonance Spectroscopy Studies on Affinity Chromatographed NADPH-Cytochrome c (P-450) Reductase

ally solubilized reductase.

In view of the fact that this flavoprotein reduces one- and two-electron acceptors with equal facility and contains two electrons in the oxidized counterpart of its catalytic cycle, and yet does not exhibit a semiquinone absorption spectrum of a magnitude comparable to many other flavoprotein radicals, the following experiments were undertaken to determine its free radical nature.

The air-stable intermediate form of NADPH-cytochrome c (P-450) reductase was formed by the addition of 1 mole of NADPH per mole of flavin and allowed to reoxidize in air to the air-stable spectrum. Fig. 2 shows the average of 8 scans of the EPR spectra obtained in such an experiment. The signal appears to be symmetrical with a g value of approximately 2, but "wings" ap-

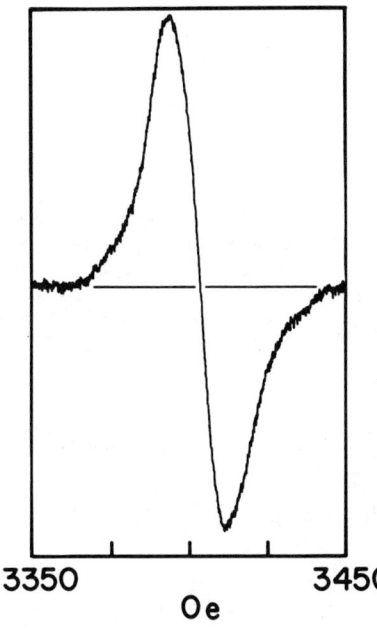

Fig. 2. The EPR signal of the air-stable form of NADPH-cytochrome c reductase. The reductase (171.6 µM) was treated with 2 equivalents of NADPH per mole of flavin in the presence of air to form the air-stable semiquinone. The EPR signal shown is the average of 8 scans recorded with the following instrument parameters: modulation amplitude, 6.3G; power, 2 mW; temperature, 25°; gain, 1×10^4 frequency, 9.54 GHz; scan rate, 8 min; and time constant 0.3 sec. The amount of radical present in the sample was estimated by comparing the integrated area of this signal with that obtained with a standard solution of 179 µM flavodoxin which had been photochemically reduced (16).

pear in the spectrum on either side of the g=2 signal. When this composite spectrum was integrated, the curve generated was distinctly broadened when compared with photochemically reduced flavodoxin from P. elsdenii at equal signal intensity. Although the studies of Fig. 2 were performed at 2 mW power, power saturation studies, shown in Fig. 3, indicate that experiments can be performed at higher concentrations of enzyme to exaggerate the "wings" and at increased power (up to 20 mW) to study the interaction of the flavin prosthetic groups. Preliminary data indicate that there are differences in power saturability between the "wings" and the major signal at g=2.

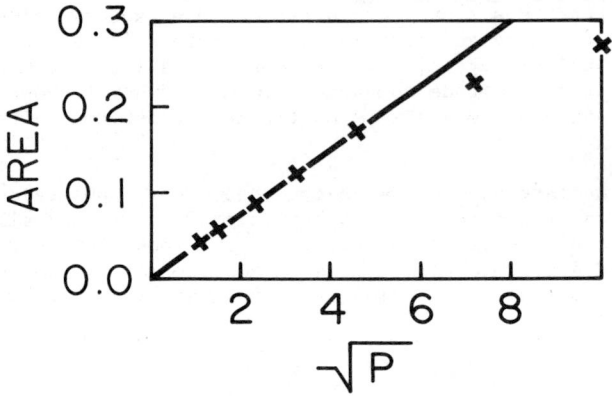

Fig. 3. Effect of power on the area of the g=2 signal of NADPH-cytochrome c (P-450) reductase. The sample of the air-stable form of NADPH-cytochrome c (P-450) reductase was prepared as described in Fig. 2.

Comparison with P. elsdenii flavodoxin semiquinone, prepared by photochemical reduction in anaerobic EDTA solutions (16), permitted the determination of the number of unpaired electrons in the air-stable intermediate. The data of Table 2 clearly show that the minimum ratio of areas is 0.68 which is substantially more than the theoretical ratio of 0.50 for 1 electron per flavin and less than 1.00 for 2 electrons per flavin. The control experiment was performed on the sample of NADPH-cytochrome c (P-450) reductase before addition of NADPH.

TABLE 2 Determination of the Number of Unpaired Electrons in NADPH-Cytochrome c (P-450) Reductase

CONDITION	AREA/FLAVIN	RATIO AREA*
Control	0.503	0.093
0.5 e-/Flavin	2.03	0.375
1.0 e-/Flavin	3.80	0.680 ± 0.03 (n=3)

*RATIO AREA The area per flavin for NADPH-Cytochrome P-450 Reductase was divided by the area per flavin for Flavodoxin

DISCUSSION

The studies reported herein were undertaken to determine the physical properties of a detergent-solubilized preparation of NADPH-cytochrome c (P-450) reductase purified to homogeneity utilizing a biospecific affinity chromatography technique developed in this laboratory (1). The apparent discrepancies in the absorption and electron paramagnetic resonance spectral properties of all of the various preparations from this laboratory compared to the recent reports from Mason's laboratory (17-19) have prompted an extensive study of the air-stable, reduced intermediate form of NADPH-cytochrome c (P-450) reductase. Although the data of Mason (17-19) were obtained with proteolytically-solubilized reductase preparations, the results obtained in the present studies with detergent-solubilized reductase are consistent with our previous studies with both the detergent-solubilized and the proteolytically solubilized reductase (1,2), indicating two electrons in the air-stable intermediate.

The nature of the half-reduced (so-called "semiquinone") species of NADPH-cytochrome c reductase has puzzled flavoprotein chemists because of its low extinction coefficient in the 500-650 nm wavelength region. It has led to the suggestion by Iyanagi and Mason (17, 19) and Iyanagi, et al. (18) that there is only one electron between the two flavin prosthetic groups in the air-stable, reduced intermediate and their experiments were designed to examine this possibility. However, recent studies by Masters, et al. (2) and Yasukochi and Masters (1) have directed our attention to the possibility that interaction of the FAD and FMN prosthetic groups occurs in such a manner that some of the flavin radicals exhibit an EPR signal and others do not, due to spin-spin interaction. The net result of such interactions is the quantitation of the EPR signal of the air-stable, half-reduced form, by comparison with P. elsdenii flavodoxin semiquinone, to give more than 1, but less than 2 electrons per mole of enzyme (more than 0.5 but less than 1 electron per flavin). The data of Figure 2 and Table 2 support the conclusion of Figure 1 that 2 electrons remain in the air-stable, reduced intermediate of NADPH-cytochrome c (P-450) reductase. In addition, there is a strong indication that the flavins are spatially contiguous to the extent that spin coupling has occurred in this free radical intermediate. Beinert and Orme-Johnson (20) reported similar findings on a proteolytically solubilized NADPH-cytochrome c reductase preparation supplied by Masters and Kamin.

Although the discrepancies between the present data and that obtained from Mason's laboratory (17-19) cannot be easily explained, the absorption spectral titrations and electron paramagnetic resonance spectra obtained in these experiments and in all our previous experiments (1, 2, 20, 21, 22) are consistent. The results strongly indicate that the more oxidized state of the flavoprotein during the catalytic cycle (21, 22), i.e., the air-stable, half-reduced enzyme (1, 2), largely consists of 2 electrons on the two distinct flavins, FAD and FMN, which are spatially arranged such that there is electron spin interaction to some degree. The significance of these results in the light of the ability of this flavoprotein to reduce 1- and 2-electron acceptors with equal facility (22) is yet to be fully understood.

REFERENCES

(1) Yasukochi, Y., and Masters, B.S.S., J. Biol. Chem. 251, 5337 (1976).
(2) Masters, B.S.S., Prough, R.A., and Kamin, H., Biochemistry 14, 607 (1975).
(3) Vermilion, J.L., and Coon, M.J., Biochem. Biophys. Res. Commun. 60, 1315 (1974).
(4) Dignam, J.D., and Strobel, H.W., Biochem. Biophys. Res. Commun. 63, 845 (1975).
(5) Satake, H., Imai, Y., and Sato, R., Seikagaku 44, 765 (1972).
(6) Dignam, J.D., and Strobel, H.W., Biochemistry, in press (1976).
(7) Imai, Y., and Sato, R., Biochem. Biophys. Res. Commun. 64, 8 (1974).
(8) Van der Hoeven, T.A., Haugen, D.A., and Coon, M.J., Biochem. Biophys. Res. Commun., 64, 569 (1974).
(9) Ryan, D., Lu, A.Y.H., Kawalek, J., West, S.B., and Levin, W., Biochem. Biophys. Res. Commun. 64, 1134 (1975).
(10) Lowry, O.H., Rosebrough, N.J., Farr, A.L., and Randall, R.J., J. Biol. Chem. 193, 265 (1951).
(11) Dulley, J.R., and Grieve, P.A., Anal. Biochem. 64, 136 (1975).
(12) Masters, B.S.S., Williams, C.H., Jr. and Kamin, H., Methods in Enzymology, X (R.W. Estabrook and M.E. Pullman, Eds.) Academic Press, New York, 565, 1967.
(13) Vermilion, J.L. and Coon, M.J., Biochem. Biophys. Res. Commun. 60, 1315 (1974).
(14) Dignam, J.D., and Strobel, H.W., Biochem. Biophys. Res. Commun. 63, 845 (1975).
(15) Kamin, H., in Reactivity of Flavins (K. Yagi, Ed.) University Park Press, Baltimore, 137, 1975.
(16) Mayhew, S.G., and Massey, V., J. Biol. Chem. 244, 794 (1969).
(17) Iyanagi, T., and Mason, H.S., Biochemistry 12, 2297 (1973).
(18) Iyanagi, T. Makino, N., and Mason, H.S., Biochemistry 13, 1701 (1974).
(19) Iyanagi, T., and Mason, H.S., in Reactivity of Flavins (K. Yagi, Ed.) University Park Press, Baltimore, 145, 1975.
(20) Beinert, H., and Orme-Johnson, W.H., in Magnetic Resonance in Biological Systems (A. Ehrenberg, B.G., Malmstrom, and T. Vanngard, Eds.) Pergamon Press, New York, 221, 1967.
(21) Masters, B.S.S., Kamin, H., Gibson, Q.H., and Williams, C.H., Jr., J. Biol Chem. 240, 921 (1965).
(22) Masters, B.S.S., Bilimoria, M.H., Kamin, H., and Gibson, Q.H., J. Biol. Chem. 240, 4081 (1967).

RELATIONSHIP BETWEEN THE MEMBRANE LIPIDS AND SUBSTRATE-CYTOCHROME P-450 BINDING REACTION IN BOVINE ADRENO-CORTICAL MICROSOMES

Shakunthala Narasimhulu

University of Pennsylvania, Philadelphia, PA 19174 USA

ABSTRACT

In the present study, lipid involvement in the substrate-cytochrome P-450 binding reaction is investigated by comparing thermally-induced changes in the fluidity of the membrane lipids and changes in the apparent substrate-dissociation constant ('K_s'). The 'K_s' increases with increase in temperature. Van't Hoff's plot of the 'K_s' shows breaks at $21°C$ and at $31°C$. These temperatures correlate well with the lipid phase transition temperatures as determined by the technique of fluorescence polarization using 1,6 diphenyl 1,3,5-hexatriene as the probe. The results indicate that the affinity of the substrate to the cytochrome depends upon the temperature. In addition the results suggest that changes in the physical state of the lipids is a factor responsible for the temperature dependency of the reaction.

INTRODUCTION

It is suggested that biological membrane lipids may form a diffusion field for the interaction of non-polar substrates and membrane-bound enzymes (1).

Cytochrome P-450 is membrane-bound in mammalian species. All of the known substrates of the cytochrome are lipid soluble. In addition there is evidence suggesting that structural integrity and enzymatic activity of the cytochrome depends upon its association with phospholipids. It is possible that the increased hydrophobicity provided by the lipid environment enhances the binding of non-polar substrates to the cytochrome. Reports regarding the role of the lipids in substrate-cytochrome binding reaction have been on studies using detergent solubilized and purified cytochrome from hepatic microsomes (2) and phospholipase C digested microsomes (3) and in adrenal microsomes depleted of their lipids by organic solvent extraction (4). In the present study the lipid involvement in the binding reaction is investigated in a more intact microsomal preparation by comparing thermally induced changes in the fluidity of the membrane lipids and changes in the apparent substrate dissociation constant using 17-hydroxyprogesterone as the substrate.

METHODS

The adrenocortical microsomes prepared as previously described (5) was further fractionated by sucrose density gradient centrifugation as follows: 2.2 ml of the microsomal fraction (4-5 mg protein per ml) were centrifuged over sucrose gradient of 5.0 ml each of 0.4M, 0.9M and 1.2M at 40,093 x gav for 2 hours. The band between 0.9M and 1.2M sucrose was used in the experiments. Preparations stored no longer than a week at $-20°C$ were used for all of the experiments with the exception of the experiment shown in Section B of

Fig. 1. For this experiment, preparation stored at -70°C was used.

Lipid-Depleted Microsomal Preparation

Aqueous suspension of microsomes were lyophylized and depleted of their lipids by extracting them with anhydrous n-butanol followed by anhydrous acetone as previously described (4). The lipid depleted powder (30 mg) was suspended in 0.12M sucrose by homogenization. The suspension was centrifuged over 1.2M sucrose. The supernatant was again centrifuged at 75,000xg for 1 hour. The supernatant after second centrifugation containing 0.33nmoles P-450 and 250 nmoles of phosphorus per mg of protein was used in the experiment reported in Fig. 3. The microsomes contain 1,000 nmoles of phosphorus per mg of protein.

Phospholipase Treatment of the Microsomal Preparation

A mixture of 0.5 ml of Band II (1.5 mg protein), 1.4 ml of 0.05 M Tris buffer pH 7.4 and phospholipase C (1 unit per mg protein) was incubated at 25°C for 10 minutes. The incubated mixture was cooled to 2°C and centrifuged over 0.9 M and 1.2 M sucrose gradient at 40,693 x gav for 1 hour. The band between 0.9 M and 1.2 M sucrose was used for the experiments.

Extraction of Lipids from the Microsomal Preparation (Band II)

Nine milliliters of the microsomal suspension (4-5 mg protein per ml) in 0.25 M sucrose were centrifuged at 50,000 x g for 15 minutes. The extraction of the sediment with a mixture of chloroform and methanol (2:1), washing and evaporation of the extract was according to the procedure described by Folch (6). The residue after evaporation is the total lipids used in the experiments. There were no proteins present in the residue in amounts detectable by the biuret procedure as described by Gornall (7).

Determination of the Apparent Substrate Dissociation Constant 'K_s'

Substrate-produced Type I difference spectrum of cytochrome P-450 which is characterized by a trough at 421 nm, peak at 388-390 nm and an isosbestic point at 407 nm was taken as criterion for binding of the substrate to the cytochrome. A spectrophotometric cuvette of 12 mm light path containing 3.0 ml of the enzyme-buffer mixture (0.2 to 0.4 mg microsomal protein per ml) (11) was placed in a dual wavelength filter photometer* fitted with interference filters with maximum transmittance at 421 nm ($\lambda 1$) and 407 nm ($\lambda 2$) of 1 nm half band width to isolate the two wavelengths of the transmitted light. A wide band filter having a transmittance range (320 nm to 470 nm) was used to filter the incident light. The enzyme in the cuvette was constantly stirred with a magnetic stirring attachment. The temperature was regulated by means of a thermostated circulator. The temperature was measured after equilibration at the beginning and end of each titration by means of a digital thermometer with a copper constantan thermocouple with an accuracy of 0.5°C. A methanolic solution of the steroid was added in 0.2 μl aliquots using a 10 μl syringe attached to an electrically operated Hamilton repeating dispenser. After each addition, the absorbance difference between 421 nm and 407 nm was recorded. The effect of methanol was tested up to 5 μl per ml of the enzyme-buffer mixture and was found to be without effect. The

---

*from Technical Consulting Service (TCS), P. O. Box 141, Southampton, PA 18966

maximum concentration of methanol ever reached in the titration experiments was 0.67 μl per ml. The concentration of the substrate required to produce half maximum absorbance change was calculated either from relative absorbance plot or from double reciprocal plots. The reliability of the $\Delta A_{407-421}$ nm values obtained with the filter photometric procedure was tested by measuring the difference spectrum with a Perkin Elmer split beam spectrophotometer Model 571.

The degree of variability of 'K_S' values obtained by the above procedure was determined as follows: 4 to 5 titrations were performed using equal aliquots of a batch of the microsomal preparation, at each of the following temperatures: 16°C, 25°C and 35°C. Coefficient of variation calculated from these 4 to 5 'K_S' values obtained at each of these temperatures are 1.35; 1.4 and 3.67% respectively.

Measurement of Fluidity of the Microsomal Membrane Lipids

Shnitizky and Inbar (8) have shown that fluorescence polarization can be used to approximate fluidity of membrane lipids. According to their procedure, fluorescence polarization of the hydrophobic probe 1,6 diphenyl 1,3,5 hexatriene (DPH, 0.33μM) added to a microsomal suspension (0.4 to 0.5 mg protein per ml) was determined as a function of temperature. Fluorescence measurements were made using a Perkin Elmer fluorescence spectrometer. Polacoat filters were used to polarize both exciting (366 nm) and the emitting (435 nm) beams. Fluorescence intensities were measured with polarizers parallel and perpendicular to the vertically polarized exciting beam. The degree of polarization was calculated as described by Shnitzky and Inbar (8).

RESULTS

Temperature Dependence of the Apparent Substrate-Dissociation Constant 'K_S'

Fig. 1 (section A) shows the substrate dissociation constant as a function of reciprocal temperature for the substrate 17-hydroxyprogesterone. The insert shows the Van't Hoff's plot of the association constant ($1/K_s$). The plot is triphasic with discontinuities at 21°C and at 31°C. The discontinuity at the lower temperature and the slopes of the three segments of the Van't Hoff's plot depended upon the temperature at which the microsomal preparations had been stored. An experiment using a preparation stored at -70°C instead of at -20°C is shown in section B of Fig. 1. The discontinuity at 21°C is shifted to 18°C. In addition the discontinuities are more clearcut.

The values for enthalpy (ΔH) for the three phases in the two cases are indicated in the figure. It is seen that ΔH for the three phases in the case of the microsomes which had been stored at -70°C (Section B: -3.7; -6.2 and -11.4 kcal mole^{-1}) are lower than for those which had been stored at -20°C (Section A: -6.4; -9.15 and -14.53 kcal mole^{-1}). Unlike the enthalpy the free energy change (ΔG = -8.8 to -8.9 kcal mole^{-1}) is nearly the same for the three phases and is also independent of the temperature of storage of the microsomal preparation. The reason for the differences between the microsomal preparations stored at the two different temperatures is not known. It is possible that the effect of the two temperatures of storage upon the preservation of the membrane structure with respect to lipid-protein interactions are different.

The effects of temperature on maximum spectral change observed at saturating

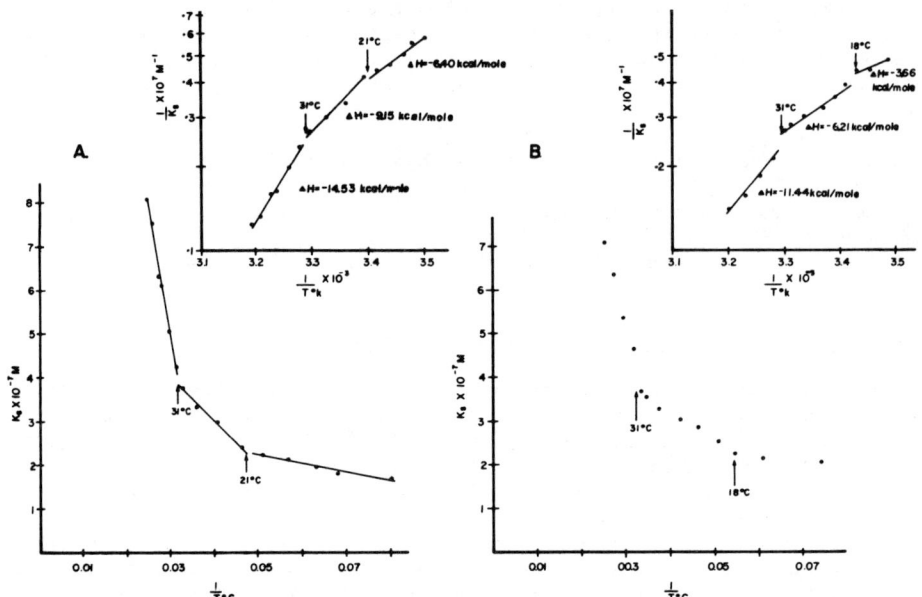

Fig. 1. The Effect of Temperature on the 'K_S' of 17-Hydroxy-Progesterone. A mixture of 2.6 ml of 1.6% albumin-containing glycyl-glycine buffer medium (5) and 0.4 ml (1.2mg protein) of the microsomal preparation (Band II, i,e between 0.9 and 1.2M sucrose described under methods) was used in these experiments. The details of the procedure is described under methods. Van't Hoff's plots (inserts) were constructed as follows: 'K_S' values at different temperatures were divided into three groups after recognizing the pattern in 'K_S' verses 1/T°C plots in which the transition points are more clearcut. Linear regression and correlation coefficients (r) were calculated by a computer. The values for 'r' are nearly 1.0 ranging between 0.99 to 0.97 with the exception of phase below 18°C (Section B) which is 0.93.

concentrations (10 x K_S) of the substrate were not significant. Since albumin-containing buffer was used in these experiments, the effect of eliminating albumin was tested and was also found to be insignificant. Since pH of the buffer varied from 7.43 at 12°C to 7.08 at 40°C, the effect of these two pHs' on the dissociation constant was tested at 17°C and 36°C. The binding constant was not altered.

Similar discontinuities observed with other membrane enzymes have been attributed to phase change in membrane lipids (9,10). Therefore, the effect of depleting the microsomes of their lipids on the temperature dependency of the substrate-cytochrome P-450 binding reaction was studied. The results of an experiment using a preparation which is 75% devoid of the microsomal lipids is shown in Fig 2. Lipid depletion eliminated the discontinuity at 20°C resulting in a biphasic Van't Hoff's plot with a single discontinuity at 30.5°C.

Although the position of the discontinuity at the higher temperature is not significantly altered, the sudden decrease in the substrate affinity is much more pronounced than that observed in the case of the microsomal suspension (Fig. 1).

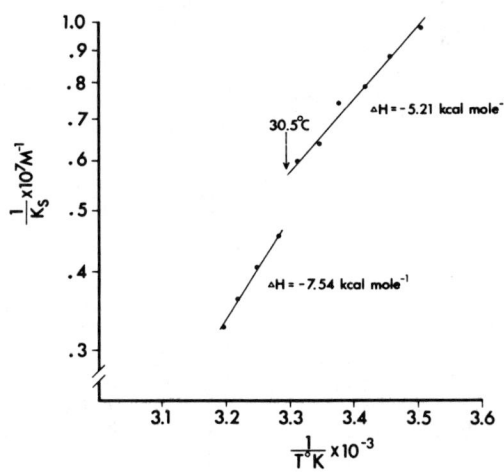

Fig. 2. Temperature Dependency of the 'K_S' in Lipid Depleted Microsomal Preparation. The enzyme preparation and the procedure for titration with 17-hydroxyprogesterone are described under Methods. 3.0 ml of the enzyme buffer mixture containing 1 mg of the microsomal protein was used for titration.

Similar to lipid depletion by organic solvent extraction, the addition of the lipid perturbing agent Triton X-114 (0.288 mg per mg protein) also eliminated the discontinuity at the lower temperature while preserving that at the higher temperature.

Temperature Dependency of Fluidity of Microsomal Lipids

Steady state fluorescence polarization of the hydrophobic dye DPH added to a suspension of the microsomal preparation was determined as described under Methods. A plot of the degree of fluorescence polarization (p) as a function of reciprocal absolute temperature is shown in Fig. 3. The plot is triphasic and exhibits discontinuities at 22°C and 31°C.

Treatment of the microsomal preparation with phospholipase C eliminated both the discontinuities (Fig. 4A). Sonicated dispersion of lipids extracted from the microsomal preparation as described under Methods exhibit only the discontinuity at the lower temperature (Fig. 4B). The enthalpies for the two

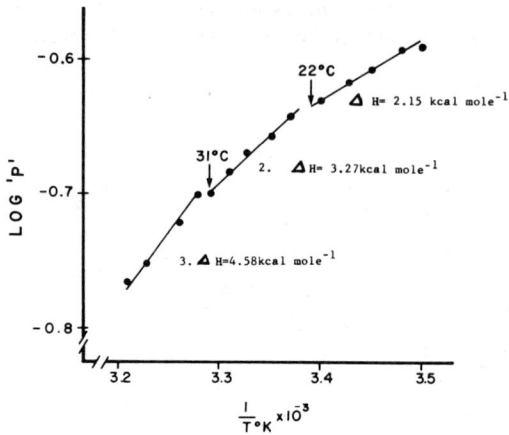

Fig. 3. Fluorescence Polarization of DPH in the Microsomal Preparation (Band II). 3.0 ml of 0.05M Tris buffer pH 7.4 containing 0.4 mg protein per ml and 0.33 µM DPH was used in this experiment. The procedure is described under Methods.

phases in the case of the lipids (Fig. 4B) are not too different from those for the 2nd and 3rd phase in the case of the microsomal preparation (Fig. 3).

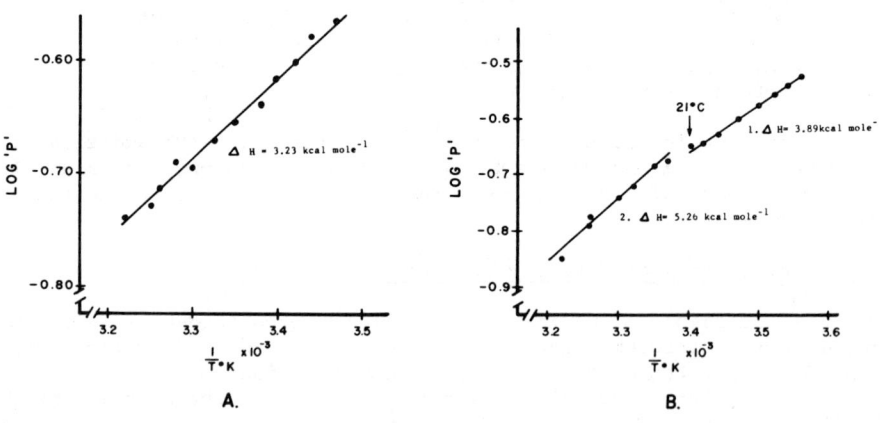

Fig. 4. Fluorescence Polarization of DPH in A. Phospholipase C Treated Band II, B. Lipids Extracted (1mg per ml) sonicated in 0.05M Tris buffer pH 7.4

These results are consistent with association of the discontinuities with transitions in lipid phase of the membrane as has been reported for other membrane systems (9, 10). The lipid phase transition temperatures correlate well with the temperatures at which discontinuities occur in Van't Hoff's plot of the substrate association constant (Fig. 1A). Furthermore, similar to the case of the association constant, the addition of Triton X-114 to the microsomal suspensions eliminated the phase transition at the lower temperature while preserving that at the higher temperature. Whether or not storage of the microsomal preparation at -70°C decreases the transition temperature at 21°C similar to that observed in the case of the substrate association constant (Fig. 1B) remains to be investigated.

DISCUSSION

Discontinuities observed in Van't Hoff's plot of the substrate association constant is analogous to the discontinuities in Arrhenius plots of reaction rates reported for various membrane enzymes (9, 10, 11). It has been suggested that discontinuities in Arrhenius plots occur at temperatures where lipids undergo a phase transition (9, 10). In the present study, the discontinuities (21°C and 31°C) in Van't Hoff's plots of the substrate association constant correlate well with the phase transition temperatures (22°C and 31°C) of the microsomal membranes as detected by the hydrophobic fluorescent probe DPH. In addition, the effects of the detergent on the substrate association constant are similar to those on the lipid phase transitions, i.e. in eliminating the transition point at the lower temperature while preserving that at the higher temperature. These observations are consistent with association of the discontinuities with lipid phase transitions.

Since lipids extracted from the microsomal membranes exhibited the transition at 22°C but not at 31°C, the transition at the lower temperature appears to be largely dependent on the membrane lipids. Since the transition at 31°C was observed only with the microsomal membranes, this transition may depend largely on protein-lipid interactions. From these results, it would appear that the discontinuity in Van't Hoff's plot of the substrate association constant at 31°C is brought about by changes in the lipids in intimate association with cytochrome P-450. This interpretation is supported by the result that a large part of the microsomal lipids could be removed preserving the discontinuity at 31°C while that at 22°C is eliminated.

The present results indicate that the affinity of the substrate to cytochrome P-450 is dependent on temperature. The results further suggest that changes in the physical state of the microsomal lipids is a factor responsible for the temperature dependency of the substrate-cytochrome binding reaction. But thermally-induced changes may not be significant for an organism which maintains constant temperature. However, physical state of lipids in biological membranes can be altered in many ways, e.g. by variation in lipid composition, by hydrophobic interaction with membrane proteins and interactions with lipophyllic substances such as cholesterol and possibly prostaglandins and steroids. Alteration in the physical state of membrane lipids offers an interesting possibility for regulation of substrate affinity which in turn may control the activity of the cytochrome P-450. In regard to substrate-dependent control of the activity, the surprisingly high value for free energy change (ΔG) of -8.9 kcal mole^{-1}

found for the substrate-cytochrome P-450 binding reaction is interesting because it is within the same range of values as reported for ΔG of hydrolysis of ATP to ADP.

ACKNOWLEDGMENTS:

supported by N.I.H grant AM 18545 and ONR contract No. N00014-75-C-0322. The author thanks Dr. J. M. Vanderkooi for the use of the Perkin-Elemr fluorescence spectrometer and for the polarizers used in the initial experiments; Ms. Mei-Feng Huang and Ms. Nancy Pratt for their excellent technical assistance

REFERENCES

1. P. Devaux and H. M. McConnell, Lateral Diffusion in Spin-Labeled Phosphatidylcholine Multilayers, J. Am. Chem. Soc. 94, 4475 (1972)
2. F. P. Guengerich and M. J. Coon, Role of Phospholipids in Reconstituted Liver Microsomal Enzyme System Containing Highly Purified Cytochrome P-450 Fed. Proc. 34, Abst. 2282 (1975)
3. M. D. Chaplin and G. J. Mannering, Role of Phospholipids in Hepatic Drug Metabolizing System, J. Mol. Pharmacol. 6, 631 (1970)
4. S. Narasimhulu, Role of Phospholipids in Adrenocortical Microsomal Hydroxylation Reaction. Adv. Exptl. Med. & Biol. 58, 271 (1975)
5. O. Rosenthal and S. Narasimhulu, Adrenal Steroid Hydroxylases, Methods in Enzymology, Ed. R. B. Clayton, Academic Press, New York, 15, 596 (1969)
6. J. Folch, M. Lees and Sloane-Stanley, A Simple Method for the Isolation and Purification of Total Lipids from Animal Tissues, J. Biol. Chem. 226, 497 (1957)
7. A. G. Gornall, C. J. Bardwell and M. M. David. Determination of the Serum Proteins by Means of the Biuret Reaction, J. Biol. Chem. 177, 751 (1949)
8. M. Shinitzky and M. Inbar, Difference in Microviscosity Induced by Different Cholesterol Levels in the Surface Membrane Lipid Layer of Normal Lymphocytes and Malignant Lymphoma Cells, J. Mol. Biol. 85, 603 (1974)
9. J. K. Raison, J. M. Lyons, R. J. Mehlhorn and A. D. Keith, Temperature Induced Phase Changes in Mitochondrial Membranes Detected by Spin Labelling, J. Biol. Chem. 246, 4036 (1971)
10. S. Eletr, D. Zakin and D. A. Vessey, A Spin Label Study of the Role of Phospholipids in Regulation of Membrane-Bound Microsomal Enzymes, J. Mol. Biol. 78, 351 (1973)
11. J. K. Raison, The Influence of Temperature-Induced Phase Changes on the Kinetics of Respiratory and Other Membrane Associated Enzymes, Bioenergetics, 4, 285 (1973)

KINETIC AND SPECTRAL EVIDENCE FOR MULTIPLE SPECIES OF CYTOCHROME P-450 IN LIVER MICROSOMES

Garth Powis, Ronald E. Talcott and John B. Schenkman

Department of Pharmacology, Yale University School of Medicine, New Haven, Connecticut 06510

Since the report of Sladek and Mannering (1), in which spectrophotometric evidence was presented for the appearance of a new cytochrome P-450 hemoprotein, called P_1-450, in animals pretreated with 3-methylcholanthrene, evidence has been rapidly accumulating indicating that P-450 is actually a family of similar but different terminal oxidases. P_1-450 differs both in its catalytic activity (2,3) and in its ability to form the type I spectral complex with its substrate benzpyrene (4,5), as well as in the absorption peak of its reduced, carbon monoxide complex (6). More recently, P_1-450 has been shown to have a different electrophoretic mobility in sodium dodecylsulfate (SDS) gel electrophoresis (7,8). SDS gel electrophoresis has provided evidence (9), as has differences in catalytic activities of the enzyme (10,11,12), for several species of cytochrome P-450 monooxygenases in liver microsomes, after treatment of rats with different inducer compounds.

In the study reported here, spectrophotometric data is provided which supports earlier enzymatic differences in the mixed function oxidases in liver microsomes of rats treated with several compounds known to cause elevation of the enzyme system. The compounds used were pregnenolone 16 α-carbonitrile, or PCN (13,14), the polychlorinated biphenyl, Aroclor 1254, or ARO (15), phenobarbital, or PB (16,17), and 3-methylcholanthrene, or MC (1,18).

MATERIALS AND METHODS

Liver microsomes were prepared from adult male rats by a rapid calcium aggregation procedure (19). PCN induced animals received eight oral doses (in 1% Tween 80) of 100 mg/Kg body weight at 12 hrs intervals. MC induced animals received five daily i.p. injections of 20 mg/Kg in corn oil, and ARO treated animals were treated similarly with 25 mg/Kg body weight. PB induction was with single daily i.p. injections of 80 mg/Kg in saline for four days. Animals were killed 48 hrs after the last treatment. Spectral studies (19,20) were performed with an Aminco DW2 spectrophotometer, with 4 mg microsomal protein/ml, unless stated otherwise. Enzyme assay conditions used are described elsewhere (dealkylations, ref. 21; benzpyrene hydroxylation, ref. 22; aniline hydroxylation, ref. 19; phenacetin, ref. 23). Destruction of cytochrome P-450 with linoleic acid hydroperoxide was by the method of Jeffery and Mannering (24,25).

RESULTS

In order to assess differences in the mixed function oxidase after various inducers by spectral means, it was necessary to ascertain that the often reported differences in elevation of metabolism could be obtained. The results are shown in Table I where activities are shown per mg microsomal protein and per n mole of P-450 hemoprotein; animals were treated with the inducers for 4 days and killed 48 hrs after the last treatment to minimize the amount of inducer remaining in the microsomes. The results indicate that PB increases the total amount of demethylase activity for ethylmorphine (278%), aminopyrine (323%), benzphetamine (847%) and caffeine (284%), and benzpyrene hydroxylase (153%). The extent of the induction was very variable, and, if expressed per n mole P-450 revealed no change for benzpyrene hydroxylation, minimal (20-40%) increase in most of the demethylases, but a real (3.6 fold) enhancement of benzphetamine demethylation. Thus it appears that PB specifically induced this latter enzyme, rather than acting as a non-specific inducer as has previously been suggested.

TABLE 1 Effect of Inducers on Mixed Function Oxidase Activity

Substrate		Control	PB	PCN	MC	ARO
ethylmorphine	V_m	13.7±0.8	16.8±4.3	*24.9±3.5	*6.4±0.5	*9.5±1.2
	V_T	10.6±0.6	*29.5±6.3	*39.3±4.2	*6.7±0.2	*18.9±3.4
aminopyrine	V_m	9.9±0.8	*13.9±1.7	9.7±1.3	7.6±1.8	*13.7±1.2
	V_T	7.7±0.5	*24.9±2.7	*15.4±1.2	7.8±1.5	*27.0±0.2
benzphetamine	V_m	12.5±1.2	*45.7±14.0	*6.6±0.7	*5.7±1.1	15.8±2.7
	V_T	9.7±0.8	*82.2±23.2	10.5±0.7	*5.8±0.9	*30.0±2.9
caffeine	V_m	0.48±0.12	0.65±0.07	ND	0.52±0.06	0.64±0.09
	V_T	0.33±0.01	*0.94±0.11	ND	0.49±0.07	*1.24±0.13
benzpyrene	V_m	0.14	0.14	0.14	0.33	-----
	V_T	0.15	0.23	0.19	0.43	-----

All values are means ± SEM. * = $P < 0.05$. V_m = n moles of product/min/n mole P-450; V_T = n moles of product/min/mg microsomal protein. Activities were taken from Lineweaver-Burk plots.

PCN acted quite different from PB, in that the activity of aminopyrine demethylation and ethylmorphine demethylation were both increased per mg protein, but benzphetamine demethylase, and benzpyrene hydroxylase were unchanged; caffeine demethylase activity was absent. When expressed relative to cytochrome P-450 content in the microsomes it is at once apparent that ethylmorphine demethylase is specifically induced by PCN (181%) while benzphetamine demethylase is not induced--the latter activity is decreased to 52% when expressed per n mole P-450. Since the activity of aminopyrine demethylase per P-450 content is unchanged, it would appear that this enzyme is increased in proportion to the total P-450 content. Thus from results with these two inducers, PCN and PB, aminopyrine demethylase, ethylmorphine demethylase and benzphetamine demethylase would appear to represent three different enzymes, based upon their differential induction. Since benzpyrene hydroxylase activity/n mole of P-450 remains constant during all three conditions, it would be expected that all of the P-450 species are capable of metabolizing this substrate, perhaps non-specifically.

Using MC as an inducer further differentiated the three above mentioned enzymes from each other and from the benzpyrene hydroxylase: ethylmorphine demethylase and benzphetamine demethylase were both diminished per mg of microsomal protein as well as per n mole of P-450. However, the microsomal content of aminopyrine demethylase activity was unchanged--only its activity per nmole of P-450 was lowered. Unlike the effect with the other inducers, benzpyrene hydroxylase activity per n mole of P-450 hemoprotein was elevated by MC pretreatment.

ARO differed from the other inducers and further aided in discriminating between the different enzymes. For example, like PB it elevated the total microsomal activity toward the substrates examined, but not all of the activities were increased to the same extent, resulting in a decrease in ethylmorphine demethylase activity, an increase in caffeine and aminopyrine demethylase activities and no change in benzphetamine demethylase activity.

Based upon the catalytic activities, then, it would be expected that spectrophotometrically it should be possible to distinguish between four different P-450's, aminopyrine demethylase, ethylmorphine demethylase, benzphetamine demethylase and benzpyrene hydroxylase and perhaps even a fifth, caffeine demethylase. This latter activity is different from the rest in that it exhibits a Km for caffeine of 5-10 mM normally, and has an activity as low as that of aniline hydroxylation.

When CO-binding spectra of the microsomes, after different inducers, were recorded on the same day and on the same chart, the absorption maxima were readily seen to differ: MC, 448 nm; ARO, 448 nm; PB, 450.5 nm and PCN, 449 nm. Such effects have been observed before by a number of laboratories (6,15,26) and, based upon enzyme activities must be considered average values (complexes of P-450 mixtures) rather than absorption maxima for specific P-450 species. Similarly, one can look at the absolute spectra of cytochrome P-450 after the different inducers. In Figure 1 is seen off-balance spectra of the different microsomes after addition of 150 uM linoleic acid hydroperoxide to the microsomal suspension in the sample cuvet. This treatment is known to cause destruction (20,21) of cytochrome P-450 but not cytochrome b_5, and thus will reveal the absolute spectrum of the sensitive hemoprotein. The data reveal differences exist in the absolute spectra of ferric cytochrome P-450 induced by the different compounds (Fig. 1), as would be suspected from the differences in catalytic activities.

Examination of substrate binding spectra with the different microsomal preparations also revealed differences in the newly induced P-450 species. One of the most striking differences was in the ability to form a type I spectral change (21) when benzpyrene is added to the microsomal suspension; addition of benzpyrene does not normally cause a spectral change (4). When added to

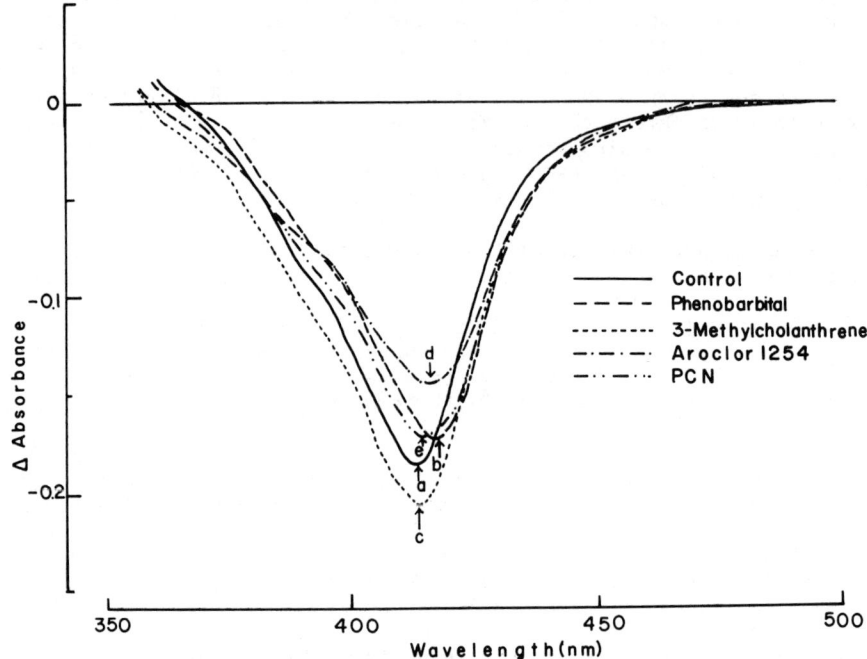

Fig. 1. Off-balance absolute spectra of P-450 after different inducers. P-450 content = 2 nmole/ml.

the microsomes of the MC treated animal (dashed line, Fig. 2) or to the microsomes of the ARO treated animal benzpyrene did cause a type I spectral change. This would be in agreement with the known induction by these agents of benzpyrene hydroxylase activity, and their ability to shift the CO-complex of the terminal oxidase to 448 nm (2,4,6,15), and would agree with a suggestion that ARO induces, among other things, P_1-450 (15).

Another striking result pointing out differences in P-450 species was seen in difference spectrum when caffeine or theophylline was added to the microsomal suspension (Fig. 3). A clear type I spectral change was seen only with microsomes of animals pretreated with PCN (note: ΔA values should be X2 for this spectrum). A reverse type I (RI) spectral change was seen with microsomes from both MC and ARO treated rats. The animals treated with PB gave an indication of containing some of the species giving rise to the type I spectral change (note peak at 390 nm), but the 420 nm trough was obliterated by an overlapping peak. [Not all of the methylated xanthines are capable of evoking the type I spectral change. For that, N-methyl groups are required on both position 1 and 3 of the xanthine ring. 1-methyl and 3-methyl xanthine, 1,9-dimethyl-, 3,9-dimethyl-, 1,7-dimethyl- and 3,7-dimethyl-xanthine were all incapable of causing the type I spectral change.]

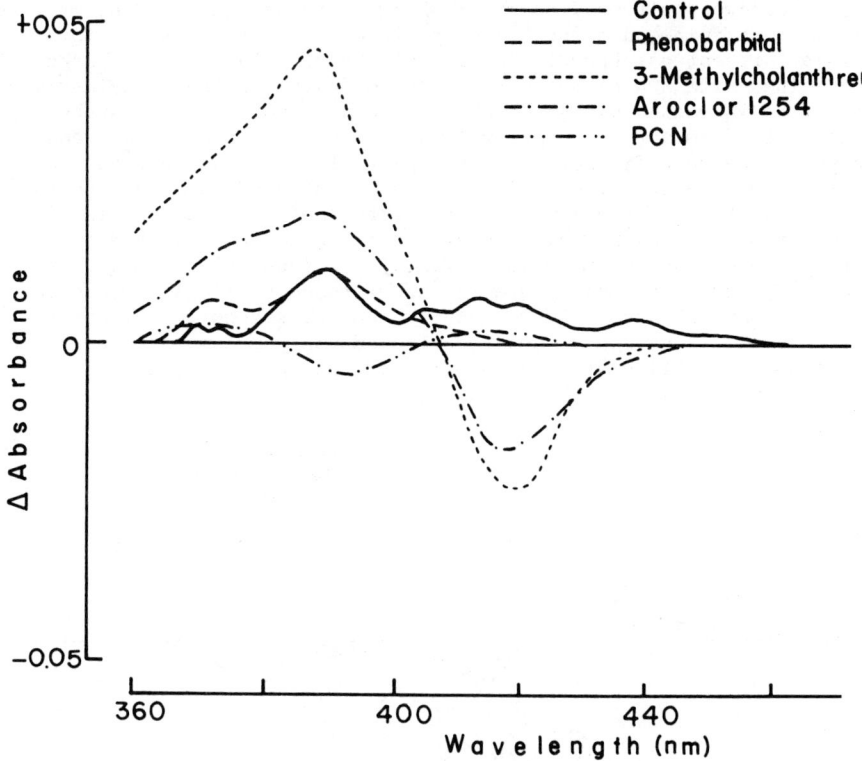

Fig. 2. Spectral changes on addition of 4 uM benzpyrene.

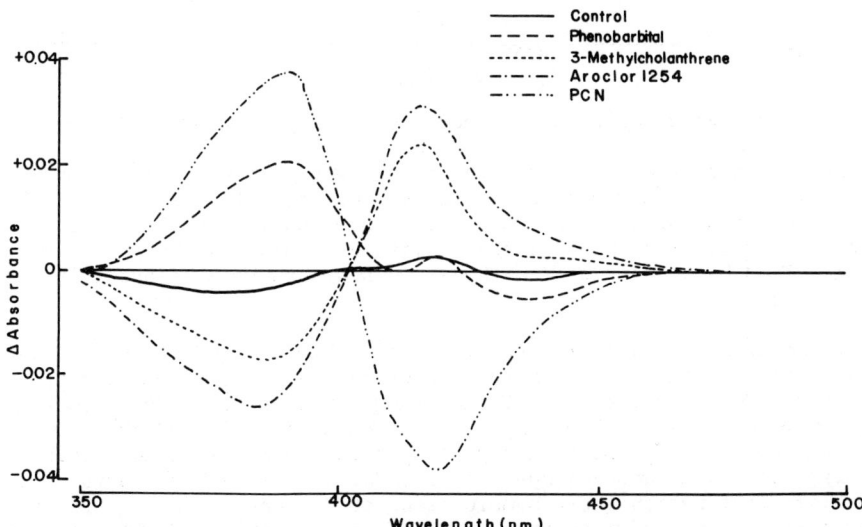

Fig. 3. Spectral change on addition of 3 mM caffeine.

The ability to evoke a type I spectral change with caffeine did not relate to metabolism of caffeine, and the P-450 specie which is responsible for the reaction is at present not known. In view of the very large type I spectral change obtained with PCN microsomes and the observation that ethylmorphine demethylase is increased almost 4-fold in these microsomes, it may be that this ability to form a type I spectral change with caffeine is a property of this enzyme. The peak at 390 nm and the partially overlapped trough with PB microsomes may reflect the smaller induction of ethylmorphine demethylase by PB. The addition of ethylmorphine to microsomes revealed the specificity of PCN induction of ethylmorphine demethylase. Note (Fig. 4) how much greater is the response in the PCN microsomes (_..._). Although PB increases the activity of ethylmorphine demethylase in the microsomes almost three fold, it does not cause an enhanced type I spectral change, suggesting either that the enhanced activity is by a more non-specific enzyme, or, as we suspect, the spectral change is obliterated by spectral overlap of an RI binding with another species of P-450 in the microsomes.

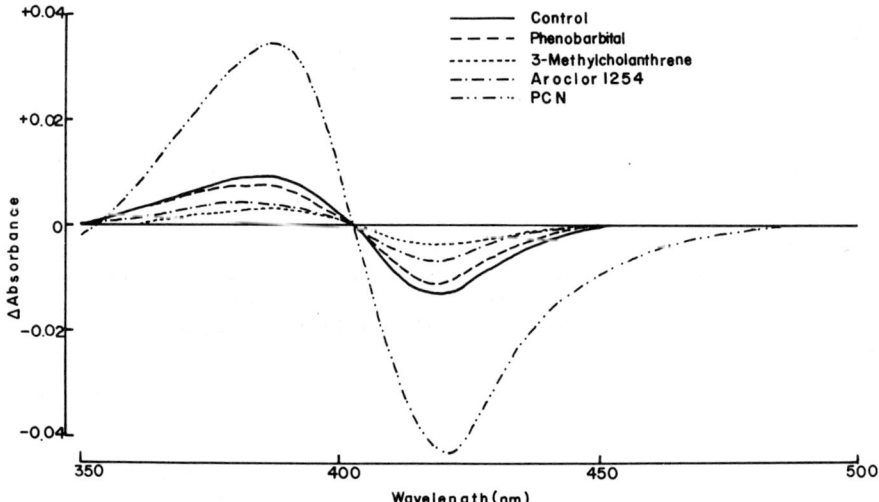

Fig. 4. Spectral change on addition of 0.3 mM ethylmorphine.

The spectral interaction of benzphetamine with the different microsomes is seen in Fig. 5. This compound caused appearance of a type I spectral change related closer than other compounds to the observed (Table I) microsomal demethylase activity; the spectral change with PB microsomes is with 2 mg microsomes/ml. Thus benzphetamine demethylase appears to be a hemoprotein induced by both PB and ARO.

One last spectrum to consider is that shown in Fig. 6, and evoked by aminopyrine. Note that only with control microsomes is there a complete type I spectral change. From the part of the trough visible from 430 nm to higher wavelengths, and from the 390 nm peak it is clear that PB induces a P-450 hemoprotein capable of forming a type I spectral change with aminopyrine. It is equally clear that another species is induced by PB which reacts to form a peak in the Soret region (note double peaks at 390 and 405 nm). From the much smaller trough with PCN microsomes, despite an aminopyrine demethylase activity similar to PB microsomes, either a more non-specific P-450 metabolizes

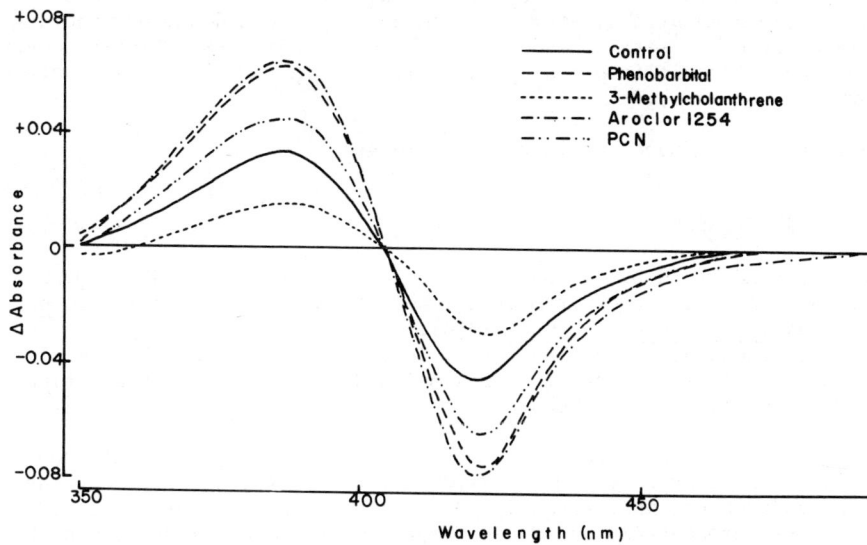

Fig. 5. Spectral change on addition of 0.3 mM benzphetamine.

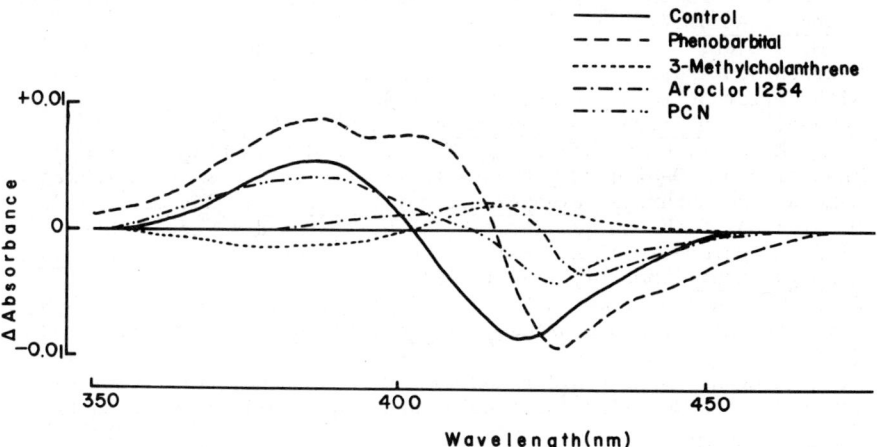

Fig. 6. Spectral change on addition of 0.3 mM aminopyrine.

aminopyrine in ARO microsomes, or spectral overlap is more complete in these microsomes. It is suspected that the latter is the case in ARO microsomes, partly from the ready formation of a large RI spectral change in these microsomes with caffeine.

Based upon differences in spectral changes seen on addition of substrates of the mixed function oxidase to liver microsomes of rats after pretreatment with different inducers, as well as upon differential induction of metabolism of the different substrates, we conclude that PCN, MC and PB, at least, induce some distinctly different forms of cytochrome P-450. If named after substrates metabolized, one would have to conclude that specific benzpyrene hydroxylase, benzphetamine demethylase, ethylmorphine demethylase and aminopyrine demethylase enzymes exist. In addition, the data indicates at least some substrates can be metabolized non-specifically by other P-450 enzymes (e.g., benzpyrene, caffeine).

Acknowledgement: Supported in part by NIH grant GM17021.

REFERENCES

1. N. E. Sladek and G. J. Mannering, Evidence for a new P-450 hemoprotein in hepatic microsomes from methylcholanthrene treated rats, Biochem. Biophys. Res. Comm. 24, 668 (1966).
2. N. E. Sladek and G. J. Mannering, Induction of drug metabolism II. Qualitative differences in the microsomal N-demethylating systems stimulated by polycyclic hydrocarbons and by phenobarbital, Mol. Pharmacol. 5, 186 (1969).
3. A. Y. H. Lu, R. Kuntzman, S. West and A. H. Conney, Reconstituted liver microsomal enzyme system that hydroxylates drugs, other foreign compounds and endogenous substrates. I. Determination of substrate specificity by the cytochrome P-450 and P-448 fractions, Biochem. Biophys. Res. Comm. 42, 1200 (1971).
4. J. B. Schenkman, H. Greim, M. Zange and H. Remmer, On the problem of possible other forms of cytochrome P-450 in liver microsomes, Biochim. Biophys. Acta 171, 23 (1969).
5. F. M. Goujon, D. W. Nebert and J. E. Gielen, Genetic expression of arylhydrocarbon hydroxylase induction IV. Interaction of various compounds with different forms of cytochrome P-450 and the effect on benzo(a)pyrene metabolism in vitro, Mol. Pharmacol. 8, 667 (1972).
6. A. P. Alvares, G. R. Schilling, W. Levin and R. Kuntzman, Studies on the induction of CO-binding pigments in liver microsomes by phenobarbital and 3-methylcholanthrene, Biochem. Biophys. Res. Comm. 29, 521 (1967).
7. D. Ryan, A. Y. H. Lu, J. Kawalek, S. B. West and W. Levin, Highly purified cytochrome P-448 and P-450 from rat liver microsomes, Biochem. Biophys. Res. Comm. 64, 1134 (1975).
8. A. F. Welton and S. D. Aust, Multiplicity of cytochrome P-450 hemoproteins in rat liver microsomes, Biochem. Biophys. Res. Comm. 56, 898 (1974).
9. D. A. Haugen, T. A. van der Hoeven and M. J. Coon, Purified liver microsomal cytochrome P-450. Separation and characterization of multiple forms, J. Biol. Chem. 250, 3567 (1975).
10. A. H. Conney, W. Levin, M. Ikeda, R. Kuntzman, D. Y. Cooper and O. Rosenthal, Inhibitory effect of carbon monoxide on the hydroxylation of testosterone by rat liver microsomes, J. Biol. Chem. 248, 3912 (1968).
11. R. Kato and J. R. Gillette, Effect of starvation on NADPH-dependent enzymes in liver microsomes of male and female rats, J. Pharmacol. Exp. Therap. 150, 279 (1965).
12. P. J. Creaven and D. V. Parke, The stimulation of hydroxylation by carcinogenic and non-carcinogenic compounds, Biochem. Pharmacol. 15, 7 (1966).
13. B. Solymoss, J. Werringloer and S. Toth, The influence of pregnenolone-16 α-carbonitrile on hepatic mixed function oxygenases, Steroids 17, 427 (1971).
14. A. Y. H. Lu, A. Somogyi, S. West, R. Kuntzman and A. H. Conney, Pregneno-

lone-16α-carbonitrile: a new type of inducer of drug-metabolizing enzymes, Arch. Biochem. Biophys. 152, 457 (1972).
15. A. P. Alvares, D. R. Bickers and A. Kappas, Polychlorinated biphenyls: a new type of inducer of cytochrome P-448 in the liver, Proc. Nat. Acad. Sci. U.S.A. 70, 1321 (1973).
16. H. Remmer, Der beschleunigte Abbau von Pharmaka in der Lebermikyosomen unter dem einfluss von Luminal, N.-S. Arch. exp. Path. Pharmak. 235, 279 (1959).
17. S. Orrenius and L. Ernster, Phenobarbital-induced synthesis of the oxidative demethylating enzymes of rat liver microsomes, Biochem. Biophys. Res. Comm. 16, 60 (1964).
18. A. H. Conney, E. C. Miller and J. A. Miller, The metabolism of methylated aminoazo dyes. V. Evidence for induction of enzyme synthesis in the rat by 3-methylcholanthrene, Cancer Res. 16, 450 (1956).
19. D. L. Cinti, P. Moldeus and J. B. Schenkman, Kinetic parameters of drug-metabolizing enzymes in Ca^{+2}-sedimented microsomes from rat liver, Biochem. Pharmacol. 21, 3249 (1972).
20. J. B. Schenkman, D. L. Cinti, S. Orrenius, P. Moldeus and R. Krasnitz, The nature of the reverse type I (modified type II) spectral change in liver microsomes, Biochemistry 11, 4243 (1972).
21. J. B. Schenkman, H. Remmer and R. W. Estabrook, Spectral studies of drug interaction with hepatic cytochrome, Mol. Pharmacol. 3, 113 (1967).
22. K. M. Robie, Y. N. Cha, R. E. Talcott and J. B. Schenkman, Kinetic studies of benzpyrene and hydroxybenzpyrene metabolism, Chem. Biol. Interact. 12, 285 (1976).
23. K. Krisch and H. Staudinger, Untersuchungen zur enzymatischen Hydroxylierung. Hydroxylierung von Acetanilid und deren Beziehungen zur mikrosomalen Pyridinnucleotidoxydation, Biochem. Z. 334, 312 (1961).
24. E. G. Hrycay and P. J. O'Brien, Cytochrome P-450 as a microsomal peroxidase utilizing a lipid peroxide substrate, Arch. Biochem. Biophys. 147, 14 (1971).
25. E. Jeffery, A. Kotake and G. J. Mannering, A study of the hepatic microsomal of an active terminal electron acceptor, Federation Proc. 34, 730A (1975).
26. J. Werringloer, Stimulation of hepatic microsomal drug metabolism by pregnenolone-16 α-carbonitrile and phenobarbital, Federation Proc. 31, 641A (1972).

DIFFERENCES IN THE INTERACTIONS OF R AND S WARFARIN WITH HEPATIC MICROSOMAL CYTOCHROME P-450

Michael J. Fasco, Lawrence J. Piper and
Laurence S. Kaminsky

Division of Laboratories and Research, New York State Department of Health, Albany, New York 12201

ABSTRACT

The enantiomers of the anticoagulant, warfarin, are metabolized to a series of monohydroxylated products by rat hepatic cytochrome P-450 (P-450). The patterns of products vary for R and S warfarin and following induction of P-450 by phenobarbital (PB) and cytochrome P-448 (P-448) by 3-methylcholanthrene (MC). We studied the binding of R and S warfarin to cytochrome P-450 by difference spectrophotometry to determine how the product patterns produced are influenced by the heterogeneity of P-450. Uninduced P-450 yielded modified type II difference spectra with R and S warfarin with binding constants, K_S = 1.60 mM. PB induced microsomes also yielded modified type II spectra which produced biphasic plots with $K_S(S)$ = 0.25, 0.07 mM and $K_S(R)$ = 0.83, 0.11 mM. MC induction with R warfarin yielded type I spectra with K_S = 0.23 mM. With S warfarin modified type II spectra were observed with K_S = 0.12 mM. In all cases prior binding of R or S warfarin influenced the binding of the opposite enantiomer. We conclude that R and S warfarin bind to two separate forms of cytochrome P-450 and two separate forms of cytochrome P-448, all of which differ from uninduced cytochrome P-450. The variety of monohydroxylated metabolites of R and S warfarin are a consequence of the interactions with these different forms of cytochrome P-450.

INTRODUCTION

A major unresolved problem associated with the elucidation of the mechanisms of cytochrome P-450 catalyzed drug metabolism involves the heterogeneity of the enzyme. Induction of experimental animals with 3-methylcholanthrene (MC) results in the synthesis of cytochrome P-448 which differs from cytochrome P-450 in many respects (e.g., 1). Other investigations have indicated that three distinct forms of cytochrome P-450 can be chromatographically separated from microsomes isolated from untreated rats (2). In the case of phenobarbital (PB) induced microsomes, immunochemical assays also indicate the presence of at least three different forms of cytochrome P-450 (3).

Warfarin [3-(α-acetonylbenzyl)-4-hydroxycoumarin] is extensively used clinically as an oral anticoagulant and is also utilized as a rodenticide. The drug is metabolized by the hepatic mixed function oxidase system to yield a variety of monohydroxylated products (4). We have recently reported that the individual warfarin enantiomers yield different patterns of metabolites and are metabolized at different rates by rat liver cytochrome P-450 (5). Prior induction of the microsomal mixed function oxidase system with PB or MC results in marked increases in the rates of R and S warfarin metabolism and alterations in the product patterns of both enantiomers (6).

These metabolic characteristics make warfarin an ideal probe for the mechanisms of hepatic drug metabolism and particularly the heterogeneity of cytochrome P-450. We have consequently investigated the binding of R and S warfarin to untreated and PB or MC induced rat liver cytochrome P-450 in an effort to probe the heterogeneity of cytochrome P-450 and to relate this heterogeneity to the patterns of metabolic products of R and S warfarin.

EXPERIMENTAL PROCEDURE

Racemic warfarin was purchased from Calbiochem and resolved by the method of West et al. (7) to yield optically pure R warfarin ($[\alpha]_D = +149$) and S warfarin ($[\alpha]_D = -149.7$).

Male Wistar rats (250-350 g) from the New York State Health Department, Griffin Laboratories, were used exclusively. Induction of hepatic microsomal cytochrome P-450 was achieved with PB (100 mg/kg, i.p., in 0.9% saline once daily for three days) or MC (25 mg/kg, i.p., in corn oil once daily for three days). Groups of five similarly pretreated rats were used in the preparation of each batch of microsomes which were prepared by the method of Tangen et al. (8). The protein (9) and cytochrome P-450 (10) concentrations were determined by published methods.

All difference spectra were determined with an Aminco DW-2 spectrophotometer in the split beam mode. Freshly prepared microsomal suspensions were used in the determination of all difference spectra following the method of Schenkman et al. (11). Binding constants, K_S, were determined from double reciprocal plots of the absorbance difference versus the corresponding warfarin concentration.

RESULTS AND DISCUSSION

The correlation of microsomal difference spectra with the binding of added compounds to the cytochrome P-450 component of the microsomes was first proposed by Remmer et al. (12).

The difference spectra arising from the addition of R or S warfarin to untreated rat liver microsomes are modified type II (Fig. 1) based on the wavelength of the absorbance maximum at 423 nm (Fig. 1), the failure of warfarin to displace CO from microsomal ferrocytochrome P-450 (13) and the effect of the type II compound, imidazole, (14) in modifying the binding spectrum of R and S warfarin to cytochrome P-450 (Fig. 1). The modified type II spectra imply that R and S warfarin ligand to the heme iron of cytochrome P-450 (15). These difference spectra also conceal a type I spectral component as evidenced by the residual spectra which closely approximate type I obtained when the modified type II spectra were eliminated by addition of (a) imidazole to sample and reference cuvettes (Fig. 1) and (b) warfarin of the opposite enantiomeric form to that being studied to both sample and reference cuvettes (Fig. 2). The latter result suggests that the binding of the warfarin enantiomers occurs initially at the modified type II site which becomes saturated. The subsequently added opposite enantiomer then binds at the type I site yielding a type I spectrum. Addition of the second enantiomer to microsomes after the first enantiomer was added at levels below that required to saturate the modified type II site permits further heme binding by the second enantiomer sufficient to overcome the type I spectrum arising from any type I binding (Fig. 2). The fact that neither R nor S warfarin individually exhibited type I spectra even at the lowest concentrations compatible with the observation of spectra supports the suggestion that initial binding occurs at the modified type II site of cytochrome P-450 followed by binding at the type I site.

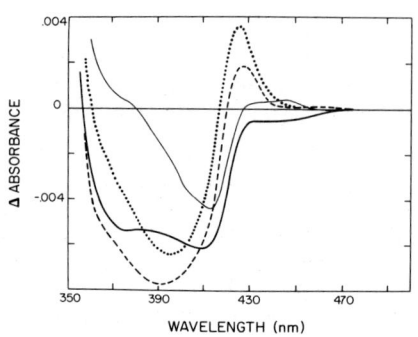 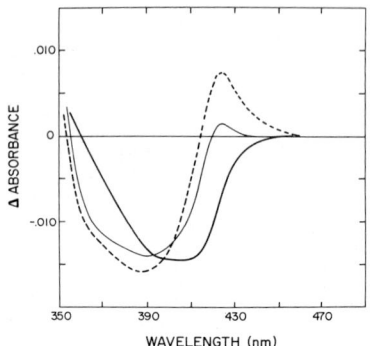

Fig. 1: The effect of bound imidazole on the difference spectra of R or S warfarin with untreated rat liver microsomes. For all spectra the two cuvettes contain equal volumes of 2.0 mg/ml microsomal protein. Spectra of (---) S warfarin (2.5 mM); (···) R warfarin (2.5 mM); and (——) S warfarin (2.5 mM) or (——) R warfarin (2.5 mM) in the sample cuvette and imidazole (1 mM) in the sample and reference cuvettes. All spectra recorded at 25° in 0.02 M tris HCl buffer, pH 7.4. Cytochrome P-450, 1.08 nmole/mg protein.

Fig. 2: The effect of bound R warfarin on the difference spectra of S warfarin with untreated rat liver microsomes. Spectra of S warfarin with R warfarin in the sample and reference cuvettes: (——) S warfarin (2.5 mM), R warfarin (1.88 mM); (——) S warfarin (1.88 mM), R warfarin (0.75 mM); and (---) S warfarin (1.50 mM), R warfarin (0.125 mM). Cytochrome P-450, 1.16 nmole/mg protein.

The effect of prior hexobarbital saturation of type I sites in enhancing the magnitude of the modified type II spectra of R and S warfarin confirms the presence of type I component in the observed warfarin spectra. A similar effect of hexobarbital was noted with the modified type II difference spectrum of agroclavine and was proposed as confirmation of a type I spectral component (16) for this compound.

The K_m values for the in vitro metabolism of R and S warfarin by untreated microsomes have been determined for each metabolite (5). All of these values are significantly lower than the K_s values obtained in this study (Table 1) for the binding of R and S warfarin to untreated microsomes. This difference is consistent with other studies of modified type II compounds (17) and probably reflects the fact that binding to the cytochrome P-450 heme is apparently not directly related to the microsomal metabolism of the substrate.

The K_s values for the binding of R and S warfarin to untreated cytochrome P-450 are not differentiated (Table 1), which implies that both enantiomers bind to the same form of cytochrome P-450.

R and S warfarin bind to PB induced microsomes to yield difference spectra which, by the same arguments previously presented, are modified type II spectra (Fig. 3). The magnitude of the spectral changes as a function of warfarin concentration is biphasic which leads to the determination of two K_S values, the low affinity values of which show variation between R and S warfarin binding (Table 1).

TABLE 1 Difference Spectral Data and Binding Constants (K_S) for the Binding of R and S warfarin to Variously Induced Rat Hepatic Microsomes

Induction	Warfarin Enantiomer	Wavelength		K_S
		Peak nm	Trough nm	mM
Untreated	R	423	394	1.69 ± .67
	S	423	394	1.29 ± .99
PB	R	423	394	0.12 ± .02, 0.79 ± .08
	S	423	394	0.07 ± .03, 0.24 ± .02
MC	R	384	415	0.238
	S	423	394	0.107

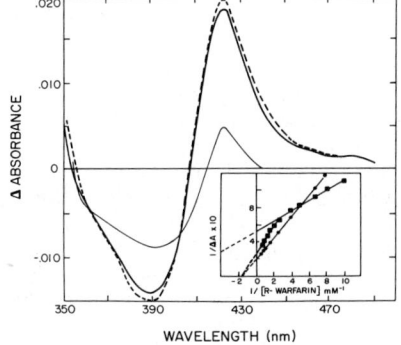 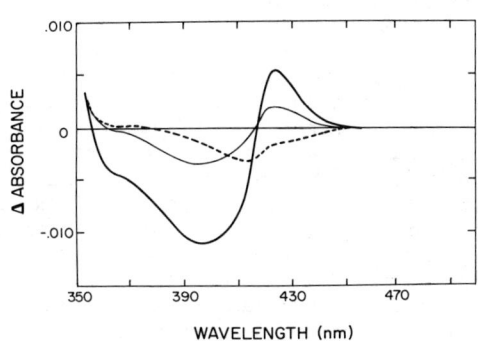

Fig. 3: The effect of bound hexobarbital on the difference spectra of R or S warfarin with phenobarbital induced rat liver microsomes. Spectra of (———) S warfarin (0.625 mM); (———) R warfarin (0.625 mM) or (- - -) S warfarin (0.625 mM) in the sample cuvette and hexobarbital (1 mM) in the sample and reference cuvettes. Insert: Double reciprocal plots of the absorbance differences between 425 and 394 nm of (■) R warfarin and (●) R warfarin with bound hexobarbital vs. R warfarin concentration. Cytochrome P-450, 2.72 nmole/mg protein.

Fig. 4: Effect of the blanking out of the R or S warfarin spectral component of the difference spectrum of racemic warfarin with phenobarbital induced rat liver microsomes. Spectra of (———) racemic warfarin (1.25 mM); racemic warfarin (1.25 mM) in the sample cuvette and (- - -) R warfarin (0.625 mM) or (———) S warfarin (0.625 mM) in the reference cuvette. Cytochrome P-450, 2.83 nmole/mg protein.

Presaturation of type I sites of PB induced microsomes with hexobarbital increases the magnitude of the difference spectra of R and S warfarin (Fig. 3) and indicates the presence of some type I binding for both enantiomers. In the case of R warfarin, the previous binding of hexobarbital primarily alters the modified type II spectrum at the lower levels of warfarin concentration (high affinity binding) (see insert, Fig. 3) by elimination of the type I component from this phase. The R warfarin induced modified type II spectrum corresponding to the low affinity site is relatively unaltered and thus elimination of the type I spectral component of R warfarin binding results in a monophasic modified type II spectral interaction (Fig. 3). These results are interpreted to indicate that R warfarin binds to PB induced cytochrome P-450 at the type I site simultaneously with its initial binding to the modified type II site and rapidly saturates the type I site. Further addition of warfarin results in binding to the modified type II site only. The binding of R warfarin to PB induced cytochrome P-450 differs from the binding of S warfarin as indicated by an experiment where the spectral contribution of each of the enantiomers in turn was excluded from the spectrum of racemic warfarin (Fig. 4). This experiment demonstrated a type I spectral component for S warfarin binding to PB induced cytochrome P-450 but not for R warfarin (Fig. 4). The fact that S warfarin yielded a type I difference spectrum in the presence of R warfarin but not in its absence (where a modified type II spectrum was obtained) indicates that R and S warfarin compete for and influence the nature of each others binding sites on cytochrome P-450. Thus the binding of R warfarin to PB induced cytochrome P-450 induces conformational changes and possibly spin state changes in the heme iron, such that subsequent binding of S warfarin to cytochrome P-450 is predominantly to type I sites. In contrast, initial binding of S warfarin to PB induced cytochrome P-450 does not induce similar changes in cytochrome P-450 to produce enhancement of type I binding, and thus subsequent binding of R warfarin still yields a modified type II spectrum. A possible explanation for these effects is that separate cytochrome P-450 molecular forms exist, each of which has modified type II and type I binding sites for one of the warfarin enantiomers. R warfarin must then bind to the modified type II and type I sites on its form of cytochrome P-450 as well as to the modified type II site on the other form of cytochrome P-450. This latter binding induces the conformation/spin state change which facilitates the subsequent binding of S warfarin to the type I site on its cytochrome P-450 molecule. In contrast, S warfarin must bind to the modified type II and type I sites of its form of cytochrome P-450 but not to the modified type II site of the other form of cytochrome P-450, and thus subsequent addition of R warfarin to the microsomes yielded a modified type II spectrum following its binding to both modified type II and type I sites. These results could also be interpreted by the use of a model incorporating a single cytochrome P-450 molecule with multiple binding sites, but the model incorporating two forms of cytochrome P-450 provides a better explanation of all of the data.

There are no available K_m values for the metabolism of R and S warfarin by PB induced microsomes, although studies to this end are being actively pursued in our laboratories. Comparison of K_m and K_s values are thus not possible.

The binding of R warfarin to MC induced microsomes to yield a type I difference spectrum (Fig. 5B) is unusual, particularly since S warfarin yields a modified type II spectrum under identical conditions (Fig. 5A). There are two possible explanations for the differentiation in the binding spectra of R and S warfarin to cytochrome P-448. These are (a) the possibility that R warfarin binds to the low spin component of cytochrome P-448 yielding type I spectrum, while S warfarin binds to the high spin component yielding a modified type II spectrum and (b) R and S warfarin bind to different molecular forms of cytochrome P-448. The first possibility appears unlikely in view of the greatly diminished low spin component of cytochrome P-448 (18). Thus it appears possible that the R and S warfarin enantiomers bind to different types of cytochrome P-448 in MC induced microsomes. The in vitro metabolism of R warfarin by MC induced microsomes presents an outstanding feature in that the production of 8-hydroxywarfarin is increased to a very much greater extent than is any of the other metabolites, relative to untreated controls, and that S warfarin metabolism does not yield a similar enhancement of this metabolite (6). It is thus not unreasonable to propose that the previously described binding of R warfarin to the specific cytochrome P-448 type results in the production of the 8-hydroxywarfarin metabolite, but this can only be confirmed when the K_m value for the metabolic reaction becomes available. The enhancement of the magnitude but not the form of the modified type II binding spectrum of S warfarin with cytochrome P-448 potentiated by the presence of R warfarin (a type I compound in this situation) in both sample and reference cuvettes (Fig. 6) indicates that there is a type I component concealed in the modified type II spectrum of S warfarin. The reverse situation where S warfarin in both reference and sample cuvettes produces a reduction in the magnitude of the subsequently added R warfarin type I binding spectrum (Fig. 6) indicates (a) that there is no modified type II component concealed in the type I spectrum of R warfarin with MC induced cytochrome P-448 and (b) that the type I binding component of S warfarin must be competing for the same binding site as R warfarin thereby diminishing the magnitude of the R warfarin type I spectrum. These results are compatible with a model of two molecular forms of cytochrome P-448, if we assume that (a) R warfarin could bind to the type I site of both proteins but to neither modified type II site, and thus subsequently added S warfarin would be excluded from its type I site only being able to interact with its modified type II site and that (b) S warfarin could also bind to the type I sites of both proteins together with a modified type II site on one, and thus subsequent addition of R warfarin could not bind to fully occupy its type I site because of its partial saturation by S warfarin and the resultant type I spectrum would thus be diminished in magnitude. These results further tend to preclude the possibility of R and S warfarin binding to different spin state forms of the same molecular type of cytochrome P-448 since, if this were the case, S warfarin binding (modified type II) should shift the equilibrium of the cytochrome spin states over to the low spin form thus facilitating subsequent R warfarin binding (type I) -- the opposite of the observed results.

In summary, our studies with warfarin enantiomers appear to indicate that both PB induced cytochrome P-450 and MC induced cytochrome P 448 bind the enantiomers to different molecular forms of the cytochrome which all differ from untreated cytochrome P-450. It appears probable that the variety of monohydroxylated products produced by hepatic microsomal metabolism of R and S warfarin are a consequence of binding to these different forms of cytochrome P-450.

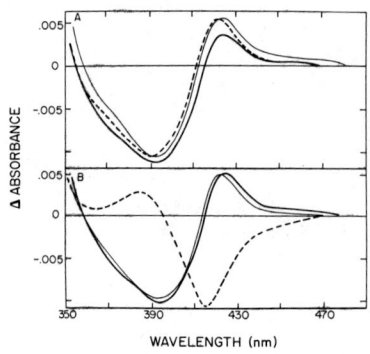 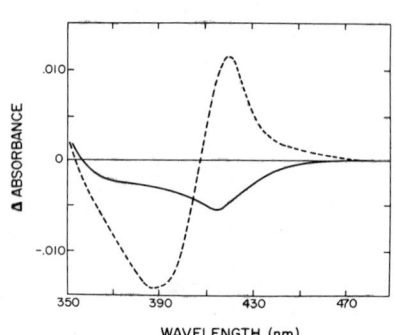

Fig. 5: The difference spectra of R or S warfarin with 3-methylcholanthrene induced rat liver microsomes. Spectra of (——) racemic warfarin (1.25 mM) and A: (– – –) S warfarin (0.625 mM); (——) R warfarin (0.625 mM) added to the S warfarin in the sample cuvette. B: (– – –) R warfarin (0.625 mM); (——) S warfarin (0.625 mM) added to R warfarin in the sample cuvette. Cytochrome P-448, 2.30 nmole/mg protein.

Fig. 6: Effect of bound opposite enantiomer on the difference spectra of R or S warfarin with 3-methylcholanthrene induced rat liver microsomes. Spectra of (——) R warfarin (0.625 mM) or (– – –) S warfarin (0.625 mM) in the sample cuvette with the opposite enantiomer (0.625 mM) in the sample and reference cuvettes. Cytochrome P-448, 2.30 nmole/mg protein.

REFERENCES

(1) A. Y. H. Lu, R. Kuntzman, S. West, M. Jacobson and A. H. Conney, Role of the cytochrome P-450 and P-448 fractions in drug and steroid hydroxylations, J. Biol. Chem. 247, 1727 (1972).
(2) K. Comai and J. L. Gaylor, Existence and separation of three forms of cytochrome P-450 from rat liver microsomes, J. Biol. Chem. 248, 4947 (1973).
(3) P. E. Thomas, A. Y. H. Lu, D. Ryan, S. B. West, J. Kawalek and W. Levin, Multiple forms of rat liver cytochrome P-450, J. Biol. Chem. 251, 1385 (1976).
(4) W. M. Barker, M. A. Hermodson and K. P. Link, The metabolism of $4\text{-}C^{14}$ warfarin sodium by the rat, J. Pharmacol. Exp. Ther. 171, 307 (1970).
(5) L. R. Pohl, S. D. Nelson, W. R. Porter, W. F. Trager, M. J. Fasco, F. D. Baker and J. W. Fenton, II, Warfarin: Stereochemical aspects of its metabolism by rat liver microsomes, Biochem. Pharmacol. in press (1976).
(6) L. R. Pohl, W. R. Porter, W. F. Trager, M. J. Fasco and J. W. Fenton, II, The stereochemical biotransformation of warfarin as a probe of the homogeneity and mechanism of microsomal hydroxylases, Biochem. Pharmacol. in press (1976).

(7) B. D. West, S. Preis, C. H. Schroeder and K. P. Link, The resolution and absolute configuration of warfarin, J. Am. Chem. Soc. 83, 2676 (1961).

(8) O. Tangen, J. Jonsson and S. Orrenius, Isolation of rat liver microsomes by gel filtration, Anal. Biochem. 54, 597 (1973).

(9) G. R. Schacterle and R. L. Pollack, A simplified method for the quantitation of small amounts of protein in biological material, Anal. Biochem. 51, 654 (1973).

(10) T. Omura and R. Sato, The carbon monoxide-binding pigment of liver microsomes, J. Biol. Chem. 239, 2370 (1964).

(11) J. B. Schenkman, H. Remmer and R. W. Estabrook, Spectral studies of drug interaction with hepatic microsomal cytochrome, Mol. Pharmacol. 3, 113 (1967).

(12) H. Remmer, J. Schenkman, R. W. Estabrook, H. Sasame, J. Gillette, S. Narasimhulu, D. Y. Cooper and O. Rosenthal, Drug interaction with hepatic microsomal cytochrome, Mol. Pharmacol. 2, 187 (1966).

(13) J. B. Schenkman, D. L. Cinti, P. W. Moldeus and S. Orrenius, Newer aspects of substrate binding to cytochrome P-450, Drug Metab. Disp. 1, 111 (1973).

(14) J. W. Gorrod and D. J. Temple, The demonstration of both type I and type II components in the difference spectra produced by aniline and N-ethylaniline with cytochrome P-450, Chem. Biol. Interact. 6, 203 (1973).

(15) Y. Yoshida and H. Kumaoka, Studies on the substrate-induced spectral change of cytochrome P-450 in liver microsomes, J. Biochem. 78, 455 (1975).

(16) B. J. Wilson and S. Orrenius, A study of the modified type II spectral change produced by the interaction of agroclavine with cytochrome P-450, Biochim. Biophys. Acta 261, 94 (1972).

(17) S. Orrenius, B. J. Wilson, C. von Bahr and J. B. Schenkman, Biological Hydroxylation Mechanisms, Academic Press, London, 1972.

(18) C. R. E. Jefcoate, J. L. Gaylor and R. L. Calabrese, Ligand interactions with cytochrome P-450, Biochemistry 8, 3455 (1969).

MULTIPLE FORMS OF HOUSEFLY CYTOCHROME P-450[1]

Jorge Capdevila and Moises Agosin

Department of Zoology, University of Georgia, Athens, Ga. 30602

ABSTRACT

Three different species of cytochrome P-450 are partially purified from a susceptible, <u>noninduced</u>, housefly strain. The hemoprotein species appear to be free of NADPH cytochrome c reductase and cytochrome b_5 and they are distinguished by column chromatography, SDS-polyacrylamide gel electrophoresis and spectral properties. Cytochromes $P-450_{HF}$, $P-452_{HF}$ and $P-453_{HF}$ are purified to specific contents of 3.5, 7.5 and 13.9 nmols of hemoprotein per mg of protein and their apparent molecular weights are 53,000; 48,000 and 43,000, respectively.

INTRODUCTION

The cytochrome P-450-containing enzyme system of the insect microsomal fractions has been shown to be as versatile as that of the hepatic endoplasmic reticulum. The insect system hydroxylates endogenous as well as exogenous substrates including drugs, insecticides and even carcinogens (1). The exact mechanism by which this and other similar systems hydroxylate a variety of structurally unrelated substrates is still unknown; however, it is becoming an increasingly accepted view that specificity is imposed upon the system by the existance of multiple forms of cytochrome P-450. Two species of P-450[2] have been solubilized and partially purified from microsomes of the insecticide-resistant Fc housefly strain (2). One, $P-450_{HF}$, absorbs maximally at 450 nm in a CO-difference spectrum while the other one, $P-448_{HF}$ at 448 nm. The latter appears to be induced by phenobarbital or naphthalene preferentially. Two P-450s of the P-452 type have been described in the endoplasmic reticulum of the insecticide-susceptible NAIDM strain (3). Microsomes from a diazinon-resistant housefly strain (Rutgers) show the presence of cytochrome P-448, which has similar spectral characteristics as $P-448_{HF}$ of the Fc strain (4). An oversimplified analysis of these observations would suggest that resistance is related to the presence of a P-450 of

---

[1] Supported by NIH grant AI-12244 and the University of Georgia, Office of General Research.

[2] Abbreviations: P-450, cytochrome P-450; $P-450_{HF}$, housefly cytochrome P-450 with an absorption maximum at 450 nm in a CO-difference spectra. The different species of housefly cytochrome P-450 are named under similar bases; $P-450_{LM}$, liver microsomes cytochrome P-450; P-420, cytochrome P-420; TX-100, triton X-100; SDS, sodium dodecyl sulfate.

the P-448$_{HF}$ type; whereas, susceptibility is related to the predominance of a P-452$_{HF}$ type of P-450. To substantiate this view from the biochemical standpoint, this laboratory has been engaged in the solubilization, purification and characterization of P-450 from three housefly strains differing in insecticide susceptibility (Fc, Rutgers and NAIDM). In the present communication we report the partial purification of three P-450 species from the NAIDM strain.

METHODS

Solubilization and Purification of P-450

All operations were performed at 2-4°. Microsomes prepared as described previously (2) are suspended in 0.1 M KH$_2$PO$_4$ buffer, pH 7.25, 30% glycerol (v/v), 0.1 mM DTT and 0.1 mM EDTA (5 mg protein/ml). Sodium cholate is added at the final concentration of 0.25% (w/v). After 30 min under gentle stirring, the solubilized material is collected by centrifugation at 212,000 g x 90 min; the phosphate buffer concentration is decreased to 0.05 M; and the suspension is loaded onto a 30 x 2.5 cm ω—amino-n-octyl Sepharose-4B column equilibrated with the same buffer. The ω-amino-n-octyl Sepharose-4B is prepared according to Jennissen and Heilmeyer (5). A fraction containing P-451$_{HF}$ is not retained by the column and elutes during the loading step. The column is washed with the equilibration buffer but containing no sodium cholate and then the retained hemoprotein (P-452$_{HF}$) is eluted by 0.1 to 0.2% (v/v) triton X-100 prepared in the equilibration buffer but containing no sodium cholate. About 50% of cytochrome $\underline{b}_5$ and about 10% of P-450 remained absorbed in the column.

The P-451$_{HF}$ fraction is concentrated in a 1 x 5 cm hydroxylapatite column equilibrated with 50 mM KH$_2$PO$_4$ buffer, pH 7.25, 30% glycerol, 0.1 mM EDTA, 0.1 mM DTT. P-451$_{HF}$ is retained in the column. The column is washed with 5 mM KH$_2$PO$_4$ buffer, pH 7.25, 30% glycerol, 0.1 mM EDTA, 0.1 mM DTT, 0.075% TX-100 (v/v) to remove remaining sodium cholate; and then, elution is carried out with the same buffer but at a phosphate concentration of 0.15 M. A turbid yellow eluate, containing P-451$_{HF}$, is clarified by centrifugation at 12,000 g x 30 min. The pH, the phosphate and glycerol concentrations are changed to 7.4, 5 mM and 20%, respectively by passage through a sephadex G-50 column and the excluded material is fractionated in a DEAE-sephadex A-50 column. One fraction is not retained by the column and contains P-453$_{HF}$; a second fraction containing P-450$_{HF}$ is eluted at a salt concentration of 0.15 M KCl. Both P-453$_{HF}$ and P-450$_{HF}$ are further purified by a hydroxylapatite column equilibrated with 5 mM KH$_2$PO$_4$ buffer, pH 7.0, 20% glycerol (v/v), 0.05% TX-100 (v/v), 0.1 mM DTT and 5 µM EDTA. P-453$_{HF}$ elutes at 0.15 M phosphate buffer concentration and P-450$_{HF}$ at a concentration of 0.075 M. P-453$_{HF}$ is further rechromatographied in a similar hydroxylapatite column.

The P-452$_{HF}$ fraction is passed through a sephadex G-50 column to adjust the phosphate concentration to 5 mM, the pH to 7.4 and glycerol to 20% and then loaded onto a DEAE-sephadex A-50 column. The P-452$_{HF}$ is not retained by the column, but cytochrome $\underline{b}_5$ and NADPH cytochrome $\underline{c}$ reductase are. The pH of the hemoprotein fraction is adjusted to 7.0 and the EDTA concentration, to 5 µM and the solution is then loaded onto a hydroxylapatite column. Elution is done by raising the phosphate concentration of the buffer to 0.15 M. Further purification is done by column chromatography in CM-sephadex at pH 6.8, hydroxylapatite at pH 7.0 and sephadex LH-20 at pH 7.25.

Electrophoresis of the Purified P-450 Fractions

SDS-polyacrylamide gel electrophoresis of the purified fractions is done by the procedure of Welton and Aust (6). Proteins are stained with Coomassie Brilliant Blue R-250 according to Fairbanks et al. (7) and heme by the benzidine stain (6). The protein bands with a positive benzidine test and in the range of 40,000 to 60,000 molecular weight are numbered from 1 to 3 according to decreasing electrophoretic mobility and increasing molecular weight. The 3 hemoprotein fractions obtained are named according to the above described electrophoretic mobilities of the heme containing peptides present. Fraction 1 corresponds to the P-453$_{HF}$ preparation and fractions 2 and 1-3, to the P-452$_{HF}$ and P-450$_{HF}$ preparations, respectively. Isoelectric focusing is done according to Behnke et al. (8) using a 5 to 30% (w/v) sucrose gradient containing 0.1% TX-100 and LKB ampholites, pH range 3.5 to 10; anode, 3% H_2SO_4 (w/v); cathode, 3% ethanol amine (v/v). The gradients are run at a constant voltage of 200 V at 2ºC for 24 hrs and in the dark.

RESULTS AND DISCUSSION

Table 1 summarizes some properties of the purified P-450 fractions. The three

TABLE 1 Properties of Different Forms of Cytochrome P-450 Purified from Housefly Microsomes

Hemoprotein Fraction	P-450$_{HF}$	P-452$_{HF}$	P-453$_{HF}$
Electrophoretic Identification	Fraction 1-3	Fraction 2	Fraction 1
Soret Band Position (nm)[1]			
Fe^{+3}	417	418.5	417
Fe^{+2}	420	419	418
Fe^{+2}·CO	448.5	450.2	451.5
Fe^{+2}·CO minus Fe^{+2}	450	452	453
Ratio "430/455" Forms[2]	1.74	0.30	0.20
Specific content of P-450 (nmols/mg protein)	3.5	13.9	7.5
Purification (Times microsomal concentration)	17.5	69.5	37.5
Apparent Molecular Weight[3]	43,000 53,000	48,000	43,000

[1]For different samples the variation in these values was within 0.5 nm in a calibrated Aminco DW-2 spectrophotometer.

[2]Measured in an Ethylisocyanide difference spectra, obtained at pH 7.25 and in 0.1 M phosphate buffer, Ethylisocyanide 80 mM.

[3]Estimated by SDS-polyacrylamide gel electrophoresis with standardization by proteins of known molecular weight. The values shown are the average of 3 determinations.

hemoprotein fractions differ in their maximum absorbance in the absolute oxidized, reduced and reduced CO-bound form, as well as in the CO-difference spectra. In all cases the absolute spectra of the reduced CO-bound form showed a small shoulder between 420 to 423 nm. The preparations also differ in the characteristics of their ethylisocyanide difference spectra. Fraction 1-3 or P-450$_{HF}$ is clearly distinguishable from the other two fractions by its "430/455" nm ratio. All three fractions appeared to be free of NADPH cytochrome c reductase activity and of cytochrome b_5. Only P-452$_{HF}$ is purified extensively to a specific P-450 content of 13.9 nmol/mg protein which corresponds to a purification factor of around 70 times when compared with the

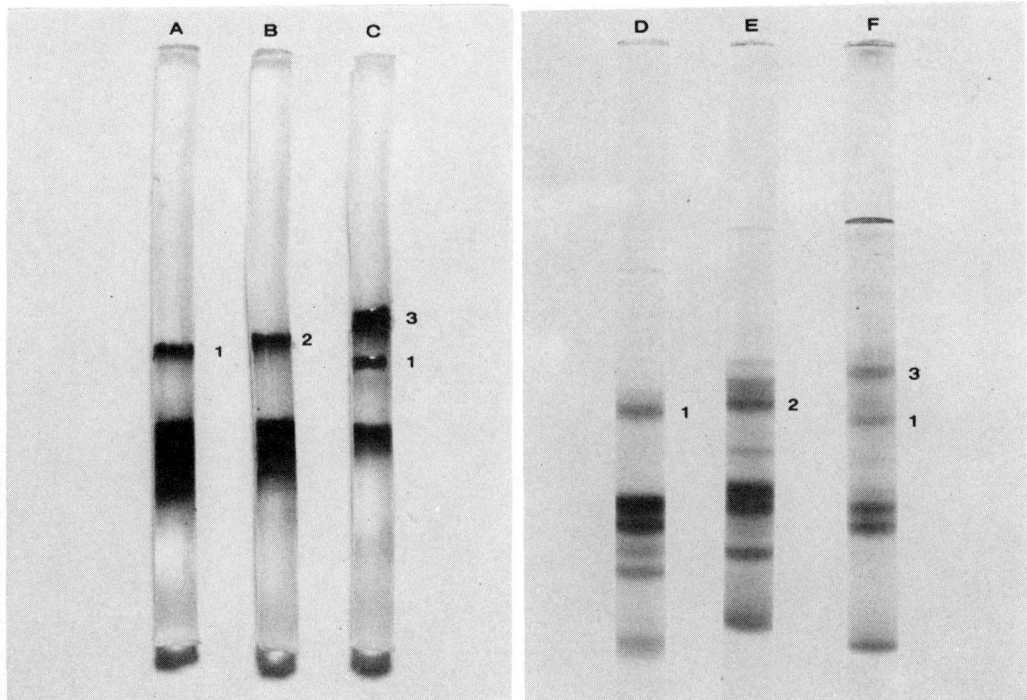

Fig. 1. SDS-polyacrylamide gel electrophoresis of the partially purified cytochrome P-450 fractions. The different fractions were analyzed after the first hydroxylapatite purification step (see Methods). The specific P-450 contents at that stage were: 4.9, 5.2 and 3.5 nmols/mg protein for the P-453$_{HF}$, P-452$_{HF}$ and P-450$_{HF}$ preparations, respectively. For each fraction, a photograph of the gels stained for heme (gels A, B and C) and one of the gels stained for protein (gels D, E and F) are included. The samples were treated with SDS (1% final concentration) at 5° for 15 min and then submitted to electrophoresis. Migration was from top to bottom, and the electrophoresis was carried out until the tracking dye was within 1 cm of the bottom of the gel. Because the benzidine stain faded out in a short time, the position of the bands was marked with drafting ink in the gel. The samples were analyzed at the amounts indicated for detection of protein bands and heme bands, respectively: P-453$_{HF}$ fraction, 40 and 90 ug (gels A and D); P-452$_{HF}$ fraction, 50 and 80 ug (gels B and E); P-450$_{HF}$ fraction, 35 and 70 ug (gels C and F).

specific content of the hemoprotein in the microsomal membrane. Gel electrophoresis indicates that P-450$_{HF}$ contains a heme-positive band with a molecular weight equal to that of P-453$_{HF}$ (Fig. 1). The three P-450s differ in their apparent molecular weight, ranging from 43,000 to 53,000.

The electrophoretic bands 1, 2 and 3 correspond to P-453$_{HF}$, P-452$_{HF}$ and P-450$_{HF}$ respectively (Fig. 1). Additional heme containing bands are seen with molecular weights ranging from approximately 12,000 to 35,000. Whether these smaller peptides correspond to additional P-450 species or represent heme binding to contaminant peptides is not clear at present. Similar observations have been reported for lung P-450 (9) and for P-450$_{LM}$ (10).

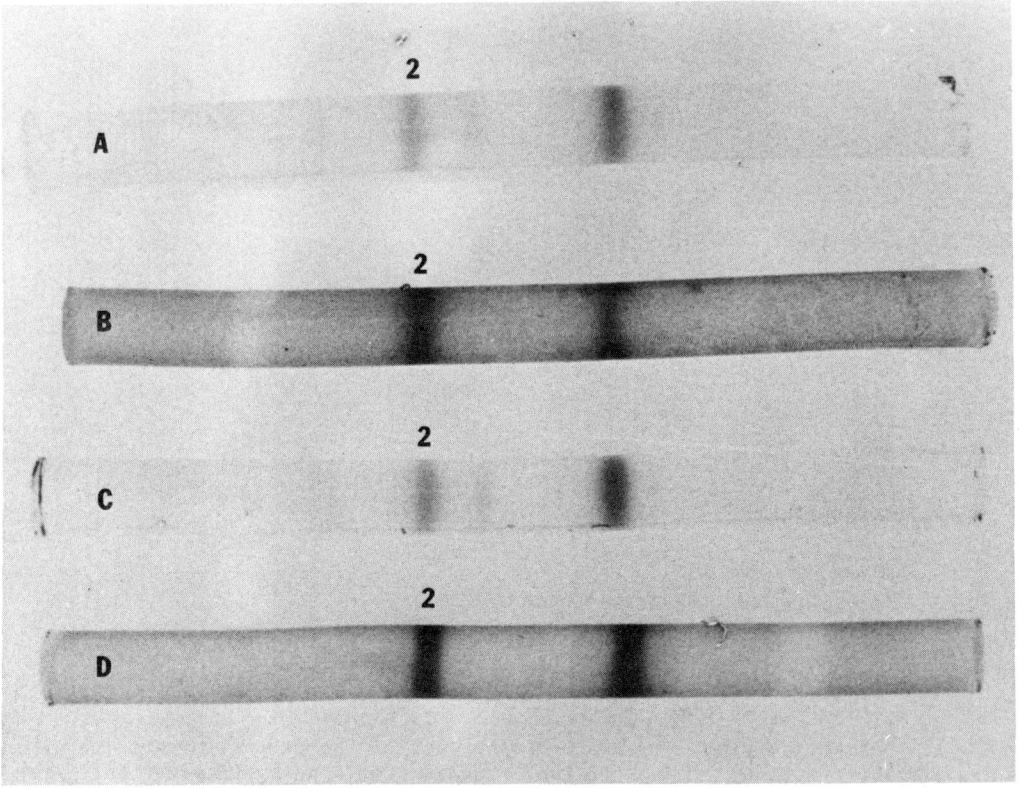

Fig. 2. SDS-polyacrylamide gel electrophoresis of the purified P-452$_{HF}$ fraction. The purified P-452$_{HF}$ preparation contained 13.9 nmols of P-450 per mg of protein. The samples were applied either directly (in 0.1 M sodium phosphate buffer, pH 7.4; 20% glycerol (v/v); 0.1 mM EDTA) to the gels (A and B) or after treatment with SDS (final concentration 1%) for 15 min at 4°C (C and D). Migration was from left to right and the electrophoresis was carried out until the tracking dye was within 2 cm of the end. Gels A and C contained 20 ug of protein each and were stained with Coomassie blue for proteins. Gels B and D contained 50 ug of protein each and were stained by the benzidine method for heme. Because the benzidine stain faded out in a short time, the position of the bands was marked with drafting ink in the gels.

SDS-gel electrophoresis of the highly purified P-452$_{HF}$ shows that pretreatment of the sample with 1% SDS does not affect the mobility of the hemoprotein (Fig. 2). This suggests that the apparent molecular weight obtained for this hemoprotein (48,000) is the minimum molecular weight of cytochrome P-452$_{HF}$. The major benzidine positive protein band that contaminates this preparation has a molecular weight of approximately 20,000. Similar experiments done with the P-450$_{HF}$ and P-453$_{HF}$ fractions resulted in a heme containing band present at the top of the gels whenever the 1% SDS pretreatment was omitted. It should be pointed out here that as discussed by Arinc and Philpot (9), the fact that SDS does not completely disrupt the non-covalent binding between the heme group and the P-450 apoprotein is rather unusual. The possibility exists that SDS-P-450 complexes, with or without bound heme or with or without component(s), non-covalently bound but SDS resistant, might differ in size and shape resulting in an erroneous molecular weight estimation.

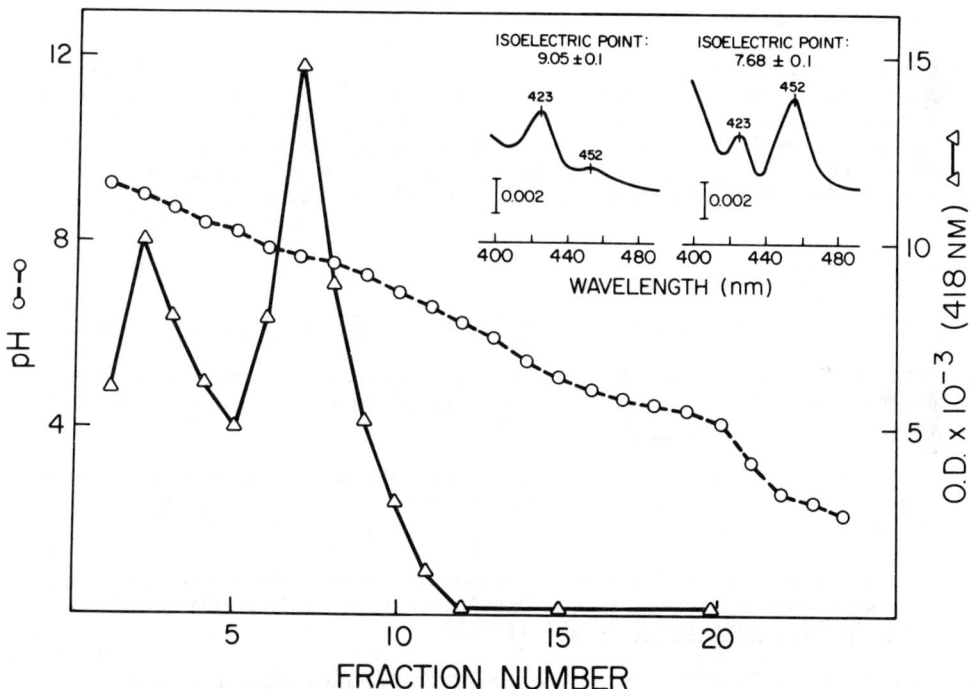

Fig. 3. Isoelectric focusing of the purified P-450$_{HF}$ preparation. The P-450$_{HF}$ aliquot (0.3 ml, 100 ug of protein; 1.39 nmol P-450) was mixed with the light sucrose solution prior to the formation of the gradient. The gradient volume was approximately 6.3 ml and it was separated from both electrodes by adding 0.3 ml of the heavy sucrose solution at the bottom and 0.3 ml of the light one at the top. After equilibrium was obtained, 25 equal volume fractions were collected from the bottom of the columns and their absorbance at 418 nm (△—△) as well as their pH at 2° (o- -o) were measured. In the insert of the figure, a CO-difference spectrum of the fractions corresponding to the two absorbance peaks eluted is given.

Isoelectric focusing of P-452$_{HF}$ shows two peaks absorbing at 418 nm. The major one, with an isoelectric point of 7.68 ± 0.1 corresponds to P-452$_{HF}$ contaminated with small amounts of cytochrome P-420. The other smaller peak with an isoelectric point of 9.05 ± 0.1 contains P-420 (Fig. 3 insert). These observations may be interpreted as follows. Firstly, more than one form of P-450 is present in our P-452$_{HF}$ and they are resolved by electrofocusing. Due to the high pH in the focusing zone of one of them, it is denaturated to cytochrome P-420 (2). Secondly, P-450 and P-420 have different isoelectric points, the denaturated protein being more basic (11). If this second possibility is correct, it would suggest that denaturation of P-450 to P-420 involves a structural change of such magnitude that allows charged groups not normally exposed to water to come in contact with the aqueous phase. Probably this structural change would involve not only the heme environment region but also other regions of the protein molecule. Thirdly, and as a consequence of the above discussion, there exists the possibility that none of the isoelectric points found correspond to the true isoelectric point. This parameter may be measured with certainty only when no P-420 is present in the focused band.

Our results indicate that at least three P-450 species exist in the non-induced NAIDM housefly strain. The two major species are P-452$_{HF}$ and P-453$_{HF}$, while P-450$_{HF}$ appears to be the minor component; however, differences in stability of the various P-450s may alter the relative proportions of each hemoprotein. It is also possible that more than three hemoproteins are normally present in our preparations and only three of them are selectively purified.

REFERENCES

1. Agosin, M., Insect cytochrome P-450, Mol. Cell Biochem., (1976, in press).

2. Capdevila, J., N. Ahmad and M. Agosin, Soluble cytochrome P-450 from housefly microsomes: Partial purification and characterization of two hemoprotein forms, J. Biol. Chem. 250, 1048 (1975).

3. Schonbrod, R. D. and L. C. Terriere, The solubilization and separation of two forms of microsomal cytochrome P-450 from the housefly, Musca domestica L., Biochem. Biophys. Res. Comm. 64, 829 (1975).

4. Capdevila, J., A. S. Perry and M. Agosin, Spectral and catalytic properties of cytochrome P-450 from a diazinon resistant housefly strain, Chem. Biol. Interactions 9, 105 (1974).

5. Jennissen, H. P. and L. M. G. Heilmeyer, Jr., General aspects of hydrophobic chromatography. Absorption and elution characteristics of some skeletal muscle enzymes, Biochemistry 14, 754 (1975).

6. Welton, Ann F. and Steven D. Aust, Multiplicity of cytochrome P-450 hemoproteins in rat liver microsomes, Biochem. Biophys. Res. Comm. 56, 898 (1974).

7. Fairbanks, G., Theodore L. Steck and D. F. H. Wallach, Electrophoretic analysis of the major polypeptides of the human erythrocyte membrane, Biochemistry 10, 2606 (1971).

8. Behnke, J. N., S. M. Dagher, T. H. Massey and W. C. Deal, Jr., Rapid

multisample isoelectric focusing in sucrose density gradients using conventional polyacrylamide electrophoresis equipment: A two peak transient in the approach-to-equilibrium, Anal. Biochem. 69, 1 (1975).

9. Arinc, Emel and Richard M. Philpot, Preparation and properties of partially purified pulmonary cytochrome P-450 from rabbits, J. Biol. Chem. 251, 3213 (1976).

10. Haugen, David A., Theodore A. Van der Hoven and Minor J. Coon, Purified liver microsomal cytochrome P-450. Separation and characterization of multiple forms, J. Biol. Chem. 250, 3567 (1975).

11. Conney, A. H., A. Y. H. Lu, W. Levin, A. Somogyi, S. West, M. Jacobson, D. Ryan and R. Kuntzman, Effect of enzyme inducers on substrate specificity of the cytochrome P-450s, Drug Metabol. and Dispos. 1, 199 (1973).

CATALYTIC HYDROXYLATION ACTIVITIES AND SOME OTHER PROPERTIES OF PARTIALLY PURIFIED RABBIT PULMONARY MICROSOMAL CYTOCHROME P-450

Emel Arinç and Richard M. Philpot

Pharmacology Branch, National Institute of Environmental Health Sciences, P.O. Box 12233, Research Triangle Park, N.C. 27709

ABSTRACT

We have purified rabbit pulmonary cytochrome P-450 to 8.9 nmoles per mg protein and have successfully reconstituted rabbit pulmonary mixed-function oxidase activity using components solubilized exclusively from pulmonary microsomes. Partially purified pulmonary cytochrome P-450 has been obtained free from cytochrome b_5 and NADPH-cytochrome c reductase but not epoxide hydrase. Epoxide hydrase activity co-purifies with the cytochrome P-450. NADPH-cytochrome c reductase fractions prepared from rabbit pulmonary or hepatic microsomes by solubilization with sodium deoxycholate supported hydroxylation activity when combined with hepatic but not pulmonary cytochrome P-450. Pulmonary NADPH-cytochrome c reductase solubilized by sodium cholate was active when combined with the cytochrome from either rabbit liver or lung.

INTRODUCTION

The lungs contain a microsomal mixed-function oxidase (mfo) system which can metabolize a large number of substrates (1-8). Studies comparing the rabbit pulmonary and hepatic mfo systems have demonstrated that the substrate specificities of the mfo systems from these two tissues are different (e.g., 4-7). We are interested in determining what factors are responsible for the observed substrate specificity differences between rabbit liver and lung. We have been able to reconstitute the rabbit pulmonary mfo system from solubilized cytochrome P-450, NADPH-cytochrome c reductase, and lipid (9) and have purified the pulmonary cytochrome P-450 to 7.4 nmoles per mg protein (10). In the course of these studies we have observed that the activity of the reconstituted rabbit pulmonary mfo system is dependent on the procedure used to obtain the NADPH-cytochrome c reductase fraction. The present study shows that the requirement for a specific reductase containing fraction is not observed when pulmonary cytochrome P-450 is replaced in the reconstituted system by hepatic cytochrome P-450.

METHODS

Pulmonary cytochrome P-450 was partially purified using the methods recently described (10). Further purification to 8.9 nmoles per mg protein was achieved by chromatography on CM-cellulose using 0.2% Emulgen 913 as the eluent. Pulmonary NADPH-cytochrome c reductase fractions were prepared by solubilization of microsomes using either sodium cholate or sodium deoxycholate followed by precipitation with ammonium sulfate. The method using sodium cholate has been reported (10).

When deoxycholate was used, precipitates containing reductase were collected between 30 and 45% and 45 and 55% saturation with ammonium sulfate. NADPH-cytochrome c reductase from rabbit hepatic microsomes was prepared by solubilization with sodium deoxycholate followed by precipitation with ammonium sulfate between 40 and 50% saturation. (The reductase fractions will be referred to as lung-cholate, lung-deoxycholate [30-45%], lung-deoxycholate [45-55%] and liver-deoxycholate [40-50%].) Fractions containing lipid were prepared as previously described (9). Hepatic cytochrome P-448 from male rats treated with 3-methylcholanthrene and hepatic cytochrome P-450 from untreated male rabbits were prepared by published methods (11, 12).

Cytochrome P-450 was assayed by the method of Omura and Sato (13) and cytochrome P-420 by the method of Nishibayashi and Sato (14). Spectra were recorded with an Aminco DW-2 spectrophotometer (American Instrument Co., Silver Springs, Md.). NADPH-cytochrome c reductase activity, 7-ethoxycoumarin O-deethylation activity, benzphetamine N-demethylation activity, and benzo(a)-pyrene hydroxylation activity were determined by the methods of Masters et al. (15), Ullrich and Weber (16), Lu et al. (17), and Wattenberg et al. (18), respectively. Protein was determined by the method of Lowry et al. (19). Incubation conditions used for the analysis of enzymic activities in reconstituted systems have been described in detail (9-12). The reductase preparations used all contained about 500 units (nmoles cytochrome c reduced per min) per mg protein. Cytochrome concentrations were 8.9, 5.6 and 2.3 nmoles per mg protein for rabbit pulmonary cytochrome P-450, rabbit hepatic cytochrome P-450, and rat cytochrome P-448, respectively.

The procedures of Weber and Osborn (20) and Alvares and Siekevitz (21) were used for analysis of samples by sodium dodecyl sulfate (SDS)-polyacrylamide gel electrophoresis. Gels stained with Coomassie blue were scanned at 550 nm using a Gilford 2400-2 spectrophotometer equipped with a linear transport attachment (Gilford Instrument Laboratories, Oberlin, Ohio). The sources and purities of all chemicals used have been described in detail elsewhere (9-12).

RESULTS AND DISCUSSION

We have reported the purification of rabbit pulmonary cytochrome P-450 to 7.4 nmoles per mg protein with 10% yield (10). Attempts to further purify the hemoprotein have been only partially successful. Chromatography of the partially purified cytochrome on CM-cellulose has resulted in preparations containing up to 8.9 nmoles cytochrome per mg protein (table 1). The recovery of cytochrome from the CM-cellulose was low--1% of the original microsomal P-450 and only about 10% of the cytochrome from the previous purification step (Table 1). One major and several minor bands of protein were resolved with Coomassie blue following SDS-polyacrylamide gel electrophoresis of the preparation containing 8.9 nmoles cytochrome per mg protein (Fig. 1). The amount of protein present in the area of the gel corresponding to a molecular weight of about 57,000 daltons was significantly reduced in comparison to that formed in preparations containing about 7 nmoles P-450 per mg protein. The protein band in the region corresponding to about 50,000 daltons has previously been found to contain peroxidase activity and was assumed to be derived from cytochrome P-450 (10).

TABLE 1 Purification of Rabbit Pulmonary Cytochrome P-450[a]

FRACTION	CYTOCHROME CONCENTRATION (nmoles/mg protein)	YIELD (%)	PURIFICATION (fold)
Microsomes	0.23	100	1
Solubilized Microsomes	0.23	90	1
Ammonium Sulfate Precipitate	0.75	42	3
DEAE Eluate	2.35	27	10
Hydroxylapatite Eluate	5.70	12	25
CM Cellulose Eluate	8.90	1	39

[a]Detailed methods for the purification of rabbit pulmonary cytochrome P-450 through the hydroxylapatite have been reported (10).

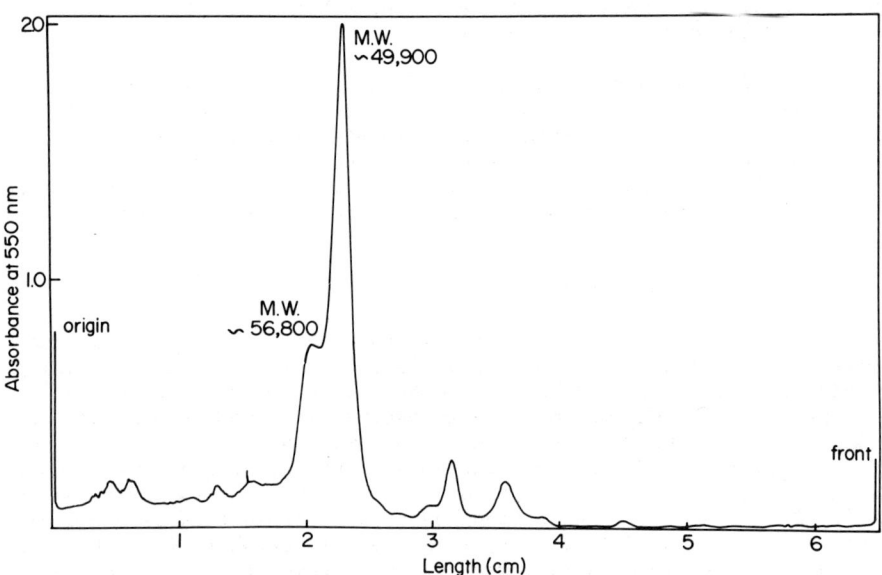

Fig. 1. SDS-polyacrylamide gel electrophoresis of partially purified pulmonary cytochrome P-450.
Subsequent to electrophoresis the gel was stained with Coomassie blue and the profile shown was obtained by scanning the gel at 550 nm. 15 µgm of protein (8.9 nmoles cytochrome per mg protein) were applied to the gel.

The oxidized, reduced, and reduced-carbon monoxide spectra of partially purified pulmonary cytochrome P-450 were similar to those of hepatic P-450 and the same as those reported for less pure preparations of pulmonary cytochrome P-450 (19). The carbon monoxide difference spectrum of dithionite-reduced pulmonary cytochrome P-450 recorded in the presence of 0.3 mg lipid per ml shows that between 5 and 7% of the hemoprotein is in the form of P-420 (Fig. 2). The apparent amounts of P-450 and P-420 were dependent on the concentration of added lipid (Fig. 2). The apparent extinction coefficient for rabbit pulmonary cytochrome P-450 has been calculated from the carbon monoxide difference spectrum (ΔA, 450-490 nm) to be 88 min^{-1} cm^{-1} (10).

Hydroxylase activity of the reconstituted pulmonary mfo system was totally dependent on both the cytochrome P-450 and NADPH-cytochrome c reductase fractions (Table 2). The effect of added lipid fraction varied with the substrate; the rate of O-deethylation of 7-ethoxycoumarin was only 6% of maximum in the absence of added lipid while the rates of N-demethylation of benzphetamine and hydroxylation of benzo(a)pyrene were 57 and 87% of maximum, respectively, when added lipid was omitted from the incubations (Table 2).

The different pulmonary NADPH-cytochrome c reductase preparations were investigated for their ability to support 7-ethoxycoumarin O-deethylation when combined with cytochrome P-450 from various sources (Fig. 3). The lung-cholate reductase was active when combined with cytochrome P-450 from rabbit

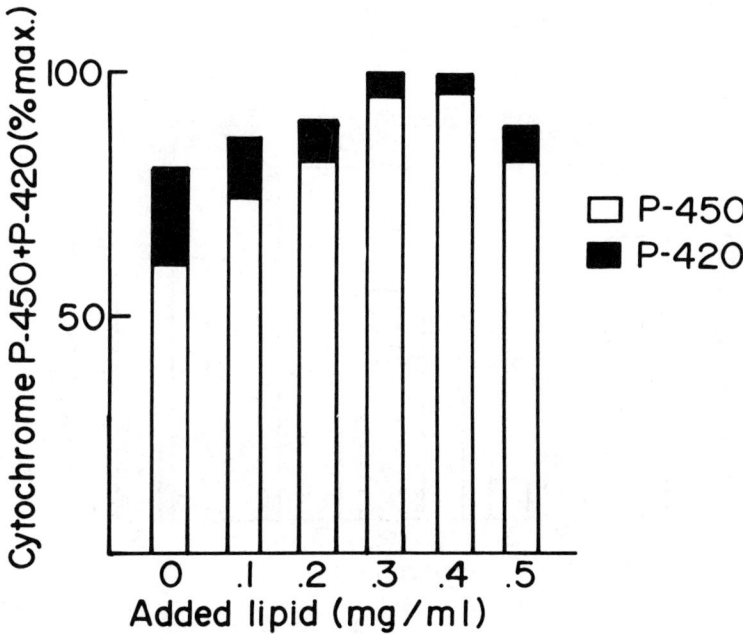

Fig. 2. Determination of cytochrome content in partially purified pulmonary cytochrome P-450 preparations.
The indicated amounts of lipid were added to the samples and the spectra recorded. The open portion of each bar represents apparent cytochrome P-450 content and the closed portion of each bar represents apparent cytochrome P-450 content.

TABLE 2 Rates of Metabolism in the Reconstituted Rabbit Pulmonary MFO System (% Maximum)[a]

FRACTIONS PRESENT	SUBSTRATE		
	7-ETHOXY-COUMARIN	BENZPHETAMINE	BENZPYRENE
Reductase[b]	4	3	0
Cytochrome[c]	0	0	0
Cytochrome[c] + Reductase[b]	6	57	87
Cytochrome[c] + Reductase[b] + Lipid[d]	100	100	100

[a]Maximum rates (nmoles product/nmoles cytochrome P-450/min) were: 2.08 for the O-deethylation of 7-ethoxycoumarin; 10.8 for the N-demethylation of benzphetamine; and 0.15 for the hydroxylation of benzo(a)pyrene.

[b]200 units of pulmonary NADPH-cytochrome c reductase prepared with sodium cholate were used.

[c]0.1 nmoles pulmonary cytochrome P-450 per ml was used.

[d]0.1 mg lipid per ml was used.

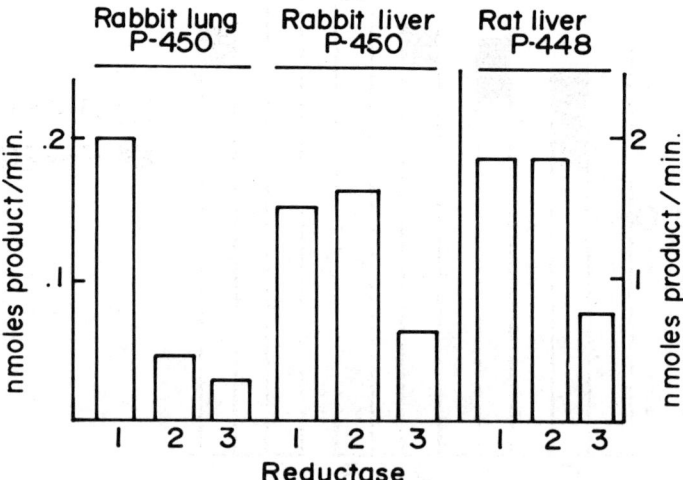

Fig. 3. The O-deethylation of 7-ethoxycoumarin in reconstituted systems containing different rabbit pulmonary reductase preparations.
The following reductase preparations were used: 1, lung-cholate; 2, lung-deoxycholate (30-45%); 3, lung-deoxycholate (45-55%). Reconstituted systems contained 0.1 nmole pulmonary P-450 and 200 units reductase per ml or 0.2 nmoles hepatic P-450 and 200 units reductase per ml.

lung or liver or cytochrome P-448 from rat liver (Fig. 3). Similar activities were observed with the lung-deoxycholate (45-55%) reductase plus rat or rabbit pulmonary cytochrome (Fig. 3). The lung-deoxycholate (35-45%) reductase was the least active with all three cytochromes but showed significantly greater activity with the hepatic cytochromes as compared to the pulmonary cytochrome. The pulmonary reductase preparations were about 80% as active as the hepatic reductase when combined with rabbit hepatic cytochrome P-450 plus lipid (Fig. 4). On the other hand, the hepatic reductase showed little activity when combined with pulmonary cytochrome P-450 (Fig. 4).

The results shown in Fig. 3 and 4 indicate that the interaction between rabbit pulmonary cytochrome P-450 and pulmonary reductase is different from the interaction between the corresponding hepatic components. The reason for this difference is not clear but it is clear that both the pulmonary and hepatic cytochrome fractions and the pulmonary cholate and deoxycholate reductase fractions are different. The similar activities of the lung-cholate and lung-deoxycholate (45-55%) fractions with hepatic cytochrome P-450 suggest that the reductases in each fraction are the same. Storage of the lung-cholate reductase results in loss of activity when the fraction is combined with the pulmonary cytochrome but retention of cytochrome c reductase activity. Thus, the lung-cholate reductase preparation appears to contain a labile factor which is required for the activity of the reconstituted mfo system. The lung-deoxycholate (45-55%) fraction either does not contain this factor or it contains an inhibitor to pulmonary but not hepatic activity.

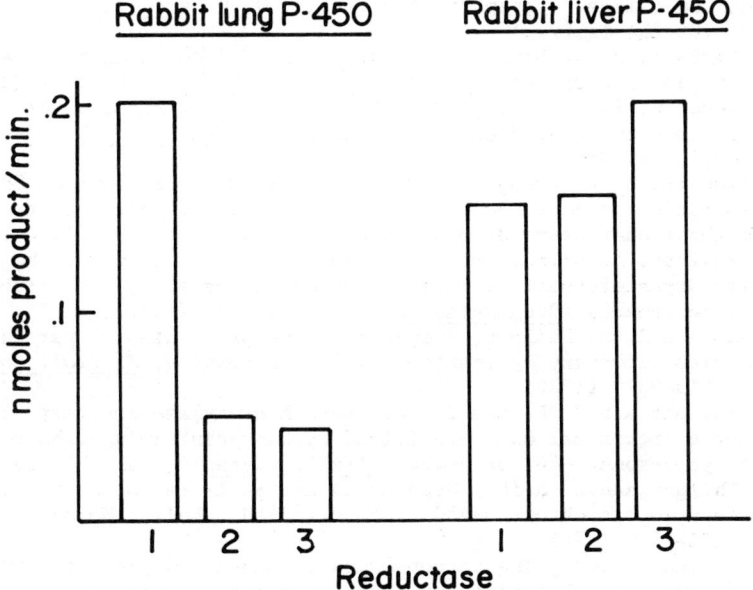

Fig. 4. The effect of reductase fractions from rabbit liver and lung on the O-deethylation of 7-ethoxycoumarin catalyzed by rabbit pulmonary or hepatic cytochrome P-450.
The following reductase fractions were used: 1, lung-cholate; 2, lung-deoxycholate (45-55%); 3, liver-deoxycholate.

Several reports concerning rat pulmonary NADPH-cytochrome c reductase suggest that the problems we have encountered are not unique to the rabbit. Buege and Aust (22) have shown that rat hepatic and pulmonary NADPH-cytochrome c reductases are either very similar or identical. Their conclusion was reached from the results of immunological and SDS-polyacrylamide gel electrophoresis studies. However, Jernstrom et al. (23) were unable to elicit mixed-function oxidase activity in a system containing rat pulmonary cytochrome P-450 and pulmonary reductase solubilized from microsomes by sodium deoxycholate.

REFERENCES

(1) H. Uehleke, Extrahepatic drug metabolism, Excerpta Med. Int. Congr. 181, 94-100 (1968).
(2) W. W. Oppelt, M. Zange, W. E. Ross, and H. Remmer, Comparison of microsomal drug hydroxylation in lung and liver of various species, Res. Commun. Chem. Pathol. Pharmacol. 1, 43-56 (1970).
(3) T. Matsubara and Y. Tochino, Electron transport systems of lung microsomes and their physiological functions. I. Intracellular distribution of oxidative enzymes in lung cells, J. Biochem. (Tokyo) 70, 981-991 (1971).
(4) G. E. R. Hook, J. R. Bend, D. Hoel, J. R. Fouts, and T. E. Gram, Preparation of lung microsomes and a comparison of the distribution of enzymes between subcellular fractions of rabbit lung and liver, J. Pharmacol. Exp. Ther. 182, 474-490 (1972).
(5) J. R. Bend, G. E. R Hook, R. E. Easterling, T. E. Gram, and J. R. Fouts. A comparative study of the hepatic and pulmonary microsomal mixed-function oxidase systems in the rabbit, J. Pharmacol. Exp. Ther. 183, 206-216 (1972).
(6) G. E. R. Hook, J. R. Bend, and J. R. Fouts, Mixed-function oxidases and the alveolar macrophage, Biochem. Pharmacol. 21, 3267-3277 (1972).
(7) J. R. Bend, G. E. R. Hook, and T. E. Gram, Characterization of lung microsomes as related to drug metabolism, Drug Metab. Dispos. 1, 358-367. (1973).
(8) T. Matsubara, R. A. Prough, M. D. Burke, and R. W. Estabrook, The preparation of microsomal fractions of rodent respiratory tract and their characterization, Cancer Res. 34, 2196-2203 (1974).
(9) R. M. Philpot, E. Arinc, and J. R. Fouts, Reconstitution of the rabbit pulmonary microsomal mixed-function oxidase system from solubilized from components, Drug Metab. Disp, 3, 118-126 (1975).
(10) E. Arinc and R. M. Philpot, Preparation and properties of partially purified pulmonary cytochrome P-450 from rabbits, J. Biol. Chem. 21, 3213-3220 (1976).
(11) R. M. Philpot and J. R. Bend, Benzpyrene hydroxylase activity in hepatic microsomal and solubilized systems containing rabbit or rat cytochrome P-448 or P-450, Life Sciences 16, 985-997 (1975).
(12) R. M. Philpot and E. Arinc, Separation and purification of two forms of hepatic cytochrome P-450 from untreated rabbits, Molec. Pharmacol. 12, 483-493 (1976).
(13) T. Omura and R. Sato, The carbon-monoxide binding pigment to liver microsomes. II. Evidence for its hemoprotein nature, J. Bio. Chem. 239, 2379-2385 (1964).
(14) H. Nishibayashi, and R. Sato, Preparation of hepatic microsomal particles containing P-450 as the sole heme constituent, J. Biochem. 63, 766-779 (1968).

(15) B. S. S. Masters, C. H. Williams, and H. Kamin, in Methods in Enzymology (R. W. Estabrook and M. E. Pullman, eds.) Vol. 10, p. 565, Academic Press, New York (1967).
(16) U. Ullrich and P. Weber, The O-deethylation of 7-ethoxycoumarin by liver microsomes, Hoppe-Seyler's Z. Physiol. Chem. 353, 1171-1177 (1972).
(17) A. Y. H. Lu, R. Kuntzman, S. West, M. Jacobson, and A. H. Conney, Reconstituted microsomal enzyme system that hydroxylates drugs, other foreign compounds and endogenous substrates. II. Role of the cytochrome P-450 and P-448 fractions in drug and steroid hydroxylations, J. Biol. Chem. 247, 1727-1734 (1972).
(18) L. W. Wattenberg, J. L. Leong, and P. S. Strand, Benzpyrene hydroxylase activity in the gastrointestinal tract, Cancer Res. 22, 1120-1125. (1962).
(19) O. H. Lowry, N. J. Rosebrough, A. L. Farr, and R. J. Randall, Protein measurement with Folin phenol reagent, J. Biol. Chem. 193, 265-275 (1951).
(20) K. Weber and M. Osborn, The reliability of molecular weight determinations by dodecyl sulfate-polyacrylamide gel electrophoresis, J. Biol. Chem. 244, 4406-4412 (1969).
(21) A. P. Alvares and P. Siekevitz, Gel electrophoresis of partially purified cytochrome P-450 from liver microsomes of variously treated rats, Biochem. Biophys. Res. Commun. 54, 923-929 (1973).
(22) J. A. Buege and S. D. Aust, Comparative studies of rat liver and lung NADPH-cytochrome c reductase, Biochem. Biophys. Acta 385, 371-379 (1975).
(23) B. Jernstrom, J. Capdevila, S. Jakobsson, and S. Orrenius, Solubilization and partial purification of cytochrome P-450 from rat lung microsomes, Biochem. Biophys. Res. Commun 64, 814-822.

HEPATIC MICROSOMAL AND SOLUBILIZED MIXED-FUNCTION OXIDASE SYSTEMS FROM THE LITTLE SKATE, *Raja erinacea*, A MARINE ELASMOBRANCH

John R. Bend, Roberta J. Pohl, Emel Arinç, and Richard M. Philpot

Pharmacology Branch, National Institute of Environmental Health Sciences, NIH, Research Triangle Park, N.C. 27709 and Mount Desert Island Biological Laboratory, Salsbury Cove, Maine 04672 U.S.A.

ABSTRACT

Pretreatment of little skates with 3-methylcholanthrene (3-MC), or 2,3,7,8-tetrachlorodibenzo-p-dioxin (TCDD) caused significant increases in hepatic microsomal benzo[a]pyrene hydroxylase activity with no concomitant increase in cytochrome P-450 content or apparent formation of cytochrome P-448.

The hepatic microsomal mixed-function oxidase (MFO) system of little skates was solubilized and separated into its component parts. Cytochrome P-450, NADPH-cytochrome c reductase, and lipid fractions were required for reconstitution of 7-ethoxycoumarin O-deethylase activity. Skate hepatic cytochrome P-450 was purified 12-fold from microsomes by solubilization with sodium cholate and chromatography on DEAE-cellulose and then hydroxylapatite. NADPH-cytochrome c reductase was purified to a specific activity of about 500 units (nmoles cytochrome c reduced/min) per mg protein.

INTRODUCTION

The sea is a repository for numerous xenobiotics that occur as environmental contaminants. Many sources of these pollutants are known including dumping of industrial and urban wastes, atmospheric fallout of airborne substances, runoff from agricultural areas treated with herbicides or insecticides, shipping spillages (both accidental and intentional), and natural seepage of hydrocarbons from underwater oil deposits. Since some xenobiotics are metabolically activated to products toxic (e.g., mutagenic or carcinogenic) to mammals, including man, the ability of marine species to biotransform and excrete foreign compounds is of considerable interest, particularly in those species that serve as direct or indirect human food sources.

Microsomes from little skate liver have relatively high MFO activity toward several substrates when compared to other marine species of coastal Maine (1, 2, 3). In the present paper the effect of pretreating little skates with 3-MC or TCDD on hepatic microsomal MFO activity and/or cytochrome P-450 content is reported, as well as the solubilization, partial purification, and reconstitution of the hepatic microsomal MFO system from untreated skates.

METHODS

Fish Collection and Maintenance

Male and female little skates (650-1300g) were collected by trawl or drag net near Mt. Desert Island, Maine. They were stored in aquaria or pools equipped with continuously circulating sea water or in live cars immersed in Frenchman Bay until use.

Microsome Preparation, Enzyme Assays, and Cytochrome P-450 Determination

All skates were sacrificed between 5:00 and 8:00 A.M. Washed hepatic microsomes were utilized for assay of MFO activities and for quantitation of cytochrome P-450. Freshly frozen samples of liver were shipped to NIEHS (in dry ice) for measurement of cytochrome P-450 content. All assays were carried out within 2 weeks of sacrifice by the dithionite difference technique due to hemoglobin contamination of the liver microsomes. The procedures utilized have been described elsewhere in detail (3). All spectra were recorded with an Aminco DW2 spectrophotometer (American Instrument Co., Silver Springs, Md.).

Statistical comparisons were made with the nonparametric Mann-Whitney U test, and P values of < 0.05 were taken to represent significant differences between means.

Solubilization, Partial Purification and Reconstitution of the Skate Hepatic MFO System

Unwashed microsomes (3) from the livers of 40 skates were suspended in 0.25 M sucrose (in batches) and stored under nitrogen at -5 to -10°C. These samples were transported to NIEHS (in dry ice) and stored at -20°C until use.

Cytochrome P-450 fraction. Thawed microsomes (about 2 g protein in 80 ml) were solubilized by suspension in a final volume of 200 ml 0.1 M phosphate buffer (pH 7.7) containing 20% glycerol, 1 mM dithiothreitol (DTT), and 1 mM EDTA, and addition of sodium cholate to a final concentration of 1 mg detergent per mg protein followed by stirring at 4° for 30 min. The clear supernatant (yellow) fraction resulting after centrifugation at 176,000 g for 2 hr was diluted 2-fold with a 20% glycerol solution containing 1 mM DTT and 1 mM EDTA and was applied to a column of DEAE-cellulose (2.5 x 45 cm) previously equilibrated with buffer I (50 mM phosphate, pH 7.7, containing 1 mM DTT, 1 mM EDTA, and 0.5% sodium cholate). Cytochrome P-450 was eluted from the column with 0.3% Emulgen 913 in buffer 1.

Cytochrome P-450 containing fractions eluted from the DEAE-cellulose column were combined, diluted 2-fold with a 20% glycerol solution, and applied to a hydroxylapatite column (2.5 x 6 cm) directly or following concentration by ultrafiltration (Amicon, PM-100 membrane). The hydroxylapatite column was equilibrated with 25 mM phosphate buffer (pH 7.7) containing 0.5 mM DTT, 0.5 mM EDTA, and 0.25% sodium cholate (buffer II). The cytochrome P-450 was eluted by increasing the phosphate concentration to 140 mM and adding Emulgen 913 (to 0.2%). Cytochrome P-450 concentrations in solubilized microsomes, and in unwashed microsomes used for solubilization, were determined by the method of Omura and Sato (4).

NADPH-Cytochrome c reductase fraction. NADPH-cytochrome c reductase was solubilized from thawed microsomes by digestion with sodium deoxycholate (1 mg/mg microsomal protein) in 0.1 M phosphate buffer (pH 7.7) containing 0.1 mM DTT and 0.1 mM EDTA. The clear supernatant fraction of the centrifuged digest was applied to a DEAE-cellulose column (2.5 x 45 cm) that was equilibrated with 0.1 M phosphate buffer (pH 7.7) containing 0.1 mM DTT, 0.1 mM EDTA, and 0.05% sodium deoxycholate (buffer III). The column was washed with buffer III (100 ml) and NADPH-cytochrome c reductase activity was eluted with a linear gradient of 0-0.6 M KCl in buffer III. The fractions containing eluted reductase activity were pooled and concentrated by ultrafiltration (Amicon, PM-30 membrane). The final preparation contained about 500 units/mg protein. NADPH-cytochrome c reductase activity was measured according to Williams and Kamin (5).

RESULTS

The administration of single doses of 3-MC (50 mg/kg P.O. or 20 mg/kg I.P.) or TCDD (1 µg/kg I.P.) to little skates did not significantly increase benzpyrene hydroxylase activities (16). However, in certain individual treated skates, benzpyrene hydroxylase activities were higher than in any control animals. Consequently, skates were given higher and/or multiple doses of these two chemicals to test for induction of MFO activities.

Table 1 shows that benzpyrene hydroxylase activities were increased about 8-fold after two oral skate treatments with 3-MC (50 mg/kg). Statistically significant increases in aniline hydroxylase and benzphetamine N-demethylase activities (about 2-fold) were also observed.

TABLE 1 Effect of 3-Methylcholanthrene Administration[*] on Several Mixed-Function Oxidase Activities of Little Skate Hepatic Microsomes

	Control	3-MC Treated	P
Benzpyrene Hydroxylase[†]	0.84 ± 0.14[‡]	6.50 ± 0.12	<0.01
7-Ethoxycoumarin Deethylase[**]	0.36 ± 0.08	0.56 ± 0.08	N.S.
Benzphetamine Demethylase[**]	0.96 ± 0.08	1.99 ± 0.10	<0.01
Aniline Hydroxylase[**]	0.43 ± 0.05	0.85 ± 0.11	<0.05

[*] Treated orally with 50 mg/kg 3-MC dissolved in DMSO on days 1 and 3, sacrificed on day 11.
[†] Fluorescence units/min/mg protein.
[‡] Mean ± S.E.M., N = 3.
[**] Nmoles product formed/min/mg protein.

As shown in Table 2, similar results were obtained following the I.P. administration of two separate TCDD doses (4.5 µg/kg) on day 1 and day 3. Benzpyrene hydroxylase activity was markedly increased (15- to 18-fold) 4 or 9 days after the second injection of TCDD. Although aniline hydroxylase, benzphetamine N-demethylase, and 7-ethoxycoumarin O-deethylase activities tended to increase, none of these changes were significant. In the skates sacrificed on day 7, the hepatic microsomal cytochrome P-450 content (per mg microsomal protein) was identical to control values, with no shift in absorption maximum. Although there was a slight increase in cytochrome P-450 content of the day 12 skates (not significant), there was also no apparent wavelength shift in the absorption maximum of the CO-bound form of reduced cytochrome P-450.

TABLE 2 Effect of 2,3,7,8-Tetrachlorodibenzo-p-dioxin Administration* on Mixed-Function Oxidase Activities and Cytochrome P-450 Content of Little Skate Hepatic Microsomes

	Control	TCDD (Day 7)	TCDD (Day 12)
Benzpyrene Hydroxylase[†]	1.22 ± 0.17 (6)[‡]	18.4 ± 3.0 (5)**	22.7 ± 5.5 (3)**
7-Ethoxycoumarin Deethylase[††]	0.36 ± 0.06 (6)	0.56 ± 0.10 (5)	0.74 ± 0.23 (3)
Benzphetamine Demethylase[††]	1.00 ± 0.20 (6)	1.38 ± 0.20 (5)	1.44 ± 0.42 (3)
Aniline Hydroxylase[††]	0.30 ± 0.06 (6)	0.58 ± 0.15 (5)	0.51 ± 0.20 (3)
Cytochrome P-450[†††]	0.24 ± 0.02 (6)	0.25 ± 0.03 (5)	0.34 ± 0.07 (3)

*Treated by I.P. injection of 4.5 μg/kg TCDD in corn oil:acetone (5:1) on days 1 and 3. Sacrificed on day 7 or day 12.
[†]Fluorescence units/min/mg protein.
[‡]Mean ± S.E.M. (N).
**Significantly different from control, $P < 0.05$.
[††]Nmoles product formed/min/mg protein.
[†††]Nmoles/mg protein; absorption maxima at 450 nm for control and treated skates.

The elution profile of cytochrome P-450 from a sodium cholate-solubilized skate hepatic microsomal preparation applied to a column of DEAE-cellulose is shown in Fig. 1. Two separate pools (Peak I and Peak II) of cytochrome P-450 were obtained, having almost identical specific P-450 concentrations (2.00 and 1.90 nmoles/mg protein, respectively). Peak II also contained a yellow pigmented impurity in addition to cytochrome P-450.

The concentrated cytochrome P-450 preparations from Peaks I and II of the DEAE-cellulose eluate were chromatographed separately on columns of hydroxylapatite (Fig. 2). The cytochrome P-450 fractions obtained were free of cytochrome b_5. The yellow pigment was also almost completely removed from Peak II. The cytochrome P-450 containing fractions (Peaks I and II) eluted from hydroxylapatite were combined and passed through a Porapak Q column to remove unbound detergent.

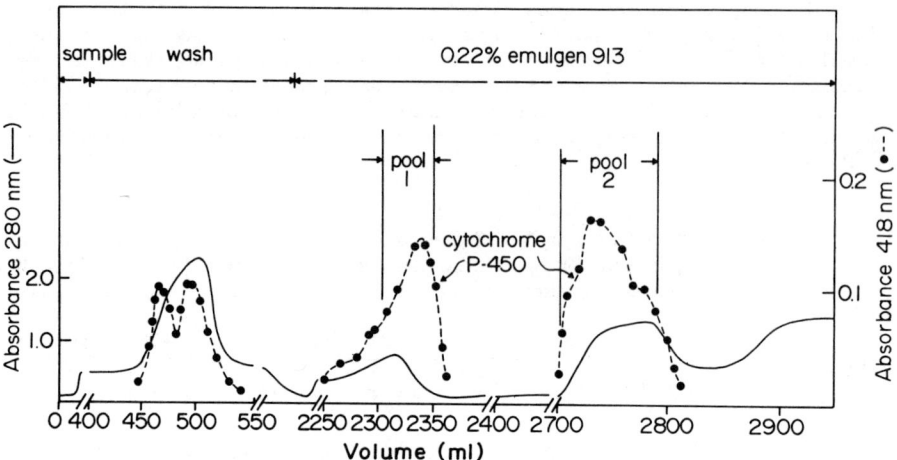

Fig. 1. Elution profile of partially purified cytochrome P-450 from DEAE-cellulose column

For experimental procedure see <u>Methods</u>. Both fractions, Pool I and Pool II, contained cytochrome P-450.

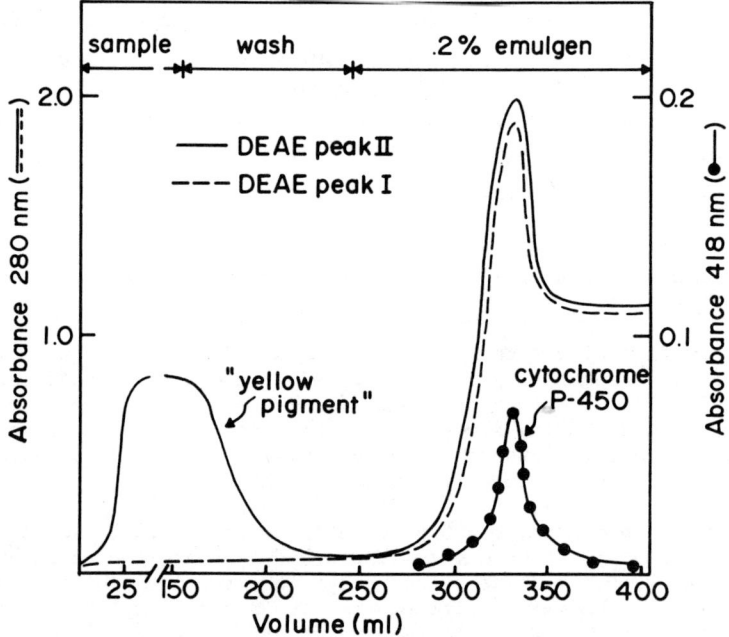

Fig. 2. Chromatography on hydroxylapatite of Peaks I and II eluted from DEAE-cellulose

The peaks were applied to separate columns of hydroxylapatite (2.5 x 6 cm) equilibrated with buffer II. For experimental details see <u>Methods</u>. The yellow pigment of Peak II appeared in the void volume.

The stepwise, partial purification of skate hepatic cytochrome P-450 is outlined in Table 3. The final preparation contained 3.00 nmoles cyto-P-450 per mg protein as compared to 0.26 nmole/mg protein in unwashed skate microsomes.

TABLE 3 Partial Purification of Cytochrome P-450 From Little Skate Hepatic Microsomes

	CYTOCHROME P-450		
	Total Amount (nmoles)	Yield (%)	Specific Content (nmoles/mg protein)
Microsomes	538	100	0.26
Solubilized Microsomes	484	90	0.28
DEAE-Cellulose Eluate	122.7	23	1.93
Peak I	42.7	8	2.00
Peak II	80.0	15	1.90
Hydroxylapatite Eluate	68.2	12	2.19
Peak I	28.2	5	2.17
Peak II	40.0	7	2.20
Porapak Q Eluate	59.4	11	3.00

The carbon monoxide difference spectrum of the dithionite reduced cytochrome P-450 preparation (3.0 nmoles/mg protein) had its absorption maximum at 450 nm. Absolute spectra of the same preparation are shown in Fig. 3 and are similar to published spectra of partially purified mammalian cytochrome P-450.

The elution of NADPH-cytochrome c reductase activity from a column cf DEAE-cellulose is shown in Fig. 4. The final preparation contained about 500 units of activity per mg protein as compared to 60 units per mg protein in hepatic microsomes. The lipid fractions used for reconstitution of MFO activity were eluted from the same column with 0.6 M KCl in buffer III (eluted mainly after the cytochrome b_5 peak). Lipid fractions used in the enzyme reconstitution experiments were free of cytochrome b_5.

The requirement of cytochrome P-450, NADPH-cytochrome c reductase, and lipid fractions for maximal 7-ethoxycoumarin O-deethylase activity in the reconstituted system is illustrated in Table 4. Individually, none of the fractions supported the oxidative metabolism of 7-ethoxycoumarin. Although there was some activity with cytochrome and reductase fractions in the absence of exogenous lipid, it was only about 25% maximal activity. Saturation of the 7-ethoxcoumarin deethylase system with respect to cytochrome P-450 was achieved in the presence of 130 units of reductase per ml and 0.08 mg per ml of lipid (at 0.12 nmole P-450/ml).

DISCUSSION

Although an early report (7) suggested that fish lacked the ability to oxidatively metabolize foreign compounds, subsequent studies (3, 8-12) have demonstrated that many freshwater and marine fish do indeed have hepatic MFO activity.

In general, such studies have emphasized the similarity of the fish MFO system to that of mammals. Thus, in the little skate, activity is localized in the microsomal fraction, requires reduced NADPH and molecular oxygen for maximal activity and can be inhibited with carbon monoxide, SKF-525A, metyraprone, and cytochrome c (1, 3, 13). However, in hepatic microsomes from this poikilotherm, maximal MFO activities (with several substrates) occur at incubation temperatures near 30°C as compared to about 45°C for mouse or rabbit (3).

The response of fish MFO systems to xenobiotic administration has been poorly characterized. Hepatic benzpyrene hydroxylase activity was induced slightly in winter flounder (3) and in rainbow trout (14) by administration of polycyclic aromatic hydrocarbons. In a field study, Payne (15) demonstrated that fish from petroleum polluted areas had higher benzpyrene hydroxylation activities in liver and gill than similar fish from unpolluted areas. The present study (Tables 1 and 2) showed that 3-MC and TCDD pretreatment caused dramatic increases in benzpyrene hydroxylase activity of hepatic microsomes from little skates. The concomitant lack of an increase in cytochrome P-450 content or of a shift in the maximum of the carbon monoxide-bound form of the reduced cytochrome is interesting since type II induction

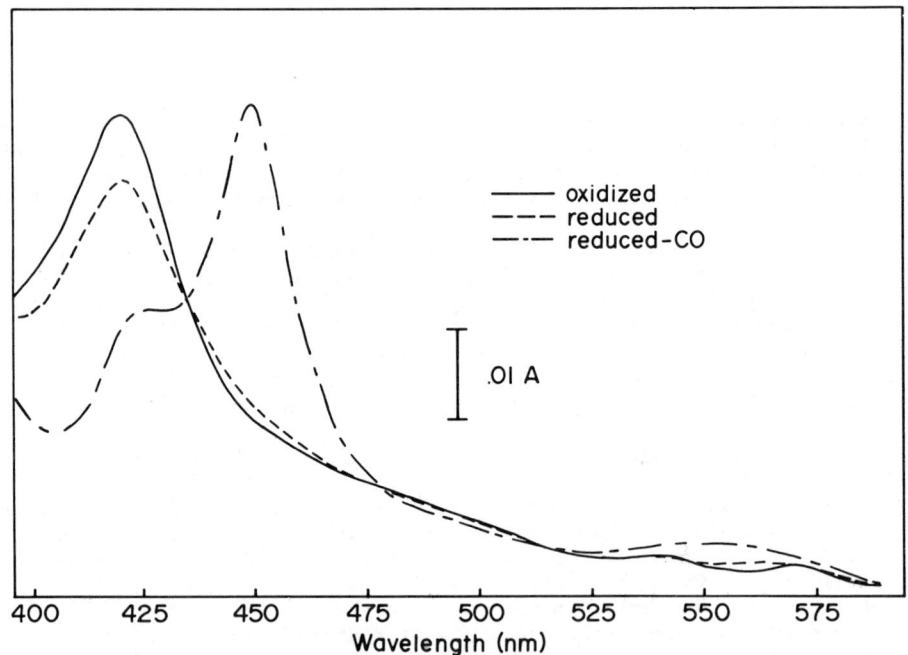

Fig. 3. Absolute spectra of partially purified skate hepatic microsomal cytochrome P-450.

The cytochrome P-450 was eluted from a Porapak Q column (3.0 nmoles P-450/mg protein). The oxidized, dithionite-reduced, and dithionite-reduced plus carbon monoxide spectra were recorded at a cytochrome P-450 concentration of 0.30 nmole/ml in 0.1 M phosphate buffer, pH 7.7, containing 30% glycerol and 0.1 M EDTA.

(i.e., polycyclic aromatic compounds, TCDD) of mammalian liver is usually associated with the formation of cytochrome P-448 (16-18). It is possible that in the skate a catalytically different form of cytochrome P-450 is synthesized in response to 3-MC or TCDD exposure which has the same wavelength maximum (in reduced microsomes bound to carbon monoxide) as the control cytochrome. It is also peculiar that benzphetamine demethylase activities should increase in hepatic microsomes from 3-MC- or TCDD-treated fish.

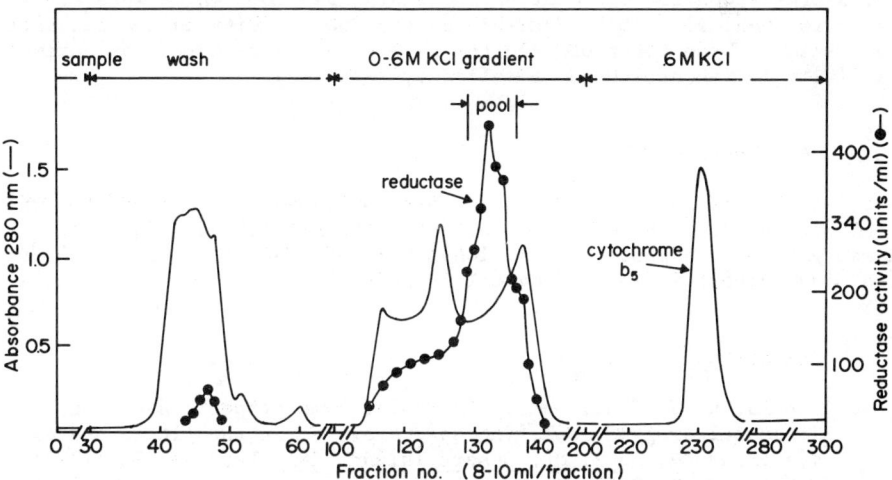

Fig. 4. Elution profile of NADPH-cytochrome c reductase and cytochrome b_5 from a column of DEAE-cellulose

The sample was prepared by digestion of hepatic microsomes (about 1.5g protein) with sodium deoxycholate followed by centrifugation at $150,000g$ for 1.5 hr. The experimental procedure is described in Methods.

TABLE 4 Reconstitution of 7-Ethoxycoumarin Deethylase Activity in Fractions Prepared from Skate Hepatic Microsomes

FRACTION(S)	7-ETHOXYCOUMARIN DEETHYLASE ACTIVITY (nmoles/min/ml)
Cytochrome P-450*	0
Reductase†	0
Lipid‡	0
Cytochrome P-450* + Reductase†	0.10
Cytochrome P-450* + Reductase† + Lipid‡	0.39
Cytochrome P-450** + Reductase† + Lipid‡	0.52
Cytochrome P-450†† + Reductase† + Lipid‡	0.52

Amount of fractions used:
*0.07 nmole/ml; †130 units/ml; ‡0.08 mg/ml; **0.12 nmole/ml; ††0.14 nmole/ml.

Currently the liver MFO system is being purified from solubilized microsomes of DBA-induced animals, and these studies should enable us to better understand these observations.

To our knowledge this is the first report of the solubilization and partial purification of the MFO system from fish liver. The data (Figs. 1-4, Tables 3-4) emphasize the similarity of the little skate system to the mammalian system. Thus, hepatic microsomes from skate were separated into cytochrome P-450, NADPH-cytochrome c reductase and lipid fractions which were all required for maximal 7-ethoxycoumarin deethylase activity in the reconstituted system. These fractions are the same as those required for maximal MFO activity in reconstituted mammalian systems (16).

ACKNOWLEDGMENTS

We are grateful to Dr. J. R. Fouts for his continued interest and encouragement and to Dr. J. K. Haseman who conducted the statistical analyses. This work was partially supported by National Science Foundation Grant #GB 28139 to the Mount Desert Island Biological laboratory.

REFERENCES

(1) J. R. Bend, R. J. Pohl, and J. R. Fouts, Some properties of the microsomal drug-metabolizing enzyme system in the little skate, *Raja erinacea*, Bull. Mt. Desert Island Biol. Lab. 12, 12 (1972).
(2) R. J. Pohl, J. R. Bend, T. R. Devereux, and J. R. Fouts, Hepatic chemical and drug metabolizing enzymes in coastal Maine marine species, Bull. Mt. Desert Island Biol. Lab. 13, 94 (1973).
(3) R. J. Pohl, J. R. Bend, A. M. Guarino, and J. R. Fouts, Hepatic microsomal mixed-function oxidase activity of several marine species from coastal Maine, Drug Metab. Disposition 2, 545 (1974).
(4) T. Omura and R. Sato, The carbon monoxide binding spectrum of liver microsomes. I. Evidence for its hemoprotein nature, J. Biol. Chem. 239, 2370 (1964).
(5) C. H. Williams and H. Kamen, Microsomal triphosphoryridine nucleotide-cytochrome "c" reductase of liver, J. Biol. Chem. 237, 587 (1962).
(6) J. R. Bend, R. J. Pohl, N. P. Davidson, and J. R. Fouts, Response of hepatic renal mixed-function oxidases in the little skate, *Raja erinacea*, to pretreatment with 3-methylcholanthrene or TCDD (2,3,7,8-tetrachlorodibenzo-p-dioxin, Bull. Mt. Desert Island Biol. Lab. 14, 7 (1974).
(7) B. B. Brodie and R. P. Maickel, Comparative biochemistry of drug metabolism, Proc. First Internat. Pharmacol. Meet. 6, 299 (1962).
(8) P. J. Creaven, W. H. Davies, and R. T. Williams, Dealkylation of alkoxybiphenyls by trout and frog liver preparations, Life Sci. 6, 105 (1967).
(9) D. R. Buhler and M. E. Rasmusson, The oxidation of drugs by fishes, Comp. Biochem. Physiol. 25, 223 (1968).
(10) J. H. Dewaide, P. T. Henderson, Hepatic N-demethylation of aminopyrine in rat and trout, Biochem. Pharmacol. 17, 1901 (1968).
(11) R. H. Adamson, Drug metabolism in marine vertebrates, Fed. Proc. 26, 1047 (1967).
(12) R. H. Stanton and M. A. Q. Khan, Mixed-function oxidase activity toward cyclodiene insecticides in bass and bluegill sunfish, Pest. Biochem.

Physiol. 3, 351 (1973).
(13) J. R. Bend, R. J. Pohl, and J. R. Fouts, Further studies of the microsomal mixed-function oxidase system of the little skate, *Raja erinacea*, including its response to some xenobiotics, Bull. Mt. Desert Island Biol. Lab. 13, 9 (1973).
(14) M. G. Pedersen, W. K. Hershberger, and M. R. Juchau, Metabolism of 2,4-benzpyrene in rainbow trout (*Salmo gairdneri*), Bull. Environ. Contam. Toxciol. 12, 481 (1974).
(15) J. F. Payne, Field evaluation of benzopyrene hydroxylase induction as a monitor for marine petroleum pollution, Science 191, 945 (1976).
(16) A. Y. H. Lu and W. Levin, The resolution and reconstitution of the liver microsomal hydroxylation system, Biochim. Biophys. Acta 344, 205 (1974).
(17) A. P. Alvares, G. Schilling, W. Levin, and R. Kuntzman, Studies on the induction of CO-binding pigments in liver microsomes by phenobarbital and 3-methylcholanthrene, Biochem. Biophys. Res. Commun. 29, 521 (1967).
(18) R. M. Philpot and J. R. Bend, Benzpyrene hydroxylase activity in hepatic microsomal and solubilized systems containing rabbit or rat cytochrome P-448 or P-450, Life Sci. 16, 985 (1975).

AMINIMIDE POLYMERS AS SUPPORTS FOR AFFINITY CHROMATOGRAPHY, UNDER LOW AND HIGH PRESSURE, OF MICROSOMAL ENZYMES*

J. C. M. Tsibris, J. E. Eppert, C. A. Ketchum, W. C. Sherrill, R. D. Safian, and C. M. West

Department of Biochemistry, University of Florida, Gainesville, Florida 32610, U.S.A.

ABSTRACT

Certain aminimide copolymers have been prepared and used to immobilize benzo-[a]pyrene, or N-phenylimidazole or 11-deoxycortisol analogs and thus purify rat liver microsomal enzymes such as cytochrome P-448, epoxide hydrase and NADH-ferricyanide reductase. These polymers have some hydrophobic properties since they bind non-ionic detergents, steroids, polycyclic hydrocarbons, and polyamines but they do not have any appreciable anion-exchange properties. Since these polymers are initially soluble in organic solvents, they can be coated on special glass beads and show great promise as supports for high-pressure affinity chromatography of biomolecules.

INTRODUCTION

Affinity chromatography on Sepharose and Sephadex polymers has become a very useful method to purify biomolecules.(1) To our knowledge, microsomal enzymes which metabolize hydrophobic molecules such as carcinogens and other xenobiotics have not been purified by substrate or inhibitor affinity polymers (Yasukochi and Masters, this Symposium, report purification of P-450

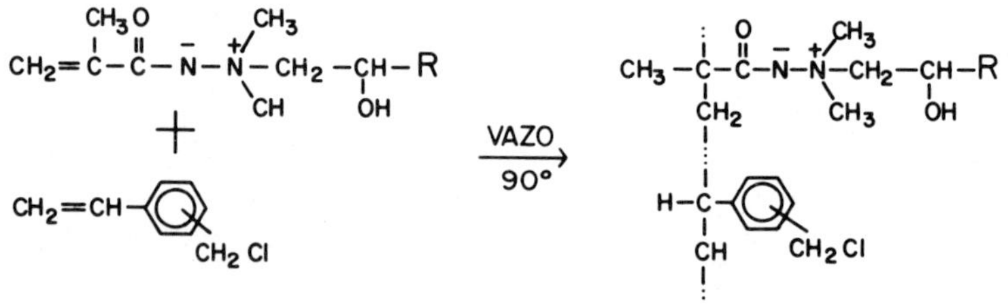

Scheme I. Preparation of "precursor" aminimide copolymers (AID-CH$_2$Cl).

---

*Supported by grant No. CA-17617 from National Cancer Institute and Research Career Development Award (GM-42386) to J.C.M.T.

†Present address: Biology Department, California Institute of Technology, Pasadena, CA.

reductase on ADP-Sepharose). Cytochrome P-450 has been partially purified on Sepharose containing n-octylamine as first reported by Imai and Sato(2) and to homogeneity by other multistep procedures (3,4,5). CNBr-activated Sepharose has anion-exchange properties due to the formation of positively charged isourea groups (6) and is not ideally suited for affinity chromatography of such acidic microsomal proteins. In addition to enzyme purification, this new class of affinity support may be useful in the isolation of different functional forms of microsomal enzymes and may improve our knowledge of their arrangement and interactions in vivo.

MATERIALS AND METHODS

Aminimide copolymer (Scheme I) and the [^{14}C]benzo[a]pyrene(BP) analog (Fig.1A) were prepared as described by Tsibris et al (7). The N-phenylimidazole analog (Fig. 1B) was prepared from p-fluorobenzonitrile and [2-^{14}C]imidazole (8) and subsequent reduction with Vitride (Eastman Kodak Co.) (9). The 11-deoxycortisol derivative (Fig. 1C) was prepared from [^{14}C]-11-deoxycortisol by first protecting the 17-substituents with formaldehyde, alkylation (10) with 6-bromocapronitrile, protection of the 3-keto group with ethanol-thiethyl orthoformate, reduction with LiAlH$_4$ and hydrolysis (11). Attachment of these amine analogs to the AID-8-CH$_2$Cl ("precursor") polymer was done in benzene (7) and addition of monoethanolamine to react with excess "benzyl" chloride groups gave a polymer insoluble in organic solvents and wettable in water.

To prepare high-pressure supports the AID-8-CH$_2$Cl polymer which was reacted with desired amount of compound A (Fig. 1) was dissolved in chloroform and mixed for 2-3 hours with Zipax glass beads (2-3% polymer to beads) in a rotating flask. The solvent was removed under reduced pressure, the beads were gently pressed with mortar and pestle to give a fine powder, and were added to a stirred solution of diethyl ether containing excess monoethanolamine and triethylamine. After stirring overnight, the beads were washed with organic solvents, buffer, ether, and dried. The washed dry beads were packed in a stainless steel column (0.8 x 43 cm), which was connected to a DuPont 830 high-pressure liquid chromatograph. The temperature was kept at 8° with circulating water, the inlet pressure was 150-200 psi and flow rates of 0.5-1.0 ml·min^{-1} were attained; samples and eluting buffers (all in 50mM potassium phosphate pH 7.4 containing 20% glycerol) were applied from a specially designed reservoir.

Microsomes from Long-Evans male rats, induced with 3-methylcholanthrene (3-MC), were sonicated and passed through Sephadex G-10 column (7) before extraction with cholate (12). The cholate extract was applied on the affinity columns and 10-13 ml fractions were collected. Enzyme activities (Fig. 2) were measured by standard methods (7), protein by a modification of the Lowry method (13), epoxide hydrase with styrene oxide (14) and UDP-glucuronosyl transferase with p-nitrophenol by a modification of the method of Henderson and Kersten (15).

RESULTS AND DISCUSSION

Aminimide copolymer AID-8 (Scheme 1) containing equimolar amounts of the aminimide and chloromethylstyrene moieties has been prepared (7) and used to immobilize three amine analogs (Fig. 1) of a substrate (A), inhibitor (B) (8), and a possible effector (C) of cytochrome P-448 monooxygenase from rat liver.

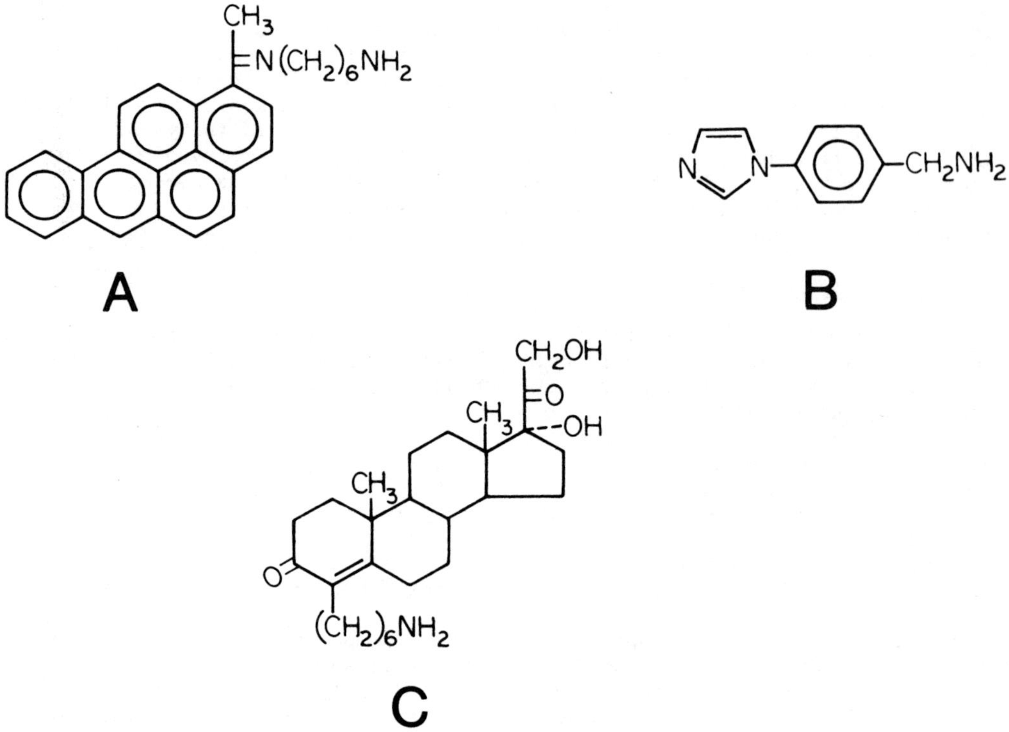

Figure 1. Amine analogs of benzo[a]pyrene (A), N-phenylimidazole (B), and 11-deoxycortisol (C), coupled to aminimide polymers.

Figure 2A shows the elution of 64 nmoles P-448, applied to a 1.5 x 13 cm a-minimide column which contained 6 μmoles of the benzo[a]pyrene analog per g dry resin (1 g gives 4 ml wet resin). P-420 was eluted with buffer (50mM potassium phosphate pH 7.4 in 20% glycerol) slightly ahead of 25-30% of the applied P-448. Most of cytochrome c reductase and 50% of cytochrome b_5 reductase are also eluted with buffer (Fig. 2B) as well as 20-30% of applied epoxide hydrase and 80-90% of UDP-glucuronosyl transferase (data not shown).

KCl elutes very little P-448 demonstrating the expected negligible anion-exchange properties of the resin since the pK_a's of the anionic aminimide nitrogen, of the "benzylimine" groups, and of the imino group on the benzo[a]pyrene analog (Fig. 1A) are expected to be 3.5-5.7 (7) for the former and below 7 for the others (16,17).

Triton N-101 in buffer elutes very pure cytochrome b_5 reductase (18) and epoxide hydrase (ca. 20% of applied), where a similar non-ionic detergent Renex-690 (3) in buffer, elutes in two peaks the remaining P-448 (specific content: 8-13 nmoles P-448 per mg protein). Thus, with a single passage through the AID-8-BP column we can achieve more than a 10-fold purification of P-448 from microsomes at an overall yield of 17% (although the recovery off the AID-8-BP column is quantitative, ca. 75% of microsomal P-448 is lost prior to affinity chromatography). The buffer elution of P-448 is not due to overloading of the column and reveals a different form of P-448 which upon storage at -20° loses its activity toward benzo[a]pyrene oxygenation and

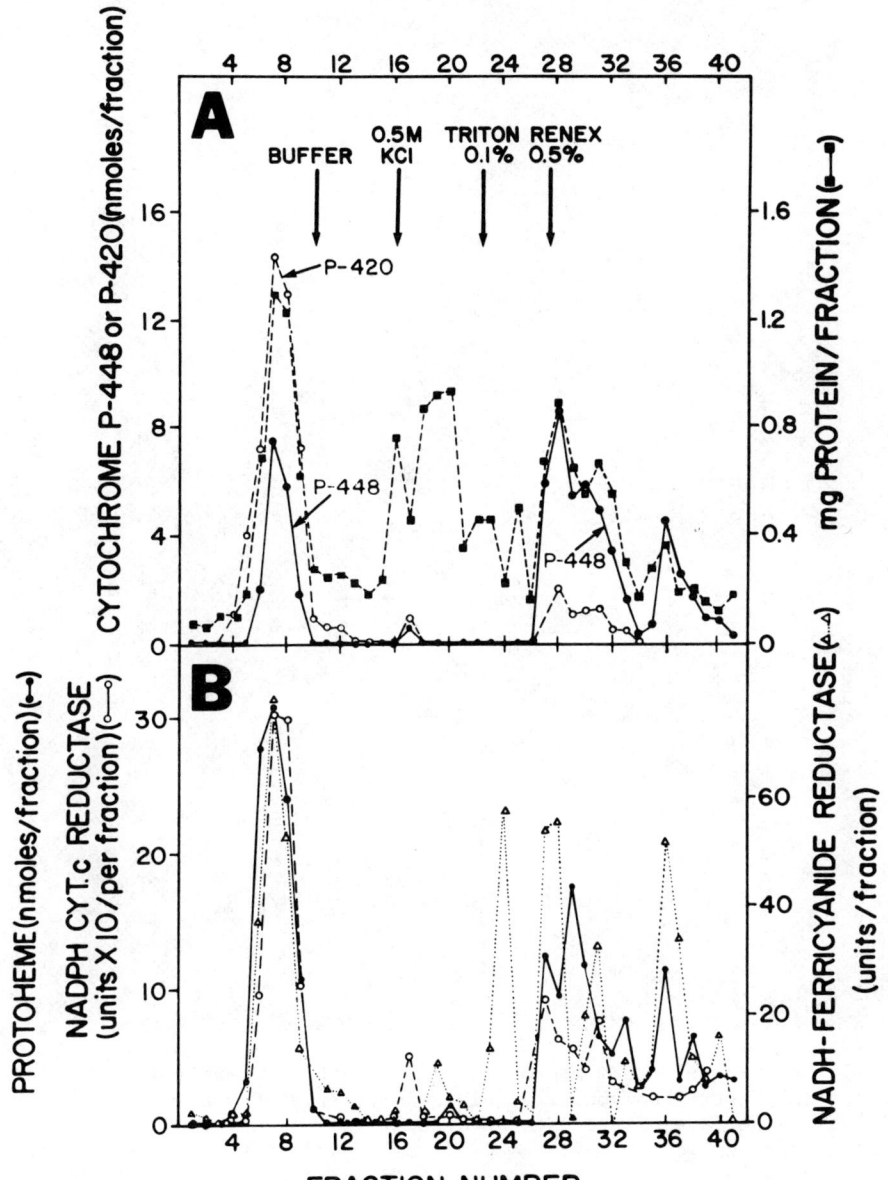

Figure 2. Elution of some microsomal enzymes from the AID-8-BP column. A and B show results from the same experiment.

the main peak of its remaining $Fe^{++} \cdot CO$ spectrum shifts to 450nm. The Renex-eluted P-448 is also active with BP and much more stable during storage.

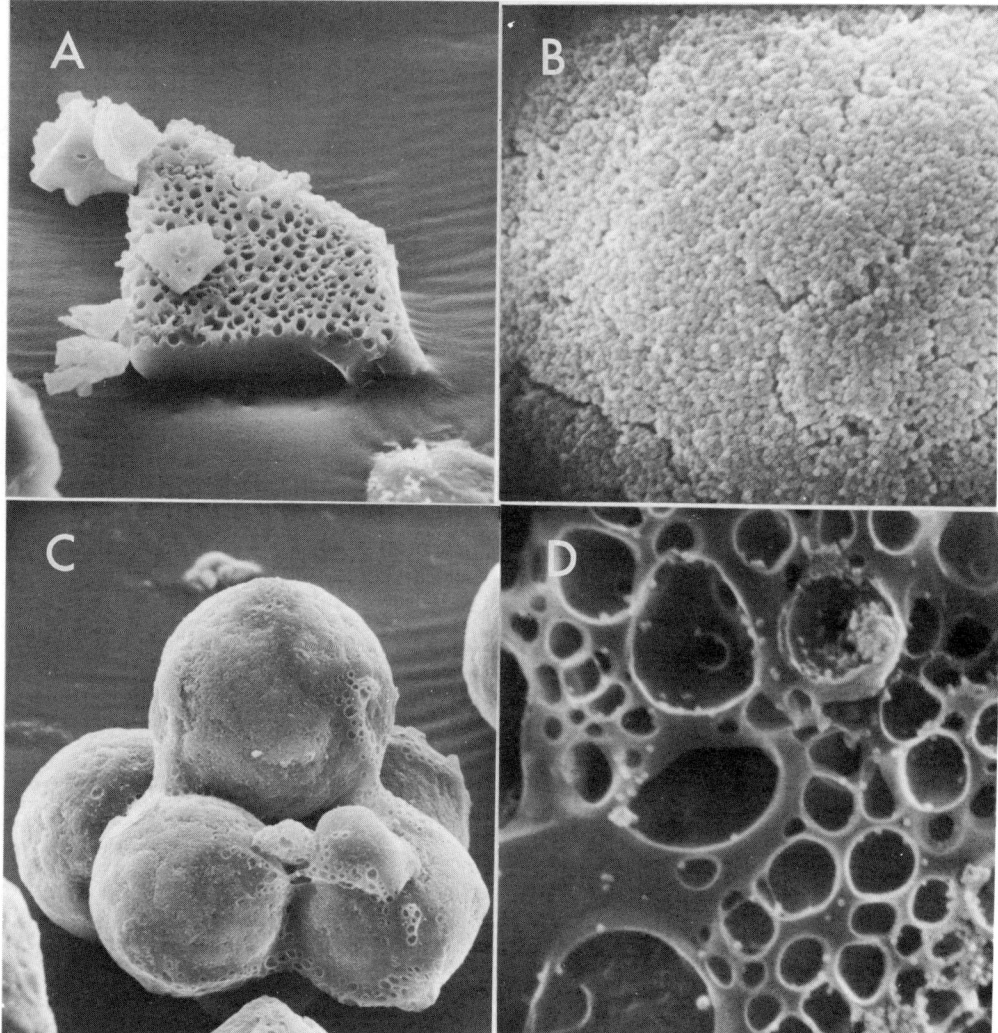

Figure 3. Scanning electrom microscope photographs of aminimide affinity supports. (A) Control polymer i.e. AID-8-CH$_2$Cl treated with monoethanolomine only, (B) surface of the untreated Zipax glass bead (DuPont Co.) showing smaller spheres (ca. 0.2-0.5 μm diameter) fused on the bead to increase surface area, (C) an aggregation of several Zipax beads coated with the AID-8-BP polymer (average bead diameter is 30 μm), and (D) surface of a Zipax bead coated with AID-8-BP. Magnification of B and D is five times greater than A and C. (Photos courtesy of Dr. Stanley R. Bates).

Renex also elutes the above reductases (Fig. 2B), epoxide hydrase and UDP-glucoronsyl transferase. It is possible that different forms of the last two enzymes are separated since a multiplicity of activity peaks are eluted by KCl and detergents.

A similar elution pattern of P-448 was observed with another AID-8 polymer containing 2 μmoles per g dry resin of the 11-deoxycortisol analog (Fig. 1C). Buffer eluted 75% of the applied P-448 and Triton N-101 25%; the latter peak had 13-16 nmoles P-448 per mg protein. Further work with this and the N-phenylimidazole column is in progress.

When P-448 is applied to a control column, made by reacting AID-8-CH_2Cl with monoethanolamine only, approximately 40% of it is eluted with buffer but the remaining is denatured to P-420. Thus the substrate (BP) seems to stabilize P-448 on the AID-8-BP column. Pure NADH-ferricyamide reductase is eluted from this column with Triton N-101. In aqueous media, both control and AID-8-BP columns bind non-ionic detergents, 17β-estradiol, spermine and sperminidine, but not putrescine, and a variety of polycyclic hydrocarbons.

Figure 3 shows scanning electron microscope photographs of the control polymer (A), of the surface of a Zipax bead (B), of AID-8-BP-coated glass beads (C) and of the surface of a coated Zipax bead (D). The beads are uniformly coated with the polymer as shown also by fluorescence microscopy, and the coat does not "peal-off" after washing with organic solvents or after an elution cycle as shown on Fig. 2. Preliminary high-pressure chromatography experiments of P-448 on the AID-8-BP-coated beads show quantitative recovery of P-448 and elution patterns similar to those under gravitational flow.

Because of the chemical stability and compatibility of the polymers with organic solvents, it is easy (synthetically) to modify the components of the polymers to change their general hydrophobic properties and/or the affinity of the enzyme for the immobilized "ligand". Moreover, copolymerization of detergent aminimides (now in progress) may allow affinity chromatography of enzymes from intact microsomes or whole cells.

REFERENCES

1. Jacoby, W. B. and Wilchek, M., eds., (1974), Methods in Enzymology, 34, (B), 1-755.
2. Y. Imai and Sato R., (1974) J. Biochem. 75, 689-697.
3. Haugen, D. A., van der Hoeven, T. A., and Coon, M. J., (1975) J. Biol. Chem. 250, 3567-3570.
4. Ryan, D., Lu, A. Y. H., Kawalek, J., West, S. B., and Levin, W., (1975) Biochem. Biophys. Res. Commun. 64, 1134-1141.
5. Hashimoto, C., Imai, Y., (1976), Biochem. Biophys. Res. Commun. 68, 831-827.
6. Nishikawa, A. H. and Bailey, P., (1975), Arch. Biochem. Biophys. 168, 574-584.
7. Tsibris, J. C. M., Ketchum, C. W., Ketchum, C. A., and Safian, R. D. and Williams, A. G., (1976) submitted for publication.
8. Johnson, A. L., Kauer, J. C., Sharma, D. C., and Dorfman, R. I., (1969), J. Med. Chem. 12, 1024-1028.
9. Bazant, B., Capka, M., Cerny, M., Chralovsky, V., Kachloefl, K., Kraus, M., and Malek, J., (1968), Tetr. Lett., 29, 3303-3306.
10. Atwater, N. W. (1960), J. Amer. Chem. Soc., 82, 2847-2452.
11. Beyler, R. E., Hoffman, F., Moriarty, R. M., and Sarett, L. H., (1961), J. Org. Chem., 26, 2421-2425.
12. Lu, A. Y. H., Kuntzman, R., West, S., Jacobson, M., and Conney, A. H., (1972) J. Biol. Chem., 247, 1727-1734.
13. Wang, C. S., and Smith, R. L., (1975) Anal. Biochem. 68, 414-417.

14. Dansette, R. M., Yagi, H., Jerina, D. M., Daly, J. W., Levin, W., Lu, A. Y. H., Kuntzman, R., and Conney, A. H., (1974) <u>Arch. Biochem. Biophys.</u> 164, 511-517.
15. Henderson, P. T., Kersten, K. J., (1970), <u>Biochem. Pharmacol.</u>, 19, 2343-2351.
16. Perrin, D. D., (1965), Dissociation Constants of Organic Bases in Aqueous Solution, Butterworth, S., London, pp. 117-118.
17. Hine, J., Craig, J. C., Underwood, II, J. G., and Via, F. A., (1970) <u>J. Amer. Chem. Soc.</u> 92, 5294-5299.
18. Ozols, J., (1976), Abstracts of 10th International Congress of Biochemistry, Hamburg, p. 238.

SEX DIFFERENCES IN THE PATTERN OF CYTOCHROMES P-450 IN RAT LIVER MICROSOMES

Regine Kahl, Michael Buecker and Karl Joachim Netter

Department of Pharmacology, University of Mainz, D 6500 Mainz, Germany

ABSTRACT

A number of sex differences in the spectral and enzymic properties of rat liver microsomes have been observed which may reflect differences in the population of hepatic cytochromes P 450 of male and female rats:
1. a blue shift in the spectrum of the reduced P 450-CO complex in females as compared to males,
2. lower ΔA_{max} values in the binding of metyrapone to reduced microsomes in females as compared to males,
3. a higher proportion of 2-hydroxylation in the metabolism of biphenyl in females as compared to males,
4. preferential inhibition of ethoxycoumarin deethylation, benzpyrene hydroxylation and biphenyl-4-hydroxylation by α-naphthoflavone in females but by metyrapone in males.

Sex differences are not abolished by a 36 h starvation period but level off after pretreatment of the animals with either 3-methylcholanthrene or phenobarbital.

Since a blue shift in the CO spectrum, high proportion of 2-hydroxylation in the metabolism of biphenyl and preferential inhibition of monooxygenase activity by α-naphthoflavone are also characteristic for animals pretreated with polycyclic hydrocarbons we conclude that similarities exist in the population of hepatic microsomal cytochromes P 450 in untreated female rats and rats of both sexes pretreated with polycyclic hydrocarbons.

INTRODUCTION

Quantitative sex differences in drug metabolism by the rat have been known since many years and have been reviewed extensively (1, 2). However, qualitative sex differences probably reflecting a different population of cytochromes P 450 in rat liver microsomes of male and female rats have also been reported. First evidence was provided by the observation that the affinity of hepatic microsomal cytochrome P 450 for type I substrates was higher in males than in females (3). Later, sex differences in the pH intercept of the reduced ethyl isocyanide spectrum (4), in the sensitivity of benzpyrene hydroxylation to the inhibitor α-naphthoflavone (5) and in the sensitivity of ethoxycoumarin deethylation to the inhibitor metyrapone (6) have been described. Moreover, it has been demonstrated that the proportions of cytochromes P 450 with short half life and cytochromes P 450 with long half life in CO-binding particles obtained from rat liver microsomes are different in males and females (7).

The aim of the present study was to add further evidence to the hypothesis that male and female rats differ in their population of hepatic cytochromes P 450 and to examine if the male and female patterns of spectral and enzymic behaviour of liver microsomes can be compared to the well known patterns which are observed after treatment with either polycyclic hydrocarbons or with phenobarbital.

METHODS

Sprague-Dawley rats of approximately 55 days of age were used. They had free access to Altromin[R] pellet diet except in starvation experiments and to tap water. The state of the estrus cycle was determined and was found to have no influence on the parameters tested.

For 3-methylcholanthrene pretreatment the animals received 3 i.p. injections of 20 mg/kg dissolved in peanut oil in 12 h intervals. The last injection was given 36 h prior to sacrifice of the animals. For phenobarbital pretreatment the animals received 3 i.p. injections of 80 mg/kg dissolved in saline in 24 h intervals. The last injection was given 24 h prior to sacrifice.

Microsomes were prepared as described previously (8) and used immediately after preparation

Spectral measurements were performed in a Perkin Elmer 356 two wavelength double beam spectrophotometer. Metyrapone binding was studied as decribed previously (9).

Arylhydrocarbon hydroxylase activity was determined by a modification of the method of Gielen et al. (10) using a protein concentration of 0.2 mg/ml and a benz(a)pyrene concentration of 50 μM. Ethoxycoumarin deethylation was measured by a modification of a method of Ullrich and Weber (11) using a protein concentration of 1 mg/ml and a substrate concentration of 0.5 mM. Biphenyl hydroxylation was measured as described by Creaven et al. (12) using a protein concentration of 1 mg/ml and a biphenyl concentration of 1.25 mM.

RESULTS

Sex differences in the binding of carbon monoxide and metyrapone to dithionite reduced liver microsomes are demonstrated in Table 1.

TABLE 1   Sex differences in CO and metyrapone binding to reduced rat liver microsomes

	Untreated males	Untreated females
Cytochrome P 450 content (nmoles × mg prot.$^{-1}$)	0.78 ± 0.06	0.62 ± 0.02
λ_{max} of CO complex (nm)	449.4 ± 0.1	448.4 ± 0.1 *
K_s for metyrapone (μM)	29.3 ± 4.4	29.4 ± 5.7
ΔA_{max} of metyrapone complex (per nmole P 450)	0.048 ± 0.005	0.023 ± 0.004 *

Values are means ± S.E.M. (n = 4 - 7)
* p against males < 0.001

It is obvious from the table that there is a marked difference of 1 nm in the location of the absorbance maximum of the CO complex. The peak is found at 448.4 nm in the females and thus is close to the peak found to be located at 447.0 - 447.5 nm in MC-stimulated microsomes.

Differences are also found in the binding of metyrapone to the reduced cytochromes. While the affinity for this ligand is virtually the same in males and females the maximal spectral changes elicited by the ligand are twice as high in the males than in the females.

Figure 1 shows the total activity of hepatic microsomal ethoxycoumarin deethylase and its inhibition by α-naphthoflavone (ANF) and metyrapone (MP) in males and females.

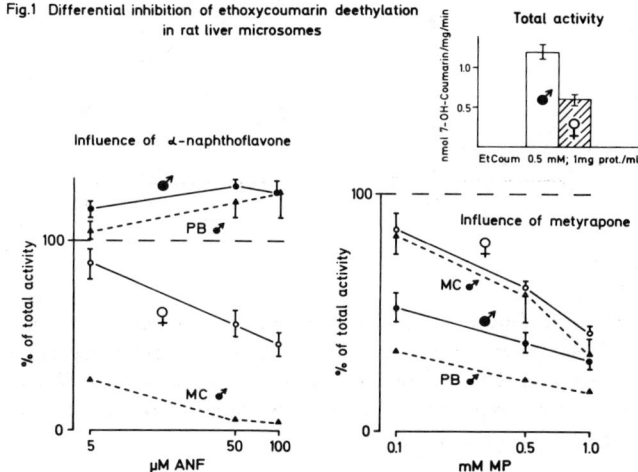

Fig.1 Differential inhibition of ethoxycoumarin deethylation in rat liver microsomes

While the female shows the usually observed lower monooxygenase activity it seems from the data on preferential inhibition that this activity in the female has similar properties to the MC-induced activity. MC-induced activity is effectively blocked by the addition of ANF, while both the phenobarbital (PB)-induced activity and the activity found in microsomes of untreated males are not inhibited but even slightly enhanced by this agent. In contrast, the ethoxycoumarin deethylation in microsomes from untreated females is clearly inhibited by ANF although not as effectively as that in MC-stimulated microsomes. Thus, PB-treated males and untreated males seem to possess one species of ethoxycoumarin deethylase and MC-treated males and untreated females seem to possess another species of ethoxycoumarin deethylase.

This is further supported by the data obtained with MP (right part of Fig. 1). While both untreated females and MC-treated males show a rather low sensitivity of ethoxycoumarin deethylation to MP, this reaction is readily inhibited by MC both in untreated males and PB-treated males.

Figure 2 demonstrates that the sex differences in the inhibition pattern of ethoxycoumarin deethylation are partially lost after induction with either PB or MC. While the PB-treated female retains part of its susceptibility to ANF, differences between males and females in the sensitivity of the reaction to MP can no longer be detected after induction with either PB or MC.

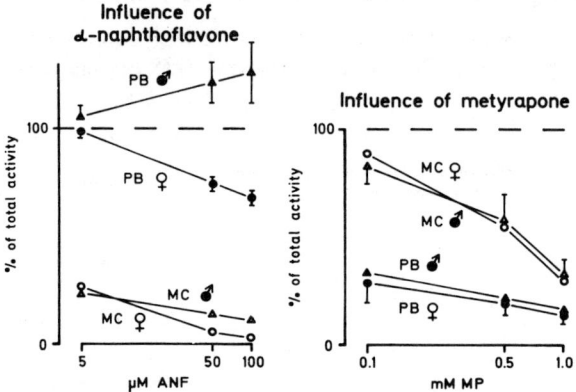

Fig.2 Differential inhibition of ethoxycoumarin deethylation after induction with PB and MC in male and female rats

In search for possible explanations for the differences between males and females one might suspect that females are more susceptible to an inducer of cytochrome P 448 which may occur ubiquitously in the environment of the animals. To test if such a factor is present in the diet starvation experiments were undertaken to remove the suspected factor at least during the last 36 h before death.

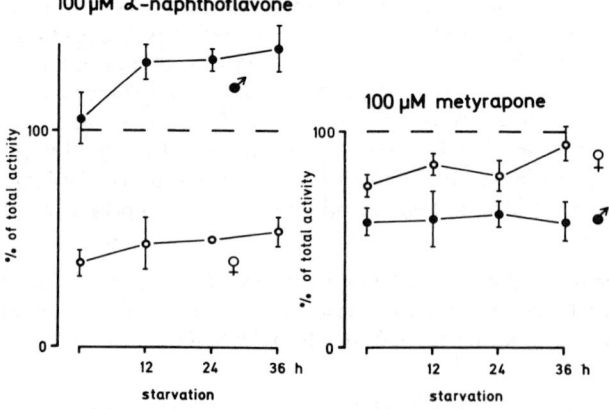

Fig.3 Lack of effect of starvation on sex differences in differential inhibition of ethoxycoumarin deethylation

No influence of this treatment on the sex-dependent inhibition pattern of ethoxycoumarin deethylation was found as shown in Fig. 3 for one selected concentration of ANF or MP. The spectral characteristics of male and female microsomes did also not change. It should, however, be taken into account that the starvation period may by far not have been long enough to effectively counteract the inducing action of a dietary factor.

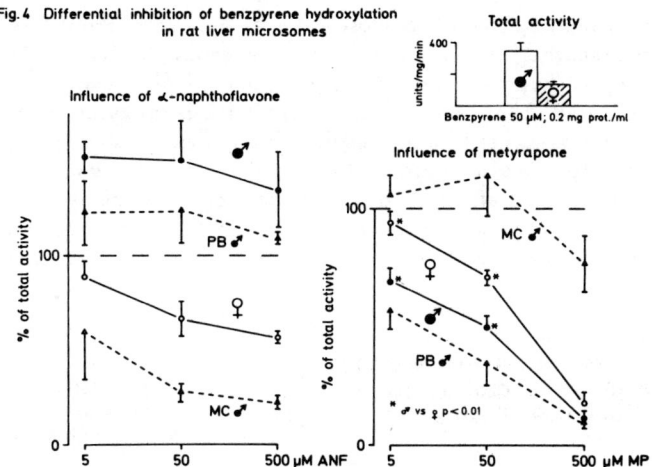

Fig. 4 Differential inhibition of benzpyrene hydroxylation in rat liver microsomes

Similar sex differences as that described in Fig. 1 for ethoxycoumarin deethylation can also be found for benz(a)pyrene hydroxylation (Fig. 4). Here again untreated females and MC-treated males on one hand and untreated males and PB-treated males on the other hand are similar in their response to ANF and to MP.

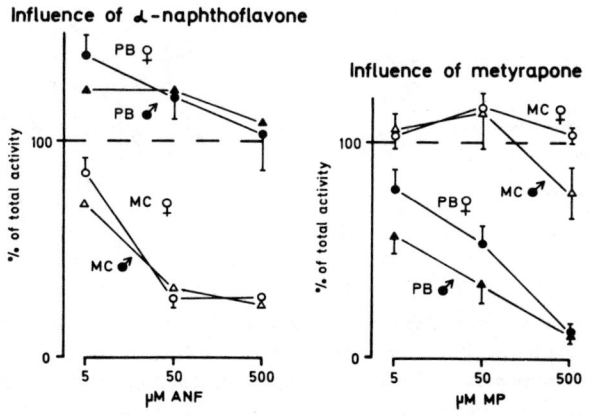

Fig. 5 Differential inhibition of benzpyrene hydroxylase after induction with PB and MC in male and female rats

Fig. 5 shows that in the case of benzpyrene hydroxylation sex differences in the inhibition pattern are completely abolished after induction with MC or with PB.

Sex differences in the sensitivity to ANF can also be observed in biphenyl-4-hydroxylation (Fig. 6). This reaction responds poorly to the inhibitor in microsomes from untreated males and from PB-treated males while it is effectively inhibited in microsomes from MC-treated males. Again, intermediate sensitivity is found in microsomes from untreated females.

The total activity of biphenyl-4-hydroxylase is lower in females than in males (Fig. 7) as could be expected from the findings on other monooxygenase activities. This is, however, different for biphenyl-2-hydroxylation. No significant sex difference in the total activity of this enzyme was found. Therefore, the proportion of 2-hydroxylation in the oxidative metabolism of biphenyl is rather high in females, a metabolic pattern which although in a more pronounced manner, is also achieved by pretreatment of the animals with polycyclic hydrocarbons. The lower part of Fig. 7 gives the ratio of 4-hydroxylation to 2-hydroxylation for different conditions, and again it can be seen that the untreated female is intermediate between the untreated male and the PB-treated male on one hand and the MC-treated male on the other hand.

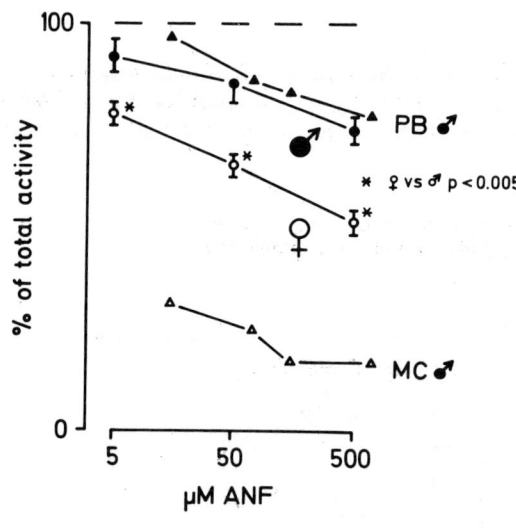

Fig. 6 Sex difference in the response of biphenyl-4-hydroxylation to α-naphthoflavone

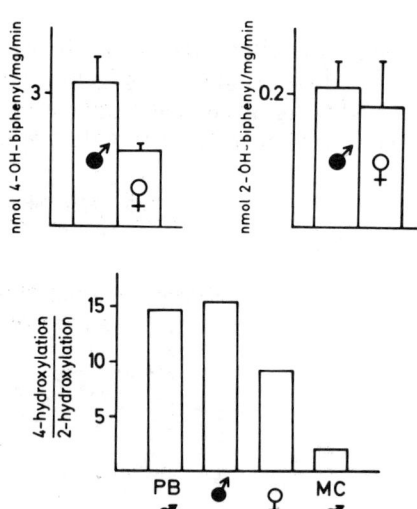

Fig. 7 Sex difference in the ratio of 2-hydroxylation to 4-hydroxylation in the metabolism of biphenyl

DISCUSSION

The results show that a number of differences exist in the spectral and enzymic properties of liver microsomes from male and female rats. These include different peak location in the CO spectrum, different maximal absorbance in metyrapone binding, different response of monooxygenase activity to preferential inhibitors and a different metabolite pattern in the biotransformation of biphenyl.

Moreover, it has been shown that these spectral and enzymic characteristics are virtually the same in microsomes from untreated males and from PB-treated males. In contrast, microsomes from untreated females display a spectral and enzymic pattern which can be compared to that found in MC-stimulated microsomes.

The appearance of certain spectral and enzymic properties as for instance the location of the CO peak at 447 - 448 nm or a preferential inhibition of monooxygenase activity by agents like ANF after pretreatment with polycyclic hydrocarbons is usually taken as indirect evidence for the MC-specific alteration of the cytochrome P 450 population characterized by the predominance of cytochrome P 448. If this interpretation is applied to our findings the conclusion may be drawn that male and female rats differ in their population of hepatic microsomal cytochromes P 450 and that the female possesses a relatively higher proportion of cytochrome P 448 or cytochrome P 448-related proteins.

This is, however, in some contrast to the finding that monooxygenase activities like benzpyrene hydroxylation and ethoxycoumarin deethylation, which are known to be specifically induced by polycyclic hydrocarbons and are therefore thought to be preferentially mediated by cytochrome P 448, show much lower levels in the untreated female than in the untreated male. Moreover, direct evidence for a different composition of the hemoprotein family in liver microsomes of males and females is still lacking since SDS-polyacrylamide gel electrophoresis in our hands fails to lead to the appearance of a distinct cytochrome P 448-related band in female microsomes. Hence, it is not yet clear if the results presented here are indeed indicative for the existence of cytochrome P 448 in the liver microsomes of female rats or if cytochrome P 448-like monooxygenase activity and spectral characteristics are due to a sex-specific arrangement of the microsomal membranes, e.g. differences in phospholipid content and composition. The reason why the female rat exhibits this cytochrome P 448-like monooxygenase activity has not yet been studied. One may speculate that inducers of cytochrome P 448 exist in the environment of the animals, e.g. in the food or in the bedding, to which the female is more sensitive. A 36 h starvation period was not sufficient to allow for the disappearance of the effect of any inducer suspected to be present in the diet.

It has been shown for a number of sex differences in the rat that the level of the circulating sex hormones must play an important role in maintaining these differences (see 2). However, it has also been shown that complete loss of sex differences can only be achieved by treatment of the animals in the newborn period, and a neonatal imprinting has been claimed to determine the cytochrome P 450 pattern of the adult male and female rat (7). From our results it seems unlikely that the actual level of female sex hormones controls the cytochrome P 450 pattern since it was not influenced by the state of the estrus cycle.

One might speculate that the constitutive pattern of monooxygenase activity described here renders the female rat more susceptible to cancer than the male, because the metabolic formation of ultimate carcinogens is assumed to be mediated by cytochrome P 448. It should, however, be noted that the sex differences are at least partially lost by induction. If an animal is exposed to a carcinogen marked induction of cytochrome P 448 will usually occur at least under experimental conditions. The pattern of cytochromes P 450 is then no longer controlled by sex, and therefore sex differences in the sensitivity to carcinogens should not be expected. Examination of the influence of sex on constitutive and induced monooxygenase activity in the target organs of carcinogens, e.g. in lung and intestine, will however be necessary to finally evaluate the role of sex in the susceptibility of the animal to cancer.

ACKNOWLEDGEMENTS

This study was supported by the Deutsche Forschungsgemeinschaft, Bonn-Bad Godesberg, Germany.
The able technical assistance of Mrs. U. Wulff is gratefully appreciated.

REFERENCES

1. A.H. Conney, Pharmacological implications of microsomal enzyme induction, Pharmacol. Rev. 19, 317 (1967)
2. R. Kato, Sex-related differences in drug metabolism, Drug Metab. Rev. 3, 1 (1974)
3. J.B. Schenkman, I. Frey, H. Remmer and R.W. Estabrook, Sex differences in drug metabolism by rat liver microsomes, Mol. Pharmacol. 3, 516 (1967)
4. S. El Defrawy El Masry and G.J. Mannering, Sex-dependent differences in drug metabolism in the rat. I. Temporal changes in the microsomal drug-metabolizing system of the liver during sexual maturation, Drug Metab. Disposition 2, 267 (1974)
5. F.J. Wiebel and H.V. Gelboin, Aryl hydrocarbon (benzo(a)pyrene) hydroxylases in liver from rats of different age, sex and nutritional status. Distinction of two types by 7,8-benzoflavone, Biochem. Pharmacol. 24, 1511 (1975)
6. V. Ullrich, P. Weber and P. Wollenberg, Tetrahydrofurane - an inhibitor for ethanol-induced liver microsomal cytochrome P 450, Biochem. biophys. Res. Commun. 64, 808 (1975)
7. W. Levin, D. Ryan, R. Kuntzman and A.H. Conney, Neonatal imprinting and the turnover of microsomal cytochrome P-450 in rat liver, Mol. Pharmacol. 11, 190 (1975)
8. K.J. Netter, Eine Methode zur direkten Messung der O-Demethylierung in Lebermikrosomen und ihre Anwendung auf die Mikrosomenhemmwirkung von SKF-525-A, Naunyn-Schmiedeberg's Arch. exp. Path. Pharmak. 238, 292 (1960)
9. H.G. Jonen, B. Hüthwohl, R. Kahl and G.F. Kahl, Influence of pyridine and some pyridine derivatives on spectral properties of reduced microsomes and on microsomal drug metabolizing activity, Biochem. Pharmacol. 23, 1319 (1974)
10. J. Gielen, F.M. Goujon and D.W. Nebert, Genetic regulation of aryl hydrocarbon hydroxylase induction. II. Simple mendelian expression in mouse tissues in vivo, J. biol. Chem. 247, 1125 (1972)
11. V. Ullrich and P. Weber, The O-dealkylation of 7-ethoxycoumarin by liver microsomes: A direct fluorimetric test, Hoppe-Seyler's Z. physiol. Chem. 353, 1171 (1972)
12. P.J. Creaven, D.V. Parke and R.T. Williams, A fluorimetric study of the hydroxylation of biphenyl in vitro by liver preparations of various species, Biochem. J. 96, 879 (1965)

CHARACTERIZATION OF MULTIPLE FORMS OF HIGHLY PURIFIED CYTOCHROME P-450 FROM THE LIVER MICROSOMES OF RATS, MICE AND RABBITS

W. Levin, D. Ryan, M.-T. Huang, J. Kawalek,
P. E. Thomas, S. B. West and A. Y. H. Lu

Department of Biochemistry and Drug Metabolism, Hoffmann-La Roche Inc., Nutley, New Jersey 07110

INTRODUCTION

The liver microsomal mixed function oxidase system is a multicomponent system consisting of two proteins, cytochrome P-450 and NADPH-cytochrome c reductase and a lipid, phosphatidylcholine. In the last few years, methods have been developed in several laboratories for the isolation and purification of liver microsomal cytochrome P-450 and NADPH-cytochrome c reductase (1-9). Our laboratory has developed procedures for the purification of multiple forms of cytochrome P-450 from animals treated with different inducers as well as from several different animal species (4,6,8,9). In this paper, we report that cytochromes P-450 and P-448 purified from rats, rabbits and mice are different hemeproteins based on their catalytic, spectral and immunological properties as well as their electrophoretic profile on SDS-polyacrylamide gels. In addition to the purification of multiple forms of cytochrome P-450 from different species treated with different inducers two forms of liver microsomal cytochrome P-450 have been separated and highly purified from $B6D2F_1/J$ mice treated with phenobarbital (8).

METHODS

Male Long-Evans rats weighing 50-60 gm were treated intraperitoneally with sodium phenobarbital (75 mg/kg/day) or 3-methylcholanthrene (25 mg/kg/day) for 4 days. Adult male $B6D2F_1/J$ mice weighing 20-25 gm were maintained on a commercial diet and water containing 0.5 mg/ml of sodium phenobarbital ad libitum for 9 days. 3-Methylcholanthrene was administered subcutaneously to male New Zealand rabbits (3 kg) at a dose of 25 mg/kg/day for 4 days. The purification procedures for cytochrome P-450 and P-448 from rats (4), rabbits (6) and mice (8) have been described elsewhere. The procedures for determining the metabolism of benzphetamine (10), benzo(a)pyrene (11), ethoxycoumarin (10), coumarin (8), zoxazolamine (12) and testosterone (11) by the reconstituted system have been described. Antibodies were prepared in rabbits against purified rat cytochrome P-450 and P-448 (10,13).

RESULTS AND DISCUSSION

The hemeproteins which will be discussed and compared in this report are: 1) cytochrome P-450 purified from phenobarbital-treated rats (designated rat cytochrome P-450), 2) cytochrome P-450 from phenobarbital-treated B6D2F$_1$/J mice (designated mouse cytochrome P-450 A$_2$ and mouse cytochrome P-450 C$_2$), 3) cytochrome P-448 from 3-methylcholanthrene-treated rats and 4) cytochrome P-448 from 3-methylcholanthrene-treated rabbits.

SDS-gel electrophoresis of the various purified preparations of cytochrome P-450 and P-448 reveals that all of the hemeprotein preparations have a major protein component with a minimum molecular weight in the region of 47,000-56,000 daltons (Figure 1). Although the differences in molecular weight are small,

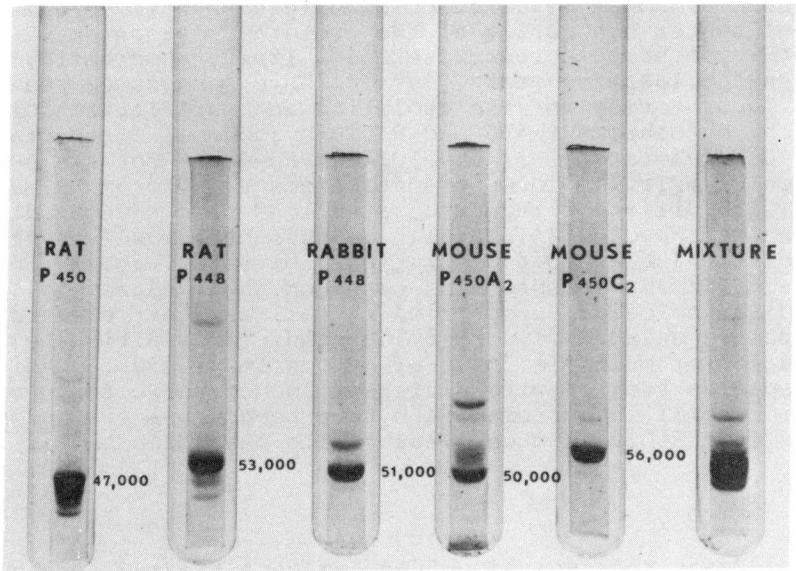

Figure 1. SDS-polyacrylamide gels of purified cytochrome P-450 and P-448 from rats, (4), mice (8) and rabbits (6). The various hemeproteins (10 µg) were treated with SDS and electrophoresis was performed as described by Laemmli (14). In the mixing experiment, 5 µg of each protein was mixed together and subjected to electrophoresis. The gels were stained for protein by the method of Fairbanks et al. (15).

some of the hemeproteins can be separated from each other electrophoretically. The following cytochrome mixtures can be separated on SDS-polyacrylamide gels: 1) rat cytochrome P-450 and P-448 (4), 2) rat cytochrome P-448 and rabbit cytochrome P-448 (6) and 3) mouse cytochrome P-450 A_2 and P-450 C_2 (8). However, as shown in Figure 1 a mixture of all of the purified cytochromes does not yield 5 major protein components. These results suggest that SDS-gel electrophoresis is probably not capable of separating all possible forms of liver microsomal cytochrome P-450.

The cytochrome P-450 preparation from phenobarbital-treated rats (Figure 1) has been shown on SDS-polyacrylamide gels to contain four proteins, all of which are hemeproteins based on their peroxidase activity using tetramethylbenzidine-H_2O_2. Immunological studies with antibody prepared against rat cytochrome P-450 or P-448 (10,13) have also indicated the presence of at least 4 hemeproteins in the rat cytochrome P-450 preparation. These immunological studies as well as studies with SDS-polyacrylamide gels have also revealed the presence of at least two hemeproteins in the rat cytochrome P-448 preparation. Thus, although rat cytochrome P-450 and P-448 are highly purified they are not "homogeneous" since they contain more than a single cytochrome component.

Table 1. Spectral Properties of Highly Purified Cytochrome P-450 and P-448 from Phenobarbital or 3-Methylcholanthrene-Treated Rats, Mice and Rabbits

Parameters	Rat[a]		Rabbit[a]	Mouse[a]	
	P-450	P-448	P-448	P-450 A_2	P-450 C_2
Specific content (nmoles/mg protein)	15-17	18-22	18-22	14-17	15-18
CO-difference spectrum, absorption maximum (nm)	450	447	447	451	450
Ethyl isocyanide difference spectrum:					
455/430 ratio at pH 7.4	0.4	2.0	1.4	0.3	1.1
pH intercept	7.9	6.9	7.3	8.1	7.3
Absolute spectrum, absorption maximum (nm)					
oxidized	418	417	395	417	417
	535	535	414	536	536
	568	568	534	570	568
			566		
			644		
reduced	414	411	410	415	419
	542	542	540	544	544
CO + reduced	450	447	447	451	450
	550	550	551	551	552

[a] Cytochrome P-450 was purified from phenobarbital-treated rats (4) and mice (8). Cytochrome P-448 was purified from 3-methylcholanthrene-treated rats (4) and rabbits (6).

Table 1 summarizes the spectral properties of the purified cytochromes. 3-Methylcholanthrene treatment of rats or rabbits results in a marked shift from 450 nm to 447 nm in the CO-reduced absorption maximum of the cytochrome. These two hemeproteins and cytochrome P-450 C_2 isolated from phenobarbital-treated B6D2F$_1$/J mice have the highest ethyl isocyanide ratios (455/430 nm) at pH 7.4. Interestingly, rat cytochrome P-450 and mouse cytochrome P-450 C_2 have identical CO spectra but differ markedly in their ethyl isocyanide spectral properties, while mouse cytochrome P-450 A_2 and rat cytochrome P-450 have similar ethyl isocyanide spectral properties but differ in their CO spectra (Table 1). All of the various isolated forms of cytochrome P-450 and P-448 have very similar absolute oxidized spectra with the exception of rabbit cytochrome P-448 which exists as a mixture of high- and low-spin forms. As was seen in the CO-reduced spectra of purified cytochrome P-448 from rats and rabbits, the absolute reduced spectral maxima of these hemeproteins also show the greatest shift to the blue (410-411 nm). The high absorption maximum of mouse cytochrome P-450 C_2 in the reduced spectrum (419 nm) is unusual but apparently it is not due to the presence of contaminating cytochrome b_5 or cytochrome P-420.

Table 2. Substrate Specificity of Highly Purified Cytochrome P-450 and P-448 Isolated from Rats, Rabbits and Mice

Substrate	Rat[a]		Rabbit[a]	Mouse[a]	
	P-450	P-448	P-448	P-450 A_2	P-450 C_2
	(nmoles/min/nmole hemeprotein)				
Benzphetamine	52	2.5	1.0	5	8
Benzo[a]pyrene	0.19	3.9	0.1	0.26	0.08
Ethoxycoumarin	4.1	56	N.D.[b]	3.7	0.3
Coumarin	0.00	0.00	0.02	0.13	0.02
Zoxazolamine	1.4	9.4	1.5	4.3	1.4
Testosterone					
6β-OH	0.12	0.25	N.D.[b]	0.01	0.88
7α-OH	0.69	0.97	N.D.[b]	0.73	0.15
16α-OH	1.50	0.17	N.D.[b]	0.03	0.06

[a] Cytochrome P-450 was purified from phenobarbital-pretreated rats (4) and mice (8). Cytochrome P-448 was purified from 3-methylcholanthrene-pretreated rats (4) and rabbits (6).

[b] N.D. = not determined.

A comparison of the substrate specificities of the various cytochrome P-450 preparations shows that rat cytochrome P-448 has the highest activity for benzo(a)pyrene and zoxazolamine hydroxylation (Table 2). Benzphetamine N-demethylation is 6-10 times higher with rat cytochrome P-450 than with mouse cytochromes P-450 A_2 or P-450 C_2. The cytochrome P-448 preparations from rats and rabbits have very low benzphetamine N-demethylase activity. Rat cytochrome P-450 has the highest activity for the 16α-hydroxylation of testosterone while mouse cytochrome P-450 C_2 has the highest activity for the 6β-hydroxylation of testosterone. Rat cytochromes P-450 and P-448 and mouse cytochrome P-450 A_2 all have good catalytic activity toward the 7α-hydroxylation of testosterone. These results clearly show the marked differences in substrate specificity of different forms of cytochrome P-450 although all of the cyrochrome preparations have some activity towards most substrates. Two striking observations are the different substrate specifities of mouse cytochromes P-450 A_2 and P-450 C_2 (two highly purified cytochromes isolated from the same animal) and the poor overall catalytic activity of rabbit cytochrome P-448 toward all substrates tested. However, Atlas et al. (16) have recently shown that the liver microsomal N-hydroxylation of 2-acetylaminofluorine is markedly enhanced by pretreatment of rabbits with 3-methylcholanthrene, emphasizing the need to investigate the catalytic activity of purified cytochrome P-450 and P-448 preparations with a large number of substrates to determine if a particular form of cytochrome P-450 or P-448 has a unique substrate specificity.

Antibodies produced in rabbits against purified rat liver cytochrome P-450 and P-448 are highly specific (10,13). As seen by double diffusion analysis (Figure 2), the immune γ-globulin produced against rat cytochrome P-450 or P-448 forms a single distinct precipitin band with its homologous antigen. On the other hand, both antibodies react poorly and incompletely with their heterologous antigens. When a high concentration of purified rat liver cytochrome P-450 cross reacts with anti-cytochrome P-448, three weak precipitin bands are observed (Figure 2), all of which are hemeproteins based on their peroxidase activity using tetramethylbenzidine-H_2O_2. However, the low intensity of these precipitin bands indicates that not all of the hemeprotein is reacting with the antibody. Using both antibodies in the same immunodiffusion plate (data not shown), there is at least one form of cytochrome P-450 that does not cross react with anti-cytochrome P-448 but does react with anti-cytochrome P-450. Similarly, using both antibodies, there is at least one hemeprotein in the rat cytochrome P-448 preparation that anti-rat cytochrome P-450 does not precipitate but which is precipitated by anti-rat cytochrome P-448. These results indicate the presence of at least 4 forms of cytochrome in the rat liver cytochrome P-450 preparation and 2 forms of cytochrome in the rat liver cytochrome P-448 preparation (13). The differential effects of these antibodies on the catalytic activity of the reconstituted monoxygenase system also suggest a multiplicity of cytochrome P-450's in the purified preparations obtained from phenobarbital- and 3-methylcholanthrene-treated rats (10,13).

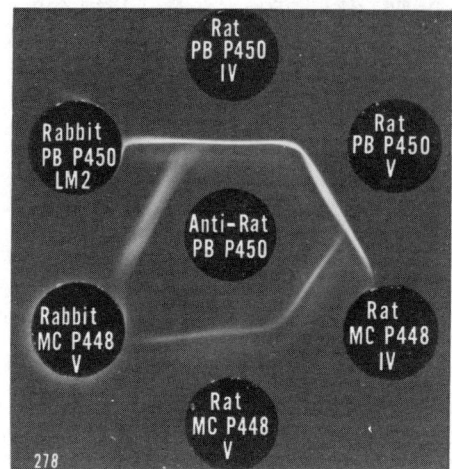
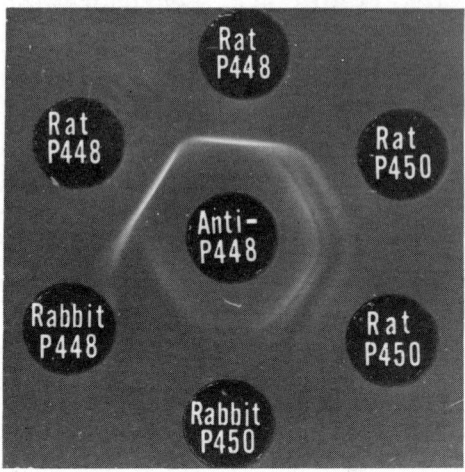
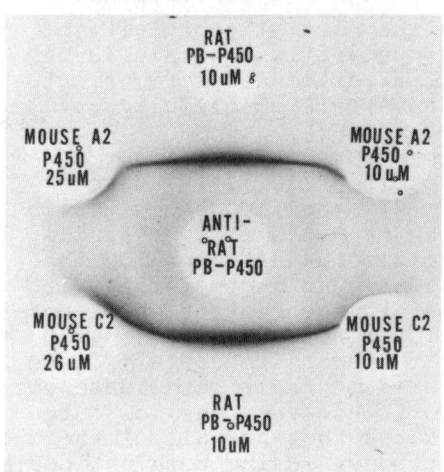
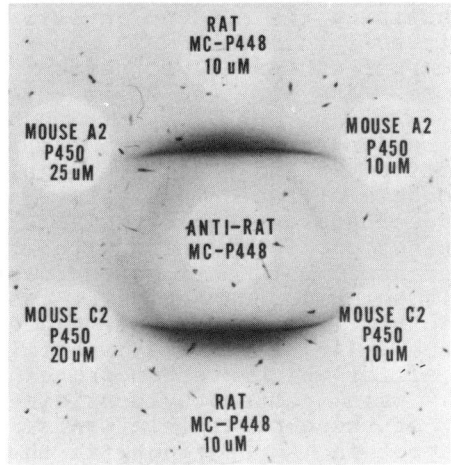

Figure 2. Ouchterlony immunodiffusion plates with anti-rat cytochrome P-450 or anti-rat cytochrome P-448 in the center well. <u>Upper left</u>. Clockwise from the top, wells 1-5 contained 20 μM of the indicated hemeproteins purified through Step IV or V. Well 6 contained 14 μM rabbit cytochrome P-450 LM$_2$ (5). <u>Upper right</u>. Wells 1-6 contained the indicated hemeproteins at a concentration of 10, 37, 28, 33, 30 and 10 μM, respectively. <u>Lower left</u> and <u>lower right</u>. The immunoprecipitin bands were stained for peroxidase activity (17).

The specificity of these antibody preparations is also demonstrated by the poor cross reactivity of mouse cytochromes P-450 A_2 and P-450 C_2 and rabbit cytochrome P-448 and P-450 LM_2 (5) with the rat antibodies (Figure 2).

In summary, the present report describes the characterization of highly purified cytochrome P-450 and P-448 from rats, mice and rabbits. These highly purified cytochrome P-450 and P-448 preparations differ in their spectral, catalytic and immunological properties as well as in their minimum molecular weights on SDS-polyacrylamide gels.

Acknowledgements

We thank Ms. Candace Caso for her assistance in the preparation of this manuscript.

References

1. Imai, Y., and Sato, R., Biochem. Biophys. Res. Commun. 60, 8-14 (1974).
2. van der Hoeven, T. A., Haugen, D. A., and Coon, M. J., Biochem. Biophys. Res. Commun. 60, 569-575 (1974).
3. Dignam, J. D., and Strobel, H. W., Biochem. Biophys. Res. Commun. 63, 845-852 (1975).
4. Ryan, D., Lu, A. Y. H., Kawalek, J., West, S. B., and Levin, W., Biochem. Biophys. Res. Commun. 64, 1134-1141 (1975).
5. Haugen, D. A., van der Hoeven, T. A., and Coon, M. J., J. Biol. Chem. 250, 3567-3570 (1975).
6. Kawalek, J. C., Levin, W., Ryan, D., Thomas, P. E., and Lu, A. Y. H., Mol. Pharmacol. 11, 874-878 (1975).
7. Philpot, R. M., and Arinc, E., Mol. Pharmacol. 12, 483-494 (1976).
8. Huang, M.-T., West, S. B., and Lu, A. Y. H., J. Biol. Chem. 251, 4659-4665 (1976).
9. Ryan, D., Thomas, P. E., Lu, A. Y. H., West, S., and Levin, W., The Pharmacologist 18, 241 (1976).
10. Thomas, P. E., Lu, A. Y. H., Ryan, D., West, S. B., Kawalek, J., and Levin, W., J. Biol. Chem. 251, 1385-1391 (1976).
11. Lu, A. Y. H., Kuntzman, R., West, S., Jacobson, M., and Conney, A. H., J. Biol. Chem. 247, 1727-1734 (1972).
12. Tomaszewski, J. E., Jerina, D. M., Levin, W., and Conney, A. H., Archiv. Biochem. Biophys., in press (1976).
13. Thomas, P. E., Lu, A. Y. H., Ryan, D., West, S. B., Kawalek, J., and Levin, W., Mol. Pharmacol. 12, 746-758 (1976).
14. Laemmli, U. K., Nature 227, 680-685 (1970).
15. Fairbanks, G., Steck, T. L., and Wallach, D. F. H., Biochemistry 10, 2606-2617 (1971).
16. Atlas, S. A., Thorgeirsson, S., Boobis, A. R., Kumaki, K., and Nebert, D. W. Biochem. Pharmacol. 24, 2111-2116 (1975).
17. Thomas, P. E., Ryan, D., and Levin, W., Anal. Biochem. 75, 168-175 (1976).

THE MECHANISM OF CYTOCHROME P450 ACTION

V. Ullrich

Department of Physiological Chemistry, University of the Saarland, 665 Homburg-Saar, GFR

INTRODUCTION

The elucidation of enzyme mechanisms is generally considered to be a rather academic task, since direct application of the results to problems in chemistry, pharmacology or medicine is not readily discernible. However, studies involving the mechanisms of drug oxidations have been an exception. They provide the theoretical basis for pharmacokinetics and for the stimulation or impairment of drug elimination by other drugs. More recently the field of toxicology and especially that part dealing with chemical carcinogenesis has profited greatly from research on the basic mechanisms of drug metabolism. The catalysis involved is so fascinating that chemists have considered developing model systems which could lead to new technical oxidation processes using molecular oxygen. Thus the combined efforts of organic and inorganic chemists as well as biochemists, physicists, pharmacologists and microbiologists have promoted our understanding of drug oxidations to such an extent that the main outlines of the corresponding enzyme mechanisms can be presented today.

It has now been 20 years since the membranes of the endoplasmic reticulum of the liver cell have been shown to contain a monooxygenase system that could oxidize lipophilic drugs and foreign compounds by the introduction of an oxygen atom derived from molecular oxygen. Besides liver, other organs like small intestine, lung, skin or kidney were also found to contain monooxygenase activities but to date only liver microsomes have been used for mechanistic studies. A major breakthrough occurred with the isolation and reconstitution of the monooxygenase system (1). Two proteins were found to be required for the enzymatic activity: a NADPH-dependent flavoprotein and a hemoprotein called cytochrome P450 because of the unusual position, at 450 nm, of the absorption band of its carbon monoxide complex. Together with phospholipids these two proteins fully reconstituted the microsomal monooxygenase activity towards a variety of drugs (2). With improved chromatographic separation it has recently been shown that different forms of cytochrome P450 exist (3), as had been postulated earlier by using different inducers of drug monooxygenase activity (4). Each species of cytochrome P450 seems to exhibit a slightly different but overlapping substrate specificity. So far up to six or seven forms have been detected in rat liver (5).

The question arises why nature has developed several enzymes with a certain degree of specificity instead of one completely unspecific enzyme. The answer is speculative but two advantages of several species become immediately apparent. With increasing specificity the affinity of an enzyme towards its substrates also increases and high affinity is required to efficiently remove low levels of toxic environmental compounds like polycyclic hydrocarbons from the body. A second advantage concerns the regulation of enzyme synthesis by specific inducers. It is more economical for the cell to rapidly build up a small steady-state concentration of an enzyme which effectively metabolizes a drug present in the body, instead of continuously synthesizing large amounts of a less efficient enzyme.

Fortunately, the mechanism of action seems to be identical for all cytochrome P450 species since only the heme iron in its obviously unusual coordination is involved in the activation process of the oxygen molecule. All mechanistic considerations therefore have to start with an elucidation of the structure of cytochrome P450.

THE STRUCTURE OF CYTOCHROME P450 AND ITS REACTIONS WITH SUBSTRATES

Cytochrome P450 has unique spectroscopic properties which can be simulated completely in model systems consisting of heme complexes to which one mercaptide ligand has been attached (6,7). Therefore, in the case of oxidized cytochrome P450 the fifth coordination position is likely to be occupied by the S^--group of a cysteinyl residue. In the absence of substrate there is also a sixth ligand of unknown structure present. This six-coordinated oxidized cytochrome P450 has a characteristic low-spin epr signal which is converted to a high-spin signal after addition of substrates. The transformation has been studied in more detail with a cytochrome P450 isolated from camphor-grown bacteria (8,9). In the presence of substrate the Soret absorption shifts from about 418 nm to about 394 nm. This complex corresponds to a five-coordinated heme with one mercaptide ligand which suggests that the sixth ligand was removed during formation of the enzyme-substrate complex. Normally, the substrate is unable to interact electronically with the iron, unless it contains heteroatoms like nitrogen, phosphorus, sulphur, selenium or oxygen. In these cases, six-coordinated, low-spin complexes can be formed with Soret bands in the region between 415 and 470 nm (Table).

TABLE 1   Ligands of Microsomal Cytochrome P450

Ligand		Soret Absorption Band	
		nm (oxidized)	nm (reduced)
$IC \equiv O$	Carbon monoxide	—	445 – 455
$I\overset{\ominus}{C} \equiv \overset{\oplus}{N}-R$	Isocyanides	430	455
$IN{\overset{\displaystyle \nearrow C...}{\searrow C...}}$	Pyridines Imidazoles	420 – 435	445 – 450
INH_2-R	Amines	~425 – 435	445 – 450
$IP{\overset{\nearrow R_1}{\underset{\searrow R_3}{- R_2}}}$	Phosphines	380, 460	459
$^{\ominus}I\underline{S} - R$	Thiols	470	(450)
$ISI{\overset{\nearrow R}{\searrow R_1}}$	Thioethers	435	450
$IOI{\overset{\nearrow R}{\searrow R_1}}$	Ethers, Alcohols	413 – 416	—

After reduction by sodium dithionite or NADPH the Soret bands of these low-spin ligand complexes are centered around 450-460 nm, similar to that of the well-known CO complex. Again the presence of a mercaptide group at the fifth ligand position could be responsible for the unusually far red shifts in these

Soret absorptions. Interestingly, organic thiol compounds form ligand spectra only when their anions are lipophilic enough to overcome the electrostatic repulsion by the negative charges of two mercaptide ligands at the ferrous ion (10). Neither undissociated mercaptans nor alcohols produce ligand spectra at the reduced cytochrome. This is of interest with regard to the still unknown nature of the sixth endogenous ligand of cytochrome P450. On the basis of the ligand spectra in the table it could well be a hydroxyl ligand since the 418 nm Soret absorption of the oxidized cytochrome and the lack of a spectrum of the reduced state would agree with this assumption.

As soon as a lipophilic compound containing a heteroatom interacts with the iron of cytochrome P450 the formation of the high-spin enzyme-substrate complex is prevented and the low-spin ligand complex is formed instead (11). Depending on the strength of the coordinate bonding to the iron and the hydrophobic interaction with the active site such compounds are more or less efficient inhibitors of cytochrome P450 and the monooxygenation process.

THE REDUCTION OF THE CYTOCHROME P450-SUBSTRATE COMPLEX

The formation of the enzyme-substrate complex at cytochrome P450 is the triggering event for the monooxygenation process. Only the enzyme-substrate complex seems to be reduced at a rate fast enough to ensure the observed hydroxylation rate (12,13). It is not yet clear whether a conformational change induced by the addition of substrate and/or an increase in the redox potential is the actual determinant for the initiation of the reduction to the ferrous cytochrome. There is, however, no doubt that the electrons from NADPH are transferred from the FMN-FAD-dependent flavoprotein to cytochrome P450 without participation of an iron-sulphur protein which is required by mitochondrial and bacterial cytochrome P450-dependent monooxygenases. NADH can also serve as an electron donor although with much lower efficiency. The use of antibodies against NADH-cytochrome b_5-reductase and cytochrome b_5 has provided clear evidence that the pathway of NADH electrons involves these two microsomal components and not cytochrome P450 reductase (14) (Fig. 1 a + b).

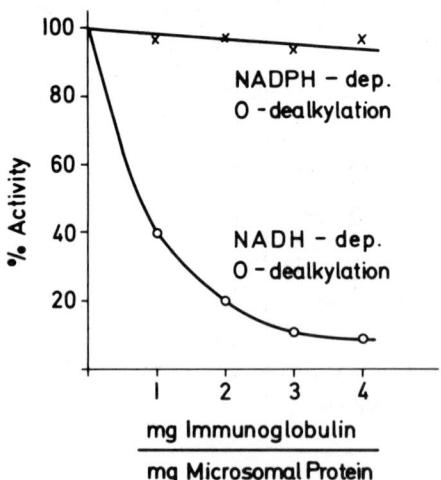

Fig. 1 a. Effect of an antibody against microsomal NADH cytochrome b_5-reductase on the NADPH- and NADH-dependent O-dealkylation of 7-ethoxycoumarin

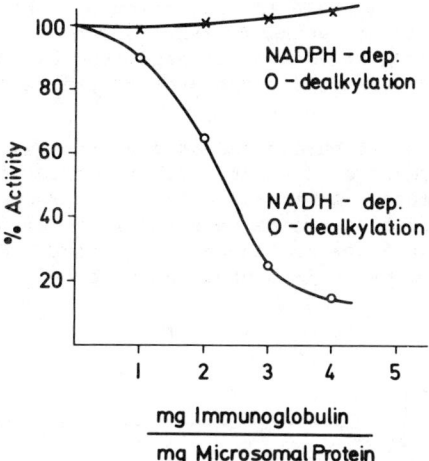

Fig. 1 b. Effect of an antibody against cytochrome b_5 on the NADPH and NADH-dependent O-dealkylation of 7-ethoxycoumarin

THE FORMATION OF THE ACTIVE OXYGEN COMPLEX

The next step in the reaction sequence is the addition of oxygen to the reduced cytochrome. With the bacterial, camphor-hydroxylating system the formation of an oxy complex of the soluble cytochrome P450 has been reported (15), but clear evidence is lacking for the microsomal drug monooxygenase system. Its electronic structure corresponds to that of oxyhemoglobin or Compound III of peroxidase. Such a complex cannot be considered as an active oxygen species unless a second electron from the reductase is transferred to yield the socalled "oxenoid complex". Earlier, in 1967, two possible structures, either $Fe^{III}O_2^{2-}$ or $[FeO]^{3+}$, had been proposed by us for this active oxygen complex (16). Recently, the surprising fact became evident that cytochrome P450 can catalyze monooxygenations without molecular oxygen and NADPH, provided certain oxidizing agents are present. Cumene hydroperoxide has been reported to be such an oxidant (17), but no decision concerning the structure of the intermediate could be made since cumenehydroperoxide could release an oxygen atom as well as a hydroperoxide anion. Looking for an oxidant with only one oxygen atom we found that iodosobenzene can also support cytochrome P450-dependent hydroxylations in anaerobic microsomes (18). Iodosobenzene can obviously act as an oxene donor using cytochrome P450 as an oxene transferase. This finding favors the structure of the active oxygen complex as an $[FeO]^{3+}$ species. The possible electronic structures of this complex can be depicted as follows:

$$-S^- \cdots Fe^{III} + \cdot \overline{O} \cdot \longrightarrow \left[-S^- \cdots Fe^{IV} - \overline{O}^\ominus \right]$$

$$\text{or} \quad \left[-S^- \cdots Fe^{V} = \overline{O}\ ^{2\ominus} \right]$$

The mercaptide ligand in trans position to the catalytically active sixth coordination site may play a dual role in the oxygen activation process. It may be required to facilitate the breaking of the O-O bond in the intermediate peroxide complex by providing an additional negative charge and secondly, it can stabilize the electron deficient oxene atom at the ferric iron by the same process.

According to the possible electronic structures of the oxenoid complex an unpaired electron at the oxygen could exist, although extensive relaxation by the Fe^{IV}-ion could be expected. An investigation by epr spectroscopy revealed that indeed paramagnetic species can be detected when microsomal cytochrome P450 is exposed to oxene donating compounds like cumene hydroperoxide, iodosobenzene, t-butyl hydroperoxide or 3-chloroperbenzoic acid (Fig. 2).

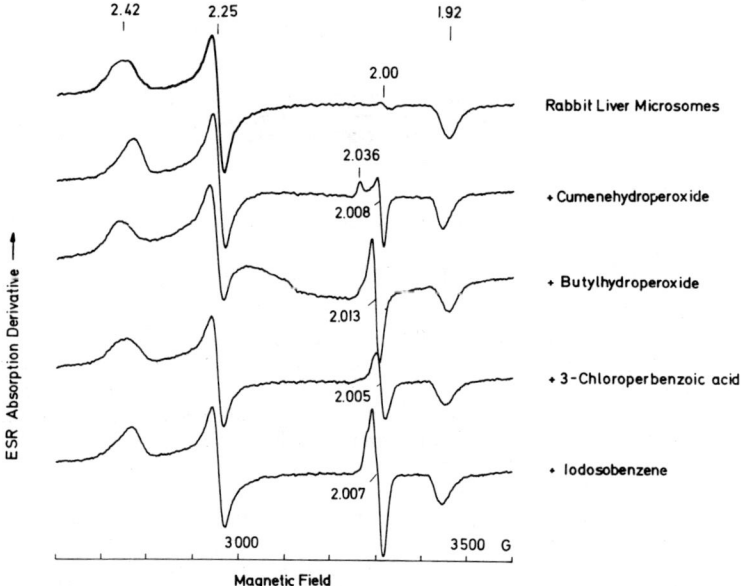

Fig. 2. Epr spectra of rabbit liver microsomes with various oxene donors

The signal with cumene hydroperoxide had been reported earlier (19). It qualitatively differs in shape from the other spectra in Fig. 2. Microsomes in which cytochrome P450 had been destroyed do not produce these epr spectra. A possible correlation between the hydroxylating activity and spectral changes in microsomes induced by various oxene donors is presented in a later section of this book (20).

THE MONOOXYGENATION OF THE SUBSTRATE

As the last step in the reaction sequence the oxygen atom is transferred to the substrate. The formation of hydroxyl groups largely occurs with retention of configuration, indicating that a direct attack of the C-H bond is involved (21). In the case of substrates containing double bonds the formation of epoxides is preferred over insertion into a C-H bond. Aromatic systems form arene oxides which can rearrange to phenols, a process which is accompanied by migration of adjacent substituents ("NIH-shift") (22).

When hydroxylation takes place at the α-carbon of an alkyl group adjacent to a heteroatom, dealkylation is usually observed. If this occurs with a secondary amine the primary product amine may bind as a ligand to cytochrome P450, thereby inhibiting the monooxygenase reaction.

Such an inhibition by the product has been reported when methylenedioxy compounds are used as substrates in microsomal monooxygenations. When these compounds are incubated in the presence of NADPH and oxygen and subsequently sodium dithionite is added, a Soret band at 455 nm in the difference spectrum emerges (23). This complex can be dissociated by light, indicating the ligand nature of this complex (24). A chemical study with a variety of methylenedioxy derivatives indicated that the free methylene group is necessary for the complex formation (D. Mansuy, W. Nastainczyk, V. Ullrich, unpublished) and that a hydroxylation represents the first step of the following postulated reaction sequence:

$$\text{R}_1\text{-benzodioxole-CH}_2 \xrightarrow{[O]} \left[\text{R}_1\text{-benzodioxole-C(OH)H} \right] \longrightarrow \text{Diphenol} + \text{HCO}_2\text{H}$$

$$\text{Fe}^{2+}_{P450} \downarrow -H_2O$$

$$\text{R}_1\text{-benzodioxole-C}\cdots\text{Fe}^{2+}$$

Carbene - Complex

The hydroxylated intermediate is unstable and may either decompose or can be stabilized, after elimination of water, as a carbene complex at reduced cytochrome P450. The high stability of this complex could be the explanation for the synergistic action of methylenedioxy compounds in insecticidal activity.

UNCOUPLED MONOOXYGENATIONS

A special situation arises if the substrate cannot accept the active oxygen to form an oxidized product. This could occur in perfluorinated hydrocarbons where the C-F bonds resist oxidative attack. With these compounds one can observe the phenomenon of uncoupling which is defined as the process of NADPH-oxidation without a concomitant monooxygenation of substrate (25).

Uncoupling may also occur if the stereochemical fixation of the substrate at the active site does not allow an appropriate transition state between the active oxygen and the substrate. This can be considered as a consequence of the broad specificity of the cytochrome P450 enzymes and probably is more the rule than an exception among microsomal monooxygenations. Since the oxene complex is a powerful oxidant it may attack the porphyrin ring or the protein if it is not transferred to a substrate. With 3-chloroperbenzoic acid as the oxene donor, parallel to the monooxygenation reaction a very destruction of cytochrome P450 is observed (Fig. 3).

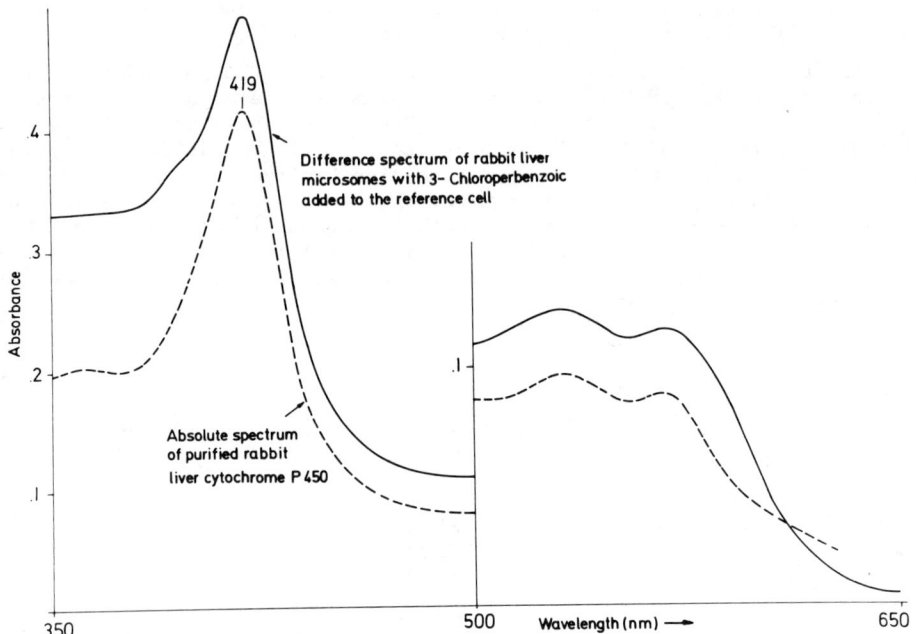

Fig. 3. Difference spectrum of rabbit liver microsomes in presence of 3-chlorobenzoic acid

10^{-3}M 3-chloroperbenzoic acid were added to the reference cell

The observed difference spectrum closely resembles the absolute spectrum of cytochrome P450. The heme chromophore of cytochrome P450, but not of cytochrome b_5, is completely bleached by this treatment indicating that the peracid acts specifically at cytochrome P450. Similarly the uncoupled active oxygen intermediate at cytochrome P450 could lead to its selfdestruction.

Reduction of the oxene complex to water requires two electrons which are provided by cytochrome P450 reductase if NADPH is the sole electron source. However, in the presence of NADH, which reflects the conditions in vivo, a much more rapid electron flux from cytochrome b_5 to the uncoupled active oxygen is observed (25). Thus, under the conditions of uncoupling, NADH can save NADPH electrons which normally are needed for oxygen activation. This also provides an alternative explanation of the synergistic effect of NADH on microsomal monooxygenases (26) and adds a physiological function to the various proposed roles of cytochrome b_5 in microsomal electron transport. The complete sequence of events can be summarized as follows:

$$Fe^{III} \underset{-RH}{\overset{+RH}{\rightleftharpoons}} \begin{bmatrix} Fe^{III} \\ | \\ RH \end{bmatrix}^{3+}$$

$$ROH \uparrow \qquad \text{"uncoupling"} \qquad \downarrow +e$$

$$\begin{bmatrix} FeO \\ | \\ RH \end{bmatrix}^{3+} \quad NADH + H^+ \quad \begin{bmatrix} Fe^{II} \\ | \\ RH \end{bmatrix}^{2+}$$

$$-OH^- \uparrow +H^+ \qquad \qquad +O_2 \updownarrow -O_2$$

$$\begin{bmatrix} Fe^{III}O_2^{2-} \\ | \\ RH \end{bmatrix}^+ \xleftarrow{+e} \begin{bmatrix} FeO_2 \\ | \\ RH \end{bmatrix}^{2+}$$

Many of the individual steps of the postulated mechanism of cytochrome P450 need further proof and support by quantitative data.

Should the proposed general structure and mechanism of cytochrome P450 action be confirmed, then this would have implications on the nomenclature of these enzymes. The obvious fact that the various forms of cytochrome P450 behave as typical enzymes makes it questionable whether they should be named cytochromes. Being enzymes, they could suitably be called monooxygenases although their typical substrates are not yet clearly defined. In order to characterize the unusual binding of the heme prosthetic group it may be appropriate to introduce a new category for electron-transfer proteins of the "heme-sulphur" type. Such a nomenclature would be analogous to that for the iron-sulphur proteins and would best account for the essential role of sulphur in these monooxygenase reactions.

REFERENCES

1) A.Y.H. Lu, K.W. Junk and M.J. Coon, Resolution of the cytochrome P450-containing -hydroxylation system of liver microsomes into three components,
   J. Biol. Chem. 244, 3714 (1969).

2) A.P. Autor, R.M. Kaschnitz, J.K. Heidema and M.J. Coon, Sedimentation and other properties of the reconstituted liver microsomal mixed-function oxidase system containing cytochrome P450, reduced triphosphopyridine nucleotide-cytochrome P450 reductase and phosphatidylcholine,
   Mol. Pharmacol. 9, 39 (1973).

3) A.F. Welton, F.O. O'Neal, L.C. Chaney and St. D. Aust, Multiplicity of Cytochrome P450 hemoproteins in rat liver microsomes,
   J. Biol. Chem. 250, 5631 (1975).

4) V. Ullrich, U. Frommer and P. Weber, Characterization of cytochrome P450 species in rat liver microsomes. I. Differences in the O-dealkylation of 7-ethoxycoumarin after pretreatment with phenobarbital and 3-methylcholanthrene,
   Hoppe Seyler's Z. Physiol. Chem. 354, 514 (1973).

5) D.A. Haugen, T.A. van der Hoeven and M.J. Coon, Purified liver microsomal cytochrome P450. Separation and characterization of multiple forms,
   J. Biol. Chem. 250, 3567 (1975).

6) E. Bayer, H.A.O. Hill, A. Röder and R.J.P. Williams, The interaction between haem-iron and thiols,
   Chem. Commun. 109 (1969).

7) S. Koch, S.C. Tang, R.H. Holm, R.B. Frankel and J.A. Ibers, Ferric porphyrin thiolates. Possible relationship to cytochrome P450 enzymes and the structure of (p-nitrobenzenethiolato) iron(III) protoporphyrin IX dimethyl ester,
   J. Amer. Chem. Soc. 97, 916 (1975).

8) R. Tsai, C.A. Yu, I.C. Gunsalus, J. Peisach, W. Blumberg, W.H. Orme-Johnson and H. Beinert, Spin-state changes in cytochrome P450 on binding of specific substrates,
   Proc. Natl. Acad. Sci. USA 66, 1157 (1970).

9) J.A. Peterson, Camphor binding by pseudomonas putida cytochrome P450,
   Arch. Biochem. Biophys. 144, 678 (1971).

10) W. Nastainczyk, H.H. Ruf and V. Ullrich, Binding of thiols to microsomal cytochrome P450,
    Chem. Biol. Interactions 14, 251 (1976).

11) V. Ullrich and K.H. Schnabel, Formation and binding of carbanions by cytochrome P450 of liver microsomes,
    Drug Metabolism and Disposition 1, 176 (1973).

12) Ph. L. Gigon, Th. E. Gram and J.R. Gillette, Studies on the rat of reduction of hepatic microsomal cytochrome P450 by reduced nicotinamide adenine dinucleotide phosphate: effect of drug substrates,
    Molec. Pharmacol. 5, 109 (1969).

13) H. Diehl, J. Schädelin and V. Ullrich, Studies on the kinetics of cytochrome P450 reduction in rat liver microsomes,
    Hoppe-Seyler's Z. Physiol. Chem. 351, 1359 (1970).

14) G.J. Mannering, S. Kuwahara and T. Omura, Immunochemical evidence for the participation of cytochrome b_5 in the NADH synergism of the NADPH-dependent Mono-oxidase system of hepatic microsomes,
    Biochem. Biophys. Res. Commun. 57, 476 (1974).

15) Y. Ishimura, V. Ullrich and J.A. Peterson, Oxygenated cytochrome P450 and its possible role in enzymic hydroxylation,
    Biochem. Biophys. Res. Commun. 42, 140 (1971).

16) V. Ullrich and Hj. Staudinger, Aktivierung von Sauerstoff in Modellsystemen in Biochemie des Sauerstoffs (Hess, B. and Staudinger Hj., eds), p. 229,
    Springer Berlin, Heidelberg, New York, 1969.

17) A.D. Rahimtula and P.J. O'Brien, Hydroperoxide catalyzed liver microsomal aromatic hydroxylation reactions involving cytochrome P450,
    Biochem. Biophys. Res. Commun. 60, 440 (1974).

18) F. Lichtenberger, W. Nastainczyk and V. Ullrich, Cytochrome P450 as an oxene transferase,
    Biochem. Biophys. Res. Commun. 70, 939 (1976).

19) A.D. Rahimtula, P.J. O'Brien, E.G. Hrycay, J.A. Peterson and R.W. Estabrook, Possible higher valence states of cytochrome P450 during oxidative reactions,
    Biochem. Biophys. Res. Commun. 60, 695 (1974).

20) F. Lichtenberger and V. Ullrich, Cytochrome P450 catalyzed oxene transfer from various donors,
    This book

21) R.E. McMahon and H.R. Sullivan, Microsomal hydroxylation of ethylbenzene, stereospecificity and the effect of phenobarbital induction,
    Life Sci. 5, 921 (1966).

22) G. Guroff, J.W. Daly, D. Jerina, J. Renson, S. Udenfriend and B. Witkop, Hydroxylation-induced intramolecular migrations: The NIH-shift,
    Science 157, 1524 (1967).

23) M.R. Franklin, The formation of a methylenedioxyphenyl derivative exhibiting an isocyanide-like spectrum with reduced cytochrome P450 in hepatic microsomes,
    Xenobiotica 1, 581 (1971).

24) International Symposium on Active Intermediates
    Turku, Finland, 1975.

25) H. Staudt, F. Lichtenberger and V. Ullrich, The role of NADH in uncoupled microsomal monoxygenations,
    Eur. J. Biochem. 46, 99 (1974).

26) A. Hildebrandt and R.W. Estabrook, Evidence for the participation of cytochrome b_5 in hepatic microsomal mixed-function oxidation reactions,
    Arch. Biochem. Biophys. 143, 66 (1971).

OXYGEN REACTIONS OF THE P450 HEME PROTEIN

S. G. Sligar, B. S. Shastry and I. C. Gunsalus

University of Illinois, Urbana 61801 U.S.A.

Studies using the camphor P450 hydroxylase have provided much insight into the molecular mechanisms of electron transfer (1-3) and mixed function oxidation (2,4,5). The present discussion will deal with the intermediate states of P450 during O_2 cleavage and oxygenation, in terms of chemistry and reactivity. The triplet ground state of O_2 and the singlet configuration of substrate and product introduces an inherent difficulty into monooxygenase catalysis (6). The slowness of this spin forbidden process is avoided in biological systems through the formation of singlet complexes with transition metals, heme iron in the case of P450. An understanding of the reactivity of the P450 heme with O_2 and other oxygen adducts is thus directly related to the mechanism and ordering of mixed function oxidation events.

A ferrous, oxygen bound intermediate in the reaction cycle of cytochrome P450 was discovered in 1970 by Gunsalus and others (7,8) using the purified camphor hydroxylase, and Estabrook et al. (9) employing the drug metabolizing system in the liver microsomes. Subsequent investigation of this intermediate by Mössbauer spectroscopy using the microbial protein, showed its great similarity to the oxygenated form of ferrous hemoglobin and myoglobin (10). Thus, the crucial differences between the monooxygenases and proteins that function only to bind O_2 reversibly centers on the input of the second reducing equivalent required for dioxygen cleavage. The chemistry of oxygenated cytochrome P450, therefore, contains the information necessary for the elucidation of a general reaction scheme of O_2 cleavage and energy segregation. The pathways for the decay of this intermediate species are schematically illustrated in Fig. 1. In bacterial P450, oxygen dissociation by reaction I proceeds at a rate of 1 sec^{-1} (11), while autoxidation, represented by reaction III occurs at only 0.008 sec^{-1} with the liberation of superoxide anion (12). Charge separation can be made competitive with O_2

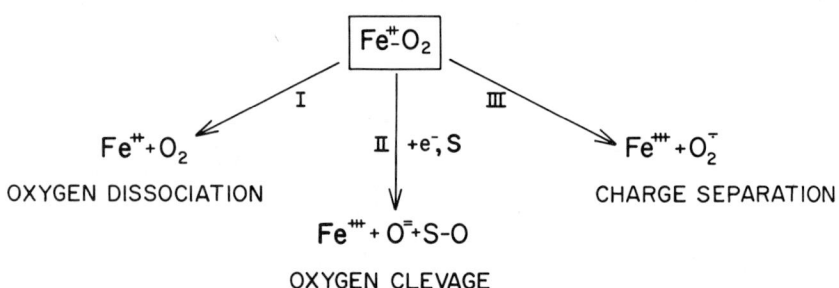

Fig. 1. Decay reactions of oxygenated cytochrome P450.

dissociation by the presence of nucleophiles, which act by a proton assisted displacement on bound oxygen, with the second order rate constants, Table 1,

TABLE 1 Nucleophile Assisted Autoxidation of Oxygenated P450

NUCLEOPHILE[1]	RATE ($M^{-1}sec^{-1}$)
CN^-	13.7
SCN^-	8.8
N_3^-	5.2
OCN^-	4.5
I^-	2.8
Br^-	2.5

(1) Ordered in decreasing nucleophilicity

scaling in order of relative nucleophilicity. A third channel for decay, corresponding to O_2 cleavage and hydroxylation, is opened when an additional redox donor is present (reaction II in Fig. 1). This rate depends on the nature of the electron source, and ranges from 17 sec^{-1} for native putidaredoxin to less than 0.02 sec^{-1} for small molecules such as dihydrolipoic acid (4). The chemistry of P450 monooxygenase hydroxylations follows reaction II which will be the center of discussion for the remainder of this article.

Implication of Putidaredoxin Carboxylate Group in Hydroxylation
The camphor hydroxylase system generates product (5-exo-hydroxy camphor) from the quasi-stable oxygenated intermediate by electron donation from reduced putidaredoxin (Pd) in a complex with P450 (2,4,13). Many other electron donors are also found to be active, although at much lower rates, including dihydroquinones, reduced flavin dyes, and several dithiol redox compounds. The much greater activity of putidaredoxin in product formation induced us to begin a systematic exploration of selective and specific protein modifications to elucidate the amino acid residue(s) of redoxin and cytochrome central to the catalytic event. The first results, a carboxypeptidase A (CPA) catalyzed stoichiometric removal of the putidaredoxin carboxy-terminal tryptophan and penultimate glutamine residues have been reported (14,15). The fluorescence maxima of native Pd at 355 nm, corresponds to a highly hydrophilic environment of the single tryptophan and shows a completely depolarized emission (rotational Stokes radius of 6.5 A) indicative of total free movement of the C-terminal residue (2,16). In addition, the accessible charge distribution of the protein, as quantitated by elution patterns under anion exchange chromatography, is drastically altered by CPA treatment. These results, together with the easy accessibility of CPA digestion, suggested a mobile, extended configuration or "tail" structure of the Gln-Trp-COOH groups. All resonance, electrochemical, and gross hydrodynamic properties of the modified redoxin, termed des-Gln-Trp-Pd, are identical to native Pd. However, the redoxin binding to P450 required for hydroxylase activity was markedly less

(15) and, with more precise assays of activity (2,4,13) a startling decrease in the velocity of product formation was observed even at saturating levels of the protein components, Fig. 2. Loss of affinity between the proteins is possibly due to removal of the hydrophobic indole side chain. With regard to velocity, as the carboxylic acid moiety is perhaps the only strong catalytic group removed during the modification, we postulate a role for this species in maintaining maximal hydroxylation activity. Discussion of a possible mechanistic interpretation for COOH assisted catalysis will be postponed following presentation of data supporting a single oxygen containing Compound I type intermediate in camphor hydroxylation. Possible models for mixed function oxidation chemistry will then be explored in detail.

$$e^- + Pd^o + m_{O_2}^{rs} \xrightleftharpoons[]{K_M} Pd^r \cdot m_{O_2}^{rs} \xrightarrow{V_{MAX}} Pd^o + m^o + S\text{-}O + H_2O$$

Redoxin	K_M	$V_{MAX}^{(a)}$
Pd	28 μM	15,300
des-gln-trp-Pd	80 μM	250

(a) nmoles hydroxycamphor formed per minute at 20°C

Fig. 2. Kinetic parameters of redoxin-P450 catalyzed camphor hydroxylation (2).

P450-Compound I as a Possible Hydroxylating Intermediate

The reaction of peracids, peroxides or other oxidants such as chlorite with peroxidase enzymes generates a monoxygenated intermediate, termed Compound I, carrying two oxidizing equivalents above the ferric resting state (17). A structure, with a formal charge of $[FeO]^{+3}$, can be written as several resonance forms including a porphyrin cation radical (18). Estabrook (19), Coon (20), and Ernster (21) have reported substrate oxygenation in the reaction of liver microsomal P450 with alkyl peroxides. In view of the strong resemblance of bacterial P450 to chloroperoxidase (22), and with the wealth of data on the kinetics and thermodynamics of these two proteins, we undertook a search for similar reactivity and intermediate states in the highly purified and recrystallized microbial hydroxylase.

Oxidized (ferric) substrate bound P450 when mixed with various peroxide, peracid and halogen containing oxidants generates 5-exo-hydroxy camphor with quantifiable velocity and affinity parameters. These reactions are neither inhibited by anaerobiasis nor by carbon monoxide and are consistent with earlier demonstrations that the oxygen atom inserted into the substrate originates from the oxidant without equilibration with water (20). Table 2 gives the maximal velocity of product formation and apparent Michaelis constant with each reagent. This latter parameter does not solely reflect affinity, for in each instance the product generation is accompanied by a loss

TABLE 2 Kinetic Parameters for the Formulation of 5-Exo-Hydroxy Camphor by Ferric P450 and Various Oxidants

	K_m(App) mM	V_{MAX} n moles/min
Peroxides:		
H_2O_2	13	4
t-butyl	1	1
Cumene	4	2
Chlorite	2	1
Periodate	20	20
Peracetate	2	400

of viable protein. Since the rate of this "suicidal reaction" is significantly less at room temperature than the rate of product formation, a side pathway not coupled stoichiometrically to hydroxylation is involved. At lower temperatures, to -40°C in mixed hydroorganic solvents, the destruction rate is drastically decreased while still allowing significant product formation. The maximal velocities for either hydrogen or alkyl peroxide supported hydroxylation are slow and nearly identical, Table 2. In contrast the rate of a peracid catalyzed reaction is several hundred times faster. These rate differences are comparable to those observed with chloroperoxidase for the formation (k_1) and breakdown (k_2) of the Compound I intermediate, Fig. 3 (23).

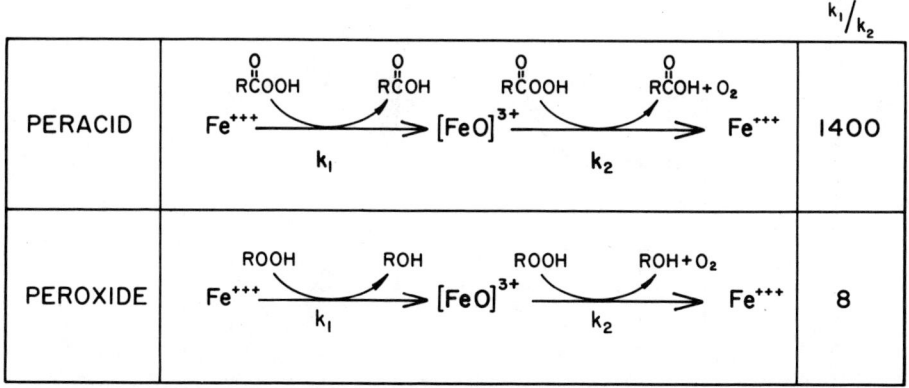

Fig. 3. Oxidant catalyzed formation and decay of Compound I in chloroperoxidase. (k_1 represents the rate of [FeO]$^{+3}$ formation and k_2 the regeneration of ferric protein. This latter reaction would be competitive with substrate hydroxylation by [FeO]$^{+3}$.)

With peroxide and peracid supported oxygenation of the camphor skeleton by cytochrome P450, it was natural to ask whether an oxygen containing intermediate could be quasi-statistically isolated. The "steady state" concentration of Compound I generated and depleted by the oxidant reactions shown in Fig. 3 is proportional to the ratio of formation and breakdown rates, k_1/k_2. With chloroperoxidase, k_1/k_2 is much larger for peracids than peroxides (23). The close similarity of chloroperoxidase and bacterial P450 suggests that peracids should be the best oxidizing species to use in looking for an $[FeO]^{+3}$ state. Furthermore, this intermediate should accumulate to a higher concentration if the substrate free P450 is used since the breakdown via camphor hydroxylation is eliminated. Figure 4 shows the optical spectrum of an intermediate formed on reacting peracetic acid with ferric P450 at -35°C in a hydroorganic solvent. Although our facilities at present did not permit monitoring in the characterization of α-β region, the spectrum in Fig. 4 is very similar to the diminished extinction and general broadening of the Soret band seen in a peroxidase Compound I (24).

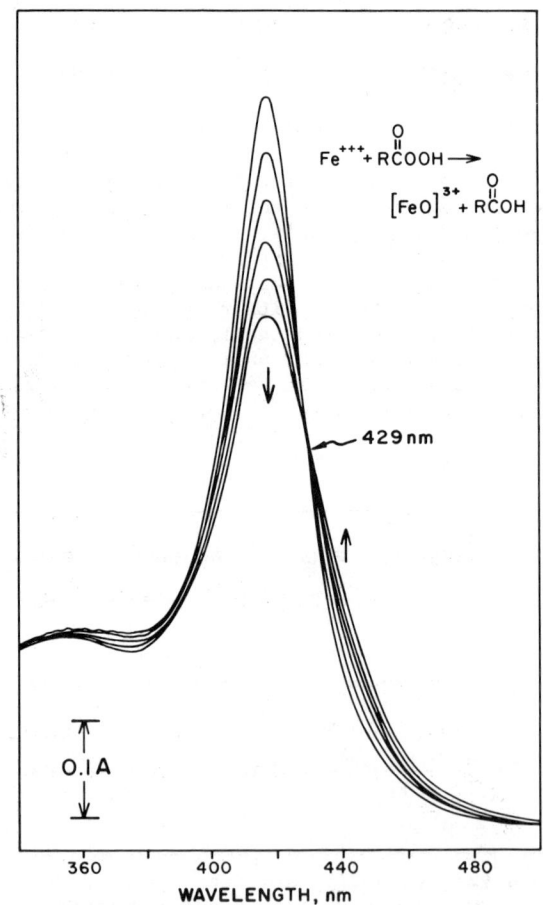

Fig. 4. Soret absorption spectrum of P450-Compound I. Peracetic acid (870 μM) is added to substrate free cytochrome P450 (6.5 μM) in 50:50 v/v ethylene glycol-50 mM potassium phosphate buffer, pH 7 at -35°C.

When camphor is added to P450 in this state, a decrease in absorption at 440 nm is observed, concomitant with generation of hydroxylated substrate and a shift of the Soret to the 391 nm band of ferric camphor bound cytochrome. Dr. T. Pederson, also reporting in this volume, has observed the same spectral species at room temperature after stopped flow mixing of peracid and substrate bound cytochrome. Maximum change in 440 nm absorbance was observed in 30 msec after which time product formation and destruction of the protein progressed.

Chemistry of P450 Catalyzed Hydroxylations

We have shown that several oxidants are active in stereo specific hydroxylation of camphor by bacterial P450, and that peracid treatment of the substrate free protein generates a spectral intermediate closely resembling Compound I of peroxidases. In addition, the carboxylate residue of putidaredoxin was implicated in maintaining the high hydroxylation velocities and product yield of oxygenated P450 when the native redox donor-effector system is used. If the observed spectral intermediate indeed represents an $[FeO]^{+3}$ state which is common for all species supporting hydroxylation, then the role of the Pd-COOH terminus must be in the rate limiting formation of this intermediate. A possible explanation of this activity, Fig. 5, involves acylation of bound dioxygen resulting from nucleophilic attack of the electron rich oxygen on the COOH carbon. Two reaction pathways are represented, corresponding to

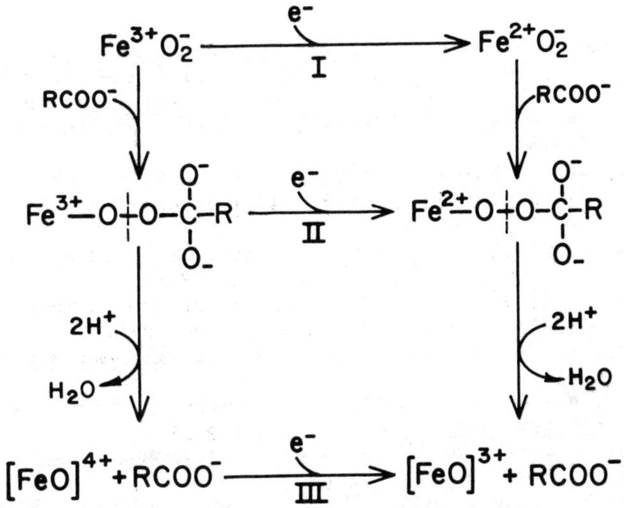

Fig. 5. Formation of P450-Compound I by acylation of bound dioxygen.

whether the second reducing equivalent is transferred from Pd^r before (I) or after (II) acylation. The former could be slightly favored due to the stronger nucleophilicity of $Fe^{2+}O_2^-$ over $Fe^{3+}O_2^-$ (6). A third alternative (III) with electron transfer as a last step seems the most uncertain due to the $[FeO]^{+4}$ species formally required. Acylation of bound dioxygen would certainly weaken the O-O bond, greatly favoring cleavage, water release, and Compound I formation. Such a mechanism utilizes all the features of peroxide and peracid catalysis schemes, and explains the facilitation of the oxygenation with an exogenous carboxylic acid moiety. The role of putidaredoxin (Pd) in camphor hydroxylation is not only as a reductant, but also as an "effector" for product generation. The Pd-P450 complex required for substrate hydroxylation from oxygenated P450 is then responsible for bringing the terminal COOH functional group from Pd into close juxtaposition to the oxygenated P450 heme. Putidaredoxin binding to P450 is known to drastically perturb the spin state of the heme group as quantitated by optical (2,13,25, 26) and electron paramagnetic resonance (27) spectroscopy, and is consistent with a close proximity of the Pd carboxy terminus. The possibility of an "effector" role in mixed function oxidation has also been advanced for cytochrome b_5 in hepatic benzphetamine metabolism (28). The lack of an accessible COOH terminus of adrenodoxin could also explain the much lower absolute activity of the adrenal 11β hydroxylase system, which in all other respects is quite homologous to putidaredoxin (29). In fact, the turnover numbers for these mammalian oxygenases are very close to that of the microbial P450 with des-Gln-Trp-Pd as redox donor-effector.

Acknowledgment. The expert technical assistance of Ms. M. J. Namtvedt and Mr. R. V. Ellis is gratefully acknowledged. The consultation and advice of Dr. J. C. Martin and Dr. T. C. Pederson was invaluable. This work was supported in part by National Institute of Health Grant AM 00562 and GM 21161.

REFERENCES

(1) Sligar, S. and I. C. Gunsalus, Proc. Nat. Acad. Sci. USA 73, 1078 (1976).
(2) Sligar, S. (1975) Ph.D. Thesis, University of Illinois, Urbana.
(3) Pederson, T., R. Austin and I. C. Gunsalus, Fed. Proc. 35, 1535 (1976).
(4) Lipscomb, J., S. Sligar, M. Namtvedt and I. C. Gunsalus, J. Biol. Chem. 251, 1116 (1976).
(5) Gunsalus, I. C., R. Meeks, J. Lipscomb, P. Debrunner and E. Münck (1974) in Molecular Mechanisms of Oxygen Activation (Hayaishi, O., ed.), Academic Press, New York.
(6) Hamilton, G. A. (1974) in Molecular Mechanisms of Oxygen Activation (Hayaishi, O., ed.), Academic Press, New York.
(7) Gunsalus, I. C. (1970) Reported at the Wenner-Gren Symposium Structure Function Oxidation Reduction Enzymes, Stockholm.
(8) Ishimura, Y., V. Ullrich and J. A. Peterson, Biochem. Biophys. Res. Comm. 42, 140 (1971).
(9) Estabrook, R. W., A. G. Hildebrandt, J. Baron, K. Netter and K. Labman, Biochem. Biophys. Res. Comm. 42, 132 (1971).
(10) Sharrock, M, P. Debrunner, C. Schulz, J. Lipscomb, V. Marshall and I. C. Gunsalus, Biochem. Biophys. Acta 420, 8 (1976).
(11) Peterson, J, Y. Ishimura and B. Griffin, Arch. Biochem. Biophys. 149, 197 (1972).
(12) Sligar, S., J. Lipscomb, P. Debrunner and I. C. Gunsalus, Biochem. Biophys. Res. Comm. 61, 290 (1974).
(13) Gunsalus, I. C., S. Sligar and P. Debrunner, Biochem. Soc. Trans. 3, 821 (1975).
(14) Sligar, S., P. Debrunner, J. Lipscomb and I. C. Gunsalus (1973) Int. Union of Biochem. No. 61, Stockholm, Sweden.
(15) Sligar, S., P. Debrunner, J. Lipscomb, M. Namtvedt and I. C. Gunsalus, Proc. Nat. Acad. Sci. USA 71, 3906 (1974).

(16) Sligar, S., P. Debrunner, J. Lipscomb and I. C. Gunsalus (1973) 9th Int. Congress of Biochemistry, Stockholm, Sweden.
(17) Thomas, J., D. Morris and L. Hager, J. Biol. Chem. 245, 3129 (1970).
(18) Dolphin, D., D. Forman, J. Borg, J. Fajer and R. Felton, Proc. Nat. Acad. Sci. USA 68, 614 (1971).
(19) Rahimtula, A., P. O'Brien, E. Hrycay, J. Peterson and R. Estabrook, Biochem. Biophys. Res. Comm. 60, 695 (1974).
(20) Nordblom, G., R. White and M. Coon (1976) Arch. Biochem. Biophys. In press.
(21) Hrycay, E., J. Gustafsson, M. Ingelman-Sundberg and L. Ernster, Biochem. Biophys. Res. Comm. 66, 209 (1975).
(22) Hollenberg, P. and L. Hager, J. Biol. Chem. 248, 2630 (1973).
(23) Hager, L., D. Doubek and P. Hollenberg (1972) in Molecular Basis of Electron Transport (Schultz, J. and Cameron, B., eds.), Academic Press, New York.
(24) Hager, L., P. Hollenberg, R. Tsafrira, R. Chiang and D. Doubek, Ann. N. Y. Acad. Sci. 244, 80 (1975).
(25) Sligar, S. (1976) Biochemistry, In press.
(26) Sligar, S. and I. C. Gunsalus, (1976) 10th Int. Congress of Biochemistry, Hamburg, Germany.
(27) Lipscomb, J. (1975) Ph.D. Thesis, University of Illinois, Urbana.
(28) Guengerich, F., D. Ballou and M. Coon, Biochem. Biophys. Res. Comm. 70, 951 (1976).
(29) Takemori, S., K. Suhara, S. Hashimoto, M. Hashimoto, H. Sato, T. Gomi and M. Katagiri, Biochem. Biophys. Res. Comm. 63, 588 (1975).

THE PEROXIDASE NATURE OF CYTOCHROME P450

Anver D. Rahimtula & Peter J. O'Brien

Dept. of Biochemistry, Memorial University of Newfoundland, St. John's, Newfoundland, Canada

ABSTRACT

It has been shown that hydroperoxides can replace NADPH, molecular oxygen, and the flavoenzyme NADPH-cytochrome P450 reductase requirements of liver microsomal cytochrome P450. Presumably the hydroperoxides can directly oxidize cyt. P450 to the same active hydroxylating species involved in carrying out cyt. P450 function. Cytochrome P450 is acting as a peroxidase with the peroxide serving as the oxidizing agent and the various drugs, steroids and carcinogens acting as electron donors. In accordance with this hypothesis it was found that cytochrome P450 has a very wide specificity and is capable of oxidizing peroxidase donors like TMPD, α-naphthol, hydroquinone etc. as well as various antioxidants including BHA, BHT and ethoxyquin. Moreover these donors effectively inhibit benzpyrene hydroxylation.

Reduced cytochrome b_5 is rapidly oxidized by hydroperoxides if cytochrome P450 is present. Reduced cytochrome b_5 may thus be the physiological donor and could play an important role in discharging the active oxygen species which in the absence of substrate would cause peroxidative destruction of cytochrome P450 and membrane lipids.

INTRODUCTION

The ability of cytochrome P450 to accept oxygen from donors other than molecular oxygen is now well established (1-5). Kadlubar et al (1) showed that the addition of organic hydroperoxides to rat liver microsomes resulted in the C-oxidation of several amine substrates. Subsequently, Rahimtula and O'Brien (2,3) showed that the organic hydroperoxide cumene hydroperoxide (CHP) when added to liver microsomes or to purified cytochrome P450 could carry out aromatic hydroxylation and O-dealkylation reactions quite efficiently. More recently, sodium periodate (4) and iodosobenzene (5) have been used as oxygen donors to hydroxylate various substrates with cytochrome P450. For all these reactions, NADPH, the flavoprotein NADPH-cytochrome P450 reductase, and molecular oxygen are not required. The specificity of the CHP dependent reaction agrees well with that of the NADPH dependent reaction. Thus, Burke and Meyer (6) found that both NADPH and CHP will hydroxylate biphenyl to a much greater extent in the 2-position with 3-MC induced microsomes or with cytochrome P448. Also, both NADPH and CHP showed comparable 'NIH' shifts with liver microsomes or cytochrome P450 during the para hydroxylation of $[4]^3$H acetanilide (7). This would indicate the similarity of the two systems and suggests that hydroxylation occurs via a common intermediate, presumably an arene oxide. Using phenanthrene as a substrate we were able to isolate phenanthrene-9,10-oxide when microsomes were incubated with CHP in the presence of cyclohexene oxide, an inhibitor of the enzyme epoxide hydrase (7). It seems reasonable to infer that CHP is capable of oxidizing cytochrome P450 to a higher oxidation state, probably a perferryl species. Hydroxylation of the substrate occurs by a peroxidase type mechanism with the hydroperoxide serving as the oxygen donor and the various drugs, carcinogens, and xenobiotics serving as oxygen acceptors.

MATERIALS & METHODS

Microsomes were prepared by differential centrifugation of rat liver homogenates as described previously (8). CHP was obtained from Matheson, Coleman & Bell. NADPH and benzpyrene were purchased from Sigma Chemical Corporation. All other chemicals and reagents were of the highest grade commercially available. Ethoxyquin was a gift of the Monsanto Company, St. Louis. Difference spectra were obtained wsing an Aminco DW-2 spectrophotometer. Benzpyrene 3-hydroxylase was assayed according to Cantrell et al (9) except that the concentration of benzpyrene used was 70 µM. Incubations with CHP and NADPH were terminated after 5 and 10 min. respectively.

RESULTS

The addition of CHP to a suspension of liver microsomes prepared from 3-MC pretreated rats results in the appearance of a transient spectral species as illustrated in Fig. 1. After addition of CHP a rather intense absorbance change at about 440 nm in the difference spectrum occurs concomitant with the loss of absorbance at about 390 nm. The intensity of the peak at 440 nm has recently been shown to be linked with the ability of cytochrome P450 to hydroxylate various substrates with CHP and sodium periodate (4).

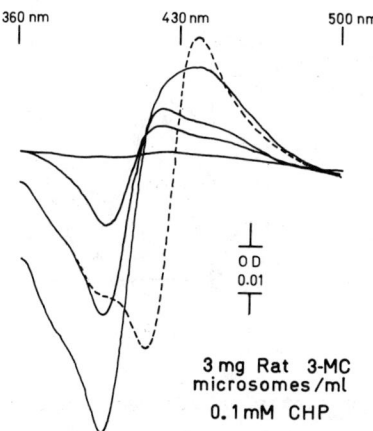

FIGURE 1. Repetitive scanning of the difference spectra obtained on the addition of 0.1 mM CHP (final conc.) to the sample cuvette containing 3 mg/ml 3-MC microsomes. The reference cuvette contained only microsomes. The broken line shows the spectra obtained with microsomes from PB pretreated animals.

Addition of dithionite reduced cytochrome b_5 to an anaerobic mixture of CHP and purified rabbit liver cytochrome P450-LM$_2$ results in the rapid oxidation of reduced cytochrome b_5. In the absence of cytochrome P450, CHP oxidizes reduced cytochrome b_5 only slowly (Fig. 2). This suggests that reduced cytochrome b_5 can effectively discharge the higher oxidation states of cytochrome P450 and must do so via one electron reduction steps. In Fig. 3 the formal oxidation states of cytochrome P450 are shown.

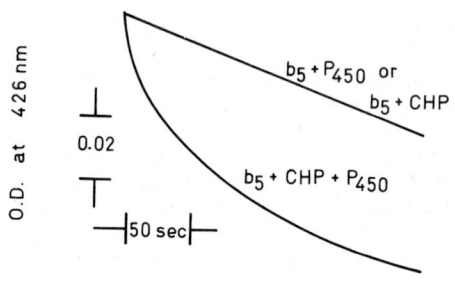

FIGURE 2. Reduced cytochrome b_5 as an electron donor for cytochrome P450 higher oxidation states. 1 μM dithionite reduced cytochrome b_5 was mixed with 0.2 uM cytochrochome P450 and to this mixture was added 90 μM CHP. Anaerobic conditions were maintained throughout the reaction time. The reaction was performed in a stopped flow apparatus; oxidation of reduced cytochrome b_5 was monitered by recording the loss in absorbance at 426 nm.

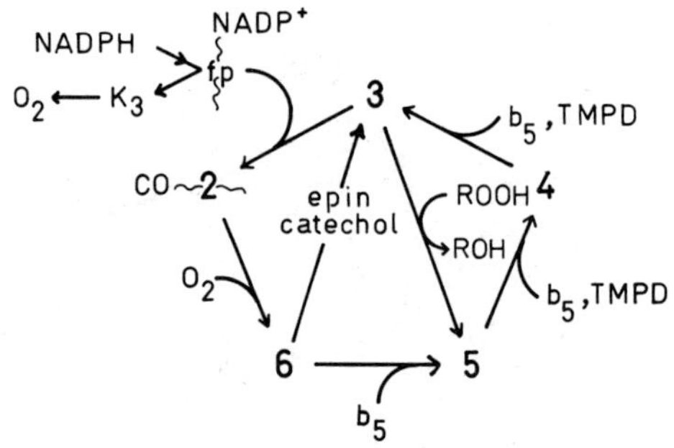

FIGURE 3. Formal oxidation states in the redox cycle of cytochrome P450.

It is well established that NADPH via the flavoprotein reduces ferric cytochrome P450 to the ferrous form which then binds molecular oxygen to give the oxyferrous form with a formal oxidation state of 6 (10,11). State 5 can result either from a further 1 electron reduction of 6 or by the direct oxidation of ferric cytochrome P450 (state 3) with CHP, sodium periodate, iodosobenzene etc. State 5 is most probably a ferric cytochrome P450 - oxene complex and the active hydroxylating species (5,7). In the absence of a substrate, state 5 probably reverts to the ferric form by two 1 electron reduction states. The 440 nm complex may represent the state 4 intermediate. From Fig. 3 it follows that agents that divert electrons away from cytochrome P450 (eg menadione) or that react specifically with ferrous cytochrome P450 (eg CO) or with the oxyferrous form (eg catechol, epinephrine) should inhibit the NADPH dependent reaction specifically but have a minimal effect on the CHP dependent reaction. Agents like reduced cytochrome b_5 that can facilitate the conversion of state 6 into state 5 should enhance the NADPH dependent reaction due to more rapid formation of the active hydroxylating species 5. Furthermore reduced cytochrome b_5 should have no effect or should inhibit the CHP dependent reaction due to discharge of the active complex via state 4 (Fig. 2). Finally, agents that act exclusively on states 3 and/or 5 should inhibit both the NADPH and the CHP dependent reactions.

TABLE I. THE INHIBITION OF CYTOCHROME P450 FUNCTION BY THE DIVERSION OF ELECTRONS OR OXIDIZING SPECIES

		Benz(o)pyrene 3-Hydroxylation	
		NADPH dependent	HYDROPEROXIDE dependent
No inhibitor		100	100
1.	Anaerobic	0	100
	CO	10	100
2.	$NADP^+$ (1 mM)	10	100
	Mechanism:	Inhibition of flavoprotein	
3.	NADH (10 mM)	140	28
	Mechanism:	6 →5	5 →4 →3
4.	K_3 (50 μM)	9	90
	Mechanism:	NADPH →fp →K_3 →O_2	
5.	α-naphthoflavone (0.2 mM)	40	28
	Metyrapone (0.1 mM)	47	60
	SKF 525A (0.6 mM)	22	26
	Androstenedione (0.1 mM)	47	10
	Mechanism:	5 →3	5 →3
6.	Hematin (10 μM)	60	0
	Cytochrome c (2 μM)	75	27
	Mechanism:	NADPH →fp →Hematin →O_2 cyt. c	Competetion for peroxide.

TABLE II. THE INHIBITION OF CYTOCHROME P450 FUNCTION BY ANTIOXIDANTS AND PEROXIDASE DONORS

	Benz(o)pyrene 3-Hydroxylation	
	NADPH dependent	HYDROPEROXIDE dependent
No inhibitur	100	100
1. Tetramethyl-p-phenylenediamine (20 μM)	50	36
Butylated Hydroxytoluene (125 μM)	72	53
2,6-Di-t-butylphenol (125 μM)	35	36
p-Cresol (1 mM)	35	36
Ethoxyquin (25 μM)	35	55
Mechanism:	5 → 4 → 3	5 → 4 → 3
2. Nordihydroguaiaretic Acid (25 μM)	35	95
Nordihydroguaiaretic Acid (100 μM)	0	14
Butylated Hydroxyanisole (25 μM)	54	85
Butylated hydroxyanisole (100 μM)	12	55
p-Benzoquinone (100 μM)	0	45
Naphthol (100 μM)	16	88
Phenol (1 mM)	45	97
Catechol (1 mM)	31	105
Mechanism:	6 → 3 ; 5 → 4 → 3	5 → 4 → 3
3. Glutathione (1 mM)	100	160
Propyl Gallate (25 μM)	82	135
Propyl Gallate (125 μM)	29	117
Epinephrine (100 μM)	37	120
Hydroquinone (100 μM)	79	137
Ascorbate (1 mM)	69	140
Pyrogallol (0.1 mM)	30	350
Mechanism:	6 → 3	Protection of cytochrome P450

Table I shows that anaerobiosis or CO inhibits the NADPH dependent but not the CHP dependent benzpyrene hydroxylation. Similarly, $NADP^+$ which inhibits the flavoenzyme NADPH-cytochrome P450 reductase or menadione which interrupts the flow of electrons from the flavoprotein to cytochrome P450, inhibits only the NADPH dependent benzpyrene 3-hydroxylation. Agents like α-naphthoflavone, metyrapone, SKF 525A, etc that bind to cytochrome P450 or are metabolised inhibit both the NADPH and the CHP dependent hydroxylations. Hematin and cytochrome c which compete effectively with cytochrome P450 for the added hydroperoxide inhibit primarily the CHP dependent hydroxylation.

Table II lists the effect of various antioxidants and peroxidase donors on benzpyrene 3-hydroxylation. These substances fall into three groups. The first group which includes tetramethyl-p-phenylenediamine, butylated hydroxytoluene etc inhibits both the NADPH and the CHP dependent activity about equally thereby indicating that these agents affect a common point in the redox cycle of cytochrome P450, presumably states 3 and 5. Indeed, tetramethyl p-phenylene diamine is known to discharge the active complex of cytochrome P450 and CHP (12) while BHT is known to bind to cytochrome P450 giving rise to a Type I spectrum (13). The second catagory which includes the antioxidants nordihydroguaiaretic acid and butylated hydroxyanisole and the phenols phenol, α-naphthol and catechol inhibits the NADPH dependent activity much more than the CHP dependent activity. The greater inhibition by these agents on the NADPH dependent activity may be due to their ability to discharge the oxyferrous complex (state 6) either directly to the ferric form (state 3) or via the other intermediate states. Glutathione, pyrogallol, propyl gallate, epinephrine, hydroquinone and ascorbate fall into the third catagory. These agents inhibit the NADPH dependent activity while they stimulate the CHP dependent activity. The inhibition of the NADPH dependent activity may well be due to the discharge of the oxyferrous complex directly to the ferric form. Rotilio et al (14) found that epinephrine accelerated the decay of oxymyeloperoxidase to ferric myeloperoxidase with concomitant release of superoxide anion. The stimulation of the CHP dependent activity may be due to the antioxidant protection these agents afford to cytochrome P450 from the destructive effects of free radicals generated during hydroperoxide destruction.

TABLE III. EVIDENCE FOR THE INVOLVEMENT OF A $^1O_2$ COMPLEX IN CYT P450 FUNCTION

	Benz(o)pyrene 3-Hydroxylation	
	NADPH dependent	HYDROPEROXIDE dependent
No inhibitor	100	100
1. $^1O_2$ Traps		
Diphenylfuran (100 uM)	36	21
Diphenylfuran (500 uM)	18	8
2. Physical Quenchers		
Bilirubin (50 uM)	31	27
Azide (1 mM)	65	12
3. $^1O_2$ Charge transfer Complex		
Triethylenediamine (10 mM)	78	90

The extensive work of Wattenberg has shown (15,16,17) that the in vivo feeding of antioxidants like butylated hydroxytoluene, butylated hydroxyanisole, ethoxyquin, disulfiram etc. has an inhibitory effect on some types of cancer like neoplasia of the forestomach and large bowel as well as mammary tumour formation. Antioxidants react poorly with various benzpyrene radicals formed during benzpyrene metabolism (18). Our observations that some of these antioxidants (Table II) inhibit benzpyrene metabolism into potentially toxic metabolites may explain the in vivo protection these agents afford against cancer.

The nature of the activated oxygen in states 6 and 5 has been suggested to be superoxy (19) and hydroperoxy (20) respectively. The active hydroxylating species formed by the elimination of water has been suggested to be a ferryl ion $Fe^{4+}O^-$ by analogy with peroxidase compound I (21) or a ferric complex $Fe^{3+}O$ (5). The dissociation of state 6 resulting in superoxide production has been demonstrated (22). We also found that cytochrome P450 in the presence of hydroperoxides can catalyze the oxygenation of singlet oxygen quenchers like diphenylisobenzofuran, diphenylfuran and bilirubin to the same products formed with singlet oxygen alone. Oxygen was required and CO inhibited. Although these quenchers inhibited benzpyrene 3-hydroxylation with NADPH and CHP, microsomes incubated with NADPH instead of hydroperoxides did not carry out these singlet oxygen reactions. This suggests that singlet oxygen complexes are not important in the formation of the active hydroxylating species (Table III). However, such a complex may be involved in the cytochrome P450 - hydroperoxide catalyzed lipid peroxidation.

REFERENCES

1. Kadlubar, F.F., Morton, K.C., and Ziegler, D.M.(1973) Biochem. Biophys. Res. Commun. 54, 1255-1261.
2. Rahimtula, A.D. and O'Brien, P.J. (1974) Biochem. Biophys. Res. Commun. 60, 440-447.
3. Rahimtula, A.D. and O'Brien, P.J. (1974) Biochem. Biophys. Res. Commun. 62, 268-275.
4. Hrycay, E.G., Gustafsson, J-A., Ingelman-Sundberg, M., and Ernster, L. (1976) Eur. J. Biochem. 61, 43-52.
5. Lichtenberger, F., Nastainczyk, W., and Ullrich, V. (1976) Biochem. Biophys. Res. Commun. 70, 939-946.
6. Burke, D.M. and Meyer, R.T. (1975) Drug Metab. & Disp. 3, 245-253.
7. Rahimtula, A.D., O'Brien, P.J., Seifried, H.E., and Jerina, D.M. (1976) manuscript in preparation.
8. Remmer, H., Greim, H., Schenkman, J., and Estabrook, R.W. (1967) in Methods in Enzymology, Vol X, pp. 703-708 (R.W.Estabrook & M. Pullman, eds.) Acad. Press, New York.
9. Cantrell, E., Abreu, M., and Busbee, D. (1976) Biochem. Biophys. Res. Commun. 70, 474-479.
10. Coon, M.J., Autor, A.P., Boyer, R.F., Lode, E.J., and Strobel, H.W. (1973) "Oxidases and Related Redox Systems" Vol 2, pp. 529-553 (T.E. King, H.S. Mason, and M. Morrison, eds.) Univ. Park Press, Baltimore.
11. Peterson, J.A., Ishimura, Y., Baron, J., and Estabrook, R.W. (1973) "Oxidases and Related Redox Systems" Vol 2, pp. 565-581 (T.E. King, H.S. Mason, and M. Morrison, eds.) Univ. Park Press, Baltimore.
12. Hrycay, E.G., and O'Brien, P.J. (1972) Arch. Biochem. Biophys. 153, 480-494.
13. Yang, C.S., Strickhart, F.S., and Woo, G.K. (1974) Life Sci. 15, 1497-1503.
14. Rotilio, G., Falcioni, G., Fioretti, E., and Brunori, M. (1975) Biochem. J.

145, 405-407.
15. Wattenberg, L.W. (1974) J. Nat. Cancer Inst. 52, 1583.
16. Wattenberg, L.W. (1975) J. Nat. Cancer Inst. 54, 1005.
17. Wattenberg, L.W., Loub, D.W., Lam, L.K., and Speier, J.L. (1976) Fed. Proc. 35, 1327-1331.
18. Menger, E.M., Spokane, R.B., Sullivan, P.D. (1976) Biochem. Biophys. Res. Commun. 71, 610-616.
19. Coon, M.J., Strobel, H.W., and Boyer, R.F. (1973) Drud Metab. and Dispos. 1, 92-97.
20. Estabrook, R.W., Werringloer, J., Hrycay, E.G., O'Brien, P.J., Rahimtula, A.D., and Peterson, J.A. (1975) Biochem. Soc. Trans. 3, 811-813.
21. Rahimtula, A.D., O'Brien, P.J., Hrycay, E.G., Peterson, J.A., and Estabrook, R.W. (1974) Biochem. Biophys. Res. Commun. 60, 695-702.

ACKNOWLEDGEMENTS

This work was supported by funds from the National Reaearch Council of Canada and the Canadian Cancer Society. We wish to thank Dr. David Ballou, Univ. of Michigan, Ann Arbor, for his help in performing the experiment shown in Fig. 2 and Dr. M.J. Coon for providing us with a sample of cytochrome P450-LM$_2$.

CYTOCHROME P450 CATALYZED OXENE TRANSFER FROM VARIOUS DONORS

Fritz Lichtenberger and Volker Ullrich

Department of Physiological Chemistry, University of the Saarland, Homburg/Saar, GFR

ABSTRACT

Iodosobenzene, 3-iodosobenzoic acid and 3-chloroperbenzoic acid as well as the well-known oxidant, cumene hydroperoxide, can substitute for NADPH and molecular oxygen in microsomal monoxygenations. Concomitant with the O-dealkylation activity for 7-ethoxycoumarin a transient peak formation at around 440 nm was observed with these oxene donors. A close correlation with temperature between activity and peak height suggests that the difference spectrum reflects the formation of the active oxygen intermediate.

INTRODUCTION

It has been reported recently that in microsomal monoxygenation reactions the physiological cosubstrates NADPH and molecular oxygen can be replaced by organic peroxides, in particular cumene hydroperoxide (1,2), but also by periodate, chlorite or hydrogen peroxide (3). From these findings it was postulated that the active oxygen species at cytochrome P450 may have the structure of a ferryl ion, $Fe^V O^{2-}$. More clear-cut evidence was presented by showing that iodosobenzene can act as an oxene donor (4), since the iodoso group can provide only one oxygen atom for the cytochrome P450-catalyzed monoxygenations. The present paper describes the results with some other oxene donors and focuses on the spectral changes associated with the interaction of these compounds at cytochrome P450. With cumene hydroperoxide a spectral change with an absorption maximum at 440 nm has been reported (5), but its significance for the function of cytochrome P450 is not yet known.

METHODS

Male Sprague-Dawley rats (80-110 g) were injected with phenobarbital (80 mg/kg) i.p. for three days and microsomes were prepared as described previously (6). Iodosobenzene was prepared according to Willgeroth (7). 3-iodosobenzoic acid, 3-chloroperbenzoic acid and cumene hydroperoxide were purchased from commercial sources and purified by chromatography.

Spectra were recorded on an Aminco DW-2 spectrophotometer and the O-dealkylation of 7-ethoxycoumarin was measured as described earlier (8).

RESULTS

The initial velocities of umbelliferone formation from 7-ethoxycoumarin in the presence of various oxene donors are listed in Table 1.

TABLE 1 O-Dealkylation Activity and Difference Spectra of Microsomes in the Presence of Various Oxene Donors

Oxene Donor	Spec. Activity* For 7-Ethoxycoumarin		Peak Formation at 440 nm
	$t = 25°C$	$t = 0°C$	
3-Chloroperbenzoic Acid	70	20	+
Iodosobenzene	35	0	+
3-Iodosobenzoic Acid	8	0	+
Cumene Hydroperoxide	3	0	+
$NaIO_4$	0	0	−

*nmol/mg protein × min

The oxene donors were added as 0.1 M methanolic or aqueous solutions to give a final concentration of 5×10^{-4}M. Difference spectra were recorded with a microsomal suspension containing 3 mg of protein/ml. The specific activity is expressed as nmol umbelliferone formed/mg protein/min. The formation of the peak at 440 nm was followed by measuring the absorbance difference between 440 and 468 nm.

The highest activity at 25°C was obtained with 3-chloroperbenzoic acid, followed by iodosobenzene, 3-iodosobenzoic acid and cumene hydroperoxide. Sodium periodate was inactive. With decreasing temperature the activities also decreased and at 0°C only 3-chloroperbenzoic acid still retained a considerable activity.

The difference spectrum of a microsomal suspension showed time dependent changes upon addition of the oxidants listed in Table 1. First a transient absorption around 440 nm was seen with all compounds except with periodate. In the case of 3-chloroperbenzoic acid this band was detectable only for a short period of time even at temperature around −20°C (Fig. 1).

At higher temperatures cytochrome P450 was rapidly destroyed resulting in a trough at 420 nm in the difference spectrum. With iodosobenzene at −20°C only a substrate binding spectrum was observed. Upon increasing the temperature the spectrum gradually changed and at about −13°C a peak at 440 nm emerged shortly after mixing. At 0°C it was fully developed but disappeared with time. Concomitantly, a trough at 420 nm was formed (Fig. 2).

The half life time of cytochrome P450 was strongly dependent on the temperature and varied from 10 min. to 30 sec. in the range from −15 to +25°C. Similar results were obtained with 3-iodosobenzoic acid, but the peak at 440 nm was not as pronounced.

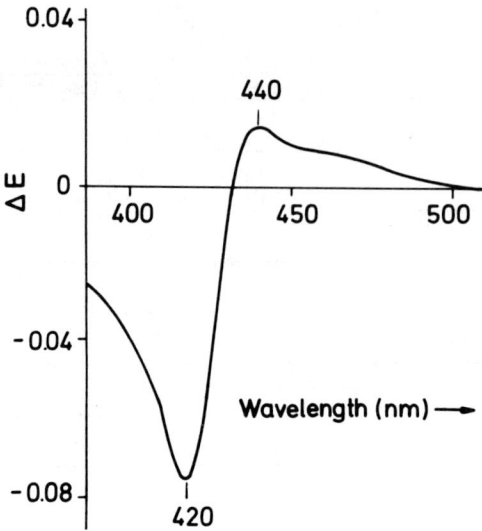

Fig. 1. Difference spectrum of 3-chloroperbenzoic acid in microsomes at $-20°C$

The incubation mixture contained 1.8 ml of 0.1 M Tris-HCl buffer pH 7.6, 1.2 ml glycerol and 9 mg of microsomal protein. 3-chloroperbenzoic acid was dissolved in methanol (0.1 M). 15 μl were added to the microsomal suspension and 15 μl of methanol to the reference cell and the spectrum was immediately recorded.

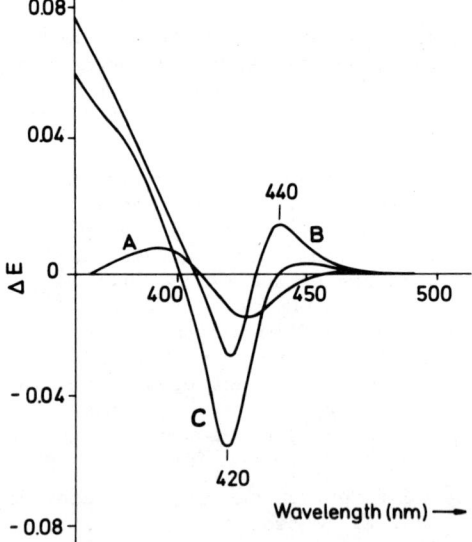

Fig. 2. Difference spectra of iodosobenzene in microsomes from phenobarbital pretreated rats at -20 and $0°C$

Assay conditions as described in legend to Fig. 1. Curve A was taken at $-20°C$, curve B at $0°C$ immediately after addition of iodosobenzene and spectrum C was obtained after 5 min.

The highest stability of the peak was observed in the presence of cumene hydroperoxide, especially when oxygen was removed to stop lipid peroxidation. At 0°C the peak was still measurable after about 20 min (Fig. 3).

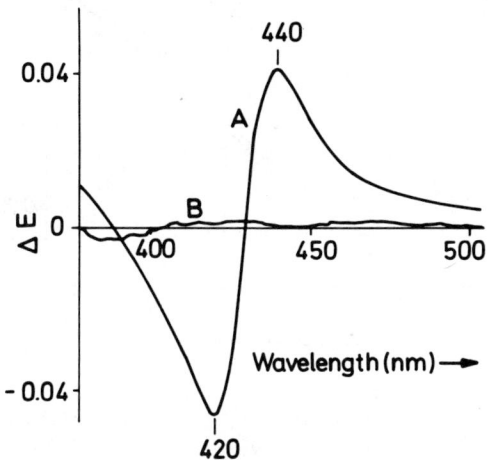

Fig. 3. Difference spectra of cumene hydroperoxide in microsomes at -20 and 10°C

Assay conditions as described in legend to Fig. 1, Curve A: 10°C, curve B: -20°C.

Since both the O-dealkylation of 7-ethoxycoumarin and the transient absorption band at 440 nm were strongly temperature dependent we looked for a correlation of both parameters. This was found for cumene hydroperoxide and iodosobenzene which both initiated the O-dealkylation as well as the peak formation at a temperature of about -10°C and then increased linearly (Fig. 4 a, b).

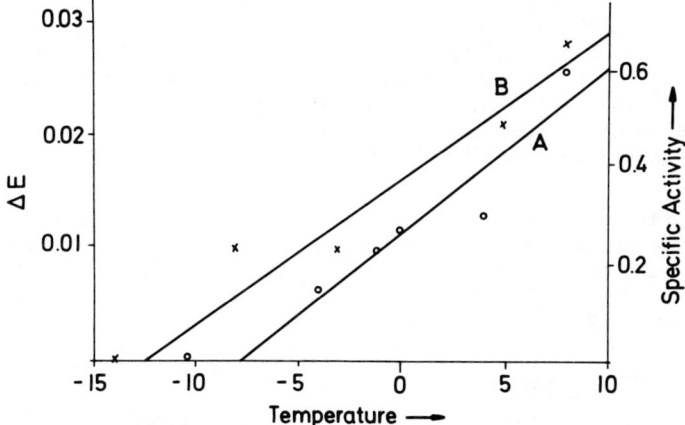

Fig. 4 a. Correlation between temperature-dependent O-dealkylation (A) and 440 nm peak formation (B) with cumene hydroperoxide.

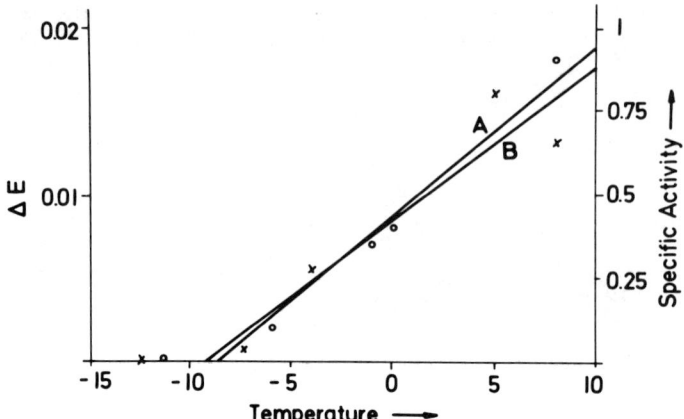

Fig. 4 b. Correlation between temperature dependent O-dealkylation (A) and 440 nm peak formation (B) with iodosobenzene

With 3-chloroperbenzoic acid the 440 nm absorption was already present at $-20°C$ and also an O-dealkylation activity of 2 nmol/mg protein/min was determined.

DISCUSSION

With regard to the monoxygenase activity, 3-chloroperbenzoic acid was the most active oxene donor for microsomal cytochrome P450. Periodate was inactive, although for microsomal steroid hydroxylations positive results had been reported (3). Correlated with the O-dealkylating activity with the various oxene donors a transient 440 nm absorption in the difference spectrum was found. It cannot be excluded that this band arises from a complex formation between the intact oxene donor and oxidized cytochrome P450, but in this case slightly different bands with each donor could be expected. Since all compounds can donate an oxygen atom to substrates via cytochrome P450 it could be postulated that the spectral species absorbing at 440 nm in the difference spectrum is identical with the active oxygen complex. This would also explain the nature of the spectrum with a maximum around 440 nm recently found by Guengerich et al. (9) during the interaction of reduced cytochrome P450 with molecular oxygen.

The O-dealkylation activity as well as the formation of the peak at 440 nm with the different oxene donors started at different temperatures, which indicates that the formation of the active oxygen complex is dependent on the oxene transferring potential of the donor and represents the rate-limiting step of the monooxygenation process.

ACKNOWLEDGEMENT

This work was supported by the Deutsche Forschungsgemeinschaft, Sonderforschungsbereich 38, Teilprojekt L.

REFERENCES

1) A.D. Rahimtula, P.J. O'Brien, Hydroperoxide catalyzed liver microsomal aromatic hydroxylation reactions involving cytochrome P450,
   Biochem. Biophys. Res. Commun. 60, 440 (1974).

2) A.D. Rahimtula, P.J. O'Brien, Hydroperoxide dependent O-dealkylation reactions catalyzed by liver microsomal cytochrome P450,
   Biochem. Biophys. Res. Commun. 62, 268 (1975).

3) E.G. Hrycay, J.A. Gustafsson, M. Ingelman-Sundberg and L. Ernster, Sodium periodate, sodium chlorite, organic hydroperoxides, and H_2O_2 as hydroxylating agents in steroid hydroxylation reactions catalyzed by partially purified cytochrome P450,
   Biochem. Biophys. Res. Commun. 66, 209 (1975).

4) F. Lichtenberger, W. Nastainczyk and V. Ullrich, Cytochrome P450 as an oxene transferase,
   Biochem. Biophys. Res. Commun. 70, 939 (1976).

5) A.D. Rahimtula, P.J. O'Brien, E.G. Hrycay, J.A. Peterson and R.W. Estabrook, Possible higher valence states of cytochrome P450 during oxidative reactions,
   Biochem. Biophys. Res. Commun. 60, 695 (1974).

6) V. Ullrich, On the hydroxylation of cyclohexane in rat liver microsomes,
   Hoppe-Seyler's Z. Physiol. Chem. 350, 357 (1969).

7) C. Willgeroth, Zur Kenntnis aromatischer Jodidchloride des Jodoso- und Jodobenzols,
   Ber. 25, 3495 (1892).

8) V. Ullrich and P. Weber, The O-dealkylation of 7-ethoxycoumarin by liver microsomes,
   Hoppe-Seyler's Z. Physiol. Chem. 353, 1171 (1972).

9) P. Guengerich, D. Ballou and M.J. Coon, Spectral intermediates in the reaction of oxygen with purified liver microsomal cytochrome P450,
   Biochem. Biophys. Res. Commun. 70, 951 (1976).

ASSOCIATION ON TYPE I, TYPE II, AND REVERSE TYPE I DIFFERENCE SPECTRA WITH ABSOLUTE SPIN STATE OF CYTOCHROME P-450 IRON

Daniel W. Nebert,* Kenji Kumaki,* Mitsuo Sato, + and Hideo Kon +

*Developmental Pharmacology Branch, National Institute of Child Health and Human Development,
+ Laboratory of Chemical Physics, National Institute of Arthritis, Metabolism and Digestive Diseases, National Institutes of Health, Bethesda, Maryland 20014 U.S.A.

ABSTRACT

Room-temperature spectrophotometry and EPR low-temperature spectroscopy were compared in liver microsomes from five species of animals. Two factors are most important in determining Type I, Type II, and Reverse Type I difference spectra--and the magnitude of the peak-to-trough absorption: (a) the hydrophobic character and/or the availability of lone-pair electrons of the substrate; and (b) the spin state of P-450 iron at the time the test compound interacts with the cytochrome. By comparisons of Type I, Type II, and Reverse Type I compounds in the sample and reference cuvettes, we conclude that the 410-nm trough represents a "Reverse Type I character" for *in vivo* low spin P-450 iron. We therefore suggest that the sixth ligand for the *in vivo* low spin P-450 ferric iron is a hydroxyl group (or moiety of similar ligand field strength) from an adjacent amino acid residue.

INTRODUCTION

The technique of difference spectrophotometry has been widely used for ten years to study substrate-induced spectral changes in hepatic microsomes (1). Typically one adds the test compound (sometimes dissolved in a solvent) to microsomes in the sample cuvette and nothing (or solvent alone) to microsomes in the reference cuvette (Fig. 1). Differences in absorbance between the two cuvettes are then recorded between 350 and 450 nm. This technique, however,

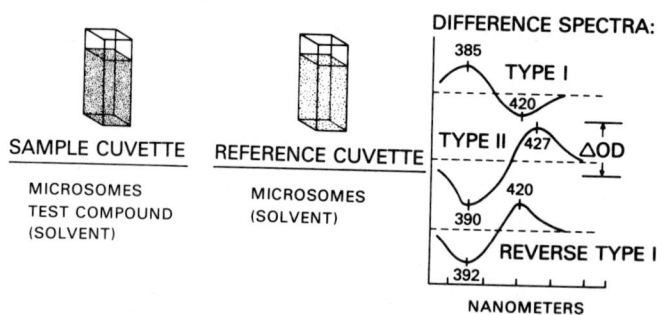

FIG. 1. Technique of difference spectrophotometry and the three fundamental types of spectra observed empirically.

reveals nothing in *absolute* terms about the changes occurring at or near the
enzyme active-site of cytochrome P-450 when the substrate interacts with the
cytochrome. Three basically empirical types of spectra have been well
described and characterized but are poorly understood: Type I, with spectral
maxima about 385 or 390 nm and minima about 420 nm; Type II, with minima about
390 or 410 nm and maxima about 430 nm; and Reverse Type I, which is almost
the mirror image of Type I spectra (1).

METHODS

All chemicals, other materials, and procedures were purchased or performed,
according to previous reports (2-4). New Zealand White rabbits, C57BL/6N
mice, Golden Syrian hamsters, guinea pigs (NIH Outbred Strain), and Sprague-
Dawley rats were obtained from the Veterinary Resources Branch, NIH Animal
Supply, and remained approximately one week before experimentation in our
animal quarters with a rigidly controlled environment, 14:10 day-night light
cycle, and food and water *ad libitum*. All animals were sexually immature
males. 3-Methylcholanthrene (MC) or β-naphthoflavone (BNF) in corn oil was
given intraperitoneally as a single dose (200 mg kg^{-1}) 48 hours before
sacrifice. Phenobarbital in 0.85% NaCl was given intraperitoneally as daily
doses (60 mg kg^{-1}) for 5 days, the first day's dose being divided; animals
were sacrificed 24 hours after the last dose. No significant differences
were found in animals receiving corn oil, 0.85% NaCl, or no injection at all;
however, control animals always routinely received one of these vehicles,
usually corn oil.

The salient features of the experimental protocol included the facts that (i)
liver microsomes were always examined within 1-2 hours after they had been
freshly prepared; (ii) a 1-mm cuvette for room temperature spectrophotometry
was used in order to ensure the same drug/cytochrome P-450/protein ratio as
had been used for EPR spectrometry; (iii) low spin P-450 was measured at 77°K
and high spin at 9.6°K (2); (iv) liquids of known relatively "pure" Type I,
Type II, or Reverse Type I "character" were added to the sample cuvette without addition of any solvents to the reference cuvette (changes in volume were
1% or less); (v) following addition of the test compound, the microsomes were
gently resuspended with two strokes of a glass-Teflon homogenizer at 20°C in
order to ensure thorough mixing and contact of the compound with P-450; and
(vi) liver microsomes from five species were examined, following control, MC
or BNF, or phenobarbital treatment *in vivo*.

RESULTS AND DISCUSSION

Introduction to Theory

The ferric iron of a hemoprotein such as P-450 has five d electrons (Fig. 2).
The strength of a ligand depends to a large degree on the magnitude of the
σ-donating power and the π-accepting power--of substrates or residues of the
protein itself--and the resultant degree with which these electrons overlap
with the d orbitals of ferric iron. A *strong ligand* will cause the electrons
to be paired in each of the two lowest d orbitals, with one unpaired electron
in the next highest orbital, and this is called *low spin*. A *weak ligand* will
result in all five electrons being unpaired, and this is called *high spin*.

Ligands have been empirically classified, with regard to their strength, or
capacity to overlap with d electrons (Table 1) (5). Halogens are notably
weak: Fe^{III}Cl$_3$, for example, is high spin. CN$^-$ is among the strongest:

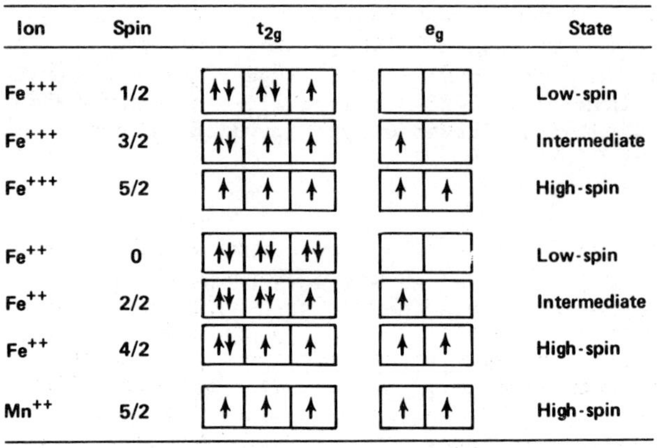

Ion	Spin	t_{2g}	e_g	State
Fe^{+++}	1/2	↕ ↕ ↑		Low-spin
Fe^{+++}	3/2	↕ ↑ ↑	↑	Intermediate
Fe^{+++}	5/2	↑ ↑ ↑	↑ ↑	High-spin
Fe^{++}	0	↕ ↕ ↕		Low-spin
Fe^{++}	2/2	↕ ↕ ↑	↑	Intermediate
Fe^{++}	4/2	↕ ↑ ↑	↑ ↑	High-spin
Mn^{++}	5/2	↑ ↑ ↑	↑ ↑	High-spin

FIG. 2. Spin distributions of d electron configurations in octahedral complexes of iron and manganese. The e_g orbitals are of higher energy than the t_{2g} orbitals (5).

$Fe^{III}(CN)_6^{\equiv}$, for example, is low spin. It should be particularly noted in Table 1 that alcohol, phenol, and water are considerably *weaker* ligands than nitrogen-containing moieties such as ethylenediamine.

Previous Empirical Observations with Room Temperature Spectra

Different spectral changes, peak-to-trough heights, or the appearance or disappearance of a spectral change, are found when the same concentration of the same test compound is added to microsomes from different species (2,4,6), strains (7), tissues (8-10), or the same strain following different *in vivo* pretreatment regimens (11-14), or when increasing amounts of test compound are added to the same microsomal preparation (15). For example, we find with C57BL/6N mouse liver microsomes that cyclohexane causes a Type I spectrum following phenobarbital treatment and essentially no spectral change after BNF or MC treatment; butanol-1 causes a Reverse Type I spectrum following BNF or MC treatment and essentially no spectral change after phenobarbital treatment (data not illustrated). All these data are consistent with the hypothesis that the spin state of iron of cytochromes P-450 *in vivo* is determined by inherent properties of the protein molecule itself in the proximity of the enzyme active-site and that the percentage of high or low spin iron can be changed by *in vivo* pretreatment with inducers, prior to addition of any test compound during the *in vitro* experiment.

Table 1 *Tsuchida's Spectrochemical Series of Increasing Ligand Field Strength* (5)

weakest	I^-
↓	Br^-
	Cl^-
	F^-
	R-OH
	H_2O
	NH_3
	Ethylenediamine
	Phenanthroline
	NO_2^-
strongest	CN^-

Room Temperature Spectral Changes by in vitro Additions

Difference spectra represent the *difference* between the sample and reference cuvettes (Table 2). Whether there is no addition (Experiment #1) or addition of cyclohexane (Experiment #2), octylamine-1 (Experiment #3), or butanol-1 (Experiment #4) to both cuvettes--the *apparent* spectrum is the same: simply a balanced baseline. Cyclohexane causes a Type I spectrum (Experiment #5), octylamine-1 produces a Type II spectrum (Experiment #6), and butanol-1 causes a Reverse Type I spectrum (Experiment #4). When cyclohexane and octylamine-1 are added to the sample cuvette (Experiment #7), the spectrum is the same as that with octylamine-1 alone. In other words, octylamine-1 apparently binds more strongly to the P-450 than cyclohexane. The same is true of octylamine-1 binding more strongly than butanol-1 (Experiment #10).

When octylamine-1 by itself (or in the presence of either cyclohexane or butanol-1) is in the sample cuvette--and either microsomes alone or butanol-1 is in the reference cuvette (Experiments #6, 7, 10, & 11)--the 410-nm trough (denoted by *arrows*) is seen in the apparent difference spectrum. This trough is *not* seen, however (Experiments #8 & 9), when cyclohexane is in the reference cuvette. We conclude that the *maximal rate of change* in absorbance (from *below* to *above* the baseline) at about 410 nm (*i.e.* between 408 and 412 nm), produced by the Reverse Type I spectrum in the reference cuvette, accounts for the 410-nm trough in the difference spectrum. We therefore believe that a ligand with "Reverse Type I character"--similar to butanol-1 *in vitro*--exists normally in untreated microsomes freshly prepared from the animal.

Correlation between Spin State of P-450 Iron Determined by EPR Spectrometry and Peak-to-trough Height of Room Temperature Difference Spectra

As increasing concentrations of cyclohexane or chloroform are added to microsomes (Fig. 3), the low spin (g $\cong$ 2.24) EPR signal height decreases, and the high spin (g $\cong$ 8.0) EPR signal height increases. Very closely parallel to these changes are increases in the peak-to-trough height of the Type I spectrum determined simultaneously by room temperature spectrophotometry. Just the opposite can be seen after addition of a Type II or Reverse Type I compound (data not illustrated), *i.e.* the g $\cong$ 8.0 signal decreases and the g $\cong$ 2.24 signal increases proportionate to increases in the peak-to-trough height of the room temperature difference spectrum. These data confirm earlier work (16-20), indicating that Type I compounds are capable of converting low spin P-450 iron to high spin and that Type II compounds convert high spin P-450 iron to low spin.

By measuring changes in EPR signal height before and after addition of Type I compounds, we solved simultaneous equations (manuscript in preparation) for estimating the percentage of high spin P-450 iron.* Among liver microsomes from all five species examined after no treatment or following BNF or MC or phenobarbital treatment, there was generally good agreement for estimating

---

*In this study with intact liver microsomes, the "spin state of P-450 iron" always denotes the *sum* of spin states of all forms of P-450 iron. Clearly, in the same microsomal preparation one form of P-450 may be 90% high spin, a second form 50%, and a third form 10%, etc. The g $\cong$ 8.0 signal height therefore reflects both the relative concentration of each form of P-450 and the per cent high spin of each form. In these equations, we would also take into account the shift in equilibrium from high to low spin P-450 iron (if there is any) (21), as the temperature is lowered from 25°C to 77°K or 9.6°K.

Table 2 Illustration of spectra observed in sample cuvette and in reference cuvette, and the resultant difference spectrum, in liver microsomes from phenobarbital-treated mice.

The curve shapes are emphasized, and no attempt is made to quantitate the peak-to-trough heights.

EXPERIMENTAL CONDITION	CUVETTES	IN VITRO ADDITION	SPECTRA SAMPLE	SPECTRA REFERENCE	APPARENT DIFFERENCE (SAMPLE) − (REFERENCE)	COMMENTS (EPR CHANGES)
1	SAMPLE	NONE	········	········	BASELINE	SAMPLE: *IN VIVO* SPIN STATE INTACT
	REFERENCE	NONE				REFERENCE: *IN VIVO* SPIN STATE INTACT
2	SAMPLE	CYCLOHEXANE	⌒⌣	⌒⌣	BASELINE	SAMPLE: *IN VIVO* LOW SPIN → *IN VITRO* HIGH SPIN
	REFERENCE	CYCLOHEXANE				REFERENCE: *IN VIVO* LOW SPIN → *IN VITRO* HIGH SPIN
3	SAMPLE	OCTYLAMINE-1	⌣⌒	⌣⌒	BASELINE	SAMPLE: *IN VIVO* HIGH SPIN AND LOW SPIN → *IN VITRO* LOW SPIN
	REFERENCE	OCTYLAMINE-1				REFERENCE: *IN VIVO* HIGH SPIN AND LOW SPIN → *IN VITRO* LOW SPIN
4	SAMPLE	BUTANOL-1	⌣⌒	⌣⌒	BASELINE	SAMPLE: *IN VIVO* HIGH SPIN → *IN VITRO* LOW SPIN
	REFERENCE	BUTANOL-1				REFERENCE: *IN VIVO* HIGH SPIN → *IN VITRO* LOW SPIN
5	SAMPLE	CYCLOHEXANE	⌒⌣	········	TYPE I	SAMPLE: *IN VIVO* LOW SPIN → *IN VITRO* HIGH SPIN
	REFERENCE	NONE				REFERENCE: *IN VIVO* SPIN STATE INTACT
6	SAMPLE	OCTYLAMINE-1	⌣⌒	········	TYPE II WITH 410-NM TROUGH	SAMPLE: *IN VIVO* HIGH SPIN AND LOW SPIN → *IN VITRO* LOW SPIN
	REFERENCE	NONE				REFERENCE: *IN VIVO* SPIN STATE INTACT
7	SAMPLE	CYCLOHEXANE + OCTYLAMINE-1	⌣⌒	⌒⌣	TYPE I WITH 410-NM TROUGH	SAMPLE: *IN VIVO* HIGH SPIN AND LOW SPIN CYCLOHEXANE-INDUCED *IN VITRO* HIGH SPIN } → *IN VITRO* LOW SPIN
	REFERENCE	NONE				REFERENCE: *IN VIVO* SPIN STATE INTACT
8	SAMPLE	CYCLOHEXANE + OCTYLAMINE-1	⌣⌒	⌒⌣	TYPE II	SAMPLE: *IN VIVO* HIGH SPIN AND LOW SPIN CYCLOHEXANE-INDUCED *IN VITRO* HIGH SPIN } → *IN VITRO* LOW SPIN
	REFERENCE	CYCLOHEXANE				REFERENCE: *IN VIVO* LOW SPIN → *IN VITRO* HIGH SPIN
9	SAMPLE	OCTYLAMINE-1	⌣⌒	⌒⌣	TYPE I	SAMPLE: *IN VIVO* HIGH SPIN AND LOW SPIN → *IN VITRO* LOW SPIN
	REFERENCE	CYCLOHEXANE				REFERENCE: *IN VIVO* LOW SPIN → *IN VITRO* HIGH SPIN
10	SAMPLE	BUTANOL-1 + OCTYLAMINE-1	⌣⌒	⌣⌒	TYPE II WITH 410-NM TROUGH	SAMPLE: *IN VIVO* HIGH AND LOW SPIN BUTANOL-INDUCED *IN VITRO* LOW SPIN } → *IN VITRO* LOW SPIN
	REFERENCE	BUTANOL-1				REFERENCE: *IN VIVO* HIGH SPIN → *IN VITRO* LOW SPIN
11	SAMPLE	OCTYLAMINE-1	⌣⌒	⌣⌒	TYPE II WITH 410-NM TROUGH	SAMPLE: *IN VIVO* HIGH SPIN AND LOW SPIN → *IN VITRO* LOW SPIN
	REFERENCE	BUTANOL-1				REFERENCE: *IN VIVO* HIGH SPIN → *IN VITRO* LOW SPIN

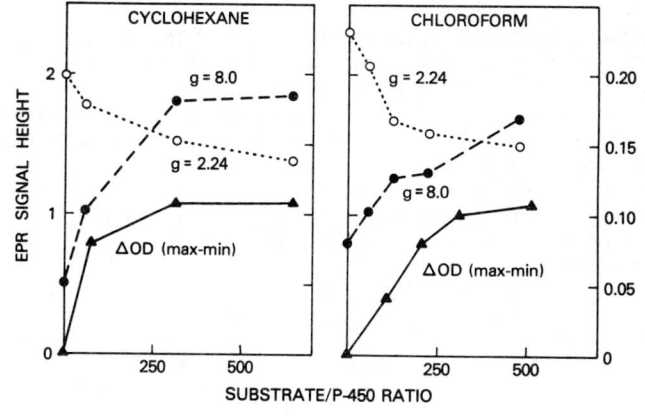

FIG. 3. EPR signal height and peak-to-trough optical density (O.D.) of difference spectrum as a function of substrate/cytochrome P-450 ratio in liver microsomes from phenobarbital-treated mice.

per cent of high spin P-450 iron per total P-450 content as a function of the ratio of the $g \cong 8.0/g \cong 2.24$ signal heights. This relationship was not zero-order or first-order (manuscript in preparation). BNF or MC treatment of all five species always caused more high spin P-450 iron than that of control animals, whereas phenobarbital treatment caused more or less high spin P-450 iron, dependent on which species was examined. Even if our estimates of per cent high spin P-450 iron were in error by 10% to 20%, such an error would not detract from the following experiments. We chose four representative liver microsomal samples having the widest range of differences: BNF-treated rabbit (57%), MC-treated guinea pig (38%), phenobarbital-treated mouse (17%), and phenobarbital-treated rat (8%). An excellent agreement in all cases was found (Table 3) between the per cent high spin *in vivo* P-450 iron and the peak-to-trough height of the room temperature difference spectrum (22,23). The *lesser* the high spin P-450 iron content at the start of the experiment, the *greater* the magnitude of the peak-to-trough height when cyclohexane (or chloroform) was added *in vitro*. The *greater* the high spin P-450 iron content at the start of the experiment, the *greater* the magnitude of either the Type II peak-to-trough height when octylamine-1 was added or the Reverse Type I peak-to-trough height when butanol-1 was added *in vitro*. If a drug (*e.g.* metyrapone or ethylmorphine) having both hydrophobic character and availability of lone pair electrons is used, this relationship becomes more complex.

Table 3 *Relationship Between Starting in vivo Spin State of P-450 Iron and Changes in Room Temperature Difference Spectra*

		Peak-to-trough spectral height		
	% High spin	Type I (cyclohexane or chloroform)	Type II (octylamine-1)	Reverse Type I (butanol-1)
BNF-treated rabbit	57	±	++++	++
MC-treated guinea pig	38	+	+++	+
PhBarb-treated mouse	17	++	++	±
PhBarb-treated rat	8	+++	+	0

Conclusions

The addition *in vitro* of a Type I compound changes the *in vivo* low spin (and perhaps some *in vivo* high spin that we cannot detect by optical methods) to *in vitro* high spin P-450 iron. Addition of a Reverse Type I compound changes the *in vivo* high spin (and perhaps some *in vivo* low spin that we cannot detect by optical methods) to *in vitro* low spin iron. Addition of a Type II compound changes both *in vivo* high and low spin to *in vitro* low spin forms of P-450 iron. The 390-nm trough represents the pure change from *in vivo* high spin to *in vitro* low spin iron. The 410-nm trough represents *in vivo* low spin iron with a sixth ligand having field strength similar to that of butanol-1 addition *in vitro*, as illustrated in Fig. 4.

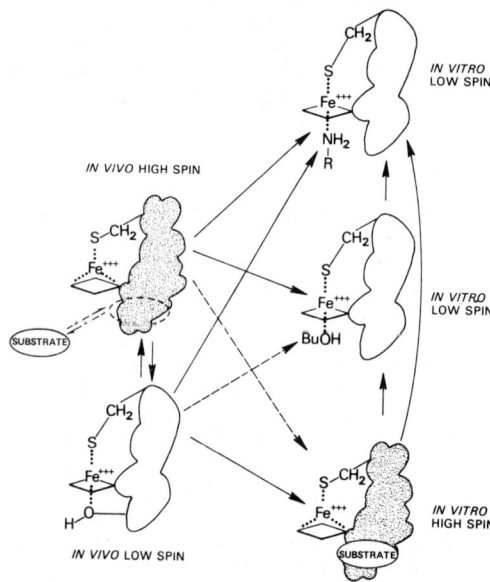

FIG. 4. Hypothetical scheme of relationship between *in vivo* and *in vitro* high (iron out of plane of porphyrin ring) and low (iron in plane of porphyrin ring) spin states of P-450 and illustrating a weak ligand, such as hydroxyl from an adjacent amino acid residue, in the sixth position of *in vivo* low spin P-450 iron.

At *lower left*, *in vivo* low spin P-450 is shown. In the presence of substrate or because of inherent properties in the P-450 protein, *in vivo* high spin P-450 iron (*upper left*) is formed. The three *in vitro* forms at *right* include: (*top*) low spin caused by Type II compounds of strong ligand field strength; (*middle*) low spin caused by Reverse Type I compounds of lesser ligand field strength; and (*bottom*) high spin caused by Type I compounds which may be substrates that bind with specificity and therefore are capable of displacing the sixth ligand. The fact that the peak for Type II compounds (about 427-435 nm) is further to the red than that for Reverse Type I compounds (about 418-420-nm) is also evidence that the ligand field strength of Type II compounds is greater than that for Reverse Type I compounds.

ACKNOWLEDGMENTS

We appreciate the critical review of this manuscript by Drs. Louise Karle Hanson and Larry M. Hjelmeland. The expert secretarial assistance of Ms. Ingrid Jordan is also acknowledged.

REFERENCES

1. Remmer, H., Schenkman, J., Estabrook, R. W., Sasame, H., Gillette, J., Narasimhulu, S., Cooper, D. Y., and Rosenthal, O. (1966) *Mol. Pharmacol.* 2: 187-190.

2. Nebert, D. W., and Kon, H. (1973) *J. Biol. Chem.* 248: 169-178.

3. Nebert, D. W., Considine, N., and Kon, H. (1973) *Drug Metab. Disp.* 1: 231-238.

4. Nebert, D. W., Robinson, J. R., and Kon, H. (1973) *J. Biol. Chem.* 248: 7637-7647.

5. Orgel, L. E. (1960) In: *An introduction to transition-metal chemistry: Ligand-field theory*. Wiley, New York, 180 pages.

6. Schenkman, J. B., Cinti, D. L., Orrenius, S., Moldeus, P., and Kraschnitz, R. (1972) *Biochemistry* 11: 4243-4251.

7. Goujon, F. M., Nebert, D. W., and Gielen, J. E. (1972) *Mol. Pharmacol.* 8: 667-680.

8. Kupfer, D., and Orrenius, S. (1970) *Mol. Pharmacol.* 6: 221-230.

9. Orrenius, S., Kupfer, D., and Ernster, L. (1970) *FEBS Lett.* 6: 249-252.

10. Ellin, A., Orrenius, S., Pilotti, A., and Swahn, C.-G. (1973) *Arch. Biochem. Biophys.* 158: 597-604.

11. Jefcoate, C. R. E., Gaylor, J. L., and Calabrese, R. L. (1969) *Biochemistry* 8: 3455-3463.

12. Jefcoate, C. R. E., and Gaylor, J. L. (1969) *Biochemistry* 8: 3464-3472.

13. Schenkman, J. B. (1970) *Biochemistry* 9: 2081-2091.

14. Jefcoate, C. R. E., Calabrese, R. L., and Gaylor, J. L. (1970) *Mol. Pharmacol.* 6: 391-401.

15. Jansson, I., Orrenius, S., Ernster, L., and Schenkman, J. B. (1972) *Arch. Biochem. Biophys.* 151: 391-400.

16. Mitani, F., and Horie, S. (1969) *J. Biochem. (Tokyo)* 65: 269-280.

17. Mitani, F., and Horie, S. (1969) *J. Biochem. (Tokyo)* 66: 139-149.

18. Whysner, J. A., Ramseyer, J., and Harding, B. W. (1970) *J. Biol. Chem.* 245: 5441-5449.

19. Peterson, J. A., Ullrich, V., and Hildebrandt, A. G. (1971) *Arch. Biochem. Biophys.* 145: 531-542.

20. Gunsalus, I. C., Tyson, C. A., Tsai, R., and Lipscomb, J. D. (1971) *Chem.-Biol. Interactions* 4: 75-78.

21. Jefcoate, C. R., Orme-Johnson, W. H., and Beinert, H. (1976) *J. Biol. Chem.* 251: 3706-3715.

22. Kumaki, K., Sato, M., Kon, H., and Nebert, D. W. (1975) *Pharmacologist* 17: 216.

23. Nebert, D. W., Kumaki, K., Sato, M., and Kon, H. (1976) *Hoppe-Seyler's Z. Physiol. Chem.* 357: 1044-1045.

NMR RELAXATION TIME STUDIES OF SUBSTRATE INTERACTIONS WITH P450 AND OTHER HEMOPROTEINS

R. F. Novak, I. M. Kapetanovic and J. J. Mieyal

Department of Pharmacology, Northwestern University Medical School, Chicago, Illinois/U.S.A.

ABSTRACT

^1H nuclear magnetic resonance longitudinal relaxation time (T_1) measurements were used to study the interaction of xylidine with solubilized rat liver microsomal cytochrome P450, myoglobin and hemoglobin, and to study the binding of aniline and imidazole to human hemoglobin. Since the carboxyferrous derivatives of all three hemoproteins are diamagnetic, they were used in all cases to give control values ($1/T_1^0$) which allowed calculation of the paramagnetic relaxation rate values ($1/T_{1p}$) where $1/T_{1p}$ represents the specific paramagnetic effect of Fe^{3+} upon the relaxation rate of the substrate. Upon addition of ferricytochrome P450, ferrimyoglobin, or ferrihemoglobin to solutions of xylidine, the T_1 values for the methyl and phenyl moieties of the xylidine molecule decreased markedly relative to the control conditions. The observed changes showed that xylidine was much more sensitive to cytochrome P450 than to myoglobin or hemoglobin; and whereas myoglobin produced essentially the same effect upon the relaxation rates of the phenyl and methyl protons, hemoglobin and cytochrome P450 produced differential changes, phenyl > methyl. Such T_1 data in conjunction with values (obtained separately) of the dissociation constants (K_S) and estimates of the correlation times (τ_c) for the various complexes allowed calculations of distances between the heme iron atoms and specific portions of the complexed substrate molecules. These estimates suggested that xylidine may not be an inner sphere ligand of solubilized P450. Similar results obtained for aniline and imidazole with hemoglobin would suggest that these substrates are directly coordinated. Further distinctions among the complexes were obtained when CN^- or F^- were added along with the substrates and ferrihemoproteins; these ligands apparently displaced imidazole from hemoglobin, but they did not abolish the interaction of xylidine or of aniline with hemoglobin.

INTRODUCTION

The hemoprotein cytochrome P450 is the terminal oxidase enzyme localized to the liver endoplasmic reticulum, which catalyzes the oxidative metabolism of a wide variety of substrates (1,2). Although this enzyme system has been widely studied, the detailed mechanism by which it carries out the selective transfer of activated oxygen to specific moieties on substrate molecules has not been fully elucidated. We found recently that hemoglobin and myoglobin can catalyze the hydroxylation of aniline about as effectively as cytochrome P450 (3,4). Thus all three of these hemoproteins have the potential to act as oxidase enzymes, at least under certain conditions.

UV difference spectroscopy has been used extensively to investigate the binding of drug molecules to the ferric form of cytochrome P450, and such complex formation has been presumed to be the first step in the overall catalytic mechanism of hydroxylation (1,2,5,6). On the contrary, we reported (7) that the aniline-ferrihemoglobin complex, although it was qualitatively the same as the aniline-ferricytochrome P450 complex (i.e. "Type II" UV difference spectrum(5,7), cannot be important in the hemoglobin-catalyzed hydroxylation reaction, because the interaction was cooperative and too weak compared to the catalytic K_M for aniline. We have therefore set out to compare the substrate complexes of P450, hemoglobin and myoglobin by the technique of NMR spectroscopy which can provide more detailed information about the relative nature of the complexes in solution. NMR spectroscopy is ideally suited for investigation of the interaction of substrates with proteins that contain paramagnetic centers (such as the hemoproteins), because marked changes may be observed in the NMR properties of those substrate moieties which most closely approach the paramagnetic site in the complex. In particular, NMR longitudinal relaxation time (T_1) measurements can be used to observe differential relaxation effects on various parts of a substrate molecule and thereby allow evaluation of the substrate orientation and distance from the paramagnetic center. In this way it might be possible eventually to predict the site of hydroxylation of a drug molecule based upon its closeness of approach to the heme oxygen-binding site. NMR techniques analogous to those described in this article have been recently employed to study certain ligand interactions with the hemoproteins myoglobin, horseradish peroxidase, catalase and cytochrome P450 (8-11).

There are two aspects to the comparative study reported here. Firstly, xylidine (2,6 dimethylaniline) was chosen for investigation of the substrate interactions with P450, hemoglobin and myoglobin, because it is hydroxylated by all three and its NMR spectrum is good for differential T_1 studies. Secondly, the interactions of aniline and imidazole with hemoglobin were compared because the former binds cooperatively while the latter does not. Parts of this study have been reported previously (12,13).

EXPERIMENTAL

Ferrimyoglobin (sperm whale skeletal muscle, Type II, 2-times recrystallized, Sigma Chemical Company) was treated with Chelex resin (Biorad Chelex 100, 50-100 mesh, sodium form) and passed over Sephadex G-25 equilibrated with 0.02 M KPi, pH_{obs} 7.5 in D_2O.

Hemoglobin (human, Type IV, 2-times recrystallized, Sigma Chemical Company) was pretreated with $K_3Fe(CN)_6$ and passed over a Sephadex G-25 column previously equilibrated with 0.02 M KPi, pH_{obs} 7.5 in D_2O. The concentrations of hemoglobin and myoglobin in solution were determined according to the assay of Van Kampen and Zijlstra (14). The spin states of Mb and Hb in solution were determined from the absolute UV/visible spectra of the solutions as described previously (7). At 35°C aquoferrimyoglobin and aquoferrihemoglobin exist at ~88% in the high spin (S = 5/2) state.

Cytochrome P450 was solubilized from the liver microsomes of phenobarbital-treated rats essentially according to the ammonium sulfate extraction procedure of Autor, et al. (15), as described previously (12). The concentrations of P450 and P420 were determined from the dithionite-reduced carbonmonoxide minus dithionite-reduced difference spectrum method of Omura and Sato (16). P450 content was obtained from $\Delta\varepsilon_{450-490\,cm}$ = 91 $mM^{-1}cm^{-1}$. P420 content was obtained from $\Delta\varepsilon_{420-490\,nm}$ = 111 $mM^{-1}cm^{-1}$ after correcting for the contribution of $P450^{2+}$ at 420 nm, i.e. $\Delta\varepsilon_{420-490\,nm}$ = -41 $mM^{-1}cm^{-1}$ (16). Total heme content was determined from the dithionite reduced pyridine oxidized difference spectra using $\Delta\varepsilon_{555-575\,nm}$ = 32.4 $mM^{-1}cm^{-1}$ (17). The concentration of the solubilized cytochrome P450 in our preparations after the 2-step ammonium sulfate precipitation ranged from 15.4 to 18.8 nmoles/ml with a specific content of 2.3-2.8 nmoles/mg protein. Protein determinations were made according to the method of Lowry (18). Cytochrome P420 content ranged from 2% for freshly prepared cytochrome P450, to < 15% for samples which were stored frozen. It was confirmed that no significant change in the relative concentrations of P450 and P420 would occur at 37° during the time course of a typical NMR experiment. The sum of the P450 and P420 content accounted for > 95% of the total heme content. Xylidine (2,6-dimethylaniline, Eastman Chemical), D_2O (99.8%, Aldrich), KCN (Fisher) were used as purchased. Tetramethylammonium phosphate was prepared by neutralization of tetramethylammonium hydroxide (Eastman) with H_3PO_4 (Fisher). All solutions were treated with chelex resin (Biorad Chelex 100, 50-100 mesh, sodium form) prior to use, in order to remove contaminating paramagnetic ions. The cyanometmyoglobin and cyanomethemoglobin derivatives were prepared directly in the NMR tubes by addition of μl volumes of 0.1 M KCN in D_2O. The corresponding carbonmonoxyferrous derivatives were prepared by reduction with dithionite (Fisher) under a stream of carbonmonoxide (Matheson).

UV spectra were recorded using an Aminco DW-2 dual beam spectrophotometer operating in the split beam mode. Spectra were recorded under ambient conditions unless otherwise noted.

^1H NMR spectra were recorded using a Perkin-Elmer R-32 spectrometer operating at 90 MHz and interfaced for Fourier Transform technique with a Nicolet TT-7 computer package; the ^1H longitudinal relaxation times, T_1, of the signals were measured using the standard inversion-recovery sequence, 180°-τ-90° (see Fig. 1). ^{13}C NMR spectra were recorded with a Varian CFT-20 Fourier Transform NMR spectrometer.

Treatment of Data--The effect of electron spin-nuclear spin interaction upon the longitudinal relaxation time (T_1) is described by the dipolar portion of the Solomon-Bloembergen equation (19,20).

$$\frac{1}{T_{1_M}} = \frac{2\hbar^2 \gamma_I^2 \gamma_S^2 (S)(S+1)}{15 \langle r^6 \rangle} \left[\frac{3\tau_c}{1 + \omega_I^2 \tau_c^2} + \frac{7\tau_c}{1 + \omega_S^2 \tau_c^2} \right] \quad [1]$$

T_{1_M} is the relaxation time of a nucleus bound in the proximity of a paramagnetic metal ion. Here γ_I is the nuclear gyromagnetic ratio, γ_S is the electron gyromagnetic ratio, $\hbar$ is Planck's constant divided by 2π, S is the total electron spin, r is the length of the vector between the nuclear-spin dipole and the electron-spin dipole, τ_c is the correlation time which describes molecular motions that modulate the electron-nuclear dipolar coupling, and ω_I and ω_S are the nuclear and electron precession frequencies, respectively. For high spin (S = 5/2) and low spin (S = 1/2) forms of the ferrihemoprotein, equation 1 reduces to

$$r(\text{Å}) = 812 \, (T_{1_M} f(\tau_c))^{1/6} \quad [2a]$$

$$r(\text{Å}) = 540 \, (T_{1_M} f(\tau_c))^{1/6} \quad [2b]$$

where $f(\tau_c)$ refers to the expression in brackets in the first term of equation 1.

The correlation time τ_c, for the dipolar term of equation 1 is given by

$$\frac{1}{\tau_c} = \frac{1}{\tau_r} + \frac{1}{\tau_s} + \frac{1}{\tau_M} \qquad [3]$$

where τ_r is the rotational correlation time, τ_s is the electron-spin relaxation time and τ_M is the mean residence time of the species complexed to the metal ion (i.e. the reciprocal of the pseudo-first order dissociation rate constant of the metal-ion complex). It can be seen that the fastest process (shortest correlation time) will contribute most significantly to τ_c. In general, τ_s is on the order of 10^{-11}-10^{-10} sec for high and low spin ferric hemoproteins (21-23), while $\tau_r \sim 10^{-9}$-10^{-8} sec and $\tau_M \sim 10^{-4}$-10^{-6} sec (21-23). Since τ_r and τ_M are 1 to 7 orders of magnitude greater than τ_s, they would be expected to contribute negligibly to the value of $1/\tau_c$. Equation 3 then reduces to $1/\tau_c \simeq 1/\tau_s$. Hence, for the substrate-ferrihemoprotein complexes under study, τ_c can be expected to lie in the range $10^{-11} < \tau_c < 10^{-10}$.

For a given substrate moiety, the contribution to the observed longitudinal relaxation rate due to the paramagnetic ion, T_{1_p}, is given by

$$\frac{1}{T_{1_p}} = \frac{1}{T_{1_{obsd}}} - \frac{1}{T_1^0} = \frac{\alpha_M}{T_{1_M} + \tau_M} \qquad [4]$$

where $1/T_1^0$ is the relaxation rate in the absence of the paramagnetism and α_M is the mole fraction of a given species interacting with the paramagnetic ion. When "fast exchange" prevails, $\tau_M \ll T_{1_M}$; then $T_{1_M} \simeq \alpha_M T_{1_p}$.

RESULTS AND DISCUSSION

The ^{1}H magnetic resonance spectrum of the substrate molecule aniline, with the phenyl protons giving rise to a multiplet at 7 ppm is shown in Fig. 1. The NMR spectrum of xylidine is shown in Fig. 2. The multiplet (δ = 7 ppm) is ascribed to the phenyl ring protons, while the methyl groups of xylidine give rise to the singlet at 2.2 ppm. The internal reference $(CH_3)_4N^+$, which gives a singlet signal at 3.2 ppm, was used as a control for viscosity changes as well as experimental varia-

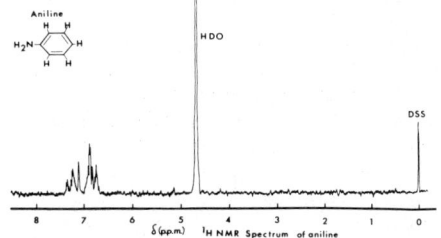

Figure 1

tion. A partially relaxed Fourier Transform series obtained from application of the inversion-recovery sequence and used in obtaining the relaxation times T_1 of the xylidine groups is given in Fig. 2 also. Here τ represents the delay time between the two pulse sequence. The signal amplitudes are then plotted semilogarithmically versus the delay time τ using the equation $\ln(A_\infty - A_\tau) = -1/T_1$. A_∞ refers to the maximum peak amplitude and A_τ is the signal amplitude (positive or negative) obtained at time τ. The slope of the line gives the ^{1}H longitudinal relaxation rate, $1/T_1$.

The changes in the relaxation rates ($\Delta(1/T_1)$) of the methyl and phenyl groups of xylidine upon addition of hemoglobin or myoglobin are shown in Fig. 3.

Here the change in relaxation rate $\Delta(1/T_1)$ is defined as $\Delta(1/T_1) = 1/T_{1obs} - 1/T_{1blank}$, where blank refers to the absence of hemoprotein. For myoglobin, both phenyl and methyl relaxation rates were identical to within experimental error. For hemoglobin (dotted lines, Fig. 3), a significant differential change in relaxation rate was observed for the methyl and phenyl protons of the xylidine molecule. The change in relaxation rates of the methyl and phenyl protons of the xylidine molecule upon addition of cytochrome P450 is shown in Fig. 3 insert. In this case the ratio of substrate to

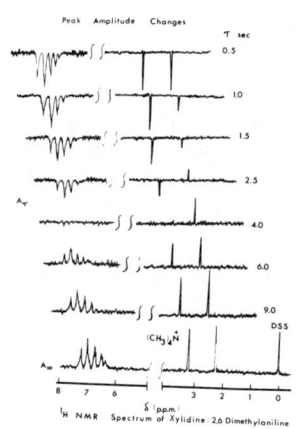

Figure 2

cyt P450 ranges from ∿ 15000 to ∿ 5000. This means that the substrate is much more sensitive to additions of cyt P450 than to Mb or Hb. A significant differential change in relaxation rate exists between the methyl and phenyl protons of xylidine (Fig. 3, insert). No significant change was observed in the relaxation rate of the internal reference in any case (see bottom line, Fig. 3). Fig. 4 shows T_1 data for the interaction of aniline and imidazole with hemoglobin.

In order to estimate distances of approach of the substrate to the paramagnetic heme iron atom using T_1 data, the rate of exchange of the substrate from the binding site should be rapid in comparison to the NMR time scale. The Arrhenius temperature dependence of the relaxation rate of the xylidine methyl groups in the presence of cyt P450 and the aniline phenyl protons in the presence of Hb appeared as region "a" of Fig. 5 (12, 13). This suggests that these complexes are in fast exchange with respect to the NMR time scale, and the resultant energy of activation (E_{act}) of ∿ 2.6 Kcal/mole obtained from this data supports this conclusion (12). Imidazole-Hb, however, appeared in the chemical control region with E_{act} = 4.4 Kcal/mole (upper region "b", Fig. 5) (13).

Distances calculated (using equation 2a,2b above) for xylidine-P450, xylidine-ferrihemoglobin, aniline-Hb and imidazole-Hb, based on a range

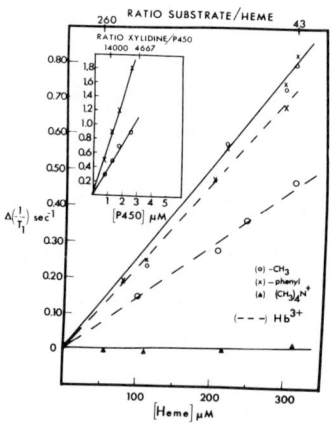

Figure 3

of values of τ_c, (the correlation time parameter, expected for these hemoproteins are given in Table I. Xylidine-P450 distances calculated for correlation time values and possible spin states range from 2.5 to 5.5 Å. Additional information on the exact spin states of the different forms of cyt P450 under ambient conditions in the presence of substrate and determination of the correlation time, τ_c, for the complex will allow calculation of specific distances for substrate-cyt P450 complexes as well as allow elucidation of the nature of the complexes in solution, for example direct or indirect coordination. The results for xylidine-Hb^{3+}, however, strongly suggest that xylidine does not directly coordinate to the heme iron atom in that complex.

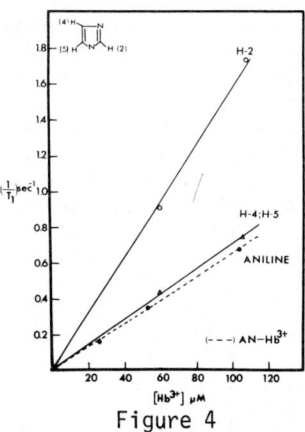

Figure 4

Further characterization of the nature of these substrate-hemoprotein interactions was obtained through experiments in which other ligands were added in the presence of the substrates. Since the observed changes in the relaxation

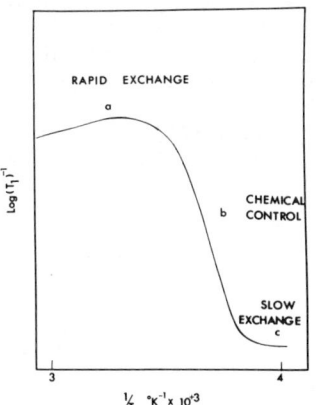

Figure 5

times of the substrate moieties may be ascribed to the effect of the paramagnetism of the iron atom in the hemoprotein, any experiment which modifies (i.e. increases or decreases) the paramagnetism should have an effect upon the relaxation time of the substrate moieties.

The system shown in Fig. 6 is aniline in the presence of ferrihemoglobin ($\sim$ 100 μM), and the effects upon the observed T_1 relaxation time in the presence of the ligands CN^- and F^-. The addition of CN^- which directly coordinates to the heme iron atom also decreases the paramagnetism by changing the spin state of the heme iron atom from high spin to low spin, and therefore it should decrease the relaxation rate of aniline either by displacing aniline completely or by virtue of the spin-state change, with aniline still capable of interacting with the Fe^{3+} atom. This is observed as shown by the lower line in Fig. 6. It should be noted that the decrease in the relaxation rate is not sufficient to return it to the control state (i.e. that found for CO-Hb^{2+}). Hence it appears that aniline may continue to associate with Hb even when CN^- is bound. Ferrihemoglobin under ambient conditions, (37°) exists as $\sim$88% high spin (7). The addition of F^- converts the hemoprotein to 100% high spin and hence the addition of fluoride ion should cause an increase in the relaxation rate only if aniline continues to interact with the six-coordinated Fe^{3+} atom. This is also observed as shown by the upper line in Fig. 6. Results analogous to these were observed when the xylidine-myoglobin

XYLIDINE-P450			
Xylidine	τ_c sec^{-1}	r (Fe---H) Å	
		S = 5/2	S = 1/2
methyl phenyl	10^{-10}	555 509	368 338
methyl phenyl	10^{-11}	381 351	253 232
XYLIDINE-Hb^{3+} S = 5/2			
Xylidine	τ_c sec^{-1}	r (Fe---H) Å	
methyl phenyl	3×10^{-10}	9.07 8.43	
ANILINE-Hb^{3+} (S = 1/2)			
τ_c	r(H-Fe^{3+}) Å		
10^{-11}	3.95		
10^{-10}	5.78		
IMIDAZOLE-Hb^{3+} S = 1/2			
τ_c	r(H2---Fe^{3+}) Å	r(H4:5-Fe^{3+}) Å	
10^{-11}	4.21	4.77	
10^{-10}	6.13	6.97	

Table I

interaction was challenged with CN^- and F^-. However, when similar T_1 studies were performed for imidazole, which does not bind cooperatively to Hb^{3+}, it was found in contrast to the case for aniline, that both F^- and CN^- caused a decrease in relaxation rate. Thus these results indicate that (a) the observed changes in relaxation rate reflect specific interactions between substrate and heme iron atom; (b) that the nature of binding of imidazole, which is probably directly coordinated to the heme iron atom, can be distinguished by this procedure from aniline and xylidine binding; (c) that the substrate molecules aniline and xylidine continue to interact with the heme iron atom even though it is liganded; and (d) that this technique may be useful in the investigation of ternary substrate-hemoprotein-ligand complexes.

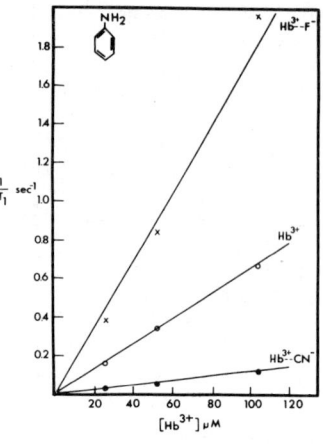

Figure 6

Application of this technique to other molecules such as antipyrine also demonstrated the observation of differential relaxation rates, with the relative relaxation rates in the presence of cytochrome P450 being phenyl > N-methyl > C-methyl. These studies may be used to determine specific substrate orientations which in turn might be important in predicting the sites of hydroxylation and closeness of approach to the oxygen binding site. An opportunity to pursue this question directly is available through the comparison of different purified forms of cyt P450 which show marked differences in substrate specificity. The P450 isozymes LM II and LM IV, whose effectiveness in dealkylation of benzphetamine is $\sim$ 20:1, respectively, have been purified from rabbit liver microsomes by Dr. M.J. Coon's group of the University of Michigan (24). In collaboration, we are presently examining the interaction of benzphetamine (Fig. 7) with those two forms using ^1H and ^{13}C T_1 relaxation time techniques which should allow us to accurately determine the mutual orientation of the substrate and heme in the benzphetamine-P450 complexes. The object of this study is to determine whether such differential metabolic activity can be predicted from the nature of the substrate interaction with some form of each P450 isozyme.

Acknowledgements. This work was carried out during the tenure of NIH postdoctoral fellowship (GM 05225) to R.F.N., a predoctoral fellowship (GM 11062) to I.M.K. and grants to J.J.M. from NIH (GM 20050) and Chicago Heart Association (B76-84).

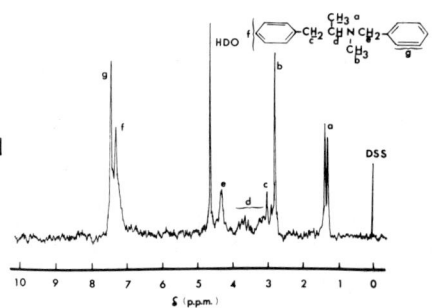

Figure 7

The ^1H spectrum of benzphetamine. The assignments of signals identifies the various moieties of the benzphetamine molecule. The ^{13}C spectrum (not shown) gives rise to signals which reflect thirteen different carbon environments.

REFERENCES

1. The Second International Symposium on Microsomes and Drug Oxidations, (R.W. Estabrook, J.R. Gillette, and K.C. Liebman, eds). (1972), 486 pages, Williams and Wilkins Co., Baltimore; and the Third International Symposium (this volume).
2. Gunsalus, I.C., Pederson, T.C.,and Sligar, S.G., Ann. Rev. Biochem. 44, 377 (1975).
3. Mieyal, J.J., Ackerman, R.S., Blumer, J.L., and Freeman, L.S., J. Biol. Chem. 251, 3436 (1976).
4. Mieyal, J.J.,and Blumer, J.L., J. Biol. Chem. 251, 3442 (1976).
5. Mannering, G.J., in Drugs and Cell Regulation, Academic Press, New York, 197 (1971).
6. Orrenius, S., Wilson, B.J., vonBahr, C.,and Schenkman, J.B., Biochem. Soc. Symp. 34, 55 (1972).
7. Mieyal, J.J.,and Freeman, L.S., Biochem. Biophys. Res. Comm. 69, 143 (1976).
8. Hershberg, R.D.,and Chance, B., Biochem. 14, 3885 (1975).
9. Vuk-Pavlovic, S.,and Benko, B., Biochem. Biophys. Res. Comm. 66, 1154 (1975).
10. Griffin, B.W.,and Peterson, J.A., J. Biol. Chem. 250, 6445 (1975).
11. Ruckpaul, K., Maricic, S., Janig, G.R., Benko, B., Vuk-Pavlovic, S. and Rein, H., Croatica Chemica Acta 48, 69 (1976).
12. Novak, R.F., Kapetanovic, I.M., and Mieyal, J.J., Mol. Pharmacol. 13 (1), in press (1977).
13. Novak, R.F., and Mieyal, J.J., The Pharmacologist 18 (2), 242 (1976).
14. Van Kampen, E.J.,and Zijlstra, W.G., Clin. Chim. Acta 6, 538 (1961).
15. Autor, A.P., Kaschnitz, R.M., Heidema, J.K., and Coon, M.J., Mol. Pharmacol. 9, 93 (1973).
16. Omura, T., and Sato, R., Methods Enzymol. 10, 556 (1967).
17. Schenkman, J.B., Cha, Y.N., Moldeus, P.,and Cinti, D.L., Drug Metabolism Disposition 1, 516 (1973).
18. Lowry, O.H., Roberts, N.R., and Kapphahn, J.I., J. Biol. Chem. 224, 1047 (1957).
19. Solomon, I., J.Chem. Phys. 99, 559 (1955).
20. Bloembergen, N., J. Chem. Phys. 27, 572 (1957).
21. Asakura, T., Ann. N.Y. Acad. Sci. 222, 68 (1973).
22. Gupta, R.K., and Koenig, S.H., Biochem. Biophys. Res. Comm. 45, 1134 (1971).
23. Gupta, R.K., and Mildvan, A.S., J. Biol. Chem. 250, 246 (1975).
24. Haugen, D.A., Van Der Hoeven, T.A., and Coon, M.J., J. Biol. Chem. 250, 3567 (1975).

THE INTERACTION OF POLYHALOGENATED METHANES WITH FERROUS CYTOCHROME P450

C. Roland Wolf, Daniel Mansuy, Wolfgang Nastainczyk and Volker Ullrich

Department of Physiological Chemistry, University of the Saarland, Homburg/Saar, GFR

ABSTRACT

The interaction of a series of halogenomethanes, CCl_4, CBr_4, CCl_3Br, CCl_3F, $CHCl_3$, $CHBr_3$, CHI_3, CCl_3CN, CH_2Cl_2 and CH_2Br_2 with ferrous cytochrome P450 was investigated in anaerobic rat liver microsomal preparations in the presence of sodium dithionite or NADPH. Apart from CH_2Cl_2 and CH_2Br_2 these compounds exhibited complexes with ferrous cytochrome P450 having absorption maxima in the difference spectra between 450 and 470 nm. The proposal that these complexes result from a cytochrome P450 mediated two electron reduction resulting in the formation of a carbene intermediate was strengthened by the finding that carbon monoxide was a product of the reaction. Carbon monoxide is a known hydrolysis product of dihalocarbenes.

INTRODUCTION

The physiological function of cytochrome P450 from various monoxygenase systems is associated with the oxygen activation process. This involves a two-electron reduction of molecular oxygen to yield an oxenoid complex which is able to transfer an oxygen atom to a substrate (1). It has been suggested, however, that not only oxygen can be reduced at ferrous cytochrome P450 but also lipophilic organic nitro- and azo compounds yielding the corresponding amines (2, 3). A similar reductive conversion has also been reported for N-oxides (4).

There are various reports indicating that polyhalogenated hydrocarbons are also subject to reductive elimination of halogen by the cytochrome P450 system. In the case of carbon tetrachloride a one-electron reduction yielding chloride and the trichloromethyl radical has been proposed in order to explain the carbon tetrachloride-mediated destruction of biological membranes (5). It is not certain however, whether this one electron reduction is mediated by cytochrome P450 or the FMN and FAD-dependent cytochrome P450 reductase. Anaerobic conditions can lead to the formation of chloroform from carbon tetrachloride and this reaction can be blocked by carbon monoxide (6). Under similar conditions carbon tetrachloride and cytochrome P450 form a complex with a Soret absorption band at 454 nm in the difference spectrum (7). This interaction has been interpreted by us as the formation of a carbene complex (8). Experimental evidence for an analogous reduction of a polyhalogenated compound has been obtained with halothane ($CF_3CHClBr$). The corresponding complex of cytochrome P450 and halothane has an absorption peak at 470 nm. An identical spectrum was formed on the addition of 1,1,1-trifluorodiazoethane to reduced microsomes suggesting that the 1,1,1-trifluoromethyl carbene was the liganding species (9).

$$CF_3\overset{H}{\underset{|}{C}} = N_2 \xrightarrow[-N_2]{Fe^{II}P450} CF_3\overset{H}{\underset{|}{C}} \mid Fe^{II}P450$$

The present study should further substantiate possible carbene formation during the reduction of polyhalogenated hydrocarbons by cytochrome P450. In order to simplify the system we have selected halogenated methane derivatives for the experiments.

METHODS

Male Sprague-Dawley rats (100-150 g) pretreated with sodium phenobarbital, (80 mg per kg body weight, i.p.) one injection per day for three days, were used. Liver microsomal fractions were prepared according to the method of Frommer (10). Protein determinations were made using the biuret method (11), and cytochrome P450 determined by the method of Omura (12). Rat hemoglobin was prepared according to the method of Rossi-Fanelli (13) and was used as an aqueous solution.

The carbon monoxide formed during anaerobic microsomal incubations was determined by measurement of the carbon monoxide-hemoglobin difference spectrum. Quantitative values for carbon monoxide concentrations were obtained from calibration curves determined by titration of a sodium dithionite reduced microsomal suspension, containing hemoglobin (25 μM), with μl quantities of an aqueous solution saturated with carbon monoxide. Microsomes were suspended in 0.1 M tris HCl buffer pH 7.6, final volume 3 ml, 1 mg protein per ml. The difference absorption between 419 and the isosbestic point at 425 nm was measured in 1 cm glass cuvettes, using an Aminco DW-2 spectrophotometer, and was plotted against carbon monoxide concentration. The carbon monoxide concentration of a saturated aqueous solution was taken as 1.0 mM.

All chemicals used were commercially available.

Substrate-induced difference spectra were measured in liver microsomal suspensions, suspended in 0.1 M tris-HCl buffer, pH 7.6, reduced with sodium dithionite or NADPH. Spectra were recorded in stoppered 1 cm glass cuvettes, containing 3 ml of microsomal suspension (1 mg protein/ml) using an Aminco DW-2 spectrophotometer. In experiments under anaerobic conditions involving NADPH the microsomal suspension was first bubbled with nitrogen for 10 min followed by the addition of substrate (4 mM) and NADPH (100 μM).

RESULTS

On the addition of carbon tetrachloride to sodium dithionite reduced microsomes from phenobarbital pretreated rats the typical CCl_4-ferrous cytochrome P450 difference spectrum was obtained (Fig. 1). With the exception of CH_2Cl_2 and CH_2Br_2 all halogenomethanes tested gave similar peaks with absorption maxima ranging between 452 and 468 nm. As in the case of carbon tetrachloride the formation of these peaks was time dependent. The stability of the spectra with time varied according to the substrate used and was in the order: $CHCl_3 \gtrsim CCl_3CN \gtrsim CCl_3F \gtrsim CCl_3Br > CCl_4 > CHBr_3 \gtrsim CBr_4 \gtrsim CHI_3$.

As shown in Fig. 1 the position of the absorption peak shifted with time, in the case of CCl_4 from 460 nm to 454 nm. On the addition of hemoglobin to test and reference cuvettes, after peak formation, the position of the absorption maximum shifted back to 460 nm, and a concomitant formation of the typical carbon monoxide-hemoglobin spectrum was observed. The shift in the absorption maximum was most marked for $CHBr_3$ and CBr_4. From these spectra it was apparent that the shift was due to the formation of a second peak absorbing around

450 nm suggesting that carbon monoxide may have been formed during the incubation.

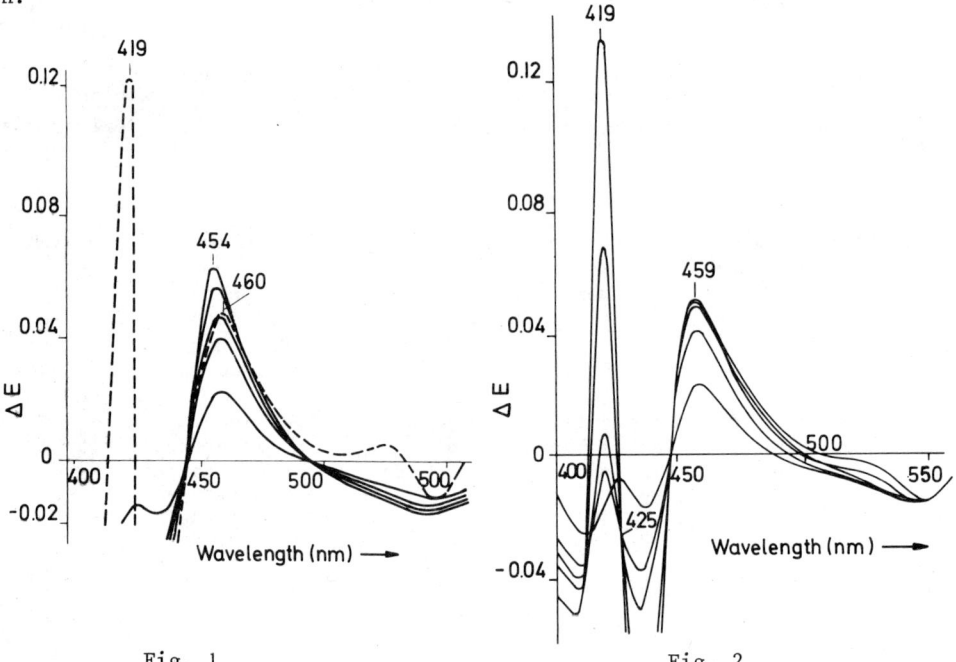

Fig. 1　Difference spectrum obtained after the addition of CCl_4 to sodium dithionite reduced liver microsomal preparations. The two cuvettes contained 3 mg of microsomal protein (2.1 nmol cyt. P450/mg protein) suspended in 3 ml 0.1 M tris-HCl buffer, pH 7.6. After reduction with sodium dithionite (2 mg) CCl_4 (4 mM) was added to the test cuvette. The scans shown were taken at 0.3, 1.0, 1.7, 4.4, and 10.0 min. Incubation temperature 25°. The dashed line represents the difference spectrum obtained after the addition of hemoglobin (25 μM) to test and reference cuvettes after 10 min. incubation.

Fig. 2　Difference spectrum obtained after the addition of CCl_4 to sodium dithionite reduced rat liver microsomes after prior addition of hemoglobin (25 μM) to test and reference cuvettes. Experimental conditions as in Fig. 1. The spectra shown were taken at 0.3, 1.0, 1.6, 4.5 and 10.0 min after CCl_4 addition.

Figure 2 shows the result of a typical experiment using CCl_4 in which hemoglobin was included in the incubation medium. In this case no shift in the Soret band was observed and the hemoglobin-carbon monoxide difference spectrum formed with time. Table 1 shows the results obtained for the series of halogenomethanes tested. In contrast to all other compounds the spectrum of $CHCl_3$ formed very slowly, 30-40 min were required before the complex reached a maximum absorption value. The spectrum observed using CCl_3Br was very small and had a maximum absorption value which was only 17 % of that obtained using CCl_4. The incorporation of hemoglobin into the incubation medium had no effect on the initial rate of complex formation and the shift in the λ_{max} value with time was inhibited. In some cases the spectrum obtained in the presence of hemoglobin was smaller than that obtained when hemoglobin was omitted from the incubation.

TABLE 1 Summary of the Spectral Data for the Halogenomethane-Mediated Complex Formation with Ferrous Cytochrome P450

Compound	Position of Peak (nm) a		b	Initial Rate of Complex Formation		ΔE_{max}/nmol P450	
	0.3 min	10 min		a	b	a	b
CCl_4	460	454	459	0.026	0.026	0.035	0.028
CBr_4	465	455	455	0.030	0.030	0.021	0.019
CCl_3F	453	452	453	0.012	0.011	0.040	0.030
CCl_3Br	454	453	(455)	0.010	0.008	0.006	0.005
CCl_3CN	468	468	468	0.024	0.022	0.020	0.021
$CHCl_3$	464	464	464	0.005	0.005	0.014	0.013
$CHBr_3$	465	455	464	0.023	0.020	0.022	0.017
CHI_3	465	463	464	0.025	0.023	0.026	0.022

a) experiment carried out in the absence of hemoglobin, b) experiment carried out in the presence of hemoglobin. Rate of complex formation expressed as ΔE/min/nmol P450. Experimental conditions are given in Figs. 1 and 2.

The rate of halogenomethane induced carbon monoxide formation is shown in Fig. 3.

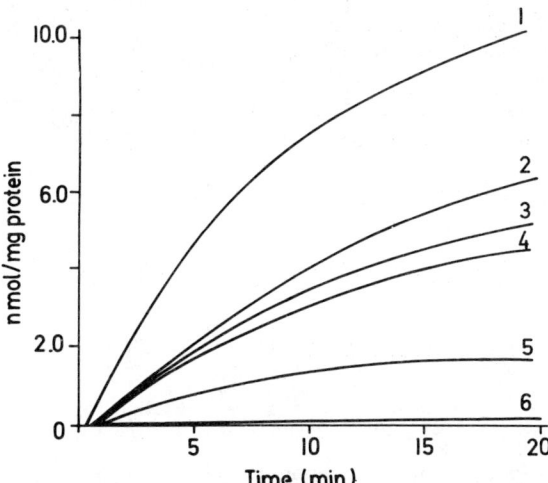

Fig. 3 Rate of carbon monoxide formation after the addition of various halogenomethanes to sodium dithionite reduced liver microsomal preparations. 3.0 ml of a microsomal protein suspension in tris buffer, (1 mg protein/ml), containing hemoglobin (25 µM), were reduced with sodium dithionite (2 mg). After the addition of substrate (4 mM) the spectral difference between 419 and 425 nm was scanned with time. Absolute values for carbon monoxide concentration were obtained from a calibration curve. 1 = CCl_4, 2 = CCl_3F, 3 = $CHBr_3$, 4 = CBr_4, 5 = CHI_3, 6 = $CHCl_3$ and CCl_3Br.

Comparison of Fig. 3 and Table 1 show that the reduction in the magnitude of the cytochrome P450 complex measured in the presence of hemoglobin, and the shift in λ_{max} was only significant for the compounds where carbon monoxide was formed. In the case of $CHCl_3$ and CCl_3CN little or no carbon monoxide was formed and no shift in λ_{max} was observed. Although the magnitude of the CCl_3CN cytochrome P450 complex was comparable to most the the compounds tested, no carbon monoxide-hemoglobin type spectrum was observed, which indicates that under these conditions the intermediates which give rise to the complex with cytochrome P450 do not bind to hemoglobin.

Experiments carried out in anaerobic microsomal suspensions reduced with NADPH gave very similar results to those obtained in the presence of sodium dithionite. The spectra formed at a comparable rate and a concomitant formation of carbon monoxide was recorded. Metyrapone (5 mM) inhibited both carbon monoxide formation and complex formation by approx. 80 %. In experiments using resuspended acid-denatured microsomes no carbon monoxide formation or complex formation could be observed. These results indicate that cytochrome P450 is involved in the reaction pathway.

DISCUSSION

With the exception of CH_2Cl_2 and CH_2Br_2 all the halogenomethanes tested gave difference spectra with ferrous cytochrome P450 with maxima ranging between 452 and 468 nm. In most cases carbon monoxide appeared as a product of the reaction and obscured the spectrum of the original complex by formation of the 450 nm-absorbing, carbon monoxide complex. This could be avoided when the spectra were recorded in the presence of hemoglobin, which trapped the carbon monoxide. The remaining spectra of the halogenomethanes resembled the ligand spectra of amines (14), phosphines (15), thiols (16), sulphides (17) or isocyanides (12) with reduced cytochrome P450. We therefore suggest that the halogenomethanes are reduced at cytochrome P450 to a species with a lone electron pair capable of liganding to the reduced heme of the cytochrome. This could either be a carbanion or a carbene formed in the following reaction sequence:

$$CX_4 + cyt\ Fe^{II} \longrightarrow [cyt\ Fe^{III} \cdot CX_3^-] + X^\ominus$$

$$\downarrow e^-$$

$$[cyt\ Fe^{III} {}^-CX_3] \rightleftharpoons [cyt\ Fe^{II} \cdot CX_3^-]$$

$$\downarrow e^-$$

$$[cyt\ Fe^{II} {}^-CX_3] \longrightarrow [cyt\ Fe^{II} |CX_2] + X^\ominus$$

$$\downarrow$$

$$cyt\ Fe^{II} + CO \xleftarrow[-2HX]{+H_2O} cyt\ Fe^{II} + |CX_2$$

As hydrolysis of halogenocarbenes is known to yield carbon monoxide, it is likely that the observed complexes involve a carbene, in analogy to the recently postulated carbene obtained on the reduction of halothane (9). The scheme shown also provides a mechanism for the reductive formation of $CHCl_3$ and $CHCl_2F$ from CCl_4 and CCl_3F respectively (6,18). It is difficult to explain why complex and carbon monoxide formation were so low in the case of CCl_3Br as it is unlikely that this compound is not reduced effectively by this system. That no spectrum was produced by CH_2Cl_2 or CH_2Br_2 can be explained by the greater difficulty in reducing these compounds. No carbon monoxide was formed in the case of CCl_3CN which seems reasonable because of the stability of the $CClCN$ carbene to hydrolysis. In all other cases the formation of carbon monoxide was established, but was first measurable about 1 min after substrate addition, whereas the cytochrome P450 carbene complex was immediately visible. This points to a relatively slow hydrolysis of the carbene complex.

Our results further establish the unusual reactivity and coordination chemistry of cytochrome P450.

ACKNOWLEDGEMENTS

Financial support for CRW from the Royal Society and the Deutsche Forschungsgemeinschaft, Sonderforschungsbereich 38, is gratefully acknowledged.

REFERENCES

1) V. Ullrich, Enzymic hydroxylations with molecular oxygen,
Angewandte Chemie 11, 702 (1972).

2) J.R. Gillette, J.J. Kamm, H.A. Sasame, Mechanism of p-nitrobenzoate reduction in liver: possible role of cytochrome P450 in liver microsomes,
Mol. Pharmacol. 4, 541 (1968).

3) P.H. Hernandez, P. Mazel, J.R. Gillette, Studies on the mechanism of mamalian hepatic azo reductase. II. The effects of phenobarbital and 3-methylcholanthrene on carbon monoxide sensitive and insensitive azo reductase activities,
Biochem. Pharmacol. 16, 1877 (1967).

4) M. Sugiura, K.I. Wasaki, R. Kato, Reduction of tertiary amine N-oxides by liver microsomal cytochrome P450,
Mol. Pharmacol. 12, 322 (1976).

5) Z.T. Wirtschafter, M.W. Cronyn, Free radical mechanism for solvent toxicity,
Arch. Environ. Health 9, 186 (1964).

6) H. Uehleke, K.H. Hellmer, S. Tarbarelli, Binding of ^{14}C-CCl_4 to microsomal proteins in vitro and formation of $CHCl_3$ by reduced liver microsomes,
Xenobiotica 3, 1 (1973).

7) O. Reiner, H. Uehleke, Bindung von Tetrachlorkohlenstoff an reduziertem mikrosomalem Cytochrom P450 und an Häm,
Hoppe-Seyler's Z. Physiol. Chem. 352, 1048 (1971).

8) V. Ullrich, K.H. Schnabel, Formation and binding of carbanions by cytochrome P450 of liver microsomes,
Drug Met. Disp. 1, 176 (1972).

9) D. Mansuy, W. Nastainczyk, V. Ullrich, The mechanism of halothane binding to microsomal cytochrome P450,
Naunyn-Schmiedeberg's Arch. Pharmacol. 285, 315 (1974).

10) U. Frommer, V. Ullrich, Hj. Staudinger, Hydroxylation of aliphatic compounds by liver microsomes,
Hoppe-Seyler's Z. Physiol. Chem. 351, 903 (1970).

11) A.G. Gornall, C.J. Bardawill, M.M. David, Determination of serum albumins by the biuret reaction,
J. Biol. Chem. 177, 751 (1949).

12) T. Omura, R. Sato, The carbon monoxide binding pigment of liver microsomes. I. Evidence for heme protein nature,
J. Biol. Chem. 239, 2370 (1964).

13) A. Rossi-Fanelli, E. Antonini, Purification and crytallization of the myoglobin in salt water fish,
    Arch. Biochem. Biophys. 58, 498 (1955).

14) J.A. Peterson, V. Ullrich, A.G. Hildebrandt, Metyrapone interaction with Pseudomonas putida cytochrome P450,
    Arch. Biochem. Biophys. 145, 532 (1971).

15) D. Mansuy, W. Duppel, H.H. Ruf, V. Ullrich, Phosphines as ligands to microsomal cytochrome P450,
    Hoppe-Seyler's Z. Physiol. Chem. 355, 1341 (1974).

16) W. Nastainczyk, H.H. Ruf, V. Ullrich, Binding of thiols to microsomal cytochrome P450,
    Chem.-Biol. Interac. 14, 251 (1976).

17) W. Nastainczyk, H.H. Ruf, V. Ullrich, Ligand binding of organic sulfides to microsomal cytochrome P450,
    Eur. J. Biochem. 60, 615 (1975).

18) C.R. Wolf, L.J. King, D.V. Parke, Anaerobic dechlorination of trichlorofluoromethane in vitro,
    Biochem. Soc. Trans. 3, 175 (1975).

SUBSTRATE-ELICITED DISSOCIATION OF THE ISOSAFROLE METABOLITE-CYTOCHROME P-450 COMPLEX AND THE CONSEQUENTIAL REACTIVATION OF MONOOXYGENATION

Clifford R. Elcombe,[*] Maurice Dickins,[†] Brian C. Sweatman,[†] and James W. Bridges[‡]

[*] Department of Pharmacology, University of Mainz 6500, W. Germany
[†] Dept. Drug Metabolism, Wellcome Research Laboratories, Beckenham, Kent, U.K.,
[‡] Dept. Biochemistry, University of Surrey, Guildford, U.K.

ABSTRACT

The present study was initiated to determine whether the substrate-elicited dissociation of the isosafrole metabolite-cytochrome P-450 complex resulted in increased monooxygenase activity. It was found that the substrates p-nitroanisole and ethoxycoumarin would elicit dissociation, furthermore this dissociation was accompanied by increased dealkylation of the substrates. Benzo(a)pyrene was found not to effect the dissociation of the complex, while biphenyl, which undergoes a similar hydroxylation, was effective.

INTRODUCTION

It has been established that methylenedioxyphenyl (1,3-dioxole) compounds react with hepatic microsomal cytochrome P-450 in the presence of O_2 and NADPH to produce characteristic difference spectra (1, 2). The expression of this spectrum is due to the generation of a methylenedioxyphenyl (MDP) metabolite-cytochrome P-450 complex. The ferrocytochrome complex has two Soret absorption maxima, that is at 427 and 455 nm, while the ferricytochrome complex exhibits a single Soret maximum at 438 nm (3).

Once produced, this complex is not functional in monooxygenation reactions, hence a dead end type of inhibition is seen (4). These complexes are also formed in vivo. After the pretreatment of animals with isosafrole and subsequent isolation of the hepatic microsomes (referred to as isosafrole-microsomes), an absorption maximum at 455 nm may be seen in the $Na_2S_2O_4$-reduced difference spectrum (5, 6).

On the addition of certain alternative substrates of the monooxygenase system to oxidized isosafrole-microsomes a time-dependent intensification of that substrate's binding spectrum (7) and a red shift in the absorption minimum (from 420 to 438 nm) of the binding spectrum are observed (6). This intensification of the binding spectrum is accompanied by the expulsion of an isosafrole-related material from the microsomes and the binding of more substrate to the cytochrome P-450 (8). This phenomenon has been termed "displacement" and compounds eliciting these changes are called "displacers" (6, 7).

The present work was initiated to determine whether this substrate-elicited dissociation of the isosafrole metabolite-ferricytochrome P-450 complex was accompanied by a corresponding increase in the monooxygenase activity of the microsomes.

MATERIALS AND METHODS

Male Wistar albino rats (180 - 200 g) were obtained from the Zentralinstitut für Versuchstierzucht (Hannover, W.-Germany). They were housed at 22 - 24° and a 12 hour (0800-2000 hr) light cycle was operated. The animals were allowed food and water ad libitum.

The rats were pretreated with three daily i.p. doses of isosafrole (150 mg/kg) in peanut oil (2.5 ml/kg). Control animals received only peanut oil. In some experiments (e.g. Fig. 4) only a single i.p. injection of isosafrole was administered. The animals were killed by cervical dislocation at 0800 hr, livers excized into ice-cold 1.15 % KCl, scissor chopped and washed three times with 1.15 % KCl buffered to pH 7.4 with 20 mM Tris-HCl. Microsomes were prepared by the method of Netter (9), and the washed microsomal pellets were finally resuspended in 20 mM Tris-HCl (pH 7.4) containing 0.25 M sucrose and 5.4 mM EDTA to a protein concentration of 20 - 30 mg/ml. Microsomes were stored at -20°C for a maximum of 2 weeks. Protein was measured by the method of Lowry et al. (10) using bovine serum albumin standards.

Biphenyl-4-hydroxylase was determined according to the method of Creaven, Parke and Williams (11). Ethoxycoumarin-O-deethylation was measured by the method of Ullrich and Weber (12), p-nitroanisole-O-demethylation was determined by a modified method of Netter and Seidel (13) and arylhydrocarbon (benzo(a)pyrene) hydroxylase was quantified by the method of Nebert and Gelboin (14) as modified by Oesch (15). All assays were carried out at 37° C.

Since substrate-elicited dissociation of the isosafrole metabolite-cytochrome P-450 complex only occurs in the oxidized state and is prevented by the addition of NADPH, the experiments were designed so that the substrates were incubated with oxidized isosafrole-microsomes for certain time periods before initiation of the monooxygenase reactions by NADPH.

The concentration of the 455 nm absorbing complex in the isosafrole-microsomes was determined as previously described (6) and the time-dependent intensifications of binding spectra were measured in the split beam repetitive scanning mode using either an Aminco DW 2 or a Perkin-Elmer 356 spectrophotometer.

RESULTS AND DISCUSSION

As demonstrated previously for other substrates, the addition of p-nitroanisole (200 μM) to oxidized isosafrole microsomes resulted in the time-dependent intensification of the p-nitroanisole type I binding spectrum (Fig. 1). The initial (30 sec) binding spectrum was barely visible but had a minimum at 420 nm, while with increasing time the trough shifted to 438 nm.

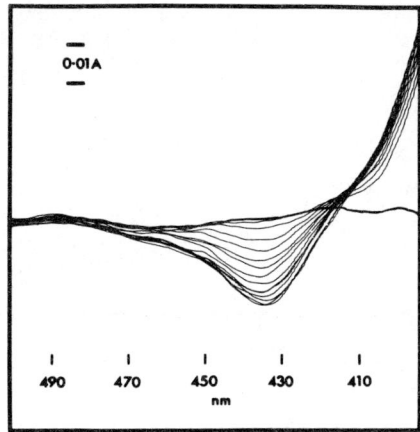

Fig. 1 The time-dependent intensification of the p-nitroanisole binding spectrum in isosafrole-microsomes.

Isosafrole-microsomes (1 mg/ml) suspended in 66 mM Tris-HCl (pH 7.4) were divided between two cuvettes. A baseline of equal light absorbance was obtained and 200 μM p-nitroanisole was added to the sample cuvette. The spectrum between 350 and 490 nm was repetitively scanned every minute.

This shift is due to the displacement of an isosafrole-related material from the microsomes (8, and Bridges, Dickins, Elcombe, Nimmo-Smith, unpublished observations).

The activity of p-nitroanisole-O-demethylase was measured in parallel to the displacement process. Since the oxidized isosafrole metabolite-cytochrome P-450 complex absorbs light at 438 nm, displacement was measured by the rate of decrease in absorbance at 438 nm. Figure 2 demonstrates that the increase of demethylase activity parallels the dissociation of the isosafrole metabolite-ferricytochrome P-450 complex. A final increase of about 250 % was observed in the O-demethylation reaction.

We have observed a similar occurrence during the dissociation of the complex by ethoxycoumarin (Fig. 3). However in this instance only a 40 % increase in ethoxycoumarin-O-deethylation was observed.

The increase in the magnitude of a displacer's binding spectrum, the shift in the position of the spectral minimum or the increase in monooxygenase activities were not observed in microsomes obtained from animals pretreated with peanut oil alone.

We have suggested (6) that both lipophilicity and molecular size are important parameters in determining the ability of a compound as a displacer. This suggestion is supported by our observation that benzo(a)pyrene does not undergo a time-dependent intensification of its binding spectrum in isosafrole-microsomes, nor does the arylhydrocarbon (benzo(a)pyrene)

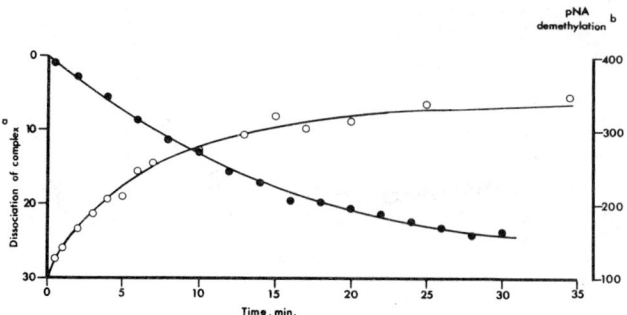

Fig. 2 Dissociation of the <u>in vivo</u> produced isosafrole metabolite cytochrome P-450 complex and the consequential increase of p-nitroanisole-O-demethylation.
a) Dissociation of complex as measured by the decrease in absorption at 438 nm (●——●).
b) O-demethylation expressed as % of activity at "zero time", i.e. no pre-incubation of the isosafrole microsomes with substrate (o——o).

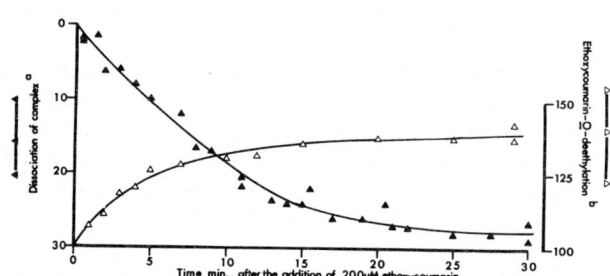

Fig. 3 Dissociation of the <u>in vivo</u> produced isosafrole metabolite-cytochrome P-450 complex and the consequential increase of ethoxycoumarin-O-deethylase.
Conditions as Fig. 2 except the substrate was 200 μM ethoxycoumarin.
▲——▲——▲ Dissociation of complex
△——△——△ O-deethlyation

hydroxylase activity increase after pre-incubation of benzo(a)pyrene with oxidized isosafrole-microsomes. Although benzo(a)pyrene is very lipophilic, it is probably too large to effect displacement of the isosafrole metabolite.

We have previously illustrated that the degree of intensification of a displacer's own binding spectrum is proportional to the amount of cytochrome P-450 which is complexed with the isosafrole metabolite (6). We have now shown that the increase of monooxygenase activity after the dissociation of the isosafrole metabolite-cytochrome P-450 complex is dependent on the amount of complex present. In order to obtain different amounts of the isosafrole metabolite-cytochrome P-450 complex, microsomes were prepared from animals at different times after the administration of a single dose (150 mg/kg) of isosafrole. Biphenyl-4-hydroxylase activity (before and after dissociation of the complex by biphenyl) and the amount of metabolite complex were measured in the various isosafrole-microsome preparations. As is seen in Fig. 4, initially inhibition and then induction of biphenyl hydroxylation occurred as previously described (16). The amount of isosafrole metabolite-cytochrome P-450 complex, as measured by the 455 nm absorbance in the $Na_2S_2O_4$-reduced difference spectrum, gradually increased with time after pretreatment.

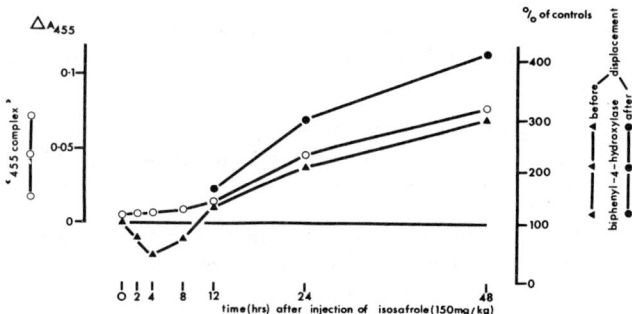

Fig. 4 The relationship between biphenyl-4-hydroxylase, appearance of isosafrole metabolite-cytochrome P-450 complex and dissociation of complex after administration of a single dose of isosafrole to rats.

o—o—o Amount of metabolite complex, measured by $Na_2S_2O_4$-reduced difference spectrum and expressed as $\Delta A_{455-490}$ mg microsomal protein.

▲—▲—▲ Biphenyl-4-hydroxylase activity before dissociation of complex by biphenyl.

●—●—● Biphenyl-4-hydroxylase activity after dissociation of complex by biphenyl.

Results for the 4-hydroxylation are expressed as % of values obtained with microsomes from control rats killed at corresponding time points.

When biphenyl was pre-incubated for 30 min with the oxidized isosafrole-microsomes (obtained at various times after dosing with isosafrole), prior to initiation of the hydroxylation reaction by NADPH, an increase of 4-hydroxylation was observed. The magnitude of increase in the 4-hydroxylation (above the levels found without pre-incubation) was dependent upon the amount of complex present.

Hence in conclusion, it is apparent that the formation of the isosafrole metabolite-cytochrome P-450 complex in vivo does not lead to irreversible inactivation of cytochrome P-450 and the substrate-elicited dissociation of the complex leads to the release of a metabolically competent cytochrome P-450.

ACKNOWLEDGEMENTS

C.R.E. gratefully acknowledges a Royal Society European Fellowship and M.D. and B.C.S. thank the Wellcome Foundation Ltd. for financial assistance in the form of Research Studentships. Part of this work was carried out under an exchange arrangement between the Universities of Surrey and Mainz.

REFERENCES

1. R.M. Philpot and E. Hodgson, The effect of piperonyl butoxide concentration on the formation of cyt. P-450 difference spectra in hepatic microsomes from mice, Mol. Pharmacol. 8, 204 (1972)

2. M.R. Franklin, The enzymic formation of a methylenedioxyphenyl derivative exhibiting an isocyanide-like spectrum with reduced cytochrome P-450 in hepatic microsomes, Xenobiotica 1, 581 (1971)

3. C.R. Elcombe, J.W. Bridges, R.H. Nimmo-Smith and J. Werringloer, Cumene hydroperoxide-mediated formation of inhibited complexes of methylenedioxyphenyl compounds with cytochrome P-450, Biochem. Soc. Transac. 3, 967 (1975)

4. M.R. Franklin, Inhibition of hepatic oxidative xenobiotic metabolism by piperonyl butoxide, Biochem. Pharmacol. 21, 3287 (1972)

5. D.V. Parke and H. Rahman, Induction of a new microsomal haemoprotein by safrole and isosafrole, Biochem. J. 123, 9P (1971)

6. C.R. Elcombe, J.W. Bridges, T.J.B. Gray, R.H. Nimmo-Smith and K.J. Netter, Studies on the interaction of safrole with rat hepatic microsomes, Biochem. Pharmacol. 24, 1427 (1975)

7. T.J.B. Gray and D.V. Parke, Interactions of safrole with hepatic cytochrome P-450, 9th Int. Congress Biochem. (Stockholm) Abs. p. 341 (1973)

8. C.R. Elcombe, J.W. Bridges and R.H. Nimmo-Smith, Substrate-elicited dissociation of a complex of cytochrome P-450 with a methylenedioxyphenyl metabolite, Biochem. Biophys. Res. Commun. in press (1976)

9. K.J. Netter, Eine Methode zur direkten Messung der O-Demethylierung in Lebermikrosomen und ihre Anwendung auf die Mikrosomenhemmwirkung von SKF 525-A, Naunyn-Schmiedebergs Arch. exp. Path. Pharmak. 238, 292 (1960)

10. O.H. Lowry, N.J. Rosebrough, A.L. Farr and R. Randall, Protein measurement with Folin phenol reagent, J. Biol. Chem. 193, 265 (1951)

11. P.J. Creaven, D.V. Parke and R.T. Williams, A fluorimetric study of the hydroxylation of biphenyl in vitro by liver preparations of various species, Biochem. J. 96, 879 (1965)

12. V. Ullrich and P. Weber, The O-dealkylation of 7-ethoxycoumarin by liver microsomes. A direct fluorometric test, Hoppe-Seyler's Z. Physiol. Chem. 353, 1171 (1972)

13. K.J. Netter and G. Seidel, An adaptively stimulated O-demethylating system in rat liver microsomes and its kinetic properties, J. Pharmacol. exp. Ther. 146, 61 (1964)

14. D.W. Nebert and H.V. Gelboin, Substrate-inducible microsomal aryl hydroxylase in mammalian cell culture. 1. Assay and properties of induced enzyme, J. Biol. Chem. 243, 6242 (1968)

15. F. Oesch, Differential control of rat microsomal "aryl hydrocarbon" monooxygenase and epoxide hydratase, J. Biol. Chem. 251, 79 (1976)

16. H. Jaffe, K. Fujii, H. Guerin, M. Sengupta and S.S. Epstein, Bi-modal effect of piperonyl butoxide on the o- and p-hydroxylations of biphenyl by mouse liver microsomes, Biochem. Pharmacol. 18, 1045 (1969)

THE INTERACTION OF HALOTHANE WITH LIVER MICROSOMAL CYTOCHROME P450

Wolfgang Nastainczyk and Volker Ullrich

Department of Physiological Chemistry, University of the Saarland, Homburg/Saar, GFR

ABSTRACT

Optimal photodissociation occurred at 470 nm of the postulated trifluoromethylcarbene cytochrome P450 complex obtained with halothane and reduced rat liver microsomal cytochrome P450, which supports the hypothesis that the carbene is a ligand to reduced cytochrome P450. The complex is formed with liver microsomal P450 in the presence of an oxygen pressure below 25 torr. It is assumed that two species of liver microsomal cytochrome P450 take part in the carbene cytochrome P450 complex formation. The carbene complex formation can be prevented in vitro by oxygen pressures greater than 25 torr and at oxygen pressures below 25 torr by 1 mM metyrapone in the NADPH containing microsomal suspension.

INTRODUCTION

Halothane has been reported to bind covalently to liver proteins when administered in vivo to mice (1) or in perfused rat liver (2). Uehleke (3) reported that under reducing conditions halothane forms a complex with cytochrome P450 which suggested a correlation between this complex formation and covalent binding. We have interpreted the chemical nature of this cytochrome P450 halothane complex as a ligand binding of the trifluoromethylcarbene (4).

The main evidence for this assumption was the observation that 2.2.2-trifluorodiazoethane reacts with reduced cytochrome P450 by formation of a difference spectrum identical to that produced by halothane. The chemical routes for formation of the cytochrome P450 carbene complex and the biochemical pathway are depicted in the following scheme:

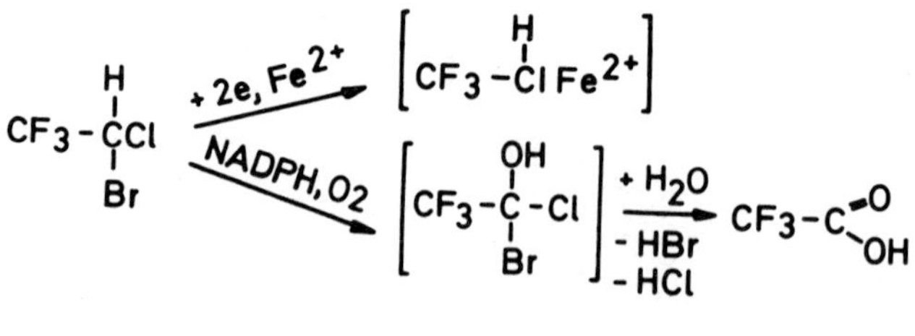

$$CF_3-\underset{Br}{\underset{|}{\overset{H}{\overset{|}{C}}}}Cl \xrightarrow[\text{NADPH}, O_2]{+2e, Fe^{2+}} \left[CF_3-\overset{H}{\overset{|}{C}}l\, Fe^{2+}\right]\; \left[CF_3-\underset{Br}{\underset{|}{\overset{OH}{\overset{|}{C}}}}-Cl\right] \xrightarrow[-HBr\\-HCl]{+H_2O} CF_3-C\overset{O}{\underset{OH}{\diagdown}}$$

It was the aim of the present study to characterize the conditions under which this halothane carbene complex can be formed in vivo.

METHODS

Male Sprague-Dawley rats (100-150 g) were used. Phenobarbital was administered intraperitoneally over 3 days by one daily injection of 80 mg/kg body weight. The microsomal fraction was isolated (5) and the protein content was determined by the biuret reaction according to Gornall et al. (6). Difference spectra were recorded on an Aminco DW-2-spectrophotometer using 10 mm cuvettes containing 3 ml of a microsomal suspension in 0.1 M Tris-HCl pH 7.6.

For the photodissociation spectra, a high pressure xenon lamp (Osram XBO, 2000 W) served as a light source. The light beam passed a condensor lens, a water bath with copper sulfate solution as a heat filter and an appropriate interference filter in the diffuse focus of the light beam. The relative energy distribution at the various wavelenghts was measured by a compensated thermophile. Reference cuvettes were always run in parallel. The irradiation time was 50 sec.

RESULTS

Metal carbon monoxide complexes contain a metal carbon bond which can be photodissociated by irradiation with light of appropriate wavelengths (7). This technique was used in the experiment shown in Fig. 1 in order to establish the ligand nature of the postulated cytochrome P450 trifluoromethylcarbene complex.

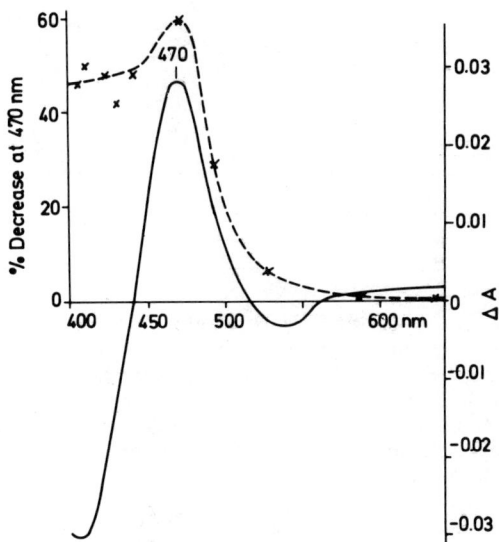

Fig. 1. Photodissociation spectrum of the cytochrome P450 halothane complex in reduced rat liver microsomes (dashed line). Each cuvette contained 7.5 mg protein (2.1 nmol cytochrome P450/mg) and 2 mg of sodium dithionite (room temperature). The sample cuvette contained in addition 2 µM halothane.

Difference spectra obtained from dithionite reduced rat liver microsomes with halothane (solid line). The two cuvettes contained 5 mg of protein (2.6 nmol cytochrome P450/mg protein) and 2 mg of dithionite. 4 µl of 10^{-3}M halothane in methanol added.

Light of about 470 nm was most efficient and an action spectrum was obtained (dashed line) which resembled the absorption spectrum of the cytochrome P450 halothane complex. The photodissociation could only be observed if the halothane concentration was in the same range as cytochrome P450. With higher concentrations the carbene complex was reformed from the excess of halothane in the medium.

In order to answer the question whether the halothane carbene complex could also be formed under oxygen pressure of in vivo conditions a cuvette was designed which allowed the simultaneous recording of the oxygen pressure and the absorbance change between 471 and 490 nm. NADPH was used as the electron donor and as can be seen from Fig. 2 the peak formation at 471 nm started at an oxygen pressure of about 25 torr in the presence of 3 mM halothane. At 0.5 mM halothane the formation occurred at about 10 torr (experiment not shown here).

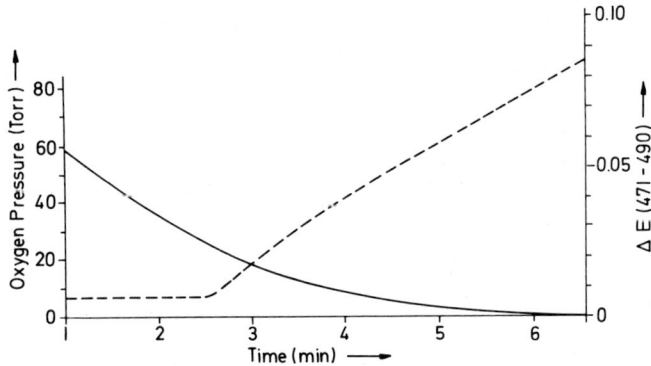

Fig. 2. Simultaneous recording of oxygen uptake and 470 nm peak formation in rat liver microsomes in the presence of 4 mM halothane.
A special cuvette with stirrer and a Clark-type oxygen electrode was placed in an Aminco-Chance DW-2 spectrophotometer.
After addition of NADPH to 3 ml of a liver microsomal suspension (3 mg of protein/ml, 1.9 nmol cytochrome P450/mg) the oxygen uptake and the increase in extinction between 470 and 490 nm were recorded.

Readmission of oxygen resulted in a rather slow disappearance of the peak and flushing with nitrogen or addition of dithionite reestablished the maximal peak height as shown in Figure 3. Not all of the absorbance at 471 nm disappeared in the presence of oxygen but about 25 % remained unaffected.

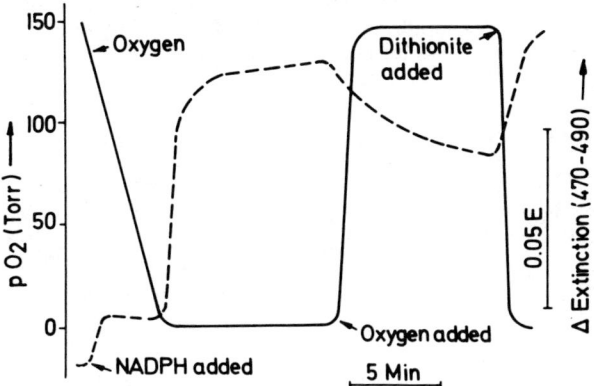

Fig. 3. The same conditions as in the Fig. 2.

A 10-fold molar excess of metyrapone over halothane were needed to displace the carbene ligand by forming the reduced cytochrome P450 metyrapone complex, but a part of the 471 nm absorption was still present (Fig. 4).
Other ligands like 1-phenylimidazole, organic sulfides (8) or phosphines (9) gave similar results.

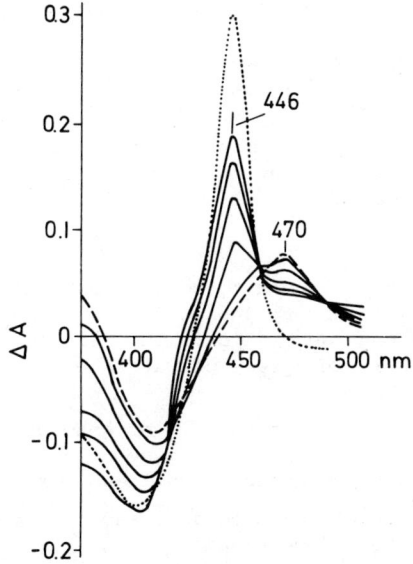

Fig. 4. Effects of increasing concentrations of metyrapone on the halothane induced spectrum of $Na_2S_2O_4$ reduced rat liver microsomes. Each cuvette contained 7.5 mg of microsomal protein (2 nmol cytochrome P450/mg protein). The dotted line corresponds to the addition of 100 μM metyrapone in the cuvette without halothane. The dashed curve was obtained by addition of 10 μM halothane. The solid lines correspond to additions of 0.3, 2, 7, 25 and 100 μM metyrapone.

Adding metyrapone to reduced liver microsomes prior to the addition of halothane considerably slowed down the formation of the carbene complex. About 20 % of the complex was formed during 3 hours after which the reaction ceased (Fig. 5).

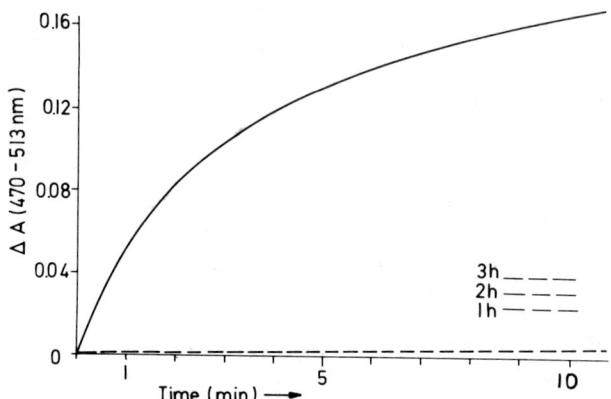

Fig. 5. Effect of metyrapone on the kinetics of the 470 nm peak formation of halothane in NADPH reduced, anaerobic rat liver microsomes.
The cuvette contained 7.6 mg of microsomal protein (2 nmol cytochrome P450/mg) and 2 mg of NADPH.
The solid curve corresponds to addition of 1 mM halothane. The dashed line was obtained in presence of 1 mM metyrapone after addition of 1 mM halothane. The increase in extinction between 470-513 nm was recorded.

Thus most of the carbene complex formation can be prevented in the presence of metyrapone. This may be important in view of the potential toxicity of this carbene complex in the liver cell.

DISCUSSION

The photodissociation of the reduced cytochrome P450 halothane complex leaves little doubt that the 471 nm absorption band corresponds to a carbon ligand complex (4). The free carbene generated during irradiation disappears in a subsequent reaction which is in agreement with the known high reactivity of these species (10).

It is evident from our results that the 471 nm absorption consists of at least two components. One component, representing about 25-30 % of the absorption band, is very stable and is not affected by light, oxygen and ligands.

The second component can be photodissociated and converted to a nitrogen ligand complex by the addition of metyrapone.

Molecular oxygen can decompose this carbene species. The existence of two species could be explained as a consequence of multiple forms of cytochrome P450 (11,12,13) with different affinities for the carbene intermediate.

In view of the potential activity of a carbene complex in covalent binding to cell constituents it was important to establish the conditions of its formation, especially with regard to the oxygen concentration. At a partial pressure of about 25 torr the formation of the 471 nm absorption started and was completed before anaerobiosis was achieved. Thus, it is quite certain that even under in vivo conditions the formation of the carbene complex can proceed although hypoxia will greatly increase its formation.

Reports in literature indicate a higher degree of covalent binding of ^{14}C-labeled halothane under hypoxic conditions (14). From this we would suggest that a reductive pathway of halothane metabolism must be involved in generating the reactive species and a carbene complex of cytochrome P450 could be a likely candidate for such an intermediate.

Sufficiently high oxygen concentrations and the presence of a strong ligand for cytochrome P450 would minimize the possibility of halothane reduction to the reactive intermediate. Theoretically any compound that could compete with the formation of the cytochrome P450 halothane substrate complex in the oxidized state, would also be effective in inhibiting a reduction at cytochrome P450.

ACKNOWLEDGEMENTS

This work was supported by the Deutsche Forschungsgemeinschaft, Sonderforschungsbereich 38.

REFERENCES

1) E.N. Cohen, Metabolism of halothane-2^{14}C in the mouse,
Anesthesiology 31, 560 (1969)

2) R.A. van Dyke, C.L. Wood, Binding of radioactivity from 14-C-labeled halothane in isolated perfused rat livers,
Anesthesiology 38, 328 (1973)

3) H. Uehleke, K.H. Hellmer and S. Tabarelli-Poplawski, Metabolic activation of halothane and its covalent binding to liver endoplasmic proteins in vitro,
Naunyn-Schmiedeberg's Arch. Pharmacol. 279, 39 (1973)

4) D. Mansuy, W. Nastainczyk and V. Ullrich, The mechanism of halothane binding to microsomal cytochrome P450,
Naunyn-Schmiedeberg's Arch. Pharmacol. 285, 315 (1974)

5) U. Frommer, V. Ullrich, Hj. Staudinger, Hydroxylation of aliphatic compounds by liver microsomes,
Hoppe-Seyler's Z. physiol. Chem. 351, 903 (1970)

6) A.G. Gornall, C.J. Bardawill, M.M. David, Determination of serum albumins by means of the biuret reaction,
J. Biol. Chem. 177, 751 (1949)

7) Antonini, E., Brunori, M., Hemoglobin and Myoglobin im Their Reactions with Ligands,
Frontier of Biology 21, North-Holland Publishing Comp., Amsterdam, 1971.

8) W. Nastainczyk, H.H. Ruf, V. Ullrich, Ligand binding of organic sulfides to microsomal cytochrome P450,
Eur. J. Biochem. 60, 615 (1975)

9) D. Mansuy, W. Duppel, H.H. Ruf, V. Ullrich, Phosphines as ligands to microsomal cytochrome P450,
Hoppe-Seyler's Z. Physiol. Chem. 355, 1341 (1974)

10) J.H. Atherton, R. Fields, Reaction of trifluoromethylcarbene with cis- and trans-but-2-ene,
    J. Chem. Soc. (London), 1450 (1967)

11) Y. Ishimura, V. Ullrich, J.A. Peterson, Oxygenated cytochrome P450 and its possible role in enzymatic hydroxylation,
    Biochem. Biophys. Res. Commun. 42, 140 (1971)

12) A.F. Welton, F.O. O'Neal, C.L. Chaney and St. D. Aust, Multiplicity of cytochrome P450 hemoproteins in rat liver microsomes,
    J. Biol. Chem. 250, 5631 (1975)

13) P.F. Guengerich, D.P. Ballou, M.J. Coon, Purified liver microsomal cytochrome P450 electron-accepting properties and oxidation-reduction potential,
    J. Biol. Chem. 250, 7405 (1975)

14) R.A. van Dyke, C.L. Wood, In vitro studies on irreversible binding of halothane metabolite to microsomes,
    Drug Metabolism and Disposition 3, 51 (1975)

THE FORMATION OF HYDROGEN PEROXIDE DURING HEPATIC MICROSOMAL ELECTRON TRANSPORT REACTIONS+

Jurgen Werringloer

Department of Biochemistry, University of Texas Health Science Center, Dallas, Texas 75235

ABSTRACT

The production of hydrogen peroxide by hepatic microsomal fractions from phenobarbital treated rats accounts for approximately 50% of the NADPH oxidized and oxygen utilized in the absence of exogenous substrates. Properties analogous to those of the mixed function oxygenase system were observed for this particular reaction by comparing the affinity for NADPH, the activation energy required, and the sensitivity to carbon monoxide inhibition. The failure of NADH to elicit a synergistic effect tends to support the hypothesis that hydrogen peroxide is formed by the dismutation of superoxide anions which are generated upon decomposition of the oxygenated complex of ferrous cytochrome P-450.

INTRODUCTION

The formation of hydrogen peroxide during the aerobic incubation of liver microsomes with NADPH, as originally described by Gillette, Brodie and La Du in 1957 (1), has been established in recent years by Thurman (2) as well as Hildebrandt (3,4) and their coworkers. This phenomenon has gained particular importance in view of the controversy over the existance of an independent microsomal ethanol oxidizing system, MEOS, (5,6) and the analysis of the reactions involved bears possible significance to further the understanding of the complexity of the microsomal electron transport reactions implicated in the activation of molecular oxygen for substrate hydroxylation. The two components of the endoplasmic reticulum suggested to be responsible for the formation of hydrogen peroxide are cytochrome P-450 (2,4) as well as the NADPH dependent flavoprotein functioning in the reduction of this pigment (7). Experiments directed to more clearly define the source of hydrogen peroxide were carried out by comparing this reaction with the N-demethylation of ethylmorphine. The results obtained are the subject of the present communication.

METHODS

Male Charles River CD outbred albino rats (150 to 200 g body weight) received daily intraperitoneal injections of sodium phenobarbital (80 mg/kg) over a period of 4 days. The animals were fasted for 12 hours prior to sacrifice on the fifth day. The liver microsomal fractions were prepared by differential centrifugation in 0.25 M sucrose containing 50 mM tris-chloride, pH 7.5, as described previously (8). The experimental conditions and the analytical procedures used are specified in the figure legends.

+Supported by a grant (NIGMS-16488) from the USPHS.

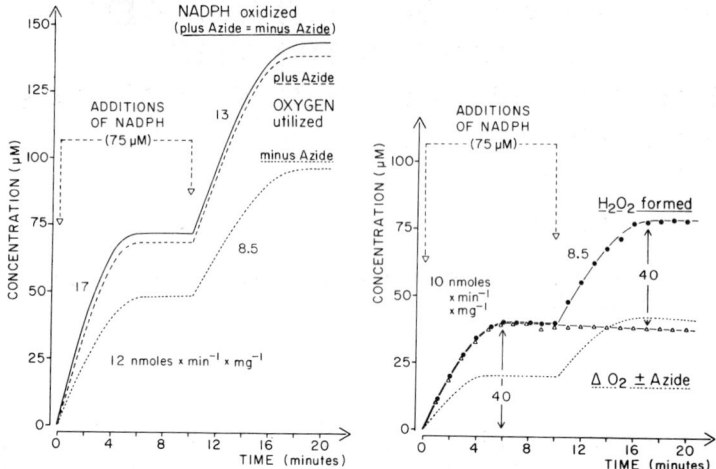

Fig. 1. Oxygen utilization and hydrogen peroxide formation during the oxidation of NADPH by liver microsomal fractions. Liver microsomes were suspended to a final concentration of 1 mg protein per ml in 50 mM tris-chloride buffer, pH 7.5, containing 10 mM $MgCl_2$, 150 mM KCl and 2 mM 5'AMP. Two types of experiments were carried out at room temperature in the presence or absence of 1 mM azide to either monitor the oxidation of NADPH at 340 nm using an Aminco DW 2 spectrophotometer (split beam mode, light path of cuvettes: 4.38 mm) or to record polarographically the utilization of oxygen. A third series of experiments were performed in the presence of 1 mM azide to follow the formation of hydrogen peroxide. The samples withdrawn from the incubation medium during the reaction were assayed colorimetrically using ferrous ammonium sulfate and potassium thiocyanate as reactants (3). Two consecutive additions of NADPH were made to initiate the reactions.

RESULTS AND DISCUSSION

Quantitative relation between NADPH oxidation, oxygen utilization and hydrogen peroxide formation.

The significance of the contribution of hydrogen peroxide formation to the NADPH dependent utilization of oxygen by liver microsomal fractions in the absence of exogenous substrates has been analyzed using incubation media which were supplemented with azide and 5'AMP. The presence of these two components is required to inhibit the breakdown of hydrogen peroxide by adventitious catalase contaminating the microsomal preparation (3) and to prevent the destruction of pyridine nucleotides by the microsomal pyrophosphatase activity (9). A close correlation between the rate and the extent of NADPH oxidation and oxygen utilization is observed under such conditions, as demonstrated in Fig. 1, and about 50% of the oxygen utilized and NADPH oxidized can be accounted for by the production of hydrogen peroxide. The omission of azide from the reaction media, which did not influence the pyridine nucleotide oxidation, resulted in a marked decrease of the rate as well as the extent of oxygen consumption consistent with the regeneration of oxygen by the catalase dependent destruction of hydrogen peroxide. The observed relationship between the amount of oxygen generated, e.g. the difference

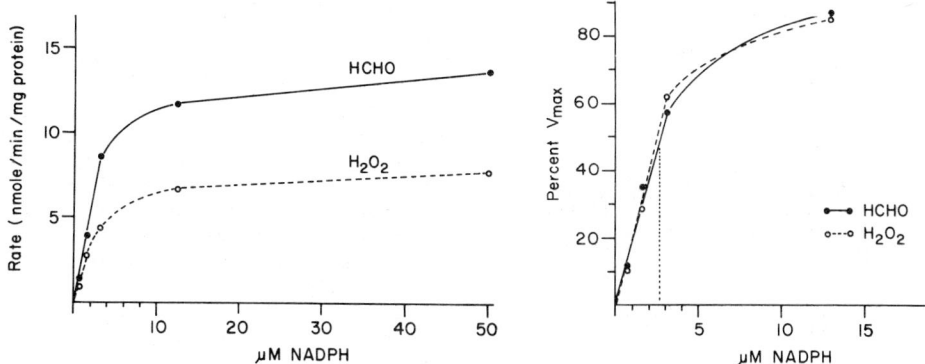

Fig. 2. The influence of varying steady state concentrations of NADPH on the initial rates of formation of hydrogen peroxide and the N-demethylation of ethylmorphine. Liver microsomes were diluted to a concentration of 1 mg protein per ml in a buffer mixture as described in the legend to Fig. 1. These incubation media were supplemented with 1 mM azide, 5 mM sodium isocitrate, 0.5 units isocitrate dehydrogenase per ml and 5 mM ethylmorphine. Varying concentrations of NADPH were added to initiate the reactions, which were followed by withdrawing samples from the incubation media at time intervals of 20 sec over at least 3 min. These samples were divided to determine both the concentration of hydrogen peroxide, as described in the legend to Fig. 1, and the amount of formaldehyde liberated from ethylmorphine using the Nash reagent (10). The actual rates of both these reactions (left portion of the Figure) have been replotted as the percent of the rates observed with 200 µM NADPH (right portion of the Figure).

between oxygen uptake in the presence and absence of azide ($\Delta O_2 \pm$ AZIDE as plotted in Fig. 1), and the concentration of hydrogen peroxide determined in the presence of azide is in agreement with the stoichiometry of the catalatic reaction of catalase, e.g. 1/2 mole of oxygen formed per mole of hydrogen peroxide uitlized. The overall stoichiometry of one mole of NADPH oxidized per mole of oxygen consumed and the requirement of only two electrons per mole of hydrogen peroxide formed suggests that the amount of cofactors consumed in excess of the amount of hydrogen peroxide generated is utilized in the oxidation of endogenous substrates - reactions which are most likely related to the function of cytochrome P-450.

<u>Comparison of certain properties of the reactions involved in the formation of hydrogen peroxide and the N-demethylation of ethylmorphine.</u>

Attempts to compare mixed function oxidation reactions with the pathway resulting in hydrogen peroxide formation have centered on measurements of the influence of varying steady state concentrations of NADPH. For this purpose, microsomal fractions were incubated in the presence of an NADPH generating system with ethylmorphine as as substrate for the mixed function oxygenase system. Relatively low concentrations of NADPH proved to be sufficient, as demonstrated in Fig. 2, to support both the N-demethylation of ethylmorphine to generate formaldehyde as well as the production of hydro-

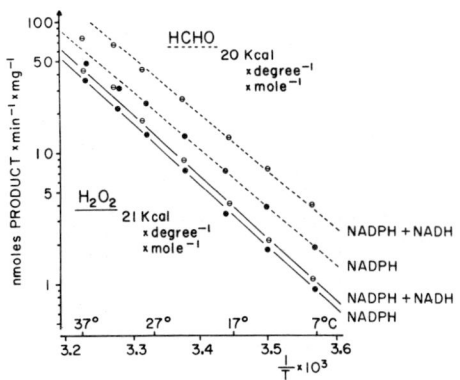

Fig. 3. Temperature dependence of the rates of hydrogen peroxide formation and the N-demethylation of ethylmorphine. A series of experiments were carried out at various temperatures using microsomes at a concentration of 1 mg protein per ml. The experimental conditions were identical to those described in Fig. 2, except that 0.5 µM rotenone was added to the incubation medium. The reactions were initiated with 200 µM NADPH or with 200 µM NADPH plus 200 µM NADH. Samples were taken up to every 15 sec at higher temperatures and were analyzed as described in Fig. 2.

gen peroxide. When the actual rates of these reactions are replotted as percent of the maximal activity determined with 200 µM NADPH it is observed that the results for formaldehyde liberation and hydrogen peroxide formation are superimposable. This failure to demonstrate significant differences in the requirements for NADPH suggests the involvement of a common electron transport carrier, e.g. the NADPH cytochrome P-450 reductase, in both the N-demethylation of ethylmorphine and the formation of hydrogen peroxide.

Further attempts to distinguish the pathways of pyridine nucleotide oxidation resulting in the formation of hydrogen peroxide from those associated with substrate hydroxylation included the determination of the activation energies required. Incubations similar to those described in Fig. 2 were performed over a temperature range from 7 to 37°C. The results obtained indicate, as illustrated in Fig. 3, that an activation energy of approximately 20 to 21 kcal per mole is required for either reaction, whether NADH - serving as a synergist for microsomal mixed function oxidation reactions (11,12) - is absent or present. The activation energy determined in these experiments for the N-demethylation of ethylmorphine is similar to values reported elsewhere (13,14) and agrees well with the activation energy determined by Peterson et al. (15) for the rate constant of the fast phase of cytochrome P-450 reduction in the presence of ethylmoprhine. It should be noted, however, that the Arrhenius plots in Fig. 3 as well as the plots obtained by Peterson et al. for the fast phase of cytochrome P-450 reduction lack any significant discontinuity at around 20°C, a temperature range at which fluidity changes within the microsomal membrane are presumed to occur (16). Since an activation energy of 33 kcal is required for the slow phase of cytochrome P-450 reduction at temperatures above 20°C (15), it may be assumed that those P-450 molecules, which are slowly reduced, do not contribute significantly to either the formation of hydrogen peroxide or the N-demethylation of ethylmorphine. The apparent discrepancy between the rate constants reported for the fast phase of cytochrome P-450 reduction (15) and the actual rates of the NADPH dependent formation of formaldehyde might be interpreted to suggest that the availability of electrons is rate determining for the donation of the second electron leading to the activation of oxygen (17). Consistent with such an interpretation is the twofold stimulation of the rate of ethylmorphine N-demethylation in the added presence of NADH without concomitant changes of the activation energy required (Fig. 3). The failure, however, to demonstrate a similar stimulation by NADPH of the NADPH dependent generation of hydrogen

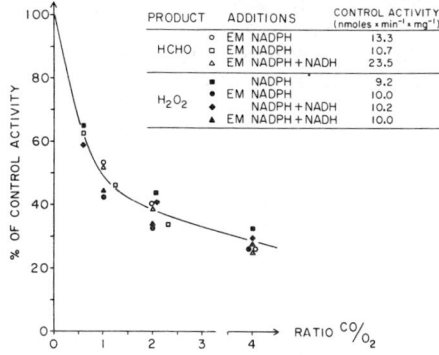

Fig. 4. Comparison of the inhibitory effect of carbon monoxide on the rates of hydrogen peroxide formation and ethylmorphine N-demethylation. The experimental conditions as well as the analytical procedures were identical to those described in legend to Fig. 3, except that the temperature was kept constant at 25°C. A gas mixing pump was used to vary the relative atmospheric concentrations of carbon monoxide and nitrogen and to maintain an oxygen level of 20% (EM = ethylmorphine).

peroxide (Fig. 3) appears to be inconsistent with the implication of an analogous reaction sequence, despite the fact that the similarity of temperature dependence of the formation of formaldehyde and hydrogen peroxide is indicative of the involvement of a common reaction step.

As a further means to evaluate possible differences between the pathways of hydrogen peroxide formation and substrate hydroxylation the sensitivity to carbon monoxide inhibition of both these reactions was examined. Summarized in Fig. 4 are the results of a series of such experiments using varying mixtures of carbon monoxide, oxygen and nitrogen and varying conditions for the formation of hydrogen peroxide by supplementing the incubation media with either ethylmorphine or NADH or with both these reagents. As demonstrated in Fig. 4, the percent inhibition of hydrogen peroxide production by carbon monoxide is essentially identical to the inhibition of the N-demethylation of ethylmorphine. These results indicate a higher degree of carbon monoxide sensitivity of the NADPH dependent generation of hydrogen peroxide than observed in studies reported from other laboratories (2,5,18), using different experimental conditions and indirect methods to estimate the rates of this particular reaction. Of interest is the observation that the added presence of NADH did not alter the inhibitory patterns observed.

Considering the evidence gathered so far, it is apparent that the NADPH cytochrome P-450 reductase as well as cytochrome P-450 itself are implicated in the production of hydrogen peroxide as they are known to function in the liberation of formaldehyde from ethylmorphine. The different response of the rates of both these reactions to the added presence of NADH indicates however, that the reaction mechanisms involved in the formation of both these products are not identical.

Possible mechanisms of cytochrome P-450 dependent hydrogen peroxide formation.

Examining the involvement of the terminal oxidase of the microsomal mixed function oxygenase system in the formation of hydrogen peroxide, the following possibilities have to be considered (Fig. 5):
1) The dissociation of the oxygenated complex of ferrous cytochrome P-450 resulting in the release of oxygen as the superoxide anion which in turn may dismutate to yield hydrogen peroxide, and
2) the breakdown of the peroxide complex of cytochrome P-450, which is formed after a second one-electron reduction, the electron being derived from either NADH or NADPH (19).

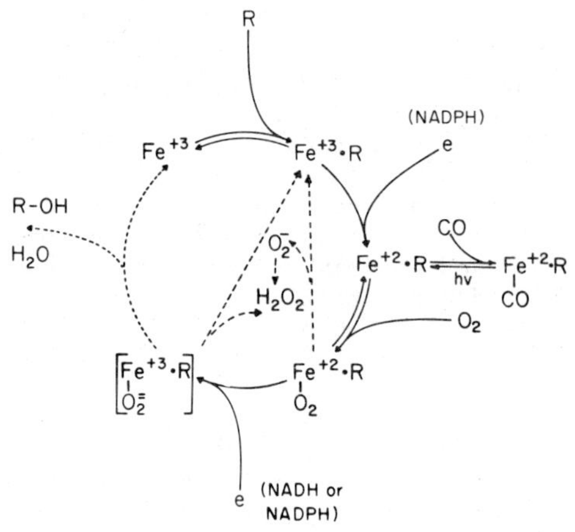

Fig. 5. Schematic representation of the possible mechanisms of cytochrome P-450 dependent hydrogen peroxide formation.

The previously discussed failure of NADH to stimulate the rate of the NADPH dependent generation of hydrogen peroxide tends to support the first hypothesis. Since the "synergistic" effect of NADH, as first described by Cohen and Estabrook (11), is most readily observed when the extent of product formation is analyzed upon complete oxidation of limiting amounts of pyridine nucleotides, additional experiments were carried out. Varying concentrations of NADH - up to 75 µM - were added together with 75 µM NADPH to incubation media supplemented with rotenone to prevent the utilization of NADH by contaminating fragments of mitochondrial membranes. As demonstrated in Fig. 6, NADH proved to be no more potent under these conditions to increase the extent of hydrogen peroxide production than it was found to promote the rate of the reaction.

When similar experiments were performed to compare the effect of such NADH concentrations on the extent of the NADPH dependent N-demethylation of ethylmorphine increases of greater than 80% in the final amount of formaldehyde liberated were obtained (Fig. 6). Critical for the interpretation of these results is the observation, that about 48 µM NADH per 72 µM NADPH were sufficient to obtain maximal stimulation of the extent of formaldehyde formation. Concentrations of NADH approaching a 1:1 ratio with NADPH proved to be an excess, as evidenced by the continued slow rate of formaldehyde production after completion of the rapid phase of substrate hydroxylation in a "synergistic" manner. These results were found to be consistent with quantitative estimations of the oxidation of NADH during NADPH dependent electron transport reactions (20).

Since the 1:1 stoichiometry of pyridine nucleotide oxidation to oxygen utilization is maintained in the presence of NADPH plus NADH (21), it is concluded that a two-electron requiring pathway is operative deriving its reducing equivalents solely from NADPH, which offsets a possible ratio of 1:1 for the relative amounts of NADPH and NADH oxidized. Considering the lack of a classic "synergistic" response of the NADPH dependent hydrogen peroxide generation to the added presence of NADH it appears likely that the particular reaction depending solely on NADPH is identical with the reaction of the microsomal electron transport system responsible for the formation of

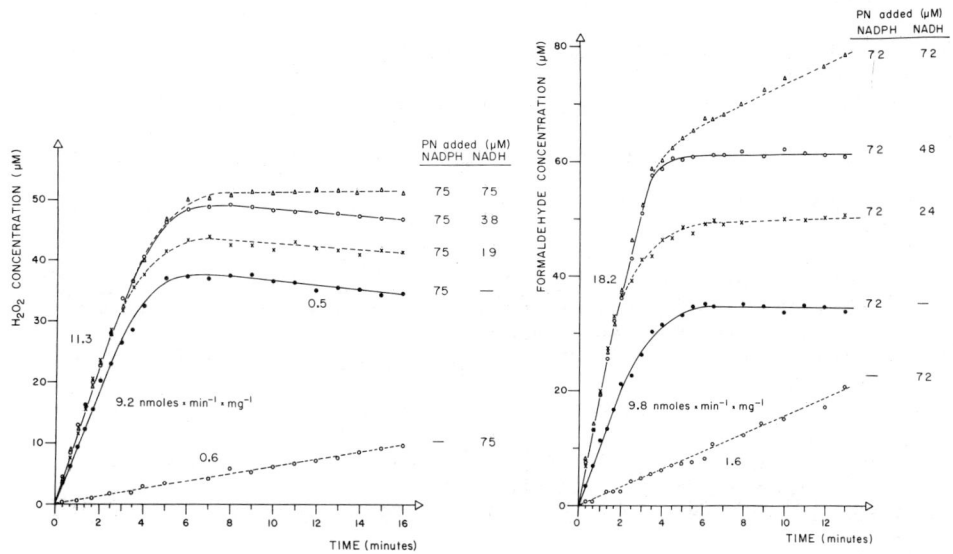

Fig. 6. Effect of NADH on the NADPH dependent formation of hydrogen peroxide and the N-demethylation of ethylmorphine. Microsomes were suspended to a final concentration of 1 mg protein per ml in 50 mM tris-chloride buffer, pH 7.5, which was supplemented with 10 mM $MgCl_2$, 150 mM KCl, 2 mM 5'AMP and 0.5 µM rotenone. The reactions were initiated by the addition of NADPH, NADH or NADPH plus NADH at concentrations as indicated in the Figure. The samples withdrawn during the reactions were analyzed as described in legends to Fig. 1 and 2. The effect of NADH on the formation of hydrogen peroxide (left portion of the Figure) was determined in the presence of 1 mM azide, whereas the catalase inhibitor was omitted from the reaction medium when the synergistic action of NADH on the N-demethylation reaction of ethylmorphine (5 mM final concentration) was analyzed (right portion of the Figure).

hydrogen peroxide. In view of the fact that NADPH is the preferred electron donor for the reduction of ferric cytochrome P-450 whereas NADH is considered to serve as a source of electrons for the reduction of oxy-cytochrome P-450 (17), it may be suggested that hydrogen peroxide is formed according to the first hypothesis, e.g. the dismutation of superoxide anions which are generated by the dissociation of the oxygenated complex of the ferrous pigment. The further examination of the modification of the rate of hydrogen peroxide formation by certain drug substrates as well as by the induction of cytochrome P-450 by a variety of agents is expected to contribute to a better understanding of the utilization of pyridine nucleotides by the microsomal electron transport system.

Acknowledgments - The author is indebted to Dr. Ronald W. Estabrook for many stimulating discussions and valuable suggestions during the course of this investigation and to Clint Young and Michael McMillan for excellent technical assistance.

REFERENCES

(1) Gillette, J.R., Brodie, B.B. and LaDu, B.N. J. Pharmacol. Exp. Ther. 119:532-540, 1957
(2) Thurman, R.G., Ley, H.G. and Scholz, R. Eur. J. Biochem. 25:420-430, 1972
(3) Hildebrandt, A.G. and Roots, I. Arch. Biochem. Biophys. 171:385-397, 1975
(4) Hildebrandt, A.G., Tjoe, M. and Roots, I. Biochem. Soc. Transactions 3:807-811,1975
(5) Lieber, C.S. and De Carli, L.M. J. Biol. Chem. 245:2505-2512, 1970
(6) Estabrook, R.W. (1974) in Alcohol and Aldehyde Metabolizing Systems (Thurman, R.G., Yonetani, T., Williamson, J.R. and Chance, B., eds) pp. 559-563, Academic Press, New York
(7) Masters, B.S.S. and Schacter, B.A. Ann. Clin. Res. 8 suppl. 17:18-27, 1976
(8) Werringloer, J. and Estabrook, R.W. Arch. Biochem. Biophys. 167:270-286, 1975
(9) Buening, M.K. and Franklin, M.R. Mol. Pharmacol. 10:999-1003, 1974
(10) Nash, T. Biochem. J. 55:416-421, 1953
(11) Cohen, B.S. and Estabrook, R.W. Arch. Biochem. Biophys. 143:46-53, 1971
(12) Correia, M.A. and Mannering, G.J. Mol. Pharmacol. 9:470-485, 1973
(13) Schenkman, J.B. Mol. Pharmacol. 8:178-188, 1972
(14) Holtzman, J.L. and Carr, M.L. Arch. Biochem. Biophys. 150:227-234, 1972
(15) Peterson, J.A., Ebel, R.E., O'Keeffe, D.H., Matsubara, T. and Estabrook, R.W. J. Biol. Chem. 251:4010-4016, 1976
(16) Duppel, W. and Ullrich, V. Biochem. Biophys. Acta 426:399-407, 1976
(17) Estabrook, R.W., Werringloer, J., Masters, B.S.S., Jonen, H., Matsubara, T., Ebel, R.E., O'Keeffe, D.H. and Peterson, J.A. (1976) in The Structural Basis of Membrane Function (Hatefi, Y. and Djavadi-Ohaniance, L., eds) pp. 429-445, Academic Press, New York
(18) Roach, M.K., Reese Jr., W.N. and Creaven, P.J. Biochem. Biophys. Res. Commun. 36:596-602, 1969
(19) Estabrook, R.W., Werringloer, J., Hrycay, E.G., O'Brien, P.J., Rahimtula, A.D. and Peterson, J.A. Biochem. Soc. Transactions 3:811-813, 1975
(20) Werringloer, J. and Estabrook, R.W. Biochem. Biophys. Res. Commun. 71:834-839, 1976
(21) Werringloer, J. and Estabrook, R.W. Hoppe-Seyler's Z. Physiol. Chem. 357:1063-1064, 1976

OXYGEN TRANSFER IN MICROSOMAL OXIDATIVE DESULFURATION

Hugo Kexel, Eike Schmelz and Hanns-Ludwig Schmidt

Lehrstuhl für Allgemeine Chemie und Biochemie der Technischen Universität München, D-8050 Freising-Weihenstephan, BRD

ABSTRACT

Microsomal sulfoxidations and desulfurations of different substrates proceed with introduction of an oxygen atom from O_2. Cumene hydroperoxide can replace NADPH/O_2, and microsomes transfer an oxygen atom from the oxidant, too. Non-enzymic desulfuration with oxygen transfer is possible with peracetic acid, other "Oxenoid" reagents do not perform the reaction. Further correlations between the NADPH/O_2- and the cumene hydroperoxide-systems are found in the D_2O-isotope effects and in the ratio of the desulfuration and oxidant-depending dearylation reactions. A mechanism is proposed for both reactions basing on a peroxide structure for the active oxygen.

INTRODUCTION

In cytochrome P-450 catalyzed oxidations the oxygen atom transferred to the substrate derives from O_2. A special microsomal oxygenation is the oxidative desulfuration in which the sulfur atom of a double bond in the substrate is substituted by oxygen from O_2 (1). This reaction is supposed to be suitable for investigations on the nature of the oxygen transferred by microsomes. In the present experiments the microsomal reaction is compared to corresponding oxygen transfers by model systems.

METHODS

Diethyl p-nitrophenyl thiophosphate (parathion)-[^{35}S] was synthesized according to Fletcher et al. (2), cumene hydroperoxide-[$^{18}O_2$] after Armstrong et al. (3); peracetic acid-[$^{18}O_2$] was obtained from $H_2^{18}O_2$ (4) and acetic anhydride. - Sulfur containing substrates were incubated with the ^{18}O-labelled oxidants or in ^{18}O-labelled water. The microsomes used were prepared from phenobarbital treated male Sprague-Dawley-rats. The isotope content of the oxidation products purified by chromatographic methods was determined by molecular mass spectrometry. - Kinetics of hydrolyses of parathion were measured spectrophotometrically from the absorption of p-nitrophenol in alkaline solution.

RESULTS

In order to find a suitable model substrate the microsomal oxygenation of substrates with sulfur in different linkages was studied: A sulfoxide was obtained from the thioether p-thioanisidine, parathion and diphenyl thiourea were desulfurated to the corresponding oxygen analogues while thioacetamide was converted to a stable oxygen containing product (5) though the substrate contains sulfur in a double bond; oxygen from O_2 was incorporated into all products (table 1). Thioacetamide sulfoxide is regarded to be a model for intermediates occurring in oxidative desulfuration. The proposed structure is supported by the fact that the corresponding N,N-dimethyl derivative is unstable (6), probably because it cannot form bridged hydrogen bonds. Because of the stability of its desulfuration product parathion was chosen for further experiments.

TABLE 1 Incorporation of Oxygen from $^{18}O_2$ into Sulfur Containing Substrates by Microsomes

Substrate in $2 \cdot 10^{-3}$ M solution, pH 7.4, 0.3 atm O_2, microsomes from 0.125 g liver/ml solution.

Substrate	Product	^{18}O incorporated [atoms]
$H_2N-C_6H_4-\underline{S}-CH_3$	$H_2N-C_6H_4-\underline{S}-CH_3$ ↓ $\overset{\bullet}{O}$	0.9
$H_3C\diagdown \overset{\underline{S}}{\diagup} \atop N-H \atop \mid \atop H$	$H_3C\diagdown \overset{\underline{S}-\overset{\bullet}{\underline{O}}^-}{\diagup} \atop N^+-H \atop \mid \atop H$	0.8
$H_5C_2-O\diagdown \overset{\underline{S}}{\diagup}P \atop H_5C_2-O \diagup \diagdown O-C_6H_4-NO_2$	$H_5C_2-O\diagdown \overset{\overset{\bullet}{\underline{O}}}{\diagup}P \atop H_5C_2-O \diagup \diagdown O-C_6H_4-NO_2$	1.0
$H_5C_6-\overset{\overset{\underline{S}}{\parallel}}{\underset{H}{N}}-C-\overset{}{\underset{H}{N}}-C_6H_5$	$H_5C_6-\overset{\overset{\overset{\bullet}{\underline{O}}}{\parallel}}{\underset{H}{N}}-C-\overset{}{\underset{H}{N}}-C_6H_5$	0.7 ^{x}

^{x} slow exchange of oxygen with water, $^\bullet$ labelled oxygen.

According to observations of Norman et al. (7) experiments with parathion-[^{35}S] showed that the sulfur atom split off during the oxidative desulfuration was nearly completely bound to the microsomes. After treatment of the incubation medium with H_2O_2 almost no ^{35}S was found in the sulfate fraction. Therefore the sulfur atom seems to be split off in the zerovalent state, and a similar structure may be assumed for the attacking oxygen. As the desulfuration reaction proceeds without a ^{35}S-isotope effect the P=S - bond splitting is not the rate limiting step of the reaction.

$$H_5C_2-O\diagdown \overset{\underline{S}}{\diagup}P{\diagdown \atop \diagup} \atop H_5C_2-O O-C_6H_4-NO_2 \quad + \quad [\underline{\overset{-}{\overline{O}}}{}^!] \quad \longrightarrow \quad H_5C_2-O\diagdown \overset{\overset{\bullet}{\underline{O}}}{\diagup}P{\diagdown \atop \diagup} \atop H_5C_2-O O-C_6H_4-NO_2 \quad + \quad [\underline{\overline{S}}{}^!]$$

After Hamilton (8) the cytochrome P-450 activated oxygen has an "Oxenoid" structure deriving from a peroxide group. As a matter of fact cumene hydroperoxide can replace O_2 in microsomal oxygenations (9); as shown in table 2 microsomes even transfer oxygen from this reagent to the substrate. At pH 7.4

TABLE 2 Oxygen Transfer by Different Oxygenizing Systems in the Desulfuration of Parathion
Microsomal systems pH 7.4, oxidant suspended to give $4 \cdot 10^{-3}$ M, substrate $2 \cdot 10^{-3}$ M solution; peracetate in 1 M solution.

Oxygenizing system	^{18}O-Atoms transferred from		Kinetic isotope effect k_{H_2O}/k_{D_2O}
	labelled oxidant	labelled water	
NADPH, O_2, micros.	1.0	0.05	1.3 ± 0.05
$H_5C_6\text{-}\underset{CH_3}{\overset{CH_3}{C}}\text{-O-O-H}$, micros.	0.95	0.05	1.1 ± 0.1
$H_5C_6\text{-}\underline{J}^+\text{-}\underline{O}I^-$ ^{x}, micros.	0.05	0.95	1.0 ± 0.1
$H_3C\text{-}\underset{O}{\overset{\|}{C}}\text{-O-O-H}$, pH 7.4	-	0	2.0 ± 0.5
$H_3C\text{-}\underset{O}{\overset{\|}{C}}\text{-O-O-H}$, pH 2.6	0.25	0.75	1.0 ± 0.1

^{x} rapid oxygen exchange with water.

a non-enzymic oxygen transfer from peracetic acid was observed while oxidation with this reagent at pH 2.6 predominantly proceeded with introduction of oxygen from water. As an oxygen exchange of the product with water under the conditions applied was excluded two different ways for the P=S - bond splitting must exist. Diphenyl thiourea was desulfurated by peracetic acid with introduction of oxygen from the oxidant at pH 7.4 and at pH 2.6.

The formation of "Oxenoid" oxygen shall also be possible from sulfoxides, aminoxides (8), and iodosobenzene (10). Microsomes catalyzed the desulfuration of parathion with iodosobenzene; a non-enzymic desulfuration was observed with diphenyl thiourea as the substrate. A proof of an oxygen transfer from the oxidant was not possible because it rapidly exchanges oxygen with water. Dimethyl aniline oxide and thioacetamide sulfoxide could not replace oxygen in oxidative desulfuration; thioacetamide sulfoxide was only positive in the sulfoxidation of p-thioanisidine.

No D_2O-isotope effect was found with the systems catalyzed by microsomes; the small D_2O-isotope effect in the natural system is due to the reduction of cytochrome P-450 (11). Therefore water does not interfere in the rate limiting step of these oxygen transfer reactions. The mechanism with peracetic acid at pH 7.4 seems to differ from that of the microsomal reactions.

Besides the desulfuration reaction microsomes catalyze the hydrolysis of parathion under formation of p-nitrophenol and thiophosphoric acid diethyl-ester (12). This "dearylation" reaction is dependent on O_2 and NADPH and is not inhibited by EDTA like the hydrolysis of diethyl p-nitrophenyl phosphate (paraoxon) catalyzed by a microsomal hydrolase.

$$\underset{H_5C_2-O}{\overset{H_5C_2-O}{>}}P\underset{O-C_6H_4-NO_2}{\overset{\underline{S}}{\diagup}} \xrightarrow[\text{NADPH, }O_2]{\text{microsomes, EDTA}} \underset{H_5C_2-O}{\overset{H_5C_2-O}{>}}P\underset{OH}{\overset{\underline{S}}{\diagup}} + HO-C_6H_4-NO_2$$

Chemical hydrolyses of phosphoric acid esters are catalyzed by the hyperoxide anion (13); therefore a peroxide group may be the catalyst of the microsomal oxygen-dependent hydrolysis, too, and corresponding model systems should accelerate the hydrolysis of parathion in a similar manner.

TABLE 3 Dearylation of Parathion in the Presence of Oxidants
Concentrations in microsomal systems see table 1, in addition $6 \cdot 10^{-3}$ M EDTA; peracetate in $2 \cdot 10^{-2}$ M solution.

Oxygenizing system	Ratio $\dfrac{\text{Dearylation}}{\text{Desulfuration}}$	Kinetic isotope effect k_{H_2O}/k_{D_2O}
NADPH, O_2, micros.	0.65	1.7 ± 0.1
$H_5C_6-\underset{CH_3}{\underset{|}{\overset{CH_3}{\overset{|}{C}}}}-O-O-H$, micros.	0.85	1.3 ± 0.1
$H_5C_6-\underline{J}^+-\underline{\bar{O}}I^-$, micros.	1.30	1.0 ± 0.1
$H_3C-\underset{O}{\underset{\|}{C}}-O-O-H$, pH 7.4	0.30	1.0 ± 0.2
$H_3C-\underset{O}{\underset{\|}{C}}-O-O-H$, pH 2.6	1.0	1.0 ± 0.2

As shown in table 3 the ratios of the reaction velocities of the microsomal "oxidative" dearylation and the oxidative desulfuration of parathion were similar for the systems NADPH/O_2 and cumene hydroperoxide. However, the speed of the dearylation reaction with iodosobenzene exceeded that of the corresponding desulfuration reaction. The coincidence between the non-enzymic oxidant peracetic acid and the natural system seems to be accidental, because, as shown before, the oxygen trassfer from peracetic acid to the substrate at pH 2.6 proceeds with a different mechanism.

The comparison of the D_2O-isotope effects of the dearylation reaction also confirms the correlation between the cumene hydroperoxide and the NADPH/O_2 systems. The value $k_{H_2O}/k_{D_2O} = 1.7$ for the latter is well distinct from that of the EDTA-inhibited hydrolysis of paraoxon ($k_{H_2O}/k_{D_2O} = 1.25$) and therefore typical for the oxygen-dependent reaction. Because of the low isotope effects an attack of H^+ or OH^- in the rate limiting step of the reactions can be excluded, yet the attack of a catalyst forming an intermediate is more possible. Finally, the intermediate must be cleaved by an attack of OH⁻ at the phosphorous because with none of the oxygenizing catalysts an introduction of oxygen from water into p-nitrophenol was observed.

CONCLUSIONS

The results obtained are in agreement with the proposal of Hamilton (8) that the cytochrome P-450 activated oxygen is an "Oxenoid" species deriving from a hydroperoxide structure. The complete substitution of the sulfur atom of P=S - and C=S - bonds by an oxygen atom from the oxidant can only be explained by a primary attack of the active oxygen at the double bond under formation of an unstable intermediate with an epoxide-analoguous structure (1). This intermediate normally would rearrange to the oxygen analogue of the substrate because of the stronger electronegativity of the oxygen atom. In acid

$$\begin{bmatrix} \geq P=\bar{S} \\ \updownarrow \\ \geq P^+-\bar{S}|^- \end{bmatrix} \xrightarrow{X-OOH \quad X-OH} \begin{bmatrix} \geq P \overset{\bar{O}|}{\underset{\bar{S}|}{\diagdown}} \end{bmatrix} \xrightarrow{H^+} [\geq P^+-\bar{S}-\bar{O}-H] \xrightarrow{OH^-} \geq P=\bar{O} + [\bar{S}|] + H_2O$$
$$\searrow \geq P=\bar{O} + [\bar{S}|]$$

solution (peracetic acid pH 2.6) the intermediate is apparently protonated; opening of the epoxide ring would result in an incorporation of oxygen from water into the product. We do not assume that the dearylation reaction pro-

$$\overset{\bar{S}}{\underset{O-Ar}{\diagup}}P \xrightarrow{X-OOH} \begin{bmatrix} \overset{\bar{S}-H}{\underset{O-Ar}{\diagup}}P\overset{}{\underset{O-O-X}{\diagdown}} \end{bmatrix} \xrightarrow{H_2O \quad X-OOH} \overset{\bar{S}}{\underset{OH}{\diagup}}P + HO-Ar$$

ceeds through the oxygen containing intermediate postulated by Ptashne et al. (1) because diethyl thiophosphate but not diethyl phosphate is found nearly in an amount equivalent to that of p-nitrophenol. A distinction between the mechanisms discussed is expected from stoichiometric experiments.

Acknowledgment: This work was in part supported by Deutsche Forschungsgemeinschaft.

REFERENCES

(1) K. Ptashne, R.M. Wolcott and R.A. Neal, Oxygen-18 studies on the chemical mechanisms of the mixed-function oxidase-catalyzed desulfuration and dearylation reactions of parathion, J.Pharmacol.Exptl.Therap. 179, 380 (1971).

(2) J.H. Fletcher, J.C. Hamilton, J. Hechenbleikner, E.J. Hoegberg, B.J. Sertl and J.T. Cassaday, Preparation of O,O-Diethyl-O-p-nitrophenyl Thiophosphate (Parathion), J.Amer.Chem.Soc. 70, 3943 (1948).

(3) G.P. Armstrong, R.H. Hall and D.C. Quin, The Autoxidation of Isopropylbenzene, J.Chem.Soc. 1950, 666.

(4) M. Anbar, Z. Baruch and D. Meyerstein, The synthesis of O^{18}-labelled hydrogen peroxide in H_2O^{16}, Internat.J.Appl.Radiat.Isotopes 17, 256 (1966).

(5) R. Ammon, H. Berninger, H.J. Haas und J. Landsberg, Thioacetamid-sulfoxid, ein Stoffwechselprodukt des Thioacetamids, Arzneimittelforschung 17, 521 (1967).

(6) W. Walter and M. Steffen, Oxidation products of thiocarboxylic acid amides XVIII, Liebigs Ann.Chem. 712, 53 (1968).

(7) R.J. Norman, R.E. Poore and R.A. Neal, Studies of the binding of sulfur released in the mixed-function oxidase-catalyzed metabolism of diethyl p-nitrophenyl phosphorothioate (Parathion) to diethyl p-nitrophenyl phosphate (Paraoxon), Biochem.Pharmacol. 23, 1733 (1974).

(8) Hamilton, G.A., Chemical models and mechanisms for oxygenases, in: Hayaishi, O., Molecular Mechanisms of Oxygen Activation, Acad. Press, New York 1974, P. 405.

(9) F.F. Kadlubar, K.C. Morton and D.M. Ziegler, Microsomal-Catalyzed Hydroperoxide Dependent C-Oxidation of Amines, Biochem.Biophys.Res.Commun. 54, 1255 (1973).

(10) F. Lichtenberger, W. Nastainczyk and V. Ullrich, Cytochrome P-450 as an oxene transferase, Biochem.Biophys.Res.Commun. 70, 939 (1976).

(11) J.L. Holtzman and M.L. Carr, Inhibition of hepatic mixed-function oxidases by deuterium oxide, Life Sciences 9, 1033 (1970).

(12) R.A. Neal, Studies on the metabolism of diethyl 4-nitrophenyl phosphorothionate (parathion) in vitro, Biochem.J. 103, 183 (1967).

(13) J. Epstein, M.M. Demek and D.H. Rosenblatt, Reaction of paraoxon with hydrogen peroxide in dilute aqueous solution, J.Org.Chem. 21, 796, (1956).

REDOX AND LIGAND DYNAMICS IN P450$_{cam}$ - PUTIDAREDOXIN COMPLEXES*

T. C. Pederson, R. H. Austin,[†] and I. C. Gunsalus

University of Illinois, Urbana 61801 USA

The 5-exo-hydroxylation of camphor by the P450 heme protein from P. putida occurs in a cyclic series of reaction states separated by binding of substrates and addition of two reducing equivalents from putidaredoxin, an iron-sulfur protein of the $Fe_2S_2Cys_4$ class (1,2). NADH oxidation is coupled to this process by a flavoprotein which serves as a putidaredoxin reductase. Reducing equivalents are transferred to P450, as illustrated in Fig. 1, in two univalent reactions; the first following camphor binding by ferric P450, m^o, and the second after oxygen binds to form an oxyferrous intermediate, $m^{rs}_{O_2}$. Rapid reaction measurements reported in this paper describe the dynamics of substrate binding and redox transfer, establish the role of putida-

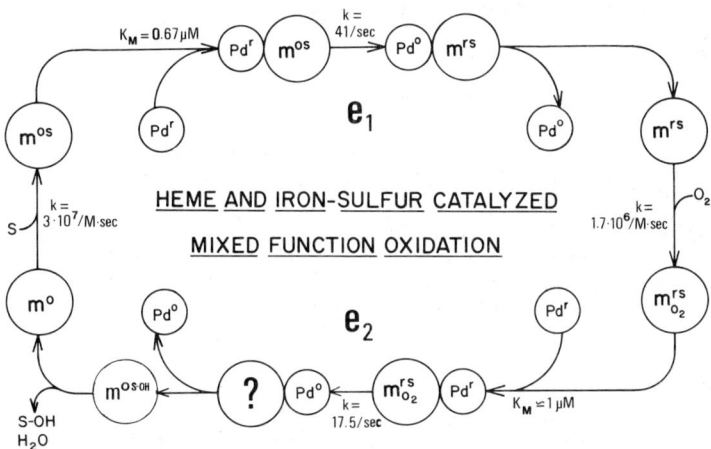

Fig. 1 Camphor hydroxylation cycle. Heme protein P450, designated m for monoxygenase and putidaredoxin, designated Pd, participate with ligand and redox states indicated: m^o and Pd^o, oxidized (ferric) proteins; m^r and Pd^r, reduced proteins; m^s and m_{O_2}, P450 bound with camphor and oxygen respectively. The reaction rates, described further in the text, were all measured at 20°C.

---

*This work supported by NIH fellowship GM 662 and grants GM21161 and AM00562
†Present address: Max Planck Institute for Biophysical Chemistry, Gottingen, West Germany

redoxin-P450 complexes in the redox reactions, and suggest the participation of other reaction intermediates between addition of the second electron and release of products.

Previous reports from this laboratory (3,4) have shown that P450 substrate binding velocities are normally much faster than the redox transfer rates, and the slowest process is the product forming reaction between reduced putidaredoxin, Pd^r, and $m_{O_2}^{rs}$. A recent subject of prominence concerns the participation of enzyme complexes and modulation of enzymatic activity coupled to the energetics of substrate-P450 or putidaredoxin-P450 binding (5). The redox dynamics described in this paper demonstrate directly the involvement of two transient dienzyme complexes, $Pd^r \cdot m^{os}$ and $Pd^r \cdot m_{O_2}^{rs}$ in the respective e_1 and e_2 redox reactions of the hydroxylation cycle. In the product forming reaction accompanying the decay of the latter complex, the appearance of ultimate products coincides with the redox transfer from Pd^r ruling out transient accumulation of reaction intermediates, but a new state of P450, perhaps analogous to peroxidase Compound I, has been observed in the product forming reaction between m^{os} and peracid.

The measurements of P450 reaction dynamics described in this paper were made possible through development of rapid reaction equipment which enhances precision and extends observation to reactions occurring in less than a millisecond. Reactions initiation by stopped-flow mixing, photoreduction and heme-ligand photodissociation, have been examined. Transient spectral changes were observed simultaneously at two wavelengths, digitalized, and stored in a transient recorder for subsequent analysis.

PUTIDAREDOXIN-P450 REDOX DYNAMICS

The two redox reactions illustrated in Fig. 1 are dependent on the binding of substrates. The initial reduction of P450 to its ferrous form by Pd^r occurs after camphor binds to P450 in its high spin form, raising the midpoint reduction potential above that of putidaredoxin (6). The second redox transfer does not occur unless oxygen binds forming the oxyferrous intermediate $m_{O_2}^{rs}$. This control by substrate addition allows the independent observation of both redox reactions between putidaredoxin and P450.

e_1: Reduction of m^{os}

The reduction of m^{os} by Pd^r has been examined following stopped-flow mixing under anaerobic conditions. At low concentrations of both proteins, <5 μM, the rate of m^{os} reduction is dependent on the concentration of Pd^r, but the reaction velocity plateaus at higher concentrations as shown in Fig. 2. The apparent K_M for this reaction, 0.63 μM, was determined from the double reciprocal plot using concentrations of free Pd^r calculated as shown in the inset of Fig. 2. A turnover number of about 50/sec is determined by this method but more precise values were derived from reactions initiated by photoreduction.

Photoreduction occurs in a reaction mixture initially containing Pd^o and a high concentration of m^{os}, assuring >90% of the Pd^o is bound in a dienzyme complex, $Pd^o \cdot m^{os}$, K_D=2.9 μM (5). The dynamics of redox reactions initiated by photoreduction are listed in Table 1. Rapid reduction of $Pd^o \cdot m^{os}$ to $Pd^r \cdot m^{os}$ occurs following the acridine sensitized photoreduction of

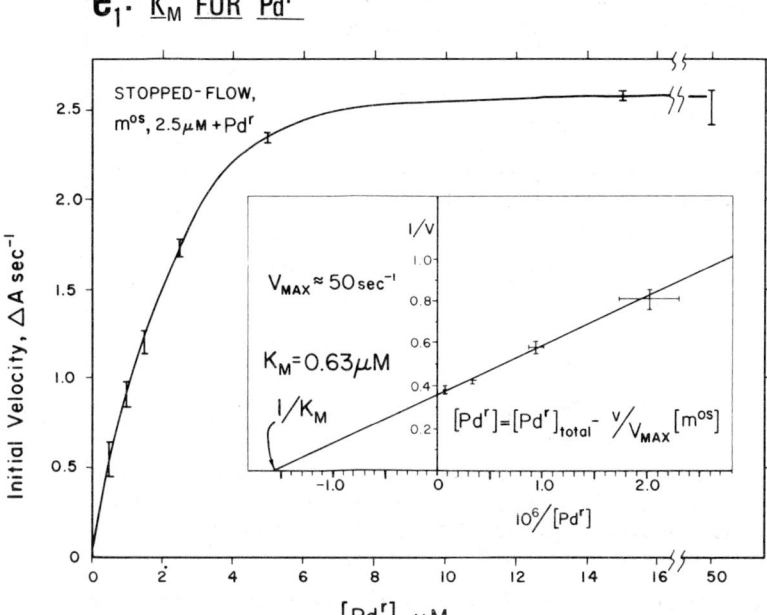

Fig. 2. Saturation and K_M determination for Pd^r in e_1 reaction. Initial velocities at 20° observed at 390 nm in 50 mM phosphate (pH 7.0), 200 mM K^+, and 0.25 mM camphor. Pd was reduced with dithionite, 100 µM.

methylviologen, MV^{2+}, by a flash of light from a 500 J Zenon flashlamp (7,8). The methylviologen radical, MV^{1+}, reduces both Pd^o and m^{os}, but reduction of Pd^o occurs about 100 times faster than m^{os} reduction, resulting in the rapid accumulation of $Pd^r \cdot m^{os}$. The subsequent redox transfer to P450 is a first order reaction.

The exponential rate of the e_1 redox transfer (Table 1) is the sum of the forward and reverse reaction rates. Measurement of the forward rate, $Pd^r \cdot m^{os} \rightarrow Pd^o \cdot m^{rs}$, was made in the presence of carbon monoxide which binds to the reduced cytochrome in competition with the reverse redox transfer reaction. Under 10 atmospheres of CO, the rate of CO binding, 600/sec, leaves the forward redox transfer reaction as rate limiting. The decrease in rate under CO corresponds to a reverse redox transfer rate of 17/sec and a redox equilibrium constant, $K_{eq}=2.3$.

e_2: Reduction of $m^{rs}_{O_2}$

The dynamics of the second redox reaction were examined following stopped-flow mixing of the reduced proteins, Pd^r and m^{rs} with oxygenated buffer. Oxygen binds very rapidly, $k=1.7 \cdot 10^6/M \cdot sec$, forming $m^{rs}_{O_2}$ and allows observation of the subsequent reaction between Pd^r and $m^{rs}_{O_2}$ resulting in camphor

TABLE 1 e_1 Transfer: Photoreduction of $Pd^o \cdot m^{os}$ Complex

$MV^{2+} \xrightarrow[\text{Acridine-EDTA}]{h\nu, 50\mu sec} MV^{1+}_{\cdot}$	
$Pd^r \cdot m^{os} \underset{k_{-1}}{\overset{k_1}{\rightleftharpoons}} Pd^o \cdot m^{rs} \xrightarrow{CO} Pd^o \cdot m^{rs}_{CO}$	
Reduction by $MV^{1+}_{\cdot}$ ($\approx 5\mu M$)	k, sec^{-1}
$m^{os} \rightarrow m^{rs}$, 50µM	80
$Pd^o \rightarrow Pd^r$, 15µM	4500
$Pd^o \cdot m^{os} \rightarrow Pd^r \cdot m^{os}$	4500
$Pd^r \cdot m^{os}$ e_1 transfer	
Argon, $k=k_1+k_{-1}$	59
CO, 10mM $k=k_1$	41

Reactions at 20° in 50 mM phosphate (pH 7.0), 200 mM K^+, 0.25 mM d-camphor, 50 mM EDTA, 50 µM proflavin, 0.5 mM methylviologen (MV^{2+}), and others as indicated.

hydroxylation. Dynamics of single-turnover reactions were measured in mixtures containing only one equivalent of camphor per equivalent of P450. When the initial rate of m^o formation is measured with varying concentrations of Pd^r, the reaction velocity saturates at higher concentrations as shown in Fig. 3, and the K_M for Pd^r is about 1 µM, similar to that observed in the first redox reaction. At saturating concentrations of Pd^r, the appearance of oxidized P450 is exponential with a rate of 17.5/sec at 20° which agrees with earlier measurements of the turnover number in P450 under steady state conditions (6,9).

Simultaneous measurements of P450 and putidaredoxin redox dynamics in the product forming reaction were best observed in a reaction limited by the amount of Pd^r but containing an excess of camphor to bind m^o forming m^{os} with its soret shifted to 391 nm. The time course of both heme and iron-sulfur center reactions are shown in Fig. 4. After the initial binding of oxygen, the oxidation of Pd^r is seen to coincide with the loss of $m^{rs}_{O_2}$ and simultaneous appearance of m^{os}. There is no evidence for the existence of an intermediate in the product forming reaction. Since Pd^r will also begin to reduce m^{os} as it accumulates, the decrease of $m^{rs}_{O_2}$ is less than the amount of Pd^r oxidized. The remaining $m^{rs}_{O_2}$ decays in the previously characterized reaction catalyzed by Pd^o yielding 1/2 mole of product (9). The reaction mixture contained less than one equivalent of Pd^r per equivalent of $m^{rs}_{O_2}$ but the rate of $m^{rs}_{O_2}$ loss accompanying Pd^r oxidation proceeds at a rate of about 15/sec, only slightly less than observed with saturating levels of Pd^r, indicating a dienzyme complex, $Pd^r \cdot m^{rs}_{O_2}$, is the enzyme intermediate involved in the second redox reaction.

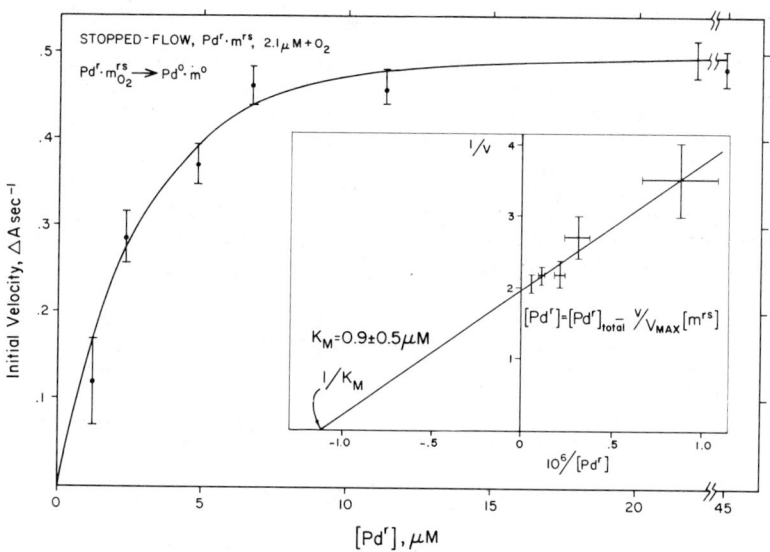

Fig. 3. Saturation and K_M determination for e_2 reaction. Reaction limited by one equivalent camphor/P450 and enzymes were reduced by NADH, 400 μM. Initial velocities were observed at 417 nm. Other conditions as described in Fig. 2

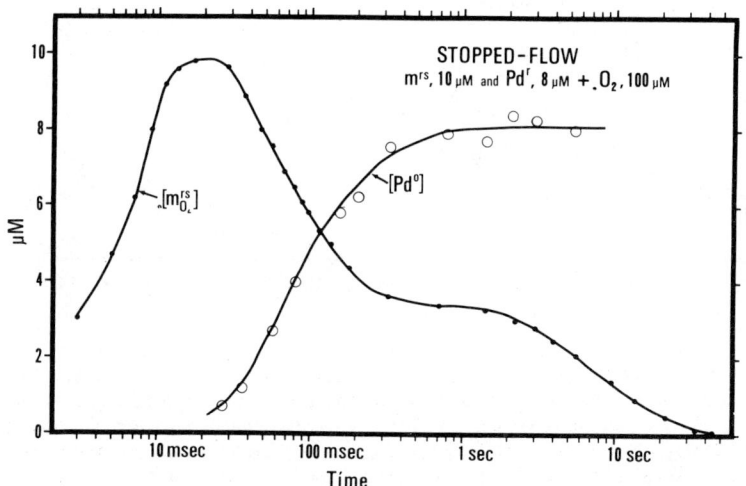

Fig. 4. Ligand and redox dynamics accompanying e_2 reaction. The formation and decomposition of $m_{O_2}^{rs}$ were determined from absorbance changes at 395 nm using $\Delta\varepsilon=-20$/mM cm for $m^r \rightarrow m_{O_2}^{rs}$ and $\Delta\varepsilon=30$ for $m_{O_2}^{rs} \rightarrow m^{os}$. Oxidation of Pd^r was observed at 525 nm using $\Delta\varepsilon=2.0$/mM cm. Other conditions as described in Fig. 3.

ROLE OF PD-P450 COMPLEX IN REDOX AND LIGAND DYNAMICS

Early evidence suggested participation of multi-enzyme complexes in the camphor hydroxylation reaction (9). The binding between putidaredoxin and P450 was later measured directly with fluorescence and spectral probes (5,10). Under equilibrium conditions, the two proteins form a dienzyme complex with K_D=2.9 µM if oxidized or K_D=0.5 µM if reduced. At high concentrations of putidaredoxin, a second equivalent is bound with K_D=2.5 µM. Redox dynamics already presented demonstrate the involvement of dienzyme complexes and the binding rates of substrate and other heme ligands have been measured with P450 complexed by putidaredoxin.

Redox Reactions: Equilibrium and Kinetic Measurements

The values derived from both kinetic and equilibrium measurements are listed in Table 2. When bound in the dienzyme complex, the midpoint potential of putidaredoxin is increased, reducing the difference in potential between the two proteins from 69 to 23 mV which decreases the redox equilibrium constant

TABLE 2 e_1: Redox Dynamics and Equilibria

$$Pd^r + m^{os} \underset{k_{-1}}{\overset{k_1}{\rightleftarrows}} Pd^r \cdot m^{os} \underset{k_{-2}}{\overset{k_2}{\rightleftarrows}} Pd^o \cdot m^{rs}$$

Kinetics:

$$K_M^{\mu M} = \frac{k_2 + k_{-1}}{k_1} = 0.63$$

$$k_1 \simeq 3 \cdot 10^8 / \text{Msec}$$

$$k_{-1} \simeq 150/\text{sec}$$

$$K_{eq} = \frac{k_2}{k_{-2}} = 2.3 \pm .3$$

Equ. Binding*:

$$K_D^{\mu M} = \frac{k_{-1}}{k_1} = 0.49$$

$$K_{eq} = e^{\Delta E_o^1 / RT} = 2.4$$

*Sligar, S. and Gunsalus, I. C., Proc. Nat. Acad. Sci 73, 1078, 1976.

to 2.4 (9). The kinetic measurement of forward and reverse redox rates are in complete agreement with this value confirming that the dienzyme complex which forms under equilibrium conditions is the redox transfer complex. The potential of P450 is unaltered in the dienzyme complex which requires that Pd^r be preferentially bound by P450 (9). The K_M for the redox reaction can therefore be compared to the K_D of the fully reduced dienzyme complex, allowing the estimation of enzyme complex formation and dissociation rates shown in Table 2. The second redox reaction also occurs in an enzyme complex with a similar binding constant, but there is no direct evidence as yet to indicate if the same binding site is involved in both reactions.

P450-Substrate Binding in Pd-P450 Complexes

The dynamics of camphor and oxygen binding to P450 have been previously measured (6,11) but the effect of enzyme-enzyme interaction on these reactions has not been examined. Thus, the binding rates of camphor, oxygen and other ligands with P450, both free and complexed by putidaredoxin, were determined following either stopped-flow mixing or photodissociation. The results recorded in Table 3, show the binding rate of substrate or other ligands is the same in the presence of putidaredoxin as in its absence. This can be contrasted to the effect of bound camphor on the binding of CO. The rate of oxygen binding is also much faster in the absence of camphor, recombination being essentially complete within the 3 msec dead time of stopped-flow mixing.

TABLE 3 P450 Ligand Dynamics: ±Pd(50 µM)

	$k_2, M^{-1} sec^{-1} \times 10^6$	
	m^o $-Pd^o/+Pd^o$	m^r $-Pd^r/+Pd^r$
$m + s \rightarrow m^s$	31/30	---
$m^s + O_2 \rightarrow m^s_{O_2}$	---	1.7/≈1.7
$m + CO \rightarrow m_{CO}$	---	8.7/8.7
$m^s + CO \rightarrow m^s_{CO}$	---	0.060/0.059
$m^s + NO \rightarrow m^s_{NO}$	2.5/2.5	2.7/3.1

Reactions at 20° in 50 mM phosphate (pH 7.0) and 200 mM K^+ observed following photodissociation except for m^o+s observed after stopped-flow mixing.

OTHER P450 INTERMEDIATES INVOLVED IN HYDROXYLATION

In the reaction between Pd^r and $m^{rs}_{O_2}$, other intermediates states cannot be observed because appearance of oxidized P450 is coincident with the oxidation of Pd^r. A P450-product complex can be demonstrated under other conditions but its rapid decomposition rate prevents the accumulation in this reaction. Another reaction state of P450 has also been observed in the camphor hydroxylating reaction between m^{os} and peracids.

The P450-Product Complex

An enzyme-product complex between m^o and hydroxy-camphor, with a dissociation constant of 80 µM, has been shown in equilibrium dialysis measurements by Dr. B. S. Shastry in our laboratory. The product competes with camphor binding as evidenced by the shift of the Soret absorbance from 391 nm back to 417 nm as hydroxycamphor displaces camphor. In the hydroxylation reaction between

Pd^r and m$^{rs}_{O_2}$, the appearance of m^O or m^{OS} would be limited by the rate of product dissociation. When the P450-product complex is mixed with 1 mM camphor, by stopped-flow, under conditions described in Table 2, m^{OS} is formed at the rate of 225/sec which presumably represents product dissociation. This rate greatly exceeds that of product formation and the P450-product complex is therefore not observed in the reaction.

P450 Intermediate in Peracid Reaction
The spectral states in the stopped-flow reaction between m^{OS} and m-chloroperbenzoic acid shown in Fig. 5 were derived from repeated observation at varied wavelengths. The soret absorption of m^{OS} at 390 nm is rapidly bleached, t½=10 msec, with the appearance of an intermediate of maximum absorbance at about 420 nm. Subsequent changes in absorbance are dominated

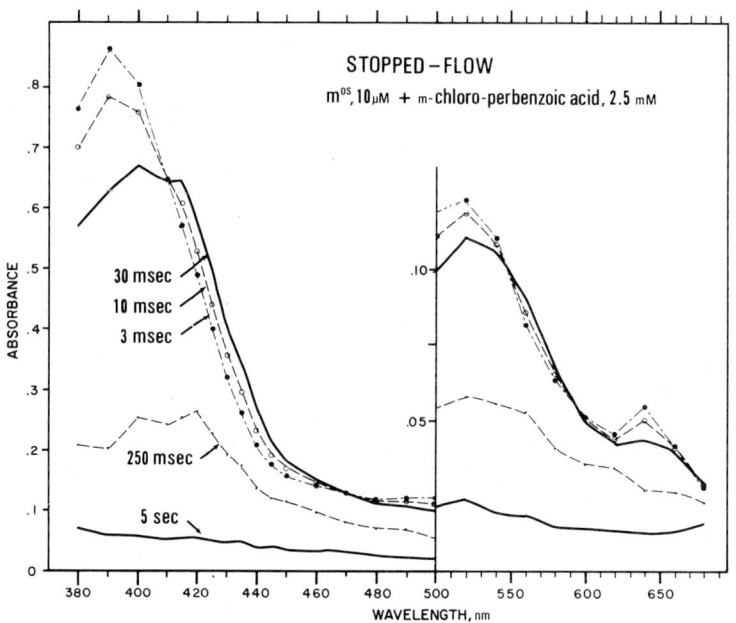

Fig. 5. Spectral states accompanying peracid promoted hydroxylation. Reaction at 20° in 50 mM phosphate (pH 7.0) 200 mM K$^+$, and 0.25 mM camphor.

by the heme destruction reaction, t½≈200 msec, which accompanies hydroxylation with peracids. The intermediate spectral state is similar in appearance to the peroxidase Compound I intermediate (12). In the accompanying paper by S. G. Sligar et al., published in these proceedings, a similar reaction intermediate is observed in the reaction between m^O and peracid at subzero temperatures. Probable mechanisms for the formation of such an intermediate and its role in hydroxylation mechanisms are discussed in the paper.

Additional P450 reaction intermediates have also been observed in liver microsomal P450 during catalysis of hydroxylation (13) or stabilized at low temperatures in mixed solvents (14), but their physical states and catalytic roles are not defined. Direct evidence for or against participation of these reaction states in hydroxylation reactions will hopefully be forthcoming.

REFERENCES

(1) Gunsalus, I. C., Meeks, J. R., Lipscomb, J. D., Debrunner, P., and Münck, E., Molecular Mechanisms of Oxygen Activation, (O. Hayaishi, ed.), Academic Press, New York, 1974.
(2) Gunsalus, I. C., Pederson, T. C., and Sligar, S. G., Ann. Rev. Biochem. 44, 377 (1975).
(3) Gunsalus, I. C., Tyson, C. A., and Lipscomb, J. D., Oxidases and Related Redox Systems, (T. E. King, H. S. Mason, and M. Morrison, eds.), University Park Press, Baltimore, 1973.
(4) Tyson, C. A., Lipscomb, J. D., and Gunsalus, I. C., J. Biol. Chem. 247, 5777 (1972).
(5) Sligar, S. G., and Gunsalus, I. C., Proc. Nat. Acad. Sci. 73, 1078 (1976).
(6) Gunsalus, I. C., Meeks, J. R., and Lipscomb, J. D., Ann. N. Y. Acad. Sci. 212, 107 (1973).
(7) Gunsalus, I. C., Lipscomb, J. D., Marshall, V., Frauenfelder, H., Greenbaum, E., and Münck, E., Biological Hydroxylation Mechanisms, Biochemical Symposia 34, (G. S. Goyd and R. M. S. Smellie, eds.), Academic Press, New York, 1972.
(8) Greenbaum, E., Austin, R. H., Frauenfelder, H., and Gunsalus, I. C., Proc. Nat. Acad. Sci 69, 1273 (1972).
(9) Lipscomb, J. D., Sligar, S. G., Namtvedt, M. J., and Gunsalus, I. C., J. Biol. Chem. 251, 1116 (1976).
(10) Sligar, S. G., Debrunner, P. G., Lipscomb, J. D., Namtvedt, M. J., and Gunsalus, I. C., Proc. Nat. Acad. Sci. 71, 3906 (1974)
(11) Peterson, J. A., and Mock, D. M., Adv. in Exp. Med. and Biol. 58, 311 (1975).
(12) Hager, L. P., Hollenberg, P. F., Rand-Meir, T., Chiang, P., and Doubek, D., Ann. N. Y. Acad. Sci. 244, 80 (1975).
(13) Guengerich, F. P., Ballou, D. P., and Coon, M. J., Biochem. Biophys. Res. Comm. 70, 951 (1976).
(14) Debey, M.-P., Leon, M., and Douzou, P., Hoppe-Seyler's Z. Physiol. Chem. 357, 1019 (1976).

THE INHIBITION OF MIXED-FUNCTION OXIDATION REACTIONS BY AMPHETAMINES IN LIVER AND LUNG MICROSOMES

Michael R. Franklin

Department of Biopharmaceutical Sciences, College of Pharmacy, University of Utah, Salt Lake City, Utah 84112

ABSTRACT

The inhibition of microsomal mixed-function oxidations (MFO) by amphetamines was greatest for those compounds capable of forming metabolic intermediate (MI) complexes with cytochrome P-450 at fast rates. Ethylmorphine N-demethylase was least sensitive, p-nitroanisole O-demethylase was intermediate, and benzopyrene hydroxylase was the most sensitive reaction to amphetamine inhibition. Formation of MI complexes prior to initiating MFO reactions increased the degree of inhibition by adding a noncompetitive component to the initial competitive inhibition. This increased inhibition was only evident in microsomes from phenobarbital-pretreated animals, not those from untreated or 3-methylcholanthrene-pretreated animals. Hepatic MFO reactions in rabbit were more susceptible to inhibition than those of rat. In rabbit, lung benzopyrene hydroxylase was more susceptible to inhibition than liver. The formation of MI complexes from amphetamines can be a significant contributing factor to the inhibition of mixed-function oxidase reactions.

INTRODUCTION

Amphetamines are known to interact with hepatic microsomal cytochrome P-450 (1-3) and undergo a variety of mixed-function oxidation (MFO) reactions, including aromatic ring hydroxylation, deamination, and other N-oxidations (4-9). Of recent interest is the oxidative reaction that amphetamines undergo which results in the formation of a metabolic intermediate (MI) complex which absorbs maximally around 455 nm (3, 10-12). The formation of similar complexes from methylenedioxybenzene derivatives (13-15) were found to dramatically enhance the inhibitory properties of these compounds (16). The inhibitory nature of the amphetamine MI complex has received only scant attention (10). Therefore, a comprehensive investigation was undertaken to examine the effect of MI complex formation on the amphetamine inhibition of MFO reactions. In addition to rat, hepatic microsomes from rabbit were also investigated, since species differences in amphetamine metabolism (7) and MI complex formation (17, 18) are known to exist. The <u>in vivo</u> accumulation of amphetamine in the lung (19) and the predilection of cytochrome P-450 in this organ for forming MI complexes (20) prompted examination of inhibiting effects in this organ in addition to liver. Also, since hepatic MI complex formation is extremely dependent upon prior induction (11, 21), the inhibition was investigated in microsomes isolated from untreated and 3-methylcholanthrene-treated, as well as phenobarbital-treated, rats.

METHODS

Hepatic microsomes were prepared from mature male Sprague-Dawley rats and New Zealand white rabbits by the procedure described previously (22). Lung microsomes were similarly isolated except that a sonication of the homogenate preceded centrifugation (23). Phenobarbital treatment was 40 mg/kg, i.p. for 5 days (rabbit), or 80 mg/kg, i.p., for 4 days (rat). 3-Methylcholanthrene treatment was 20 mg/kg, i.p., 3 doses at 36 hr. intervals. Ethylmorphine N-demethylation was determined by formaldehyde formation (24). p-Nitroanisole O-demethylation was either measured as formaldehyde or p-nitrophenol formation (25). Benzopyrene hydroxylase activity was determined by the aqueous alkaline fluorescence (Ex = 396 nm Fl = 522 nm) of hexane extracted metabolites (26), and related as 3-hydroxybenzopyrene equivalents to a quinine sulfate standard (27). All enzyme activity determinations were performed at 25°.

RESULTS AND DISCUSSION

The inhibition of three MFO reactions by amphetamine concentrations of 0.017 to 2.00 mM are shown in Table 1. The dashed lines show the approximate boundary of 50% inhibition. Of the three reactions investigated, benzopyrene hydroxylase was found to be the most sensitive and ethylmorphine demethylase the least sensitive to amphetamine inhibition. N-hydroxyamphetamine was the most inhibitory and p-hydroxyamphetamine and methamphetamine the least, for all three reactions. For the other amphetamines, the relative inhibition varied with the reaction. For some, if not all the compounds, reactions other than MI complex formation obviously contribute to their initial inhibitory properties (e.g., benzphetamine demethylation). To assess the contribution of MI complex formation to the inhibitory effects, aerobic preincubation of microsomes with the amphetamine and NADPH (conditions under which MI complex formation occurs) was performed prior to determining the inhibition (Table 2). Since p-hydroxyamphetamine does not form an MI complex (3), preincubation did not increase its inhibitory effect. In contrast, preincubation with d-amphetamine markedly increased the inhibition of ethylmorphine demethylase, and to a lesser extent, benzopyrene hydroxylase. Preincubation with norbenzphetamine, one of the fastest MI complex formers (28), markedly increased the inhibition of both ethylmorphine demethylase and benzopyrene hydroxylase. By way of comparison, preincubation with the non-nitrogenous methylenedioxyphenyl compound, piperonyl butoxide, is known to enhance the

TABLE 1   % Inhibition of Phenobarbital-Treated Rat Hepatic Mixed-Function Oxidations by Amphetamines*

	Ethylmorphine N-demethylase Amphetamine concentration (mM)					p-Nitroanisole O-demethylase Amphetamine concentration (mM)					Benzopyrene hydroxylase Amphetamine concentration (mM)				
	0.017	0.05	0.10	0.20	2.00	0.017	0.05	0.10	0.20	2.00	0.017	0.05	0.10	0.20	2.00
Amphetamine															
p-Hydroxyamphetamine				0	1±1					20±1					14±2
Methamphetamine								-5		16±7					21±9
l-Amphetamine		2		3±1	13±7		12±1		16±4	51±5			30	44±3	
d-Amphetamine		-4±1			7±6		21±1	23±5		61±8	26±4	28±15	55±3	62±5	
Norbenzphetamine		-2	2±2	13±4			5±5	15±6	31±5		24±7	32±6	55±3		
Benzphetamine			47±2	71			16	9	16		43±5	58±2			
N-hydroxyamphetamine	4±3	24±3				-21	33±14	35±8			27±3	66±4			

*Values are the mean ± S.E.M. Single values indicate one microsomal preparation.

TABLE 2 Effect of Preincubation of Phenobarbital-Treated Rat Microsomes with Various Inhibitors and NADPH on Their % Inhibition of Mixed-Function Oxidation Reactions

Inhibitor	(mM)	Ethylmorphine N-demethylation Initial	With preincubation	Benzopyrene hydroxylation Initial	With preincubation
p-Hydroxyamphetamine	(2.00)	1±1	0		
d-Amphetamine	(2.00)	7±6	37±1	74	84
	(0.05)			23±1	31±8
Norbenzphetamine	(0.20)	13±4	64±6		
	(0.10)			32±6	75±5
	(0.05)			26	64
Piperonylbutoxide	(0.07)			33±5	61±4

inhibition of ethylmorphine demethylase (16) and, as shown here, has the same effect on benzopyrene hydroxylase.

To further delineate the role of MI complex formation in the inhibition of MFO reactions by amphetamines, the inhibition kinetics were examined before and after preincubation (Table 3). The inhibition of p-nitroanisole demethylase and ethylmorphine demethylase by norbenzphetamine, and ethylmorphine demethylase by N-hydroxyamphetamine, was largely competitive under initial conditions, with a three-fold increase in K_m by norbenzphetamine and

TABLE 3 Inhibition of O- and N-Demethylation Reactions by Norbenzphetamine and N-Hydroxyamphetamine With and Without Preincubation

Reaction	Amphetamine	K_m(mM)	V_{max}	After preincubation* K_m(mM)	V_{max}
p-Nitroanisole	None	.062	1.30	.067	1.13
O-demethylation	Norbenzphetamine(33µM)	.217	1.03	.186	.38
Ethylmorphine	None	.28	23	.26	16
N-demethylation	Norbenzphetamine(167µM)	.75	16	.91	13
	None	.50	20	.38	16
	N-hydroxyamphetamine(50µM)	7.16	18	12.28	10

*Preincubation of microsomes with NADPH, O_2 and amphetamine for 5 min. (norbenzphetamine) or 1 min. (N-hydroxyamphetamine) before substrate addition.

Values are the mean of 5 (ethylmorphine N-demethylation) and 2 (p-nitroanisole demethylation) rat liver microsomal preparations. V_{max} values are nmoles HCHO/mg protein/min.

over ten-fold increase in Km by N-hydroxyamphetamine. After preincubation, in addition to the changes in Km, larger decreases in Vmax were observed with both norbenzphetamine inhibition of p-nitroanisole demethylase and N-hydroxyamphetamine inhibition of ethylmorphine demethylase. Although not conclusive, these results again suggest that MI complex formation is a reaction partially responsible for the inhibition of MFO reactions, and that the complex, once formed, is sufficiently stable to prevent cytochrome P-450 participation in other MFO reactions. As further proof, the effect of aerobic preincubation of microsomes with norbenzphetamine and NADPH on benzopyrene hydroxylase was investigated in microsomes from untreated animals and those treated with 3-methylcholanthrene (Table 4). Since MI complex formation is an

TABLE 4 Effect of Animal Treatment on Norbenzphetamine Inhibition of Benzopyrene Hydroxylase

Animal treatment	Benzopyrene hydroxylase (nmoles/mg protein/min)	50 μM Norbenzphetamine inhibition %	
		Initial	With preincubation
None	0.22 (2)	20	24
Phenobarbital	0.78 (2)	14	61
3-Methylcholanthrene	3.00 (1)	-16*	0

*Negative inhibition indicates activation.

almost exclusive property of phenobarbital-induced microsomes (21) preincubation should not increase the amphetamine inhibition in other than these microsomes. As shown, preincubation produced an additional 47% increase in inhibition in phenobarbital microsomes compared with only a 4% in untreated and a possible (since only one determination was performed in which an activation upon norbenzphetamine addition was seen) 16% in 3-methylcholanthrene-treated animals. These effects were independent of the absolute rate of benzopyrene hydroxylation which was four-fold higher in 3-methylcholanthrene than phenobarbital-treated animals, which, in turn, was four-fold higher than untreated animals. Investigation of the amphetamine inhibition of MFO reactions in rabbit (Table 5), as compared to rat (Table 1), showed that in contrast to rat, p-nitroanisole demethylase was no more sensitive to inhibition than ethylmorphine demethylase. Like rat, however, benzopyrene hydroxylase was the most sensitive to inhibition. In the comparison of the animals, rat appeared more sensitive to inhibition in only one reaction, p-nitroanisole demethylase. Ethylmorphine demethylase and benzopyrene hydroxylase were more sensitive to inhibition in the rabbit. Thus, in considering inhibitory properties of amphetamines, it is necessary to consider both the species as well as the MFO reaction being used as a model. Also, it is necessary to consider the organ from which the microsomes were isolated. Benzopyrene hydroxylase in lung microsomes was equal to or more susceptible to amphetamine inhibition compared to that from livers of phenobarbital-treated rabbits (Table 5).

TABLE 5   % Inhibition of Phenobarbital-Treated Rabbit Liver and Lung Mixed-Function Oxidations*

Amphetamine		p-Nitroanisole O-demethylase Rabbit (liver)	Ethylmorphine N-demethylase Rabbit (liver)	Benzopyrene hydroxylase Rabbit (liver)	(lung)
p-Hydroxyamphetamine	2.0 mM	14±12	43±3	29±15	68±24
Methamphetamine	2.0 mM	17±5	--	--	--
l-Amphetamine	2.0 mM	42±9	42±8	65±6	88±5
d-Amphetamine	2.0 mM	--	64±8	60±4	94±3
Norbenzphetamine	0.1 mM	8±7	26±7	54±9	61±6
Benzphetamine	0.05mM	-3±11	--	63±8	77±6
N-hydroxyamphetamine	0.05mM	22±12	29±2	63±7	94±2

*Values are the mean of 3 to 5 microsomal preparations.

This correlates with the similar ability of cytochrome P-450 of these two organs to form MI complexes, despite the fact that phenobarbital does not induce lung microsomal cytochrome P-450. Thus, in assessing amphetamine inhibitory effects, the knowledge of the accumulation of amphetamines in the lung, their ability to effectively inhibit benzopyrene hydroxylase in this organ, and the exposure to polycyclic hydrocarbons and benzopyrene from the environment and inhalation of tobacco smoke that the lung receives, the formation of MI complexes may have far-reaching consequences on carcinogenesis in the human environment.

ACKNOWLEDGEMENTS

This investigation was supported by Grant No. CA 15760 awarded by the National Cancer Institute, DHEW.

REFERENCES

1. D.S. Hewick and J.R. Fouts, Metabolism in vitro and hepatic microsomal interactions of some enantiomeric drug substrates, Biochem. J. 117, 833 (1970).
2. I. Hoffstrom and S. Orrenius, The interaction of various N-substituted amphetamines with cytochrome P-450 of rabbit liver microsomes, FEBS. Letters 31, 205 (1973).
3. M.R. Franklin, Complexes of metabolites of amphetamines with hepatic cytochrome P-450, Xenobiotica 4, 133 (1974).
4. J. Axelrod, The enzymatic deamination of amphetamine (Benzedrine), J. Biol. Chem. 214, 753 (1955).
5. A.H. Beckett and P.M. Belanger, Metabolic N-oxidation of secondary and primary aromatic amines as a route to ring hydroxylation, to various N-oxygenated products, and to

dealkylation of secondary amines, Biochem. Pharmacol. 25, 211 (1976).
6. A.H. Beckett and S. Al-Sarraj. The identification, properties and analysis of N-hydroxyamphetamine--a metabolite of amphetamine, J. Pharm. Pharmacol. 25, 328 (1973).
7. L.G. Dring, R.L. Smith, and R.T. Williams, The metabolic fate of amphetamine in man and other species, Biochem. J. 116, 425 (1970).
8. J.A. Johnsson, Hydroxylation of amphetamine to parahydroxyamphetamine by rat liver microsomes, Biochem. Pharmacol. 23, 3191 (1974).
9. H. Rommelspacher, H. Honecker, G. Schulze, and S.M. Strauss, The hydroxylation of d-amphetamine by liver microsomes of the male rat, Biochem. Pharmacol. 23, 1065 (1974).
10. J. Werringloer and R.W. Estabrook, Evidence for an inhibitory product·cytochrome P-450 complex generated during benzphetamine metabolism by liver microsomes, Life Sci. 13, 1319 (1973).
11. J. Werringloer and R.W. Estabrook, Heterogeneity of liver microsomal cytochrome P-450: The spectral characterization of reactants with reduced cytochrome P-450, Arch. Biochem. Biophys. 167, 270 (1975).
12. M.R. Franklin, The formation of a 455 nm complex during cytochrome P-450-dependent N-hydroxyamphetamine metabolism, Molec. Pharmacol. 10, 975 (1974).
13. R.M. Philpot and E. Hodgson, A cytochrome P-450-piperonyl butoxide spectrum similar to that produced by ethyl isocyanide, Life Sci. 10 (II), 503 (1971).
14. M.R. Franklin, The enzymic formation of a methylenedioxyphenyl derivative exhibiting an isocyanide-like spectrum with reduced cytochrome P-450 in hepatic microsomes, Xenobiotica 1, 581 (1971).
15. E. Hodgson and R.M. Philpot, Interaction of methylenedioxyphenyl (1,3-Benzodioxole) compounds with enzymes and their effects on mammals, Drug Metab. Rev. 3, 231 (1974).
16. M.R. Franklin, Inhibition of hepatic oxidative xenobiotic metabolism by piperonyl butoxide, Biochem. Pharmacol. 21, 3287 (1972).
17. R.C. James and M.R. Franklin, Cytochrome P-450-455 nm complex formation in hepatic microsomes from mouse, rat, guinea pig, rabbit, and dog, Pharmacologist 17, 185 (1975).
18. R.C. James and M.R. Franklin, Comparisons of the formation of cytochrome P-450 complexes absorbing at 455 nm in rabbit and rat microsomes, Biochem. Pharmacol. 24, 835 (1975).
19. T.C. Orton, M.W. Anderson, R.O. Pickett, T.E. Eling and J.R. Fouts, Xenobiotic accumulation and metabolism by isolated perfused rabbit lungs, J. Pharmacol. Exp. Ther. 186, 482 (1973).
20. M.K. Buening and M.R. Franklin, The formation of cytochrome P-450-455 nm complexes during amine metabolism in rabbit liver, lung and small intestine, Fed. Proc. 34, 729 (1975).
21. M.R. Franklin, The influence of cytochrome P-450 induction on the metabolic formation of 455-nm complexes from amphetamines, Drug Metab. Disp. 2, 321 (1974).

22. M.R. Franklin and R.W. Estabrook, On the inhibitory action of mersalyl on microsomal drug oxidation: A rigid organization of the electron transport chain, Arch. Biochem. Biophys. 143, 318 (1971).
23. J.R. Fouts and T.R. Devereux, Use of 10-sec sonication of homogenates to increase microsomal protein yield in liver and lung from young and adult Dutch Belt rabbits, Biochem. Pharmacol. 22, 1393 (1973).
24. T. Nash, The colorimetric estimation of formaldehyde by means of the Hantzsch Reaction, Biochem. J. 55, 416 (1953).
25. K.J. Netter and G.Seidel, An adaptively stimulated O-demethylating system in rat liver microsomes and its kinetic properties, J. Pharmacol. Exp. Ther. 146, 61 (1964).
26. D.W. Nebert and H.V. Gelboin, The in vivo and in vitro induction of aryl hydrocarbon hydroxylase in mammalian cells of different species, tissues, strains, and developmental and hormonal states, Arch. Biochem. Biophys. 134, 76 (1969).
27. D.E. Rickert and J.R. Fouts, Benzpyrene pretreatment changes the kinetics and pH optimum for aniline hydroxylation in vitro, but not those for benzphetamine demethylation in vitro by rat liver microsomes, Biochem. Pharmacol. 19, 381 (1970).
28. M.R. Franklin, Inhibition of the metabolism of N-substituted amphetamines by SKF 525-A and related compounds, Xenobiotica 4, 143 (1974).

THE PHYSIOLOGICAL IMPLICATIONS OF DRUG OXIDATIONS TO THE CELL

Sten Orrenius, Peter Moldéus, Hjördis Thor and Johan Högberg

Department of Forensic Medicine, Karolinska Institutet, S-104 01 Stockholm, Sweden

The conversion of various lipophilic drugs to hydrophilic and excretable products occurs predominantly in the liver as a multistep metabolic process which most often involves cytochrome P-450-linked monooxygenation and subsequent conjugation reactions. During this process, the hepatocyte is exposed not only to the possible toxic effects of the unmetabolized drug but also to those of its metabolites which may sometimes be even more deleterious. Moreover, drugs may compete with endogenous substrates for common enzymes, cofactors and oxygen and may thus also indirectly affect cellular intermediary and energy metabolism. It is therefore obvious that drug metabolism may have pronounced physiological implications to the cell. Some of these are outlined in Fig. 1.

During recent years most studies on drug oxidations have utilized isolated liver microsomes or solubilized and purified enzymes. As a result of such studies critical information has been obtained about a number of important aspects of this process including the reaction mechanism and substrate and reaction specificity of the enzyme system (cf refs 1 and 2). Considerably less has been revealed, however, about the interaction of drug metabolism and other metabolic pathways and on the effects of drugs and their metabolites on the various functions of the intact cell. For obvious reasons, other and more complex experimental systems are required to obtain this information.

For some time the isolated, perfused liver has been frequently employed in various metabolic studies including those on drug biotransformation. This experimental tool, which was early used to investigate patterns of drug metabolites, has more recently also been successfully employed to study other aspects of drug biotransformation including effects on oxygen consumption (3), pyridine nucleotide levels (4) and certain cellular biosynthetic pathways (5,6).

During the last few years suspensions of isolated hepatocytes have been introduced as an experimental model in studies on drug biotransformation and also proven to be a valuable tool in elucidating some of the physiological implications of this process to the cell (7-9). The availability of viable hepatocytes in high yields by enzyme perfusion of the liver and the possibility

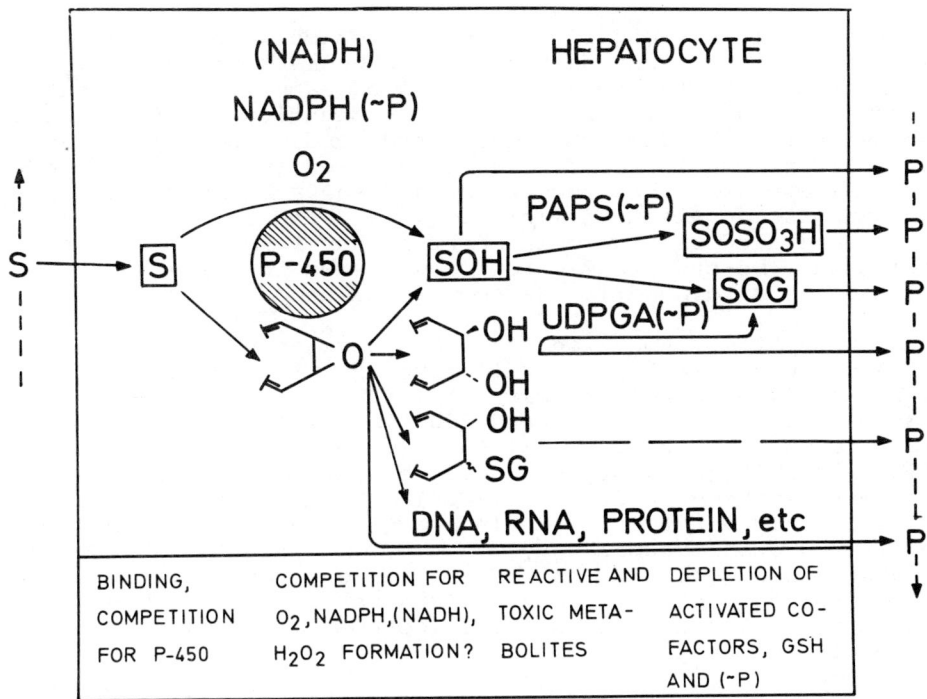

Fig. 1. Physiological implications of drug oxidations to the liver cell. S, substrate; P, product; PAPS, 3'-phosphoadenosine-5'-phosphosulphate; UDPGA, uridine-5'-diphosphoglucuronic acid; ($\sim$P), energy requiring.

to continuously monitor drug biotransformation, as well as a number of other important cellular functions, contribute to the advantage of this model and make it almost ideal for the investigation of various effects of drugs and drug metabolites, as well as of the process of drug metabolism per se, on the cell.

Isolation and some properties of rat hepatocytes

Liver cells are isolated by perfusion of the rat liver with an oxygenated, collagenase containing medium. The method, which is a modification of that described by Berry and Friend (10), may be outlined as follows: (a) Perfusion with Locke's medium, EGTA and albumin, 2% (4 min); (b) Perfusion with Hank's medium, collagenase and Ca^{2+} (6 min); (c) Disruption of liver; (d) Incubation in Hank's medium with or without collagenase and albumin, 2% (5 min); (e) Filtration (100 mesh); (f) Sedimentation by centrifugation at 80 g for 2 min or by gravity and suspension in Krebs-Henseleit's buffer, pH 7.5, containing 2% albumin (see refs 11 and 12 for detailed discription).

As a result of this modified, shorter method we obtain higher yields of cells as well as more viable cells containing high levels of e.g. glutathione. The yield of each liver preparation is about $5-6 \times 10^8$ cells, as measured by counting the final cell suspension in a Buerker chamber, corresponding to approximately 5 g of liver. The isolated liver cell fraction consists mainly of normal-appearing hepatocytes as judged by light microscopy and transmission electron microscopy (7). The cells are essentially separated and the cell membrane has a smooth, rounded appearance. The endoplasmic reticulum shows the usual arrangement and the mitochondria are normal and uncondensed.

Immediately after isolation the cells exclude both trypan blue and NADH (90-95%) which is indicative of an intact cell membrane. The endogenous cell respiration is about 15 nmoles O_2 consumed per min per 10^6 cells and addition of succinate increases this rate only by about 0-10%. Addition of ADP has no effect on cellular respiration but the uncoupler carbonyl cyanide-p-trifluoromethoxyphenyl hydrazone (FCCP) is stimulatory. Oxygen uptake is mainly mitochondrial since antimycin A, an inhibitor of the mitochondrial respiratory chain, at a concentration of 10 µM causes an almost 75% inhibition. Various metabolic functions of the isolated cells are also unimpaired as exemplified by a high rate of gluconeogenesis from lactate, 140 nmoles glucose formed per 10^6 cells per hour (7).

When the liver cells are kept at $4°$ in Krebs-Henseleit's buffer containing albumin, they retain all functions at maximal for up to two hours, but then membrane leakage can be observed and after five hours only about 60-70% of the cells are viable as judged by trypan blue exclusion. However, if a more complete medium is used (Waymoth medium (MB 752/9), 7.5% horse serum, 2.5 mM Hepes buffer supplemented with heparin and penicillin) the cells are still viable after four to eight hours (Fig. 2 and ref. 12). Using this complete medium it is also possible to keep the cells at room temperature under carbogen atmosphere.

The metabolism of a number of lipophilic drugs is conveniently measured in suspensions of isolated hepatocytes. This is illustrated in Fig. 3 using 7-ethoxycoumarin as a substrate. Upon addition of the drug to the medium it is rapidly bound to cytochrome P-450 as revealed by the appearance of a type I spectral change, i.e. the conversion of part of the hemoprotein from the low spin to the high spin form. Cytochrome P-450-linked monooxygenation, which in this case involves an oxidative deethylation reaction to yield the fluorescent product 7-hydroxycoumarin (13), then proceeds linearly with time until the reaction is stopped by the addition of the selective inhibitor, α-naphthoflavone. The subsequent decrease in fluorescence is due to the formation of nonfluorescent conjugates - glucuronides and sulphates - from the hydroxylated product. These reactions are in turn inhibited by the addition of the competitive inhibitor salicylamide. As also shown in Fig. 3, monooxygenation of 7-

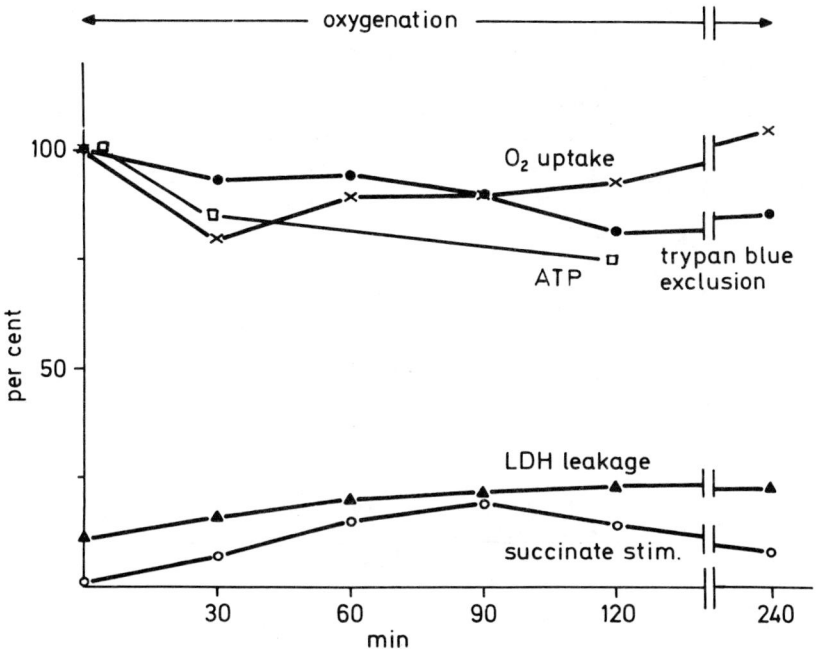

Fig. 2. Endogenous and succinate-stimulated oxygen uptake, ATP level, extracellular lactate dehydrogenase (LDH) activity and trypan blue exclusion frequency in hepatocytes during incubation at 37° in the presence of 95% O_2-5% CO_2

ethoxycoumarin is a rather slow process which does not appreciably affect cellular oxygen uptake or under these conditions - hepatocytes isolated from fed rats - the level of reduced NADPH. However, by the use of selective inhibitors monooxygenation as well as subsequent conjugation of this drug is easily monitored in isolated hepatocytes which makes it a valuable model substrate for the study of cellular drug metabolism and of various factors that affect this process.

A protective effect of drug binding and metabolism by the cytochrome P-450 system.

A large portion of cytochrome P-450 in isolated hepatocytes, and in the isolated perfused liver, is thus present in the oxidized, non-substrate bound state and readily binds entering drug molecules (4,7,8). Drugs may in fact compete with endogenous substrates like steroid hormones for binding to the cytochrome although there is still little experimental evidence to support a physiological importance of such a competition.

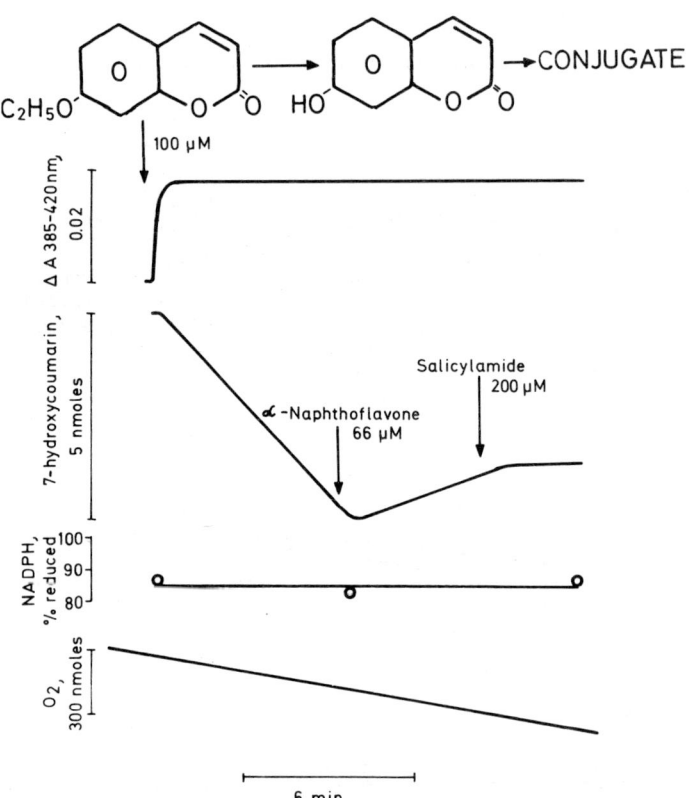

Fig. 3. <u>7-Ethoxycoumarin metabolism in isolated hepatocytes from 3-methylcholanthrene-treated rat.</u> 10^6 cells per ml were incubated at $25°$.

Endogenous substrates are generally favored by higher binding affinities for the cytochrome and accumulating evidence for a heterogeneity of cytochrome P-450 may also imply that different species of the hemoprotein are often responsible for the binding and metabolism of drugs and endogenous substrates.

In contrast, there is little doubt that drug binding, and metabolism by the cytochrome P-450 system, offers a protective mechanism of physiological significance against toxic effects of drugs on vital cellular functions. This may be illustrated by comparing the effects of certain drugs on mitochondrial respiration in different experimental systems. As outlined in Fig. 4, drugs may act as inhibitors of the mitochondrial NADH oxidase or uncouplers of phosphorylation from respiration. The inhibitory effect was early demonstrated with barbiturates, and the site of attack characterized (14). More recently a variety of other drugs including the β-receptor blocking agents alprenolol and propranolol have also been shown to inhibit the NADH oxidase at a similar site (15).

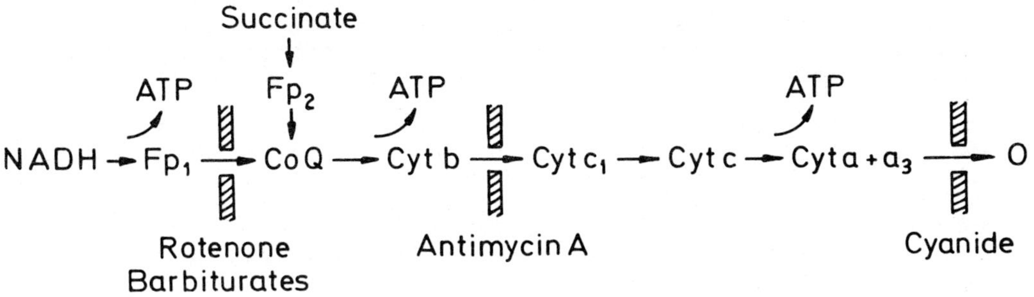

$$NADH \rightarrow Fp_1 \xrightarrow{ATP} CoQ \xrightarrow{Succinate \downarrow Fp_2} Cyt\,b \xrightarrow{ATP} Cyt\,c_1 \rightarrow Cyt\,c \xrightarrow{ATP} Cyt\,a+a_3 \rightarrow O$$

Rotenone / Barbiturates — Antimycin A — Cyanide

Inhibitors		Uncouplers	
relative effects			
Rotenone	+++	FCCP	+++
Alprenolol	++	Benzphetamine	++
Hexobarbital	+(+)	Norbenzphetamine	+(+)
Propranolol	+	Biphenyl	+
		4-OH-Biphenyl	+
		o,o'-Biphenol	++(+)

Fig. 4. **Schematic representation of effects of drugs on the mitochondrial NADH oxidase.** FCCP, carbonyl cyanide p-trifluoromethoxyphenyl hydrazone.

The observed extent of drug inhibition of respiration is, however, markedly affected by the experimental system employed. Thus, Brauser et al (3) early observed that the inhibitory effect of amobarbital on oxygen uptake in the isolated perfused rat liver changed into a stimulatory effect when livers from phenobarbital-treated rats were used. Similarly, we observed that considerably higher concentrations of hexobarbital, propranolol and alprenolol were required to obtain inhibition of oxygen uptake in isolated hepatocytes as compared to liver or heart mitochondria or heart slices, and that the addition to the medium of low concentrations of SKF 525-A, which efficiently competes with the other drugs for binding to cytochrome P-450, made the hepatocytes as sensitive as the isolated mitochondria to the inhibitory effects of these drugs on oxygen uptake (15). Since alprenolol and propranolol have been found to bind extensively to cytochrome P-450 (16), we concluded that this process may provide a protective mechanism against toxic effects caused by high concentrations of free drug in the liver cells. Binding of the drug to cytochrome P-450, rather than its metabolism via this system to less toxic products, seemed to be of major importance for the protective effect observed, since the rate of metabolism was too low, and the added drug concentration too high, to allow any appreciable decrease in the concentration of unchanged drug during the course of the recording.

When certain drugs are present in higher concentrations, however, binding to cytochrome P-450 does not seem to offer enough protection against toxic effects on vital cellular functions. Amphetamine derivatives like benzphetamine do for example, when added to hepatocytes, cause a rapid decrease in cellular ATP level - probably due to uncoupling of phosphorylation from respiration - followed by a disintegration of the plasma membrane (17). In this case, the main metabolite, norbenzphetamine, is almost as efficient as an uncoupler as the parent drug and the N-demethylation of benzphetamine by the cytochrome P-450 system would therefore not decrease the mitochondrial toxicity of this drug much. This is in contrast, however, to the mitochondrial toxicity of fatty acids undergoing cytochrome P-450-linked ω-hydroxylation, whose metabolites exhibit a dramatically decreased toxicity as compared to the parent fatty acids - a phenomenon which may have physiological significance in conditions of enhanced levels of circulating fatty acids (18).

Cytochrome P-450-linked drug monooxygenation - physiological implications

Cytochrome P-450-linked drug monooxygenation utilizes molecular oxygen and NADPH and may thus compete with other O_2- and NADPH-dependent metabolic processes. Moreover, even in the presence of an excess of O_2 and NADPH, drugs and endogenous substrates, such as steroids and fatty acids have been shown to be competitive inhibitors of each other's metabolism by the cytochrome P-450 system in a number of investigations (cf refs 19 and 20). Yet, little is known about the physiological importance of this competition. High drug concentrations, which do probably not occur in vivo, have namely been used to demonstrate this effect in microsomal incubations and in the perfused liver steroid hormones must be added to the perfusate in rather high concentrations in order to obtain measurable inhibition of drug oxidation (21). Little is therefore known about the importance of metabolic interaction between drugs and endogenous substrates of cytochrome P-450 under physiological conditions.

Since cytochrome P-450 has a high affinity for oxygen, rapid drug metabolism by this system may act to decrease the tissue concentration of oxygen and may thus, at least in theory, affect other oxygen requiring processes. Under special conditions, like in the perfused liver isolated from a fed, phenobarbital-treated rat, a significant part of the total oxygen uptake is in fact utilized for drug monooxygenation when a rapidly metabolised substrate like hexobarbital is added in a rather high concentration to the perfusate (4). On the other hand, drug monooxygenation is more often less rapid and when 7-ethoxycoumarin is added to hepatocytes there is no, or only a marginal, effect on oxygen consumption. Furthermore, oxygen appears to be readily accessible under normal conditions and there is really no evidence that oxygen utilization by cytochrome P-450-linked drug monooxygenation may affect other oxygen requiring processes.

TABLE 1 Effect of 7-ethoxycoumarin on O_2 uptake in antimycin A-inhibited rat hepatocytes[x]

Addition	Control	PB-treated	MC-treated
	nmoles O_2/min/10^6 cells		
None	18.1	27.1	15.2
+ Antimycin A, 25 µM	6.5	13.5	5.1
+ 7-Ethoxycoumarin, 100 µM	6.5	13.5	7.1
ΔO_2	0	0	2.0
7-Hydroxycoumarin formed, nmoles/min	0.02	1.1	2.1

[x] Hepatocytes were obtained from control rats, or rats injected with 80 mg of sodium phenobarbital, or 20 mg of 3-methylcholanthrene, per kg bodyweight once daily for three days. Incubations were performed at 25°.

Competition for cytosolic NADPH seems to be a more important physiological implication of drug monooxygenation. The dependence of this process on a high concentration of cytosolic NADPH was early indicated by the observation of Sies and Brauser (4), of an almost complete inhibition of hexobarbital metabolism in livers isolated from fasted, phenobarbital-treated rats which had been preperfused with antimycin A. More recently we observed that the cytochrome P-450-linked metabolism of alprenolol was markedly inhibited in hepatocytes isolated from fasted, but not from fed, rats (7,8). This observation could be correlated to differences in the level of reduced NADPH, which was always lower in hepatocytes from fasted rats and further decreased in the presence of respiratory inhibitors like antimycin A or rotenone. This was subsequently shown to be due to a decreased energy level under these conditions and supported the previous observation that energy requiring shuttle mechanisms are of major importance for the generation of cytosolic NADPH in the liver of the fasted rat (5). Under the same conditions, rapid NADPH utilization by drug monooxygenation has been found to inhibit biosynthetic pathways like gluconeogenesis from lactate (5) and fatty acid synthesis (6). Thus, we may conclude that drug monooxygenation, fatty acid synthesis and probably also glutathione reductase compete for cytosolic NADPH which in the fasted, glycogen-depleted state is mainly generated via energy requiring shuttle mechanisms including the α-ketoglutarate-isocitrate shuttle and the shuttle involving malic enzyme (Fig. 5). This competition is easily observable in experimental models and may well have physiological implications.

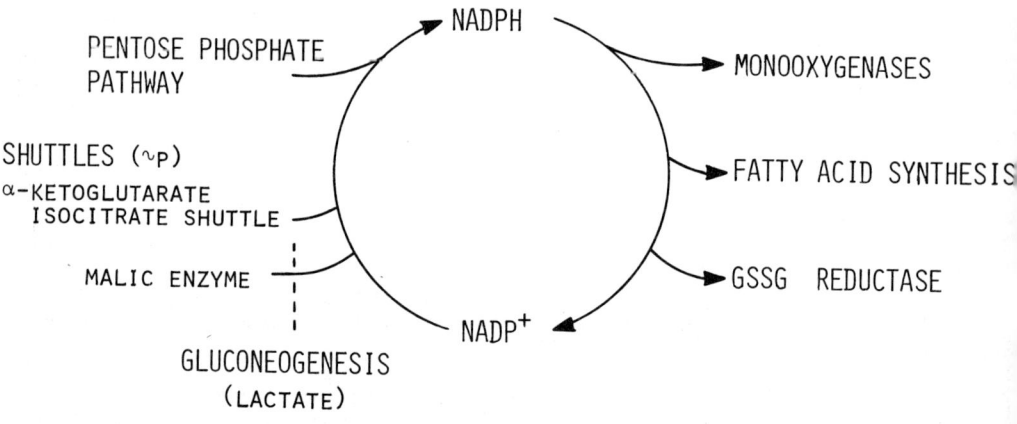

Fig. 5. Schematic representation of the generation and utilization of cytosolic NADPH in the liver.

In studies with isolated liver microsomes NADH has been shown to affect drug monooxygenation (22,23), stimulating drug metabolism in synergism with NADPH. Several possible mechanisms have been proposed to explain this phenomenon, such as increased hepatic reduction level leading to increased production of NADPH (24), transfer of electrons from reduced cytochrome b_5 to the reduced, oxygenated, substrate-bound cytochrome P-450 complex (25), or that electrons from NADH are utilized for the reduction of active oxygen during "uncoupling" of the monooxygenase, resulting in the formation of H_2O and H_2O_2 (26).

It is too early to decide, however, whether reducing equivalents from NADH are also utilized during drug monooxygenation under more physiological conditions than those represented by a fortified microsomal fraction. The unability to observe changes in the $NADH/NAD^+$ ratio, or an enhanced H_2O_2 production, during drug monooxygenation in the perfused liver or isolated hepatocytes would suggest that this is not the case. On the other hand, Grundin (27) has reported that the presence of alprenolol stimulates ethanol oxidation in isolated hepatocytes under conditions when the reoxidation of cytosolic NADH is rate-limiting for this process and causes a corresponding decrease in the lactate/pyruvate ratio. This was interpreted as an indication of utilization of reducing equivalents from NADH by cytochrome P-450-linked drug monooxygenation. However, further experimental evidence is needed to decide one way or the other.

Reactive and toxic metabolites of drug oxidations

The products of cytochrome P-450-linked drug monooxygenation most often undergo further metabolism before being excreted. During this second phase of metabolism activated cofactors like uridine-5'-diphosphoglucuronic acid (UDPGA) and 3'-phosphoadenosine-5'-phosphosulphate (PAPS) are utilized and, furthermore, the oxygenated, primary drug products may compete with endogenous substrates for certain enzymes, e.g. glucuronyl transferase, sulphotransferase, glutathione-S-epoxide transferase, aldehyde dehydrogenase, etc. The physiological implications of such competition are, however, yet largely unknown and at the present time the formation of reactive and toxic drug metabolites through the cytochrome P-450 system seems to present a more important hazard to the cell.

During recent years a number of compounds including amphetamine derivatives have been found to yield microsomal metabolites that bind to cytochrome P-450 (cf ref. 28). The binding, which is spectrophotometrically easily distinguishable, often causes inactivation of the cytochrome, and many of these compounds are in fact known as efficient inhibitors of drug metabolism. On the other hand, the formation of the complex seems to be more easily obtained in microsomal incubations than in the cell, where the reactive product may undergo inactivation by further metabolism (29). Thus, under physiological conditions the formation of catalytically inactive cytochrome P-450 - drug metabolite complexes may be of less significance.

The formation of toxic metabolites by cytochrome P-450-linked drug monooxygenation has been known for quite some time but has received increased attention during recent years. This is due in particular to the observations that cytochrome P-450-dependent reactions, such as N-hydroxylation and epoxide formation, may be critical steps in the metabolic activation of chemical carcinogens. Such metabolites may, however, also exert acute toxic effects and cause cell death and tissue necrosis. Also in this case, the resulting effect is dependent on the balance between the formation of the toxic metabolite and its inactivation by further metabolism. The latter may proceed by way of formation of glucuronides and sulphates from hydroxylated metabolites, or transdihydrodiols from epoxides, but in both instances inactivation by the formation of glutathione conjugates appears to be of major importance (30,31). This is also obvious when hepatocytes are incubated with either bromobenzene (Fig. 6) or paracetamol (Fig. 7). In both cases the loss of viability is preceeded by a dramatic decrease in the level of reduced glutathione. Whether the disappearance of glutathione per se constitutes the toxic mechanism, e.g. by permitting uncontrolled lipid peroxidation (32), or if the damage is attributed to the binding of the reactive drug metabolites to other and vital cell constituents in the glutathione-depleted cell, cannot yet be decided.

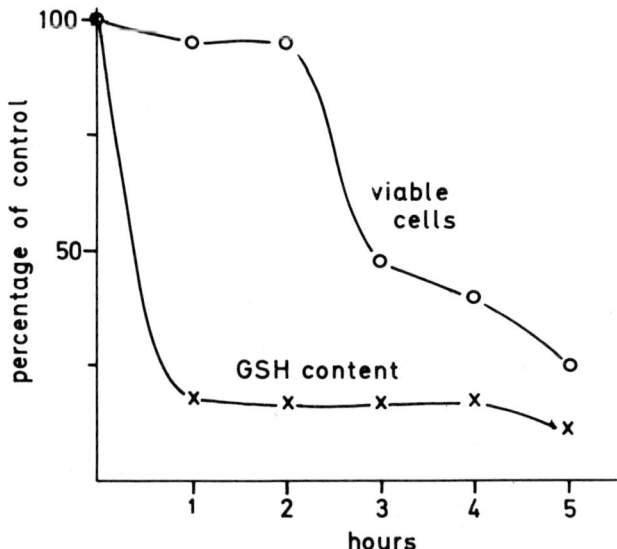

Fig. 6. Effect of bromobenzene on isolated hepatocytes from phenobarbital-treated rats. 10^6 cells per ml were incubated in complete medium (12) at 37° in the presence of 4 mM bromobenzene. Latency of lactate dehydrogenase activity served as an index for cell viability (12). GSH content was measured as previously described (12). Incubation without added bromobenzene served as control.

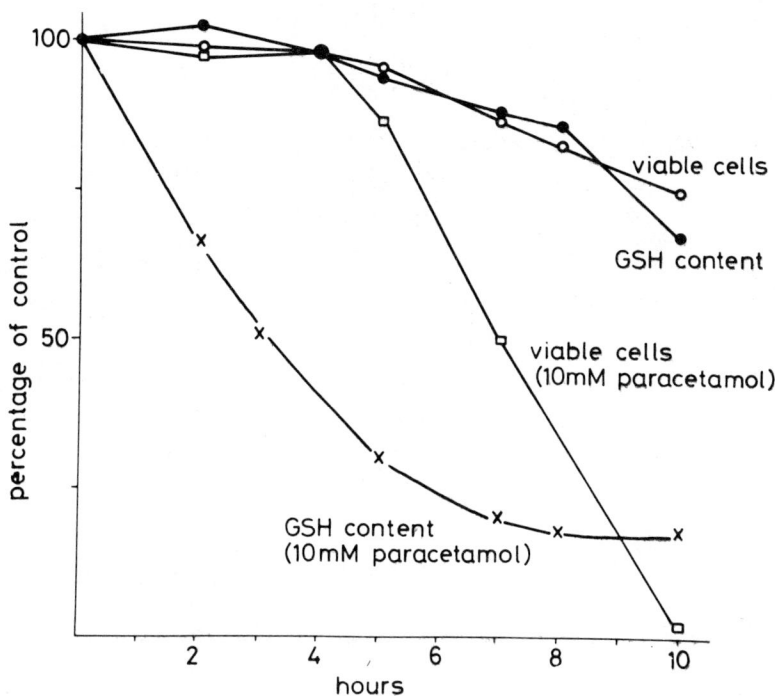

Fig. 7. Effect of paracetamol on isolated hepatocytes from phenobarbital treated rats. 10^6 cells per ml were incubated in complete medium (12) at 37°. Latency of lactate dehydrogenase activity served as an index for cell viability (12). GSH content was measured as previously described (12).

Concluding remarks

It has been the purpose of this overview to point out some of the physiological implications of drug oxidations to the cell. A number of possible implications have been mentioned, like protection against drug toxicity by binding and metabolism through the cytochrome P-450 system, competition between drugs and endogenous substrates for common enzymes, oxygen and cofactors, and the formation of reactive and toxic drug metabolites (cf Fig. 8). Most of these phenomena are, however, only partly elucidated and little is known about their physiological importance. Thus, this is another field of drug metabolism which calls for further, active research.

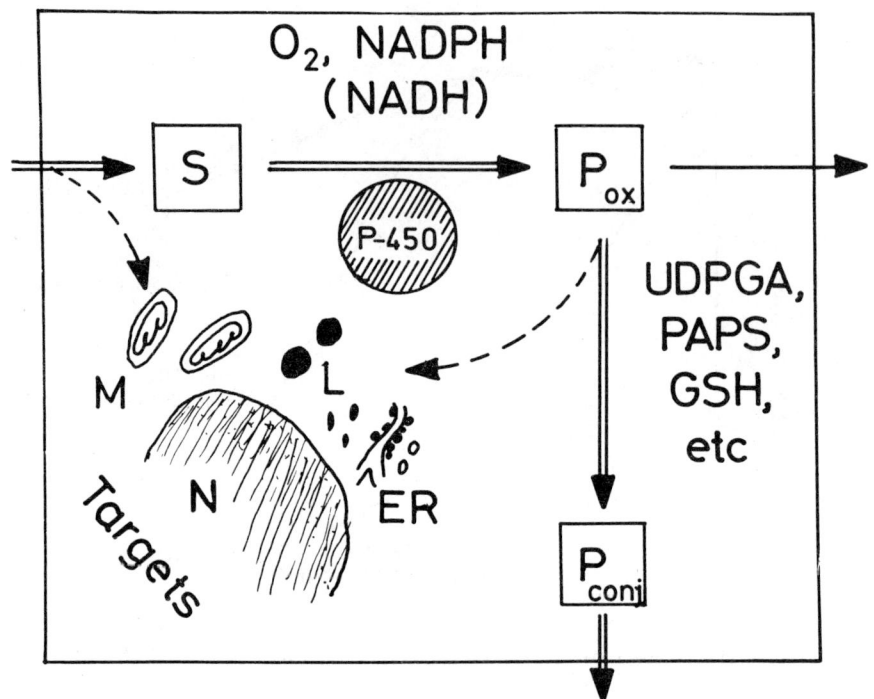

Fig. 8. Drug oxidations - inactivation, competition and activation.

ACKNOWLEDGEMENTS

The work from the authors' laboratory was supported by the Swedish Medical Research Council.

REFERENCES

(1) S. Orrenius and L. Ernster, Microsomal cytochrome P-450-linked monooxygenase systems in mammalian tissue. In <u>Molecular Mechanisms of oxygen activation</u>, O. Hayaishi, ed, Academic Press, New York, 1974, p. 215.
(2) S. Orrenius, Reaction mechanisms of cytochrome P-450. In Proceedings Sixth Intern. Congr. Pharmacol., vol. 6, <u>Mechanisms of toxicity and metabolism</u>, N. T. Kärki, ed, Helsingfors, 1975, p. 39.
(3) B. Brauser, H. Sies and Th. Bücher, Action of amobarbital on microsomal and mitochondrial respiratory state in perfused rat liver with and without phenobarbital, <u>FEBS Letters</u>, 2, 170 (1969).

(4) H. Sies and B. Brauser, Interaction of mixed function oxidase with its substrates and associated redox transitions of cytochrome P-450 and pyridine nucleotides in perfused rat liver, Eur. J. Biochem. 15, 531 (1970).
(5) R. Scholz, W. Hansen and R. G. Thurman, Interaction of mixed-function oxidation with biosynthetic processes. 1. Inhibition of gluconeogenesis by aminopyrine in perfused rat liver, Eur. J. Biochem. 38, 64 (1973).
(6) R. G. Thurman and R. Scholz, Interaction of mixed-function oxidation with biosynthetic processes. 2. Inhibition of lipogenesis by aminopyrine in perfused rat liver, Eur. J. Biochem. 38, 73 (1973).
(7) P. Moldéus, R. Grundin, H. Vadi and S. Orrenius, A study of drug metabolism linked to cytochrome P-450 in isolated rat liver cells, Eur. J. Biochem. 46, 351 (1974).
(8) R. Grundin, P. Moldéus, H. Vadi, S. Orrenius, C. von Bahr, D. Bäckström and A. Ehrenberg, Drug metabolism in isolated rat liver cells. In Adv. Exp. Med. Biol. 58, Cytochromes P-450 and b_5: Structure, function and interaction, D. Y. Cooper, O. Rosenthal, R. Snyder and C. Witmer, eds, Plenum Press, New York, 1975, p. 251.
(9) P. Moldéus, H. Thor, J. Högberg and S. Orrenius, Drug metabolism and toxicity studies in isolated rat liver cells, Proc. Eur. Soc. Toxicol. in press.
(10) M. N. Berry and D. S. Friend, High-yield preparation of isolated rat liver parenchymal cells, J. Cell Biol. 43, 506 (1969).
(11) J. Högberg, S. Orrenius and P. J. O'Brien, Further studies on lipid-peroxide formation in isolated hepatocytes, Eur. J. Biochem. 59, 449 (1975).
(12) J. Högberg and A. Kristoferson, A correlation between glutathione levels and cellular damage in isolated hepatocytes, Manuscript in preparation.
(13) V. Ullrich and P. Weber, The O-dealkylation of 7-ethoxycoumarin by liver microsomes, Hoppe-Seyler's Z. Physiol. Chem. 353, 1171 (1972).
(14) O. Jalling, O. Lindberg and L. Ernster, On the effect of substituted barbiturates on mitochondrial respiration, Acta Chem. Scand. 9, 198 (1955).
(15) R. Grundin, H. Thor and S. Orrenius, The inhibition of the mitochondrial NADH oxidase by alprenolol and the protective effect of cytochrome P-450, Acta Pharmacol. et Toxicol. 37, 154 (1975)
(16) R. Grundin, P. Moldéus, S. Orrenius, K. O. Borg, I. Skånberg and C. von Bahr, The possible role of cytochrome P-450 in the liver "first pass elimination" of a β-receptor blocking drug, Acta Pharmacol. et Toxicol. 35, 242 (1974).
(17) M. Hirata, H. Thor, J. Högberg, P. Moldéus and S. Orrenius, Manuscript in preparation.
(18) Å. Ellin, Fatty acid hydroxylation in rat kidney and liver microsomes, Dissertation thesis, University of Stockholm, 1975.

(19) S. Orrenius and B. P. Lisboa, Metabolic interactions of drugs and steroid hormones in rat liver microsomes, In Biochemical aspects of antimetabolites and of drug hydroxylation, S. Shugar, ed, Academic Press, London, 1969, p. 143.
(20) S. Orrenius and H. Thor, Fatty acid interaction with the hydroxylating enzyme system of rat liver microsomes, Eur. J. Biochem. 9, 415 (1969).
(21) C. von Bahr, F. Sjöqvist and S. Orrenius, The inhibitory effect of hydrocortisone and testosterone on the plasma disappearance of nortriptyline in the dog and the perfused rat liver, Eur. J. Pharmacol. 9, 106 (1970).
(22) A. H. Conney, R. R. Brown, J. A. Miller and E. C. Miller, The metabolism of methylated aminoazo dyer. VI. Intracellular distribution and properties of the demethylase system, Cancer Res. 17, 628 (1957).
(23) B. S. Cohen and R. W. Estabrook, Microsomal electron transport reactions. II. Use of reduced triphosphopyridine nucleotide and/or reduced disphosphopyridine nucleotide for the oxidative N-demethylation of aminopyrine and other drug substrates, Arch. Biochem. Biophys. 143, 46 (1971).
(24) J. R. Williamson, R. Scholz, E. T. Browning, R. G. Thurman and M. H. Fukami, Metabolic effects of ethanol in perfused rat liver, J. Biol. Chem. 244, 5044 (1969).
(25) A. Hildebrandt and R. W. Estabrook, Evidence for the participation of cytochrome b_5 in the hepatic microsomal mixed-function oxidase reaction, Arch. Biochem. Biophys. 143, 66 (1971).
(26) H. Staudt, F. Lichtenberger and V. Ullrich, The role of NADH in uncoupled microsomal monooxygenations, Eur. J. Biochem. 46, 99 (1974).
(27) R. Grundin, Metabolic interaction of ethanol and alprenolol in isolated liver cells, Acta Pharmacol. et Toxicol. 37, 185 (1975).
(28) J. Werringloer and R. W. Estabrook, Heterogeneity of liver microsomal cytochrome P-450: The spectral characterization of reactants with reduced cytochrome P-450. Arch. Biochem. Biophys. 167, 270 (1975).
(29) M. Hirata, H. Thor and S. Orrenius, Manuscript in preparation.
(30) J. R. Mitchell, D. J. Jollow, W. Z. Potter, J. R. Gillette and B. B. Brodie, Acetaminophen-induced hepatic necrosis. IV. Protective role of glutathione, J. Pharmacol. Exp. Therap. 187, 211 (1973).
(31) W. D. Reid, G. Krishna, J. R. Gillette and B. B. Brodie, Biochemical mechanism of hepatic necrosis induced by aromatic hydrocarbons, Pharmacology 10, 193 (1973).
(32) J. Högberg, Regulation and effects of lipid peroxidation in isolated hepatocytes. In Active intermediates. Formation, toxicity and inactivation, R. Snyder and H. Vainio, eds, Plenum Press, New York, in press.

NICOTINAMIDE NUCLEOTIDE SYSTEMS AND DRUG OXIDATION IN THE LIVER CELL

Helmut Sies and Klaus Weigl

Institut für Physiologische Chemie und Physikalische Biochemie der Universität München, 8-München-2, Goethestr. 33, Germany

ABSTRACT

1. The redox states of NADPH and NADH systems and of cytochrome P-450 have been studied in rat liver using hemoglobin-free perfusion and isolated hepatocytes from phenobarbital-pretreated rats.
2. The results indicate that in the intact cell control of electron flow through the monooxygenase electron transport chain of the endoplasmic reticulum is exerted by properties intrinsic to the system. Thermodynamically, a plausible description is afforded by a positive increase of the midpoint potential of cytochrome P-450 upon interaction with its substrate, thus facilitating electron flow. In accordance with this, in the intact cell the steady state degree of reduction of cytochrome P-450 is substantially higher in the presence of monooxygenase substrate(30-50 %) than in its absence(about 6 %).
3. The marked decrease in the NADPH/NADP$^+$ ratio of total tissue levels from 4.0 to 2.3 upon addition of monooxygenase substrate is accompanied by a similar decrease in the isocitrate dehydrogenase indicator couple, isocitrate/2-oxoglutarate. Also, the citrate levels decrease significantly due to equilibration by aconitate hydratase.
4. In hepatocytes subfractionated with the digitonin method into cytosolic and mitochondrial fractions, it is shown that not only in the cytosol but also in the mitochondrial compartment there is a decrease of isocitrate and citrate and an increase of 2-oxoglutarate levels upon addition of monooxygenase substrate, e.g. aminopyrine.
5. The cytosolic citrate concentration decreases from 0.17 mM to 0.10 mM during aminopyrine demethylation. This probably explains the inhibition of lipogenesis reported in the literature. Furthermore, in livers from fasted rats, the process of ureogenesis from ammonia, requiring mitochondrial NADPH for amination of 2-oxoglutarate, is partially inhibited; this further illustrates mitochondrial-cytosolic interrelationships occurring during monooxygenation reactions.

INTRODUCTION

Control of hepatic drug monooxygenase, i.e. the regulation of electron flow through the microsomal electron transport chain from NADPH to oxygen and substrate with cytochrome P-450 as the terminal oxidase, is exerted at or near the level of the terminal oxidase. It has early been postulated by Narasimhulu et al(1,2) that "substrate binding is associated with transformation of a pigment from an inactive form to an active state which permits increased electron flow to the oxidase P-450". Thermodynamically, this can be described by a positive increase of the midpoint potential of cytochrome P-450 upon interaction with the substrate, as we have suggested in a study on the redox properties in the perfused rat liver(3), and by a low-to-high spin transition as detected with isolated liver microsomes(4). The extensive studies on the soluble camphor methylene hydroxylase from P.putida provided further evidence for the control of electron flow by substrates(see recent review by Gunsalus et al(5)). Kinetically, this is expressed by an enhanced reducibility of cytochrome P-450 upon substrate binding(6,7).

A steady state oxidation of NADPH during drug oxidation by the monooxygenase system has been identified and characterized in our studies with perfused rat liver(3,8). The implication of the steady state oxidation of NADPH to the cell is further investigated in the present paper. For this purpose, mitochondrial and cytosolic concentrations of citrate and 2-oxoglutarate were determined with and without enhanced flux in the monooxygenase system. This was achieved by application of the digitonin fractionation procedure of Zuurendonk & Tager(9) to isolated hepatocytes from male fed rats pretreated with phenobarbital.

METHODS

Hemoglobin-Free Liver Perfusion

Livers from male fed Wistar rats of 130-180 g body wt, pretreated with phenobarbital(80 mg/kg for 3 days, were perfused in an open(non-recirculating)system(10) with a flow rate of 4 ml/min per g liver at $37°C$. The bicarbonate-buffered saline(10) contained L-lactate(2.1 mM) and pyruvate(0.3 mM).

Isolation of Hepatocytes and Incubation and Fractionation Procedure

Hepatocytes were isolated(11) and incubated and subfractionated with a modification(12) of the digitonin method into mitochondrial and cytosolic fractions. A final concentration of digitonin of 1.67 mM in the separation medium was found to give better separation than the 2 mM used with cells from untreated animals; this may be due to a lower cholesterol content in the liver plasma membrane of the pretreated animals. Fractions called cellular were obtained by centrifugation of cells suspended shortly in a medium(12) at $-2°C$, layered on top of a silicone oil mixture, into perchloric acid. Fractions called mitochondrial(containing particulate structures) were obtained in a similar way, but the medium contained digitonin in this case to disrupt the plasma membrane. Metabolite assays were performed in neutralized extracts of these samples. Cytosolic metabolite levels were calculated by difference of cellular <u>minus</u> mitochondrial levels. Concentrations were calculated with the water contents of 2.0 and o.21 ml/g dry wt for the cytosolic and mitochondrial spaces, respectively(13). Detailed analytical data will be published elsewhere.

Redox State of Cytochrome P-450 in Perfused Liver

This was measured essentially as described(8) except that a newly designed organ spectrophotometer unit was employed(see (14)).

Reduced Nicotinamide Nucleotides in Isolated Hepatocytes

This was performed by dual-wavelength photometry of suspended cells at 350 <u>minus</u> 380 nm as described(11).

RESULTS AND DISCUSSION

Redox State of NADPH and of Cytochrome P-450 in Liver During Monooxygenase Activity

The total tissue level of NADPH decreases and that of $NADP^+$ increases in perfused liver from phenobarbital-pretreated rats when a substrate for monooxygenation, e.g. hexobarbital, is added(8). The ratio, $NADPH/NADP^+$, decreases from 4.0 to 2.3 in presence of hexobarbital; in other words, the degree of re-

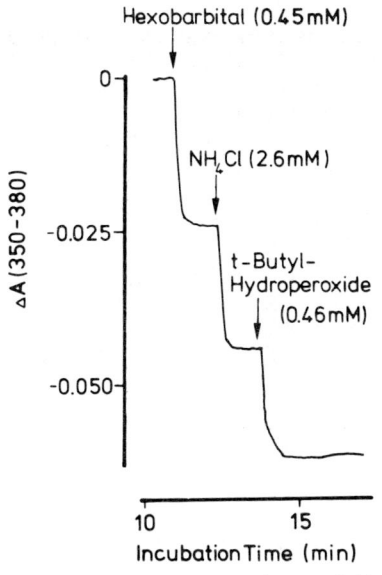

Fig. 1. Time course of the oxidation of NADPH in isolated hepatocytes upon addition of hexobarbital, ammonium chloride, and t-butyl hydroperoxide. Dual-wavelength recording at 350-380 nm.

duction of NADPH, i.e. $(NADPH)/(NADPH+NADP^+)$, decreases from 80 % to 70 % in presence of hexobarbital. These findings have been confirmed in a similar study with isolated hepatocytes(15).

The oxidation in the NADPH system occurs rapidly upon addition of monooxygenase substrate, as can be shown in the intact organ by surface fluorometry at nicotinamide nucleotide-specific wavelengths or by dual-wavelength spectrophotometry through a lobe of perfused liver(8,14). The experiment shown in Fig. 1 was performed with isolated hepatocytes. A new steady state is reached within less than 1 min after addition of hexobarbital, and furthermore the oxidation in the mitochondrial NADPH system subsequently initiated by ammonium chloride is additive to the oxidation obtained with hexobarbital. The further oxidation of NADPH upon addition of t-butyl hydroperoxide is due to the activity of GSSG reductase which re-reduces the GSH peroxidized by GSH peroxidase (10,11) present in both the cytosolic and the mitochondrial compartments(16).

The degree of reduction of cytochrome P-450 is about 6 % in the endogenous steady state of perfused rat liver and is substantially increased upon addition of substrate(8), as is shown in the upper trace of Fig.2 for aminopyrine. A perfused liver was given antimycin A(8 nmol/ml of perfusate for 3 min)prior to the addition of aminopyrine to inhibit mitochondrial O_2 uptake, thereby raising effluent O_2 concentration towards 0.63 mM. Thus, extra O_2 uptake by aminopyrine(lower trace in Fig.2) is unlikely to cause a hypoxic condition which can lead to a reduction of cytochrome P-450. With the midpoint potential of -410 mV for cytochrome P-450(17), redox potentials of -335 mV and -390 mV are calculated for the states in absence and presence of substrate, respectively. A comparison with the data for free NADPH(Table 1) shows that apparently cytochrome P-450 would have a more negative redox potential than its reductant NADPH in presence of substrate, obviously an unreality. Assuming a midpoint potential of 210 mV more positive in presence of substrate(18), the redox potential in presence of aminopyrine is about -180 mV. The value calculated without the change in midpoint potential is given in parentheses. The midpoint potential of cytochrome b_5 is reported to be around 0 mV. Since cy-

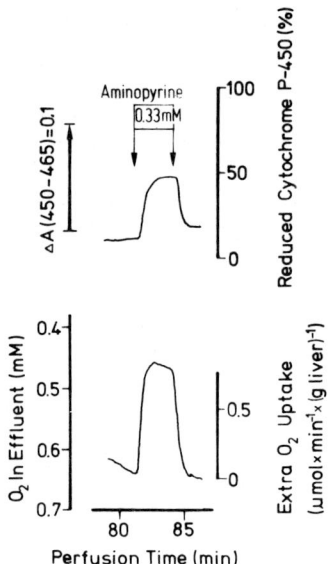

Fig. 2. Effect of aminopyrine on the degree of reduction of cytochrome P-450(upper trace) and on oxygen uptake(lower trace) in perfused liver of a phenobarbital-pretreated rat.
Earlier in this experiment(before antimycin A was present) the addition of aminopyrine had a similar effect on the redox state of cytochrome P-450; O_2/CO then was approx. 3-fold lower than during the period between 80-85 min shown here.

TABLE 1   Redox Potentials of Free Cytosolic Nicotinamide Nucleotides and of Cytochrome P-450 in Perfused Rat Liver from Phenobarbital-Pretreated Male Fed Rats

Addition	NADPH	NADH	Cytochrome P-450
		mV	
None	-391	-241	-335
Aminopyrine(1 mm)	-380	-244	-180(-390)

The indicator systems are cytosolic isocitrate dehydrogenase(see Table 2), lactate dehydrogenase, and the reduced cytochrome P-450-CO complex, respectively. Midpoint potentials are -433, -215, and -410(or -200 with aminopyrine) mV, respectively.

tochrome b_5 is almost fully reduced within intact liver(11), it is difficult at present to estimate with any certainty the redox potential of this cytochrome in rat liver. The donor system for reduction of cytochrome b_5, cytosolic free NADH, does not show any oxidation but rather a slight reduction, indicated by the redox potentials of -241 and -244 mV in the absence and presence of aminopyrine, respectively(Table 1).

Tricarboxylates and 2-Oxoglutarate in Mitochondria and Cytosol During Monooxygenation

Employing the mitochondrial and cytosolic NADPH-linked isocitrate dehydrogenases(coupled to aconitate hydratases) as indicator metabolite systems for the redox potentials of free NADPH in these subcellular compartments(12,19), it is revealed that even in cells obtained from fed--as opposed to fasted--rats and in presence of 10 mM glucose in the incubation medium there is a substantial participation of the mitochondrial compartment in the metabolite changes. As

TABLE 2 Citrate and 2-Oxoglutarate Concentrations and Redox Potentials of Free NADPH in Mitochondrial and Cytosolic Spaces in Hepatocytes Isolated From Phenobarbital-Pretreated Male Rats in Absence and Presence(5 min) of 1 mM Aminopyrine

Fraction	Addition	Citrate	2-Oxoglutarate	Citrate / 2-Oxoglut.	Redox Potential of Free NADPH
		mM	mM		mV
Mitochondrial	-	3.06	1.06	2.90	-407
	Aminopyrine	2.56	1.54	1.66	-400
Cytosolic	-	0.17	0.20	0.87	-391
	Aminopyrine	0.10	0.26	0.38	-380
M/C Gradient	-	18.0	5.3		
	Aminopyrine	25.6	5.9		

Data were obtained as described in METHODS. Redox Potentials were calculated with $E'= -433$ mV for NADP-dependent isocitrate dehydrogenase, assuming constant CO_2 of 1.2 mM and equilibration of the reactants of aconitate hydratase.

shown in Table 2, a significant decrease of citrate concentration occurs in the mitochondrial and cytosolic fractions when aminopyrine is present. Concomitantly, there is an increase in the concentration of 2-oxoglutarate. Assuming equilibration of the aconitate hydratase reactants, the ratio of citrate/2-oxoglutarate may be taken as an indicator of the free $NADPH/NADP^+$ ratio. The citrate/2-oxoglutarate ratio in the cytosol decreases to 44 % of the control value, whereas in the mitochondria it decreases to 57 % of the control value. Accordingly, the citrate/2-oxoglutarate ratio in the total cellular extracts decreases to an intermediate value, 55 %(not shown), quite similar to the changes in total $NADPH/NADP^+$(8) in perfused liver, mentioned above. Redox potentials calculated from these data are also given in Table 2. The cytosolic potential for free NADPH changes from -391 to -380 mV, and the corresponding numbers for the mitochondrial space are -407 and -400 mV. Thus, in liver mitochondrial NADPH is more reduced than cytosolic NADPH, as predicted(20) from experiments with a reconstituted system using isolated mitochondria.

Aminopyrine was selected because of the lack of effects on the ATP/ADP ratio (to be publ. elsewhere). Further, the addition of aminopyrine in presence of SKF 525A did not elicit changes in the metabolites compared to the controls, indicating that the metabolism of aminopyrine led to the observed metabolite changes shown above.

Metabolic Significance of the Metabolite Changes

Probably the most striking result is the substantial decrease of the cytosolic concentration of citrate from 0.17 to 0.10 mM. This result suggests that one of the major implications of the oxidation in the NADPH system by high rates of monooxygenation to the metabolism of the cell is transmitted by a key metabolite for many processes, citrate. For example, it appears evident that the

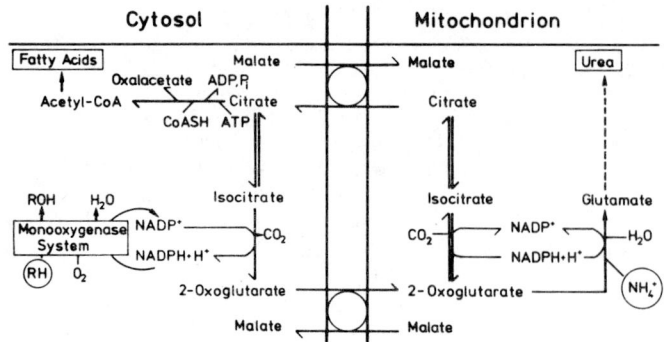

Fig. 3. Metabolic scheme depicting interaction in tricarboxylate utilization by monooxygenase and fatty acid synthase systems.

decrease of citrate concentration may account for the inhibition of lipogenesis(21) observed in presence of aminopyrine. A direct competition of the two NADPH-requiring processes for NADPH(21) is unlikely(22). A metabolic scheme depicting citrate diversion from the ATP-citrate lyase reaction towards oxidation by isocitrate dehydrogenase is given in Fig. 3. Since the K_m(citrate) of the tricarboxylate carrier for export from the mitochondria is in the range of 3-6 mM(23), a rate-limitation could also occur at the translocation step; and, in addition, by a restricted allosteric activation by citrate of acetyl-CoA carboxylase. These possibilities of citrate-linked effects of monooxygenation on lipogenesis shall be further investigated.

Mitochondrial-Cytosolic Interactions

A dependence of the monooxygenation process on reducing equivalents generated from the mitochondrial compartment has been deduced from the lack of an extra O_2 uptake upon addition of a drug to perfused livers from fasted rats in presence of antimycin A(24,8) as was confirmed with isolated hepatocytes(15). Under more physiological conditions(without added antimycin A) a functional interdependence between the two compartments can be demonstrated by activating a mitochondrial NADPH-requiring process, ureogenesis from ammonia(14,25). Aminopyrine addition leads to a small but significant steady state inhibition of ureogenesis from ammonia by 15 % in perfused livers from fasted but not from fed rats. The pentosephosphate shunt is capable of providing sufficient supply of reducing equivalents in the glycogen-supplemented state in the fed rat(26).

A further interaction between mitochondria and endoplasmic reticulum has been revealed as a "shielding effect" of the ER-membranes in a comparison of barbiturate titration curves of the inhibition of O_2 uptake in the intact organ and in isolated mitochondria(8,27).

ACKNOWLEDGEMENTS

Expert technical assistance was provided by Annegret Marklstorfer and Ingrid Linke. This investigation was supported by Deutsche Forschungsgemeinschaft, Sonderforschungsbereich 51, Medizinische Molekularbiologie und Biochemie.

REFERENCES

1 S.Narasimhulu, D.Y.Cooper, O.Rosenthal, Steroid-Induced Changes in Absorption Spectrum and Electron Transport of Adrenocortical 21-Hydroxylase Preparations, Fed.Proc. 25, Abs. 523(1966)

2 S.Narasimhulu, Significance of the Steroid-Induced Type I Spectral Change in Steroid C-21-Hydroxylase System of Bovine Adrenocortical Microsomes, Arch.Biochem.Biophys.147, 391(1971)

3 H.Sies, M.Kandel, Positive Increase of Redox Potential of the Extramitochondrial NADP(H) System by Mixed Function Oxidations in Hemoglobin-free Perfused Rat Livers, FEBS Letters 9, 205(1970)

4 M.Waterman, V.Ullrich, R.W.Estabrook, Effect of Substrate on the Spin State of Cytochrome P-450 in Hepatic Microsomes, Arch.Biochem.Biophys. 155, 355(1973)

5 I.C.Gunsalus, T.C.Pederson, S.G.Sligar, Oxygenase-Catalyzed Biological Hydroxylations, Ann.Rev.Biochem. 44, 377(1975)

6 P.L.Gigon, T.E.Gram, J.R.Gillette, Studies on the Rate of Reduction of Hepatic Microsomal Cytochrome P-450 by Reduced Nicotinamide Adenine Dinucleotide Phosphate: Effect of Substrate, Mol.Pharmacol. 5, 109(1969)

7 B.Brauser, H.Sies, Th.Bücher, Reduction Kinetics and Content of Cytochrome P-450 by Application of Dual Wavelength Techniques to Hemoglobin-free Perfused Rat Liver, FEBS Letters 2, 167(1969)

8 H.Sies, B.Brauser, Interaction of Mixed Function Oxidase with Its Substrates and Associated Redox Transitions of Cytochrome P-450 and Pyridine Nucleotides in Perfused Rat Liver, Eur.J.Biochem. 15, 531(1970)

9 P.F.Zuurendonk, J.M.Tager, Rapid Separation of Particulate Components and Soluble Cytoplasm of Isolated Rat-Liver Cells, Biochim.Biophys.Acta 333, 393(1974)

10 H.Sies, K.H.Summer, Hydroperoxide-Metabolizing Systems in Rat Liver, Eur. J.Biochem. 57, 503(1975)

11 H.Sies, M.Grosskopf, Oxidation of Cytochrome b_5 by Hydroperoxides in Rat Liver, Eur.J.Biochem. 57, 513(1975)

12 H.Sies, T.P.M.Akerboom, J.M.Tager, Mitochondrial and Cytosolic NADPH Systems and Isocitrate Dehydrogenase Indicator Metabolites During Ureogenesis from Ammonia in Isolated Rat Hepatocytes, Eur.J.Biochem., submitted

13 J.R.Williamson, Calculation of Metabolic Concentrations in the Cytosol and Mitochondria of Rat Liver, in: The Energy Level and Metabolic Control in Mitochondria(S.Papa, J.M.Tager, E.Quagliariello, E.C.Slater, eds.) Adriatica Editrice, Bari, p. 385(1969)

14 H.Sies, D.Häussinger, M.Grosskopf, Mitochondrial Nicotinamide Nucleotide Systems: Ammonium Chloride Responses and Associated Metabolic Transitions in Hemoglobin-free Perfused Rat Liver, Hoppe-Seyler's Z.physiol.Chem. 355, 305(1974)

15 P.Moldéus, R.Grundin, H.Vadi, S.Orrenius, A Study of Drug Metabolism Linked to Cytochrome P-450 in Isolated Rat-Liver Cells, Eur.J.Biochem. 46, 351(1974)

16 L.Flohé, W.Schlegel, Glutathion-Peroxidase, IV.: Intrazelluläre Verteilung des Glutathion-Peroxidase-Systems in der Rattenleber, Hoppe-Seyler's Z. Physiol.Chem. 352, 1401(1971)

17 M.J.Waterman, H.S.Mason, The Redox Potential of Liver Cytochrome P-450, Biochem.Biophys.Res.Comm. 39, 450(1970)

18 V.Ullrich, Enzymatic Hydroxylations with Molecular Oxygen, Angew.Chem. Int.Ed. 11, 701(1972)

19 Th.Bücher, H.Sies, Mitochondrial and Cytosolic Redox States in Perfused Rat Liver: Methods and Problems in Metabolic Compartmentation, in: Use of Isolated Liver Cells and Kidney Tubules in Metabolic Studies(J.M.Tager, J.R.Williamson, H.D.Söling, eds.)p.41, North-Holland, Amsterdam(1976)

20 J.B.Hoek, L.Ernster, Mitochondrial Transhydrogenase and the Regulation of Cytosolic Reducing Power, in: Alcohol and Aldehyde Metabolizing Systems (R.G.Thurman, T.Yonetani, J.R.Williamson, B.Chance, eds.)p.351, Academic Press, New York(1974)

21 R.G.Thurman, R.Scholz, Interaction of Mixed-Function Oxidation with Biosynthetic Processes. 2. Inhibition of Lipogenesis by Aminopyrine in Perfused Rat Liver, Eur.J.Biochem. 38, 73(1973)

22 R.L.Veech, R.W.Guynn, The Control of Hepatic Fatty Acid Synthesis in vivo, in: Regulation of Hepatic Metabolism(F.Lundquist, N.Tygstrup, eds.)p.337, Munksgaard, Copenhagen(1974)

23 J.Kleineke, H.D.Söling, Heterogenicity of Tricarboxylate Antiport in Liver Mitochondria, 10th IUB Congress Abstracts, p.344, Hamburg(1976)

24 R.G.Thurman, R.Scholz, Mixed Function Oxidation in Perfused Rat Liver. The Effect of Aminopyrine on Oxygen Uptake, Eur.J.Biochem. 10, 459(1969)

25 H.Sies, K.H.Summer, Th.Bücher, A Process Requiring Mitochondrial NADPH: Urea Formation from Ammonia, FEBS Letters 54, 274(1975)

26 O.Junge, K.Brand, Mixed Function Oxidation of Hexobarbital and Generation of NADPH by the Hexose Monophosphate Shunt in Isolated Rat Liver Cells, Arch.Biochem.Biophys. 171, 398(1975)

27 B.Brauser, H.Sies, Th.Bücher, Action of Amobarbital on Microsomal and Mitochondrial Respiratory State in Perfused Rat Liver With and Without Phenobarbital Induction, FEBS Letters 2, 170(1969)

THE ROLE OF REDUCING EQUIVALENTS GENERATED IN MITOCHONDRIA IN HEPATIC MIXED-FUNCTION OXIDATION[1]

Ronald G. Thurman, Marie Lurquin,* Roxanne Evans and Frederick C. Kauffman†

*Dept. Biochemistry and Biophysics, University of Pennsylvania, Philadelphia, Pa. 19174
†Dept. of Pharmacology and Experimental Therapeutics, University of Maryland, Baltimore, MD 21201

INTRODUCTION

A number of years ago (1) we demonstrated that oxygen uptake by the hemoglobin-free perfused rat liver could be stimulated markedly by addition of aminopyrine or hexobarbital, two substrates for mixed-function oxidation. It has subsequently been demonstrated that mixed-function oxidation is accompanied by oxidation of the hepatocellular NADP:NADPH system (2). Although microsomal mixed-function oxidations utilize reducing equivalents available in the cytoplasm, mitochondrial oxidations may also contribute reducing equivalents in certain metabolic states. Mitochondria might provide reducing equivalents for drug metabolism either by oxidation of fatty acids (3) or by succinate-stimulated reversed electron transfer (4). Previous conclusions regarding interactions between mitochondrial and microsomal electron transfer systems based on data obtained from mixtures of these systems in vitro or of measurements of oxygen uptake only in perfused organs may have serious shortcomings. For example, mixtures of mitochondria and microsomes are greatly diluted under in vitro situations and important intermediates necessary to shuttle reducing equivalents between the cytosol and the mitochondrial space may be missing. Estimation of rates of drug metabolism from stimulation of oxygen uptake alone by drugs could lead to erroneous conclusions because other oxygen requiring reactions may be stimulated or uncoupling of oxidative phosphorylation by the drug or its product may occur.

We have attempted to overcome these difficulties by developing a method to study the mixed-function oxidation of a model drug substrate, p-nitroanisole, in the hemoglobin-free perfused rat liver (5). O-demethylation of this compound as well as oxygen uptake may be monitored continually in the perfused organ. Moreover, cytosolic NADPH:$NADP^+$ ratios may be calculated from metabolites measured in freeze-clamped livers, and compared with the kinetics of drug metabolism. To identify major pathways for the generation of reducing equivalents for drug metabolism, maximal rates of mixed-function oxidation have been studied in a variety of nutritional states, as well as in the presence of inhibitors of the pentose phosphate shunt and mitochondrial oxidation.

---

[1]. Support: USPHS Grant AA-00288 and the Pangborn Fund of the University of Maryland.

METHODS

The technique of hemoglobin-free liver perfusion (6,7) and the simultaneous determination of p-nitrophenol from p-nitroanisole have been described in detail elsewhere (5).

RESULTS AND DISCUSSION

NADPH as controlling factor of mixed-function oxidation in the intact cell

Netter and his co-workers showed a number of years ago that p-nitroanisole O-demethylation by microsomes incubated with NADPH and oxygen was linear with time (19). Moreover, when p-nitroanisole is infused into the perfused liver of a normal, well-fed rat, a small but linear production of p-nitrophenolate is observed (Fig. 1). In livers from barbiturate-treated rats, however, two striking differences from the normal pattern were observed (Fig. 1): a) the maximal rate is markedly enhanced, and b) the reaction is only linear for a minute or two and is followed by a steady decline. Conjugation reactions may only account for a small portion, about 20% (5), of

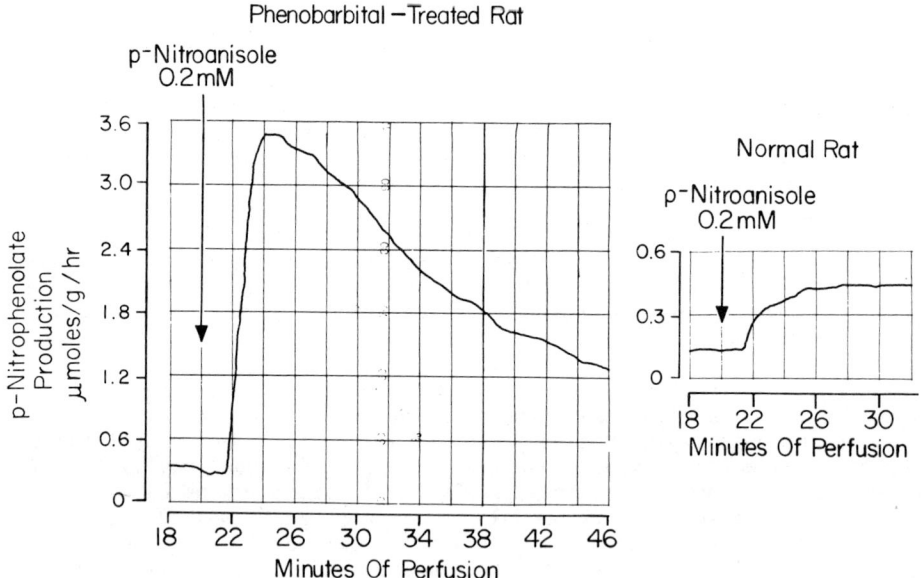

Figure 1. Rates of p-nitrophenolate production by perfused rat liver of normal (right panel) and phenobarbital-treated (left panel) rats. Rates were calculated from changes in optical density of the perfusate at 436 nm (5). Addition of p-nitroanisole (0.2 mM) is indicated by arrows. Time of perfusion is denoted on the horizontal axis.

the declining phase of the drug kinetic ("off-kinetic"). Moreover, infusion of the product, p-nitrophenol, was not associated with an "off-kinetic". Since substrate and oxygen were supplied in excess, it seemed likely that the decline in rate was due to a diminished supply of reduced cofactor (NADPH). This postulate is supported by activation of p-nitrophenolate production in livers from fasted animals perfused with substrates for glucose synthesis (e.g. dihydroxyacetone) (5,8). Studies described below were designed to determine whether or not mixed-function oxidation in livers from barbiturate-treated animals is regulated by the availability of reduced cofactor.

Relationship between NADPH redox state and rates of drug metabolism

The redox state of the NADP:NADPH system may be altered by fasting and refeeding or by treatment with 6-aminonicotinamide. We have compared the maximal rate of drug metabolism with the basal cytosolic NADPH/NADP ratio calculated from either the malic enzyme or isocitrate dehydrogenase equilibrium (Table 1). The maximal rate of p-nitrophenolate production varied considerably with the metabolic state. For example, fed, phenobarbital-treated livers oxidized the drug at rates 2 to 3 fold less than in livers

TABLE 1

BASAL CYTOSOLIC NADPH:NADP RATIOS AND RATES OF
p-NITROANISOLE O-DEMETHYLATION IN DIFFERENT METABOLIC STATES

Metabolic State	p-nitrophenolate production μmole/g/h	NADPH:NADP from malic enzyme	from isocitrate dehydrogenase
Fed	3.7 ± 0.6	53.9 ± 14.6	100.1 ± 16.5
Fasted (24 h)	4.6 ± 1.0	132.4 ± 11.3	
Fed + 6-aminonicotinamide	9.7 ± 0.9	205.7 ± 66.4	217.9 ± 48.0

NADPH:NADP ratios were calculated (9) from concentration of malate, pyruvate, isocitrate, and α-ketoglutarate measured in extracts of freeze-clamped livers (10). Values are averages obtained from 3 to 6 livers per experimental group ± S.E.M. 6-aminonicotinamide (70 mg/Kg) was injected i.p. 6 to 8 hours before perfusion experiments.

treated with the potent inhibitor (11) of 6-phosphogluconate dehydrogenase, 6-aminonicotinamide (Table 1). The rationale for selecting this inhibitor is discussed below. Intermediate values were obtained in livers from fasted rats. The NADPH:NADP$^+$ ratio varied four-fold under the different metabolic perturbations employed in these studies, and a direct correlation was observed between the basal NADPH:NADP redox state and maximal rates of drug metabolism (Table 1).

In accord with earlier studies (1,12), oxygen consumption was increased in all of the above metabolic states by addition of drug. However, a direct correlation did not exist between increases in oxygen uptakes and rates of O-demethylation of p-nitroanisole. It is concluded that oxygen uptake measurements alone do not adequately reflect changes in rates of mixed-function oxidation.

Role of the pentose phosphate shunt in providing reducing equivalents for mixed-function oxidation

Since pentose shunt flux is stimulated by the addition of aminopyrine (13) and by barbiturate-pretreatment (14), reducing equivalents for mixed-function oxidations have been considered to arise via this pathway. Furthermore, a large substrate reserve for this pathway is present in well-fed animals in the form of glycogen. It was, therefore, of interest to examine the effect of an inhibitor of pentose shunt activity on the kinetics of p-nitrophenol production from p-nitroanisole. For this purpose we have employed 6-aminonicotinamide, which is converted into a compound closely resembling NADP. The 6-aminonicotinamide analog of NADP does not participate in hydride transfer reactions (11), and is an exceptionally potent inhibitor of 6-phosphogluconate dehydrogenase ($K_i = 1 \times 10^{-7}$M) (11).

Six to eight hours following the administration of 6-aminonicotinamide to rats, 6-phosphogluconate levels were elevated approximately 700-fold (Table 2).

TABLE 2

EFFECT OF 6-AMINONICOTINAMIDE PRETREATMENT
ON HEPATIC 6-PHOSPHOGLUCONATE CONTENTS

	CONTROL	6-AMINONICOTINAMIDE
	μmoles/kg wet tissue	
6-phosphogluconate	7.3 ± 2.8* (6)	5,196 ± 486 (4)

*Mean ± S.E.M.   number of livers in parenthesis.

Although 6-aminonicotinamide appears to be a relatively specific inhibitor of the pentose shunt, its administration to phenobarbital-treated rats did not diminish maximal rates of mixed-function oxidation. On the contrary, it stimulated p-nitrophenolate production from p-nitroanisole over 2-fold (Table 1).

If the pentose shunt is inhibited by 6-aminonicotinamide treatment, as is suggested by the enormous increase in 6-phosphogluconate (Table 2), other sources must provide reducing equivalents for mixed-function oxidation under these conditions.

TABLE 3

EFFECT OF ANTIMYCIN A AND ETHANOL ON p-NITROANISOLE O-DEMETHYLATION IN PERFUSED RAT LIVER

Metabolic State	Addition	p-Nitrophenolate Production μmoles/g/h	% of control
Fasted	none	3.0	100
Fasted	antimycin A	0.5	16
Fed	none	1.6	100
Fed	ethanol	1.2	75
Fasted	none	2.9	100
Fasted	ethanol	0.4	14
Fasted	ethanol + pyruvate	1.8	62

Each set of data was obtained from one representative liver in the presence or absence of indicated additions.
antimycin A, 30 μM; ethanol, 76 mM; pyruvate, 16 mM.

Role of mitochondrial oxidations in providing reducing equivalents for mixed-function oxidation

Previous work from this laboratory demonstrated that mixed-function oxidation of aminopyrine in glycogen-depleted livers was markedly impaired by inhibitors of the mitochondrial respiratory chain such as potassium cyanide and antimycin A (1). Similar inhibition of p-nitroanisole oxidation occurred in the presence of antimycin A (Table 3), suggesting that mitochondria serve a major role in supporting mixed-function oxidation.

In addition to the absence of glycogen and reduced pentose shunt activity, livers from fasted rats also contain more fat and have higher rates of ketogenesis. Therefore, we have examined the effect of fatty acids on rates of p-nitrophenolate production from p-nitroanisole. For these experiments the fasted-refed rat was employed because the "off-kinetic" was much sharper in this metabolic state (Fig. 2).

When octanoate (0.5 mM) was infused at or near the peak of the maximal rate of drug metabolism, the sharp "off-kinetic" observed in fasted-refed animals was abolished (Fig. 2, center panel). Since fatty acids inhibit microsomal p-nitroanisole O-demethylation *in vitro* (3), it is concluded that fatty acid oxidation via the β-oxidizing system in the mitochondria provides reducing equivalents for extramitochondrial processes such as NADPH-dependent mixed-function oxidation of p-nitroanisole. For comparison, the

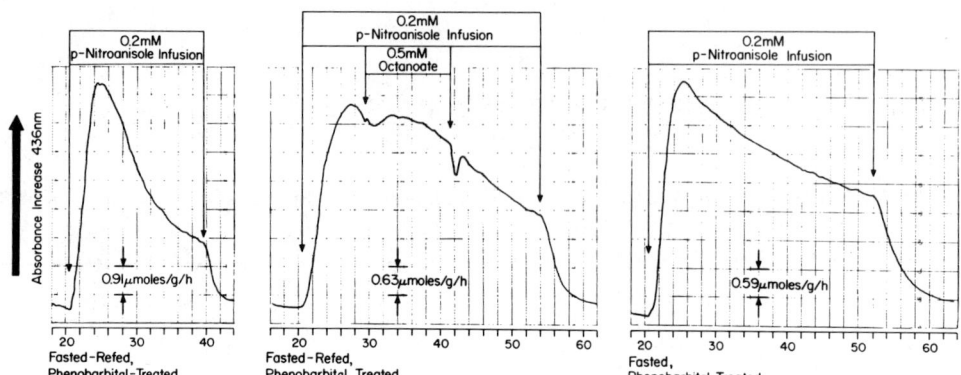

Figure 2. Effect of endogenous and exogenous fatty acids on mixed-function oxidation of p-nitroanisole by perfused rat liver. Conditions as in Fig. 1. Animals were either fasted for 24 h or fasted for 48 h and refed for 24 h prior to perfusion experiments. octanoate = 0.5 mM

kinetics of p-nitroanisole O-demethylation in a liver from a fasted rat is also shown in Fig. 2. In this metabolic state, the high rate of drug metabolism is maintained for a much longer period of time and resembles that obtained by the addition of exogenous fatty acids to the perfusate.

It has been known for a number of years that both glucose-6-phosphate dehydrogenase and malic enzyme are induced by the fasting-refeeding procedure (15). Increased levels of malic enzyme in the fasted-refed animal could be related to the faster "off-kinetic" in this metabolic state if it served as a component of a substrate shuttle mechanism (see below).

Possible involvement of a malic enzyme shuttle in transporting mitochondrial reducing equivalents into the cytosol

Since enzymes responsible for the oxidation of acyl-CoA compounds are located in the mitochondrial space and because pyridine nucleotides cannot penetrate the mitochondrial membrane (16,17), a substrate shuttle mechanism must move mitochondrial hydrogen into the cytosolic space for drug metabolism. There are at least two shuttle mechanisms which could serve such a purpose: one involving mitochondrial and cytosolic isocitrate dehydrogenases, the other involving mitochondrial malate dehydrogenase and malic enzyme (18,1).

The possible involvement of a malic enzyme shuttle in supplying reducing equivalents for p-nitroanisole O-demethylation in livers from fed and fasted rats may be examined by depleting pyruvate, a key shuttle intermediate, with ethanol. When ethanol was administered to livers from fed rats, a 25% inhibition of p-nitrophenol production from p-nitroanisole was observed

(Table 3). In contrast, mixed-function oxidation of this drug was diminished by 80 to 90% in livers from fasted animals following addition of ethanol. Addition of pyruvate greatly attenuated the inhibition produced by ethanol in livers from fasted rats. Thus, it is concluded that an ethanol-sensitive enzyme shuttle mechanism is involved in moving hydrogen from mitochondrial fatty acid oxidation into the cytosol to support mixed-function oxidation.

In support of this conclusion, the ATP-ADP ratio decreased from around 4.0 to 2.0 upon the addition of p-nitroanisole, presumably as a result of the uncoupling action of the product, p-nitrophenol (5,10). Slower activation of fatty acids to acyl-CoA brought about by the lower ATP/ADP ratio might be responsible for the observed "off-kinetic" of mixed function oxidation.

REFERENCES

(1) Thurman, R.G., and Scholz, R., Eur. J. Biochem., 10, 459-467, (1969).

(2) Sies, H., and Kandel, M., FEBS Letters, 9, 205-208, (1970).

(3) Lurquin, M., Kauffman, F., and Thurman, R.G., The Pharmacologist, 11, 264, (1976).

(4) Cinti, D.L., Ritchie, A., and Schenkman, J.B., Molecular Pharmacology, 8, 339-344, (1972).

(5) Thurman, R.G., Marazzo, D.P., Jones, L.S., and Kauffman, F.C., Molecular Pharmacology, in press.

(6) Scholz, R., Hansen, W., and Thurman, R.G., Eur. J. Biochem., 38, 64-73, (1973).

(7) Scholz, R., Thurman, R.G., Williamson, J.R., Chance, B., and Bücher, Th., J. Biol. Chem., 244, 2317-2324, (1969).

(8) Thurman, R.G., Marazzo, D.P., and Scholz, R., Adv. Exp. Med. and Biology, 58, 355-368, (1975).

(9) Krebs, H.A., in Davies, D.D. (ed.), Symposiums of the Society for Exp. Biology, 27, 299-318, University Press, Cambridge, (1973).

(10) Kauffman, F.C., Evans, R., and Thurman, R.G., Biochem. J., submitted.

(11) Kohler, E., Barrach, H.-J., and Neubert, D., FEBS Letters, 6, 225-228, (1970).

(12) Sies, H., and Brauser, B., Eur. J. Biochem., 15, 531-540, (1970).

(13) Busch, U., Thesis, Medical Faculty of the University of Munich, (1975).

(14) Kunz, W., Schaude, G., Schimassek, H., Schmid, W., and Siess, M., Medical International Congress Series, 115, 138-153, (1966).

(15) Tepperman, H.M., and Tepperman, J., Am. J. Physiol., 206, 357-361, (1964).

(16) Lehninger, A.L., J. Biol. Chem., 190, 345-354, (1951).

(17) Purvis, J.C., and Lowenstein, J.M., J. Biol. Chem., 236, 2794-2803, (1961).

(18) Hoek, J.B., and Ernster, L., in Thurman, R.G., Yonetani, T.Y., Chance, B., and Williamson, J.R. (eds.), Alcohol and Aldehyde Metabolizing Systems, Academic Press, New York, 351-364, (1973).

(19) Netter, K.J., and Seidel, G., J. Pharmacol. Exp. Ther., 146, 61-66, (1964).

DESTRUCTION OF CYTOCHROME P-450 BY LINOLEIC ACID HYDROPEROXIDE

E. H. Jeffery, D. Nerland, R. el-Azhary and
G. J. Mannering

Department of Pharmacology, University of Minnesota, Minneapolis, Minnesota 55455

ABSTRACT

Increasing concentrations of linoleic acid hydroperoxide (10-250 μM) destroyed increasing amounts of P-450 hemoprotein in hepatic microsomes from untreated, phenobarbital-treated and 3-methylcholanthrene-treated rats. Loss of P-450 hemoprotein in microsomes from untreated and phenobarbital-treated rats was biphasic; 50% was destroyed at low concentrations of linoleic acid hydroperoxide (50 μM or less) and only about half of that which remained was destroyed at relatively high concentrations (150 μM or more). The labile P-450 hemoprotein was designated P-450$_{(11)}$, the stable, P-450$_{(1s)}$. The loss of P-450 hemoprotein in microsomes from 3-methylcholanthrene-treated rats was not biphasic and most of the hemoprotein was in the stable form. Almost all of the monooxygenase activity (ethylmorphine N-demethylase; aniline hydroxylase) and type I binding (hexabarbital) is associated with P-450$_{(11)}$. Microsomal NADPH-cytochrome c reductase activities and cytochrome b$_5$ levels were not affected by linoleic acid hydroperoxide until high concentrations were reached (150 μM or more). NADH-cytochrome c reductase activity was similarly unaffected by linoleic acid hydroperoxide in microsomes from untreated rats, but was greatly elevated in microsomes from phenobarbital-treated rats. NADPH oxidase activity was unaffected by linoleic acid hydroperoxide in microsomes from untreated rats. These results led to the conclusion that P-450$_{(11)}$ functions in the oxidation of exogenous substrates, that P-450$_{(1s)}$ functions in the oxidation of endogenous substrates, and that the two functions are essentially independent of each other.

INTRODUCTION

The monooxygenase system of the hepatic endoplasmic reticulum is comprised of at least four components: P-450 hemoprotein, cytochrome b$_5$, NADPH-cytochrome c reductase, and NADH-cytochrome b$_5$ reductase. The composition and physical state of the membrane to which these components are bound are thought to influence the activities of the monooxygenase systems to such a degree that the membrane itself may be thought of as a component of the system (Reviewed by Stier, 1). Certain studies of the monooxygenase systems would be facilitated if it were possible to alter the relative concentrations of these components. Although this can be achieved *in vivo* to some degree by administering agents which induce the synthesis of P-450 hemoprotein and NADPH-cytochrome c reductase, it has not been possible to change the composition of the monooxygenase system *in vitro* without drastically disturbing the integrity of the membrane. A method is described in this presentation which employs linoleic acid hydroperoxide to remove cytochrome P-450 in a step-wise manner without disturbing other components of the system. The bleaching action of linoleic acid hydroperoxide also revealed the existence of two forms of membrane-bound P-450 hemoprotein, one relatively labile, the other relatively stable.

METHODS

Fresh hepatic microsomes, from male, Holtzman strain rats (180-220 g) injected with saline, sodium phenobarbital (40 mg/kg for 3 days) or 3-methylcholanthrene, (20 mg/kg for 3 days) were suspended in a solution of 1.15% KCl, 0.1 M phosphate buffer, pH 7.4, to a concentration of 5 mg microsomal protein/ml. Amounts of linoleic acid hydroperoxide were added to these suspensions to give graded concentrations between 10 and 250 µM. The preparations were allowed to stand at 24°C for 20 min, after which reduced glutathione (final concentration, 2.0 mM) was added to reduce unreacted linoleic acid hydroperoxide. Linoleic acid hydroperoxide was synthesized by air oxidation of linoleic acid as described by O'Brien (2) and quantified by the use of both the extinction coefficient (25.25 $mM^{-1}cm^{-1}$ at 233 nm) and a radioactive tracer. It was stored in ethanol at -10°C.

RESULTS

Effect of Linoleic Acid Hydroperoxide on P-450 Hemoprotein

Figure 1 shows the biphasic destruction of cytochrome P-450 with increasing concentrations of linoleic acid hydroperoxide. About 50% of the hemoprotein was destroyed at a concentration of linoleic acid hydroperoxide of 50 µM or less and only half of the remaining hemoprotein was destroyed at a concentration of 150 µM. The least stable hemoprotein is designated $P-450_{(11)}$, the more stable $P-450_{(1s)}$.

The relative stabilities of $P-450_{(11)}$ and $P-450_{(1s)}$ to linoleic acid hydroperoxide are shown in Fig. 2. It is seen that at a concentration of 35 µM all of the $P-450_{(11)}$ had been destroyed while only 90% of the $P-450_{(1s)}$ had disappeared.

When P-450 hemoprotein of microsomes from rats treated with phenobarbital was destroyed in this manner a similar distribution of $P-450_{(11)}$ and $P-450_{(1s)}$ was seen (Fig. 3). However, 3-methylcholanthrene treatment produced a type of P-450 hemoprotein similar to $P-450_{(11)}$; it was relatively resistant to destruction by linoleic acid hydroperoxide, but not divisible into $P-450_{(11)}$ and $P-450_{(1s)}$ phases (Fig. 3). We have no explanation for the apparent increase in P-450 hemoprotein observed after the addition of low concentrations of linoleic acid hydroperoxide (Fig. 3).

Stern and Peisach (3), provided evidence that the sixth ligand of hepatic cytochrome P-450 is a sulphur-containing ligand, probably a mercaptide ion. If linoleic acid hydroperoxide destroys cytochrome P-450 by reacting with the sulphur ligand, as has been proposed by O'Brien and associates (4,5), the question arises as to whether cytochrome $P-450_{(1s)}$ possesses a sulfur ligand, and if it does, how it is prevented from reacting with linoleic acid hydroperoxide. Conceivably, the difference between cytochrome $P-450(11)$ and

P-450$_{(1s)}$ may be that the sulfur ligand of the former may be free to react with linoleic acid hydroperoxide, whereas that of the latter may be bound to membrane components which prevent the ligand from reacting readily with lipid peroxides.

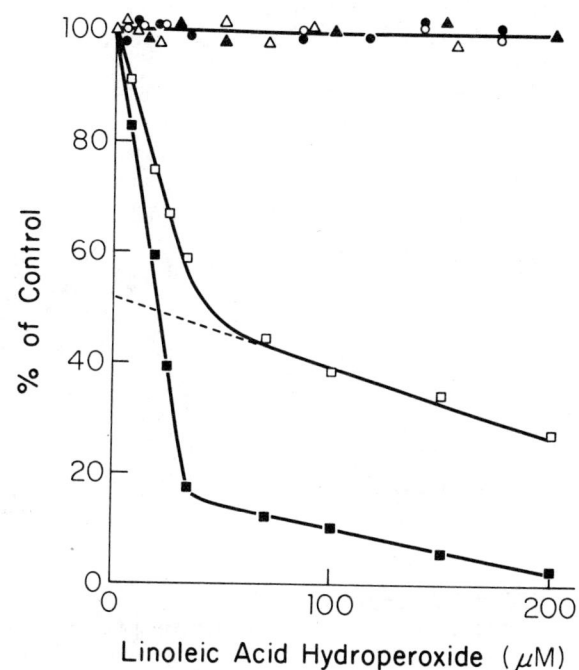

Fig. 1. Effect of linoleic acid hydroperoxide on microsomal monooxygenase systems. ▲-▲ NADPH oxidation (6), △-△ cytochrome b5 (7); ●-● NADH-cytochrome c reductase (8); ○-○ NADPH cytochrome c reductase (8); □-□ cytochrome P-450 (7); ■-■ ethylmorphine N-demethylase (9). Respective 100% values (nmole or nmole/min per mg of protein) are: 7.5; 0.37; 1536; 244; 0.56; 4.0.

Effect of Linoleic Acid Hydroperoxide on Monooxygenase Components Other Than Cytochrome P-450

Cytochrome b5 levels were not affected by linoleic acid hydroperoxide concentrations below 200 µM (Fig. 1). Integrity of the flavoproteins was monitored by measuring NADH- and NADPH-cytochrome c reductase activities and found to be unaffected by concentrations of linoleic acid hydroperoxide up to 200 µM in microsomes from untreated rats (Fig. 1). NADH-cytochrome c reductase activity in microsomes from rats treated with phenobarbital was about 40-50% lower than that observed in microsomes from untreated rats. Increasing concentrations of linoleic acid hydroperoxide gradually restored this activity (data not shown). The mechanisms of depression and restoration of NADH-cytochrome c reductase activity probably involve cytochrome b5 because NADH-cytochrome b_5 reductase activity was not lowered in microsomes from phenobarbital-treated rats when ferricyanide replaced cytochrome c as the electron acceptor. The addition of linoleic acid hydroperoxide did not alter this activity.

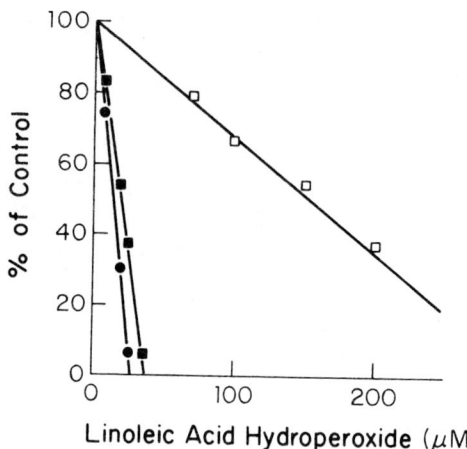

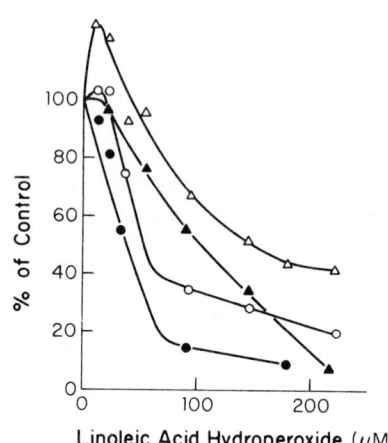

Fig. 2. Effect of linoleic acid hydroperoxide on cytochrome P-450(11); cytochrome P-450(1s) and ethylmorphine N-demethylase activity. □ - □ P-450(1s); ■ - ■ P-450(11); ● - ● P-450(11)-dependent ethylmorphine N-demethylase activity. Disappearance of P-450(1s) was extrapolated to zero as shown by the dotted line in Fig. 1. The disappearance of P-450(1s) was then plotted as a percentage of P-450(1s) at 0 μM linoleic acid hydroperoxide rather than as a percentage of total P-450 hemoprotein. P-450(11) was calculated from total P-450 hemoprotein less the calculated P-450(1s) and replotted as a percentage of P-450(11) at 0 μM linoleic acid hydroperoxide. The ethylmorphine N-demethylase activity that disappeared rapidly (Fig. 1.) was replotted in the manner used to plot P-450(11) disappearance.

Fig. 3. Effect of linoleic acid hydroperoxide on microsomal monooxygenase systems from rats treated with phenobarbital or 3-methylcholanthrene. △ - △ P-450 hemoprotein and ▲ - ▲ ethylmorphine N-demethylase from rats treated with 3-methylcholanthrene. ○ - ○ P-450 hemoprotein and ● - ● ethylmorphine N-demethylase from rats treated with phenobarbital. 100% values for phenobarbital and 3-methylcholanthrene-treated rats were 1.02 and 0.87 nmole P-450/mg of protein and 23.0 and 5.3 nmole HCHO/min/mg of protein, respectively.

Effect of Linoleic Acid Hydroperoxide on the Metabolism of Drug Substrates

Ethylmorphine N-demethylase activity disappeared almost entirely with the destruction of P-450$_{(11)}$ (Figs. 1 and 2). Type I binding (hexobarbital) and aniline hydroxylase activity were also almost entirely associated with P-450$_{(11)}$ (data not shown). In the absence of drug substrate the rate of NADPH oxidation was not affected by linoleic acid hydroperoxide (Fig. 1). The increased rate of NADPH oxidation seen in the presence of ethylmorphine disappeared with increasing addition of linoleic acid hydroperoxide; this was concomitant with the disappearance of N-demethylase activity, hexobarbital binding and P-450$_{(11)}$. These results suggest that only P-450$_{(11)}$ is involved in drug metabolism, but leave unanswered the question of the function of P-450$_{(1s)}$. With respect to loss of N-demethylase activity with increasing additions of linoleic acid hydroperoxide, microsomes from phenobarbital-treated rats responded much like microsomes from untreated rats (Fig. 3). As was the case with the loss of P-450 hemoprotein from microsomes from 3-methylcholanthrene-treated rats, the loss of ethylmorphine N-demethylase activity was monophasic rather than biphasic (Fig. 3).

Sex Difference in the Relative Amounts of P-450$_{(11)}$ and P-450$_{(1s)}$ in the Rat

It has long been known that female rats metabolize certain drugs less rapidly than male rats (10). When the destructive effect of linoleic acid hydroperoxide on hepatic P-450 hemoprotein from adult female rats was studied, a sex difference was immediately obvious (Fig. 4). Whereas 48% of the total P-450 in the male was P-450$_{(11)}$, only 22% was P-450$_{(11)}$ in the female. In preparations from rats of both sexes, most of the ethylmorphine N-demethylase activity disappeared concomitantly with P-450$_{(11)}$, (Fig. 4).

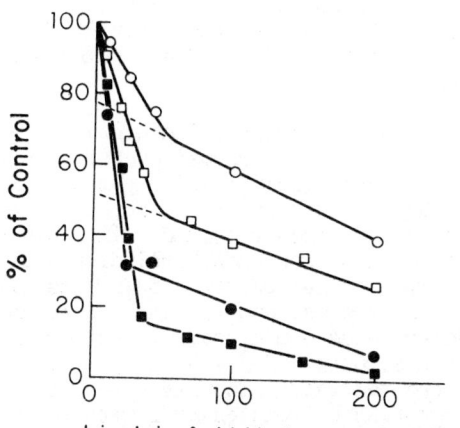

Fig. 4. Sex differences in the relative amounts of P-450$_{(11)}$ and P-450$_{(1s)}$ hemoprotein in rats. ○ - ○ P-450 hemoprotein and ● - ● ethylmorphine N-demethylase, female rats; □ - □ P-450 hemoprotein and ■ - ■ ethylmorphine N-demethylase, male rats. 100% values for male and female rats were 0.67 and 0.49 nmol of P-450 hemoprotein/mg of protein and 4.1 and 1.0 nmol HCHO/min/mg of protein, respectively.

N-demethylase activity based on total P-450 hemoprotein was 6.1 and 2.1 nmole/min/nmole P-450 in microsomes from the male and female, respectively. Activity based on the concentration of P-450$_{(11)}$ rather than total P-450 was 9.6 and 5.4 nmole/min/nmole P-450$_{(11)}$ for hepatic microsomes from male and female rats, respectively. These data suggest that the sex difference in P-450 hemoprotein reflects both quantitative and qualitative differences in P-450$_{(11)}$.

Sex differences in drug metabolism are not observed in adult mice or sexually immature rats (11); we observed no sex difference in the relative amounts of P-450$_{(11)}$ and P-450$_{(1s)}$ in mice or immature rats.

Use of Linoleic Acid Hydroperoxide to Determine the Absolute Spectra of Membrane-Bound P-450 Hemoprotein

The determination of the absolute spectrum of hepatic membrane-bound P-450 hemoprotein is hindered by the presence of an interfering chromophore, cytochrome b$_5$. To overcome this problem, Kinoshita and Horie (11) made use of the fact that phenobarbital induction increases the concentration of cytochrome P-450 in the microsome with little change in cytochrome b$_5$ concentration. By placing microsomes from untreated rabbits in the reference cuvette and microsomes from phenobarbital-treated rabbits in the sample curvette they were able to balance cytochrome b$_5$ absorption and measure the spectrum of the excess P-450 hemoprotein. The method assumes that the P-450 hemoprotein induced by phenobarbital has spectral characteristics identical to those of the P-450 hemoprotein present in microsomes from untreated animals. It is not applicable for the determination of the absolute spectrum of cytochrome P$_1$-450, the P-450 hemoprotein produced with 3-methylcholanthrene and certain other polycyclic inducing agents, because the spectrum of cytochrome P$_1$-450 is known not to be identical to that of the P-450 hemoprotein found in microsomes from untreated animals.

As shown in Fig. 1, linoleic acid hydroperoxide bleaches P-450 hemoprotein without affecting cytochrome b$_5$. A microsomal preparation was split between reference and sample cuvettes and linoleic acid hydroperoxide added to the reference cuvette only. The absolute spectrum of the unbleached P-450 hemoprotein in the sample cuvette was then recorded. The absolute spectra of membrane-bound P-450 hemoproteins from phenobarbital and 3-methylcholanthrene-treated rats exhibited maxima at 418, 535 and 568 nm and 417, 535 and 568 nm, respectively, for oxidized microsomes and 412 and 545 nm and 411 and 545 nm, respectively, for dithionite-reduced microsomes. Fujita and Mannering (12) observed similar maxima for solubilized P-450 hemoproteins from phenobarbital and 3-methylcholanthrene treated rats. The extinction coefficients of membrane-bound P-450 hemoprotein correlated well with those of purified hemoprotein for α and β bands, but extinction of the soret band was about 30% lower. This difference might be explained either by spectral interference caused by the membrane or by the binding of the P-450 hemoprotein to membrane components *per se*.

DISCUSSION

The view that not all of the P-450 hemoprotein of hepatic microsomes functions as a terminal oxidase in the biotransformation of xenobiotics has been expressed from time to time by several investigators. For example, Ullrich

(13) estimated from spectral binding studies that only 12% of the P-450 hemoprotein of microsomes from untreated rats functions in the hydroxylation of cyclohexane. Our results show that not all hepatic P-450 hemoprotein is active in the metabolism of xenobiotics. In the absence of exogenous substrate, NADPH oxidase activity remained constant regardless of the degree of destruction of P-450 hemoprotein by linoleic acid hydroperoxide (Fig. 1). If all of the NADPH oxidase activity was mediated through cytochrome P-450, one would then have to conclude that essentially all endogenous NADPH oxidation is catalyzed by cytochrome P-450$_{(1s)}$. The increase in NADPH oxidation seen in the presence of ethylmorphine disappeared with increasing concentrations of linoleic acid hydroperoxide in direct proportion to the disappearance of N-demethylase activity and cytochrome P-450$_{(11)}$. This suggests that when ethylmorphine is present, cytochrome P-450$_{(11)}$ functions only in the oxidation of ethylmorphine, and that cytochrome P-450$_{(1s)}$ is turning over at the same rate regardless of whether exogenous substrate is present or not. That NADPH oxidation is not altered by the total destruction of cytochrome P-450$_{(11)}$ suggests that cytochrome P-450$_{(11)}$ is not turning over when exogenous substrate is absent. In other words, cytochrome P-450$_{(11)}$ functions in the oxidation of exogenous substrates whereas cytochrome P-450$_{(1s)}$ functions in the oxidation of endogenous substrates.

This work was supported by U.S. Public Health Service Grant GM 15477.

REFERENCES

1. A. Stier, Lipid structure and drug metabolizing enzymes. Biochem. Pharmacol. 25, 109 (1976).
2. P. J. O'Brien, Intracellular mechanisms for the decomposition of a lipid peroxide. 1. Decomposition of a lipid peroxide by metal ions, heme compounds, and nucleophiles. Can. J. Biochem. 47, 485 (1969).
3. J. O. Stern and J. Peisach, A model compound study of the CO-adduct of cytochrome P-450. J. Biol. Chem. 249, 7495 (1974).
4. R. C. Green, C. Little and P. J. O'Brien, The inactivation of isocitrate dehydrogenase by a lipid peroxide. Arch. Biochem. Biophys. 142, 598 (1971).
5. E. G. Hrycay and P. J. O'Brien, Cytochrome P-450 as a microsomal peroxidase utilizing a lipid peroxide substrate. Arch. Biochem. Biophys. 147, 14 (1971).
6. E. Jeffery and G. J. Mannering, Discrepancy in the measurement of reduced triphosphopyridine nucleotide oxidized during ethylmorphine N-demethlation due to the presence of a nucleotide pyrophosphatase. Mol. Pharmacol. 10, 1004 (1974).
7. T. Omura and R. Sato, The carbon monoxide-binding pigment of liver microsomes. 1. Evidence for its hemoprotein nature. J. Biol. Chem. 239, 2370 (1964).
8. A. H. Phillips and R. G. Langdon, Hepatic triphosphopyridine nucleotide cytochrome c reductase: isolation, characterization and kinetic studies. J. Biol. Chem. 237, 2652 (1962).
9. M. A. Correia and G. J. Mannering, DPNH synergism of TPNH-dependent mixed function oxidase reactions. Drug Metab. Dispos. 1, 139 (1973).
10. S. el Defrawy el Masry, G. M. Cohen and G. J. Mannering, Sex dependent differences in drug metabolism in the rat. 1. Temporal changes in the microsomal drug metabolizing system of the liver during sexual maturation. Drug Metab. Dispos. 2, 267 (1974).

11. T. Kinoshita and S. Horie, Studies on P-450. III. On the absorption spectrum of P-450 in rabbit liver microsomes. J. Biochem. 61, 26 (1967).
12. T. Fujita and G. J. Mannering, Differences in soluble P-450 hemoproteins from livers of rats treated with phenobarbital and 3-methylcholanthrene. Chem-Biol. Interactions 3, 264 (1971).
13. V. Ullrich, On the hydroxylation of cyclohexane in rat liver microsomes. Hoppe-Seyler's Z. Physiol. Chem. 350, 357 (1969).

STOICHIOMETRY OF OXYGEN UPTAKE, NADPH OXIDATION AND ETHYLMORPHINE N-DEMETHYLATION BY HEPATIC MICROSOMES*

Jordan L. Holtzman, Ronald P. Mason and
Richard R. Erickson

Clinical Pharmacology Section, Veterans Administration Hospital, Minneapolis, Minnesota 55417; and Departments of Pharmacology and Medicine, University of Minnesota, Minneapolis, Minnesota 55455

ABSTRACT

The initial rate of oxygen uptake is as much as 30% greater than the initial NADPH uptake. After about one minute the rates parallel each other. The lack of stoichiometry occurs in both the presence and absence of ethylmorphine. Although there is not the expected 1:1 stoichiometry between oxygen uptake and NADPH oxidation as determined by the initial rate, when known concentrations of NADPH are added there is an excellent 1:1 stoichiometry of NADPH added to oxygen consumed indicating that in the end all the oxygen goes to peroxide which in turn disproportionates to oxygen and water. The excess oxygen uptake is only minimally affected by superoxide dismutase or NADH and, in fact, the initial oxidation of NADH by itself is also significantly less than the initial oxygen uptake. Unlike NADPH, the rate of NADH oxidation and oxygen uptake are reduced by ethylmorphine. Our data suggest formation of intramicrosomal superoxide anion radical which then disproportionates to peroxide and oxygen.

INTRODUCTION

The reaction scheme of the hepatic, microsomal, mixed function oxidases has usually been taken to be:

1. $NADPH + O_2 + sub. \xrightarrow{MFO} NADP^+ + OH^- + sub. - OH$

This scheme states that the NADPH donates two electrons to the oxygen to form an activated ("peroxyl") species which in turn attacks the substrate. Further, it predicts that the stoichiometry of the reaction should be that one pyridine nucleotide

---

*Portions of this study were presented at the meetings of the American Society for Pharmacology and Experimental Therapeutics at Atlantic City, April, 1972 (Fed. Proc. 31, 2071 (1972)) and Davis, California, August, 1975 (Pharmacology (1975)). This work was supported in part by grant GM21310-01A1 from the National Institute of Health.

molecule is oxidized for each oxygen molecule consumed and for each substrate molecule hydroxylated.

In the current study we have compared the initial rate of oxygen uptake to the initial rate of NADPH oxidation and to the rate of N-demethylation of ethylmorphine. Our findings suggest that initially there is a higher rate of oxygen uptake than NADPH oxidation both in the absence and presence of ethylmorphine. These data suggests that superoxide anion may be formed. Since this radical has a short half life, its presence can only be observed at short time intervals.

MATERIALS AND METHODS

All animals used in these studies were untreated, 180-200 g, male C-D rats obtained from Charles River Inc. (North Willmington, MA) and were maintained on standard laboratory chow with free access to water. Microsomes were prepared in KCl-tris and washed once by resuspension in this buffer and resedimenting (1). Oxygen uptake was determined with a Clark electrode at 37° in a jacketed glass chamber (1). NADPH and NADH oxidation and NADPH-cytochrome c reductase activity were determined spectrophotometrically at 37° in an Aminco DW-2 spectrophotometer (1). Ethylmorphine N-demethylase was determined by the formation of formaldehyde (1). The formation of superoxide anion radical was estimated by the formation of adrenochrome from epinephrine (2).

RESULTS

In an initial study we examined the stoichiometry of NADPH oxidation, oxygen uptake, and HCHO formation (Table 1). In this study the oxygen uptake and HCHO formation were determined in the same incubation but the NADPH oxidation was determined separately. The concentration of all components, including NADPH, were the same in all incubations. As we and other investigators have observed (1,3-9), there is an excellent 1:1 stoichiometry between the stimulation of the initial rate of NADPH oxidation and HCHO formed in 10 minutes (Table 1). Yet, both in the absence and presence of ethylmorphine, the initial rate of NADPH oxidation is significantly lower than the initial rate of oxygen uptake ($p < 0.01$ without ethylmorphine, $p < 0.001$ with ethylmorphine). This lack of stoichiometry only occurs in the first minute of the incubation (Fig. 1). Hence, there appears to be an initial burst of excessive oxygen uptake and then the uptake parallels the NADPH oxidation. The excess oxygen uptake can be estimated by extrapolating the steady state uptakes back to zero time. There is a 6 µM excess of "activated oxygen" in solution. This is 6 fold greater than the concentration of electron chain components in the microsomes.

This excess uptake of oxygen may represent the formation of superoxide anion radical. If such is the case, the addition of superoxide dismutase, which disproportionates the radical to

TABLE 1 Stoichiometry of Oxygen Uptake to the Oxidation of NADPH and the Formation of Formaldehyde in the Metabolism of Ethylmorphine by Hepatic Microsomes from Male Rats

Component	Ethylmorphine 2 mM	Metabolism nmoles/min-mg protein	Δ
Oxygen	−	$12.4 \pm 0.6^{a,b}$	
	+	29.3 ± 0.9	16.9 ± 1
NADPH	−	9.7 ± 0.2^{c}	
	+	21.6 ± 0.1^{d}	11.9 ± 0.3
HCHO	+	10.5 ± 0.2	

[a] Values are the average of triplicate incubations ± SEM. [b] The initial NADPH concentration for both the oxygen uptake and NADPH oxidation was 0.13 mM. [c] Compared to basal initial oxygen uptake $P < 0.01$. [d] Compared to initial oxygen uptake in the presence of ethylmorphine $P < 0.001$.

peroxide and oxygen, or the addition of NADH which, as shown by Hildebrandt and Estabrook (10) and Correia and Mannering (11,12), can donate the second electron to give the hydroxylating species, should significantly affect the oxygen uptake by reducing the concentration of superoxide. Indeed, superoxide dismutase does have a small, statistically significant effect on the oxygen uptake (Table 2), decreasing it seven percent in the absence of ethylmorphine, but had no effect on oxygen uptake in the presence of ethylmorphine, or when NADH was the sole pyridine nucleotide. Further, it had no effect on the formation of formaldehyde, in agreement with the results of Stroebel and Coon (13), for the metabolism of benzphetamine by intact microsomes.* In view of the stoichiometry data suggesting that superoxide anion is formed, the failure of superoxide dismutase to have a greater effect would suggest that this anion is intramicrosomal where the enzyme cannot penetrate.

Similarly, NADH did not significantly improve the stoichiometry between the stimulation of oxygen uptake and the formation of HCHO. In fact, in spite of a significant increase in formation of formaldehyde, there is no change in ΔO_2. As observed for NADPH, there is only a 0.7:1 stoichiometry between NADH oxidation and oxygen uptake in the absence of ethylmorphine (Table 3). Another point of interest is that the addition of ethylmorphine to the vessel actually decreased oxygen uptake when NADH was used as the sole source of reducing equivalents (Table 3).

*Although this study primarily concerned solubilized microsomal cytochrome P-450, they also included negative data for intact microsomes (cf. p. 7828, column 2, ref. 13).

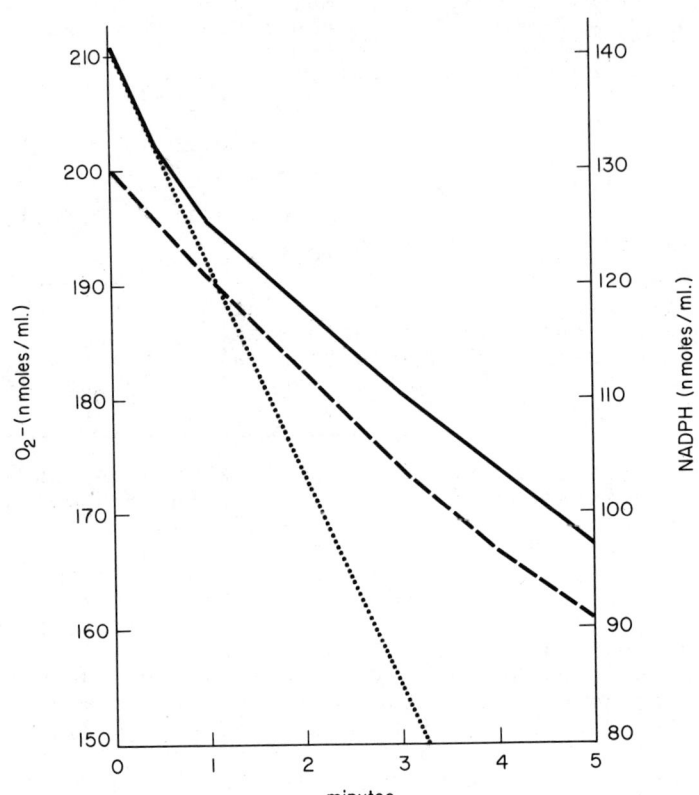

Fig. 1. Comparison of time course of NADPH oxidation (---) to the initial (···) and steady state (———) oxygen uptake. The scales of the two indices are the same except that the initial concentration of NADPH was 130 µM and oxygen was 211 µM. The graphs were plotted from basal uptakes of samples run for the study reported in Table 1. In both samples the initial concentration of NADPH was 130 µM and oxygen 211 µM.

Since this pyridine nucleotide serves as a poor source of reducing equivalents for the N-demethylation of ethylmorphine, we feel that this effect is most likely due to some indirect effect on the membrane.

Table 2 The Effect of Superoxide Dismutase and NADH on Oxygen Uptake and Formaldehyde Formation from Ethylmorphine by Hepatic Microsomes from Male Rats[a]

Ethyl-morphine 2 mM	Superoxide Dismutase 15 μg/ml (124 units/ml)	NADH 0.3 mM	nmoles HCHO[b] min-mg protein	nmoles O_2 min-mg protein	ΔO_2
-	-	-	-	13.4 ± 0.2[c,d]	-
+	-	-	7.29 ± 0.02	32.1 ± 0.9	18.7 ± 1.0
-	+	-	-	12.4 ± 0.2[f]	-
+	+	-	7.17 ± 0.00	30.7 ± 0.5[g]	18.3 ± 0.6
-	-	+[e]	-	25.1 ± 2.6[j]	-
+	-	+[e]	-	13.4 ± 0.7	-11.7 ± 2.8[h]
-	-	+	-	32.4 ± 0.9	-
+	-	+	8.99 ± 0.16	50.4 ± 1.7	17.0 ± 1.9[i]

[a]Preparations and assays are given in the text. In this study the microsomal preparation was resuspended and resedimented one time. [b]Values are for a 10 minute incubation. [c]Values are initial rates. [d]All values are the average of triplicate incubations ± SEM. [e]NADPH was not present. [f]P < 0.05 compared to control. [g]P - N.S. compared to control + ethylmorphine. [h]P < 0.02. [i]P < 0.001. [j]This value was unaffected by superoxide dismutase

Table 3 The Effect of Ethylmorphine on the Stoichiometry of NADH Oxidation and Oxygen Uptake by Hepatic Microsomes from Male Rats

Ethyl-morphine 2 mM	NADH Oxidation nmoles/min-mg protein	Oxygen uptake nmoles/min-mg protein
-	4.62 ± 0.35[a]	6.68 ± 0.06
+	2.91 ± 0.18	4.03 ± 0.13
Δ	-1.71 ± 0.40[b]	-2.65 ± 0.14[c]

[a]Values are the average of triplicate incubations ± SEM.
[b] 0.02 > P > 0.01. [c]0.001 > P

When we examined for superoxide formation by epinephrine oxidation, we found that on the addition of NADPH there was rapid adrenochrome formation which was complete in two minutes. This oxidation was totally inhibited by the addition of superoxide dismutase to both reference and sample cuvettes. Although this assay rarely quantitates the fraction of oxygen going to superoxide anion radical, it does, in line with our stoichiometry studies, confirm that the ion is produced.

Interestingly, cytochrome c reduction, although an alternative measure of superoxide formation, was not affected by the dismutase, suggesting that the direct reduction of this substrate by the NADPH-cytochrome c or P-450 reductase is more rapid than superoxide formation under the conditions of this assay.

We do not wish to imply that the overall reaction presented in scheme I is incorrect. And indeed, in a final study when we added 252 µM NADPH to a suspension of microsomes, there was an uptake of oxygen of 126 $\pm$ 2 nmoles/ml. This 2:1 stoichiometry between NADPH added and oxygen taken up indicates that the peroxide is formed and then destroyed by microsomal catalase without a significant fraction participating in lipid peroxidation. The uptake of oxygen is sufficiently slow that even the low concentrations of catalase found in the microsomes will destroy all of it.

DISCUSSION

The major finding of this study is that in the first minute of incubation there is less than a 1:1 stoichiometry between NADPH oxidation and oxygen uptake. On the other hand, the ultimate stoichiometry between these two is 2:1, as would be expected if all the oxygen goes to peroxide and then is disproportionated by catalase to water and oxygen. We feel that at the present time this result for the initial activity would best be explained on the basis of the formation of intramicrosomal superoxide anion radical. Mishin et al. (14) have, on the basis of a quantitative adrenochrome assay, come to a similar conclusion. Yet a number of problems arise from this hypothesis.

The first is that the oxygen could be reduced by species other than NADPH. Since we have observed these results in washed microsomes, it is unlikely that any of the usual soluble cofactor reductants, as reduced glutathione or tetrahydropteridines, are present. Further, the reaction was initiated by the addition of NADPH; hence, the burst of oxygen uptake is not due to the presence of a fully reduced microsomal electron chain being discharged on addition to the incubation vessel, especially since the microsomal components are fully oxidized when prepared.

Similarly, lipid peroxidation with subsequent auto-oxidation of lipids could play a significant role in providing extra reducing equivalents in this system. This seems unlikely since the shape of this curve is not that of an autocatalytic oxidation, suggesting that the high oxygen uptake is not due to lipid peroxidation. We find that in the presence of ADP-Fe^{3+} there is an initial slow rate followed by an acceleration of oxygen uptake as the number of chain promoting radical species increased. This would suggest that extra reducing equivalents are not being contributed by unsaturated fatty acids in the membrane.

Another problem with postulating the formation of superoxide anion radical is that it should have such a short half life that we should not be able to detect it by oxygen uptake studies. The fact that it takes about one minute for the oxygen to parallel the NADPH oxidation (Fig. 1) would suggest that the half life of the superoxide anion radical is between 15 and 20 seconds. The half life in water at pH 7.4 is 0.7 seconds (2).

REFERENCES

(1) J.L. Holtzman, The effect of 4,4-dideuteration of NADPH on the hepatic mixed function oxidases, Biochemistry 9, 995 (1970).
(2) McCord, J.M., Beauchamp, C.O., Goscin, S., Misra, H.P. and Fridovich, I. (1973) Oxidases and Related Redox Systems, University Park Press.
(3) E. Jeffery and G.J. Mannering, Discrepancy in the measurement of reduced triphosphopyridine nucleotide oxidized during ethylmorphine N-demethylation due to the presence of a nucleotide pyrophosphatase, Mol. Pharmacol. 10, 1004 (1974).
(4) M.K. Buening and M.R. Franklin, Limitations in the use of the 340 nm absorbance maximum of reduced nicotinamide adenine dinucleotide phosphate for the determination of oxidation rates and stoichiometry during rat hepatic microsomal metabolism, Mol. Pharmacol. 10, 999 (1974).
(5) J.L. Holtzman and M.L. Carr, Inhibition of hepatic microsomal mixed function oxidases by D_2O, Life Sci. 9, 1033 (1970).
(6) J.L. Holtzman and M.L. Carr, Inhibition by deuterated water of the mixed function oxidases of hepatic microsomes of the male rat, Mol. Pharmacol. 8, 481 (1972).
(7) J.A. Thompson and J.L. Holtzman, Kinetics of N- and O-dealkylation of ethylmorphine by hepatic microsomes from male rats, Pharmacologist 16, 407 (1974).
(8) J.L. Holtzman and M.L. Carr, The temperature dependence of the components of the hepatic microsomal mixed function oxidases, Arch. Biochem. Biophys. 150, 227 (1972).
(9) H. Staudt, F. Lichtenberger and V. Ullrich, The role of NADH in uncoupled microsomal monoxygenations, Eur. J. Biochem. 46, 99 (1974).
(10) A. Hildebrandt and R.W. Estabrook, Evidence for the participation of cytochrome b_5 in hepatic microsomal mixed function oxidation reactions, Arch. Biochem. Biophys. 143, 66 (1971).

(11) M.A. Correia and G.J. Mannering, Reduced diphospho-pyridine nucleotide synergism of the reduced triphosphopyridine nucleotide-dependent mixed function oxidase system of hepatic microsomes, I, Mol. Pharmacol. 9, 455 (1973).

(12) M.A. Correia and G.J. Mannering, Reduced diphosphopyridine nucleotide synergism of the reduced triphosphopyridine nucleotide-dependent mixed function oxidase system of hepatic microsomes, II, Mol. Pharmacol. 9, 470 (1973).

(13) H.W. Stroebel and M.J. Coon, Effect of superoxide generation on hydroxylation reactions catalyzed by liver microsomal P-450, J. Biol. Chem. 246, 7826 (1971).

(14) V. Mishin, A. Pokrousky and V.V. Lyakhovich, Interactions of some acceptors with superoxide anion radicals formed by the NADPH-specific flavoprotein in rat liver microsomal fractions, Biochem. J. 154, 307 (1976).

STRUCTURAL AND FUNCTIONAL INTERACTION OF ROUGH ENDOPLASMIC RETICULUM (RER) AND MITOCHONDRIA DURING SYNTHESIS OF HEPATIC CYTOCHROME P450

Peter J. Meier, Max A. Spycher and Urs A. Meyer*

Div. of Clinical Pharmacology, Depts. of Medicine and Pharmacology, and Dept. of Pathology, University of Zürich, CH 8091 Zürich, Switzerland

ABSTRACT

Apparent accumulation of hepatic apocytochrome P450 was observed after concomitant treatment of rats with phenobarbital and cobaltous chloride; incubation of liver homogenate from these rats with hemin or with protoporphyrin and Fe^{++} resulted in increased cytochrome P450 concentration. The magnitude of this effect was dependent on the dose of phenobarbital administered and on the amount of hemin generated in the in vitro incubation. After differential centrifugation of the homogenate, the largest hemin-mediated increase in cytochrome P450 concentration was found in a subfraction of rough endoplasmic reticulum (RER) membranes sedimenting at low speed. Ultrastructural and marker-enzyme studies suggest that these RER-membranes are associated with mitochondria. Structural contact between mitochondria (the site of heme formation) and RER-membranes (the presumed site of apoprotein synthesis) may be a prerequisite for the formation of holocytochrome P450.

INTRODUCTION

Drug mediated induction of cytochrome P450 requires de novo synthesis of apoprotein(s) and of heme (iron-protoporphyrin IX) and is associated with the biogenesis of membranes of the endoplasmic reticulum (ER). The presumed site of apoprotein synthesis is at ribosomes of the ER (rough endoplasmic reticulum, RER), while heme is formed in mitochondria. It remains unclear how the two processes of apocytochrome and heme synthesis are coordinated and at which stage of this process mitochondria and ER-membranes interact to assemble the holocytochrome.

An apparent dissociation of apocytochrome and heme synthesis was experimentally produced by Correia and Meyer (1). Concomitant treatment of rats with inducers of cytochrome P450 and compounds

---

* To whom correspondence should be addressed

that interfere with hepatic heme synthesis and/or affect its breakdown resulted in the accumulation of "free" apocytochrome P450 in the liver. This was deduced from the observation that incubation of liver homogenate in the presence of hemin resulted in increased concentration of microsomal cytochrome P450 (determined by its CO-binding spectrum) and parallel increases in microsomal oxidative function. We have interpreted these and additional findings to indicate that drug-mediated induction of apocytochrome P450 proceeds independent of heme availability and therefore may be the primary and rate-limiting event in the formation of cytochrome P450 (Refs. 1, 2). Similar conclusions have been reached by Rajamanickam, Rao and Padmanaban (3) on the basis of experiments in which the sequential incorporation of radioactive precursors into apoprotein and heme moieties of hepatic cytochrome P450 was studied.

In the present paper, we explore further biochemical and structural aspects of the in vitro reconstitution of apocytochrome P450 with hemin. Using the same approach of an experimentally produced excess in hepatic apocytochrome P450 we investigated the subcellular distribution of the hemin-mediated changes in cytochrome P450 concentration. In doing so, we discovered a subfraction of RER-membranes characterized by a strikingly higher content of apocytochrome P450. In this subfraction, RER-membranes are present in close association with mitochondria. On the basis of these findings, we propose that mitochondria-RER complexes may operate as functional units in the synthesis of cytochrome P450.

METHODS

Male Sprague-Dawley rats (150-200 g) were treated with cobaltous chloride ($CoCl_2$, 60 mg/kg subcutaneously, at 48 and 24 hours) and phenobarbital sodium (50 mg/kg at 48 hours; 80 mg/kg at 24 hours, intraperitoneally) before they were killed. All animals were fasted throughout the experiment with free access to water.

Preparation of subcellular fractions:

After decapitation of the animals, livers were perfused in situ with ice-cold 0.9% NaCl and homogenized in 2.5 volumes of medium (0.35 M sucrose, 2.5 mM Mg-acetate, 10 mM Trisacetate, pH 7,4) in a glass-Teflon Potter-Elvehjem homogenizer (radial clearance 0,18 - 0,24 mm) rotating at 1200 rpm. The homogenate was then filtered through one layer of Miracloth (Calbiochem.) and subcellular fractions prepared by differential centrifugation (Fig. 3). The fraction enriched in mitochondria-RER complexes (Mito-RER fraction) was isolated from crude nuclear pellets by the method described by Lewis and Tata (4) with minor modifications (Fig. 3). The relative purity of this fraction was evaluated by electronmicroscopy (Fig. 4) and with marker enzymes. For electronmicroscopy, the pellet of the isolated Mito-RER fraction was resuspended in homogenization medium and fixed in

2% glutaraldehyde in 0,1 M sodium cacodylate buffer (pH 7,4) containing 3.5 mM $CaCl_2$, while beeing resedimented by centrifugation. The pellet thus obtained was then postfixed by osmification with 1% osmium tetroxide in 0,1 M sodium cacodylate buffer (pH 7,4). These studies will be reported in detail elsewhere.

Incubations with hemin

Hemin (ferriprotoporphyrin hydrochloride, Fluka Chemical Co.) was dissolved in ethanolic KOH (Ethanol: 0,02 N KOH = 1:1, v/v) and added to the incubate to give a concentration of 4 μM, except where stated otherwise. For incubation with hemin, 50% homogenates were prepared directly in the incubation medium (0.25 M sucrose, 20 mM Tris-HCl, pH 7,4). Other subcellular fractions were incubated in medium beeing supplemented with 105000 x g supernatant at a concentration of 5-10 mg protein per ml of incubate. Incubations were carried out at 37°C with gentle shaking and stopped after 20 minutes by cooling on ice.

Biochemical assays

Cytochrome P450 was measured in all subcellular fractions by the $[(CO + Na_2S_2O_4) - Na_2S_2O_4]$ difference spectrum (5). In fractions containing mitochondria, corrections for the interference of cytochrome oxidase with the CO-binding spectrum of cytochrome P450 were made according to Kowal et al. (6). Protein was quantitated according to the method of Lowry et al. (7).

RESULTS AND DISCUSSION

Incubation of liver homogenate of rats treated with phenobarbital and cobaltous chloride in the presence of hemin (4 to 8 μM) produced an increase in the CO-difference spectrum between 450 and 490 nm (Fig.1). Identical results were obtained when heme was replaced by equimolar amounts of protoporphyrin and Fe^{++} (Ref. 8). Under these conditions, heme newly formed in the mitochondria of the homogenate was quantitatively incorporated into apocytochrome P450. Moreover, a linear relationship between the hemin-mediated increase in cytochrome P450 and the total dose of phenobarbital administered (100 to 200 mg/kg in 48 hours) was demonstrated

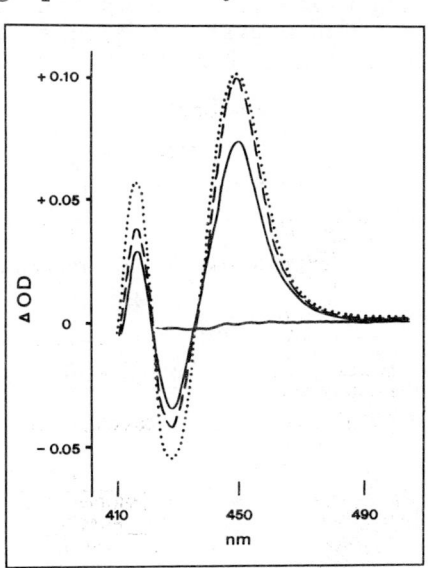

Fig. 1. Hemin-mediated change in the $[(CO+Na_2S_2O_4)-Na_2S_2O_4]$ difference spectrum in liver homogenates of rats treated with $CoCl_2$ and phenobarbital.
⎯⎯ control, no hemin; ---- hemin, 4 μM; .... hemin, 8 μM.

(Meier et al., in preparation). These data confirm and extend earlier studies, in which the cytochrome P450 was quantitated in microsomal fractions prepared from hemin-exposed homogenate of identically prepared rats. The direct demonstration of the effect of hemin on cytochrome P450 concentration in the homogenate largely excludes non-specific effects of hemin on the subsequent sedimentation behaviour of microsomes as an explanation for the effects observed in the previous studies (1). This and the dependence of the observed effect on the dose of phenobarbital lends strong support to the concept of drug-mediated accumulation of apocytochrome P450, the synthesis of which appears to proceed independent of the effects of cobalt on intracellular heme.

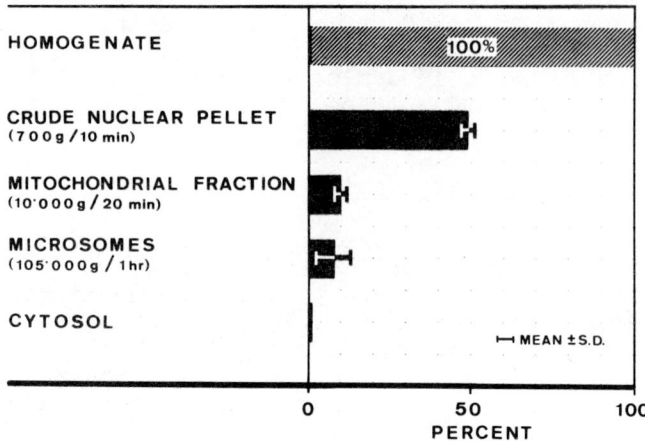

Fig. 2. Effect of hemin (4 μM) on the increase in cytochrome P450 concentration in subcellular fractions. The data are expressed as percent of the increase (nmoles/g liver) in the homogenate (= 100%).

We next approached the question at what stage of apocytochrome synthesis the heme addition may take place. For this purpose we subfractionated liver cell homogenates by differential centrifugation and incubated the subcellular fractions with hemin (Fig. 2). To our surprise, the largest proportion of the hemin-mediated increase was found in the crude nuclear pellet (sedimenting at 600 to 700 x g/10 minutes), accounting for almost 50 percent of the effect in the homogenate. Extensive studies on the relative content of subcellular organelles in this fraction (manuscript in preparation) revealed that the crude nuclear fraction indeed contains a large proportion of

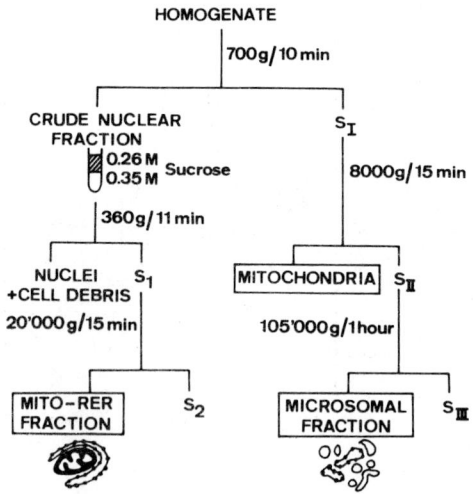

Fig. 3. Isolation of a fraction enriched with RER-membranes structurally associated with mitochondria (Mito-RER fraction). For details, see METHODS and Ref. 4.

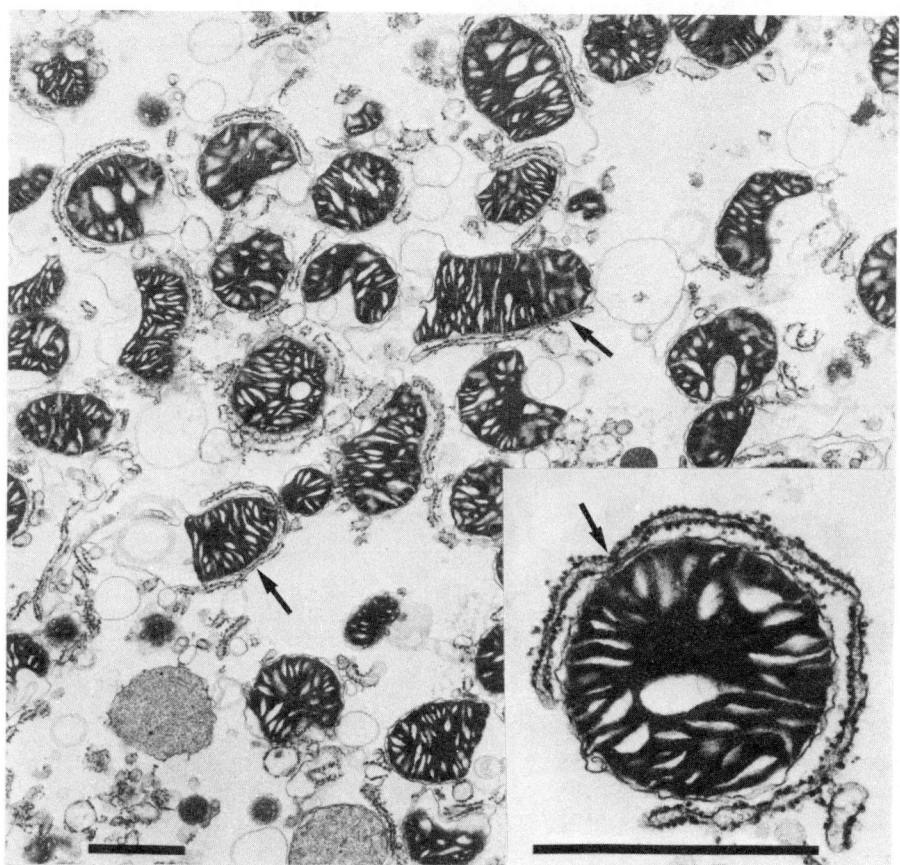

Fig. 4. Electron micrograph of Mito-RER fraction prepared as described in Fig. 3. The arrows point to areas of close association between RER-membranes and the outer mitochondrial membrane. Inset: Single Mito-RER complex at higher magnification. Bars represent 1 μm.

RER-membranes and that these RER-membranes frequently are found in close association with mitochondria, confirming the observations in nuclear fractions of normal rat liver by Lewis and Tata (4). Because of the obligatory biochemical interaction of mitochondria and RER-membranes during induction of cytochrome P450, we wondered if mitochondria-RER complexes could represent functional units involved in the regulation of microsomal cytochrome P450 synthesis, as suggested by Jones and Emans (9). The association of these two organelles could then account both for their cosedimentation at low centrifugal force and the apparent high content of apocytochrome P450. To evaluate this possibility, attempts were made to separate intact Mito-RER complexes from the contaminating nuclei, erythrocytes and other components of the crude fraction (Fig. 3) and to explore the apocytochrome content in the isolated organelle complexes by the described

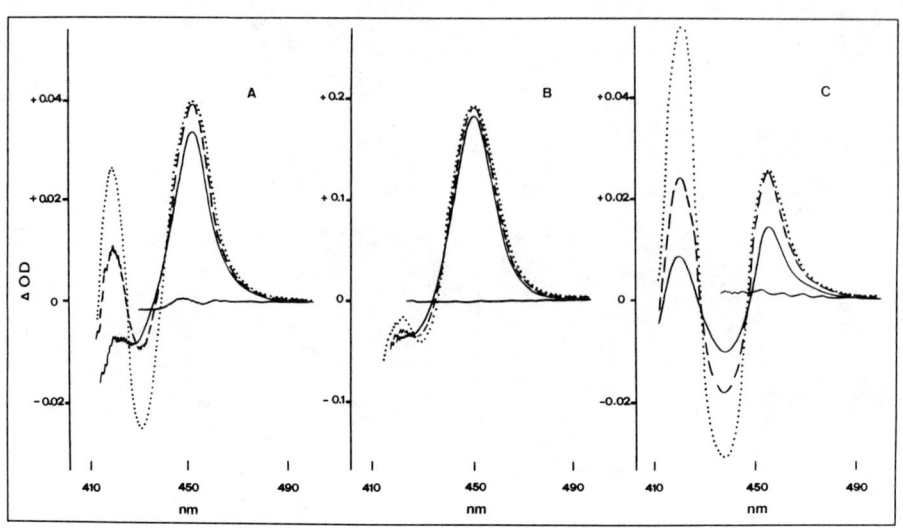

Fig. 5. Hemin-mediated change in the $[(CO+Na_2S_2O_4)-Na_2S_2O_4]$ difference spectrum in 3 subcellular liver fractions from rats treated with $CoCl_2$ and phenobarbital.
A: Mitochondrial fraction. B: Microsomal fraction. C: Mito-RER fraction.
——— control, no hemin; ---- hemin, 4 µM; .... hemin, 8 µM.

methods. The relative purity of the isolated complexes of mitochondria and RER-membranes was evaluated by electronmicroscopy (Fig. 4) and by marker enzymes. As demonstrated by the results in Fig. 5 and Table 1, a strikingly higher hemin-mediated increase in cytochrome P450 concentration was observed in this Mito-RER fraction, suggesting that RER-membranes of these organelle complexes may be responsible for the high content of apocytochrome P450 in the crude nuclear fraction.

Table 1 Hemin mediated increase in cytochrome P450 concentration in 3 subcellular fractions (mean ± S.D., 3 experiments)

	nmoles/mg protein	nmoles/g liver	percent increase
Mitochondria	0,018 ± 0,004	0,270 ± 0,050	18,7 ± 9,3
Microsomes	0,015 ± 0,008	0,215 ± 0,157	3,7 ± 1,5
Mito-RER fraction	0,031 ± 0,002	0,480 ± 0,048	60,8 ± 8,1

More definitive proof for a functional or regulatory role of Mito-RER complexes in microsomal hemoprotein synthesis obviously will depend on the direct demonstration of increased apocytochrome P450 in these membranes and its quantitative correlation with the inductive event in vivo. Nevertheless, on the basis of the indirect evidence described in this paper, we propose that the synthesis of cytochrome P450 requires the functional and structural interaction of mitochondria and membranes of the rough endoplasmic reticulum.

REFERENCES

(1) Correia, M.A., Meyer, U.A.: Apocytochrome P450: Reconstitution of functional cytochrome with hemin in vitro. Proc. Nat. Acad. Sci. USA, 72, 400 (1975).

(2) Meyer, U.A., Meier, P.J., Correia, M.A.: Interaction between Mitochondria and rough endoplasmic reticulum during induction of cytochrome P450, in: The Liver-Quantitative Aspects of Structure and Function, R. Preisig et al., editors, Editio Cantor, Aulendorf (1976), p. 172.

(3) Rajamanickam C., Rao M.R.S., Padmanaban G.: On the sequence of reactions leading to cytochrome P450 synthesis - Effect of drugs. J. Biol. Chem. 250, 2305 (1975).

(4) Lewis, J.A., Tata, J.R.: A rapidly sedimenting fraction of rat liver endoplasmic reticulum. J. Cell. Sci. 13, 447 (1973).

(5) Omura T., Sato, R.: The carbon monoxide-binding pigment of liver microsomes. J. Biol. Chem. 239, 2370 (1964).

(6) Kowal, J., Simpson, E.R., Estabrook, R.W.: Adrenal cells in tissue culture. J. Biol. Chem. 245, 2438 (1970).

(7) Lowry, O.H., Rosebrough, N.J., Farr, A.L., Randall, R.J.: Protein measurement with the Folin phenol reagent. J. Biol. Chem. 193, 265 (1951).

(8) Israels, L.G., Yoda, B., Schacter, B.A.: Heme binding and its possible significance in heme movement and availability in the cell. Ann. N.Y. Acad. Sci. 244, 651 (1975).

(9) Jones, A.L., Emans, J.B.: The effects of progesterone administration on hepatic endoplasmic reticulum: an electron microscopic and biochemical study. In: Metabolic Effects of Gonadal Hormones and Contraceptive Steroids. Plenum Press, New York (1969, p 68.

Supported by the Swiss National Foundation for Scientific Research and the Anita Saurer Foundation.

MECHANISM OF N-OXIDE FORMATION BY HEPATIC MIXED FUNCTION OXIDASES

Peter Hlavica and Maria Kehl

Pharmakologisches Institut der Universität, D-8000 München 2, Nussbaumstr. 26, West Germany

ABSTRACT

This communication establishes evidence for the existence of alternate metabolic routes of N-oxidation of N,N-dimethylaniline in rabbit liver microsomes. One pathway involves the participation of two types of cytochrome P-450 with differing sensitivity towards heat. Both types may represent distinct haemoprotein species or two physical forms of a single P-450 species. The other pathway is represented by a mixed function amine oxidase. The amine oxidase lacks NADPH diaphorase activity and is insensitive to treatment with 2-bromo-4'-nitroacetophenone and lipase.

INTRODUCTION

The metabolic route of N-oxidation of a wide variety of tertiary amine compounds to produce N-oxides has received increasing attention during the past decade.
Attempts have been made to isolate and characterize the enzyme system(s) responsible for the N-oxidation of tertiary amine substrates. Recently, a N,N-dimethylaniline monooxygenase (N-oxide forming; EC 1.14.13.8), also designated as mixed function amine oxidase, has been purified from pork liver microsomes (1). A large number of tertiary and some secondary amines are N-oxidized by this system. On the other hand, evidence has accumulated hinting at the participation of the liver microsomal cytochrome P-450 (P-450LM; EC 1.14.14.1) in the N-oxidation of tertiary amines (2, 3, 4, 5),
The further characterization of the systems supporting N-oxidation of tertiary amine compounds constitutes the focus of this presentation.

METHODS

Rabbit liver microsomes were prepared as described by Jagow et al. (6). Alkylation with 2-bromo-4'-nitroacetophenone was performed by mixing the reagent with a microsomal suspension containing 2 mg of protein per ml mixture in 0.15 M phosphate, pH 7.4. The reaction was stopped by the addition of 1.4 mmoles of histidine. The microsomal pellets were washed twice with ice cold 0.15 M KCL and resuspended in phosphate. Lipase treatment

of the microsomal preparations was accomplished by incubating the suspensions (10 mg of protein/ml mixture in 0.05 M phosphate, pH 7.7) with steapsin (2 mg/ml mixture) for five minutes at 37° C. The microsomal pellets were washed twice with ice cold 0.15 M KCL and resuspended in phosphate. $CoCl_2$ (60 mg/kg) was injected subcutaneously at 72, 48 and 24 hours prior to sacrifice of the rabbits. Pregnenolone-16 α-carbonitrile was administered to the animals five times at a dose of 20 mg/kg. The mixed function amine oxidase was purified according to Ziegler et al. (1). For routine assays, the "acid-ammonium sulfate" fraction was used. Cytochrome P-450 was partially solubilized by the method of Autor et al. (7).
The enzyme assay contained: 0.15 M phosphate, pH 7.4; 1.2 mM $NADP^+$; 10 mM glucose-6-phosphate; 6 mM $MgCl_2$; 2.5 U glucose-6-phosphate dehydrogenase (EC 1.1.1.49); 0.1 to 1.0 mM amine substrates, and varying amounts of enzyme (microsomes: 2 mg/ml; amine oxidase: 0.14 mg/ml; cytochrome P-450: 1.43 nmol/ml) in a total volume of 3.5 ml. Incubation was carried out for 10 minutes at 37° C. N,N-dimethylaniline-N-oxide was determined according to Ziegler et al. (8). Imipramine-N-oxide was measured by the method of Bickel et al. (9). Trimethylamine-N-oxide was estimated according to Fok et al. (10). NADPH-cytochrome c reductase activity was measured as indicated by Masters et al. (11). Cytochrome P-450 was determined as described by Omura et al. (12).

RESULTS AND DISCUSSION

As shown in Fig. 1, treatment of rabbit liver microsomes with 0.1 mM 2-bromo-4'-nitroacetophenone (BNAP) causes partial inhibition of the particle-bound NADPH-cytochrome c reductase. Similarly, the rate of microsomal N-oxide formation from N,N-dimethylaniline is considerably decreased. Alkylation, however, does not modify the content of microsomal cytochrome P-450. As can be furthermore seen, the purified mixed function amine oxidase is insensitive to treatment with the alkylating agent. Steapsin treatment of microsomes almost depletes the particles from NADPH-cytochrome c reductase (Fig. 2). There is also some loss of microsomal cytochrome P-450 due to conversion of the haemoprotein to the P-420 form. Solubilization of the microsomal reductase is accompanied by partial inhibition of the microsomal amine oxide formation from N,N-dimethylaniline. The amine oxidase-dependent N-oxidation of the amine, however, is insensitive to steapsin treatment.
Switching of the NADPH-cytochrome c (P-450) reductase-directed electron transport from drug oxidation to menadione almost completely blocks the cytochrome P-450-dependent N-dealkylation of N,N-dimethylaniline (Fig. 3), whereas the microsomal N-oxidation of the tertiary arylamine is inhibited by only 60 %. Partial inhibition of the N-oxidation reaction is not due to poor stability of N,N-dimethylaniline-N-oxide formed in the presence of menadione or lack of NADPH in the incubation mixtures (cf. Fig. 3). Although the naphthoquinone reduces the level of NADPH in the assay mixtures from 0.78 to 0.46 mM, the latter concen-

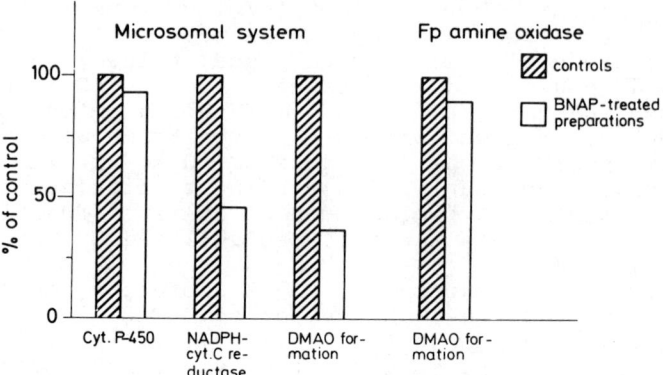

Fig. 1. Effect of treatment of microsomes and purified mixed function amine oxidase with 2-bromo-4'-nitroacetophenone (BNAP) on the N-oxidation of N,N-dimethylaniline. For experimental details see text. DMAO = N,N-dimethylaniline-N-oxide.

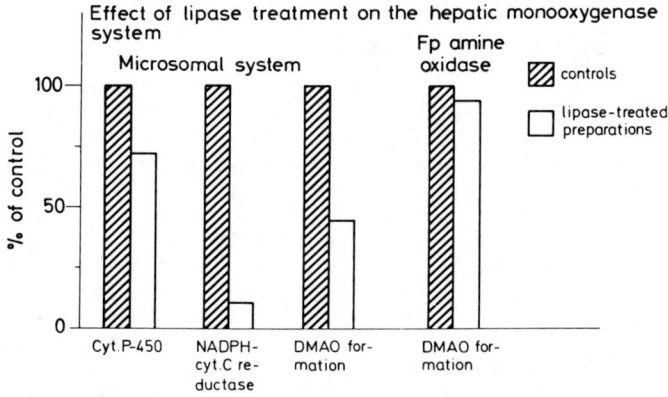

Fig. 2. Effect of treatment of microsomes and purified mixed function amine oxidase with steapsin on the N-oxidation of N,N-dimethylaniline. The values are the means of five determinations. DMAO = N,N-dimethylaniline-N-oxide.

tration of the reduced pyridine nucleotide has been found to be still within the range of saturating concentrations of the cofactor. Figure 3 illustrates that the mixed function amine oxidase-catalyzed N-oxidation of N,N-dimethylaniline is not affec-

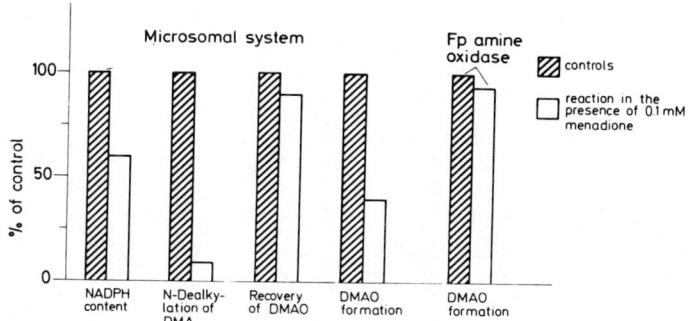

Fig. 3. Effect of menadione on the microsomal and mixed function amine oxidase-catalyzed N-oxidation of N,N-dimethylaniline. Microsomal suspensions, containing 2 mg of protein/ml mixture and NADPH-generating system were preincubated for five minutes at 37°C. Subsequently, the mixtures were supplemented with 0.1 mM menadione and 1.0 mM N,N-dimethylaniline. Incubation was performed for 10 minutes at 37° C. Solutions of purified amine oxidase were handled analoguously. The data is the mean of five to seven experiments. DMAO = N,N-dimethylaniline-N-oxide.

ted by menadione. This finding is in accord with previous reports on the porcine amine oxidase which has been shown to be devoid of detectable NADPH diaphorase activity (1).
When rabbits are injected with $CoCl_2$, a considerable decrease in the level of the hepatic microsomal cytochrome P-450 and rate of N-oxide formation from N,N-dimethylaniline occurs (Fig. 4). On the other hand, pretreatment of rabbits with pregnenolone-16α-carbonitrile increases the content of microsomal cytochrome P-450, elevates NADPH-cytochrome c reductase activity and enhances the microsomal N-oxidation of N,N-dimethylaniline to comparable extent (Fig. 4).

In order to assess the role of cytochrome P-450 in the N-oxidation of N,N-dimethylaniline more precisely, the haemoprotein was partially solubilized from liver microsomes of untreated rabbits. The preparations still contained small amounts of cytochrome b_5 and sufficient NADPH-cytochrome c reductase to support N-oxidation of N,N-dimethylaniline without the addition of exogenous reductase. When partially solubilized cytochrome P-450 was preincubated at 37° C in the absence of NADPH, a biphasic decrease in the N-oxygenating activity for N,N-dimethylaniline was observed (Fig. 5). The corrected values for the fast-phase component are given in the lower part of the figure. The ratio of the fast-phase to the slow-phase component was found to be 1.4. The pseudo first order rate constant for inactivation of the fast-phase component was determined to be 5.78 h^{-1}, the corresponding value for the slow-phase component was calculated to be 0.19 h^{-1}. Preincubation of the purified mixed function amine oxidase at

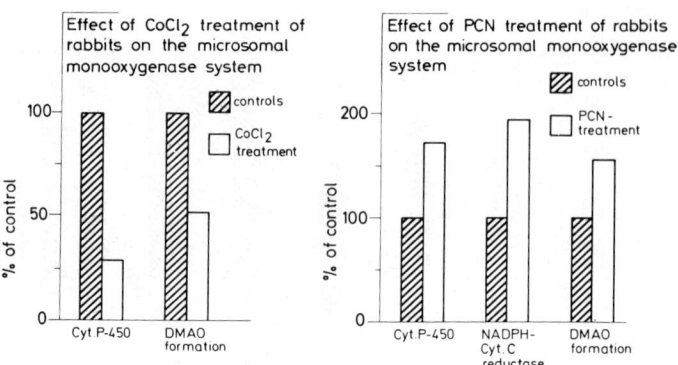

Fig. 4. Effect of pretreatment of rabbits with $CoCl_2$ and pregnenolone-16α-carbonitrile (PCN) on the microsomal N-oxidation of N,N-dimethylaniline. For details see METHODS. The values are the means of four to eight experiments. DMAO = N,N-dimethylaniline-N-oxide.

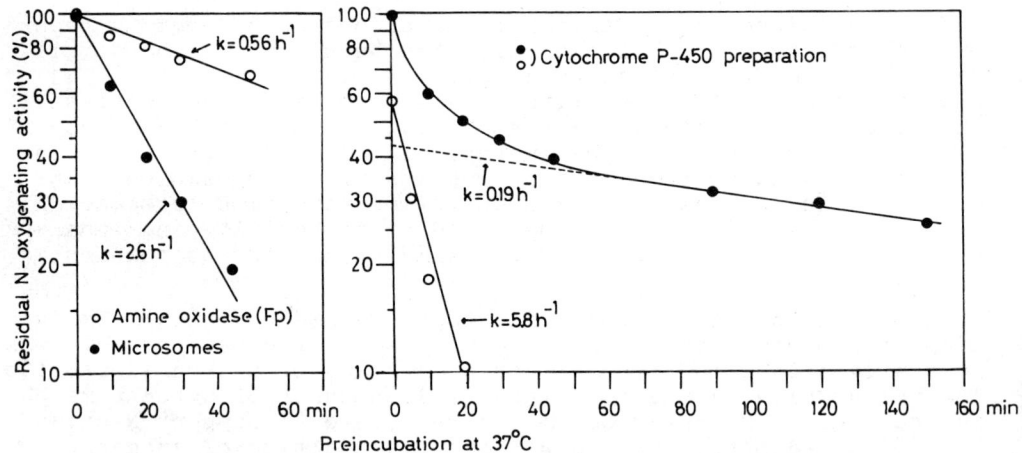

Fig. 5. Influence of temperature on the N-oxygenating activity of rabbit liver microsomes, solubilized cytochrome P-450 and mixed function amine oxidase. All preparations were preincubated in the absence of NADPH. The corrected values for the fast-phase component of the cytochrome P-450 system were obtained by subtracting the extrapolated slow-phase values from the fast-phase portion of the curve; k designates the pseudo first order rate constant for inactivation of the individual oxygenating systems.

37° C in the absence of the cofactor results in a slow loss of N-oxygenating activity for N,N-dimethylaniline, the pseudo first order rate constant for inactivation being 0.56 h^{-1} (Fig. 5). Preincubation of intact rabbit liver microsomes causes the complex system catalyzing N-oxidation of N,N-dimethylaniline to lose activity quite rapidly.(Fig. 5). The pseudo first order rate constant for inactivation was found to be 2.6 h^{-1}. Finally, comparative studies on the N-oxidation of imipramine (pK_a = 8.0), trimethylamine (pK_a = 9.7), and N,N-dimethylaniline (pK_a = 5.1) were performed, since the pK_a value of amine substrates has been recently proposed to determine by which of the N-oxygenating systems amines might be acted upon (5). Fig. 6

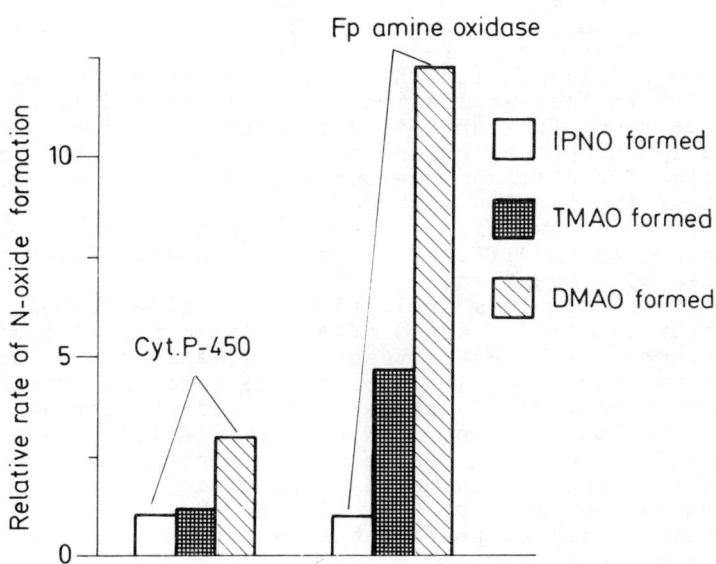

Fig. 6. N-Oxidation of imipramine, trimethylamine and N,N-dimethylaniline by partially solubilized cytochrome P-450 and mixed function amine oxidase. In these experiments the concentration of the amine substrates was 0.1 mM. IPNO = imipramine-N-oxide; TMAO = trimethylamine-N-oxide; DMAO = N,N-dimethylaniline-N-oxide.

shows that strongly basic amines, e.g. imipramine and trimethylamine, are not excluded from being N-oxidized by the cytochrome P-450 system. The ratio of the N-oxidation products formed from imipramine, trimethylamine, and N,N-dimethylaniline by cytochrome P-450-dependent reactions was calculated to be 1 : 1.2 : 3. The mixed function amine oxidase was found to catalyse N-oxidation of these compounds in the proportion 1 : 4.7 : 12.3.

In summary, our data establishes unequivocal evidence for the existence of alternate metabolic routes of N-oxidation of N,N-dimethylaniline in rabbit liver microsomes.
One pathway involves the participation of the cytochrome P-450-linked monooxygenase system. The data suggests the involvement of at least two types of the pigment in this N-oxidation process, both characterized by differing sensitivity to heat.
These might represent two distinct haemoprotein species independently synthetized in the rabbit liver or two physical forms of a single cytochrome P-450 species. Indeed, this communication supports evidence for the participation of two distinct P-450 species in the N-oxidation of N,N-dimethylaniline, namely cytochrome P-450 from untreated rabbits and a second pigment specifically induced by pretreatment of the animals with PCN. Synthesis of the former haemoprotein can be suppressed by pretreatment of the rabbits with $CoCl_2$.
The other pathway is represented by the mixed function amine oxidase, which differs from the cytochrome P-450 system in a series of properties.
From the incomplete inhibition of N-oxide formation from N,N-dimethylaniline observed after treatment of intact rabbit liver microsomes with BNAP, steapsin and menadione or pretreatment of rabbits with $CoCl_2$ it is concluded that the cytochrome P-450-linked monooxygenase system might account for at least 50 to 60 % of the total amount of N,N-dimethylaniline-N-oxide formed, the remainder depending on the mixed function amine oxidase. This view is strongly supported by the inactivation profile of liver microsomes exposed to $37°$ C in the absence of NADPH, which closely resembles that of the fast-phase component of solubilized cytochrome P-450.

ACKNOWLEDGEMENTS

The authors wish to thank Mrs. Mietaschk and Miss Heide for competent technical assistance.
This investigation was supported by Grant Hl 1/7 from the Deutsche Forschungsgemeinschaft.

REFERENCES

(1) D.M. Ziegler and C.H. Mitchell, Properties of a mixed function amine oxidase isolated from pig liver microsomes, Arch.Biochem.Biophys. 150, 116 (1972).

(2) P. Hlavica and M. Kehl, Effect of carbon monoxide on the N-oxidation of N,N-dimethylaniline, Hoppe-Seyler's Z.Physiol.Chem. 355, 1508 (1974).

(3) P. Hlavica, Interaction of oxygen and aromatic amines with hepatic microsomal mixed-function oxidase, Biochim.Biophys.Acta 273, 318 (1972).

(4) I.N.H. White and A.R. Mattocks, Some factors affecting the conversion of pyrrolizidine alkaloids to the N-oxides and to pyrrolic derivatives in vitro, Xenobiotica 1, 503 (1971).

(5) J.W. Gorrod, Differentiation of various types of biological oxidation of nitrogen in organic compounds, Chem.Biol.Interactions 7, 289 (1973).

(6) R. Jagow, H. Kampffmeyer and M. Kiese, The preparation of microsomes, Naunyn-Schmiedebergs Arch.exp.Path.Pharmacol. 251, 73 (1965).

(7) A.P. Autor, R.M. Kaschnitz, J.K. Heidema and M.J. Coon, Sedimentation and other properties of the reconstituted liver microsomal mixed-function oxidase system containing cytochrome P-450, reduced triphosphopyridine nucleotide-cytochrome P-450 reductase, and phosphatidylcholine, Mol.Pharmac. 9, 93 (1973).

(8) D.M. Ziegler and F.H. Pettit, Formation of intermediate N-oxide in the oxidative demethylation of N,N-dimethylaniline catalyzed by liver microsomes, Biochem.Biophys.Res.Commun. 15, 188 (1964).

(9) M.H. Bickel and P.L. Gigon, Metabolic interconversions and binding of imipramine, imipramine-N-oxide, and desmethylimipramine in rat liver slices, Xenobiotica 1, 631 (1971).

(10) A.K. Fok and D.M. Ziegler, Estimation of amine oxides in the presence of hepatic microsomes, Biochem.Biophys.Res.Commun. 41, 534 (1970).

(11) B.S.S. Masters, C.H. Williams and H. Kamin, Methods in Enzymology, Vol. X, Academic Press, New York, 1967.

(12) T. Omura and R. Sato, The carbon monoxide-binding pigment of liver microsomes, J.Biol.Chem. 239, 2379 (1964).

A PRELIMINARY COMPARATIVE STUDY ON THE INFLUENCE OF CYSTEAMINE AND METYRAPONE ON MIXED-FUNCTION OXYGENASE ACTIVITIES IN VARIOUSLY PRETREATED LIVER MICROSOMES FROM RATS AND MICE

Robert Mull, Wolfgang Hinkelbein, Johannes Gertz and Kurt Flemming

Institute of Biophysics and Radiobiology, University of Freiburg, Albertstr. 23, 7800 Freiburg i.Br., W. Germany

ABSTRACT

It has been found that metyrapone can inhibit both type I and type II mixed-function oxygenase reactions, while cysteamine inhibits only type I activity. Following pretreatment with phenobarbital and 3-methylcholanthrene the half-maximal inhibiting concentrations for the O-demethylation of para-nitroanisol are increased for cysteamine and decreased for metyrapone. Both cysteamine and metyrapone give type II binding spectra with oxidized cytochrome P-450. The negative and positive peaks are at 393 and 426 nm respectively for metyrapone, and 410 and 434 nm for cysteamine. Cysteamine showed no binding comparable to that of metyrapone for reduced cytochrome P-450. Metyrapone showed little or no inhibition for the NADH- or NADPH-cytochrome-c reductases while cysteamine had a more or less strong inhibiting effect depending on the pretreatment of animals.
Neither the binding to P-450 heme nor the inhibition of NADH and NADPH cytochrome-c reductase correlate well with cysteamine inhibition of total activity. It is therefore suggested that cysteamine reacts with an intermediate electron carrier of non-heme iron or glycoprotein character thus inhibiting mixed-function oxygenase activity.

INTRODUCTION

Cytochrome P-450 is the CO-binding enzyme (1) responsible for drug and steroid metabolism in liver microsomes (2,3) and corticosteroid biosynthesis in adrenal glands (4). The various substrates for this enzyme have been classified as polycyclic hydrocarbons type I, and aliphatic amines, type II (5). This system is complicated by an array of inhibitors also giving type I and type II binding spectra (6), and the apparent competition between type I and type II substrates for cytochrome P-450 (7). Further complications enter the picture with the use of P-450 inducing drugs. They also can be divided into two groups, e.g. phenobarbital-like, inducing more type I substrate activity, and 3-methylcholanthrene-like inducing more type II substrate activity (8). The so-called cytochrome P-420 (1) does not appear to have any functional importance in this system, and could be an

artifact of non-specific heme binding to liver microsomes (9).
In this laboratory it was recently found that the 11 β-hydroxylation of desoxycorticosterone in adrenal mitochondria was inhibited _in vivo_ and _in vitro_ (10,11) by cysteamine. The fact that the order of magnitude for this effect was comparable to that of metyrapone, led us to ask if cysteamine could also be a specific inhibitor of the mixed-function oxygenase system in the liver and by what mechanism this might proceed.

At least four groups have tested cysteamine or its disulfide cystamine on liver mixed-function oxygenase reactions (12,13,14,15) but none have, to the best of our knowledge, made parallel experiments with a known inhibitor of "similar type", or chosen a group of reactions and treatments large enough to answer the above questions.

METHODS

All animals were fed an Altromin diet and water _ad lib_. Rats were of the SIV-50 strain obtained from S. Ivanovas, Kisslegg/Allgäu, West Germany when 7 weeks old, with an average weight of 200 g (untreated). Phenobarbital treated animals were injected twice daily for two days with 70 mg/kg phenobarbital in physiological NaCl and sacrificed by decapitation 48 hours after the last injection; 3-methylcholanthrene was injected 1 x daily for 4 days and the animals sacrificed 24 hours after the last injection. Mice were of an inbred strain from our institute. They were also treated when 7 weeks old, having an average weight of 25-30 g. Phenobarbital and 3-methylcholanthrene doses were the same as for rats. Mice were sacrificed by breaking the neck. All other methods were as previously described (16).

RESULTS AND DISCUSSION
Cysteamine and Metyrapone Effects on Total Mixed-Function Oxygenase Activity.

Cysteamine showed an inhibiting effect on the O-demethylation of para-nitroanisol similar to that of metyrapone in both untreated and induced rats, however, the effect of metyrapone was stronger than that of cysteamine (Fig. 1). Only small quantitative differences were found between rats and mice (data not presented). If one compares the half-maximal inhibiting concentrations (I_{50}), determined from semi-log plots to the specific activity of O-demethylation in control and 3-methylcholanthrene or phenobarbital induced microsomes, it immediately becomes apparent that the I_{50} value for metyrapone decreases with induction while that of cysteamine increases (Table 1). The increase in cysteamine's I_{50} after 3-methylcholanthrene pretreatment is especially large. Another interesting variable is seen when one pretreats animals with phenobarbital, 1 x 70 mg/kg sacrificed 48 hours later, so as to induce O-demethylation to about the same specific activity as in 3-methylcholanthrene pretreated microsomes. In the case of metyrapone one then sees a decrease in I_{50} comparable to that for 3-methylcholanthrene microsomes, and about one-forth of that for microsomes pretreated 4 x 70 mg/kg phenobarbital, cysteamine's increased I_{50} after pretreatment with 1 x 70 mg/kg phenobarbital is also about one-forth of that for microsomes pretreated four times with the same dosis, but it is not comparable to the large

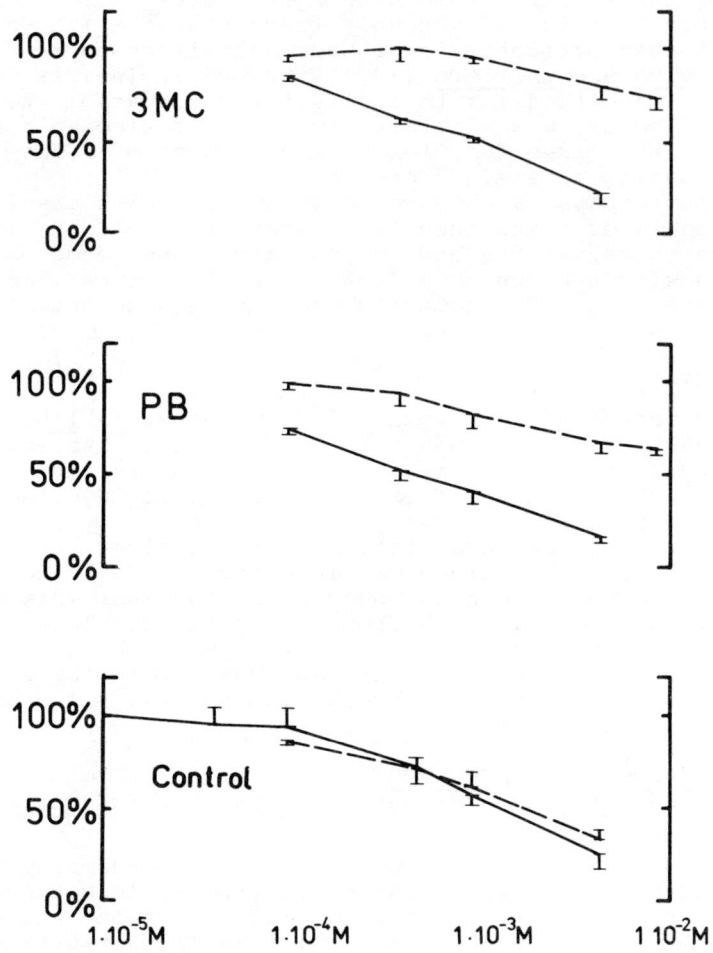

Fig. 1 The effect of cysteamine and metyrapone on the O-demethylation of p-nitrophenol in variously pretreated liver microsomes from rats.

O-demethylation was determined by a method modified from that of Netter (5). The reaction mixture was incubated for 15 minutes and contained: G-6-P = 5 µmoles; nicotinamide = 0.05 µmoles; NADP = 0.1 µmole; $MgCl_2$ = 1.0 µmole; pNA = 1.0 µmole; G-6-PDH = 0.42 units; microsomes 1-3 mg/assay and buffer 0.067 M potassium phosphate (pH 7.9) added to give a final volume of 2.0 ml.
Specific activities for controls (100 %) were: 3.0, 3.8 and 9.2 nmoles para-nitrophenol per minute and per milligram protein for untreated control, 3-methylcholanthrene, 3-MC, and phenobarbital, PB, pretreated microsomes respectively.

change after 3-methylcholanthrene pretreatment, thus demonstrating that the 3-methylcholanthrene induction-effects on cysteamine's inhibition differ qualitatively as well as quantitatively from those of metyrapone. No inhibition of acetanilide-hydroxylation was found for cysteamine.

TABLE 1 Half-maximal Inhibiting Concentrations (I_{50}) for Cysteamine and Metyrapone in Variously Treated Mice.

Treatment	Cysteamine m moles/l	Metyrapone m moles/l	O-demethylation nmole·p-nitrophenol mg protein · min
Untreated	2.0	2.0	3.0
Phenobarbital 1 x 70 mg/kg	4.0	1.2	4.0
4 x 70 mg/kg	20.0	0.5	9.2
3-methylcholanthrene	50.0	1.0	3.8

Half-maximal inhibiting concentrations were determined from semi-logarithmic plots of rest activity (x-axis, linear scale) and inhibitor concentration (y-axis, log-scale). Concentrations of inhibitors used were as in Figure 1.

Effects of Cysteamine and Metyrapone on Individual Components of Mixed-Function Oxygenase System

Hemoproteins

We and others have been able to demonstrate the presence of two CO binding cytochromes in liver microsomes from mice (17) and rats (18) respectively. As these two cytochromes were differentially induced by phenobarbital and 3-methylcholanthrene, differences in these components could explain the variations in inhibition after induction and thus reveal the locus of action for cysteamine and metyrapone. To examine this possibility we studied the binding spectra of cysteamine, ethandithiol, a sulfur containing derivative of cysteamine and metyrapone in microsomes from variously treated mice. As has been previously reported (14,15) cysteamine gives a so-called type II binding spectrum similar to that of metyrapone indicating an amine-heme iron ligand (15). Ethandithiol, however, gave a maximum at 468 nm which is similar to that for carbene ligands described by Ullrich (19). Neither cysteamine nor ethandithiol gave binding spectra with reduced cytochrome P-450 preparations; although this form of cytochrome P-450 is thought to be the active state of hydroxylation.

Spectral binding constants, K_s were also determined for these compounds (Table II). The metyrapone K_s values were strongly influenced by pretreatment and agree well with previously reported values (20) and partially with the changes in I_{50} for O-demethylation (Table I), making hemoprotein binding the most probably explanation for metyrapone inhibition.

The K_s values for cysteamine and ethandithiol were unchanged after induction. This and the absence of a reduced binding spectrum would indicate that cysteamine binding to cytochrome P-450's heme can not form the basis of any inhibition mechanism. From the above, we therefore propose that cysteamine acts on some intermediate electron carrier specific for type I activity.

TABLE 2 Microsomal Binding Constant (K_s) Values for Cysteamine and Metyrapone in Variously Pretreated Mice for Oxidized Cytochrome P-450

	Untreated	Phenobarbital	3-Methylcholanthrene
Metyrapone m molar	0.015	0.002	0.14
Cysteamine m molar	2	3	2
Ethandithiol m molar	0.15	0.27	0.22

K_s values, e.g. the concentrations which gives a half-maximal spectral development, were determined in a solubilized and partially purified cytochrome P-450 preparation. As the values for metyrapone were comparable to those of whole microsomes (20), it was assumed that binding characteristics were unchanged by this modification, while the problems of turbidity and sedimentation seen when using whole microsomes were eliminated.

Reductases
As type I and type II activities have differing synergetic relationships to NADH as an electron donor in NADPH dependent hydroxylation (21), we decided to study the effect of inhibitors on the NADH and NADPH dependent cytochrome-c reductases. In agreement with our previous studies (17), the specific activity of the NADH-dependent reductase appeared to be unchanged after pretreatment with phenobarbital and 3-methylcholanthrene while the NADPH-dependent activity was strongly induced by phenobarbital and unchanged by 3-methylcholanthrene pretreatment (Table III). Inhibition studies with cysteamine and metyrapone from the same preparations revealed several interesting effects, the largest of these being an inhibition of both NADPH- and NADH-dependent reductases by cysteamine in microsomes from untreated animals. Only 65 per cent of control activity was obtained by 10 mM cysteamine. Metyrapone, however, had little or no effect on either the NADH- or NADPH-dependent reactions.
Cysteamine's inhibiting effect was slightly decreased after phenobarbital and more so after 3-methylcholanthrene pretreatment. The decrease in inhibition was most prominate for the

NADH-dependent reductases. This reaction still demonstrated 88 per cent of control activity by 10 nM cysteamine in microsomes for 3-methylcholanthrene pretreated animals.

TABLE 3   Inhibition of Microsomal Cytochrome-c Reductases by Cysteamine and Metyrapone in Variously Pretreated Mice.

Reductase	Untreated		PB		3-MC	
	NADPH	NADH	NADPH	NADH	NADPH	NADH
Control	100 ±5.9	100 ±5.3	100 ±4.5	100 ±4.7	100 ±4.4	100 ±6.5
Cysteamine 10 mmolar	65 ±5.3	66 ±7.5	63 ±1.9	71 ±2.3	78 ±4.1	88 ±4.1
Metyrapone 10 mmolar	87 ±4.0	105 ±7.4	86 ±2.8	102 ±2.6	102 ±16.0	105 ±4.9

Experimental methods were as cited. The data represent the mean ± S.E. of five experiments, three samples per experiment, for two preparations in each group of animals.

As can be seen from the above data, the flavoprotein components of microsomal electron transport can be inhibited by cysteamine but not metyrapone. The inhibition of the cytochrome-c reductase is also changed in the same manner as the inhibition of the O-demethylation after induction with 3-methylcholanthrene and phenobarbital, e.g. induction decreases cysteamine's inhibition, but the magnitude of the effects do not correlate well with the inhibition seen in the total reaction.
This lack of correlation between cysteamine's influence on individual components and inhibition of total hydroxylase activity before and after induction could be explained by at least two hypotheses:
1) Due to the existing stoichoimetry between flavoprotein(s) and hemoprotein(s) |1:10 as reported by Estabrook (22)| and changes therein after induction, an amplification of the inhibition on any single component could occur. 2) The often suggested presence of unknown intermediate electron carriers (23), where cysteamine may react, could also explain the weak effects on known components as compared to the strong inhibition of total activity. Our experiments characterizing the protein components of microsomal electron transport before and after induction (16,17) tend to support the latter.
On the one hand, we found a ratio of NADPH dependent flavoprotein

to CO binding between 1:1.5 in controls and 1:3 after phenobarbital induction. These values do not agree with previous reports (22), and are thus not of a magnitude which could directly explain cysteamine's inhibition of P-450 dependent activities. One must, however, also consider the role of the NADH-dependent flavoprotein. We have shown that the NADH cytochrome-c reductase can be made more resistant to cysteamine than the NADPH-dependent reductases and one could suggest that 3-methylcholanthrene induction switches the flow of more electrons from the NADPH- to the NADH-dependent reductases.
On the other hand, we have found two additional components which appear to be active in the mixed-function oxygenase system (16, 17) and, therefore, could help explain the complex inhibition patterns described in the first part of this paper.

ACKNOWLEDGEMENT

This investigation was supported by Research grant SS 257 A, Bundesminister für Forschung und Technologie, Bonn.

REFERENCES

1. T. Omura and R. Sato, The CO binding pigment of liver microsomes. I. Evidence for its hemoprotein nature, J. Biol. Chem. 239, 2370 (1964)
2. A.H. Conney, Pharmacological implications of microsomal enzyme induction, Pharmacol. Rev. 19, 317 (1967)
3. N.P. Sanzari and F.G. Person, Influence of metopirone on substrate-supported hydroxylation of DOC by rat adrenal mitochondria, Steroids 8, 929 (1966)
4. D.Y. Cooper, S. Narasimhulu, A. Slade, W. Raich, O. Forhoff and O. Rosenthal, Hemoprotein content and activity of solubilized steroid 11 β-hydroxylase preparations from adrenocortical mitochondria, Life Sci. 4, 2109 (1965)
5. K.J. Netter, Eine Methode zur direkten Messung der O-Demethylierung in den Lebermikrosomen, Naunyn-Schmiedeberg's Arch. exp. Path. u. Pharmakol. 338, 292 (1960)
6. A.G. Hildebrandt, The binding of metyrapone to cytochrome P-450. In Biological Mechanisms of Oxidation (Biochem. Soc. Symposium No. 34, Univ. Edinburgh 1971) pp. 79 Academic Press New York (Boyd, G.S. and Smellie, R.M.S., eds.) (1972)
7. A.Y.H. Lu, R. Kuntzman, S.B. West and A.H. Conney, Determination of substrate specificity by cytochrome P-450 and P-448 fractions, Biochem. Biophys. Res. Comm. 42, 1200 (1971)
8. A.Y.H. Lu, W. Levin, S.B. West, M. Jacobson, D. Ryan, R. Kuntzman and A.H. Conney, Different substrate specificities of the cytochrome P-450 fractions from control and phenobarbital treated rats, J. Biol. Chem. 248, 456 (1973)
9. A.E.M. McLean and R.C. Garner, Does P-420 exist? Haem reductase activity in liver microsomes, Biochem. Pharmacol. 23, 475 (1974)
10. K. Flemming, B. Geierhaas and V. Seydewitz, Inhibition by cysteamine of steroid 11 β-hydroxylation, Biochem. Pharmacol. 22, 1241 (1973)
11. K. Flemming and V. Seydewitz, A study of the inhibition by cysteamine on the steroid 11 β-hydroxylation, Experientia 30, 989 (1974)

12. Z.M. Bacq and P. Alexander, Cysteamine and SH-protectors; in Fundamentals of Radiobiology, p. 470 Pergamon Press, New York (1961)
13. N.A. Scholan and G.S. Boyd, The cholestrol 7α-hydroxylase enzyme system, Hoppe Seyler's Z. Physiol. Chem. 319, 1628 (1968)
14. G.A. Castro and C.R. Castro, Effects of cysteamine on the mixed-function oxygenase system for rat liver microsomes and its preventive effect on destruction of cytochrome P-450 by CCI_4, Pharm. Res. Comm. 4, 185 (1972)
15. C.R.E. Jefcoate, J.L. Gaylor and R.L. Calabrese, Ligand interactions with cytochrome P-450 I. Binding of primary amines, Biochem. 8, 3455 (1969)
16. R. Mull, T. Voigt and K. Flemming, Distribution of electron transport components in ammonium sulfate fractions from mouse liver microsomes, Biochem. Biophys. Res. Comm. 64, 1098 (1975)
17. R. Mull, M. Schgaguler and K. Flemming, Resolution and quantitation of cytochromes P-450 and P_1-450 and the so-called "Factor X" in SDS-polyacrylamide gels of total microsomes, Biochem. Biophys. Res. Comm. 67, 849 (1975)
18. T.A. Van der Hoeven, D.A. Haugen and M.J. Coon, Cytochrome P-450 purified to apparent homogeneity from phenobarbital-induced rabbit liver microsomes, Biochem. Biophys. Res. Comm. 60, 569 (1974)
19. V. Ullrich, W. Mastainczyk and H.H. Ruf, Ligand reactions of cytochrome P-450, Biochem. Soc. Trans. 3, 803 (1975)
20. R. Kahl, H.G. Jonen, G.F. Kahl, The effect of phenobarbital and 3-methylcholanthrene induction and iso-octane extraction on metyrapone binding to cytochrome P-450. Biochem. Pharmacol. 23, 2305 (1974)
21. A.I. Archakov, V.M. Derichensky and A.V. Karjakin, The interaction of NADPH and NADH specific electron transfer chains in microsomes, Arch. Biochem. Biophys. 166, 308 (1975)
22. R.W. Estabrook, J. Baron, M. Franklin, I. Mason, M. Waterman, and J. Peterson, in the molecular basis of electron transport (Schultz, J. and Cameron, B.F., eds.) pp. 197, Academic Press, New York (1972)
23. Y. Ichikawa and T. Yamano, Preparation and physico-chemical properties of functional hemoproteins, Biochim. Biophys. Acta 200, 220 (1970)

INHIBITION OF RAT LIVER MICROSOMAL DT DIAPHORASE BY BENZO (a) PYRENE METABOLITE (S)

Christina Lind and Lars Ernster

Department of Biochemistry, Arrhenius Laboratory, University of Stockholm, S-104 05 Stockholm, Sweden

ABSTRACT

The increase in DT diaphorase activity after treatment of rats with 3-Meyhylcholantrene (MC) seems to be due to an induced enzyme synthesis. Not only liver but also other tissues show an increased DT diaphorase activity after MC-treatment. This increase parallels that of the aryl hydrocarbon monooxygenase (AHM) activity except in heart, where only DT diaphorase activity increases. This indicates a difference in gene expression between the inductions of DT diaphorase and the AHM system. Microsomal hydroxylation of benzo(a)pyrene (BP) leads to an inhibition of DT diaphorase with a simultanous decrease in dicoumarol sensitivity of the enzyme. Probably a product of BP metabolism is responsible for these effects.

INTRODUCTION

The enzyme DT diaphorase (E.C. 1.6.99.2) is a flavoprotein catalyzing the oxidation of NADH and NADPH by various dyes and quinones (1). This highly dicoumarol sensitive enzyme is found both in the cytosol and in the microsomes as well as in the mitochondria. As has been reported (2,3) treatment of rats with MC leads to a severalfold increase of DT diaphorase activity, in contrast to NADPH-cytochrome $\underline{c}$ reductase. The increase is parallel to that of the AHM activity (measured with BP as substrate (3)). The finding that 7,8-benzoflavone, a potent inhibitor of the AHM system in MC-treated rats (4,5), also inhibited DT diaphorase to the same degree (3), while NADPH-cytochrome $\underline{c}$ reductase was uneffected, suggested a functional relationship between DT diaphorase and the AHM system.

A plausible possibility appeared to be that DT diaphorase might function as cytochrome P-448 reductase. If so, one would expect that dicoumarol should inhibit MC-induced AHM activity parallel to its inhibition of DT diaphorase. As shown below, this does not seem to be the case. However, active hydroxylation of BP by MC-induced microsomes leads to a decreased dicoumarol sensitivity of DT diaphorase, with a simultaneous inhibition of DT diaphorase by some metabolic product(s) of BP.

METHODS

Liver microsomes were isolated as described before (3) from MC-treated rats (2 daily i.p. injections of 20 mg MC/kg). AHM activity was measured according to DePierre et al. (6) using ^{3}H-BP as substrate. DT diaphorase activity was determined as previously described (3). The reaction mixture, in a final volume of 1 ml, consisted of 50 mM Tris chloride pH 7.5, 0.5 mM NADPH, 10 μM menadione, 77 μM cytochrome c, 0.5 mM KCN and 0.08 % Triton X-100. In some experiments Triton X-100 was replaced by the cationic detergent cetyltrimethyl ammonium bromide (Cetavlon), which greatly stimulates microsomal DT diaphorase while inhibiting NADPH-cytochrome c reductase (Fig. 1). The activity of NADPH-cytochrome c reductase and the microsomal content of cytochrome P-450 (P-448) were determined according to Dallner (7).

RESULTS AND DISCUSSION

The increase in DT diaphorase activity occurs in all cell fractions that contain the enzyme and seems to be due to an induced enzyme synthesis since cycloheximide, an inhibitor of protein synthesis (8), counteracts the increase in DT diaphorase as well as AHM activities and cytochrome P-450 content after MC-treatment (Table 1). Besides liver also lung, kidney and heart but not brain show an increased DT diaphorase activity after MC-treatment. The same respons was also observed for the AHM system except in heart, which did not show an increased AHM activity (Table 2). This would indicate a difference in gene expression between the inductions of DT diaphorase and the AHM system. Recently Kumaki et al. (9) have reported similar genetic differences in the mouse. Their results suggest that there are two loci which regulate DT diaphorase induction by MC, one gene being the same as for AHM or linked to that locus, the other unlinked on another chromosome.

As shown in Fig. 2, addition of 10^{-4} M dicoumarol gave only a weak (appr. 30 %) inhibition of BP hydroxylation. When aliquots were removed from the incubation media and diluted 10-fold with an assay mixture for DT diaphorase activity, the latter was about 80 % inhibited by the dicoumarol present (10^{-5}) at time 0. The DT diaphorase activity diminished during BP hydroxylation and so did also its sensitivity to dicoumarol; after 15 minutes of incubation, the decrease in activity was appr. 60 %, and the dicoumarol inhibition about 40 %.

The decrease in DT diaphorase activity did not occur in the absence of BP or when BP hydroxylation was inhibited by 7,8-benzoflavone or by treatment of the microsomes with trypsin (10,11) (Fig. 3). Under these conditions, the dicoumarol sensitivity of the enzyme also remained constant (not shown). These data suggest that the decrease in DT diaphorase activity may be due to some metabolic product(s) of BP, which also counteracts the sensitivity of the enzyme to dicoumarol. It should be added that the NADPH-cytochrome c reductase activity of the microsomes remained unchanged during active BP hydroxylation.

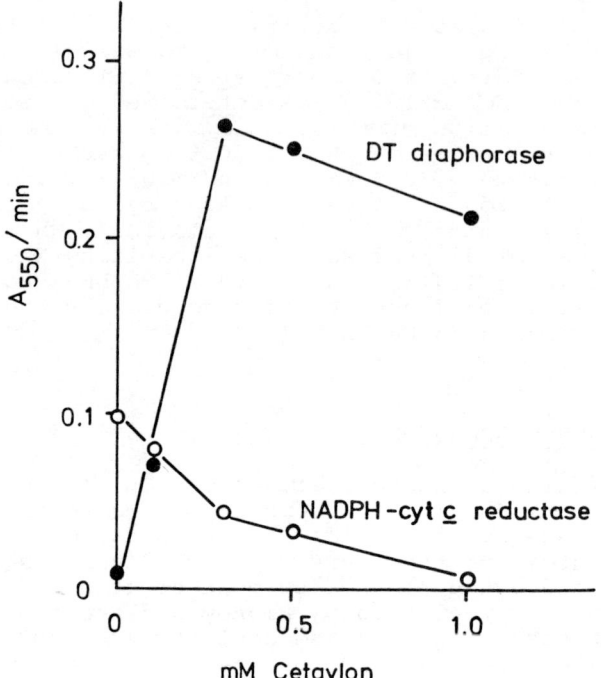

Fig. 1. Effect of Cetavlon on liver microsomal DT diaphorase and NADPH-cytochrome c reductase.

3-OH-BP, a metabolic product of BP (12), was found to inhibit microsomal DT diaphorase, but the inhibition was weaker than that observed with the endogeneously formed metabolites. This is shown in Fig. 4, which compares the effects of added 3-OH-BP and of endogenous metabolites as deduced from the data in Fig. 2. It appears from these results that the endogenous metabolite(s) responsible for the diminished DT diaphroase activity inhibit the enzyme more efficiently than does 3-OH-BP. Preliminary experiments indicate that further metabolism of 3-OH-BP by microsomes (13) leads to a diminished inhibition of DT diaphorase, and, furthermore, that 1,2-epoxy-3,3,3-trichloropropane, an inhibitor of epoxide hydrase (14), potentiates the inhibition of DT diaphorase occurring during BP hydroxylation. These data are consistent with the conclusion that a product of BP metabolism is the inhibitory metabolite. Attempts to identify this metabolite and to further characterize the relationship between DT diaphorase and the AHM system are in progress.

This work has been supported by a grant from the Swedish Medical Research Council.

TABLE 1 Effect of 3-Methylcholanthrene (MC) and Cycloheximide on DT Diaphorase and Aryl Hydrocarbon Monooxygenase Activities and Cytochrome P-450 (P-448) Content

The rats were injected i.p. with cornoil, NaCl, cycloheximide (1.5 mg/kg) and MC (20 mg/kg) 24 hours before sacrifice. A second injection of cycloheximide (0.75 mg/kg) was given 9 hours after the first one. DT diaphorase, AHM activities and cytochrome P-450 content were determined as described in METHODS. Each value is the average of those obtained with a group of 4 rats.

Treatment	DT Diaphorase					
	Cytosol	x Increase	Microsomes (nmoles/min/mg protein)	x Increase	Mitochondria	x Increase
Cornoil + NaCl	1830	1	28.1	1	462	1
Cornoil + cycloheximide	2632	1.4	28.3	1	639	1.4
MC + NaCl	4394	2.4	69.8	2.5	871	1.9
MC + cycloheximide	2478	1.3	31.5	1.1	546	1.2

Treatment	Aryl Hydrocarbon Monooxygenase		Cytochrome P-450 (P-448)	
	nmoles/min/mg protein	x Increase	nmoles/mg protein	x Increase
Cornoil + NaCl	0.27	1	0.28	1
Cornoil + cycloheximide	0.13	0.5	0.25	0.9
MC + NaCl	2.61	9.6	0.62	2.2
MC + cycloheximide	1.14	4.2	0.33	1.2

TABLE 2 Induction of DT Diaphorase and Aryl Hydrocarbon Monooxygenase in Various Tissues by Treatment of Rats with 3-Methylcholanthrene (MC)

Rats were treated for 2 days with a daily dose of 20 mg/kg MC. DT diaphorase and aryl hydrocarbon monooxygenase activities were measured in 10,000 x g supernatants of tissue homogenates. Each value is the average of those obtained with a group of 3 rats.

Expt. No.	Tissue	DT diaphorase (nmoles/min/mg protein)			Aryl hydrocarbon monooxygenase (nmoles/min/mg protein)		
		control	MC-treated	x increase	control	MC-treated	x increase
1	liver	1646	6944	4.2	0.062	1.13	18.2
	heart	65.9	202.9	3.1	0.029	0.029	1.0
	brain	186.2	191.5	1.0	0.024	0.019	0.8
2	liver	1600	8920	5.6	0.067	1.08	16.2
	lung	910	1881	2.1	0.0057	0.03	5.4
	kidney	130.4	319.8	2.5	0.026	0.111	4.3

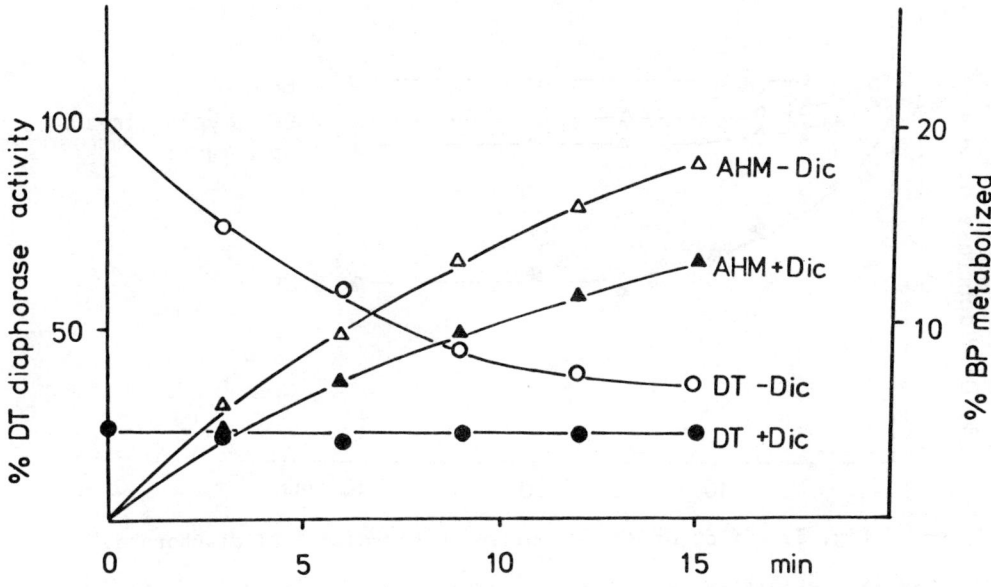

Fig. 2. Effect of dicoumarol on DT diaphorase and AHM.

The incubation mixture for BP metabolism was as described by DePierre et al. (6), except that the ^{3}H-BP used had a 10-fold higher specific activity. During incubation at 37°C, two aliquots of 100 μl were withdrawn at the times indicated in the figure, and the AHM and DT diaphorase activities were tested. In the case of DT diaphorase, the aliquot was added directly to the assay system (see METHODS), which also contained 1 mM Cetavlon, in order to inhibit NADPH-cytochrome c reductase. The dicoumarol concentration was 10^{-4} M in the incubation mixture for the measurement of AHM activity and 10^{-5} M in the DT diaphorase assay.

REFERENCES

(1) L. Ernster, L. Danielson, and M. Ljunggren, DT diaphorase I. Purification from the soluble fraction of rat-liver cytoplasm, and properties, Biochim. Biophys. Acta 58, 171 (1962).

(2) C. Huggins and R. Fukunishi, Induced protection of adrenal cortex against 7,12-dimethylbenz(a)anthracene. Influence of ethionine. Induction of menadione reductase. Incorporation of thymidine-H^3, J. Exptl. Med. 119, 923 (1964).

(3) C. Lind and L. Ernster, A possible relationship between DT diaphorase and the aryl hydrocarbon hydroxylase system, Biochem. Biophys. Res. Commun. 56, 392 (1974).

(4) L. Diamond and H.V. Gelboin, Alpha-naphtoflavone: An inhibitor of hydrocarbon cytotoxicity and microsomal hydroxylase, Science 166, 1023 (1969).

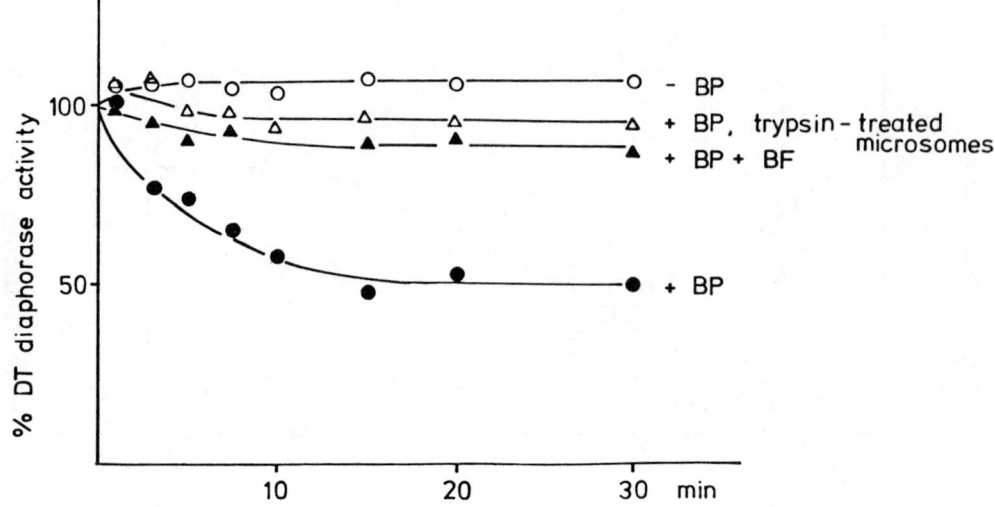

Fig. 3. Effect of BP metabolism on microsomal DT diaphorase.

Experimental conditions as in Fig. 2. The trypsin-treated microsomes had no measurable AHM activity. The 7,8-benzoflavone (BF) concentration used, 10^{-4} M, inhibited AHM 80 %.

(5) F.J. Wiebel, J.C. Leutz, L. Diamond, and H.V. Gelboin, Aryl hydrocarbon (benzo(a)pyrene) hydroxylase in microsomes from rat tissues: Differential inhibition and stimulation by benzoflavones and organic solvents, Arch. Biochem. Biophys. 144, 78 (1971).

(6) J.W. DePierre, M.S. Moron, K.A.M. Johannesen, and L. Ernster, A reliable and convenient radioactive assay for benzpyrene monooxygenase, Anal. Biochem. 63, 470 (1975).

(7) G. Dallner, Studies on the structural and enzymatic organization of the membranous elements of liver microsomes, Acta Pathol. Microbiol. Scand. Suppl. 166 (1963).

(8) S. Pestka, Inhibitors of ribosome functions, Ann. Rev. Biochem. 41, 697 (1971).

(9) K. Kumaki, N.M. Jensen, J.G.M. Shire, and D.W. Nebert, Genetic differences in induction of cytosol reduced-(NAD(P):menadione oxidoreductase and microsomal aryl hydrocarbon hydroxylase in the mouse, in press.

(10) A. Ito, and R. Sato, Proteolytic microdissection of smooth-surfaced vesicles of liver microsomes, J. Cell Biol. 40, 179 (1969).

(11) S. Orrenius, A. Berg, and L. Ernster, Effect of trypsin on the electron transport system of liver microsomes, Eur. J. Biochem. 11, 193 (1969).

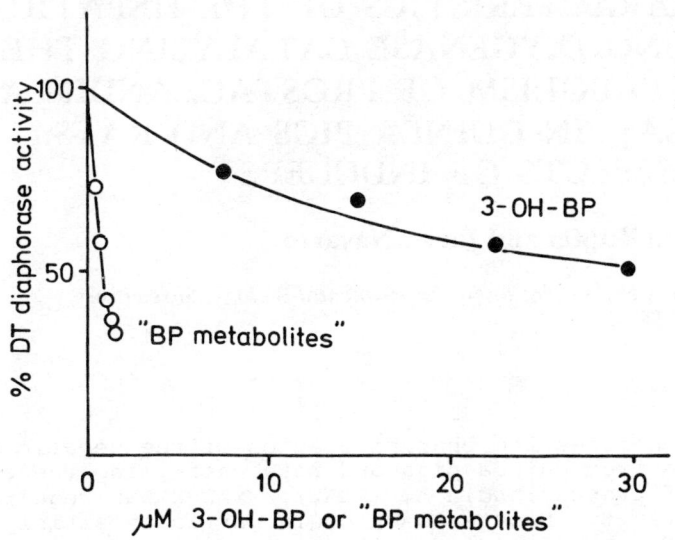

Fig. 4. Inhibition of DT diaphorase by BP metabolites.

The curve showing the inhibition of DT diaphorase by BP metabolites was deduced from Fig. 2, by plotting DT diaphorase activity against the concentration of BP metabolites found at different times, taken into account that the concentration of metabolites was diluted 10 times when DT diaphorase activity was measured. In the case of the curve marked 3-OH-BP, DT diaphorase was assayed as in Fig. 2. Varying concentrations of 3-OH-BP were added as indicated.

(12) A.H. Conney, E.C. Miller, and J.A. Miller, Substrate-induced synthesis and other properties of benzo(a)pyrene hydroxylase in rat liver, J. Biol. Chem. 228, 753 (1957).

(13) F.J. Wiebel, Metabolism of monohydroxybenzo(a)pyrenes by rat liver microsomes and mammalian cells in culture, Arch. Biochem. Biophys. 168, 609 (1975).

(14) F. Oesch, N. Kaubisch, D.M. Jerina, and J.W. Daly, Hepatic epoxide hydrase. Structure-activity relationships for substrates and inhibitors, Biochemistry 10, 4858 (1971).

CHARACTERISTICS OF THE HEPATIC MONOOXYGENASE CATALYZING THE METABOLISM OF PROSTAGLANDIN A_1 (PGA_1) IN GUINEA PIGS AND RATS. I. EFFECTS OF INDUCERS

David Kupfer and Javier Navarro

Worcester Foundation for Experimental Biology, Shrewsbury, MA 01545

SUMMARY

This study describes the characteristics of the hepatic microsomal monooxygenase from guinea pigs and rats catalyzing the oxidative metabolism of prostaglandin A_1 (PGA_1). NADPH was required for enzymatic activity, however NADH could partially satisfy this requirement. Furthermore at nonsaturating concentration of NADPH, the addition of NADH increased the oxidative metabolism of PGA_1, indicating additive effects. However, at saturating levels of NADPH, NADH had no effect, demonstrating absence of synergism. It was observed that PGA_1 stimulates the rate of NADPH oxidation by guinea pig liver microsomes and that this stimulation is inhibited by CO, pointing to the participation of cytochrome P-450 in this effect. The administration to rats of 3-methylcholanthrene (3MC), pregnenolone-16α-carbonitrile (PCN) or p,p'DDT enhanced hepatic microsomal PGA_1 metabolism; similarly, p,p'DDT enhanced the microsomal metabolism of PGA_1 in the guinea pig. Surprisingly, phenobarbital treatment only marginally enhanced the rate of PGA_1 metabolism in the rat. Kinetic studies showed that whereas treatment of rats with phenobarbital increased the apparent K_M of PGA_1, 3MC administration diminished this value. It was observed that the K_M values of PGA_1 in control rats and guinea pigs were similar. However, though the concentrations of cytochrome P-450 were similar in the two species, the rate of metabolism of PGA_1 was substantially higher in guinea pigs than in rats, suggesting that the P-450s catalyzing PGA_1 metabolism in the two species might be different or that the proportion of a "specific form" of P-450 involved in PGA_1 metabolism in guinea pigs is higher than in rats.

INTRODUCTION

Previous studies by Israelsson et al. (1) showed that liver microsomes from guinea pigs, but not from rats, catalyze the ω-1 (19-) and ω(20-) hydroxylation of prostaglandin A_1 (PGA_1). However under somewhat different conditions, we observed that rat liver microsomes also oxidize PGA_1 to polar metabolites, having the same chromatographic characteristics as metabolites derived from guinea pigs (2). Furthermore we demonstrated in rats and guinea pigs that the hepatic microsomal activity responsible for the oxidative metabolism of PGA_1 is a typical monooxygenase (2,3). The present study examines further the characteristics of the monooxygenase system catalyzing PGA_1 metabolism. We also investigated the effects of inducers of monooxygenase on the oxidative metabolism of PGA_1 in guinea pigs and rats.

MATERIALS AND METHODS

Compounds

The various compounds utilized in this study were obtained from sources previously described (2,3). PCN was a gift from Searle Laboratories. The following were purchased: Sodium phenobarbital (Mallinckrodt); 3-MC (Eastman); p,p'DDT (Aldrich).

Animals

Male rats weighing 90-100 g and male albino Hartley strain guinea pigs 350-400 g were obtained from Charles River Breeding Labs and from Elm Hill Farm, respectively.

Preparation of Microsomes, Incubations and Isolation of Metabolites

Carried out as previously reported (2), except that incubation time was usually 1/2 hour and 1 hour with guinea pig and rat liver microsomes, respectively.

Administration of Inducers

Rats were injected i.p.: sodium phenobarbital, dissolved in water, was injected at 75 mg/kg (in divided doses - bidaily) for 4 days; p,p'DDT was injected at 35 mg/kg/day in corn oil for 3 days; PCN, suspended in water with minimal amount of Tween 80, was injected at 50 mg/kg/day for 3 days; 3-MC was injected at 25 mg/kg in corn oil for 2 days. Guinea pigs were injected i.p. with p,p'DDT at 100 mg/kg/day for 7 days. Controls received the respective vehicle.

RESULTS AND DISCUSSION

The pyridine nucleotide requirement for PGA_1 metabolism by liver microsomes from guinea pigs and rats was determined (Table 1). In guinea pigs, at saturating concentrations of NADH, the activity was 31% of that observed with NADPH. Furthermore, additive effects were observed when NADPH was at nonsaturating (1 mg) levels and NADH was saturating (2.5 mg), yielding 86% activity as compared with incubations containing saturating levels of NADPH. Similarly in incubations with rat liver microsomes NADH could partially satisfy the NADPH requirement, yielding 42% activity (Table 1). Again additive effects were observed when NADPH was utilized at nonsaturating concentrations. By contrast to certain previous findings with other substrates (9,10), the addition of NADH to incubations containing saturating concentrations of NADPH did not enhance the enzymatic activity, indicating the absence of a synergistic effect. It is of interest that similarly to our findings with PGA_1, in the hydroxylation of another fatty acid

TABLE 1 Additive Effects of NADH and NADPH in Metabolism of PGA_1 by Guinea Pig and Rat Liver Microsomes

	NADPH (mg)	NADH (mg)	Activity[a] (nmoles formed/mg prot.)	% Control
Guinea pig	2.5	0	7.7	100
	1.0	0	4.4	57
	0	2.5	2.4	31
	2.5	2.5	7.9	102
	1.0	2.5	6.6	86
Rat	5	0	4.0	100
	2	0	2.7	67
	0	5	1.7	42
	5	5	4.5	112
	2	5	3.9	98

[a] Values are for 45 min and 60 min for guinea pig and rat incubations, respectively. Reactions were carried out in duplicate.

Guinea pig: 2.5 mg of NADPH or NADH was demonstrated to be saturating and 1 mg of NADPH was not saturating. The concentration of microsomal protein was 5.9 mg/ml.

Rat: 5 mg of NADPH or NADH was demonstrated to be saturating and 2 mg of NADPH was not saturating. The concentration of protein was 6.2 mg/ml.

(laurate) by rat kidney cortex microsomes, NADH was about 50% as active as NADPH and additive effects by these nucleotides were observed (4).

Previous studies demonstrated that substrates of hepatic monooxygenase stimulate the rate of oxidation of NADPH and in certain cases stoichiometry between NADPH oxidation and oxidative metabolism of a given compound was demonstrated (5,6,7,8). Thus, we examined whether PGAs stimulate NADPH oxidation by guinea pig liver microsomes (Table 2). As can be seen, PGA_1 stimulated NADPH oxidation, though to a lesser extent than did benzphetamine. Furthermore, carbon monoxide inhibited the stimulation of NADPH oxidation by either PGA_1 or benzphetamine suggesting that the stimulation of NADPH oxidation by these compounds is mediated by cytochrome P-450. When similar experiments were performed with rat liver microsomes there was no stimulation of NADPH oxidation by PGA_1. This lack of effect is presumably due to a substantially lower metabolism of PGA_1 by rat liver microsomes as compared with preparations derived from guinea pigs. The findings in the guinea pig indicated that if a stoichiometric relationship between PGA_1 metabolism and NADPH oxidation is demonstrated, it will be possible to monitor PGA_1 metabolism by merely following NADPH oxidation.

TABLE 2 Stimulation of NADPH Oxidation by
PGA_1 in Guinea Pig Liver Microsomes and the
Effect of Carbon Monoxide

	NADPH oxidat. (nmoles/min/mg prot.)	Increase (%)
Control	5.7	–
PGA_1 (1.2 x 10^{-4} M)	8.6	51
Benzphetamine (3.3 x 10^{-4} M)	19.7	246
Control + *CO	4.1	–
PGA_1 + *CO	4.8	17
Benzphetamine + *CO	9.7	136

*The gas mixture consisted of: $CO:O_2:N_2$ (5:1:4) versus $O_2:N_2$ (1:9) in control.

Each cuvette contained 2 mg of microsomal protein per 3 ml and NADPH (90 µM). NADPH was assayed at 340 nm. Experiment was carried out at 37°C.

In subsequent experiments we examined the effects of various inducers of hepatic monooxygenase on the in vitro PGA_1 metabolism. The regimen of treatment utilized has been previously established to be favorable for induction of hepatic monooxygenase, however we did not explore whether the treatment was also optimal for stimulation of PGA_1 metabolism. Results demonstrate that the various types of inducers: p,p'DDT, PCN and 3-methylcholanthrene (3MC) stimulate PGA_1 metabolism (Table 3). The most pronounced induction in the rat appeared to be with p,p'DDT. Similarly, the guinea pig which has higher basal PGA_1-metabolizing activity, exhibited a pronounced induction by p,p'DDT. Phenobarbital treatment was only marginally effective in elevating PGA_1 metabolism in the rat. The reason for the lack of activity of phenobarbital is not known. However it is possible that the various inducers utilized by us stimulate different pathways of PGA_1 hydroxylation and that phenobarbital enhances the less pronounced route; i.e. the ω-1 hydroxylation (1). That such preference might occur was shown by Orrenius (personal comm.) who observed that phenobarbital treatment of rats enhanced the hepatic microsomal ω-1, but not the ω-,hydroxylation of laurate. Whether the induction of phenobarbital or by other inducers stimulates the formation of one metabolite of PGA_1 in preference to another is currently not known. This study must await the determination whether in the rat, as is the case in the guinea pig, there are two PGA_1 metabolites (19- and 20-hydroxy). Nevertheless, despite the lack of this information it was of interest to determine whether the various inducers alter the kinetic characteristics of the microsomal metabolism of PGA_1. We observed that the treatment of rats with 3-MC lowered the K_M from 2.0×10^{-4} M to 1.0×10^{-4} M (Fig. 1 and Table 4). On the other hand, phenobarbital treatment seems to increase the apparent K_M. It is also of interest that though the K_M values for PGA_1 in liver microsomes

TABLE 3   Effect of Inducers of Hepatic Monooxygenase on PGA_1 Metabolism in Rats and Guinea Pigs

Species	Inducer	PGA_1 Metabolites (nmoles/mg/protein)[a]	Increase (%)
Rat	--	2.6 ($\pm$ 0.5)	
	p,p'DDT	5.1 ($\pm$ 0.5)	96*
Rat	--	2.9 ($\pm$ 0.3)	
	PCN	4.1 ($\pm$ 0.3)	41**
Rat	--	2.6 ($\pm$ 0.25)	
	3-MC	4.5 ($\pm$ 0.3)	73[†]
Guinea Pig	--	5.7 ($\pm$ 0.4)	
	p,p'DDT	15.3 ($\pm$ 1.7)	168[†]

*$P \leq 0.025$;   **$P \leq 0.05$;   [†]$P \leq 0.005$

Rats: Males (95-100 g) were injected i.p. with: p,p'DDT (35 mg/kg/day) or with PCN (50 mg/kg/day) for 3 days and sacrificed 24 hours later; rats receiving 3-MC (25 mg/kg/day) were injected for 2 days and sacrificed 48 hours later.

Guinea Pigs: Males (350-400 g) were injected i.p. with p,p'DDT in corn oil (100 mg/kg/day) for 7 days and sacrificed 48 hours after last dosing.

[a]Values are per 60 min for rats and per 30 min for guinea pigs.

TABLE 4   Michaelis Constants of PGA_1 in Hepatic Microsomes from Male Guinea Pigs and Rats

Treatment	Guinea Pig	Rat
	K_M (10^{-4} M)	
---	2.1	2.0*
Phenobarbital		3.6
3-MC		1.0

*The same value for K_M was obtained independently of the vehicle injected: H_2O or corn oil.

Guinea pigs weighing 350-400 grams were used. Rats (90-100 grams) were injected with phenobarbital (75 mg/kg/day in divided doses) for 4 days or with 3-methylcholanthrene (3-MC) (25 mg/kg/day) for 2 days.

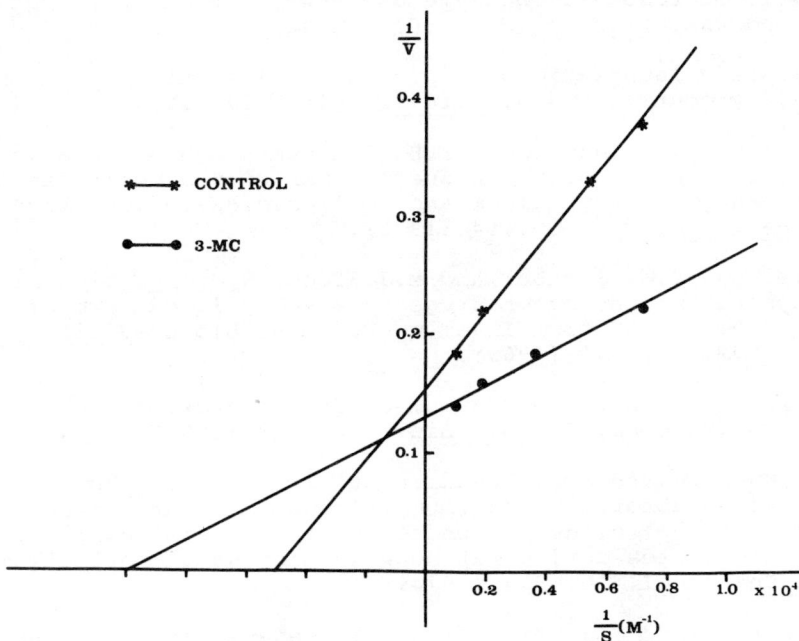

Fig. 1. Kinetic studies on the effect of 3-MC administration on the metabolism of PGA_1 by rat liver microsomes. A double reciprocal plot of the velocity of PGA_1 oxidative metabolism versus concentration of PGA_1.

from guinea pigs and rats were similar, the rate of metabolism of PGA_1 was substantially higher with microsomes from guinea pigs than from rats. Since the concentration of cytochrome P-450 in microsomes from these species was found to be similar, being 0.65 ± 0.02 and 0.74 ± 0.07 nmoles/mg protein in the guinea pig and rat, respectively, it is likely that the respective cytochrome P-450s are different and that the cytochrome in the guinea pig is more active towards PGA_1 metabolism than in the rat. These findings may also be due to a higher proportion of a "specific" form of P-450 of the total P-450s, catalyzing PGA_1 oxidation in guinea pigs than in rats.

ACKNOWLEDGEMENT

This study was supported by NIH grants GM22688 and ES00834.

REFERENCES

(1) U. Israelsson, M. Hamberg and B. Samuelsson, Biosynthesis of 19-hydroxy-prostaglandin A_1, Europ. J. Biochem. 11, 390 (1969).

(2) D. Kupfer and J. Navarro, Metabolism of prostaglandin A_1 by hepatic microsomal monooxygenase P-450 system in the guinea pig and rat, Life Sci. 18, 507, (1976).

(3) D. Kupfer, Interactions of prostaglandins with hepatic microsomal cytochrome P-450, Life Sci. 15, 657 (1974).

(4) A. Ellen, S.V. Jakobsson, J.B. Schenkman and S. Orrenius, Cytochrome P_{450K} of rat kidney cortex microsomes: Its involvement in fatty acid ω and (ω-1)-hydroxylation, Arch. Biochem. Biophys. 150, 64 (1972).

(5) A.Y.H. Lu, H.W. Strobel and M.J. Coon, Hydroxylation of benzphetamine and other drugs by a solubilized form of cytochrome P-450 from liver microsomes, Biochem. Biophys. Res. Comm. 36, 545 (1969).

(6) S. Orrenius, On the mechanism of drug hydroxylation in rat liver microsomes, J. Cell Biol. 26, 713 (1965).

(7) M.K. Buening and M.R. Franklin, Limitations in the use of the 340 nm absorbance maximum of reduced nicotinamide adenine dinucleotide phosphate for the determination of oxidation rates during rat hepatic microsomal metabolism, Mol. Pharmacol. 10, 999 (1974).

(8) E. Jeffery and G.J. Mannering, Discrepancy in the measurement of reduced TPN oxidized during ethylmorphine N-demethylation due to the presence of a nucleotide pyrophosphatase, Mol. Pharmacol. 10, 1004 (1974).

(9) B.S. Cohen and R.W. Estabrook, Microsomal electron transport reactions, Arch. Biochem. Biophys. 143, 46 (1971).

(10) A. Hildenbrandt and R.W. Estabrook, Evidence for the participation of cytochrome b_5 in hepatic microsomal mixed-function oxidation reactions, Arch. Biochem. Biophys. 143, 66 (1971).

STUDIES ON A CYTOCHROME P-450-DEPENDENT HYDROXYLASE SYSTEM ACTIVE ON STEROIDS IN *Bacillus megaterium*

Anders Berg, Kjell Carlström,* Magnus Ingelman-Sundberg, Joseph Rafter and Jan-Åke Gustafsson

Department of Chemistry, Karolinska Institutet, S-10401 Stockholm 60, Sweden

*Present address: Department of Obstetrics and Gynecology, Sabbatsberg Hospital, S-11382 Stockholm, Sweden

ABSTRACT

The steroid hydroxylase system in Bacillus megaterium that hydroxylates 3-oxo-Δ^4-steroids primarily in position 15β has been resolved into an NADPH-specific FMN-containing reductase (megaredoxin reductase), an iron-sulphur protein (megaredoxin) and cytochrome P-450$_{meg}$. Megaredoxin reductase can be replaced by rabbit liver NADPH-cytochrome P-450 reductase. In reactions supported by oxygen-donating agents as NaIO$_4$ or iodosobenzene, hydroxylation occurs in both position 15α and 15β. Using hydrophobic chromatography cytochrome P-450$_{meg}$ has been purified to 20% homogeneity. This protein has an isoelectric point of 4.9 and a molecular weight of approximately 37 000 D. Phosphatidylcholine stimulated NaIO$_4$-supported, P-450$_{meg}$-catalyzed hydroxylation reactions by a factor of two.

INTRODUCTION

A major advantage of studying mechanisms of cytochrome P-450-dependent hydroxylations in bacterial systems is that the hydroxylase systems are soluble. Two different bacterial P-450-dependent hydroxylase systems have been characterized previously: the camphor hydroxylase system in Pseudomonas putida (1,2) and the n-octane hydroxylating system in Corynebacterium (3,4). Both systems contain an NADH-specific FAD-containing reductase which donates electrons to an iron-sulphur protein that serves as electron donor to cytochrome P-450. Cytochrome P-450$_{cam}$ and its mechanism of action have been extensively studied by Gunsalus and coworkers (1,2). In addition to the bacterial systems, a soluble type of P-450 (P-450$_c$) from Rhizobium japonicum, has been purified to homogeneity and its amino acid composition determined (5).

In view of our current studies on mechanisms involved in steroid hydroxylations (6-8) we have tried to find a suitable bacterial cytochrome P-450-dependent steroid hydroxylase system as a model system for these studies. We have found it advantageous to use the steroid hydroxylase system in Bacillus megaterium and have resolved it into an NADPH-specific FMN-containing reductase (megaredoxin reductase), an iron-sulphur protein (megaredoxin) and cytochrome P-450$_{meg}$ (cf. Fig. 1) (9,10). The present paper describes the purification of cytochrome P-450$_{meg}$ to 20% homogeneity and some characteristics of this protein.

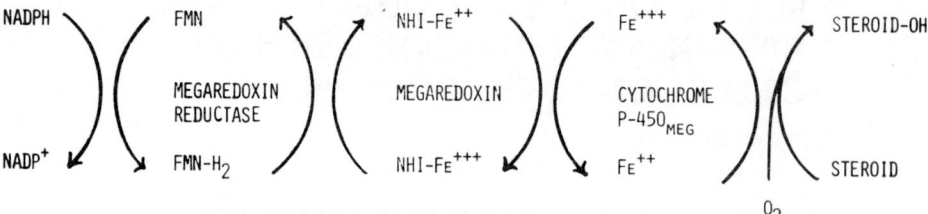

Fig. 1. Electron transport in the steroid hydroxylase system of B. megaterium.

MATERIALS AND METHODS

Sources for most materials have been given in a previous paper (10). Sephadex G-75, superfine, and Octyl-Sepharose were obtained from Pharmacia Fine Chemicals, Uppsala, Sweden. Ampholine^R was purchased from LKB-Produkter, Bromma, Sweden. Bacillus megaterium ATCC 13368 was obtained from the American Type Culture Collection, Rockville, Md., USA.

The organism was cultivated at 28 C in a medium containing 20 g Bacto-yeast extract and 1.6 g Bacto-soybean extract (Difco Laboratories) per l. In recent experiments 40 g coconut oil (Kebo AB, Spånga, Sweden), 2.5 g NH_4Cl, 15 g $Na_2HPO_4 \cdot 2H_2O$, 6 g KH_2PO_4, 0.5 g $Na_2SO_4 \cdot 10\ H_2O$, 0.2 g $MgSO_4 \cdot 7\ H_2O$ and 1 ml of Tween 80 was added per l of medium, also containing 10 g Bacto-yeast extract and 16 g Bacto-soybean extract, and the cultivation temperature was kept at 32 C. Cultivation was performed for 48-55 h in a 10 l Biotec fermentor (LKB-Biotec AB, Sundbyberg, Sweden) with 8 l of medium at a pH of 7.2 ± 0.2 with an air supply of 2 l per l of medium and min. At the end of cultivation the cells were centrifuged at 6 000 x g and stored at -20 C. Cell-free extracts were prepared by sonication as described previously (10).

Assay methods. All spectrophotometric measurements were performed with a Cary model 118 spectrophotometer. Cytochrome P-450 was assayed according to Omura and Sato (11) and hydroxylations of progesterone were assayed as described previously (10). SDS-polyacrylamide gel electrophoresis was performed essentially according to Weber and Osborn (13).

RESULTS

Purification of cytochrome P-450$_{meg}$. Cytochrome P-450$_{meg}$ was first purified from the cell-free extract of B. megaterium, using chromatography on DEAE-cellulose and gel filtration on Ultrogel ACA-54 to a concentration of 1.3 nmoles per mg of protein as described earlier (10). The cytochrome P-450 fraction from the Ultrogel column was further purified by hydrophobic chromatography on Octyl-Sepharose. An aliquot fraction corresponding to 10 nmoles of P-450$_{meg}$ was made 24% with respect to ammonium sulphate. After stirring for 30 min the fraction was centrifuged at 6 000 x g for 15 min. The supernatant containing all of the P-450$_{meg}$ was applied to an Octyl-Sepharose column equilibrated with 50 nM phosphate buffer containing 25% ammonium sulphate. Non-absorbed proteins were eluted with 10 ml of 25% ammonium sulphate containing phosphate buffer before the proteins were eluted with a linear gradient of decreasing ammonium sulphate concentration from 25 to 17% and increasing concentration of polyethylene glycol 6 000 from 0 to 8% (see Fig. 2). The procedure gave about four times purification of P-450$_{meg}$ and a concentration of about 5 nmoles of P-450$_{meg}$ per mg of protein was obtained (cf. Table I).

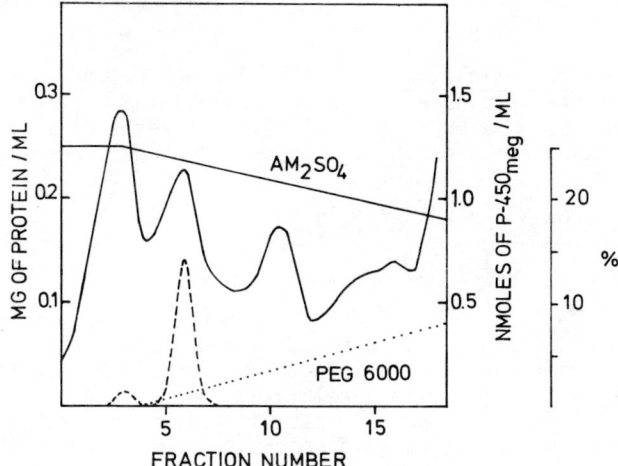

Fig. 2. Purification of cytochrome P-450$_{meg}$ by hydrophobic interaction chromatography on Octyl-Sepharose CL-4B. The supernatant obtained after fractionation of 10 nmoles of P-450$_{meg}$ (obtained from the Ultrogel chromatography) with 25% ammonium sulphate was applied to an Octyl-Sepharose column (0.3 x 3 cm) equilibrated with 50 mM phosphate buffer, pH 7.4, containing 25% ammonium sulphate. The column was eluted with linear gradients of decreasing concentration of ammonium sulphate and increasing concentration of polyethylene glycol 6 000. Protein concentration is shown by solid line, P-450$_{meg}$ concentration by dashed line.

Table I Purification of cytochrome P-450$_{meg}$

Preparation	Protein mg	nmol P-450/ mg protein	Purification factor	Yield %
Cell-free extract	3840	0.008	1	100
Cell-free extract from bacillae grown on coconut fat	2600	0.016	2	100
0.1 M KCl fraction from DEAE-cellulose	660	0.32	40	75
Fraction from Ultrogel ACA-54	150	1.3	163	53
Fraction from hydrophobic chromatography	60	5.0	625	20

Sucrose density gradient centrifugation. $P-450_{meg}$ obtained from the Ultrogel column was subjected to sucrose density gradient centrifugation against standard proteins (cytochrome c, 1.7 S, hemoglobin, 2.8 S, and bovine serum albumin, 4.8 S) in 5-20% gradients. $P-450_{meg}$ sedimented as a homogeneous zone with a sedimentation coefficient of 3.3 S.

Molecular weight determination. When the Ultrogel column was calibrated with cytochrome c (Stokes radius a = 17 Å), hemoglobin (a = 24 Å), ovalbumin (a = 27 Å) and bovine serum albumin (a = 35 Å), a Stokes radius of 27 Å was obtained for $P-450_{meg}$. Calculation of the molecular weight was performed according to the formula

$$M = \frac{6\pi N \times a \times s_{20,w} \times \eta_{20,w}}{1 - \bar{v} \, \varsigma_{20,w}}$$

where $\eta_{20,w}$ and $\varsigma_{20,w}$ is the viscosity and density, respectively, of water at 20 C, N Avogadro's number, a Stokes radius and $\bar{v}$ the partial specific volume of the protein. Different types of P-450 have been reported to have similar amino acid composition (5) with a relatively high content of hydrophobic amino acids. In view of this the partial specific volume of $P-450_{meg}$ was calculated (14) on the basis of the amino acid composition given for $P-450_{cam}$ and for $P-450_c$ from R. japonicum; thus, a $\bar{v}$ value of 0.735 was used. The molecular weight of $P-450_{meg}$ was then calculated to be 37 000 D.

Isoelectric focusing of $P-450_{meg}$. Isoelectric focusing was performed on flat beds of Sephadex G-75 superfine as described elsewhere (15). 3 Nmoles of $P-450_{meg}$ obtained from the Ultrogel column were desalted on a Sephadex G-25 column equilibrated with 5 mM phosphate buffer, pH 7.4. The sample was applied at a pH of about 7, to plates 3 x 25 cm with a 3 mm thick layer of Sephadex containing 1% Ampholine pH 3.5 - 10 that had been prefocused for 2 h. The plate was focused at 170 V/cm for 20 h. A piece of filter paper corresponding in size to the plate was applied on the gel. When the paper strip was saturated it was dried and stained with 0.2% Coomassie brilliant blue and then immediately destained in water:ethanol:HAc, 87:5:7 (by vol.), at 100 C for 5 - 10 min. pH was measured on the gel with a surface pH-electrode. As shown in Fig. 3, a zone containing all $P-450_{meg}$ applied was detected at a pH of 4.9. SDS-polyacrylamide gel electrophoresis of this zone revealed a few contaminating proteins.

Catalytic activity of $P-450_{meg}$. In NADPH-supported, $P-450_{meg}$-catalyzed reactions, progesterone and several 3-oxo-Δ^4-steroids were hydroxylated in position 15β and to a minor extent in position 6β. In oxidizing agent-supported reactions (with direct activation of P-450; no electron donors present in the incubation mixture (16)), hydroxylations occurred in both position 15β and 15α (cf. Table II). With different oxidizing agents the ratio between 15β- and 15α-hydroxylated products varied somewhat. The highest ratio was obtained in ClO_2^--supported reactions (3.5), while a significantly lower value was obtained with IO_4^- as hydroxylating agent (1.2).

Role of phospholipid in $P-450_{meg}$-catalyzed reactions. In the liver microsomal hydroxylase system it is known that addition of phospholipid to oxidizing agent-supported reactions enhance the velocity of the reaction by a factor of about two (17). Since $P-450_{meg}$ is a soluble protein it was of interest to investigate possible effects of phospholipid in this system. $P-450_{meg}$, 0.33 nmole, was incubated with 5 mM $NaIO_4$, 32 nmoles of progesterone and increasing amounts

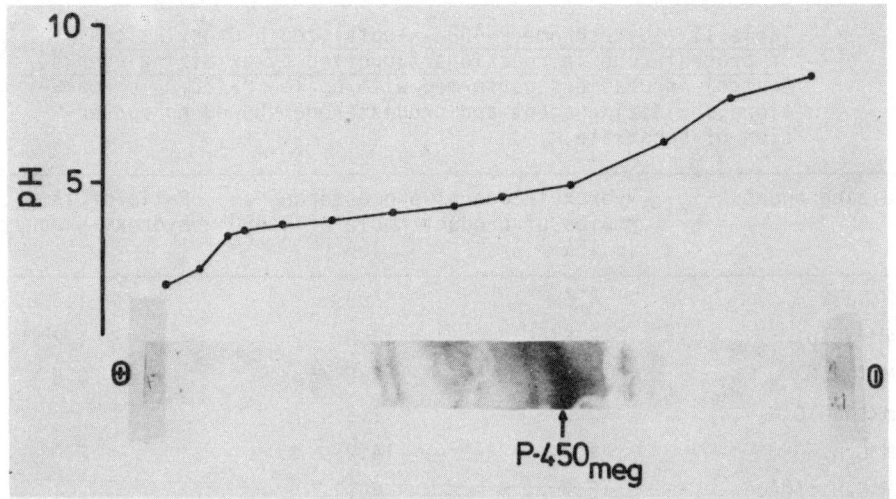

Fig. 3. Isoelectric focusing of P-450$_{meg}$ preparation obtained from the Ultrogel column. P-450$_{meg}$, 4 nmoles, was desalted on a column of Sephadex G-25 equilibrated with 5 mM phosphate buffer, pH 7.4. Electrofocusing was performed on flat beds of Sephadex G-75. The different protein zones were eluted from the gel with 50 mM phosphate buffer, pH 7.4. The eluates were assayed for cytochrome P-450.

of phosphatidylcholine. As shown in Table III a two-fold stimulation of progesterone hydroxylase activity was obtained with a saturation concentration of about 10 µg of phosphatidylcholine in the incubation mixture, i.e. with about 40 nmoles of phospholipid per nmole of P-450. A constant ratio of 15β-/15α-hydroxylated products was obtained at all concentrations of phospholipid.

Replacement of electron carriers in the cytochrome P-450$_{meg}$ hydroxylase system. In order to evaluate the specificity of various electron donors to P-450$_{meg}$, NADPH-cytochrome P-450 reductase was purified to a specific activity of 1 000 U per mg of protein from phenobarbital-induced rabbit liver according to the method of Coon (18) and experiments were performed to replace megaredoxin and megaredoxin reductase by rabbit liver reductase. P-450$_{meg}$, 0.25 nmole, was incubated with 32 nmoles of progesterone and either with 160 DCIP unit of megaredoxin reductase, megaredoxin corresponding to 34 nmoles of acid labile sulphur and 0.5 mg NADPH or with 370 units of NADPH-cytochrome P-450 reductase from rabbit liver and megaredoxin (in amounts as mentioned) or with 370 units of the rabbit liver reductase alone. Full activity, i.e. about 10 nmoles of 15β-hydroxyprogesterone formed per nmole of P-450, was obtained in the first two types of incubations, while no activity was found when the rabbit liver reductase replaced both of the natural electron carriers.

DISCUSSION

In a previous paper we have described the resolution of the steroid hydroxylase system in B. megaterium into an FMN-containing NADPH-specific reductase (megaredoxin reductase), an iron-sulphur protein (megaredoxin) and cytochrome

Table II Cytochrome P-450$_{meg}$-catalyzed hydroxylations of progesterone in reactions supported by oxidizing agents. Control incubations performed with boiled P-450$_{meg}$ preparations, oxidizing agent and progesterone showed no conversion of substrate.

Oxidizing agent	Hydroxylation of progesterone nmoles of product/nmole of P-450		Ratio of 15β-/15α- hydroxy-progesterone
	15α	15β	
2 mM NaIO$_4$	4.6	5.8	1.3
20 mM NaIO$_4$	2.6	3.0	1.2
2 mM NaClO$_2$	0.7	2.5	3.6
20 mM NaClO$_2$	0	1.5	-
5 mM C$_6$H$_5$IO	4.7	14.2	3.0
5 mM C$_6$H$_5$I(OAc)$_2$	2.9	6.1	2.1
5 mM o-NO$_2$C$_6$H$_4$IO	8.6	18.7	2.2
5 mM m-NO$_2$C$_6$H$_4$IO	4.8	11.2	2.3
5 mM o-NO$_2$C$_6$H$_4$(OAc)$_2$	7.1	17.2	2.4
5 mM m-NO$_2$C$_6$H$_4$(OAc)$_2$	2.2	3.9	1.8
5 mM C$_6$H$_5$I(OCOCF$_3$)$_2$	5.3	12.5	2.4

Table III Effect of addition of phosphatidylcholine (PC) on P-450$_{meg}$-catalyzed hydroxylations of progesterone supported by NaIO$_4$. The incubation mixture contained 0.33 nmole of P-450$_{meg}$, 5 mM NaIO$_4$, 32 nmoles of progesterone and increasing amounts of PC in a total volume of 2 ml 50 mM phosphate buffer, pH 7.4. Incubations were performed at 37 C for 5 min.

Amount of PC μg	Hydroxylation of progesterone nmoles of product/nmole of P-450		Ratio of 15β-/15α- hydroxyprogesterone
	15α	15β	
0	6.0	9.2	1.51
2	7.8	12.0	1.53
5	9.9	15.1	1.53
10	10.2	16.1	1.58
100	10.6	16.0	1.48

P-450$_{meg}$, using a two-step procedure by ion exchange chromatography on DEAE-cellulose and gel filtration on Ultrogel ACA-54 (10). By this procedure the proteins were purified between 100 - 200 times. By further use of hydrophobic chromatography, $P-450_{meg}$ has now been purified to about 5 nmoles per mg of protein, i.e. to about 20% homogeneity.

NADPH-supported, cytochrome $P-450_{meg}$-catalyzed hydroxylation of progesterone mainly resulted in the formation of 15β-hydroxyprogesterone whereas oxidizing agent-supported reactions gave rise to both 15β- and 15α-hydroxyprogesterone. This change in steric specificity which we have not seen when comparing NADPH- and oxidizing agent-supported steroid hydroxylation reactions in rat liver microsomes (16) could have several explanations. Hydroxylating agents are known to compete with the substrate for its binding site on P-450 and consequently the affinity for the steroid decreases considerably (16), which could result in changed steric specificity in the hydroxylation reaction. Putidaredoxin has recently been shown to act as an effector in hydroxylation reactions catalyzed by cytochrome $P-450_{cam}$ and an effector role of megaredoxin resulting in changed conformation and specificity of $P-450_{meg}$ cannot be ruled out. It is furthermore possible that B. megaterium, as is the case with R. japonicum, contains multiple forms of P-450. Such forms may differ in specificity but would conceivably be difficult to separate because of similar size and isoelectric point (4.9). Experiments are in progress to investigate these alternative possibilities.

A two-fold enhancement of $NaIO_4$-supported hydroxylations of progesterone was obtained by adding phosphatidylcholine to the incubation mixture at a molar ratio of P-450:PC of about 40. It is well known that the membrane-bound liver microsomal P-450 is dependent on phospholipid for activity (18,19) and a two-fold increase in hydroxylation activity has also been described by lipid addition to oxidizing agent-supported hydroxylation reactions catalyzed by highly purified liver microsomal P-450 (17). However, $P-450_{meg}$ is a soluble protein and therefore the stimulatory effect on enzyme activity by phosphatidylcholine was unexpected. It has been reported that the lipid environment affects substrate binding to adrenal cortex microsomal P-450 (20) and it is likely that the lipid in the case of $P-450_{meg}$ may act by binding to hydrophobic parts of the cytochrome, thereby facilitating interactions between the hydrophobic substrate and $P-450_{meg}$. This may be a general mechanism of lipid action also in reactions catalyzed by microsomal types of P-450.

ACKNOWLEDGEMENTS

We are indebted to Professor Anders Ehrenberg and Dr. Dan Bäckström for carrying out EPR analyses on megaredoxin and to Professor Tord Holme for advice and facilities in cultivation of B. megaterium.

This work was supported by grants from the Swedish Medical Research Council (No. 2819), Karolinska Institutets fonder and from Magnus Bergvalls Stiftelse.

REFERENCES

1. Katagiri, M., Ganguli, B.N., and Gunsalus, I.C., A soluble cytochrome P-450 functional in methylene hydroxylation, J. Biol. Chem. 243, 3543 (1968).
2. Yu, C.-A., Gunsalus, I.C., Katagiri, M., Suhara, K., and Takemori, S., Cytochrome $P-450_{cam}$. I. Crystallization and properties, J. Biol. Chem. 249, 94 (1974).
3. Cardini, G., and Jurtshuk, P., Cytochrome P-450 involvement in the oxidation of n-octane by cell-free extracts of Corynebacterium sp. Strain 7 E1C, J. Biol. Chem. 243, 6070 (1968).
4. Cardini, G., and Jurtshuk, P., The enzymatic hydroxylation of n-octane by Corynebacterium sp. Strain 7 E1C, J. Biol. Chem. 245, 2789 (1970).

5. Dus, K., Goewert, R., Weaver, C.L., Carey, D., and Appleby, C.A., P-450 hemoproteins of Rhizobium japonicum. Purification by affinity chromatography and relationship to P-450$_{cam}$ and P-450LM$_2$, Biochem. Biophys. Res. Commun. 69, 437 (1976).
6. Berg, A., and Gustafsson, J.-Å., Regulation of hydroxylation of 5α-androstane-3α,17β-diol in liver microsomes from male and female rats, J. Biol. Chem. 248, 6559 (1973).
7. Gustafsson, J.-Å. and Ingelman-Sundberg, M., Regulation and properties of a sex-specific hydroxylase system in female rat liver microsomes active on steroid sulphates. I. General characteristics, J. Biol. Chem. 249, 1940 (1974).
8. Ingelman-Sundberg, M., Rane, A., and Gustafsson, J.-Å., Properties of hydroxylase systems in the human fetal liver active on free and sulphoconjugated steroids, Biochemistry 14, 429 (1975).
9. Berg, A., Carlström, K., Gustafsson, J.-Å., and Ingelman-Sundberg, M., Demonstration of a cytochrome P-450-dependent steroid 15β-hydroxylase system in Bacillus megaterium, Biochem. Biophys. Res. Commun. 66, 1414 (1975).
10. Berg, A., Carlström, K., Gustafsson, J.-Å., and Ingelman-Sundberg, M., Characterization of a cytochrome P-450-dependent steroid hydroxylase system present in Bacillus megaterium, J. Biol. Chem. 251, 2831 (1976).
11. Omura, T., and Sato, R., The carbon monoxide-binding pigment of liver microsomes.I. Evidence for its hemoprotein nature. J. Biol. Chem. 239, 2370 (1964).
12. Lowry, O.H., Rosebrough, N.J., Farr, A.L., and Randall, R.I., Protein measurement with the folin-phenol reagent. J. Biol. Chem. 193, 265 (1951).
13. Weber, K., and Osborn, M., The reliability of molecular weight determination by dodecylsulfate-polyacrylamide gel electrophoresis, J. Biol. Chem. 244, 4406 (1969).
14. Cohn, E.J., and Edsall, J.T. in Proteins, Amino Acids and Peptides as Dipolar Ions, Reinhold, New York, 1943 p. 370-375.
15. Ingelman-Sundberg, M., and Gustafsson, J.-Å., Resolution of multiple forms of phenobarbital induced form of rabbit liver microsomal cytochrome P-450, FEBS Letters, submitted (1976).
16. Hrycay, E.G., Gustafsson, J.-Å., Ingelman-Sundberg, M. and Ernster, L., The involvement of cytochrome P-450 in hepatic microsomal steroid hydroxylation reactions supported by sodium periodate, sodium chlorite and organic hydroperoxides, Eur. J. Biochem. 61, 43 (1976).
17. Guengerich, F.P., Ballou, D.P., and Coon, M.J., Spectral intermediates in the reaction of oxygen with purified liver microsomal cytochrome P-450, Biochem. Biophys. Res. Commun. 70, 951 (1976).
18. Haugen, D.A., van der Hoeven,T.A., and Coon, M.J., Purified liver microsomal cytochrome P-450. Separation and characterization of multiple forms, J. Biol. Chem. 250, 3567 (1975).
19. Ingelman-Sundberg, M., Protein-lipid interactions in the liver microsomal hydroxylase system, Biochem. pharm., in press (1977).
20. Narasimhulu, S., Relationship between the membrane lipids and substrate-cytochrome P-450 binding reaction in bovine adrenocortical microsomes, Hoppe-Seyler's Z. physiol. chem. 357, 1019 (1976).

II PHARMACOLOGY

IMPACT OF DRUG MONOXYGENASES IN CLINICAL PHARMACOLOGY

A. Breckenridge,[*] M. R. Bending[†] and G. Brunner[‡]

[*] Department of Pharmacology and Therapeutics, University of Liverpool
[†] Westminster Hospital Medical School, London
[‡] Med. Universitatsklinik, Gottingen

We have taken as our terms of reference an overview of clinical aspects of drug metabolism which have shown significant advances since the 2nd International Symposium on Microsomes and Drug Oxidation at Stanford University in 1972. The choice of topics discussed is obviously eclectic but while some emphasis will naturally be placed on topics of special personal interest, an objective assessment of the field has been attempted. Two special constraints have been imposed - firstly, not to discuss at any length, implications of drug-drug interactions, and secondly not to deal extensively with those problems raised by studies of drug metabolism in disease states since both these topics are presented elsewhere in this volume.

I   INTER-INDIVIDUAL DIFFERENCES IN DRUG RESPONSE

Inter-individual differences in response to drug administration can often be related to differences in rates of metabolism, and this is especially true for lipid soluble drugs which are not excreted unchanged in urine to any significant extent. Although extra hepatic drug metabolism may on occasion be of great clinical importance, the liver is the most important site of drug detoxication. In this context, a number of factors in addition to specific hepatic enzyme activity may control rates of drug oxidation in vivo. Diminution of liver blood flow may decrease the elimination of those drugs whose hepatic extraction is high; and for this type of drug, binding to plasma proteins will enhance drug elimination (Ref 1). Further, dietary factors such as those outlined by Dr. Kappas elsewhere in this volume are of obvious significance too (Ref 2). Proposals that there are multiple forms of cytochrome P450 have been readily accepted by clinical pharmacologists for at least three reasons. Firstly, studies of species differences in the oxidation of ethylmorphine suggest that the reduction of the cytochrome P450 - substrate complex may be rate limiting (Ref 3). If this were so for many drugs, then in theory it ought to be possible to classify individuals according to their ability to oxidise drugs by measuring the rate of oxidation of one marker compound. In practice, this met with only limited success. While correlations between plasma half lives of structurally similar drugs such as antipyrine and phenylbutazone have been described (Ref 4) the same does not hold between antipyrine and glutethimide, amylobarbitone, diphenylhydantoin or sulphinpyrazone (Ref 5). Surprisingly, a correlation has been shown between the rates of elimination of diphenylhydantoin and amylobarbitone (Ref 6) and of acetanilide and phenytoin (Ref 7). One obvious interpretation of this non-correlation in rates of drug elimination is the existence of multiple forms of cytochrome P450.

A second strand of supporting evidence from clinical studies for this concept comes from studies of drug metabolism in patients with liver disease. One of the problems of these studies is definition of the nature and stage of the disease process. Acute viral hepatitis, however, is a reasonably circumscribed illness and over the past few years a number of well conducted studies have clearly shown discrepancies in the effect of hepatitis on the rate of elimination of different drugs. A lengthening of the plasma half life and a reduction in the plasma clearance of hexobarbitone, diazepam, pethidine and antipyrine has been reported in patients with acute viral hepatitis, but no change in the handling of phenobarbitone, diphenylhydantoin or warfarin. This topic has recently been reviewed (Ref 8). None of these drugs are highly extracted by the liver, and thus the possibility that alteration of liver blood flow or of protein binding caused by the disease do not merit consideration as the cause of these differences. Again, a likely candidate appears to be the occurrence of separate forms of cytochrome P450 responsible for the oxidation of different drugs.

The third reason for acceptance of multiple forms of cytochrome P450 arises from work done on the effects of corticosteroids on rates of drug elimination in man. It was reported to the 2nd International Symposium on Microsomes and Drug Oxidation that infusions of hydrocortisone at a rate sufficient to increase plasma steroid levels to their maximal physiological concentrations caused an immediate and significant increase in the rate of elimination of antipyrine (Fig 1, Ref 4). That this was due to a change in drug metabolism was inferred from the fact that hydrocortisone caused an alteration in neither the apparent volume of distribution of antipyrine nor in the small amount of antipyrine excreted unchanged in urine. This finding has been followed up in two ways. An infusion of hydrocortisone which raised the plasma hydrocortisone concentration from an average of 7.4 ± 1.2 µg/% to 14.1 ± 0.8 µg/% caused a shortening of the antipyrine half life in 7 out of 8 subjects. The rate of appearance of 4 hydroxyantipyrine, the principal metabolite of antipyrine was measured in urine and this followed the change in plasma half lives. These studies confirm that the change in half life of antipyrine was due to a change in the rate of drug oxidation.

Hydrocortisone has been shown to have a selective effect on drug oxidation. The rate of 4-hydroxylation of antipyrine and of side chain oxidation of phenylbutazone are increased by hydrocortisone administration, while the 3-hydroxylation of amylobarbitone and aromatic hydroxylation of phenylbutazone remain unaffected. While the underlying mechanism of these changes remain to be elucidated one reasonable hypothesis is that the steroid alters the conformation and thus activity of some forms of cytochrome P450, accounting for the immediacy and selectivity of the process observed.

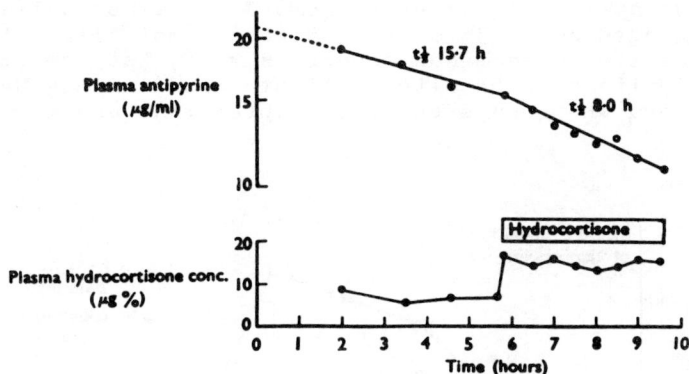

Fig. 1. Effect of hydrocortisone on elimination of antipyrine in one subject

(reproduced by permission of the publishers of the British Journal of Pharmacology)

Two further aspects of variations in drug metabolism have provoked considerable interest. At the extremes of age, there may be alterations in rate of drug oxidation. For example, Vestal and his colleagues (9) have investigated the metabolic clearance of antipyrine in 307 healthy volunteers whose ages ranged from 18 to 97 years, and showed that age and antipyrine clearance were apparently negatively correlated. But when contributing factors were examined, both caffeine (coffee) and cigarette consumption were positively correlated with antipyrine metabolism and the old both drank less coffee and smoked fewer cigarettes than their younger counterparts. Age itself, therefore, accounted for less than 3% of the variance in antipyrine clearance. Old age per se is not a therapeutically important determinant of rates of drug oxidation.

At the other extreme of age, variations in routes rather than rates of drug elimination may be important. It has been demonstrated that paracetamol (acetaminophen) is eliminated predominantly as a sulphate in the newborn and in infants, but increasingly as a glucuronide in children and in adults (Ref 10). However, the overall rate of paracetamol elimination was not affected by age. Since sulphate conjugation is clearly a capacity limited system and since other co-administered drugs may compete for available sulphate either in the gut wall or the liver, this route of drug elimination may assume importance in the young. It has been shown in experiments using an isolated loop of dog gut that some 3.6% isoprenaline is absorbed unchanged and 68% as sulphate. When the gut was pretreated with salicylamide which competes for available sulphate, some 73.7% isoprenaline was absorbed unchanged and 3.3% as the sulphate (Ref 11). It is not clear if the same phenomenon occurs in vivo, but the implication is that conditions which alter sulphate handling may be important in the newborn when drugs such as isoprenaline are administered.

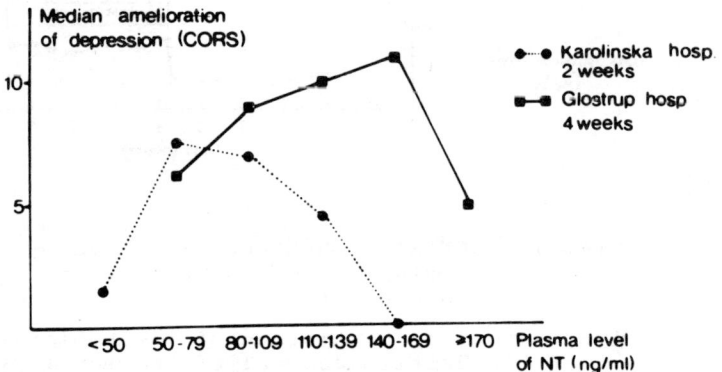

Fig. 2. Relationship between plasma concentration and antidepressant effect of nortriptyline.

Data computed from Kragh-Sørensen et al (1973) above, and Asberg et al (1971) below.

(reproduced by permission of the authors and Pitman Medical)

Differences in drug metabolism are but a contributory cause to variations in drug response. One area where significant advances continue to be made is in measurement of drug response. Even in clinical situations such as where assessment of response is difficult, a relationship between plasma concentration of tricyclic antidepressants and amelioration of the depression has been demonstrated (Ref 12). This curvilinear relation (Fig 2) implies that at higher drug levels one may produce a worsening rather than an improvement, and has now been confirmed in a second independent study (Ref 13).

In clinical practice, probably the most important sigle cause of variations in drug response is whether the patient takes his drugs as prescribed. Even in a potentially lethal disease such as tuberculosis less than 50% of patients surveyed took para-aminosalicylic acid (PAS) as prescribed (Ref 14) and for patients with rheumatoid arthritis, less than 35% complied with prescribed drug regimen (Ref 15). There seems to be little point in planning complicated therapeutic regimes based on scientific principles if it cannot be ensured that patients comply with therapeutic regimens.

II   PLASMA CONCENTRATIONS OF DRUGS

A second area of progress and change in attitude is the use of plasma concentrations of drugs in dose adjustment to optimise therapeutic response. A review in 1973 (Ref 16) listed 10 drugs where knowledge of plasma concentrations were considered to be of value (Table 1). Some modification of this is now necessary. It has been shown for two of these drugs, procainamide and lignocaine that therapeutically active metabolites accumulate in plasma and tissues. Thus to measure only unchanged drug in the plasma as a guide to dose alteration may be insufficient. N-acetyl procainamide (NAPA) is a major metabolite of procainamide, especially in patients who acetylate the drug rapidly: up to 34% of a single dose of procainamide may be recovered as NAPA in the urine of patients with normal renal function (Ref 17). Dosing with NAPA (Ref 18) has been shown to diminish ventricular premature contractions in 7 of 9 patients studied, confirming its therapeutic effectiveness. Monoethylglycinexylidide (MEGX) is the major metabolite of lignocaine in man: its plasma half life is similar to that of the parent drug; in some patients plasma levels of MEGX in excess of lignocaine have been recorded (Ref 19). There is good evidence that the central nervous system toxicity of lignocaine is due to the metabolite, and it has antiarrhythmic potency too.

Two aspects of methodology of measurement of plasma drug concentrations have come to the fore in the last few years. For a drug to be measured in routine clinical practice, methods must be inexpensive, repeatable and if possible, must lend themselves to automation. Thus immunoassay has received considerable attention as a method of measuring drug levels in plasma. Radioimmunoassay is the most widely used application, although enzyme mediated immunoassay techniques (EMIT) and free radical assay techniques (FRAT) are in some respects more attractive since they do not depend on high initial outlay for expensive equipment although the cost of the commercially available kits for drug measurement using these techniques currently remains high. For research purposes where precision of measurement and the ability to distinguish drug from metabolite at low concentrations are all important, immunoassay may present difficulties, although these have sometimes been over-emphasised. An example of the importance of applying sophisticated methodology to drug analysis is the measurement of the kinetics of the anti-inflammatory agent indomethacin (Ref 20). Using spectrophotometric methods, the plasma half life of indomethacin has been shown to be between 1.5 and 2.3 hours, which is surprising in view of a clinical effect lasting some 8 to 12 hours. Using a newly developed mass fragmentographic assay it has been shown that the half life of indomethacin may be as long as 11.2 hours, and that the previous figures for half life probably represent a distribution phase (Ref 20).

TABLE 1 Ranges of therapeutic serum concentrations of various drugs

Drug	Range
Digitoxin	14 - 30 µg/l
Digoxin	0.9 - 2.0 µg/l
Diphenylhydantoin	10 - 20 mg/l
Lignocaine	1.5 - 4.0 mg/l
Lithium	0.5 - 1.3 m-Equiv/l
Nortriptyline	50 - 140 µg/l
Procainamide	4 - 8 mg/l
Propranolol	20 - 50 µg/l
Quinidine	2 - 5 mg/l
Salicylates	150 - 300 mg/l

The other aspect of methodology worthy of comment is the increasing use of saliva as a fluid in which to monitor drug levels. It is frequently difficult or inconvenient to obtain blood samples at the appropriate time either because the patient is a child or is extremely ill or samples are required at unsociable hours. Under these circumstances measurements of drug levels in saliva have much to commend them. The ratio of drug in saliva to that in plasma depends on several factors, principally the extent of drug protein binding and the degree of ionisation of the drug, since the pH of saliva is usually lower than that of plasma. It has been shown that the ratio (R) of tolbutamide in saliva to that in plasma is:

$$R = \frac{1 + 10^{(pHs - pKa)}}{1 + 10^{(pHp - pKa)}} \times \frac{fp}{fs} \qquad (Ref\ 21)$$

where pHs = saliva

pHp = plasma

fp = fraction of drug unbound in plasma

fs = fraction of drug unbound in saliva

Correlation between plasma and salivary drug concentration has now been shown for sulphonamides, salicylic acid, paracetamol, tolbutamide, theophylline, antipyrine, amylobarbitone, phenobarbitone and phenytoin. The ratio of salivary to plasma concentration varies from 0.012 for tolbutamide to 1 for paracetamol and antipyrine.

III   CLINICAL TOXICOLOGY

Perhaps the greatest impact of drug monoxygenases in clinical pharmacology is in the field of clinical toxicology, and a potentially extremely important application comes from attempts to treat patients with hepatic failure. In liver failure, central nervous system abnormalities culminating in coma are probably due to the accumulation of toxins. The identity of these toxic substances is uncertain, but phenols, mercaptans, fatty acids and ammonia have all been implicated. Phenol accumulation for example results in further damage to liver cells inhibiting the activity of mitochondrial enzymes such as succinic dehydrogenase and monoamine oxydase and also mitochondrial respiration. Free fatty acids and mercaptans inhibit a large number of gluconeogenesis and ATP generation. As a result, a vicious circle of decrease in enzyme activity, further accumulation of toxins, further liver cell damage is created. A possible solution is to establish an extracorporeal system to remove these toxins and thus interrupt the cycle (Ref 22). Apart from physical techniques, biochemical methods are now being utilised.

UDP-glucuronyl transferase has been solubilised from microsomes by deoxycholate treatment. By use of ammonia precipitation, DEAE and sephadex G-200 chromatography the enzyme can be purified some 500-600 fold. This purified enzyme can then be bound on to artificial carriers. The best results have been achieved by binding the enzyme to copolymers of acrylamide and acrylic and acrylic acid or to Br-CN-sepharose beads. Binding of an already stable enzyme has been shown to increase its stability some 20 fold. UDPGA can be entrapped inside the beads and diffuses out at a rate which allows maximum enzyme reaction for 50-60 minutes. The beads can then be "reloaded" with UDPGA. For many toxins, glucuronidation is the second phase of detoxication and many must be hydroxylated prior to conjugation. Two enzymes are necessary for hydroxylation - cytochrome P450 and NADPH cytochrome P450 reductase. Both enzymes have been prepared according to established techniques and bound to sepharose beads (Fig 3) together with the necessary cofactors. These beads can then be incorporated into a circuit through which blood is passed. This concept of extracorporeal detoxication obviously has many implications other than those of liver failure. Binding of the enzymes covalently to a carrier overcomes the problems of immune reactions and allows multiple applications with the same enzyme preparation.

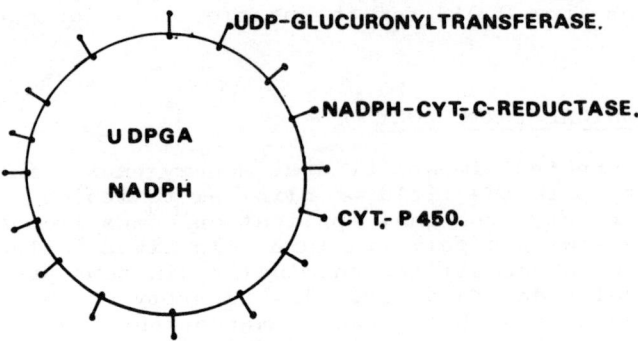

Fig. 3. Multienzyme complex. Binding of three enzymes to Br-CN-sepharose beads; cofactors included inside the beads

CONCLUSIONS

In this overview, the impact of drug monoxygenases in three areas has been presented - in the understanding of causes of inter-individual differences in drug response, in the philosophy of using plasma drug concentrations to optimise drug dosing and in clinical toxicology. From these three examples it is abundantly clear that the last few years has seen an increasing dependance of clinical pharmacology on basic concepts of drug metabolism.

ACKNOWLEDGEMENTS

We acknowledge the support of the Medical Research Council and The Deutsche Forschungesgemeinschaft.

REFERENCES

1. J.R. Gillette, Overview of Drug-protein binding, Ann.N.Y. Acad.Sci. 226, 6-17 (1974)

2. A. Kappas, Regulation of human drug metabolism by nutritional factors, This volume

3. D.S. Davies, P.L. Gigon and J.R. Gillette, Species and sex differences in electron transport systems in liver microsomes and their relation to ethylmorphine demethylation, Life Sciences. 8, 85-91 (1969)

4. D.S. Davies, S.S. Thorgeirsson, A. Breckenridge and M. Orme, Inter-individual differences in rates of drug oxidation in man, Drug Metabolism and Disposition, 1, 411-416 (1973)

5. D. Kadar, T. Inaba, L. Endrenyi, G.E. Johnson and W. Kalow, Comparative drug elimination capacity in man - glutethimide, amobarbital, antipyrine and sulfinpyrazone, Clinical Pharmacology and Therapeutics, 14, 552-560 (1973)

6. J.F. Brien, T. Inaba and W. Kalow, Comparative drug elimination in man. Diphenylhydantoin and amobarbital, Europ.J.Clin.Pharmacol. 9, 79-84 (1975)

7. J.L. Cunningham, M.F. Bullen and D.A. Price-Evans, The pharmacogenetics of acetanilide and diphenylhydantoin sodium, Europ.J.Clin.Pharmacol. 7, 461-466 (1975)

8. G.R. Wilkinson and S. Schenker, Drug disposition and liver disease, Drug Metabolism Reviews, 4, 139-176 (1975)

9. R.E. Vestal, A.H. Norris, J.D. Tobin, B.H. Cohen, N.W. Shock and R. Andres, Antipyrine metabolism in man; Influence of age, alcohol, caffeine and smoking, Clinical Pharmacology and Therapeutics, 18, 425-432 (1975)

10. R.P. Miller, R.J. Roberts and L.J. Fischer, Acetaminophen elimination kinetics in neonates, children and adults, Clinical Pharmacology and Therapeutics, 19, 284-294 (1976)

11. C.F. George, E. Blackwell and D.S. Davies, Metabolism of isoprenaline in the intestine, J.Pharm.Pharmacol. 26, 265-267 (1974)

12. M. Asberg, B. Cronholm, F. Sjöqvist and D. Tuck, Relationship between plasma level and therapeutic effect of nortriptyline, Brit.Med.J. 3, 331-334 (1971)

13. P. Kragh-Sørenson, C.E. Hansen and M. Asberg, Plasma levels of nortriptyline in the treatment of endogenous depression, Acta.Psychiatr.Scand. 49, 444-456 (1973)

14. W.M. Dixon, P. Stradling and I.D.P. Wootton, Outpatient PAS therapy, Lancet, 2, 871-872 (1957)

15. C.R. Joyce, Patient cooperation and the sensitivity of clinical trials, J.Chron.Dis. 15, 1025-1036 (1962)

16. J. Koch-Weser, Serum concentrations as therapeutic guides N.Eng.J.Med. 287, 227-231 (1972)

17. T.P. Gibson, J. Matusik, E. Matusik, H.A. Nelson, J. Wilkinson and W.A. Briggs, Acetylation of procainamide in man and its relationship to isonicotonic acid hydrazide acetylation phenotype, Clinical Pharmacology and Therapeutics, 17, 395-399 (1975)

18. J. Elson, J.M. Strong, W.K. Lee and A.J. Atkinson, Anti-arrhythmic potency of N-acetylprocainamide, Clinical Pharmacology and Therapeutics, 17, 134-140 (1975)

19. H. Halkin, P. Meffin, K.L. Melmon and M. Rowland, Influence of congestive heart failure on blood levels of lidocaine and its active monode-ethylated metabolite, Clinical Pharmacology and Therapeutics, 17, 669-676 (1975)

20. G. Alvan, M. Orme, L. Bertilsson, R. Ekstrand and L. Palmer, Pharmacokinetics of indomethacin, Clinical Pharmacology and Therapeutics, 18, 364-372 (1975)

21. S.B. Matin, S.H. Wan and J.J. Karaim, Pharmacokinetics of tolbutamide: prediction by concentration in saliva, Clinical Pharmacology and Therapeutics, 16, 1052-1058 (1974)

22. G. Brunner, Microsomal enzymes bound to artificial carriers in Artificial Liver Support, ed. R. Williams and I.M. Murray-Lyon, p.153-157, Pitman Medical, (1975)

THE ROLE OF LIVER NUCLEI IN THE FORMATION OF DNA BINDING PRODUCTS FROM BENZO(a) PYRENE

Bengt Jernström, Helena Vadi and Sten Orrenius

Department of Forensic Medicine, Karolinska Institutet, S-104 01 Stockholm, Sweden

It is generally assumed that the metabolic activation of the parent compound to electrophilic species that can interact with DNA is an early and critical event in the mutagenic and carcinogenic effect of benzo(a)pyrene (BP) and related compounds (1). Such electrophilic metabolites are produced by monooxygenation through the cytochrome P-450 system and include both primary epoxides and dihydrodiol epoxides (2-5). The latter are formed by a second monooxygenation of transdihydrodiol metabolites and include highly mutagenic species such as the 7,8-dihydrodiol-9,10-epoxide of BP (6-8).

When trying to elucidate the process of metabolic activation of BP to products which interact with DNA, one has to consider both the properties of the electrophilic species formed and their transfer from the site of generation to the target site, i.e. nuclear DNA. Thus it seems probable that the possibility for the highly electrophilic product(s) to escape various cytoplasmic defence systems and reach and interact with DNA would be increased by a proximity between the formation and target sites. The experiments described in this communication were part of a study designed to further investigate the metabolic activation of BP to DNA binding products with special regards to the site of generation of the electrophilic products.

As shown in Fig. 1, incubation of hepatocytes isolated from 3-methylcholanthrene(MC)-treated rats with [^{3}H]BP resulted in the incorporation of radioactivity into cellular DNA. Incorporation was time dependent and reached a maximum after 25 min of incubation. In contrast, when hepatocytes from control animals were incubated under the same conditions, no detectable binding of radioactive material to DNA was observed. This experiment suggested that an enhanced BP monooxygenase activity, caused by the pretreatment of the rats with MC, is required to obtain measurable incorporation of BP products into DNA in this model but did not reveal the identity or site of generation of the DNA binding metabolites. Therefore, a series of experiments was performed in which the isolated liver microsomal and nuclear fractions were incubated, alone or in combination, with BP and, in case of the nuclear fraction, microsomal metabolites of BP, and the resulting incorporation of labeled metabolites into nuclear DNA was measured. Three possible mechanisms for the metabolic activation of BP to DNA binding products were considered: a/ activation takes place

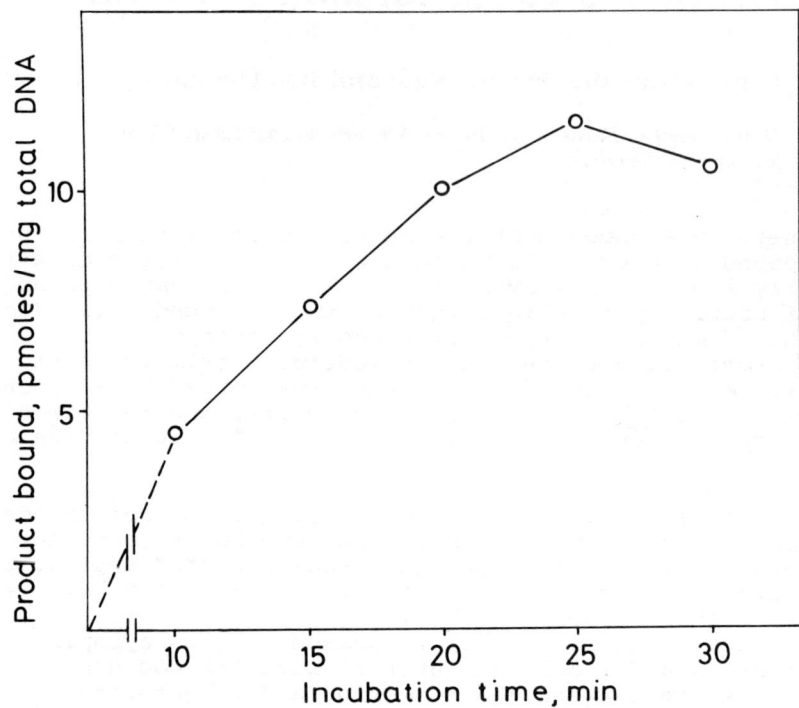

Fig. 1. Binding of BP products to DNA in hepatocytes isolated from MC-treated rats. Incubations contained 80 μM [^{3}H]BP, 2% bovine serum albumin and 5×10^6 cells/ml suspended in Krebs-Henseleit's buffer, pH 7.4. At each time point the cells were solubilized in SDS-EDTA and extracted with chloroform-phenol and ether. DNA was precipitated with ethanol, treated with RNA'ase and pronase, dialysed, banded on a cesium chloride density gradient, precipitated with TCA, collected on cellulose filters and the bound radioactivity was measured with liquid scintillation counting. Results are uncorrected for tritium loss during the incubation.

in the nuclear envelope, which has been shown to display BP monooxygenase activity (9,10), b/ activation takes place in the endoplasmic reticulum (ER), the major site for oxygenation of polycyclic hydrocarbons (11), and the electrophilic metabolites reach nuclear DNA by diffusion, and c/ microsomal metabolites are activated by a second monooxygenation in the nuclear membrane. Fig. 2 summarizes results of experiments where isolated nuclei, or nuclei in combination with microsomes, were incubated with BP under various conditions. In agreement with previous reports (12-14), pretreatment of the animals with MC enhanced the BP monooxygenase activity in the nuclear fraction and this stimulation was associated with a 5-fold increase in the amount of BP metabolites incorporated into nuclear DNA (Fig. 2B, hatched bar). Addition of the inhibitor of BP monooxygenase, α-naphthoflavone (α-NF), reduced the extent of binding by about 70% (Fig. 2B, open bar).

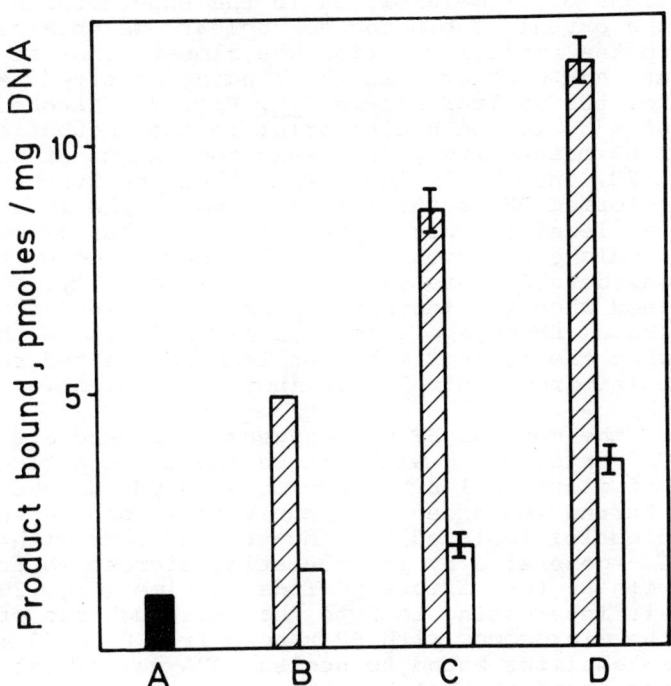

Fig. 2. Binding of BP products to nuclear DNA. (A) Control nuclei (2 mg protein/ml) incubated with 1 mM NADPH and 80 μM [^{3}H]BP for 20 min. (B) Nuclei from MC-treated rats (2 mg protein/ml) incubated with 1 mM NADPH and 80 μM [^{3}H]BP for 20 min. ▨, No addition; ▢, 0.1 mM α-NF. (C) Nuclei from MC-treated rats (2 mg protein/ml) incubated for 20 min with microsome-free supernatants derived from incubations containing 1 mM NADPH, 80 μM [^{3}H]BP and microsomes from MC-treated rats (200 μg protein/ml). Microsomes were removed by centrifugation and the remaining supernatants contained 0.7 μM BP and 3-7 μM BP-metabolites. The limit of experimental results are also shown for each bar. ▨, No addition; ▢, 0.1 mM α-NF. (D) Nuclei from MC-treated rats (2 mg protein/ml) incubated for 20 min with microsomes from MC-treated rats (200 μg protein/ml), 1 mM NADPH and 80 μM [^{3}H]BP. Microsomes were preincubated with BP and NADPH for 0-40 min prior to addition of nuclei. The limit of experimental results are shown for each bar. ▨, No addition; ▢, Preincubation of nuclei with 0.1 mM α-NF.

The possibility that BP metabolites formed in the microsomes and released into the incubation medium could act as substrates for the nuclear BP monooxygenase was tested in the following experiment: liver microsomes from MC-treated rats were incubated with BP for different periods of time, the microsomes removed by centrifugation, and the remaining supernatants were incubated with the nuclear fraction isolated from the same animals. Although, the concentrations of BP metabolites in the supernatants ranged from 3-7 µM, the extent of binding to nuclear DNA obtained upon incubation with the nuclear fraction was almost constant (Fig. 2C, hatched bar) and greater than the binding observed using BP as substrate for the nuclear enzyme (cf. Fig. 2B, hatched bar). The addition of α-NF to the nuclei prior to the incubation with the microsomal BP metabolites, decreased the extent of binding markedly (Fig. 2C, open bar), indicating that the further metabolism of the preformed BP metabolites leading to the interaction with DNA, was at least in part dependent on nuclear BP monooxygenase activity. This conclusion was further supported by the observation that considerably less DNA-bound radioactivity, only about 10%, was obtained when the incubations of the nuclear fraction with the microsomal BP metabolites were carried out in the absence of NADPH, or when the nuclear fraction from MC-treated rats was replaced by nuclei from control rats (not documented).

The capacity of the nuclear BP monooxygenase to generate BP products capable of interacting with DNA seemed to be optimal at a concentration of microsomal BP metabolites ≤3 µM. However, this rate could be further enhanced by incubating a combination of microsomes and nuclei isolated from MC-treated rats with BP and BP metabolites - generated by preincubating microsomes from MC-treated rats with BP for various periods of time (Fig. 2D, hatched bar). It is interesting to note that although the preincubation time of the microsomes with BP ranged from 0 to 40 min, the amount of BP metabolites bound to nuclear DNA was almost constant and in all cases greater than when the microsomes were omitted (cf. Fig. 2C, hatched bar). Thus, it seems that under our conditions, a direct transfer of electrophilic microsomal BP metabolites to nuclear DNA did indeed contribute to the total binding observed. However, nuclear BP monooxygenase still seemed responsible for the formation of most of the DNA-bound BP metabolites, since this amount was markedly reduced by preincubating the nuclei with α-NF, which inhibited nuclear BP monooxygenase activity by approximately 70% without affecting the microsomal activity (Fig. 2D, open bar). In a similar experiment, when liver microsomes from MC-treated rats were replaced by microsomes from control rats, the extent of binding of BP products to nuclear DNA was markedly decreased. In fact, the amount of bound BP metabolites never exceeded the extent of binding observed when nuclei were incubated with BP in the absence of microsomes. The inhibitory effect of control microsomes was probably due to a decreased availability of substrates for the nuclear enzyme because of a limited microsomal production of BP metabolites and a trapping of BP and BP metabolites by the microsomes. This is illustrated in Fig. 3 which demonstrates that the amount of BP metabolites released into the incubation medium by control microsomes was considerably less than by microsomes from MC-treated rats. The recovery of unmetabolized

BP in the incubation medium from both systems never exceeded 0.7 µM, indicating that much of the lipophilic substrate was trapped by the microsomes.

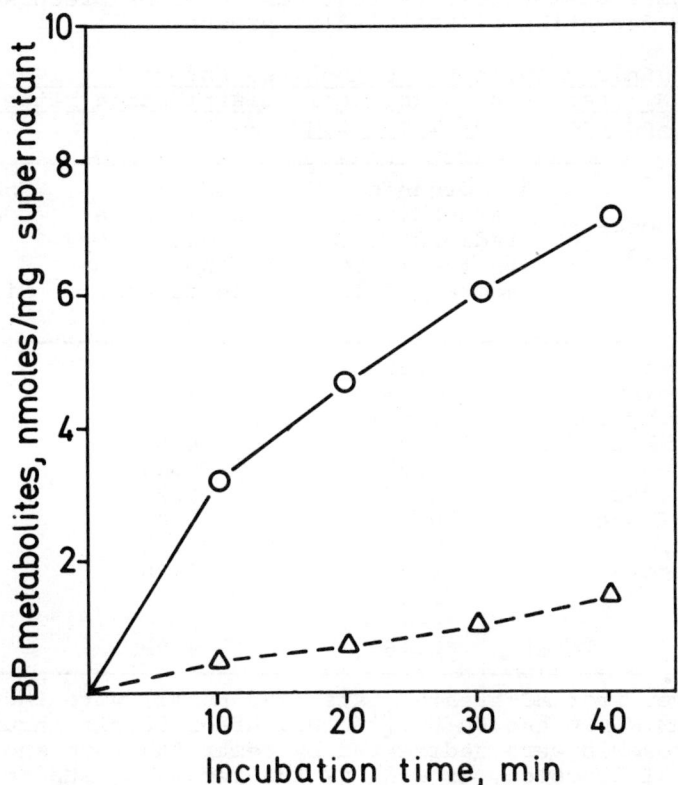

Fig. 3. <u>Presence of BP metabolites in the supernatant fraction after removal of the microsomes</u>. Incubations contained microsomes (200 µg/ml), 1 mM NADPH and 80 µM [^{3}H]BP. The microsomes were removed by centrifugation and the supernatant was analyzed for metabolites (17). ---, Microsomes from control rats; ——, microsomes from MC-treated rats.

It thus seems that the rate of BP metabolism and the distribution of BP and BP metabolites between the microsomes and the incubation medium are important factors to consider when interpreting the results of the present study. At present, it is difficult to decide which microsomal BP metabolite may serve as the prime substrate for the formation of DNA binding products in the nuclear fraction, although the distribution pattern shown in Table 1 seems to indicate that the dihydrodiols are the metabolites most easily released by the microsomes into the medium and thus accessible to the nuclear enzyme. In a previous study we have found that both phenols and dihydrodiols accumulate intracellularly, when rat hepatocytes are incubated with BP (15). However, further experimental work is needed to establish which of the microsomal BP metabolites that can serve as precursors for the nuclear formation of DNA binding products.

TABLE 1 Organic soluble BP metabolites formed by liver microsomes from MC-treated rats and their distribution between the microsomes and the incubation medium [a]/.

Peak	[b]/ BP metabolite	Organic extractable radioactivity in the microsomes: nmoles	Organic extractable radioactivity in the incubation medium: nmoles	Incubation medium / Microsomes
I	unknown	0.58	0.36	0.62
II	9,10-dihydrodiol	1.71	2.05	1.2
III	4,5-dihydrodiol	1.58	0.37	0.23
IV	7,8-dihydrodiol	1.71	0.67	0.39
V	1,6-dione	1.94	0.25	0.13
VI	3,6-dione	2.25	0.20	0.09
VII	6,12-dione	0.29	0.02	0.07
VIII	9-hydroxy	1.46	0.13	0.09
IX	3-hydroxy	4.24	0.26	0.06
X	BP	52.40	0.68	0.01
	Total	68.16	4.99	

[a]/ Microsomes from MC-treated rats (200 µg/ml) were incubated with 1 mM NADPH and 80 µM [^{14}C]BP. After 20 min incubation the microsomes were sedimented by centrifugation and the microsomal fraction, as well as the remaining supernatant, were analysed for organic-extractable metabolites by high pressure liquid chromatography (16).

[b]/ Peaks co-chromatographing with authentic BP derivatives are listed in order of increasing retention time.

In summary, this study has shown that the nuclear BP monooxygenase can convert BP, and metabolites of BP preformed in the microsomes, to DNA binding products In addition, a direct transfer of electrophilic microsomal metabolites seems to have contributed to the total incorporation of BP products into nuclear DNA observed with our model. The relative importance of these mechanisms under physiological conditions remains to be established although it is tempting to speculate that the formation of the DNA binding species in the close vicinity of the nucleus would permit them to escape trapping by the various cytoplasmic defence systems and thus facilitate their binding to DNA.

ACKNOWLEDGEMENT

This study was supported by NIH contract 1 CP 33363.

REFERENCES

(1) C. Heidelberger, Chemical carcinogenesis, Ann. Rev. Biochem. 44, 79 (1975)
(2) A. Borgen, H. Darvey, N. Castagnoli, T. T. Crocker, R. E. Rasmussen and I. Y. Wang, Metabolic conversion of benzo-(α)pyrene by syrian hamster liver microsomes and binding of metabolites to deoxyribonucleic acid, J. Med. Chem. 16, 502 (1973)
(3) D. M. Jerina and J. W. Daly, Arene oxides: A new aspect of drug metabolism, Science 185, 573 (1974)
(4) N. Kinoshita, B. Shears and H. V. Gelboin, K-region and non-K-region metabolism of benzo(α)pyrene by rat liver microsomes, Cancer Res. 33, 1937 (1973)
(5) P. Sims, P. L. Grover, A. Swaisland, K. Pal and A. Hewer, Metabolic activation of benzo(a)pyrene proceeds by a diol-epoxide, Nature 252, 326 (1974)
(6) E. Huberman, L. Sachs, S. K. Yang and H. V. Gelboin, Identification of mutagenic metabolites of benzo(a)pyrene in mammalian cells, Proc. Nat. Acad. Sci. U.S.A. 73, 607 (1976)
(7) C. Malaveille, H. Bartsch, P. L. Grover and P. Sims, Mutagenicity of non-K-region diols and diol-epoxides of benz(a)-anthracene and benzo(a)pyrene, in S. Typhimurium TA 100, Biochem. Biophys. Res. Commun. 66, 693 (1975)
(8) P. G. Wislocki, A. W. Wood, R. L. Chang, W. Levin, H. Yagi, O. Hernandez, D. M. Jerina and A. H. Conney, High mutagenicity and toxicity of a diol epoxide derived from benzo-(a)pyrene, Biochem. Biophys. Res. Commun. 68, 1006 (1976)
(9) K. Alexandrov and C. Frayssinet, Aryl and aniline hydroxylases in rat nuclear membranes after pretreatment with pregnenolone 16α-carbonitrile, phenobarbital and methylcholanthrene, Experientia 31, 778 (1975).
(10) A. S. Khandwala and C. B. Kasper, Preferential induction of aryl hydroxylase activity in rat liver nuclear envelope by 3-methylcholanthrene, Biochem. Biophys. Res. Commun. 54, 1241 (1973)
(11) A. H. Conney, E. C. Miller and J. A. Miller, Substrate-induced synthesis and other properties of benzpyrene hydroxylase in rat liver, J. Biol. Chem. 228, 753 (1957)
(12) K. Alexandrov, P. Brookes, M. W. S. King, M. R. Osborne and M. H. Thompson, Comparison of the metabolism of benzo(a)-pyrene and binding to DNA caused by rat liver nuclei and microsomes, Chem.-Biol. Interactions 12, 269 (1976)
(13) E. G. Rogan and E. Cavalieri, 3-Methylcholanthrene-inducible binding of aromatic hydrocarbons to DNA in purified rat liver nuclei, Biochem. Biophys. Res. Commun. 58, 1119 (1974)
(14) J. Vaught and E. Bresnick, Binding of polycyclic hydrocarbons to nuclear components in vitro, Biochem. Biophys. Res. Commun. 69, 587 (1976)
(15) H. Vadi, P. Moldéus, J. Capdevila and S. Orrenius, The metabolism of benzo(α)pyrene in isolated rat liver cells, Cancer Res. 35, 2083 (1975)

(16) S. J. Stohs, R. C. Grafström, M. D. Burke, P. W. Moldéus and S. G. Orrenius, The isolation of rat intestinal microsomes with stable cytochrome P-450 and their metabolism of benzo-(α)pyrene, Arch. Biochem. Biophys. In press, 1976.
(17) J. W. DePierre, M. S. Moron, K. A. M. Johannesen and L. Ernster, A reliable, sensitive and convenient radioactive assay for benzpyrene monooxygenase, Anal. Biochem. 63, 470 (1975).

MECHANISM OF MICROSOMAL ACTIVATION OF BENZO[a] PYRENE TO DIOL-EPOXIDES: THE SEPARATION AND CHARACTERIZATION OF INTERMEDIATES AND PRODUCTS

Shen K. Yang,[1] Peter P. Roller,[2] and Harry V. Gelboin[1]

Chemistry Branch[1] and Carcinogen Metabolism and Toxicology Branch,[2] National Cancer Institute, National Institutes of Health, Bethesda, Maryland 20014, U.S.A.

ABSTRACT

Benzo[a]pyrene is metabolically and stereospecifically converted by rat liver microsomal mixed-function oxidases and epoxide hydratase to the single enantiomer (-)r-7,t-8-dihydroxy-7,8-dihydrobenzo[a]pyrene. Its identification as the optically pure trans isomer was accomplished by chromatographic separation of the cis and trans isomers and the di(-)menthoxyacetates of the (+) and (-) trans isomers. The metabolically formed dihydrodiol is further metabolized highly stereoselectively by the mixed-function oxidases predominantly to a single enantiomer of the diol-epoxide, r-7,t-8-dihydroxy-t-9,10-oxy-7,8,9,10-tetrahydrobenzo[a]pyrene. Small amounts of r-7,t-8-dihydroxy-c-9,10-oxy-7,8,9,10-tetrahydrobenzo[a]pyrene and other unidentified metabolites are also formed. The diol-epoxides are unstable in aqueous medium and their identification and characterization were accomplished by the identity of their tetrahydroxy-tetrahydrobenzo[a]pyrenes hydrolysis products with those of the authentic synthetic compounds. The diol-epoxides are non-enzymatically reduced in the presence of NADPH to distinct trihydroxy-pentahydrobenzo[a]pyrenes. Since r-7,t-8-dihydroxy-t-9,10-oxy-7,8,9,10-tetrahydrobenzo[a]pyrene is very highly mutagenic in mammalian cells, and it is the major metabolically formed diol-epoxide, we suggest that this compound is the predominant ultimate carcinogenic form of benzo[a]pyrene.

INTRODUCTION

Benzo[a]pyrene (BzPy)* is the most common polycyclic aromatic hydrocarbon in the environment and exerts several biological effects (1). Early studies demonstrated that BzPy was itself biologically inert and its biological activity was dependent on its metabolism by microsomal enzymes (2,3). The biological activities included toxicity, mutagenicity, tumorigenicity, and covalent binding to DNA (3,4). A major finding was that microsomes catalyzed the binding of BzPy to DNA (2). Subsequently, Borgen, et al (5) showed that the 7,8-diol of BzPy bound more readily to DNA than BzPy in the presence of microsomes and NADPH. Sims, et al (6) showed that a synthetic 7,8-diol-9,10-epoxide of BzPy was chemically very reactive in binding to DNA and suggested that diol-epoxide may be the reactive intermediate. BzPy is converted by the microsomal mixed-function oxidases, epoxide hydratase, and various conjugases to at least 20-25 metabolites (4-13). Recently we reported that the three dihydrodiols formed metabolically from BzPy with rat liver microsomes are all optically active (14).

*See refs. 22 and 24 for the nomenclature and abbreviation of compounds described in this paper.

The central problem in BzPy carcinogenesis is the elucidation of the structure of the activated carcinogenic form and its mechanism of formation by the microsomal enzyme systems. Recent studies have indicated that a 7,8-diol-9,10-epoxide of BzPy is very active in the binding to DNA in vitro and to the DNA and RNA of mammalian cells in culture (15-16). Two stereoisomeric diol-epoxides have been synthesized (17-19) and their stereochemistry have been elucidated (18,19).

Malaveille, et al (20) reported that a BzPy 7,8-diol-9,10-epoxide exhibits similar mutagenic activity as the 4,5-epoxide in strains TA 100 of S.Typhimurium. A subsequent report by Wislocki,et al (21) showed that r-7,t-8-dihydroxy-c-9,10-oxy-7,8,9,10-tetrahydrobenzo[a]pyrene (diol-epoxide II) (22) was more mutagenic than BzPy 4,5-epoxide in S. Typhimurium strains TA 98 and TA 100, respectively. The same report (21) also showed that diol-epoxide II was more mutagenic than BzPy 4,5-epoxide in the cultured Chinese hamster V79 cells. We have reported (23) that r-7,t-8-dihydroxy-t-9,10-oxy-7,8,9,10-tetrahydrobenzo[a]pyrene (diol-epoxide I) was more highly mutagenic in cultured Chinese hamster V79 cells than either the diol-epoxide II, the 4,5-epoxide, or any other BzPy derivatives including the known BzPy metabolites. In the same report, we presented evidence showing that the trans-7,8-diol, formed metabolically from BzPy, is converted predominantly to diol-epoxide I (23). Subsequently, we reported (24) that a single enantiomer of diol-epoxide I was enzymatically and highly stereoselectively formed from the (-)trans-7,8-diol enzymatically prepared from BzPy. Findings similar to ours showing the exceptionally high mutagenic activity of diol-epoxide I was also reported by Newbold and Brookes (25). In a recent report, analysis of the RNA bound BzPy derivative formed in vivo indicated the intermediate was diol-epoxide I (16). The high metagenic activity of diol-epoxide I and the demonstration of its metabolic formation suggests that it is the major ultimate carcinogenic form of BzPy.

MATERIALS AND METHODS

The [^{3}H]BzPy (5 Ci/mmol) was purchased and purified (10). Nonradioactive synsynthetic diol-epoxide I, cis-7,8-diol, and trans-7,8-diol, [^{3}H]trans-7,8-diol (157 mCi/mmol) and [7-^{14}C]diol-epoxide I (53.9 mCi/mmol) were obtained by synthesis on National Cancer Institute Contract No. NO1-CP-33387 (17,26), and the diol-epoxide II by synthesis on National Cancer Institute Contract No. NO1-CP-33385 (18). The detailed synthesis of BzPy cis-7,8-diol can be obtained from Dr. James Engel, 425 Volker Boulevard, Midwest Research Institute, Kansas City, Mo. 64110. The metabolically formed [^{3}H]trans-7,8-diol was obtained by incubating [^{3}H]BzPy with rat liver microsomes and purified (24).

Either the metabolically formed or the synthetic [^{3}H]trans-7,8-diol was incubated with liver microsomes from 3-methylcholanthrene-treated male Sprague-Dawley rats (23,24). [^{3}H]trans-7,8-diol and its metabolites were extracted and prepared for HPLC analysis as described (23,24). Diol-epoxides I and II were each incubated under identical conditions as the [^{3}H]trans-7,8-diol. Diol-epoxides I and II (200 nmol in 100 μl tetrahydrofuran) were also treated with water (2 ml) at 37°C for 10 min. Products of diol-epoxides I and II and the metabolites of trans-7,8-diol were analyzed by HPLC as previously described (23,24). The enzymatic hydration product of the synthetic racemic 7,8-epoxide was obtained by incubating the 7,8-epoxide with partially purified rat liver epoxide hydratase prepared according to Lu, et al (27) and was isolated by HPLC. The separation of the cis and trans 7,8-diols, and the di(-)menthoxyacetates of BzPy (-) and (+) trans-7,8-diols by HPLC is described in detail elsewhere (28,29). The optical rotation was measured on a Perkin-Elmer 241-MC polarimeter.

RESULTS

The Metabolically Formed 7,8-diol Is trans-7,8-diol

The enzymatically formed dihydrodiols of BzPy although assumed to have trans configurations, have not been definitively characterized (4). We have developed HPLC conditions in which the synthetic cis and trans 7,8-diols of BzPy are separated chromatographically (Fig. 1). We then found that 7,8-diol formed metabolically from BzPy migrates identical to the synthetic trans-7,8-diol (Fig. 1). Thus we have established unequivocally that BzPy is converted to the trans-7,8-diol.

The trans-7,8-diol Enzymatically Formed from BzPy and Racemic 7,8-epoxide Are Optically Active

The optical rotation of the trans-7,8-diol formed metabolically from BzPy and that formed metabolically from the racemic 7,8-epoxide are shown in Table 1. These are compared with the HPLC resolved optically pure (-)trans-7,8-diol and indicate that BzPy is converted metabolically to an optically pure (-)trans-7,8-diol. The optical purity of the metabolically formed (-)trans-7,8-diol is further supported by the results shown in Fig. 2. The optical rotation of the

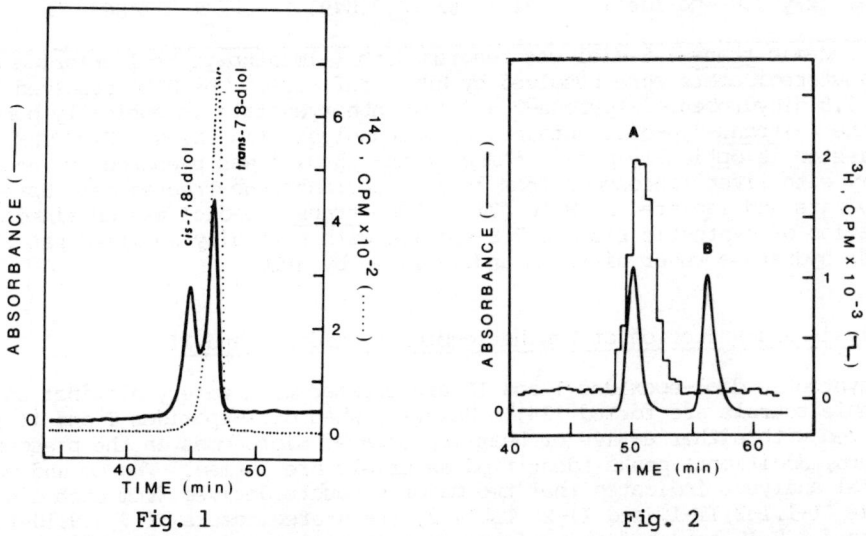

Fig. 1   Fig. 2

Fig. 1. HPLC Separation of BzPy cis and trans 7,8-diols. The [^{14}C]BzPy 7,8-diol metabolically formed from BzPy is co-chromatographed with the synthetic BzPy cis and trans 7,8-diols. HPLC was carried out as described (28).

Fig. 2. HPLC Separation of the Di(-)menthoxyacetates of BzPy (-)trans-7,8-diol (peak A) and (+)trans-7,8-diol (peak B). The metabolically formed [^{3}H]trans-7,8-diol (ca. 0.1 μg) and non-radioactive synthetic (±)trans-7,8-diol (50 μg) in 0.1 ml pyridine were reacted with (-)menthoxyacetyl chloride (10 mg) for 16 hr at room temperature (ref. 30). One ml water was added to the mixture and was extracted with 3 x 2 ml ether/ethyl acetate(1:1, v/v). The sample was analyzed as described (28).

trans-7,8-diol obtained by incubation of synthetic racemic 7,8-epoxide with partially purified epoxide hydratase indicates that the epoxide hydratase is stereoselective for the substrate preferring the form of the epoxide made metabolically from BzPy. Thus if the mixed-function oxidases converting the BzPy to the 7,8-epoxide and the subsequent epoxide hydratase catalyzing the formation of trans-7,8-diol are both highly stereoselective, the resultant product of the two enzymatic reactions would be close to optically pure. Direct analyses shown in Fig. 2 and Table 1 suggest that the oxygenation of BzPy to 7,8-epoxide by the mixed-function oxidases is stereospecific, i.e., a single enantiomer of the 7,8-epoxide is stereospecifically formed from BzPy by the mixed-function oxidases.

TABLE 1 Optical Rotation of HPLC Resolved and Metabolically Formed Benzo[a]pyrene trans-7,8-diol

Source of trans-7,8-diol	$[\alpha]_{400}^{25}$ (c, methanol)	% (−) Enantiomer*
HPLC Resolved*	−3730 ± 160 (c, 0.019)	100
BzPy Metabolism**	−3482 ± 62 (c, 0.025)	97
Racemic BzPy 7,8-epoxide†	−2701 ± 62 (c, 0.049)	86

* The racemic trans-7,8-diol was reacted with (−)menthoxyacetyl chloride and the diastereoisomers were resolved by HPLC (ref. 29). The HPLC resolved (−)-trans-7,8-dihydrobenzo[a]pyrene-7,8-di(−)menthoxyacetate is optically pure. Thus the (−)trans-7,8-diol obtained by methanolysis (in $CH_3ONa/CH_3OH/THF$) of its diester is optically pure. **The trans-7,8-diol was prepared by incubation of BzPy with liver microsomes from 3-methylcholanthrene-treated male Sprague-Dawley rats and isolated by HPLC (24). †The trans-7,8-diol was obtained by incubation of synthetic racemic 7,8-epoxide with partially purified rat liver epoxide hydratase under nitrogen and isolated by HPLC.

Hydrolysis and Reduction of Synthetic Diol-epoxides I and II

When synthetic diol-epoxides I and II are treated with water, distinct and HPLC separable tetrols are formed (24). However, when diol-epoxides I and II are incubated with either active or heat-inactivated microsomes in the presence of NADPH, additional peaks identified as triols are formed. The uv and mass spectral analyses indicated that two major products derived from each diol-epoxide (I-1, I-2, II-1, and II-2, Table 2) are stereoisomers of 7,8,9,10-tetra-hydroxy-7,8,9,10-tetrahydrobenzo[a]pyrenes (tetrols) and I-3 and II-4 are isomeric trihydroxypentahydrobenzo[a]pyrenes (triols). I-3 and II-4 are reduced forms of the diol-epoxides formed in the presence of NADPH (31). In addition to uv and mass spectral analysis (23,24,31,32), the structures of the tetrols and triols were characterized by reactions with potassium triacetylosmate which reacts irreversibly with compounds containing vicinal cis-diol (31,32), and mass spectral analysis of their acetonides (32).

Metabolism of the Metabolically Formed and Synthetic Racemic trans-7,8-diol

The HPLC system previously reported (23,24) is ideally suited for the study of trans-7,8-diol metabolism. Since diol-epoxide I or diol-epoxide II rapidly undergoes hydrolysis in aqueous medium to distinct tetrols, diol-epoxides formed metabolically from trans-7,8-diol would rapidly convert to tetrols whose detection would indicate the metabolic formation of the diol-epoxides. When synthetic racemic trans-7,8-diol was incubated with microsomes and cofactors, HPLC analysis (Table 2) showed the presence of tetrols (M-2,M-3,M-4, and M-6) and triols (M-5 and M-9) with mobilities identical to the hydrolysis and NADPH reduction products of diol-epoxides I and II. Thus the results suggest that this synthetic racemic trans-7,8-diol is metabolically converted to both diol-epoxides I and II, which are hydrolyzed subsequently to tetrols and reduced to triols. Other unidentified metabolites were also detected.

The metabolites of the metabolically formed trans-7,8-diol were prepared and analyzed under identical conditions (24). The amount of metabolites formed is compared with those of the synthetic racemic trans-7,8-diol (Table 2). The results in Table 2 indicate that the metabolically formed trans-7,8-diol is stereoselectively metabolized predominantly to diol-epoxide I.

TABLE 2 Identification and Quantification of the Metabolites Formed From Synthetic Racemic and Metabolically Formed trans-7,8-diol

Products of Synthetic Diol-epoxides*		Metabolite (m/e,M+)*	r_t on HPLC, min.[†]	Nmoles Metabolites formed from[‡]	
Compound (m/e,M+)	Identification			Metabolic (−)trans-7,8-	Synthetic (±)trans-7,8-
–	–	M-1 (320)	23.0	0.23	0.08
I-1 (320)	(7,10/8,9)-tetrol	M-2 (320)	26.7	1.51	1.25
I-2 (320)	(7/8,9,10)-tetrol	M-4 (320)	30.6	0.38	0.32
I-3 (304)	(7/8,9)-triol	M-5 (304)	34.0	0.23	0.28
II-1 (320)	(7,9/8,10)-tetrol	M-3 (320)	27.8	0.03 − 0.05[§]	0.14
II-2 (320)	(7,9,10/8)-tetrol	M-6 (320)	36.1	0.03 − 0.05[§]	ca. 0.33[¶]
–	–	M-7 (302)	37.0	0.63	ca. 0.30[¶]
–	–	M-8 (?)	38.0	0.05	0.30
II-4 (304)	(7,9/8)-triol	M-9 (304)	44.8	0.01 − 0.04[§]	0.23
–	–	trans-7,8-diol	53.5	–	–
–	–	M-10 (304)	61.1	0.05	0.06
–	–	M-11 (304)	63.2	0.05	0.06

* The products were separated on HPLC and characterized as previously described (23,24). The M-1, etc. designations refer to metabolites formed from the trans-7,8-diol and in some cases as indicated are identical to the tetrol and triol products derived from the synthetic diol-epoxides I and II. The identification of tetrols and triols are described in detail elsewhere (24, 31,32). [†]Retention time on HPLC under conditions described previously (23,24) [‡]Nmoles of metabolites per ml incubation mixture. The metabolites are those extracted by the organic solvent. [§]With different preparations of microsomes the amounts of M-3,M-6 and M-9 varied. In all cases these were considerably lower than the values for M-2,M-4 and M-5. [¶]M-6 and M-7 are not completely separated and the quantities can not be accurately determined.

DISCUSSION

Our results indicate that BzPy is metabolically converted to an optically pure (−) enantiomer of trans-7,8-diol. The latter is subsequently metabolized predominantly to diol-epoxide I. Since the trans-7,8-diol is one of two optically active forms, the diol-epoxide I derived from the (−)trans-7,8-diol must also be a single enantiomer. Hence the absolute configuration of the major metabolically formed diol-epoxide I is determined by the absolute configuration of the metabolically formed (−)trans-7,8-diol and the stereoselective nature of the oxygenation by the mixed-function oxidases at the 9,10-position. Other metabolites were also detected indicating that the formation of diol-epoxides is not the only metabolic product of the trans-7,8-diol. The metabolic events which lead to the formation and disposition of diol-epoxides are depicted in Fig. 3. It is expected that all the products derived from optically pure (−)trans-7,8-diol would possess optical activity.

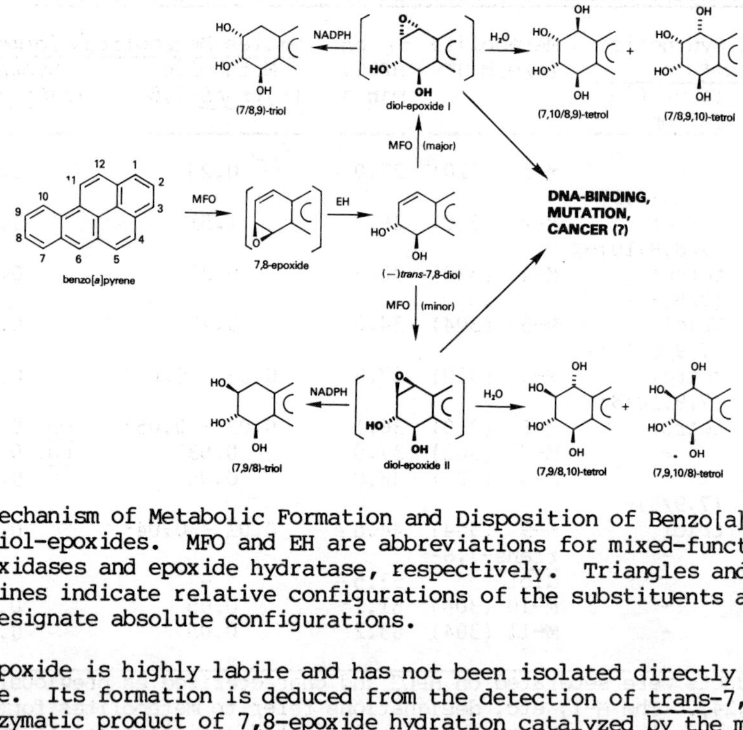

Fig. 3 Mechanism of Metabolic Formation and Disposition of Benzo[a]pyrene Diol-epoxides. MFO and EH are abbreviations for mixed-function oxidases and epoxide hydratase, respectively. Triangles and dotted lines indicate relative configurations of the substituents and do not designate absolute configurations.

The 7,8-epoxide is highly labile and has not been isolated directly as a metabolite. Its formation is deduced from the detection of trans-7,8-diol which is the enzymatic product of 7,8-epoxide hydration catalyzed by the microsomal epoxide hydratase. The conversion of BzPy to an optically pure (−)trans-7,8-diol thus requires two enzymatic steps; oxygenation by the mixed-function oxidases and hydration by the epoxide hydratase. The demonstration (Table 1) that the epoxide hydratase acts stereoselectively toward one of the two 7,8-epoxide enantiomers and the optical purity of the trans-7,8-diol metabolically formed from BzPy indicate the oxygenation of BzPy by the mixed-function oxidases is at least stereoselective and may be stereospecific. It is clear however that two sequential enzymatic reactions yield an optically pure product.

The (-)trans-7,8-diol is stereoselectively oxygenated by the mixed-function oxidases to mainly diol-epoxide I and to small amounts of diol-epoxide II. The structure of tetrols indicated that the formation of tetrols from diol-epoxides I and II follows an S_N1 reaction mechanism, i.e., the formation of carbonium ions at C_{10} (an electrophile) with subsequent nucleophilic attack by the weakly nucleophilic water molecules (and OH^-) to form tetrols (32). The results thus indicate that the major metabolically formed diol-epoxide I can undergo S_N1 reactions at C_{10} with less nucleophilic groups on cellular macromolecules such as DNA and proteins. This is consistent with the recent reports showing that an adduct is formed between C_{10} of diol-epoxide I and the 2-amino group of guanine in RNA from cultured bovine bronchial mucosa (16), and the diol-epoxide I is the form predominantly made from (-)trans-7,8-diol which binds strongly to DNA in cultured human bronchus (34).

The metabolism of BzPy is regulated by several enzymes which include the various forms of cytochrome P-450 (35,36), epoxide hydratase (37,38), glutathione S-transferase (12), and UDPGA transferase (13). Understanding the various enzymatic pathways may lead to a precise definition of factors regulating the formation of the ultimate carcinogenic form vis-a-vis the formation of detoxification forms. The relative rates of these metabolic routes may govern tissue, species and individual differences in human susceptibility to polycyclic hydrocarbon carcinogenesis. Carcinogenicity by BzPy may be initiated by the formation of a metabolite of high reactivity and yet sufficiently stable to reach the target site in the cell. This target site may indeed be the DNA and the interaction of the highly reactive BzPy metabolite with DNA would be reflected by a high mutagenic activity. Our previous report (23) showed that the diol-epoxide I was much more mutagenic than either the corresponding isomeric diol-epoxide II or the K-region 4,5-epoxide. In this study we show that a single enantiomer of the diol-epoxide I is metabolically and stereoselectively formed from (-)trans-7,8-diol. The latter is one of the major metabolites enzymatically formed from BzPy (7-10). The metabolic formation of diol-epoxide I and its extraordinarily high mutagenicity (23) indicate that this metabolite is the major ultimate carcinogenic form of BzPy. The amount of this diol-epoxide formed from BzPy relative to other inactive metabolites may be an important determinant of carcinogenesis by BzPy.

REFERENCES

1. Anonymous, Particulate Polycyclic Organic Matter (1972) National Academy of Sciences, Washington, D.C.
2. Gelboin, H.V., Kinoshita, N., and Wiebel, F.J., Fed.Proc. 31, 1298 (1972).
3. Gelboin, H.V., Cancer Res., 29, 1272 (1969).
4. Sims, P. and Grover, P.L., Advan. Cancer Res., 20, 165 (1974).
5. Borgen, A., Darvey, H., Castagnoli, N., Crocker, T.T., Rasmussen, R.E. and Wang, I.Y., J. Med. Chem., 16, 502 (1973).
6. Sims, P., Grover, P.L., Swaisland, A., Pal, K. and Hewer, A., Nature, 252, 326 (1976).
7. Kinoshita, N., Shears, B., and Gelboin, H.V., Cancer Res., 33, 1937 (1973).
8. Selkirk, J.K., Croy, R.G., Roller, P.P., and Gelboin, H.V., Cancer Res., 34, 3474 (1974).
9. Holder, G., Yagi, H., Dansette, P., Jerina, D.M., Levin, W., Lu, A.Y.H. and Conney, A.H., Proc. Nat. Acad. Sci. USA, 72, 4356 (1974).
10. Yang, S.K., Selkirk, J.K., Plotkin, E.V. and Gelboin, H.V., Cancer Res., 35, 3642 (1975).

11. Selkirk, J.K., Croy, R.G., Harvey, R.G., Engel, J.F. and Gelboin, H.V., Biochem. Pharmacol., 25, 227 (1976)
12. Nemoto, N., Gelboin, H.V., Habig, W.H., Ketley, J.N. and Jakoby, W.B., Nature, 255, 512 (1976)
13. Nemoto, N. and Gelboin, H.V., Biochem. Pharmacol., 25, 1221 (1976)
14. Yang, S.K. and Gelboin, H.V., Biochem. Pharmacol., 25, 2221 (1976)
15. Sims, P. Grover, P.L., Swaisland, A., Pal, K., and Hewer, A., Nature, 252, 326 (1974)
16. Weinstein, J.B., Jeffrey, A.M., Jennette, K.W., Blobstein, S.H., Harvey, R.G., Harris, C., Autrup, H., Kasai, H., and Nakanishi, K., Science, 193, 592 (1976)
17. McCaustland, D.J. and Engel, J.F., Tetrahedron Lett., 30, 2549 (1976)
18. Beland, F.A. and Harvey, R.G., J. C. S. Comm., 3, 84 (1976)
19. Yagi, H., Hernandez, P. and Jerina, D.M., J. Am. Chem. Soc., 97, 6881 (1975)
20. Malaveille, C., Bartsch, H., Grover, P.L., and Sims, P., Biochem. Biophys. Res. Comm., 66, 693 (1975)
21. Wislocki, P., Wood, A.W., Chang, R.L., Levin, W., Yagi, H., Hernandez, O., Jerina, D.M., and Conney, A.W., Biochem. Biophys. Res. Comm., 68, 1006 (1976)
22. Fletcher, J.H., Dermer, O.C., and Fox, R.B. (eds.) (1974) Nomenclature of Organic Compounds, in Advances in Chemistry Series, no. 126 (American Chemical Society, Washington, D.C.), 103
23. Huberman, E., Sachs, L., Yang, S.K., and Gelboin, H.V., Proc. Nat. Acad. Sci. USA, 73, 607 (1976)
24. Yang, S.K., McCourt, D.W., Roller, P.P., and Gelboin, H.V., Proc. Nat. Acad. Sci. USA, 73, 2594 (1976)
25. Newbold, F.F. and Brookes, P., Nature, 261, 52 (1976)
26. McCaustland, D.J., Duncan, W.P. and Engel, J.F., J. Labeled Compds and Radiopharmaceuticals, in press
27. Lu, A.Y.H., Ryan, D., Jerina, D.M., Daly, J.W. and Levin, W., J. Biol. Chem., 250, 8283 (1975)
28. Yang, S.K., McCourt, D.W., Leutz, J.C., and Gelboin, H.V., submitted for publication
29. Yang, S.K., Gelboin, H.V., Weber, J.D., Fischer, D.L., Sankaran, V. and Engel, J.F., submitted for pubication
30. Cook, J.W., Loudon, J.D., and Williamson, W.F., J. Chem. Soc., 911 (1950)
31. Yang, S.K. and Gelboin, H.V., Cancer Res., in press
32. Yang, S.K., Roller, P.P., McCourt, D.W., Miller, J.R., and Gelboin, H.V., submitted form publication
33. Criegee, R., Marchand, B. and Wannowius, M., Justus Liebigs Ann. Chem., 550, 99 (1942)
34. Yang, S.K., Gelboin, H.V., Trump, B.N., Autrup, H.N., and Harris, C.C., submitted for publication
35. Haugen, D.A., van der Hoeven, T.A., and Coon, M.J., J. Biol. Chem., 250, 3567 (1775)
36. Ryan, D., Lu, A.Y.H., West, S., and Levin, W., J. Biol. Chem., 250, 2157 (1975)
37. Oesch, F., Xenobiotica, 3, 305 (1973)
38. Leutz, J.C. and Gelboin, H.V., Arch. Biochem. Biophys., 168, 722 (1975)

THE METABOLISM OF BENZO[a]PYRENE IN CELL CULTURES AND HOMOGENATES FROM DIFFERENT HUMAN FETAL TISSUES

Olavi Pelkonen and Pekka Korhonen

Departments of Pharmacology and Microbiology, University of Oulu, SF 90220 Oulu 22, Finland

ABSTRACT

Benzo(a)pyrene is efficiently metabolised by cells derived from human fetal tissues into known (dihydrodiols, phenols, quinones) and unknown metabolites. Benzo(a)pyrene metabolism in cell cultures does not necessarily represent the activity of aryl hydrocarbon hydroxylase in the original tissue in vitro. The intraindividual variability of benzo(a)pyrene metabolism is only about 15 per cent in liver cell cultures and very large, several fold, in different fibroblast cultures. At least part of the intraindividual variability can be overcome by sub-culturing fetal fibroblasts. Interindividual variability is 5- to 10-fold, a figure also obtained by others in different human cell and tissue culture systems. When compared to some established cell lines, for instance BS-C-1 or BHK-21, human fetal cell cultures metabolise benzo(a)-pyrene to a comparable extent. Human fetal cells in culture seem to be promising models for studying the significance of carcinogen metabolism in terms of toxicity, mutagenicity and carcinogenicity.

INTRODUCTION

Recently considerable interest has been centred on the development of short-term screening methods for detecting late harmful effects of foreign compounds (1). One line of research has been to use cells in culture, in the hope that results obtained in cell cultures could be extrapolated to the situation in the intact organism. It is reasonable to assume that it is possible to develop screening methods using human cells with the final goal of predicting the toxicity, mutagenicity, teratogenicity, and carcinogenicity of foreign compounds in man. Benzo(a)pyrene (BP) is an ubiquitous environmental pollutant which is carcinogenic (2). Nowadays it is thought that the carcinogenicity of BP is related to some specific aspects of its metabolism (3) and thus the elucidation of its metabolism may give insight into its carcinogenicity and be the first step in developing useful cell culture systems for short-term mutagenicity and carcinogenicity testing.
Earlier we have studied the inducibility of aryl hydrocarbon hydroxylase (AHH), an enzyme oxidizing BP, in cell cultures derived from different human fetal tissues and we found that the enzyme is inducible especially in liver cell cultures (4). This prompted us to continue our studies in order to elucidate the metabolism of BP in human fetal cell cultures, to compare cell cultures with their corresponding organs and to compare fetal cells with established cell lines derived from different rodent and primate species. The basic aim has been to discover a human cell line with high ability to metabolize benzo(a)pyrene and with high inducibility of aryl hydrocarbon hydroxylase. This cell line could then

be used in attempts to develop a useful method for screening the toxicity of foreign compounds.

MATERIALS AND METHODS

Human fetuses were obtained during the hysterotomy performed for socio-medical reasons to otherwise healthy women during 9 to 14 weeks of gestation. Cell cultures were prepared as described in detail earlier (4) except that in the present studies we used a lower concentration of fetal calf serum, 10 % instead of 20 %. Nonconfluent cultures after three days of growth were exposed to BP (final concentration 0.25 uM, including about 6×10^6 cpm of ^{3}H-BP per nanomole) usually for one day and culture medium was analyzed with respect to BP metabolites and unchanged BP essentially as described by Sims (5) and Borgen et al. (6). The procedure used is described in more detail earlier (7). Authentic reference metabolites of BP including 9,10-dihydro-9,10-dihydroxy BP (BP-9,10-dihydrodiol), 7,8-dihydro-7,8-dihydroxy BP (BP-7,8-dihydrodiol), 4,5-dihydro-4,5-dihydroxy BP (BP 4,5-dihydrodiol), 3- and 9-hydroxy BP (3-OH-BP, 9-OH-BP) and several quinones were kindly donated by Harry V. Gelboin (NCI, Bethesda, Maryland).
Established cell lines were obtained from Fluka Labs (Irvine, Scotland) and propagated in the Department of Microbiology, University of Oulu. Aryl hydrocarbon hydroxylase activity in human fetal tissue homogenates was determined according to Nebert and Gelboin (8).

RESULTS AND DISCUSSION

The metabolite pattern of benzo(a)pyrene in cell cultures

The metabolite pattern of BP produced by human fetal liver homogenate in vitro and in culture medium of cells derived from the same liver after one day exposure to BP is shown in Fig. 1. In liver cell culture, a large part of the radioactivity is concentrated near the origin of thin layer plate and to the spot of authentic BP. At one day, radioactivity migrating with BP-9,10-dihydrodiol and BP-7,8-dihydrodiol is the most prominent and radioactive peaks at the origin are somewhat lower. The reverse is true after 3 days (not shown) and seems to suggest that the peak at the origin represent the metabolites derived from "primary" metabolites of BP. The low amounts of phenols and quinones at all times after the exposure of cells to BP and decreasing amounts of dihydrodiols as a function of time show that most "primary" metabolites are in vivo conditions quite efficiently metabolized further. After three days, most of the radioactivity, in liver cell cultures almost 100 per cent, remains in the water phase suggesting the importance of the further metabolism of BP when elucidating its total fate in any biologic system.
Essentially similar kind of metabolite pattern was obtained when BP was added into the cell cultures derived from other human fetal tissues, from kidney, lungs, skin, trunk, intestine or adrenals. The radioactivity migrating with reference dihydrodiols and at the origin of the TLC plate was always the most prominent and the amounts of phenols and quinones were relatively small. The water-soluble radioactivity increased steadily towards the end of incubation.
When comparing the metabolite patterns between cell cultures and homogenates, clear differences can be seen (Fig. 1). In homogenates, the fraction containing 3-OH-BP and other phenols is the predominant one, whereas in cell cultures, more polar metabolites dominate.

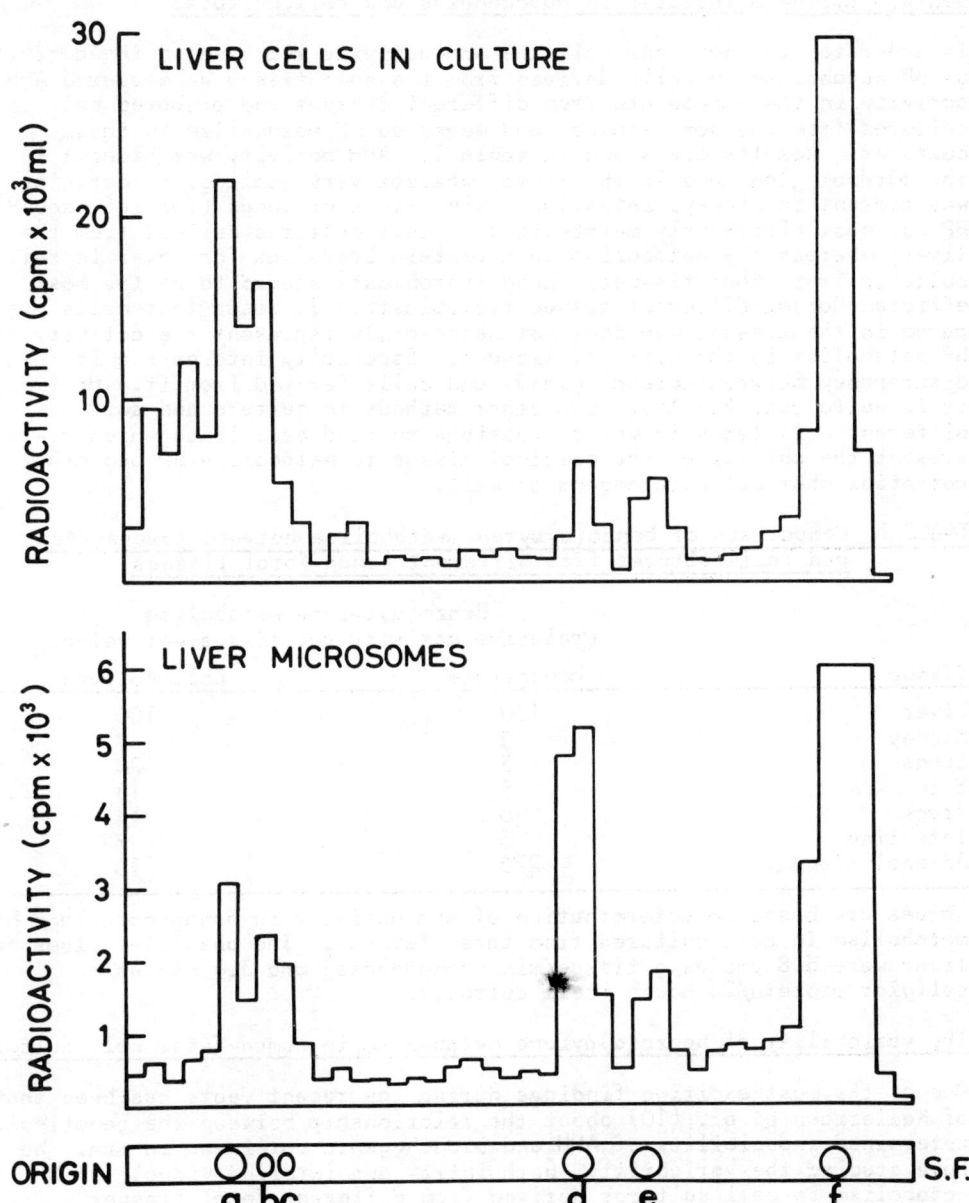

Fig. 1. The metabolite pattern of BP obtained using TLC (Silica gel 60) developed in benzene:ethanol (19:1, v/v). The spots a, b and c are dihydrodiols, d represents phenols and e quinones. f is unchanged benzo(a)pyrene.

Benzo(a)pyrene metabolism in homogenates and cell cultures

In order to elucidate how well the AHH activity represented the activity of BP metabolism in cells derived from the same tissue we measured AHH activity in the homogenate from different tissues and prepared cell cultures from the same tissues and measured BP metabolism in those cultures. Results are shown in table 1. AHH activity was highest in the adrenal gland and in the liver, whereas very small or no activity was present in kidney, intestine, skin, trunk or lungs (for ref. see 9). BP was most efficiently metabolized in cell cultures derived from the liver, whereas the metabolism on a protein basis was far lower in cell cultures from other tissues. Lung fibroblasts seemed to be the most efficient among different tissue fibroblasts. It seems that cells prepared in the present way does not necessarily represent the activity of BP metabolism in the original tissues. Especially interesting is the discrepancy between adrenal glands and cells derived from it. However, it is quite possible that with other methods to culture and sort out different cell types it may be possible to find cell lines which represent the ability of the original tissue to metabolize BP and other potential chemical carcinogens as well.

TABLE 1 Comparison of benzo(a)pyrene metabolism between homogenates and cell cultures from different human fetal tissues

Tissue	Benzo(a)pyrene metabolism (relative activity per tissue wet weight)	
	Homogenate	Cell culture
Liver	100	100
Kidney	5	8
Lungs	5	28
Skin	5	18
Trunk	5	14
Intestine	5	7
Adrenal gland	270	10

Values are based on determination of AHH activity in homogenates and BP metabolism in cell cultures from three fetuses. The absolute values for liver were 6.8 pmoles/g tissue/min (homogenate) and 3.1 nmoles/mg cellular protein/24 hours (cell culture).

The variability of benzo(a)pyrene metabolism in human fetal cell cultures

One of the most exciting findings during the recent years has been that of Kellermann et al. (10) about the relationship between the genetically determined inducibility of AHH and bronchogenic carcinoma in man. We have studied the variability, both intra- and interindividual, of BP metabolism in cell cultures derived from different fetal tissues. Findings are summarized in table 2. As seen, the intraindividual variation in fetal liver cell cultures is negligible, about 15 per cent, indicating the reproducibility of the enzyme activity during culture. Interindividual variability is much higher, about 4-fold, which is similar to our earlier results (4) and to those by others obtained in various human cell and tissue cultures (11-14). On the other hand, the intraindividual variation on fibroblast cultures is very high, about 3-fold in the case of fibroblasts derived from trunk. This may indicate that during culture different sub-populations of fibroblasts are selected in different culture bottles. This fact makes it very difficult to study the contributions of genetic background and environment to BP metabolism in human fetal fibroblasts. Our preliminary studies indicate, however, that by subculturing it is possible to obtain human fetal fibroblasts which are rather reproducible in their ability to metabolize BP.

TABLE 2 Intra- and interindividual variability of benzo(a)pyrene metabolism in human fetal cell cultures

Source	Variability between individual cultures (%)	
	Intraindividual	Interindividual
Liver	14	378
Lungs	129	539
Kidney	74	392
Trunk	334	1033

Values are based on determinations of BP metabolism in cell cultures from three fetuses, from 3 to 10 culture bottle per individual tissue.

Comparison of human fetal cell cultures with some established human, primate and rodent cell lines

Because the supply of fetuses is haphazard and limited, we have tried to find established cell lines which can be used for pilot experiments before going into studies with human fetal cell cultures. Table 3 shows semiquantitative comparison of human fetal cell cultures with established cell cultures originated from monkey, rodent and human tissues.

TABLE 3 Preliminary screening of the ability of primary and established cell cultures from human, primate and rodent species to metabolize benzo(a)pyrene

Cell line	Source	BP metabolism
Primary		
hepatocytes	fetal liver	+++
fibroblasts	fetal trunk	+ to ++
fibroblasts	fetal lungs	+ to ++
epithelial cells	fetal kidney	(+) to +
SUBCULTURED		
epithelial cells	fetal kidney (III p)	- to (+)
fibroblasts	fetal lungs (VI p)	+ to ++
ESTABLISHED		
BS-C-1	green monkey kidney	++ to +++
VERO	green monkey kidney	-
GMK	green monkey kidney	-
BHK-21	baby hamster kidney	++ to +++
RK-13	rabbit kidney	+
McC	human synovial fluid	- to (+)
Tu	human amnion fluid	- to (+)

+++ = >50 %, ++ = 25 to 50 %, + = 10 to 25 %, (+) = 5 to 10 %, and - = <5 % of the added benzo(a)pyrene metabolized in 24 hours. It has been taken into account that human fetal cell cultures contain, much less cell per bottle than established cell cultures and thus differences between fetal cells and established cells are larger when compared on a basis of cell counts or protein amounts.

It is interesting to find out that those cell lines with epithelial cell morphology (BS-C-1, RK-13, BHK-21) are far more capable of metabolizing BP than cells with fibroblast morphology (VERO, GMK). This is not in agreement with human fetal cell cultures, because epithelial-like cells from human fetal kidneys are rather poor in metabolizing BP and lose their activity during subculturing, whereas fibroblast-like cells from many fetal tissues metabolize BP quite efficiently and retain their activity during several passages. Two established human cell lines (Tu and McC) have only a very low ability to metabolize BP.
Table 3 shows that with respect to BP metabolism human fetal cells can be compared to cells from e.g. monkey, hamster or rabbit kidney and, in fact, human fetal liver cells are superior to them. Unfortunately we have not managed to subculture human fetal liver cells.
In conclusion, we have shown that cell cultures from different human fetal tissues metabolize BP efficiently and we are proceeding to study the importance of this metabolic activity in terms of the metabolism of other carcinogens, mutagenicity, toxicity and malignant transformation.

REFERENCES

(1) Workshop on Rapid Screening Tests to Predict Late Toxic Effects of Environmental Chemicals, Brussels, Belgium, June 1975.

(2) International Agency for Research on Cancer, IARC Monographs on the Evaluation of Carcinogenic Risk of the Chemical to Man, 1973, Vol. 3. Lyon, France.

(3) P. Sims and P. L. Grover, Epoxides in polycyclic aromatic hydrocarbon metabolism and carcinogenesis, Adv. Cancer Res. 20, 166 (1974).

(4) O. Pelkonen, P. Korhonen, P. Jouppila and N. T. Kärki, Induction of aryl hydrocarbon hydroxylase in human fetal liver cell and fibroblast cultures by polycyclic hydrocarbons, Life Sci. 16, 1403 (1975).

(5) P. Sims, Qualitative and quantitative studies on the metabolism of a series of aromatic hydrocarbons by rat-liver preparations, Biochem. Pharmacol. 19, 795 (1970).

(6) A. Borgen, H. Darvey, N. Castagnoli, T. T. Crocker, R. E. Rasmussen and I. Y. Wang, Metabolic conversion of benzo(a)pyrene by syrian hamster liver microsomes and binding of metabolites to deoxyribonucleic acid, J. Med. Chem. 16, 502 (1973).

(7) J. Ahokas, O. Pelkonen and N. T. Kärki, Metabolism of polycyclic hydrocarbons by a highly active aryl hydrocarbon hydroxylase in the liver of a trout species, Biochem. Biophys. Res. Commun. 63, 635 (1975).

(8) D.W. Nebert and H. V. Gelboin, Substrate-inducible microsomal aryl hydroxylase in mammalian cell culture I. Assay and properties of induced enzyme, J. Biol. Chem. 243, 6242 (1968).

(9) O. Pelkonen, The metabolism of benzo(a)pyrene in human adult and fetal tissues, in Polynuclear Aromatic Hydrocarbons: Chemistry, Metabolism, and Carcinogenesis, edited by R. I. Freudenthal and

P. W. Jones, Raven Press, New York, 1974.

(10) G. Kellermann, C. R. Shaw and M. Luyten-Kellermann, Aryl hydrocarbon hydroxylase inducibility and bronchogenic carcinoma, N. Engl. J. Med. 289, 934 (1973).

(11) E. Huberman and L. Sachs, Metabolism of the carcinogenic hydrocarbon benzo(a)pyrene in human fibroblasts and epithelial cells, Int. J. Cancer 11, 412 (1973).

(12) A. P. Alvares, A. Kappas, W. Levin and A. H. Conney, Inducibility of benzo(a)pyrene hydroxylase in human skin by polycyclic hydrocarbons, Clin. Pharmacol. Ther. 14, 30 (1973).

(13) J. P. Whitlock, H. L. Cooper and H. V. Gelboin, Aryl hydrocarbon (benzopyrene) hydroxylase is stimulated in human lymphocytes by mitogens and benz(a)anthracene, Science 177, 618 (1972).

(14) D. L. Busbee, C. R. Shaw and E. T. Cantrell, Aryl hydrocarbon hydroxylase induction in human leukocytes, Science 178, 315 (1972).

ARYL HYDROCARBON HYDROXYLASE IN CULTURED HUMAN LYMPHOCYTES[1]

B. Paigen, H. L. Gurtoo, Jun Minowad, and K. Paigen

Departments of Molecular Biology (BP, KP), Experimental Therapeutics (HLG), and Immunology (JM), Roswell Park Memorial Institute, Buffalo, N.Y. 14263

INTRODUCTION

Polycyclic aromatic hydrocarbons are metabolized by aryl hydrocarbon hydroxylase (AHH), a component of the microsomal mixed-function oxygenases. This inducible enzyme complex metabolizes steroids, drugs, insecticides and carcinogens (1). Benzo(a)pyrene, a prototype substrate, is metabolized by AHH to epoxide forms and the epoxides, through several alternative pathways, may form dihydrodiols, phenols, dihydrodiol epoxides, or conjugates with glutathione, glucuronic acid or sulfate.

Some inbred strains of mice are inducible for AHH and some are not (2). The inducible strains are more susceptible to tumor induction by 3-methylcholanthrene than are the non-inducible strains (3,4). Genetic crosses between inducible and non-inducible strains of mice indicate that AHH inducibility is determined by at least 2, and possible more, genetic loci (5).

If people also varied in the levels or inducibility of carcinogen metabolizing enzymes, such variation might be related to differential susceptibility to cancer. It has been reported that the human population does vary in AHH inducibility and that persons toward the high end of the AHH inducibility range are more susceptible to lung cancer induced by smoking (6,7). Due to the clinical importance of demonstrating that a genetically determined biochemical trait influences an individual's susceptibility to lung cancer, we have reinvestigated the studies of Kellermann and coworkers. In this paper we report that AHH activity levels and inducibility do vary in the human population, that some of the variation is genetically determined, and that some of the variation is environmentally determined by factors such as drugs and season of the year.

---

[1]Supported by NIH contract CP-55626 and by NIH grants GM-19521, CA-13038, and CA-14413.

METHODS

A 20 ml sample of heparinized blood was withdrawn from healthy volunteers and used within 4 hours. The blood was diluted with an equal volume of RPMI-1640 medium and each 20 ml aliquot layered onto a 7 ml Hypaque-Ficoll gradient solution (s.g. 1.080) and centrifuged for 30 min at 1400 rev/min at room temperature. The lymphocyte rich interphase was aspirated, the fractions for each donor combined, and the cells washed twice with RPMI-1640 medium. The cells were resuspended at 10^6/ml in RPMI-1640 medium containing 10% heat-inactivated fetal calf serum, penicillin at 100 units/ml, streptomycin at 50 µg/ml, 1:100 dilution of pokeweed mitogen and phytohemagglutinin. The cell suspension from each donor was distributed into 4 plastic Falcon T-30 flasks (25 cm^2) in 8 ml aliquots and incubated in a 5% CO_2 atmosphere at 37°C. After 2 days, 5 µl acetone was added to 2 flasks and 8 nmoles of the inducer 3-methylcholanthrene in 5 µl acetone (1.0 µM in culture) was added to the other 2 flasks. Cultures were incubated 24 additional hrs, harvested, and DNA and AHH determined. AHH was assayed using benzo(a)pyrene as substrate and measuring fluorescent products according to modifications we published earlier (8). Enzyme activity is expressed as picomole equivalents of 3-hydroxybenzopyrene formed per 30 min. The inducibility ratio is calculated as the total enzyme activity in induced culture flasks compared to control culture flasks.

RESULTS AND DISCUSSION

We tested 30 people 2-5 times each and found that AHH inducibility was reproducible with an average coefficient of variation of 0.13. Using a normal population of over 300 persons, we observed no difference in AHH inducibility between smokers or non-smokers or between male and female.

Induced AHH changes with the season of the year (Fig. 1). Induced AHH activity levels observed during the winter months are only 20% of those seen during the summer and early fall. Individuals measured repeatedly throughout this 18-month period show the same pattern of seasonal change. This seasonal change is not due to problems in the enzyme assay since cultured lymphocytes frozen during the summer and assayed during the winter are still high.

During the rise in activity from June, 1975 through September, 1975, no change in reagents, media, or procedure occurred suggesting that the variation exists in people rather than in the laboratory. Additional evidence that the variation is in people and that it is related to seasonal factors comes from AHH levels measured during the winter time on people who have just returned from a Florida vacation. Buffalo winters are cold with little sunshine but Florida is warm and sunny. Some people showed dramatic increases in induced AHH levels, characteristic of mid-summer values, just after their return from Florida, but 2-3 weeks later, their induced AHH activity decreased to values more typical of mid-winter. Not all individuals who were exposed to Florida weather showed an increase in induced AHH, and we have been unable to discover what causes some people to respond. The fact that some people do respond, however, shows that the seasonal variation occurs in people rather than in the laboratory.

We next asked whether the marked seasonal changes in AHH activity caused changes in AHH inducibility ratios. Figure 2 illustrates the distribution of AHH inducibility observed in the normal population during the summer, fall and winter. The range of AHH inducibility shifts to lower values and the presence of high inducers disappears altogether in winter. The shift to lower ratios and the disappearance of high inducers are primarily due to changes in the induced activity.

Seasonal changes in microsomal enzymes have also been reported in animals. Dr. Minor Coon reports that rabbits are more inducible for microsomal enzymes in summer than in winter (personal communication) and Beuthin and Bousquet report that metabolism of hexobarbital, aminopyrine, and p-nitroanisole in rat liver show peak inducibility by phenobarbital during the summer (10).

Thus a population distribution of AHH inducibility in the human population cannot be accumulated over a long period of time and is best done during the summer and early fall. The population distribution obtained during the high season of 1975 is depicted in Fig. 3. There is considerable variation in AHH inducibility but we do not observe a distinct trimodal distribution as reported earlier by Kellermann and coworkers (6).

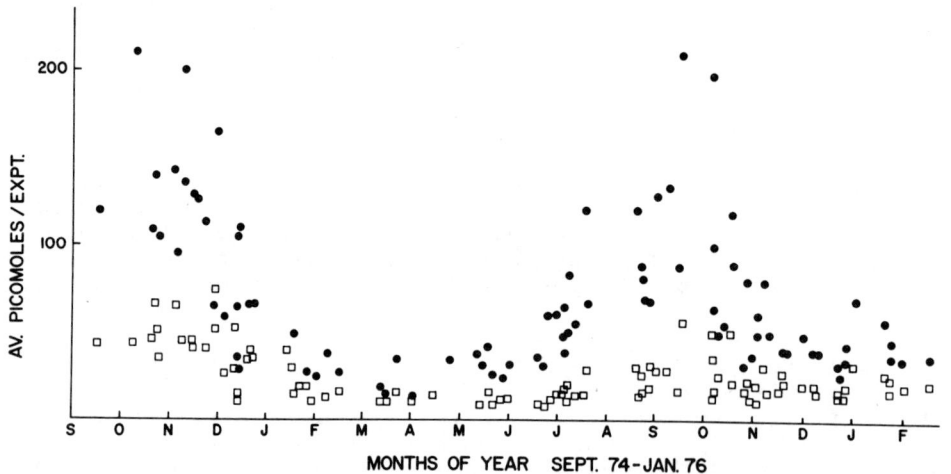

Fig. 1 Seasonal changes in AHH activity
The closed circles represent the average induced activity and the open squares represent the average basal activity of all people measured on a particular day. Values based on over 600 donors.

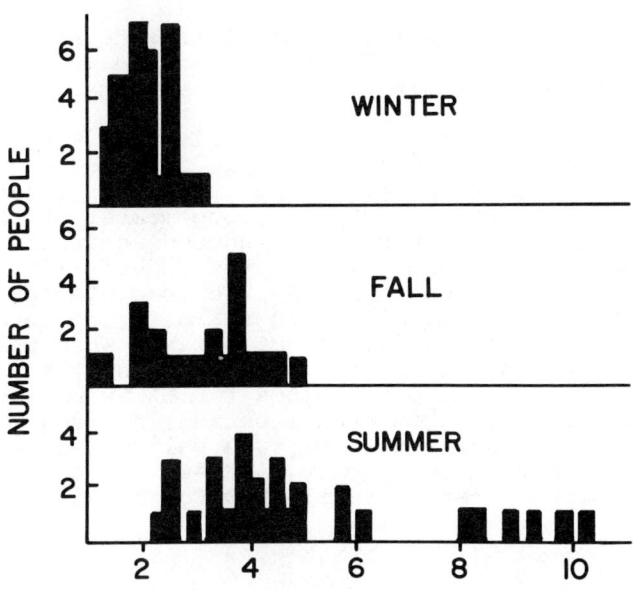

Fig. 2. Seasonal changes in AHH inducibility ratios. The distribution of AHH inducibility for 92 normal individuals is shown for summer (July-August), fall (September-October), and winter (November-December).

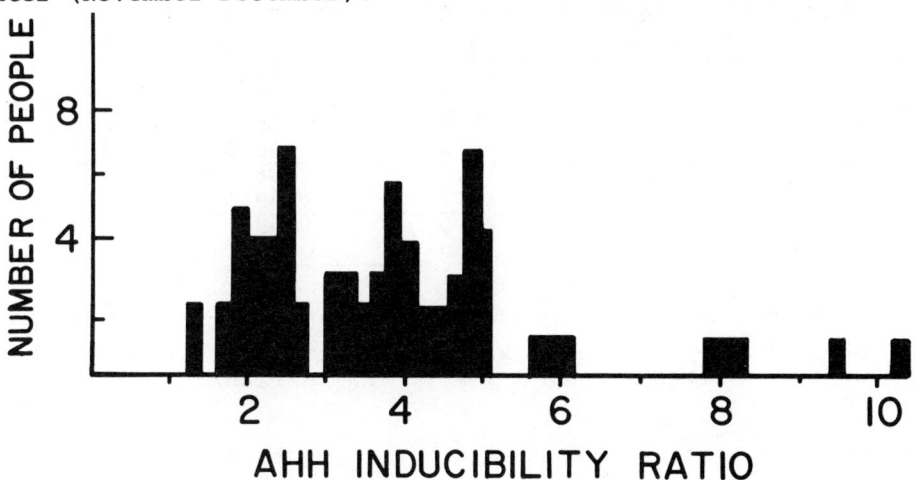

Fig. 3. Distribution of AHH inducibility. This distribution of 76 normal individuals was obtained from July-September 1975.

Since the population distribution showed considerable variation, we next asked whether any of the variation in AHH inducibility between individuals was genetically determined. To do this, we measure AHH in 18 pairs of monozygotic (MZ) and 30 pairs of dizygotic (DZ) twins. In order to control for the seasonal effect, we worked only during the high season and we always measured both members of a twin pair on the same day.

Some typical data from the twin study are given in Table 1. Most MZ twins had similar values for basal, induced, and inducibility ratio as illustrated by the twin pair 603/604. In 2 sets of MZ twins 619/620 and 698/699), one member of the twin pair was on medication. The basal and induced activities of the twin receiving medication were low, but the inducibility ratio was similar. Thus medication depresses the absolute AHH activity levels but does not appear to alter the ratios. The twin pair 692/693 was the most different of all MZ pairs. When the data for 16 pairs of MZ twins were averaged, (excluding the 2 twin pairs with 1 twin on medication), values for MZ twins were as similar as a blood sample taken from 1 person, divided in half and assayed on the same day. Most DZ twin pairs also gave very similar values as represented by pair 643/644. However, some DZ twin pairs were quite different as illustrated by pairs 696/697 and 710/711.

TABLE 1 Aryl hydrocarbon hydroxylase in twins

Twin Pair	Zygosity	Basal AHH	Induced AHH	Ratio
603/604	MZ	8.17	30.6	3.74
		7.70	26.4	3.43
619/620	MZ	10.6	54.1	5.12
		19.7	112.5	5.71
698/699	MZ	12.7	50.0	3.94
		23.5	109.0	4.64
692/693	MZ	45.0	111.0	2.46
		36.6	136.0	3.72
643/644	DZ	14.1	66.1	4.69
		14.7	69.6	4.73
696/697	DZ	20.0	203.0	10.2
		35.0	115.0	3.28
710/711	DZ	24.1	124.0	5.14
		18.3	52.5	2.87

AHH activity is pmoles product/30 min per culture flask. Culture flasks averaged 2×10^6 cells/flask.

Using the classical formula for determining heritability:

$$\text{Heritability} = \frac{\text{Variance DZ} - \text{Variance MZ}}{\text{Variance DZ}}$$

the heritability for AHH inducibility ratio is 0.68, for the basal 0.32 and for the induced 0.61. If the 2 MZ twin pairs with 1 pair on medication are excluded from the calculations, the heritability for basal and induced activity increases to 0.57 and 0.85 respectively. These heritabilities were calculated using intra pair differences normalized to the mean of each twin pair.

As mentioned previously, most DZ twins had values as similar as the MZ twins. Table 2 gives the normalized intrapair differences for AHH inducibility for MZ and DZ pairs. Of 30 DZ twin pairs, 26 fell into the range observed for MZ twins. Only 4 pairs were significantly different in phenotype. This high frequency of concordant twin pairs is expected if a trait is determined by a single gene because two sibs from the same set of parents will frequently be concordant for a given trait. The percentage of concordant pairs expected for traits determined by 1, 2, or 3 genes is given in Fig. 4. From the high degree of concordancy observed for AHH inducibility in this study, we consider it likely that the trait is determined by a single gene. However, the family studies required to prove single gene inheritance cannot be done because distinct phenotypes for AHH inducibility cannot be assigned to individuals.

TABLE 2 Normalized intrapair differences in AHH inducibility between members of twin pairs

Range	Number of Pairs	
	MZ	DZ
0 - .10	10	14
.11 - .20	6	10
.21 - .30	1	1
.31 - .40	1	1
.41 - .50	0	2
.51 - .60	0	1
.61 - .70	0	0
.71 - .80	0	1
Total	18	30

From the twin study, we conclude that AHH activity levels and inducibility ratios are genetically determined traits and that the data for the inducibility ratios are consistent with a trait determined by a single gene.

We have attempted to determine AHH inducibility in lung cancer patients but have failed to obtain measurable AHH from the cultures of at least half the patients. We found that lymphocytes from many lung cancer patients do not survive the 3-day culture period to the same degree as normals.

In summary we find that the human population varies in AHH inducibility and that the variation is genetically determined. These findings agree with an earlier report by Kellermann, Kellermann and Shaw (6). We also find that AHH activity varies with the season of the year and is affected by medication.

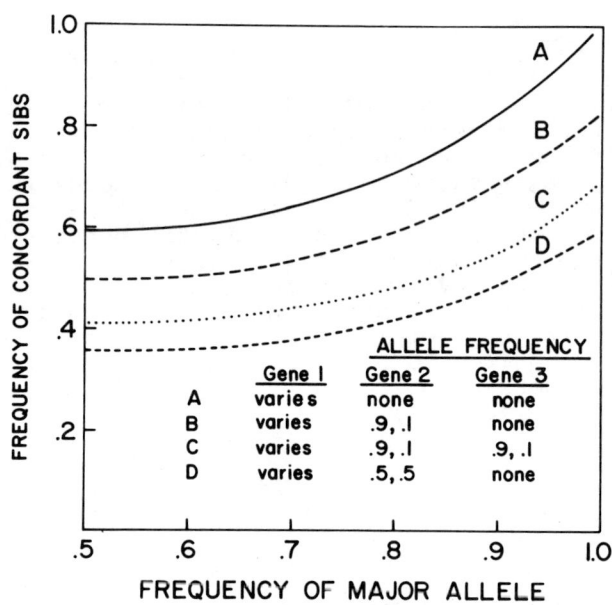

Fig. 4. Expected concordancy of sib pairs

REFERENCES

(1) A.H. Conney, Pharmacological implications of microsomal enzyme induction, Pharmacol. Rev. 19, 317 (1967).

(2) P.E. Thomas and J.J. Hutton, Genetics of aryl hydrocarbon hydroxylase induction in mice: additive inheritance in crosses between C3H/HeJ and DBA/2J. Biochem. Genet. 8, 249 (1973).

(3) R.E. Kouri, H. Ratrie, and C.E. Whitmire, Evidence of a genetic relationship between susceptibility to 3-methylcholanthrene-induced subcutaneous tumors and inducibility of aryl hydrocarbon hydroxylase, J. Nat. Cancer Inst. 51, 197 (1973).

(4) P.E. Thomas, J.J. Hutton, and B.A. Taylor, Genetic relationship between aryl hydrocarbon hydroxylase inducibility and chemical carcinogen induced skin ulceration in mice, Genetics 74, 655 (1973).

(5) D.W. Nebert, J.R. Robinson, A. Niwa, K. Kumaki, and A.P. Poland, Genetic expression of aryl hydrocarbon hydroxylase activity in the mouse, J. Cell Physiol. 85, 393 (1975).

(6) G. Kellermann, M. Luyten-Kellermann and C.R. Shaw, Genetic variation of aryl hydrocarbon hydroxylase in human lymphocytes, Amer. J. Human Genetics 25, 371 (1973).

(7) G. Kellermann, C.R. Shaw, and M. Luyten-Kellermann, Aryl hydrocarbon hydroxylase inducibility and bronchogenic carcinoma, New Eng. J. Med. 289, 971 (1973).

(8) D.W. Nebert and H.V. Gelboin, Substrate-inducible microsomal aryl hydroxylase in mammalian cell culture. I. Assay and properties of induced enzyme, J. Biol. Chem. 243, 6242 (1968).

(9) H.L. Gurtoo, N. Bejba, and J. Minowada, Properties, inducibility and an improved method of analysis of aryl hydrocarbon hydroxylase in cultured human lymphocytes, Cancer Res. 35, 1235 (1975).

(10) P.K. Beuthin and W.F. Bousquet, Long-term variation in basal and phenobarbital-stimulated oxidative metabolism in the rat, Biochem. Pharmacol. 19, 620 (1970.

ASSIGNMENT OF A HUMAN GENE FOR ARYL HYDROCARBON HYDROXYLASE EXPRESSION TO CHROMOSOME 2

F. J. Wiebel,[*] S. Brown,[†] J. D. Minna[†] and H. V. Gelboin[‡]

[*] Gesellschaft für Strahlen- und Umweltforschung, Abteilung Toxikologie, Neuherberg, Germany,
[†] NCI-VA Medical Oncology Branch and
[‡] Chemistry Branch, NCI, NIH, Bethesda, U.S.A.

ABSTRACT

Hybrid cells were formed from mouse RAG cells and fresh human bone marrow cells. The expression of the mixed-function oxidase, aryl hydrocarbon hydroxylase (AHH), in 15 out of 31 hybrid clones correlated with the presence of the human isozymes malate dehydrogenase and isocitrate dehydrogenase, known markers for chromosome 2. AHH activity was not detectable in the parental mouse cell line. Isozyme markers for 17 other chromosomes segregated independently of AHH activity in the mouse x human cell hybrids. The results suggest that the structural gene(s) for AHH are located on human chromosome 2.

INTRODUCTION

Activity and distribution of microsomal mixed-function oxidases are key factors in mammalian metabolism of xenobiotics. Genetic dissection of the regulatory mechanism(s) and identification of different forms of these cytochrome P-450 containing enzymes in man would greatly aid our understanding of the human disposition to toxic and carcinogenic chemicals.

Formation of interspecific hybrids of somatic cells has been used to map genes (1 - 3) and study the regulation of gene action (4). Specifically, the preferential segregation of human chromosomes in hybrids formed from established rodent cells and human cells provides a basis for human gene linkage studies (5 - 7).

The following describes our attempts to correlate the expression of a mixed-function oxidase, aryl hydrocarbon hydroxylase (AHH), with the presence of human chromosome(s) in hybrids formed from mouse cells that do not exhibit AHH activity and human bone marrow cells.

The presence of human chromosomes in the hybrid cells was determined by analyzing the expression of "marker" enzymes which have been assigned to specific human chromosomes (8) and which can readily be distinguished from the mouse enzymes by gel electrophoresis (9 - 11).

The results indicate that the structural gene(s) for constitutive and induced AHH are located on human chromosome 2.

Abbreviations (see also Table 1): AHH = aryl hydrocarbon hydroxylase

METHODS

Hybrid cells were formed by fusion of mouse 8-azaguanine resistent RAG cells, a heteroploid cell line derived from a renal adenoma (12), and fresh human bone marrow from a normal 24 year old female as described previously (7, 9, 11). Mouse and human isozymes in homogenates of cells (Table 1) were analyzed by vertical starch gel electrophoresis following established procedures (9-11). AHH activity was determined in parallel sets of cultures as described previously (13) except that 3 ml of the organic phase were extracted with 1 ml of 1 N NaOH. AHH specific activity is expressed as femtomol equivalents of 3-hydroxy-benzo(a)pyrene formed/min/mg protein. Heat inactivated homogenates served to determine the "background" fluorescence.

Table 1 Summary of Isozymes Tested

Human Isozymes	Symbol	Human Chromosome	E. C. Nr.	Reference
Glucose 6-Phosphate Dehydrogenase	G6PD	X	1.1.1.49	10
Peptidase C	PEP C	1	3.4	10
Adenylate Kinase-2	AK-2	1	2.7.4.3	20
Malate Dehydrogenase	MDH(MOR)	2	1.1.1.37	10
Isocitrate Dehydrogenase	IDH	2	1.1.1.42	10
Hexosaminidase B	HEX B	5	3.2.1.30	20
Malic Enzyme	ME (MOD)	6	1.1.1.40	10
ß-Glucuronidase	ß-GLU	7	3.2.1.31	21
Adenylate Kinase-1	AK-1	9	2.7.4.3	20
Glutamate Oxaloacetate Transaminase	GOT	10	2.6.1.1	10
Lactate Dehydrogenase	LDHA	11	1.1.1.27	10
Peptidase B	PEP B	12	3.4	10
Esterase D	EST D	13	3.1.1.1	22
Nucleoside Phosphorylase	NP	14	2.4.2.1	10
Mannose Phosphate Isomerase	MPI	15	5.3.1.8	10
Adenine Phosphoribosyltransferase	APRT	16	2.4.2.7	23
Peptidase A	PEP A	18	3.4	10
Glucosephosphate Isomerase	GPI	19	5.3.1.9	10
Adenosine deaminase	ADA	20	3.5.4.2	10
Superoxide dismutase-1	SOD-1	21	1.15.1.1	10

RESULTS AND DISCUSSION

The hybrid character of the cells which grew out in colonies 6 - 8 weeks after fusion of mouse RAG cells and human bone marrow cells was verified by the concomitant expression of some human isozymes and the mouse isozymes expected for mouse strain BALB. The hybrid colonies grew with a doubling time of 18 - 24 hours.

Parent RAG cells and 12 primary hybrid clones were examined for the expression of induced AHH and human isozymes. In RAG cells neither constitutive nor

Table 2 Aryl Hydrocarbon Hydroxylase¶ and Human Enzyme Activity† in Human × Mouse Hybrid Clones

Cell Line	AHH	G6PD	PEPC	AK-2	MDH	IDH	HEX B	ME	β-GLU	AK-1	GOT	LDHA	PEPB	EST.D	NP	MPI	APRT	PEPA	GPI	ADA	SOD-1
	(x)	(1)‡	(1)	(2)	(2)	(5)	(5)	(6)	(7)	(9)	(10)	(11)	(12)	(13)	(14)	(15)	(16)	(18)	(19)	(20)	(21)
RBM 6	−	+	−	(+)§	−	−	−	−	−	+	+	−	−	+	−	−	−	−	−	−	−
RBM 8	−	+	−	−	−	−	−	+	−	+	+	+	+	(+)	−	−	−	−	−	−	−
RBM 9	+	+	+	+	+	−	−	+	−	+	+	+	−	+	+	+	+	+	+	+	+
RBM 13	−	+	+	+	−	(+)	−	−	−	+	+	+	+	+	(+)	+	(+)	+	+	−	+
RBM 15	−	+	+	−	−	−	−	−	−	+	+	−	−	+	(+)	+	+	+	+	−	+
RBM 16	−	+	−	−	−	+	−	−	−	+	−	−	−	+	−	−	−	(+)	−	−	−
RBM 18	−	+	−	−	−	−	−	−	−	+	−	−	−	+	−	−	−	−	−	−	−
RBM 19	+	+	−	−	−	−	−	+	−	+	+	−	−	+	(+)	−	+	−	−	−	+
RBM 24	+	+	+	+	+	−	−	+	−	+	+	−	−	+	−	−	−	−	−	+	+
RBM 25	+	+	+	−	−	(+)	−	−	−	+	−	−	−	+	−	+	+	+	−	+	+
RBM 27	−	+	(+)	(+)	−	−	−	−	−	−	−	−	−	+	−	−	−	−	−	−	−
RBM 29	−	+	(+)	(+)	−	−	−	−	−	−	+	−	−	+	−	−	−	−	−	−	−

¶ Cultures were exposed to benz(a)anthracene for 24 hr. AHH activity was assayed in cell homogenates from duplicate cultures. Incubation time was 30 min.. Range of values from duplicate cultures was less than 20% of their mean. AHH (+) represents a specific fluorescence (398 nm ex./522 nm em.) two times greater than the fluorescence of the blanks.

† Human enzyme activities were determined by starch gel electrophoresis. Mouse enzymes were always expressed.

§ (+) = weak band of activity after incubation at 37° C for 30 min.

‡ Chromosome number.

Table 3 Correlation of Aryl Hydrocarbon Hydroxylase¶ and Human Enzyme Activities† in Subclones of RBM 24 and 25.

Clone RBM	AHH	AK-2	MDH	IDH	ME	GOT	EST-D	NP	SOD-1
24 S_1	−	−	−	−	−	+	+	+	+
S_2	−	−	−	−	−	−	+	−	+
S_3	−	+	−	−	+	+	+	+	+
S_4	1720 ± 350	+	+	+	−	−	+	−	+
S_5	−	−	−	−	−	+	+	+	+
S_6	−	−	−	−	+	−	−	−	+
S_7	315 ± 15	+	(+)§	(+)	−	−	+	−	+
S_8	200 ± 20	+	(+)	(+)	−	−	+	−	+
S_9	−	−	−	−	(+)	−	+	−	+
25 S_1	1570	+	+	+	+	+	−	+	+
S_2	152 ± 5	+	+	+	+	+	+	+	+
S_3	−	+	−	−	+	+	+	+	+
S_4	545 ± 23	+	+	+	−	+	+	+	+
S_5	199 ± 15	+	+	+	+	−	+	+	+
S_6	178 ± 70	+	+	+	+	+	+	+	+
S_7	777 ± 90	+	+	+	+	−	+	+	+
S_8	376 ± 12	+	+	+	+	+	+	+	+
S_9	−	+	−	−	+	−	+	+	+
S_{10}	−	+	−	−	−	+	+	+	+

¶   Cultures were exposed to inducers (4 μM benz(a)anthracene and 0.7 mM aminophylline) for 18 hrs. AHH activity (fmol/min/mg protein) was determined in homogenates of 2 pooled cultures. Incubation time was 1 hr. Values give the mean and range of duplicate sets of cultures. (−) represents a specific fluorescence (398 nm ex./522 nm em.) less than two fold the fluorescence of the blanks. The calculated specific activities of AHH in these "negative" samples amount maximally to 25. They were considered to be insignificant.

†, §   see legend to Table 2.

induced AHH could be detected (Fig. 1). As shown in Table 2, the presence of induced AHH in 3 hybrid clones and the absence in 9 other clones correlated only with the expression of the human isozymes, malate dehydrogenase (MDH), isocitrate dehydrogenase (IDH), and malic enzyme (ME), markers for chromosome 2 and 6, respectively (10). Seventeen other analyzed human chromosomes segregated independently of AHH induction. The possible involvement of the remaining 5 chromosomes (3, 4, 7, 17, 22) has to await further examination.

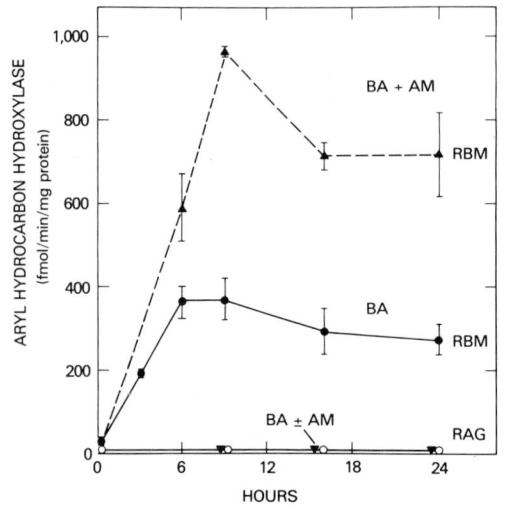

Fig. 1. Time course of induction of aryl hydrocarbon hydroxylase in RAG cells and a clone of hybrid cells.
RAG cells (=RAG) and the hybrid clone RBM 24S4 were exposed to culture medium containing 4 μM benz(a)anthracene (=BA) without or with 0.7 mM aminophylline (=BA + AM) for various time periods. AHH activity was determined in homogenates of 2 pooled dishes. Incubation time was 60 min. Mean and range of determinations from 2 sets of dishes.

Two of the inducible lines, RBM 24 and 25, were subcloned and retested for AHH and isozyme activities. The sensitivity of the test for induced AHH activity was increased by addition of aminophylline to the growth medium (14). As shown in Fig. 1 induced AHH activities in the hybrid clone RBM 24S4 after exposure to aminophylline and benz(a)anthracene are twice those after exposure to benz(a)anthracene alone. The subclones expressed fewer isozymes than the primary clones (Table 3) and a number of clones lost the expression of induced AHH. Examination of the relationship between the expression of MDH, IDH, ME and induced AHH in primary clones (cf. Table 2) and subclones (cf. Table 3) shows a strong correlation between induced AHH and the markers for chromosome 2, MDH and IDH, but not the marker for chromosome 6, ME (Table 4).

Table 4 Relationship between Selected Human Isozymes and Expression of Aryl Hydrocarbon Hydroxylase Activity

	Number of Clones				Syntenic Fraction
	+/+	+/-	-/+	-/-	
MDH/AHH	13	1	0	17	30/31
IDH/AHH	13	0	0	18	31/31
ME/AHH	9	6	4	12	21/31

Table 5 Constitutive and Induced AHH Activities and the Expression of Human Isozymes MDH and IDH in Hybrid Cells.

Clone RBM	AHH Activity[¶] Constitutive	Induced	Human Isozymes[†] MDH	IDH
16	< 2.5[‡] -	< 2.5 -	-	-
24S1	< 2.5 -	< 2.5 -	-	-
24S2	< 2.5 -	< 2.5 -	-	-
25S3	< 2.5 -	< 5.2[‡] -	-	-
25S9	< 2.5 -	< 5.5 -	-	-
24S8	< 2.5 -	15.0 $\pm$ 1	(+)[§]	-
24S7	< 2.5 -	19.0 $\pm$ 2	(+)	(+)
25S6	< 2.5 -	100.0 $\pm$ 7	+	+
18	< 2.5 -	132.0 $\pm$ 54	+	-
24	5.9 $\pm$ 1.6	127.0 $\pm$ 5	(+)	(+)
25S2	4.2 $\pm$ 0	168.0 $\pm$ 30	+	+
25S7	< 2.5 -	351.0 $\pm$ 40	+	+
25S5	5.2 $\pm$ 0.2	460.0 $\pm$ 26	+	+
25S8	11.2 $\pm$ 1.0	552.0 $\pm$ 136	+	+
25S4	11.5 $\pm$ 1.0	562.0 $\pm$ 23	+	+
24S4	40.1 $\pm$ 6.8	713.0 $\pm$ 33	+	+
25S1	64.6 $\pm$ 8.0	856.0 $\pm$ 3	+	+
25	291.0 $\pm$ 80.0	950.0 $\pm$ 90	+	+

[¶] AHH activity (fmol/min/mg protein) was determined in homogenates of 4 pooled dishes for control cultures and 2 pooled dishes for inducer-treated cultures. Incubation time was 3 hrs. Values give the mean and range from duplicate sets of cultures. Cells were exposed to inducers (cf. Table 3) or fresh medium for 18 hrs.

[‡] Specific fluorescence (398 nm ex./522 nm em.) less than 1.5 times the fluorescence of the blanks.

[†], [§] see legend to Table 2.

The single exemption to the syntenic relationship between MDH and AHH, primary clone 18 (cf. Table 2), was found to contain inducible AHH activity at a later testing period (see below).

When the sensitivity of the assay for AHH was further increased (see legend to Table 5) constitutive hydroxylase activity was detectable in 9 out of 20 clones tested (Table 5). It is apparent that the constitutive activity occured only in clones which were inducible and with the exception of clones RBM 24 and RBM 25S2 exhibited relatively high levels of induced AHH.

It is interesting to note that clone RBM 18 expresses isozymes MDH but not IDH. This dissociation could result from a break in human chromosome 2 which carries the structural genes for MDH and IDH on opposite regions, the p-arm and q-arm (15). The observation may suggest that the locus for the expression of AHH is more closely linked to the gene of MDH (p-arm) than that of IDH (q-arm).

Since the human and mouse hydroxylases are presently undistinguishable in cultured cells, we cannot exclude that AHH activity in the RBM hybrid cells arises from complementation of this multicomponent enzyme system or from the activation of a previously unexpressed function of the mouse genome by human chromosome 2. However, this appears less likely in view of observations made on hybrids from other mouse and human cells which showed the same dependency of AHH expression on the presence of human chromosome 2 (unpublished observations).

It is known that more than one form of cytochrome P-450 metabolizes benzo(a)-pyrene in mammalian cells (16 - 19) and both the constitutive and the induced AHH might consist of different forms of the enzyme. The present study does not differentiate between the various forms of the cytochrome.

The stringent correlation between induced AHH activity and the expression of isozyme markers for chromosome 2 suggests that the structural gene is carried on this chromosome rather than a regulatory gene. Otherwise some clones would be expected to evolve which lost the structural gene with another chromosome and retained the regulatory unit on chromosome 2 that can no longer be expressed. This apparently did not occur: The presence of the chromosome 2 markers was invariably associated with expression of AHH activity. Furthermore if the sturctural gene for the constitutive AHH activity were located on another chromosome than chromosome 2, this activity would be expressed independently of the induced enzyme and the expression of marker enzymes for chromosome 2. Again, this was not the case: All hybrids expressing constitutive AHH activity were also inducible and expressed human MDH and IDH. Thus, the data suggest that the structural gene(s) for both, constitutive and induced AHH are located on human chromosome 2. The regulatory gene on the other hand is not necessarily located on this chromosome but might be part of either another human chromosome which is present in all hybrid clones (e.g. 1 or 21) or a mouse chromosome.

Further studies in hybrid cells may lead to the assignment of the gene(s) involved in mono-oxygenase induction and help to unravel the genetic regulation of different forms of these enzymes and their role in the human response to toxic and carcinogenic xenobiotics.

ACKNOWLEDGEMENT

We are grateful for the excellent technical assistance of Ms. Stephenson and Mr. H. L. Waters and the skillful secretarial help of Ms. I. Steffen.

REFERENCES

1. M. Nabholz, V. Miggiano, and W. Bodmer, Genetic analysis with human-mouse somatic cell hybrids, Nature 223, 358 (1969).
2. F. H. Ruddle, Linkage analysis using somatic cell hybrids, Adv. Hum. Genet. 3, 173 (1972).
3. K. H. Grzeschik, P. W. Alderice, A. Grzeschik, J. M. Opitz, O. J. Miller, and M. Siniscalco, Cytological mapping of human X-linked genes by use of somatic cell hybrids involving an X-autosome translocation, Proc. Nat. Acad. Sci. U.S.A. 69, 69 (1972).
4. Davidson, R. L., and de la Cruz, F. (eds.) Regulation of gene expression in hybrid cells, p. 131, in Somatic cell hybridization, Raven Press, New York, 1974.
5. M. Weiss, and H. Green, Human-mouse hybrid cell lines containing partial complements of human chromosomes and functioning human genes, Proc. Nat. Acad. Sci. U.S.A. 58, 1101 (1967).
6. A. Westerveld, P. Meera Khan, R. P. L. S. Visser, and D. Bootsma, Loss of human genetic markers in man-chinese hamster somatic cell hybrids, Nature New Biol. 234, 20 (1971).
7. J. D. Minna, and H. G. Coon, Human x mouse hybrid cells segregating mouse chromosomes and isozymes, Nature 252, 401 (1974).
8. T. B. Shows, Gene markers for mapping the human genome, Cytogenetics and Cell Genet. 14, 199 (1975).
9. J. Minna, D. Glazer, M. Nirenberg, Genetic dissection of neural properties Nature New Biol. 235, 225 (1972).
10. E. A. Nichols, and F. H. Ruddle, A review of polymorphism, linkage, and electrophoretic conditions for mouse and somatic cell hybrids in starch gels, J. Histochem. and Cytochem. 21, 1066 (1973).
11. J. D. Minna, A. F. Gazdar, G. M. Iverson, T. H. Marshall, K. Stromberg, and S. H. Wilson, Oncornavirus expression in human x mouse hybrid cells segregating mouse chromosomes, Proc. Nat. Acad. Sci. U.S.A 71, 1695 (1974).
12. B. Felluga, A. Claude, and E. Mrena, Electron microscope observations on virus particles associated with a transplantable renal adenocarcinoma in BALB/cf/Cd mice, J. Nat. Cancer Inst. 43, 319 (1969).
13. F. J. Wiebel, E. J. Matthews, and H. V. Gelboin, Ribonucleic acid synthesis-dependent induction of aryl hydrocarbon hydroxylase in the absence of ribosomal ribonucleic acid synthesis and transfer, J. Biol Chem. 247, 4711 (1972).
14. H. Yamasaki, E. Huberman, and L. Sachs, Regulation of aryl hydrocarbon (benzo(a)pyrene) hydroxylase activity in mammalian cells. Induction of hydroxylase activity by dibutyryl-cyclic AMP and aminophylline, J. Biol. Chem. 250, 7766 (1975).

15. J. L. Hamerton, and P. J. L. Cook, Report of the committee on the genetic constitution of chromosomes 1 and 2, Cytogenetics and Cell Genet. 14, 173, (1975).

16. F. J. Wiebel, J. C. Leutz, L. Diamond, and H. V. Gelboin, Aryl hydrocarbon (benzo(a)pyrene) hydroxylase in microsomes from rat tissues: differential inhibition and stimulation by benzoflavones and organic solvents, Arch. Biochem. Biophys. 144, 78 (1971).

17. A. Y. Lu, R. Kuntzmann, S. West, M. Jacobsen, and A. H. Conney, Reconstituted liver microsomal enzyme system that hydroxylates drugs, other foreign compounds, and endogenous substrates, J. Biol. Chem. 247, 1727 (1972).

18. F. M. Goujon, D. W. Nebert, and J. E. Gielen, Genetic expression of aryl hydrocarbon hydroxylase induction. IV. Interaction of various compounds with different forms of cytochrome P-450 and the effect on benzo(a)pyrene metabolism in vitro, Mol. Pharmacol. 8, 667 (1972).

19. F. J. Wiebel, J. K. Selkirk, H. V. Gelboin, D. A. Haugen, T. A. van der Hoeven, and M. J. Coon, Position-specific oxygenation of benzo(a)pyrene by different forms of purified cytochrome P-450 from rabbit liver, Proc. Nat. Acad. Sci. U.S.A. 72, 3917 (1975).

20. P. A. Lalley, M. C. Rattazzi, and T. B. Shows, Human B-D-N-acetylhexosaminidases A and B: expression and linkage relationships in somatic cell hybrids, Proc. Nat. Acad. Sci. U.S.A. 71, 1569 (1974).

21. P. A. Lalley, and T. B. Shows, Cytogenetics and Cell Genet., in press.

22. V. van Heynigen, M. Bobrow, W. F. Bodmer, S. Povey, S. E. Gardiner, and D. A. Hopkinson, Chromosome assignment of some human enzyme loci: mitochondrial malate dehydrogenase to 7, mannose phosphate isomerase and pyruvate kinase to 15 and probably, esterase D to 13, Ann. Hum. Genet. 38, 295 (1975).

23. E. A. Nichols, and F. H. Ruddle, A modified technique for separating mouse and chinese hamster from human HPRT and human and chinese hamster from mouse APRT, Cytogenetics and Cell Genet. 13, 132 (1974).

OXIDATIVE METABOLISM OF CARCINOGENS BY TROUT LIVER RESULTING IN PROTEIN BINDING AND MUTAGENICITY

J. Ahokas, R. Pääkkönen, K. Rönnholm, V. Raunio, N. Kärki, and O. Pelkonen

Departments of Pharmacology and Microbiology, University of Oulu, SF-90220 Oulu 22, Finland

ABSTRACT

Protein binding of benzo(α)pyrene and mutagenicity resulting from metabolic activation by trout liver microsomal mixed function oxidase was investigated. The metabolite pattern produced by the relatively active benzo(α)pyrene hydroxylase of trout liver microsomes was established by HPLC and compared with that produced by rat liver microsomes. Three dihydrodiols, quinones and two phenols of benzo(α)pyrene were detected as well as indications of other, unidentified metabolites. The pathway of production of some of the metabolites is apparently via reactive intermediates which bind covalently to protein. Like in the mammalian system the intermediates seem to include epoxides as the inhibition of epoxide hydratase results in a relative increase in covalently bound benzo(α)pyrene. Trout liver microsomal mixed function oxidase results in formation of mutagenic intermediates from benzo-(α)pyrene, aflatoxin B_1 and 2-acetylaminofluorene.

INTRODUCTION

There is considerable amount of data on the incidence of hepatoma, epidermal and other forms of cancerous growths in trout and other species of fish (1,2). Chemicals are being implicated as causative agents in these reports. This is also in accordance with the view that majority of cancers are a result of exposure to chemical agents (3). Nevertheless, the presence of mixed function oxidase (MFO) and particularly its functional significance in fish has not gained full acceptance (4). It is well documented that various chemicals are carcinogenic to fish (5), therefore it should follow that fish have an ability to activate procarcinogens metabolically. Although benzo(α)pyrene (BP) has not been shown to cause cancer in trout, we chose it as a model substrate because it has been widely studied and much background information is available. Using BP, we have looked for what are thought to be essential requirements of carcinogen activation, e.g. metabolism, metabolically mediated protein binding and mutagenicity.

MATERIALS AND METHODS

The fish were hatchery (Oulujoki Oy, Finland) reared juvenile lake trout, *Salmo trutta lacustris*. The trout weighed between 100 and 200 g. The rats used were either untreated or Clophen pretreated (500 mg/kg i.p., 5 days before killing) male Sprague-Dawley rats (200 to 300 g). The microsomes were prepared as described previously (6). The incubations were carried out at or near previously established "optimum" temperatures (7), being 30 °C and 37°C for trout and rat, respectively.

The incubations for the high performance liquid chromatographic (HPLC) analysis of metabolites were carried out using 12.5 mg microsomal protein, 1.59 µmoles of BP and sufficient amount of NADPH generating system in a total volume of 8 ml. The incubations were carried out in wide based flasks for 15 min, after which the incubation mixture was extracted twice with 10 ml of ethyl acetate. The ethyl acetate fractions were combined and dried over anhydrous Na_2SO_4. The extracts were evaporated to dryness under nitrogen atmosphere and redissolved in 50 µl of ethyl acetate from which 10 µl fractions were used for HPLC analysis. A Du Pont 830 HPLC instrument with 1 m x 2.1 mm ODS-permaphase (DuPont) column was used. Linear gradient from 30% to 70% aqueous methanol increasing at 3%/min was used. The elution pressure was 500 psi and the column temperature was 50°C. Monitoring was by u.v. absorbance at 254 nm.

To determine the extent of metabolically mediated covalent binding to protein, 100 nmoles of $(7,10-^{14}C)$benzo(α)pyrene (0.5 µCi/µmol) was incubated in the presence of 5 mg of trout liver microsomal protein, 40 mg albumin and sufficient amount of NADPH generating system in a total of 4 ml. After 15 min incubation at 30°C the reaction was stopped by an addition of 6 ml of ethyl acetate and extracted with further 6 ml of ethyl acetate. The ethyl acetate phases were collected, combined and concentrated under a nitrogen stream. The ethyl acetate extracts were analysed for total metabolites according to the method of DePierre *et al*., 1975 (8).

The aqueous phase with the precipitated protein was analysed for covalently bound BP according to the method of Siekevitz (9).

The mutagenicity tests were conducted according to Ames *et al*. (10), using TA 1538 and TA 98 strains of *Salmonella typhimurium* (obtained from B. Ames).

RESULTS

Using BP as a model substrate it can be seen that trout liver microsomes metabolize it into numerous hydroxylated metabolites and quinones in quantities exceeding those produced by control male rat liver microsomes (Fig. 1a and 1b). A number of metabolites can be identified on the basis of their retention times using HPLC and authentic standards. The metabolites identified are 9,10-dihydrodiol, 4,5-dihydrodiol and 7,8-dihydrodiol of BP, 3-OH-BP

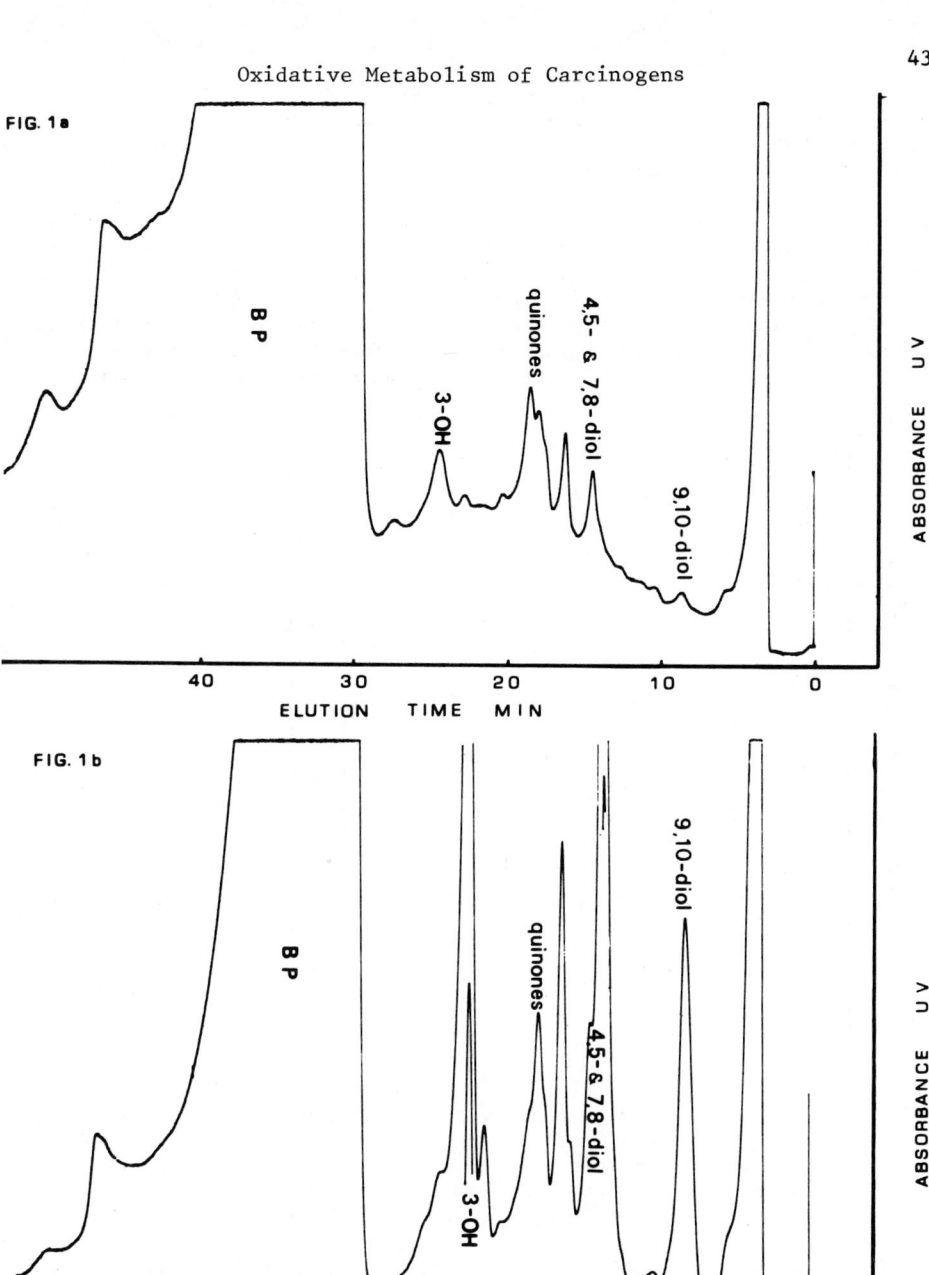

Fig. 1a. HPLC separation of BP metabolites produced by male rat liver microsomes. Fig. 1b. The metabolite pattern produced by trout liver microsomes. Detection is by u.v. absorbance at 254 nm; sensitivity 8×10^{-2} absorbance units full scale.

and 9-OH-BP (appearing immediately before 3-OH-BP on the chromatograms), as well as three quinones (1,6-, 3,6- and 6,12-quinones of BP; not well separated under present chromatographic conditions).

The production of electrophilic intermediates, such as epoxides, is reflected in the fact that ^{14}C labelled BP is extensively bound covalently to protein. Approximately 4% of total hydroxylated BP can be found to be bound to protein (Fig. 2). (With tritium labelled BP considerably higher level of binding of radioactivity to protein was encountered, but this is due to the inherent problems involved in using tritium labelled compounds in this kind of study (11)).

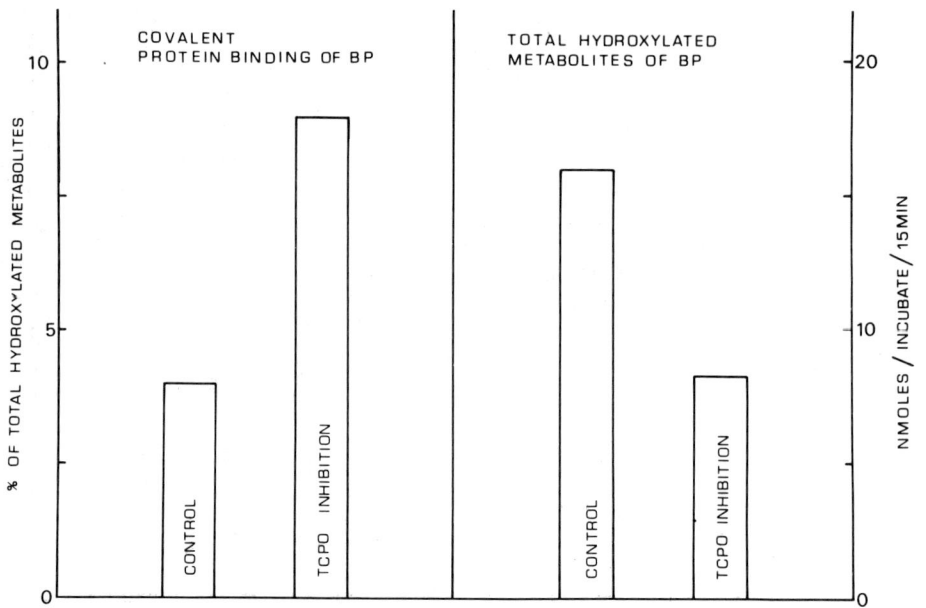

Fig. 2. Metabolically mediated binding of ^{14}C-BP covalently to protein related to the total hydroxylation and the effect of an inhibitor (TCPO, 2.5 mM) of epoxide hydratase to this binding. Trout liver microsomal protein content is 5 mg per each 4 ml incubate.

Epoxide hydratase inhibitor 1,2-epoxy-3,3,3-trichloropropane (TCPO) was found to increase the relative degree of covalent binding of BP to protein (Fig. 2). This would be an indication that epoxide intermediates are accumulated in the incubation medium as a result of inhibition of epoxide hydratase.

The mutagenicity studies were carried out using *Salmonella typhimurium* TA 1538 and TA 98 strains and 9000 x g supernatants of Clophen pretreated male rat and untreated trout liver homogenates. A significant increase in the formation of revertants can be seen using a number of procarcinogens and 9000 x g supernatant of trout liver homogenates (table 1). With strain TA 1538 the mutagenicity caused by BP and 2-acetylaminofluorene (2-AAF) is marginal. However, aflatoxin B_1 (ATB_1) which fails to cause mutagenicity with the rat liver 9000 x g supernatant does so with the trout liver 9000 x g supernatant. When TA 98 strain is used also BP causes a significant number of revertants in conjunction with trout liver MFO.

TABLE 1   Induction of mutations using *Salmonella typhimurium* TA 1538 and TA 98 strains

Strain used	TA 1538			TA 98		
	Without S9 mix	Rat S9 mix	Trout S9 mix	Without S9 mix	Rat S9 mix	Trout S9 mix
Histidine[a]	11	63	164	12	40	100
BP	10	344	250	13	330	297
ATB_1	6	52	304	22	716	304
2-AAF	13	310	268	18	222	258

[a] Histidine concentration is 10^{-7} mole/plate.
All values are means from two plates. S9 mix = required cofactors of MFO and 9000 x g supernatant of liver homogenate; 100 μl/plate. The incubation temperature was 37°C. The rats used were Clophen pretreated (500 mg/kg i.p., 5 days before killing).
BP = benzo(α)pyrene, 5 μg/plate;
ATB_1 = aflatoxin B_1, 1 μg/plate;
2-AAF = 2-acetylaminofluorene, 20 μg/plate.

DISCUSSION

In the literature there has been an obvious conflict of ideas. On one hand trout are found to be sensitive to chemical carcinogens (2) and it is now known that chemical carcinogens in many cases require metabolically mediated conversion from procarcinogens to proximate carcinogens (3). On the other hand, it has been commonly stated that fish do not have a metabolizing system capable of metabolizing lipophilic compounds (12).

In this study we have shown that BP is metabolized into numerous metabolites by trout liver microsomal MFO and that some of these metabolites are formed via intermediates capable of covalently bonding with macromolecules. This bonding is enhanced when epoxide hydratase is inhibited. From the carcinogenicity point of view another important fact is that some metabolites are formed which are mutagenic to living cells. This was shown by using bacterial mutagenicity tests. Although this does not suffice as conclusive proof that trout forms metabolites capable of inducing mutagenicity in its own cells *in vivo*, this forms a firm basis for the assumption that this may occur in trout liver.

It must be noted that trout S9 mix causes relatively high rate of "spontaneous" mutation. Whether this is of any physiological significance remains to be seen, but at present it is a problem in conducting mutagenicity tests with trout liver preparations. It has been reported that a decomposition product of peroxidized polyunsaturated fatty acids, malonaldehyde is mutagenic (13). This aspect has not been investigated with trout liver preparations in this context. Another point of interest is the fact that trout S9 mix in conjunction with ATB_1 is able to cause mutagenicity in the TA 1538 strain. This does not normally occur with rat S9 mix.

REFERENCES

(1) Neoplasms and related disorders of invertebrate and lower vertebrate animals, National Cancer Institute Monograph 31, Washington (1968).

(2) Trout hepatoma research conference papers, edited by J. E. Halver and I. A. Mitchell, Research Report 70, U.S. Bureau of Sport Fisheries and Wildlife, Washington (1967).

(3) C. Heidelberger, Studies on the mechanisms of carcinogenesis by polycyclic aromatic hydrocarbons and their derivatives, in Polynuclear Aromatic Hydrocarbons: Chemistry, Metabolism, and Carcinogenesis, edited by R. I. Freudenthal and P. W. Jones, Raven Press, New York, 1976.

(4) A. Goldstein, L. Aronow, and S. M. Kalman, Principles of Drug Action: The Basis of Pharmacology, Wiley, New York, 1974.

(5) G. B. Pliss and V. V. Khudoley, Tumor induction by carcinogenic agents in aquarium fish, J. Natl. Cancer Inst. 55, 129 (1975).

(6) J. T. Ahokas, O. Pelkonen, and N. T. Kärki, Metabolism of polycyclic hydrocarbons by a highly active aryl hydrocarbon hydroxylase system in the liver of a trout species, Biochem. Biophys. Res. Commun. 63, 635 (1975).

(7) J. T. Ahokas, O. Pelkonen and N. T. Kärki, Cytochrome P-450 and drug-induced spectral interactions in the hepatic microsomes of trout, Salmo trutta lacustris, Acta pharmacol. et toxicol. 38, 440 (1976).

(8) J. W. DePierre, M. S. Moron, K. A. Johannesen, and L. Ernster, A reliable, sensitive and convenient radioactive assay for benzpyrene monooxygenase, Anal. Biochem. 63 470 (1975).

(9) P. Siekevitz, Uptake of radioactive alanine in vitro into the proteins of rat liver fraction, J. Biol. Chem. 195, 549 (1952).

(10) B. N. Ames, W. E. Durston, E. Yamashaki and F. D. Lee, Carcinogens are mutagens: a simple test system combining

liver homogenate for activation and bacteria for detection, Proc. Nat. Acad. Sci. 70, 2281 (1973).

(11) J. Ahokas, R. Pääkkönen, K. Rönnholm, V. Raunio and O. Pelkonen, Oxidative metabolism of carcinogens by trout liver resulting in protein binding and mutagenicity, Hoppe-Seler's Z. Physiol. Chem. Bd. 357, S. 1019 (1976).

(12) B. B. Brodie and R. P. Maickel, Comparative biochemistry of drug metabolism, Proc. 1st Int. Pharmac. Meet. 6, 299 (1961).

(13) F. H. Mukai and B. D. Goldstein, Mutagenicity of malonaldehyde, a decomposition product of peroxidized polyunsaturated fatty acids, Science 191, 868 (1976).

DRUG-DRUG INTERACTIONS VIA INHIBITION OF MICROSOMAL ENZYMES INVOLVED IN METABOLISM OF EPOXIDES PRODUCED BY MICROSOMAL MONOOXYGENASE

F. Oesch, H. R. Glatt and P. Bentley

Institute of Pharmacology, Obere Zahlbacher Strasse 67 D-6500 Mainz, Federal Republic of Germany

SUMMARY

Benzo(a)pyrene was activated by liver microsomes to mutagens detected by the reversion of histidine dependent Salmonella typhimurium TA 1537. Using pure epoxide hydratase or epoxide hydratase inhibitors, comparing animal species with high and low epoxide hydratase activity, or inducing monooxygenase activity, it was shown that epoxide hydratase was a critical enzyme for the inactivation of these mutagens. Many clinically used drugs are metabolized to epoxides. Epoxides are not necessarily mutagenic, but since epoxide hydratase has a very low substrate specificity, such epoxides may competitively inhibit the hydration of mutagenic epoxides, as demonstrated in the present study for the metabolically produced epoxide from the clinically used drug cyproheptadine. Interestingly the structurally closely related epoxide derived from carbamazepine did not significantly inhibit epoxide hydratase. In a therapeutic situation it would be expected that the concentration of the epoxides metabolically produced from the drug would be much greater than that of the epoxides produced from polycyclic hydrocarbons which are present ubiquitously, but at very low levels. Thus, competitive inhibition by the former may be very effective and may potentiate adverse biological effects of the latters.

INTRODUCTION

Microsomal monooxygenases can convert unreactive aromatic or olefinic compounds to epoxides (1,2). Several epoxides bind spontaneously to cellular macromolecules, cause mutations or transform cells in culture (1-4). Metabolic formation of epoxides, some of them relatively stable, has been reported from several clinically used drugs (for references see 5), e.g. cyproheptadine, carbamazepine, protriptyline, allobarbital, secobarbital and alphenal. Moreover, terminal metabolites which implicate metabolism via an epoxide intermediate have been observed with numerous drugs such as diethylstilbestrol, diphenylhydantion, phenobarbital and mephobarbital, phensuximide, lorazapam, imipramine, acetanilide, methaqualone. The alarming biological effects of some epoxides do not automatically imply that all epoxides have similar effects (6,7). However, the enzymes involved in epoxide metabolism possess a very low substrate specificity. Thus, epoxides differing widely in structure may effectively compete with each other for the inactivating systems. In this study we have investigated whether mutagenic metabolites produced from benzo(a)pyrene by liver microsomes can be inactivated by epoxide hydratase, and if epoxides very different from those derived from polycyclic hydrocarbons may inhibit epoxide hydratase and thus potentiate the mutagenicity of benzo(a)pyrene metabolites. Such interactions would be of practical interest, if per se harmless metabolites of compounds to which we can be exposed at high concentrations, such as to clinically used drugs, would potentiate adverse effects of environmental contaminants.

INACTIVATION OF MUTAGENS PRODUCED FROM BENZO(A)PYRENE BY EPOXIDE HYDRATASE

Benzo(a)pyrene itself did not revert histidine-dependent Salmonella typhimurium TA 1537 to histidine prototrophy. However, in the presence of liver microsomes (from untreated C3H mice) and NADPH, benzo(a)pyrene was metabolized to mutagens reverting this strain. This mutagenicity was strongly decreased by homogeneous (8) epoxide hydratase (Fig. 1). Quite modest amounts of epoxide hydratase, in fact less than present in microsomes from some other species were required to reduce the mutagenicity greatly. Addition of 16 units of enzyme (which is 16 times more than present in the mouse microsomes used) prevented the reversion of TA 1537 nearly completely (Fig. 1).

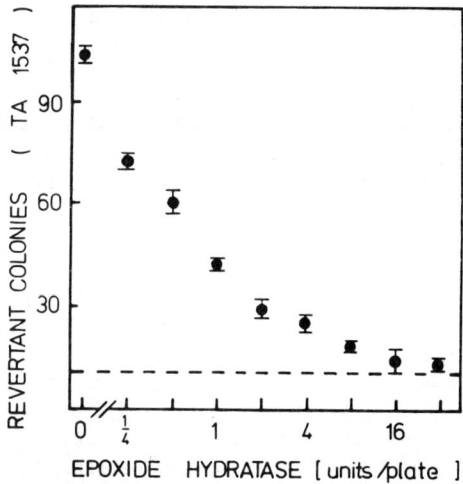

Fig. 1: Effect of pure epoxide hydratase on the mutagenicity of metabolically activated benzo(a)pyrene: Benzo(a)pyrene (10 μg per plate) was incubated with Salmonella typhimurium, liver microsomes from female C3H mice, a NADPH generating system, and various amounts of homogeneous epoxide hydratase on a histidine poor petri dish as described elsewhere (6). The figure shows the number of his$^+$ revertant colonies as a function of added epoxide hydratase (logarithmic scale). Bars represent the S.E.M. (n = 2-10). The horizontal line represents the mean of the number of spontaneous reversions.

Adult male Sprague Dawley rats, Hartley guinea pigs and humans (Caucasians) have, respectively, 7, 18 and 17 fold higher hepatic epoxide hydratase activities than C3H mice (4, 9). Indeed, the mutagenic effect of benzo(a)pyrene was quantitatively very different in the presence of liver microsomes from male Sprague Dawley rats compared to C3H mice. With rat microsomes, the mutagenic effect, although statistically significant ($p < 0.01$) was very weak (Fig. 2A). However, besides the seven fold higher

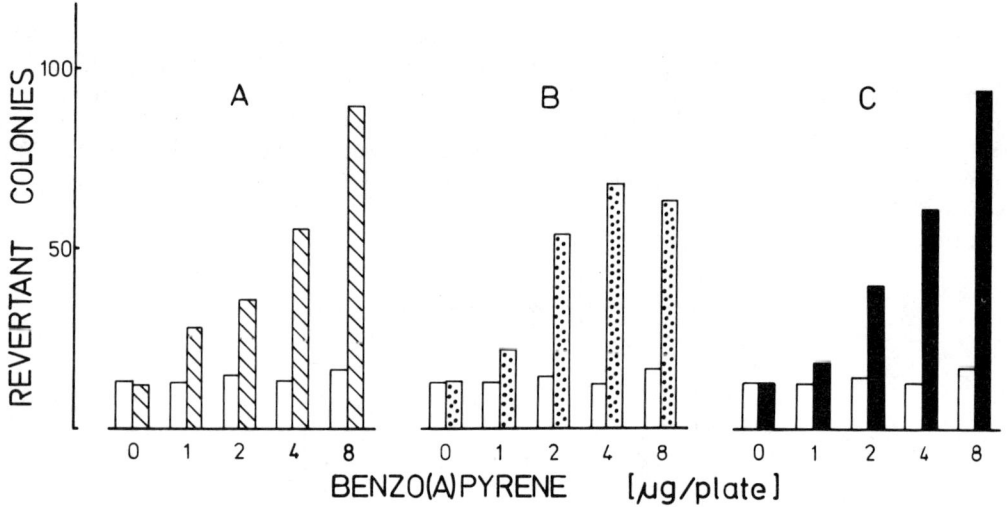

Fig. 2: Activation of benzo(a)pyrene to a mutagen reverting Salmonella typhimurium TA 1537 by various activating systems:
A) Activation by liver microsomes from male Sprague Dawley rats (□) and female C3H mice (▧).
B) Effect of the epoxide hydratase inhibitor 1,1,1-trichloropropene 2,3-oxide (930 μg/plate) (▦) on the activation by rat microsomes.
C) Activation by liver microsomes from control (□) and 3-methylcholanthrene treated (10 mg/kg, i.p. 3 days before sacrifice) (■) rats.

epoxide hydratase activity rat microsomes have also a two fold lower benzo(a)pyrene hydroxylase activity (9). If the assumption is correct that the mutagenic effect of benzo(a)pyrene in the presence of rat microsomes was much lower than in the presence of mouse microsomes because of the higher epoxide hydratase activity (or the higher ratio of epoxide hydratase towards monooxygenase activity), it should be possible to cancel this difference by appropriate enzyme modulations. Indeed, when using rat liver microsomes, inhibition of epoxide hydratase (Fig. 2 B) or induction of monooxygenase (Fig. 2 C) led to a degree of benzo(a)pyrene induced mutagenicity much more reminiscent of that with mouse than that with rat liver control preparations.

Of course, modulation of enzyme activity may affect not only the overall rate of oxidation or hydration but also the pattern of metabolism if homologous enzymes with non-identical catalytic properties are present.

DRUG-DRUG INTERACTIONS VIA INHIBITION OF EPOXIDE HYDRATASE

Studies on the substrate specificities of the active fractions obtained during purification of epoxide hydratase, the relative potencies of inhibitors to the fractions and the effect of antibodies against the homogeneous enzyme on pure and crude fractions indicated that a single enzyme in microsomes is responsible for the hydration of styrene 7,8-oxide (an alkene oxide) and benzo(a)pyrene 4,5-oxide (an arene oxide) (10).

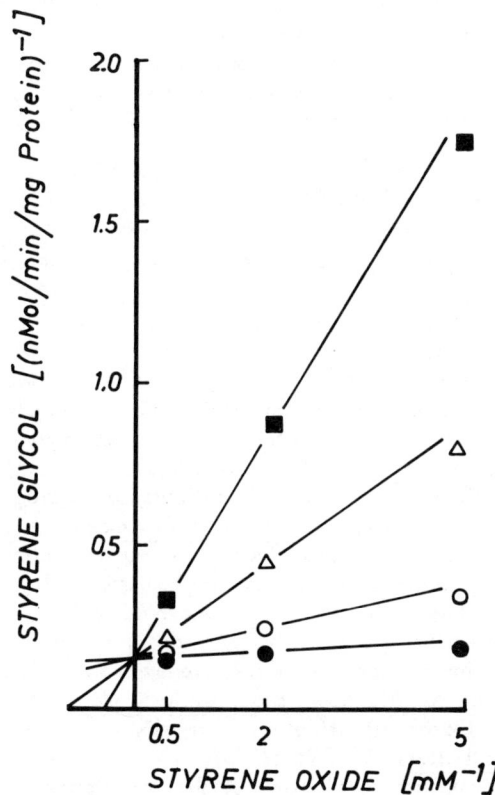

Fig. 3: Effect of cyproheptadine 10,11-oxide on epoxide hydratase activity: Styrene 7,8-oxide hydratase activity in rat liver microsomes was measured (11) in the presence of different concentrations of cyproheptadine 10,11-oxide: Control (●), 0.1 mM (○), 0.5 mM (△), 2 mM (■).

Furthermore, a great number of epoxides inhibited epoxide hydratase (4). We therefore tested two epoxides which are major metabolites of clinically used drugs, carbamazepine 10,11-oxide and cyproheptadine 10,11-oxide as inhibitors of epoxide hydratase. Carbamazepine 10,11-oxide inhibited epoxide hydratase in liver microsomes only weakly, cyproheptadine 10,11-oxide was, however, a potent inhibitor of the hydration of styrene 7,8-oxide as well as of benzo(a)pyrene 4,5-oxide (5). The epoxide hydratase inhibitors cyclohexene oxide and 1,1,1-trichloropropene 2,3-oxide potentiated strongly the mutagenicity when benzo(a)pyrene was activated by liver microsomes (Fig. 2 B and Ref. 9). Cyclohexene oxide is a non-competitive inhibitor of epoxide hydratase (4). The relative inhibition is therefore independent of substrate concentration. With 1,1,1-trichloropropene 2,3-oxide, an uncompetitive inhibitor (4), the relative inhibition decreases with decreasing substrate concentration. Cyproheptadine 10,11-oxide inhibits epoxide hydratase competitively (Fig. 3). Therefore, inhibition is very strong at low substrate concentrations. Since the concentration of epoxides produced metabolically from the very low levels of ubiquitous environmental polycyclic hydrocarbons would be smaller by several orders of magnitude than those produced from medical drugs in a therapeutic situation, competitive inhibition of the inactivation of epoxides derived from these environmental compounds may be very pronounced and potentiate their adverse biological effects.

REFERENCES

(1) D.M. Jerina, and J. Daly, Arene oxides: A new aspect of drug metabolism, Science 185, 573 (1974).
(2) P. Sims and P.L. Grover, Epoxides in polycyclic aromatic hydrocarbons metabolism and carcinogenesis, Adv. Cancer Res. 20, 165 (1974).
(3) C. Heidelberger, Chemical carcinogenesis, Ann. Rev. Biochem. 44, 79 (1975).
(4) F. Oesch, Mammalian epoxide hydrases: Inducible enzymes catalyzing the inactivation of carcinogenic and cytotoxic metabolites derived from aromatic and olefinic compounds, Xenobiotica 3, 305 (1973).
(5) F. Oesch, Metabolic transformation of clinically used drugs to epoxides: New perspectives in drug-drug interactions, Biochem. Pharmacol. 25, 1935 (1976).
(6) H.R. Glatt, F. Oesch, A. Frigerio and S. Garattini, Epoxides metabolically produced from some known carcinogens and from some clinically used drugs. I. Differences in mutagenicity. Int. J. Cancer 16, 787 (1975).
(7) H.R. Glatt, M. Metzler, H.-G. Neumann and F. Oesch, Metabolic epoxidation of trans-4-acetylaminostilbene: A protective mechanism against its activation to a mutagen, submitted to: Biochem. Biophys. Res. Commun.
(8) P. Bentley and F. Oesch, Purification of rat liver epoxide hydratase to apparent homogeneity, FEBS Letters 59, 291 (1975).
(9) F. Oesch and H.R. Glatt, Evaluation of the importance of enzymes involved in the control of mutagenic metabolites, Screening Tests in Chemical Carcinogenesis (Eds. R. Montesano, H. Bartsch, L. Tomatis) p. 255 ff, IARC Lyon (1976).
(10) F. Oesch, and P. Bentley, Antibodies against homogeneous epoxide hydratase provide evidence for a single enzyme hydrating styrene oxide and benzo(a)pyrene 4,5-oxide, Nature 259, 53 (1976).
(11) F. Oesch, D.M. Jerina and J. Daly, A radiometric assay for hepatic epoxide hydratase activity with 7-^{3}H-styrene oxide, Biochim. Biophys. Acta 227, 685 (1971).

LEVELS OF GLUTATHIONE AND GLUTATHIONE-METABOLIZING ENZYMES IN RAT LUNG

Maria S. Moron, Joseph W. DePierre, Kerstin Jacobsson, and Bengt Mannervik

Arrhenius Laboratory, Department of Biochemistry, University of Stockholm, Fack, S-104 05 Stockholm, Sweden

INTRODUCTION

There is growing evidence that glutathione, which is involved in many biological functions, plays an important role in the metabolism of polycyclic hydrocarbons. These compounds are known to be carcinogenic in several living organisms, including humans and in several organs, including the lung.

A great amount of data (Refs 1,2,3,4) has been accumulated suggesting that arene oxides, intermediates in the metabolism of polycyclic hydrocarbons, are the immediate carcinogens in cases of tumor induction.

Arene oxides can undergo a number of further metabolic steps <u>in vivo</u> to give less dangerous products (Refs 5,6); they may rearrange in water to form phenols, be hydrated by epoxide hydrase to dihydrodiols or be conjugated enzymatically or non-enzymatically with glutathione.

It is of great importance to understand the pathways for synthesis and breakdown of arene oxides, especially in the lung. We have therefore investigated the levels of glutathione and glutathione metabolizing enzymes in the rat lung and compared these with the corresponding levels in rat liver.

METHODS

<u>Reduced glutathione</u> was measured by reaction with 5,5'-dithiobis (2-nitrobenzoic acid) (DTNB) to give a compound that absorbs at 412 nm (the Ellman's method) (Ref 7).

The disadvantage of the DTNB method is its nonspecificity - the reagent will conjugate with thiols other than glutathione which may be present in the tissue investigated. However, we have checked the possibility that DTNB gives values that are too high by comparing the results obtained with this method to those obtained with a specific enzymatic assay for reduced glutathione based on the use of a glutathione-S transferase (Ref 10). We found no significant difference in the glutathione levels measured by the enzymatic and DTNB methods in the lung or in the liver. Consequently, use of the DTNB method in this study is justified.

<u>Glutathione-S-transferase</u> activity was assayed with three different substrates (Refs 8,9).

<u>Glutathione reductase</u> activity was measured by the standard spectrophotometric technique (Ref 11).

For some unexplained reason reduced glutathione in rat lung homogenates is oxidized much more rapidly than in liver or blood homogenates; so that the procedure described below has to be carried out as quickly as possible for lung tissue.

Moreover, we soon found out that the lung contains such large amounts of blood and such relatively small amounts of glutathione that correction had to be made for the glutathione content in the blood. In untreated lungs this correction was 50-75 % of the total value obtained, so that the results were unreliable. Consequently, we perfused the lung <u>in situ</u> through the right ventricle of the heart with 0,15 M NaCl before beginning determinations of reduced glutathione, a procedure which resulted in a correction for blood of only 10 % or less. Perfusion could be carried out reasonably successfully in rats whose necks had been dislocated, but it was much easier and more complete after anesthetizing the animals with nembutal. Consequently, this latter procedure was used routinely. Table 1 shows that nembutal had no effect on the levels of reduced glutathione in lung, liver and blood.

Table 1. <u>Lack of effect of nembutal anesthesia on the levels of reduced glutathione (GSH) in lung, liver and blood</u>

Tissue	Treatment	n	GSH nmoles / mg soluble prot.
Lung	Nembutal	7	71.3 ± 12.6
	Control	4	69.9 ± 9.0
Liver	Nembutal	3	92.4 ± 9.5
	Control	4	96.5 ± 7.1
Blood	Nembutal	5	15.8 ± 2.8
	Control	1	12.6

The perfused lung was homogenized vigorously and sonicated for 1 min; protein was precipitated by addition of a small volume of 25 % trichloroacetic acid to give a final concentration of 5 %. After centrifugation for 10 min. in a desk centrifuge the resulting supernatant was assayed for glutathione.

Estimation of liver glutathione content for comparison was carried out essentially in the same manner omitting the perfusion step; usually a 20 % homogenate in water was prepared.

RESULTS AND DISCUSSION

Table 2 contains a comparison of the levels of reduced glutathione in rat lung and liver.

It can be seen that the conclusions drawn will vary depending on the basis on which this comparison is made. In terms of nmoles/mg soluble protein (assuming an approximately equal concentration of cytoplasmic protein in lung and liver cells this will reflect the concentration of glutathione in the cytoplasm) there is very little difference. In terms of µmoles/g wet weight liver has more than 4 times as much glutathione as lung (assuming that the weights of lung and liver are accounted for by water to the same extent, this measure will reflect the concentration of glutatione in the cytoplasm). Finally, on a per cell basis (µmoles/mg DNA) liver contains about 10 times as much glutathione as lung tissue.

Table 2. Levels of reduced glutathione (GSH) in rat lung and liver[a]

Tissue	Lung	Liver
n	11	7
$\frac{\text{nmoles GSH}}{\text{mg soluble prot.}}$	70.8 ± 10.2	94.7 ± 11.9
$\frac{\mu\text{moles GSH}}{\text{g wet weight}}$	1.59 ± 0.26	6.67 ± 0.72
$\frac{\mu\text{moles GSH}}{\text{mg DNA}}$	0.32 ± 0.05	3.03 ± 0.33

[a] the figures represent the means ± S.D.

We have also investigated the levels of some glutathione metabolizing enzymes. It can be seen from Table 3 that rat lung contains levels of glutathione S-transferases that are approximately 5 - 20 % (on a wet weight basis) of those found in the liver. The level of lung glutathione reductase is rather high compared to that of the liver - about 40 %.

Table 3.

Levels of some glutathione metabolizing enzymes in rat lung and liver.

Enzyme	Lung			Liver	
	units[b]/mg sol. protein	units/g wet weight	units/mg DNA	units/mg sol. protein	units/g wet weight	units/mg DNA
GSH reductase	0.067	1.50	0.31	0.057	3.99	1.81
GSH S-transferase with DCNB[c] as 2:nd substrate	0.003	0.067	0.01	0.022	1.55	0.70
with CDNB[d] as 2:nd substrate	0.164	3.66	0.75	0.567	39.70	18.04
with NPEP[e] as 2:nd substrate	0.007	0.15	0.03	0.009	0.65	0.29

a) the figures in the table represent the means of two experiments
b) μmoles/min
c) 3,4-dichloro-1-nitrobenzene
d) 1-chloro-2,4-dinitrobenzene
e) 1,2-epoxy-3-(p-nitrophenoxy)propane

No definite conclusion can be drawn from the results presented here. On the one hand, the difference between lung and liver glutathione contents do not appear to be so great that arene oxides produced in the lung cannot be successfully conjugated with glutathione. In addition, maintenance of glutathione in a reduced state so that it can be conjugated with arene oxides is apparently successfully carried out in lung tissue by the active glutathione reductase present.

On the other hand, the finding that on a wet weight basis, the lung contains at least 4 times less reduced glutathionethan the liver may partially account for the difference in the susceptibility of these two organs to the carcinogenic effects of polycyclic hydrocarbons.

ACKNOWLEDGMENT

This study was supported by NCI Contract NO. I CP 33363, "Polycyclic Hydrocarbon Metabolism in the Respiratory Tract".

REFERENCES

(1) M. J. Cookson, P. Sims, and P.L.Grover, Mutagenicity of epoxides and other derivatives of polycyclic hydrocarbons, Nature, 234, 186, (1971).

(2) E. Huberman, L. Aspiras, C. Heidelberger, P. L. Grover, and P. Sims, Mutagenicity to mammalian cells of epoxides and other derivatives of polycyclic hydrocarbons, Proc. Nat. Acad.Sci.U.S.A., 68, 3195, (1971).

(3) H. Marquardt, T. Kuroki, E. Huberman, J. K. Selkirk, C. Heidelberger, P. L. Grover, and P. Sims, Malignant transformations of cells derived from mouse prostate by epoxides and other derivatives of polycyclic hydrocarbons, Cancer Res., 32, 716, (1972).

(4) C. Heidelberger, Chemical carcinogenesis, Ann. Rev. Biochem., 44, 71, (1975).

(5) E. Boyland, and L. F. Chasseaud, The role of glutathione and glutathione S-transferase in mercapturic acid biosynthesis, Adv. Enzymol., 32, 173, (1969).

(6) D. M. Jerina, and J. W. Daly, Arene oxides: a new aspect of drug metabolism, Science, 185, 573, (1974).

(7) M. Winell, and B. Mannervik, The nature of the enzymatic reduction of S-sulfoglutathione in liver and peas, Biochim. Biophys. Acta, 184, 374, (1969).

(8) W. H. Habig, M. J. Pabst, and W. B. Jacoby, Glutathione S-transferases. The firstenzymatic step in mercapturic acid formation, J. Biol. Chem., 249, 7130, (1974).

(9) P. Askelöf, C. Guthenberg, I. Jacobson, and B. Mannervik, Purification and characterization of two glutathione S-aryltransferase activities from rat liver, Biochem. J., 147,513, (1975).

(10) C. Crowley, B. Gillham, and M. B. Thorn, A direct enzymic method for the the determination of reduced glutathione in blood and other tissues, Biochem. Med., 13, 287, (1975).

(11) I. Carlberg, and B. Mannervik, Purification and characterization of the flavoenzyme glutathione reductase from rat liver, J.Biol. Chem., 250, 5475, (1975).

THE DISPOSITION OF BENZO(A)PYRENE IN ISOLATED PERFUSED RAT LUNG

Kirsi Vähäkangas, Kaisu Nevasaari, Olavi Pelkonen and Niilo T. Kärki

Department of Pharmacology, University of Oulu, SF-90 220 Oulu 22, Finland

INTRODUCTION

The development of lung tumours by polycyclic hydrocarbon carcinogens has been studied *in vivo* in different experimental animals. Other approaches in the elucidation of lung carcinogenesis have been the study of lung microsomal oxidative enzyme systems (1), which are thought to be responsible for the activation of procarcinogens to ultimate reactive forms (2). Some workers have used as a model different kinds of organ and cell cultures derived from lung tissue (3, 4). Recently Vainio et al. (5) and Cohen and Moore (6) have used the perfusion of isolated rat lung in the study of the fate of benzo(a)pyrene. We have studied the disposition of benzo(a)pyrene in short- and long-term recirculating perfusion of lungs from 3-methylcholanthrene- and phenobarbital-treated rats and have tried to analyze the metabolite pattern and covalent binding as a function of time.

MATERIAL AND METHODS

Non-fasted male rats of Sprague-Dawley strain weighing 190-260 g were used in all experiments. Phenobarbitone in the drinking water (0.5 g/l) was given to one group for 7 days, the controls drinking water ad. lib. 3-methylcholanthrene was given to the second group 40 mg/kg in vegetable oil (1 ml/kg) i.p. once a day for 3 days, controls received the same amount of the vehicle. The third group was exposed to cigarette smoke for 10 minutes by means of a self-made smoking chamber, for 2 days before the day of the perfusions, twice on the first day, once on the second day. For lung preparation rats were anaesthetized by urethane (1.5 g/kg i.p.).
Trachea was cannulated for ventilation (0.5 ml per stroke 16 times per minute) and a.pulmonalis for perfusion (20 ml/min). 12.5 umol (about 15×10^6 cpm) of ^3H-benzo(a)pyrene dissolved in 0.25 ml of dimethylsulphoxide was added into the perfusate after an equilibration period of 5 min. The temperature was maintained at 37.5 ± 0.5 °C. The volume of the perfusate was 50 ml, containing 12 ml fresh rat blood, 38 ml Krebs-Ringer phosphate buffer solution, 1.25 g bovine serum albumin and 50 mg D-glucose.
For analysis of radioactivity the lungs were homogenized in 4 volumes of 0.1 M potassium-sodium phosphate buffer, pH 7.4. Lung and perfusate samples were extracted twice with two volumes of ethyl acetate and small samples (25 ul) were taken from both aqueous and ethyl acetate phases for calculation of radioactivity. Ethyl acetate-soluble material was further resolved by thin-layer chromatography according to Sims (7) and Borgen

et al. (8). The binding of radioactivity to lung tissue was determined according to the method of Siekevitz (9). The benzo(a)pyrene hydroxylase activity determinations were carried out according to the fluorometric method of Nebert and Gelboin (10).

RESULTS

The effects of methylcholanthrene (MC) and phenobarbitone (PB) pretreatment were studied in two hours' perfusions. The half-life of BP in the perfusions of lungs from MC pretreated rats was about 15 minutes (Fig. 1). In perfusions using control and PB pretreated lungs the half-life of BP was 5-6 times longer (Fig. 2). The rapid disappearance of BP in the perfusions of MC pretreated lungs was accompanied with the appearance of water-soluble metabolites which was less marked in the perfusions of control and BP pretreated lungs (Fig. 1 and 2). In fact it seemed that the accumulation of water-soluble metabolites in the perfusions of PB pretreated lungs was retarded when compared with the control lungs.

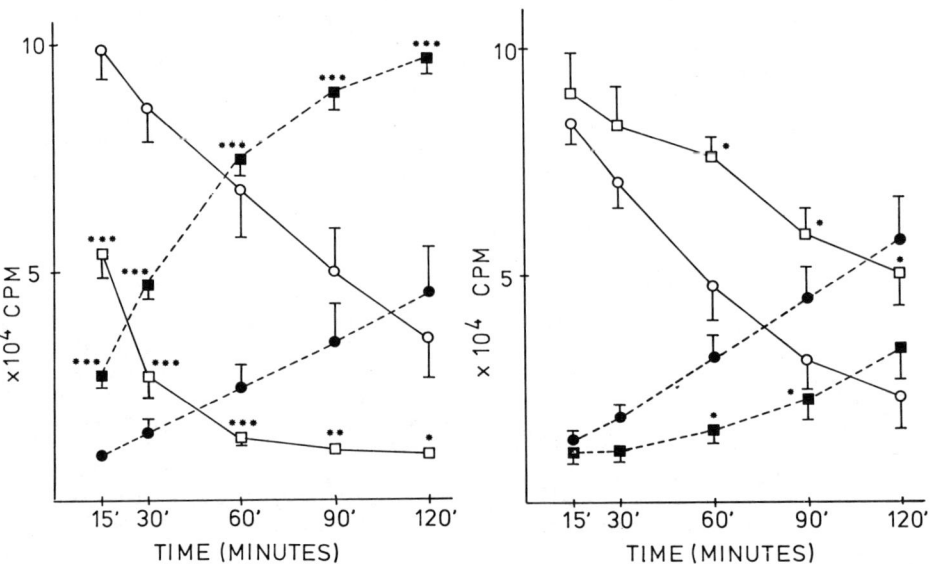

Figure 1.   Figure 2.

Fig. 1. Disappearance of ^{3}H-benzo(a)pyrene from perfusate, □—□ methylcholanthrene, pretreated, ○—○ control, and the appearance of water-soluble metabolites into the perfusate, ■- - -■ methylcholanthrene pretreated, ●- - -● control.

Fig. 2. Disappearance of ^{3}H-benzo(a)pyrene from the perfusate, □—□ phenobarbitone pretreated, ○—○ control, and the appearance of water-soluble metabolites into the perfusate, ■- - -■ phenobarbitone pretreated, ●—● controls.

Ethyl acetate-soluble metabolites were detected in the perfusion fluid of all lungs (Table 1).

TABLE 1   Ethyl acetate-soluble metabolites (pmol/ml) in the perfusion medium

	Perfusion time 15 min			Perfusion time 120 min		
	Controls	MC	PB	Controls	MC	PB
near origin	608	967	384	1074	1752	636
9,10-diol	592	1215	368	1566	810	1540
7,8-diol	37	711	259	755	488	778
4,5-diol	426	811	381	560	448	607
phenols	1297	3840	689	2255	945	1965
quinones	1202	1999	1038	1420	1039	1224
unchanged BP	8711	4767	7925	3167	913	4403

MC = methylcholanthrene pretreated group
BP = phenobarbitone pretreated group
The values represent means of 6 perfusion in each group

The amount of covalently bound radioactivity after two hours' perfusion in the lungs from MC pretreated rats was about 3.5 times higher than in the control lungs (Table 2). No significant difference existed between the lungs from PB pretreated rats and the controls. There was a positive correlation between the covalent binding and the aryl hydrocarbon hydroxylase activity in lung homogenates.

TABLE 2   Covalently bound radioactivity in the lungs

Pretreatment

3-methylcholanthrene	316 127 $\pm$ 120964[x]
phenobarbitone	81007 $\pm$ 15970
controls	89067 $\pm$ 43242

x) significantly different from controls (p   0.005).

The effect of cigarette smoke was studied in three 30 minutes' perfusions.
In the perfused lungs exposed to cigarette smoke there was two times more water-soluble metabolites than in the control lungs (Fig. 3). The amount of covalently bound radioactivity was also greater in the lungs exposed to cigarette smoke than in the controls.

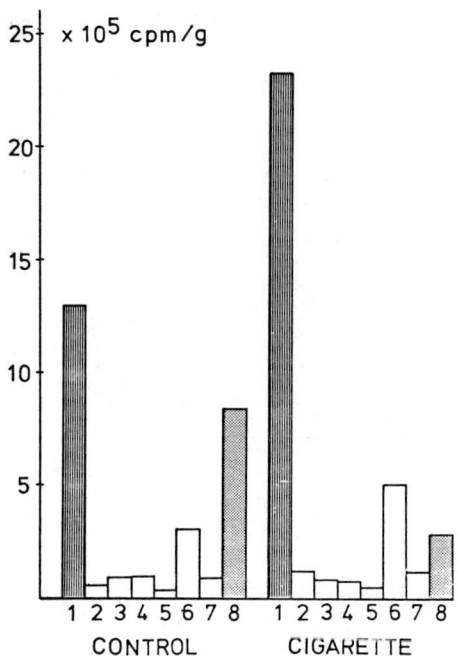

Fig. 3. ^{3}H-benzo(a)pyrene and its tritiated water-soluble metabolites and ethyl acetate-soluble metabolites in the lungs after perfusions of lungs from control and cigarette smoke pretreated rats. 1) water-soluble metabolites 2) unidentified radioactivity at the origin of TLC plate 3) 9,10-diol 4) 7,8-diol 5) 4,5-diol 6) OH-metabolites 7) quinones 8) benzo(a)pyrene

DISCUSSION

The pretreatment of rats with 3-methylcholanthrene (MC) caused several fold increase in the metabolism and binding of benzo(a)pyrene (BP) covalently in isolated, perfused rat lungs. The effect, if any, of phenobarbitone (PB) pretreatment was a decrease in the metabolism of BP. The induction of BP metabolism by polycyclic aromatic hydrocarbons is almost ubiquitous (11) and with respect to isolated organs or cells, it has been studied earlier (12, 5). In our study the induction of BP-hydroxylase in the lungs of MC pretreated rats was indicated by the rapid disappearance of BP from the perfusion fluid and the appearance of great amounts of phenols into the perfusion medium. Epoxide hydrase is known to be inducible by MC (13). Our findings are consistent with this, the concentration of dihydrodiols being two times greater at fifteen minutes in the perfusions of lungs from MC pretreated rats than in the controls. The amounts of phenols and other organic-soluble metabolites decreased during the perfusion, most notably in MC-lungs, and most probably they were converted to polyhydroxylated products and to water-soluble metabolites. The unidentified fraction of metabolites at the origin of TLC plate was not further characterized but it may represent polyhydroxylated products, because it increased during the perfusion. The further metabolism of "primary" metabolites of BP was shown in this study, especially in lung perfusions from MC-pretreated rats.

In our perfusions water-soluble radioactivity constituted about 2/3 of all BP metabolites after two hours' perfusion.

The induction of lung BP-hydroxylase by cigarette smoke has been shown

in many studies (14, 15, 16, 17). In our preliminary studies using 30 minutes' perfusions cigarette smoke had an inductive effect on BP-hydroxylase. The conjugative enzymes were also stimulated, the water-soluble radioactivity in the perfusion medium of lungs exposed to cigarette smoke being twice that found in the controls.
Pretreatment of rats with phenobarbitone has been reported to induce the metabolism of BP (18) and to stimulate the conjugation of hydroxylated BP metabolites (19). In our study it seems that lung enzymes metabolizing BP are not inducible by PB pretreatment.
All chemical carcinogens that have been thoroughly studied have been shown to bind covalently to DNA, RNA and protein of the target tissues (13). We also found that MC pretreatment greatly increases the binding of BP covalently in comparison to the controls and the PB treated ones. In short term perfusions by Vainio et al. (5) the results were similar. In our study, cigarette smoke has a similar effect.
The results of the present study suggest that the isolated perfused lung may be a useful tool in studying the metabolism of polycyclic hydrocarbon carcinogens in their target organ, free from the influences of the rest of the body, and in a more complete system than mere enzymes could provide.

ACKNOWLEDGEMENTS

The skilful technical assistance of Ms. Ritva Saarikoski is gratefully acknowledged. Reference metabolites of benzo(a)pyrene were kindly donated by Dr. Harry V. Gelboin, National Cancer Institute, Bethesda, Maryland, U.S.A.

REFERENCES

(1) J. Capdevila, S. W. Jakobsson, B. Jernström, O. Helia and S. Orrenius, Characterization of a rat lung microsomal fraction obtained by Sepharose 2B ultrafiltration, Cancer Res. 35, 2820 (1975).
(2) P. Sims and P. L. Grover, Epoxides of polycyclic aromatic hydrocarbon metabolism and carcinogenesis, Adv. Cancer Res. 20, 166 (1974).
(3) C. Leuchtenberger, R. Leuchtenberger and I. Zbinden, Gas vapour phase constituents and SH reactivity of cigarette smoke influence lung cultures, Nature 247, 565 (1974).
(4) C. C. Harris, V. M. Genta, A. L. Frank, D. G. Kaufman, L. A. Barrett, E. M. McDowell and B. F. Trump, Carcinogenic polynuclear hydrocarbons bind to macromolecules in cultured human bronchi, Nature 252, 68 (1974).
(5) H. Vainio, P. Uotila, J. Hartiala and O. Pelkonen, The fate of intratracheally installed benzo(a)pyrene in the isolated perfused rat lung of both control and 20-methylcholanthrene pretreated rats, Res. Commun. Chem. Pathol. Pharmacol. 13, 259 (1976).
(6) G. M. Cohen and B. P. Moore, Metabolism of (^{3}H)benzo(a)pyrene by different portions of the respiratory tract. Biochem. Pharmacol. 25, 1623 (1976).
(7) P. Sims, Qualitative and quantitative studies on the metabolism of a series of aromatic hydrocarbons by rat-liver preparations, Biochem. Pharmacol. 19, 795 (1970).

(8) A. Borgen, H. Darvey, N. Castagnoli, T. T. Crocker, R. E. Rasmussen and I. Y. Wang, Metabolic conversion of benzo(a)pyrene by Syrian hamster liver microsomes and binding of metabolites to deoxyribonucleic acid, J. Med. Chem. 16, 502 (1973).
(9) P. Siekevitz, Uptake of radioactive alanine in vitro into the proteins of rat liver fraction, J. Biol. Chem. 195, 549 (1952).
(10) D. W. Nebert and H. V. Gelboin, Substrate-inducible microsomal aryl hydroxylase, J. Biol. Chem. 243, 6242 (1968).
(11) A. H. Conney, Pharmacological implications of microsomal enzyme induction, Pharmacol. Rev. 19, 317 (1967).
(12) H. Vadi, P. Moldeus, J. Capdevila and S. Orrenius, The metabolism of benzo(a)pyrene in isolated rat liver cells, Cancer Res. 35, 2083 (1975).
(13) D. M. Jerina and J. W. Daly, Arene oxides: A new aspect of drug metabolism, Science 185, 573 (1974).
(14) R. K. Abramson and J. J. Hutton, Effects of cigarette smoking on aryl hydrocarbon hydroxylase activity in lungs and tissues of inbred mice, Cancer Res. 35, 23 (1975).
(15) P. G. Holt and D. Keast, Induction of aryl hydrocarbon hydroxylase in the lungs of mice in response to cigarette smoke. Experientia 29, 1004 (1973).
(16) R. U. Welch, A. Loh and A. H. Conney, Cigarette smoke: Stimulatory effect on metabolism of 3,4-benzpyrene by enzymes in rat lung, Life Sci. 10, 215 (1971).
(17) J. Marcotte, F. S. Skelton, M. G. Coté and H. Witschi, Induction of aryl hydrocarbon hydroxylase in rat lung by marijuana smoke, Toxicol. Appl. Pharmacol., 33, 231 (1975).
(18) W. G. Levine, The role of microsomal drug-metabolizing enzymes in the biliary excretion of 3,4-benzpyrene in the rat, J. Pharmacol. Exp. Ther. 175, 301 (1970).
(19) E. Schlede, R. Kuntzman and A. H. Conney, Stimulatory effect of benzo(a)pyrene and phenobarbital pretreatment on the biliary excretion of benzo(a)pyrene metabolites in the rat, Cancer Res. 30, 2898 (1970).

EPOXIDE HYDRATASE IN RAT LUNG

Janeric Seidegård, Joseph W. DePierre, Maria S. Moron,
Karin A. M. Johannesen, and Lars Ernster

*Arrhenius Laboratory, Department of Biochemistry, University of Stockholm,
Fack, S-104 05 Stockholm, Sweden*

INTRODUCTION

Polycyclic hydrocarbons may be a major cause of cancer in human beings. A number of studies have demonstrated the carcinogenic properties of these compounds in different species, including man, and in different tissues, including the lung (Refs 1,2,3,4). In addition, polycyclic hydrocarbons have been observed to cause malignant transformations (Ref 5) and mutagenesis (Ref 6) when applied to cells in culture.

There is much evidence in support of the hypothesis that epoxide metabolites of the polycyclic hydrocarbons, rather than the parent hydrocarbons themselves, are the immediate agents responsible for causing lung and skin cancer (Refs 7, 8,9). Such epoxides can rearrange in water to form phenols; be hydrated by epoxide hydratase to dihydrodiols; or be conjugated enzymatically or non-enzymatically with glutathione (Ref 10). All these products are considerably less carcinogenic than the epoxides themselves. Epoxide hydratase may have an important role to play in determining the steady state levels of arene oxides during the metabolism of polycyclic hydrocarbons in mammalian tissues and, consequently, an important role to play in etiology of cancer caused by these compounds.

Oesch and his coworkers have concluded that benzpyrene oxide hydratase and styrene oxide hydratase activities are catalyzed by the same protein (Ref 11). This conclusion is based on the findings that the relative ratio of these activities in the preparation remains essentially unchanged during purification and that an antibody to the purified enzyme inhibits both benzpyrene oxide hydratase and styrene oxide hydratase in solubilized microsomes.

We have modified a published assay procedure for styrene oxide hydratase (Ref 12) in order to increase its sensitivity. This increased sensitivity has allowed us to characterize the epoxide hydratase of rat lung in various ways. A published report suggests that the epoxide hydratase of human liver are quite similar (Ref 13); so it is reasonable to expect that an investigation of rat lung epoxide hydratase will give information relevant to the corresponding enzyme in human lung.

METHODS AND RESULTS

Assay of epoxide hydratase

The procedure used for measuring epoxide hydratase activity was essentially that developed for Oesch and his coworkers (12) for liver microsomes. Radioactive styrene oxide is incubated with the microsomes; remaining substrate is then selectively extracted into petroleum ether; and finally, the styrene glycol product is extracted into ethyl acetate and scintillation couting is performed. However, this procedure had to be modified in three ways in order to make it sensitive enough for routine use with rat lung preparations.

First, the substrate was purified before use. The most serious limitation to the sensitivity of a radioactive enzymatic assay is often the background from contaminants in the substrate and from non-enzymatic reactions. 1µl of the undiluted ^{3}H-styrene oxide was dissolved in 1.5 ml hexane and this solution was extracted 3 times with 100µl 0.5 M Tris-Cl, pH 7.5. The ^{3}H-styrene oxide was then extracted from the hexane into 1.5 ml acetonitrile, and this procedure was repeated twice.

^{3}H-styrene oxide as purchased gave a zero time background of about 1.5%, and this background rose sharply during the first minutes of incubation, and after 10 min bacame linear up to at least 30 min. On the other hand, the purified substrate gave a zero time background of 0.1-0.2%, and this background increased slowly with time in a linear manner. As a result, the background after 45 min of incubation (the standard incubation time - see below) is decreased about 2.5-fold by purification of the ^{3}H-styrene oxide.

The background with purified substrate is linear with time for at least one hour, is independent of the presence or absence of boiled microsomal protein, and is about 2.4% of the total added radioactivity after a 45 min incubation.

The second modification was to scale down the 0.4 ml incubation volume used with liver microsomes to 0.1 ml. Thus, the details of the assay were as follows: 25 µl 0.5 M Tris-Cl, pH 7.5, and 75 µl of microsomes (usually containing about 0.75 mg of protein) or other fraction are mixed in a glass-stoppered tube and preincubated at 37° for a min or two. The reaction is started by addition of 2 µl of the purified substrate solution. After incubation at 37°, the reaction is terminated by adding ice-cold petroleum ether and shaking vigorously on a mixer for 5 min. After separation of the two phases by spinning in a desk centrifuge for 5 min, the tubes are placed in a freezer to allow the lower water phase to freeze and the upper petroleum ether phase is subsequently decanted. This extraction with petroleum ether is repeated once and then the product styrene glycol is extracted using 1 ml ethyl acetate in an essentially identical manner (without the freezing step). A 100 µl aliquot of the combined petroleum ether extracts and a 400 µl aliquot of the ethyl acetate phase are counted and corrected for quenching. The yield of product is determined by calculating the fraction of the total radioactivity that is recovered in the ethyl acetate phase and then multiplying this value by th amount of substrate added to the reaction mixture.

It can be seen from Fig. 1 that epoxide hydratase activity measured using a 0.1 ml incubation volume is identical to that using a 0.4 ml volume. Fig. 1 also illustrates that this assay of epoxide hydratase with lung microsomes is linear for at least 60 min. Thus, the third modification made in the assay procedure is to incubate for 45 min instead of 5 min, the time period used with liver microsomes. However, when such long incubation times were used, a net loss of radioactivity was noticed. This loss, which is presumably due to evaporation of styrene oxide, could be prevented by carrying out the assay in glass-stoppered tubes.

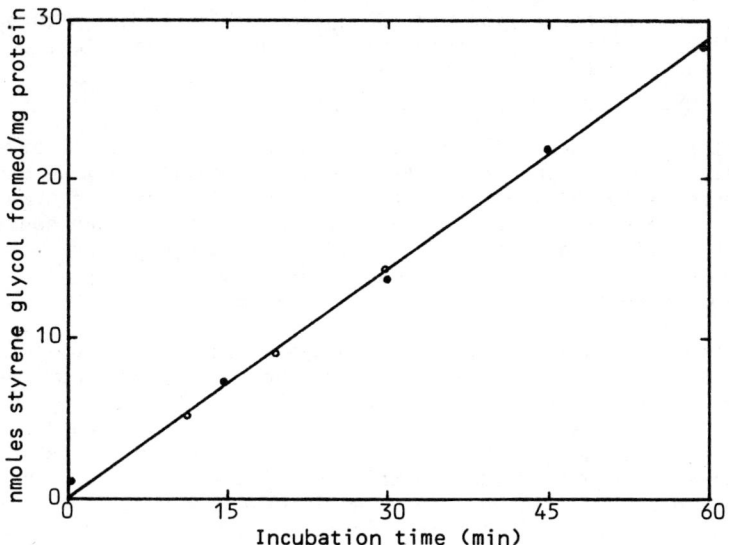

Fig. 1. Linearity of the epoxide hydratase activity of lung microsomes with time.
● = volume of the incubation mixture was 0.1ml
o = volume of the incubation mixture was 0.4ml

These three modifications-purification of the substrate, scale-down of the incubation volume, and a considerably longer incubation time - increase the sensitivity of the assay for epoxide hydratase 75- to 150-fold over the procedure developed with liver microsomes. As a result, using rat lung microsomes a 15-20 % conversion of substrate into product over a 2-3 % background is normally obtained.

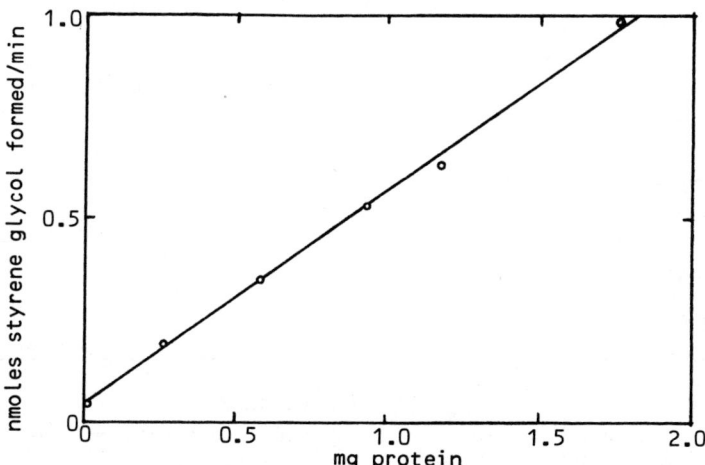

Fig. 2. Linearity of the epoxide hydratase activity of lung microsomes with protein.

The present assay procedure for epoxide hydratase is linear up to at least 1.8 mg lung microsomal protein per incubation mixture, as can be seen from Fig. 2.

Apparent K_m and V_{max} of lung epoxide hydratase

It is important that the concentration of substrate in a radioactive enzyme assay be saturating but not in great excess. The use of too much substrate involves a decrease in the specific activity of the substrate or a waste of radioactive styrene oxide. More seriously, the use of too much substrate decrease the sensitivity of the assay, since the % background remains essentially unchanged while the % yield of products decreases linearly with the increase in substrate concentration above saturation.

The dependence of lung microsomal epoxide hydratase activity on substrate concentration is shown in Fig. 3. A smooth line has been drawn, but there are indications of a preliminary plateau between 0.5 and 2.0 mM styrene oxide before the activity rises to its final apparent V_{max} of about 0.5 nmoles styrene glycol formed/min-mg lung microsomal protein. Many factors might be involved in the appearance of such a plateau in this complicated system containing at least two phases (the hydrophobic phase of the microsomal membrane and the aqueous phase) and a substrate that is only slightly soluble in water. With these reservations, if the data in Fig. 3 are arranged in a Lineweaver-Burke plot (14), an apparent K_m for styrene oxide of 0.11-0.25 mM is obtained. The 1.6 mM concentration of styrene oxide used in the present assay procedure thus seems to be appropriate.

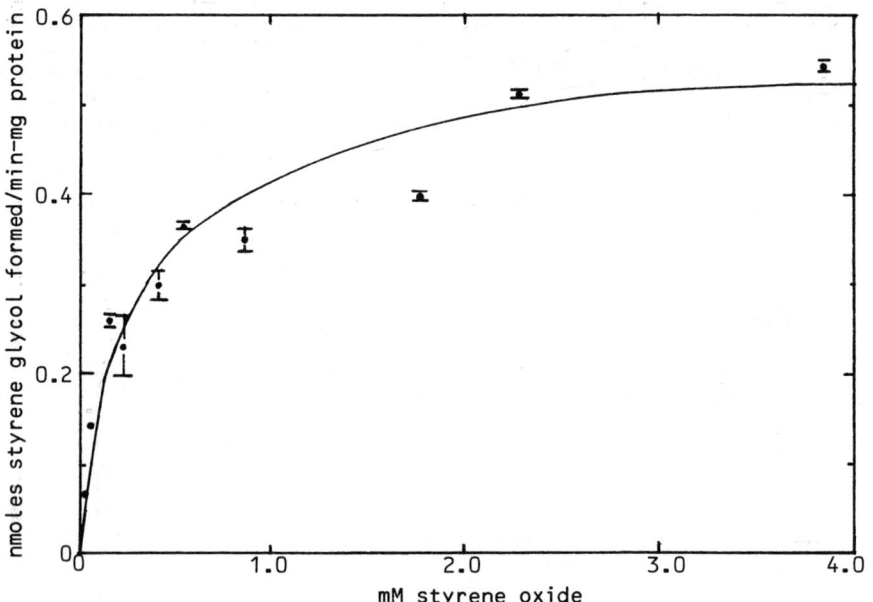

Fig. 3. Dependence of the epoxide hydratase activity of lung microsomes on substrate concentration.

pH dependence

Lung epoxide hydratase activity was investigated over a wide range of pH. Three different buffers - phosphate, Tris, and glycine - were used and no individual buffer effects were observed. The microsomal suspension, which itself has a pH of 7.26 ± 0.07, seems to have considerable buffering capacity; so the pH values were measured after mixing microsomes and buffer. The activity increases slowly up to pH 8.7, then rises more rapidly to its optimum at pH 9.7 ± 0.3, and falls off very rapidly above pH 9.8. The background in the assay procedure does not increase as the pH is raised above 7.5, but it does go up as the incubation mixture is gradually made more acidic.

Attempts to induce epoxide hydratase activity in the lung

The synthesis of arene oxides by the cytochrome P-450 system can be induced 10-20-fold in liver and lung by intraperitoneal administration of methylcholanthrene to rats (Refs 15,16). This activity is also inducible in liver, although to a much lesser, extent, by phenobarbital (Ref 17). Therefore, it is of great interest in the study of polycyclic hydrocarbon-caused lung cancer to find out if these substances can induce the activity of epoxide hydratase (which metabolizes arene oxides to less toxic products) in lung. Treatment of rats either with methylcholanthrene or with phenobarbital has no effect on lung epoxide hydratase activity, even though in these same rats methylcholanthrene induced liver benzpyrene monooxygenase activity 16-fold and phenobarbital induced liver NADPH-cytochrome c reductase activity 4-5-fold.

An attempt to bring about "substrate induction" of epoxide hydratase by intraperitoneal administration of styrene oxide in corn oil to rats has been done. This treatment had no significant effect on epoxide hydratase, NADPH-cytochrome c reductase, or DT-diaphorase activities in the lung and liver and no effect on cytochrome P-450 and b_5 in liver microsomes. Even very high doses of styrene oxide did not have any apparent toxic effects on the animals.

Inhibitors

Cyclohexene oxide and 1,1,1-trichloropropene 2,3-oxide are known to be non-competitive and uncompetitive inhibitors, respectively, of hepatic epoxide hydratase with respect to the substrate styrene oxide (Ref 18). However, it appeared that lung microsomal epoxide hydratase is affected in an identical manner by these two inhibitors. An apparent K_i for cyclohexene oxide of 80 μM could be calculated for the lung enzyme. Other experiments showed that 40-50 μM 1,1,1-trichloropropene 2,3-oxide inhibits the saturated lung epoxide hydratase approximately 50 %.

Effects of alcohols

Oesch (18) has demonstrated that liver microsomal epoxide hydratase is activated by low concentrations of 1-alkanols and inhibited by higher concentrations. This is also the case for lung microsomal epoxide hydratase. 10 % (v/v) (1.7 M) ethanol is seen to activate the lung enzyme approximately 100 %, while inhibition is seen above 18 % (v/v) ethanol. Butanol has only a small activating effect at 5 % (v/v) (0.55 M) and inhibits potently at higher concentrations.

DISCUSSION

Three modifications have been made in the standard radioactive assay procedure for epoxide (styrene oxide) hydratase: the substrate is purified before use; the volume of the incubation mixture is scaled down 4-fold; and the incubation time is lengthened 9-12-fold (to 45-60 min). These changes increase the sensitivity of the assay procedure 75-150-fold, thus providing a reliable and convenient method for characterizing epoxide hydratase in tissues where its activity is quite low, such as rat lung.

Since polycyclic hydrocarbons appear to cause cancer much more easily in lung than in liver and since epoxide hydratase may play an important role in the etiology of such cancer, it is of interest to compare the properties of epoxide hydratase in these two organs. Table 1 shows that these properties are quite

TABLE 1 Comparison of the properties of epoxide (styrene oxide) hydratase in liver and lung

Property	Liver	Rat lung[a]
Specific activity in microsomes[b]	8[c]	0.5
Apparent K_m, mM	0.25[d]	0.11-0.25
pH optimum	8.7-9[e]	9.7
Subcellular localization	endoplasmic reticulum[f]	endoplasmic reticulum
Inducibility with		
methylcholanthrene	none[g]	none
phenobarbital	300% of control[f]	none
styrene oxide	–	none
Inhibition with		
1,1,1-trichloropropene 2,3-oxide	uncompetitive[h]	uncompetitive
cyclohexene oxide	noncompetitive[h]	noncompetitive
Effect of		
ethanol	activates at low conc.[h] inhibits at high conc.	activates at low conc. inhibits at high conc.
butanol	inhibits at high conc.[h]	activates at low conc. inhibits at high conc.
Tween 80	activates[f]	none

a) Values from present study.
b) nmoles styrene glycol formed/min-mg protein.
c) A rough average of the values from Refs. 21,22,23.
d) Rat liver microsomes. See Ref. 21.
e) Epoxide hydratase purified from rat liver. See Refs. 22,23.
f) Guinea pig liver. See Ref. 10.
g) Our own results (not presented).
h) Epoxide hydratase purified from guinea pig liver. See Ref. 10.

similar. In both liver and lung the apparent K_m of the enzyme for styrene oxide is about 0.2-0.3 mM. The corresponding value for purified human liver epoxide hydratase is 0.38 mM (Ref 19). The pH optimum of epoxide hydratase in both organs is quite alkaline, as it also is for purified human liver epoxide hydratase (pH optimum = 9; Ref 19). However, the activity of lung microsomal epoxide hydratase at pH 7.5 (which is presumably close to the pH of lung cell cytoplasm in vivo) is fully 70 % of the activity at the pH optimum of 9.7.

1,1,1-Trichloropropene 2,3-oxide and cyclohexene oxide inhibits lung epoxide hydratase in the same way they inhibit the liver enzyme, and ethanol and butanol have the same effect on these two enzymes. The only differences, other than in specific activity, are that Tween 80 stimulates the liver but not the lung enzyme and that phenobarbital induces epoxide hydratase activity in the liver but not in the lung. This latter finding is consistent with previous reports that intraperitoneal injection of phenobarbital greatly increases the levels of NADPH-cytochrome c reductase, cytochrome P-450, and drugmetabolizing activity in rat liver but affects none of these components in rat lung (Ref 20).

ACKNOWLEDGMENT

This study was supported by NCI Contract No. 1 CP 33363, "Polycyclic Hydrocarbon Metabolism in the Respiratory Tract".

REFERENCES

(1) J.W.S. Blacklock, The production of lung tumours in rat by 3:4 benzpyrene, methylcholanthrene and the condensate from cigarette smoke, Br. J. Cancer, 11, 181 (1957).

(2) O. Auerbach, E.C. Hammond, D. Kirman, and L. Garfinkel, Effects of cigarette smoking on dogs, Arch. Environ. Helth, 21, 754 (1970).

(3) U. Saffiotti, R. Montesano, A.R. Sellakumar, F. Cefis, and D.G. Kaufman, Respiratory tract carcinogenesis in hamster induced by different numbers of administrations of benzo(a)pyrene and ferric oxide, Cancer Res., 32, 1073 (1972).

(4) M.C. Henry, C.D. Port, R.R. Bates, and D.G. Kaufman, Respiratory tract tumor in hamsters induced by benzo(α)pyrene, Cancer Res., 33,1585 (1973).

(5) Y. Berwald, and L. Sachs, In vitro cell transformation with chemical carcinogens, Nature, 200, 1182 (1963).

(6) E.Y.Y. Chu, E.G. Bailiff, and H.V. Malling, mutagenicity of chemical carcinogens in mammalian cells, Abstr. 10th Int. Cancer Congr.,p62,1970.

(7) P.L. Grover, P. Sims, E. Huberman, H. Marquardt, T. Kuroki,and C. Heidelberger, In vitro transformation of rodent cells by K-region derivatives of polycyclic hydrocarbons, Proc. Natl. Acad.Sci. USA,68, 1098, (1971).

(8) E. Huberman, T. Kuroki, M. Marquardt, J.K. Selkirk, C. Heidelberger, P.L. Grover, and P. Sims, transformation of hamster embryo cells by epoxides and other derivatives of plycyclic hydrocarbons, Cancer Res., 32, 1391 (1972).

(9) M. Marquardt, T. Kuroki, E. Huberman, J.K. Selkirk, C. Heidelberger, P.L. Grover, and P. Sims, Malignant transformation of cells derived from mouse prostate by epoxides and other derivatives of polycyclic hydrocarbons, Cancer Res., 32, 716 (1972).

(10) F. Oesch, Mammalian epoxide hydrases: Inducible enzymes catalysing the inactivation of carcinogenic and cytotoxic metabolites derived from aromatic and olefinic compounds, Xenobiotica, 3, 305 (1972).

(11) F. Oesch, Transplacental control of epoxide hydratase and its relationship to the control of microsomal monooxygenase, FEBS Lett., 53, 205 (1975).

(12) F. Oesch, D.M. Jerina, and J. Daly, A radiometric assay for hepatic epoxide hydrase activity with 7-^{3}H styrene oxide, Biochim. Biophys. Acta, 227, 685 (1971).

(13) F. Oesch, Purification and specificity of a human microsomal epoxide hydratase, Biochem. J., 139, 77 (1974).

(14) H. Lineweaver, and D. Burk, The determination of enzyme dissociation constants, J. Am. Chem. Soc., 56, 658 (1934).

(15) C. Harper, R.T. Drew, and J.R. Fouts, Species differences in benzene hydroxylation to phenol by pulmonary and hepatic microsomes, Drug Metab. Dispos. 3, 381 (1973).

(16) J.C. Leutz, and H.V. Gelboin, benzo(a)pyrene-4,5-oxide hydratase: Assay, properties, and induction, Arch. Biochem. Biophys., 168, 722 (1975).

(17) Y. Gnosspelius, H. Thor, and S. Orrenius, A comparative study on the hydroxylating enzyme system of rat-liver microsomes, Chem. Biol. Interactions, 1, 125 , 169/70.

(18) F. Oesch, N. Kaubisch, D.M. Jerina, and J.W. Daly, Hepatic epoxide hydrase. Structure-activity relationships for substrates and inhibitors, Biochemistry, 10, 4858 (1971).

(19) F. Oesch, and H. Thoenen, Epoxide hydrase in human liver biopsy specimens: Assay and properties, Biochem. Pharmacol., 23, 1307 (1974).

(20) T. Matsubara, R.A. Prough, M.D. Burke, and R.W. Estabrook, The preparation of microsomal fractions of rodent respiratory tract and their characterization, Cancer Res., 34, 2196 (1974).

(21) G. Belvedere, J. Pachecka, L. Cantoni, E. Mussini, and M. Salmona, A specific gas chromatografic method for the determination of microsomal styrene monooxygenase and styrene epoxide hydratase activities, J. Chromat., 118, 387 (1976).

(22) P. Bentley, and F. Oesch, Properties and amino acid composition of pure epoxide hydratase, FEBS Lett., 59, 296 (1975).

(23) A.Y.H. Lu, D. Ryan, D.M. Jerina, J.W. Daly, and W. Levin, Liver microsomal epoxide hydrase, J. Biol. Chem., 250, 8283 (1975).

MONOOXYGENASE CATALYZED ACTIVATION OF THIONO-SULFUR CONTAINING COMPOUNDS TO REACTIVE INTERMEDIATES

R. A. Neal, T. Kamataki, A. L. Hunter and G. Catignani

Center in Environmental Toxicology, Department of Biochemistry, Vanderbilt University School of Medicine, Nashville, Tennessee 37232

A number of thiono-sulfur containing compounds which find use as pesticides, drugs, and in industrial processes exhibit toxic properties. These effects include bone marrow depression (1), liver damage and the induction of neoplasia. Table 1 shows representative thiono-sulfur containing compounds which have been shown to induce cancer in experimental animals and cause liver damage in experimental animals and man. A number of the compounds listed in Table 1 have also been shown to bring about a

TABLE 1 Representative Thiono-sulfur Containing Compounds Which Have Been Shown to Cause Cancer and/or Liver Damage in Experimental Animals and/or Man

Cancer	Liver Damage
Thiourea (2)	Disulfiram[a] (3)
Thioacetamide (2)	Methimazole[a] (1,4)
Ethylene thiourea (2)	Propylthiouracil[a] (4,5)
Propylthiouracil (2)	Thiobarbital[a] (6)
Thiouracil (2)	Carbon Disulfide (7)
Methylthiouracil (2)	Methylthiouracil[a] (8)
	Thiouracil[a] (9)
	Diethylphenyl phosphorothionate (10)
	Thioacetamide (11)

[a]Have been reported to cause liver damage in man.

decrease in the level of hepatic cytochrome P-450 detectable as its carbon monoxide complex and/or decrease the ability of hepatic microsomes to metabolize substrates like aniline, benzphetamine or aminopyrine when administered *in vivo* or when incubated with hepatic microsomes *in vitro* in presence but not in the absence of NADPH (4).

The centrilobular hepatic necrosis seen with carbon disulfide can be largely prevented by administration of SKF-525A prior to the administration of carbon disulfide. The hepatic necrosis seen with thioacetamide can also be markedly decreased by administration of SKF-525A, pyrazole or cobaltous chloride prior to the administration of thioacetamide. These data suggest that the thiono-sulfur containing compounds are causing liver necrosis and inhibiting mixed-function oxidase activity by a mechanism which involves metabolic activation of these compounds. These and other data also suggest that the enzyme system involved in this activation reaction is most likely the cytochrome P-450 containing mixed-function oxidase system, although it is considered that the amine oxidase system of Poulsen and Zeigler might also be involved as concerns the effects of some of these compounds (12).

In examining the mechanism by which thiono-sulfur containing compounds cause liver necrosis and inhibit the activity of the cytochrome P-450 containing mixed-function oxidase enzyme systems, we have studied two compounds of divergent structure in some detail. These are carbon disulfide and parathion. As noted in Table 1, the compound diethylphenyl phosphorothionate causes centrilobular hepatic necrosis and reduces the level of hepatic cytochrome P-450 when administered in vivo (10). The insecticide parathion (diethyl p-nitrophenyl phosphorothionate) also decreases the level of cytochrome P-450 when incubated with hepatic microsomes in the presence of NADPH (13). It is quite likely that parathion would also cause hepatic necrosis if sufficient amounts of this compound could be administered. However, the powerful anticholinesterase properties of parathion precludes the administration of a dose large enough to produce necrosis. The most probable mechanism by which parathion is metabolized by the cytochrome P-450 containing mixed-function oxidase enzyme system to its major product, paraoxon, is shown in Fig. 1. It is thought that the initial reaction is the reaction of a singlet oxygen atom bound to cytochrome P-450 with one of the unshared electron pairs on the sulfur atom. As indicated, the resultant sulfine can assume a number of forms. One of these forms undergoes a cyclization to form a "phosphooxythirane" intermediate followed by an electron rearrangement to form paraoxon with the loss of the sulfur atom. The data supporting this mechanism has been reported previously (14-17). The initial product of the metabolism of carbon disulfide (S=C=S) by hepatic mixed-function oxidase enzyme systems is carbonyl sulfide (O=C=S) (18). A similar mechanism to that shown in Fig. 1 can be written to explain the cytochrome P-450 catalyzed metabolism of S=C=S to O=C=S.

In Fig. 1 the sulfur atom is shown to be released as singlet atomic sulfur. Atomic sulfur is a highly reactive form of sulfur. In its singlet state, it has reactivity similar to carbenes and nitrenes (19,20). One would therefore expect if the sulfur is released as atomic sulfur that it should readily covalently bind to available nucleophiles. Table 2 shows the results of experiments in which the NADPH stimulated covalent binding of radioactivity from ^{14}C and ^{35}S-labeled carbon disulfide to hepatic microsomes was compared. In the absence of NADPH

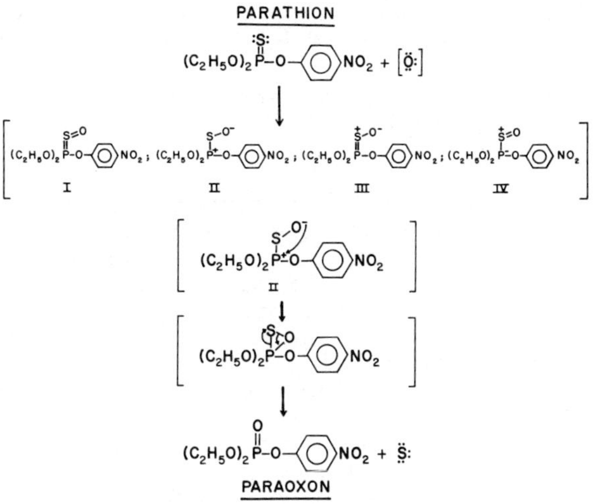

Fig. 1. The chemical mechanism for the metabolism of parathion to paraoxon by the mammalian cytochrome P-450 containing monooxygenases.

there was an approximately equal amount of binding of ^{14}C and ^{35}S-labeled carbon disulfide to the hepatic microsomes. This most likely represents the formation of dithiocarbamate deriva-

TABLE 2 Comparison of ^{14}C and ^{35}S Binding Following Incubation of ^{14}C-Carbon Disulfide and ^{35}S-Carbon Disulfide with Hepatic Microsomes from Phenobarbital-treated Adult Male Rats[d]

Substrate	NADPH	^{14}C or ^{35}S Bound (nmoles/10 min/mg protein[b])
^{14}C-Carbon Disulfide (1 x 10^{-3}M)	+	2.06 ± 0.01
	−	0.96 ± 0.05
^{35}S-Carbon Disulfide (1 x 10^{-3}M)	+	7.05 ± 0.36
	−	1.02 ± 0.02

[a] ^{14}C or ^{35}S-Carbon disulfide (1 x 10^{-} M) was incubated with aliquots of the same preparation of microsomes isolated from the pooled livers of 4 phenobarbital-pretreated rats. The incubation procedures and the methods used to wash the microsomes

free of unmetabolized carbon disulfide and its non-covalently bound metabolites have been described previously (18).

b Mean ± standard deviations of the means of 3 determinations under each incubation condition.

tives of amino groups and/or trithiocarbonate derivatives of sulfhydryl groups present on the macromolecules of the microsomes. This binding in the absence of NADPH has no effect on cytochrome P-450 concentrations or mixed-function oxidase activity toward substrates like benzphetamine. In the presence of NADPH, an approximate doubling of the amount of ^{14}C bound to the microsome is seen. In contrast, about 7 times more ^{35}S was bound in the presence than in the absence of NADPH. Thus, NADPH-stimulated binding of ^{35}S to the microsomes was about 6 times that of ^{14}C. Therefore, the majority of the sulfur bound to the microsomes in the presence of NADPH is in a form free of the carbon atom of CS_2. Similar results have been obtained comparing the NADPH stimulated binding of ^{32}P and ^{35}S-labeled parathion to hepatic microsomes (21). As was the case with carbon disulfide, the majority of the sulfur bound was free of the phosphorus containing portion of the parathion molecule. The comparison of the binding of ^{14}C (ethyl) and ^{35}S-labeled parathion to cytochrome P-450 of a reconstituted mixed-function oxidase enzyme system from rat liver is shown in Table 3. In the presence of the complete system only about 3% as much ^{14}C as ^{35}S became bound to the cytochrome P-450 of the reconstituted system. Thus, 97% or greater of the sulfur bound to the cytochrome P-450 was free of the remainder of the parathion molecule and thus must be some form of the sulfur atom released in the metabolism of parathion to paraoxon. In experiments like those described in Table 3, approximately 25% of the cytochrome P-450 present in the incubations containing the complete system was no longer detectable as its carbon monoxide complex at the end of the experiment (22). In contrast, under conditions which did not lead to the binding of appreciable amounts of sulfur (minus reductase) no significant loss of cytochrome P-450 was seen. For each nmole of apparently homogeneous cytochrome P-450 no longer detectable, approximately 4 nmoles of sulfur were bound. In contrast, under similar conditions only about 0.1 nmole of ^{14}C from ^{14}C (ethyl) parathion was bound per nmole of cytochrome P-450 no longer detectable as its carbon monoxide complex. Thus, the binding of some form of the sulfur atom must be responsible for the loss of cytochrome P-450 in the reconstituted system and most likely in intact microsomes as well.

We have also examined the nature of the sulfur bound to hepatic microsomes incubated with ^{35}S-carbon disulfide (23) and to purified cytochrome P-450 of a rat liver reconstituted system incubated with ^{35}S-parathion (22). These studies have shown that about 50% of the sulfur bound to both microsomes and to cytochrome P-450 can be released as ^{35}S-thiocyanate ($^{35}SCN^-$) on incubation with CN^- (22,23). These results indicate that the

TABLE 3 An Examination of the Binding of ^{35}S and ^{14}C (ethyl) and ^{35}S-Labeled Parathion to Rat Liver Cytochrome P-450[a]

Incubation Conditions	^{35}S or ^{14}C Binding (nmole/nmole P-450/5 min)
^{35}S-Parathion	
Complete system	1.33 ± 0.12
Minus NADPH	0.08 ± 0.02
Minus NADPH-cytochrome c reductase	0.06 ± 0.02
Minus dilauroyl phosphatidylcholine	0.52 ± 0.01
Minus deoxycholate	1.05 ± 0.03
Minus Mg	1.29 ± 0.07
^{14}C-Parathion	
Complete system	0.031 ± 0.004

[a] The incubation procedures have been described previously (22).

sulfur atom of both carbon disulfide and parathion is released as atomic sulfur on metabolism by the cytochrome P-450 containing mixed-function oxidase enzyme systems. In addition, these data indicate that at least half the sulfur released becomes bound to sulfhydryl groups of cysteine side chains in the polypeptide chains forming a hydrodisulfide. The probable mechanism of this reaction is shown in Fig. 2. R-CH$_2$-S-H represents the side chain of a cysteine molecule where R is the remainder of the polypeptide. The R-CH$_2$-S$^-$ (or R-CH$_2$S-H followed by loss of a proton) is subject to electrophilic attack by the sulfur atom to form a hydrodisulfide. The hydrodisulfide is then subject to nucleophilic attack by CN$^-$ to release the terminal sulfur as HSCN (SCN$^-$ + H$^+$). In practice the ethereal sulfur of hydrodisulfides is also subject to nucleophilic attack by CN$^-$ (24). The degree to which this latter reaction occurs with hydrodisulfides in microsomes or the cytochrome P-450 molecule is not known at this time.

With both carbon disulfide and parathion, an NADPH stimulated binding of the carbon and phosphorus containing portions of these compounds to the macromolecules of the microsomes is seen (18, 21). The requirement for NADPH for binding infers that, like the release and covalent binding of the sulfur atom, some metabolic activation reaction is required. A possible reactive intermediate is the sulfine formed in the reaction of the cytochrome P-450 activated oxygen atom with one of the unshared electron pairs on the sulfur atom of these compounds (see Fig.1).

Although some sulfines are stable enough to be isolated, those of parathion and carbon disulfide are not. However, the results of various studies (10,13,14-18) leave little doubt that they are intermediates in the metabolism of these compounds by the cytochrome P-450 containing monooxygenases. Some of the theoretical resonance forms of these sulfines are electrophilic. Therefore, it appears possible that at least a part of the NADPH stimulated binding of the carbon containing portion of carbon disulfide and the carbon and phosphorus containing portion of parathion is the result of attack by tissue nucleophiles on an electrophilic form or forms of these sulfines prior to their rearrangement leading to the loss of the sulfur atom. A possible reaction sequence involving a thioketone is shown in Fig. 3.

In summary, a number of thiono-sulfur containing compounds bring about an inhibition of cytochrome P-450 containing monooxy genases when administered <u>in vivo</u> or incubated with hepatic microsomes <u>in vitro</u> in the presence of NADPH. Some of these compounds also cause hepatic necrosis. The loss of cytochrome P-450 detectable as its carbon monoxide complex and the accompanying loss in mixed-function oxidase activity is, with at least some of these compounds, most probably the result of the binding of a sulfur atom released in a reaction catalyzed by

$$R-CH_2-S-H \underset{+H^+}{\overset{-H^+}{\rightleftharpoons}} R-CH_2-\bar{S} + S \longrightarrow R-CH_2-S-\bar{S} \underset{-H^+}{\overset{+H^+}{\rightleftharpoons}} R-CH_2-S-SH$$

$$RCH_2-S-SH + CN^- \longrightarrow R-CH_2-\bar{S} + HSCN$$

$$HSCN \rightleftharpoons H^+ + SC\bar{N}$$

Fig. 2. Scheme for the formation of a hydrodisulfide bond in a reaction between atomic sulfur and the side chain of a cysteine molecule contained in a polypeptide chain

these same cytochrome P-450 containing enzyme systems. The binding of the sulfur atom may also be responsible for the liver necrosis seen with some of these compounds. However, the binding of the intermediate sulfine derivatives of these compounds may also be responsible for this effect.

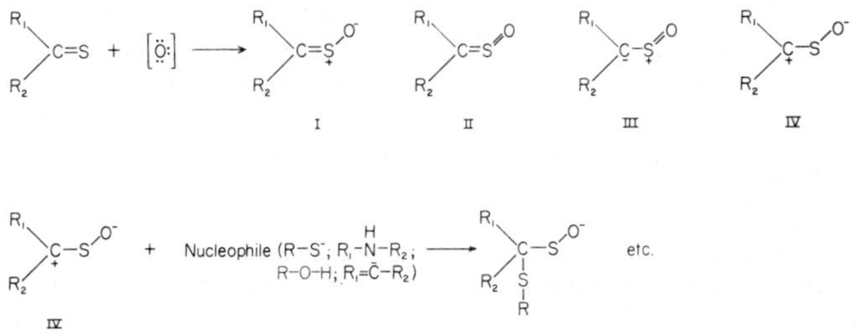

Fig. 3. Scheme for the reaction of tissue nucleophiles with a sulfine derivative of a thioketone.

REFERENCES

1. J.I. Martinez-Lopez, S.E. Greenberg, and R.R. Kling, Drug induced hepatic injury during methimazole therapy, Gastroenterology 43, 84 (1962).
2. IARC Monographs on the Evaluation of Carcinogenic Risk of Chemicals to Man, Vol. 7, International Agency for Research on Cancer, Lyon, 1974.
3. H.J. Eisen and A.L. Ginsberg, Disulfiram hepatotoxicity, Ann. Intern. Med. 83, 673 (1975).
4. A.L. Hunter and R.A. Neal, In vitro and in vivo inhibition of hepatic mixed function oxidase activity by various thiono-sulfur containing compounds, Biochem. Pharmacol. 24, 2199 (1975).
5. M.S. Fedotin and L.G. Lefer, Liver disease caused by propylthiouracil, Arch. Intern. Med. 135 (2), 319 (1975).
6. E.B. Astwood, Some observations on the use of thiobarbital as an antithyroid agent in the treatment of Graves' disease, J. Clin. Endocrinol. Metab. 5, 345 (1945).
7. E.J. Bond and F. DeMatteis, Biochemical changes in rat liver after administration of carbon disulphide, with particular reference to microsomal changes, Biochem. Pharmacol. 18, 2531 (1969).
8. R.L. Stirret, D.W. Petit and P. Starr, Therapeutic studies in hyperthyroidism: Methylthiouracil, J. Clin. Endocrinol. Metab. 12, 719 (1952).
9. J.E. Holoubek, W.R. Mathews and W.J. Hollis, Thiouracil hepatitis, Amer. J. Med. 5, 138 (1948).

10. A.A. Seawright, J. Hrdlicka and F. DeMatteis, The hepatotoxicity of O,O-diethyl, O-phenyl phosphorothionate (SV_1) for the rat, Brit. J. Exp. Path. 57(1), 16 (1976).
11. O.G. Fitzhugh and A.A. Nelson, Liver tumors in rats fed thiourea or thioacetamide, Science 108, 626 (1948).
12. L.L. Poulsen, R.M. Hyslop and D.M. Zeigler, S-oxidation of thioureylenes catalyzed by a microsomal flavoprotein mixed-function oxidase, Biochem. Pharmacol. 23, 3431 (1974).
13. B.J. Norman, R.E. Poore and R.A. Neal, Studies of the binding of sulfur released in the mixed-function oxidase-catalyzed metabolism of diethyl p-nitrophenyl phosphorothionate (parathion) to diethyl p-nitrophenyl phosphate (paraoxon), Biochem. Pharmacol. 23, 1733 (1974).
14. T. Kamataki, M.C.M.L. Lin, D.H. Belcher and R.A. Neal, Studies of the metabolism of parathion using an apparently homogeneous preparation of rabbit liver cytochrome P-450, Drug Metab. Disp. 4, 180 (1976).
15. K.A. Ptashne, R.M. Wolcott and R.A. Neal, Oxygen-18 studies on the chemical mechanisms of the mixed-function oxidase catalyzed desulfuration and dearylation reactions of parathion, J. Pharmacol. Exp. Ther. 179, 380 (1971).
16. K.A. Ptashne and R.A. Neal, Reaction of parathion and malathion with peroxytrifluoroacetic acid, A model system for the mixed-function oxidases, Biochemistry 11, 3224 (1972).
17. J.B. McBain, I. Yamamoto and J.E. Casida, Mechanism of activation and deactivation of Dyfonate(R) (O-ethyl S-phenyl ethylphosphonodithioate) by rat liver microsomes, Life Sci. 10(2), 947 (1971).
18. R.R. Dalvi, R.E. Poore and R.A. Neal, Studies of the metabolism of carbon disulfide by rat liver microsomes, Life Sci. 14, 1785 (1974).
19. K. Gollnick and E. Leppin, Direct photolysis of carbonyl sulfide in solution. Mechanism of singlet D and triplet P sulfur atom formation, J. Amer. Chem. Soc. 92, 2217 (1970).
20. E. Leppin and K. Gollnick, Direct photolysis of carbonyl sulfide in solution. Reactions of singlet D sulfur atoms in the liquid phase, J. Amer. Chem. Soc. 92, 2221 (1970).
21. R.E. Poore and R.A. Neal, Evidence for extrahepatic metabolism of parathion, Tox. Appl. Pharmacol. 23, 759 (1972).
22. T. Kamataki and R.A. Neal, Metabolism of diethyl p-nitrophenyl phosphorothionate (parathion) by a reconstituted mixed-function oxidase enzyme system: Studies of the covalent binding of the sulfur atom, Mol. Pharmacol. (In press).
23. G.L. Catignani and R.A. Neal, Evidence for formation of a protein bound hydrodisulfide resulting from the microsomal mixed-function oxidase catalyzed disulfuration of carbon disulfide, Biochem. Biophys. Res. Commun. 65, 629 (1975).

24. S. Kawamura, Y. Otsuji, T. Nakabayashi, T. Kitao and J. Tsurugi, Aralkyl hydrodisulfides. IV. The reaction of benzyl hydrodisulfide with several nucleophiles, J. Org. Chem. 30, 2711 (1965).

STIMULATORY EFFECTS OF POLYCHLORINATED BIPHENYLS (PCB) ON CYTOCHROMES P-450 AND P-448 MEDIATED MICROSOMAL OXIDATIONS

Alvito P. Alvares

The Rockefeller University, New York, N.Y. 10021

PCB are industrial chemicals which are widely distributed in the environment. The use of PCB in consumer products has been drastically curtailed in the U.S. in the past 2 years. Due to inadvertent contamination of the environment, PCB have been found in fish, cows milk, human milk as well as human adipose tissue. They are sold commercially in the U.S. under the trade name of Aroclor. In the present studies, the PCB mixture, Aroclor 1254 was used since many of the components present in this mixture are similar to those found in extracts of human adipose tissue (1). Previous studies by Alvares et al. (2) have shown that PCB possess some of the inducing properties of the two main classes of microsomal enzyme inducers, the prototypes of which are phenobarbital (PB) and the polycyclic hydrocarbon, 3-methylcholanthrene (3-MC). The CO-difference spectrum of hepatic microsomes from PCB-treated rats showed an absorption maximum at 448-449 nm compared to the absorption maximum of 450 nm observed with untreated rats (2). PCB pretreatment also resulted in an increase in the ratio of the 455:430 nm peaks in the ethyl isocyanide difference spectrum of hepatic microsomes, the increase being similar to that observed when rats are pretreated with 3-MC (2). Subsequent to these initial studies, further studies have been carried out to provide additional information on the characteristics of PCB induction and to compare the inductive effects of PCB, PB and 3-MC. In these studies, each inducer was administered at a maximal inducing dosage. As shown in Table 1, the PCB mixture, Aroclor 1254, like PB, significantly increased liver weight, microsomal protein, cytochromes b_5 and P-450 content, NADPH-cytochrome c reductase activity, the N-demethylation of ethylmorphine and benzphetamine and markedly decreased hexobarbital-induced sleeping time. Like the 3-MC class of inducers, Aroclor 1254 caused marked enhancement of benzo(a)pyrene hydroxylase activity and decreases in zoxazolamine-induced paralysis time (Table 1). These data provide further evidence that PCB possess the inducing properties of both the PB as well as the 3-MC class of inducing substances.

Besides being potent inducers of benzo(a)pyrene hydroxylase in the liver, Aroclor 1254 had a marked stimulating effect on this hydroxylase activity in non-hepatic tissues as shown in Fig. 1. Ten-15 fold induction of benzo(a)-pyrene hydroxylase activity occurred in homogenates of liver, kidney and small intestine of male rats and in the placentas of pregnant rats pretreated with Aroclor 1254. The lung homogenates showed a 5-fold increase in hydroxylase activity. Previous studies have shown that PCB also have a markedly stimulatory effect on skin hydroxylase activity, when they are painted on the skin of rats (3). The data obtained on extrahepatic metabolism of benzo(a)pyrene with PCB-treated rats are very similar to those obtained when

Stimulatory Effects of Polychlorinated Biphenyls

TABLE 1 Comparative effects of polychlorinated biphenyls (PCB), phenobarbital (PB) and 3-methylcholanthrene (3-MC) on various hepatic microsomal parameters in the male rat

Parameter	Percent Control		
	PB	Aroclor 1254	3-MC
Microsomal protein	140	140	106
Liver/body weight	132	129	118
Cytochrome b_5	138	141	103
Cytochrome P-450	302	286	237
NADPH-cytochrome c reductase	190	182	108
Ethylmorphine N-demethylase	366	321	110
Benzphetamine N-demethylase	709	455	86
Benzo(a)pyrene hydroxylase	268	739	780
Hexobarbital-induced sleeping time	12	24	120
Zoxazolamine-induced paralysis time	14	1	2

Male Sprague Dawley rats were administered i.p. Aroclor 1254, 25 mg/kg/day for 6 days, or PB, 75 mg/kg/day for 4 days or 3-MC, 25 mg/kg/day for 4 days and sacrificed 24 hours after the administration of the last dose. Each value represents the mean for 5 rats.

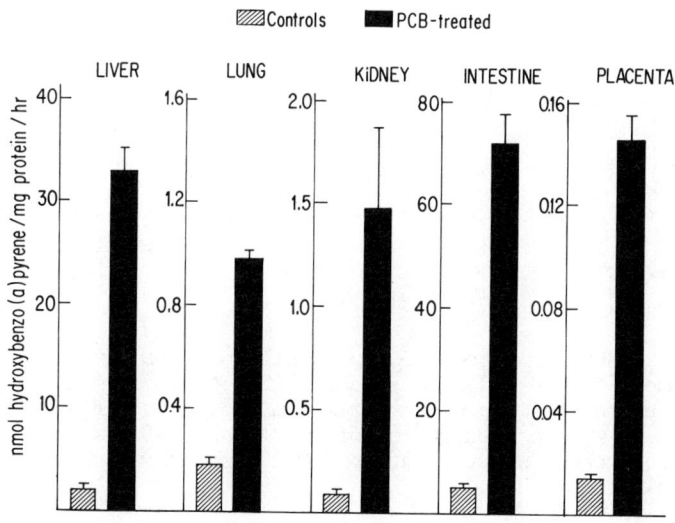

Fig. 1 Influence of Aroclor 1254 pretreatment on benzo-(a)pyrene hydroxylase activity in various tissues. Aroclor 1254, at a dosage of 25 mg/kg/day for 6 days was administered intraperitoneally. Benzo(a)pyrene hydroxylase activity was determined on homogenates of the various tissues.

rats are pretreated with 3-MC. The polycyclic hydrocarbon has a marked stimulatory effect on benzo(a)pyrene hydroxylase activity in extrahepatic tissues whereas PB is a very weak inducer (4). Thus although the liver is a major site of xenobiotic biotransformations, other tissues may play a significant role in such biotransformations when such tissues have markedly induced enzymic activities as observed following exposure to PCB.

In order to further examine the hemeprotein(s) induced by PCB, a partially purified preparation of the hemeprotein induced in rats pretreated with Aroclor 1254 was prepared by the method of Lu and Levin (5) and carried through the calcium phosphate gel step of their method. The absolute spectra of the induced hemeprotein are shown in Fig. 2. The oxidized spectrum

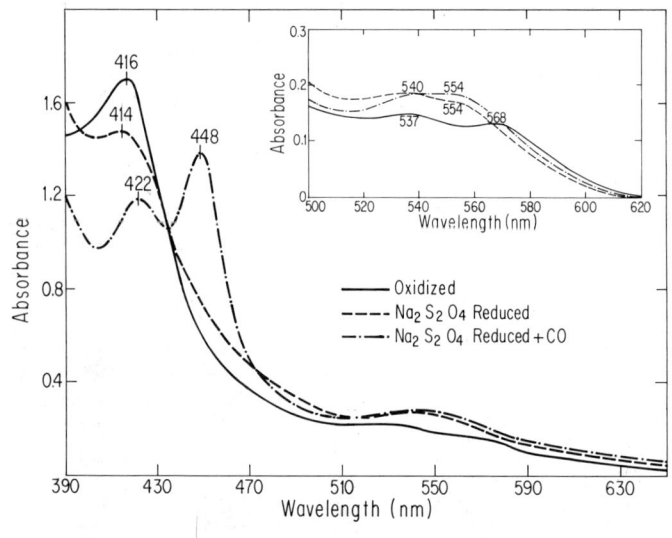

Fig. 2 Absolute spectra of partially purified cytochrome P-450 (P-448) from Aroclor 1254 treated rats. Cytochrome P-450 in 0.1 M potassium phosphate buffer (pH 7.4) at a concentration of 10.8 nmoles/ml was used. The reference cuvette contained 0.1 M potassium phosphate buffer (pH 7.4).

has a Soret band at 416 nm and absorption bands at about 537 and 568 nm. When the oxidized form was reduced with dithionite, the Soret peak shifted downward to 414 nm and a broad peak was observed at around 550 nm. These spectra are very similar to what Lu and Levin have reported with their partially purified P-450 and P-448 preparations derived from PB and 3-MC treated rats, respectively. The CO-spectrum of the reduced hemeprotein shows a small peak at 422 nm and an absorption maximum of 448 nm. The peak at 422 nm may be due to the presence of contaminating amounts of cytochrome b_5 or due to the presence of cytochrome P-420. The occurrence of absorption maximum at

448 nm in the CO-spectrum of the reduced hemeprotein from Aroclor 1254 treated rats provides further evidence that the hemeproteins induced by this PCB mixture is a mixture of cytochromes P-450 and P-448, since studies by Ryan et al. (6) show that the spectral maxima for P-450 purified from PB treated rats and P-448 purified from 3-MC treated rats occurred at 450 nm and 447.5 nm, respectively. Previous studies from this laboratory (7) have shown that when the hemeproteins partially purified from Aroclor 1254 treated rats were subjected to sodium dodecyl sulfate-polyacrylamide gel electrophoresis, the gel electrophoretic band patterns resembled a combination of the bands obtained with cytochrome P-450 from separately injected PB and 3-MC rats.

Since PCB, which are commercially available, are mixtures comprised of chemical congeners of both high and low chlorine containing biphenyls, it was of importance to determine whether PCB mixtures contain one or more isomers having the PB type of inductive properties and one or more isomers having the 3-MC type of induction. However, there was a second possibility, namely that a single component of the mixture possesses the combined properties of the two classes of inducers. Approximately 20 pure chlorinated isomers of biphenyl (obtained from Analabs, 99% pure) were tested in rats at dosages comparable to the PCB mixture Aroclor 1254. The biphenyl moiety itself and the di-, tri-, hepta-and deca-chlorine containing isomers tested were devoid of inducing capacities. The tetra- and penta-chlorine containing isomers were either not inducers or were very poor inducers in comparison with PB, 3-MC or Aroclor 1254. A comparison of the inducing properties of the hexachlorobiphenyl isomers is shown in Table 2. Of the hexachlorobiphenyls

TABLE 2 Comparative effects of hexachlorobiphenyl (HCB) isomers on microsomal enzyme activities and on cytochrome P-450 content

Compound	Percent of control values		
	Cytochrome P-450 or P-448	Ethylmorphine N-demethylase	Benzo(a)pyrene Hydroxylase
Control	100	100	100
2,3,4,2',3',4'-HCB	121	139	206
2,3,6,2',3',6'-HCB	101	101	98
2,3,4,2',4',5'-HCB	271	275	883
2,4,5,2',4',5'-HCB	324	367	922
2,4,6,2',4',6'-HCB	126	190	91
2,3,5,6,2',5'-HCB	108	111	106
2,3,4,5,2',5'-HCB	102	111	122
PB	310	373	212
3-MC	176	93	895
Aroclor 1254	275	290	898

HCB isomers and Aroclor 1254 were administered i.p. at a dose of 25 mg/kg/day for 6 days. PB and 3-MC were administered i.p. at a dose of 75 mg/kg/day for 4 days and 25 mg/kg/day for 4 days, respectively. Each value represents the mean for 5 rats.

tested, only 2,3,4,2',4',5' and 2,4,5,2',4',5' induced cytochrome P-450 (with the spectral maxima occurring at 448-449 nm), ethylmorphine N-demethylase and benzo(a)pyrene hydroxylase activities comparable to the induction observed with the PCB mixture, Aroclor 1254. The induction of demethylase by these two isomers was comparable to that observed with PB treated rats and the induction of benzo(a)pyrene hydroxylase was comparable to the induction observed with 3-MC treated rats. Thus these isomers are unique in that a single compound is capable of inducing both P-450 and P-448 associated enzymic activities.

Previous studies have demonstrated that PCB can cross the placental barrier and can be transmitted through maternal milk and induce fetal and neonatal hepatic drug-metabolizing enzymic activities (8). However, during gestation the effects of PCB on the liver enzymes of the maternal host differ markedly from those elicited by PB and 3-MC. At dosages which elicited maximal response in the male rat, PB induced ethylmorphine N-demethylase and cytochrome P-450 in the pregnant rat (Table 3). Similarly, 3-MC induced

TABLE 3 Effects of polychlorinated biphenyls (PCB), phenobarbital (PB) and 3-methylcholanthrene (3-MC) on microsomal enzymes and cytochrome P-450 in livers of 19-day pregnant rats

Measurement	Control	Aroclor 1254	3-MC	PB
Ethylmorphine N-demethylase nmol HCHO/mg protein/hr	44.26^1 ±3.34	57.51 ±8.67	55.81^2 ±2.73	194.41^2 ±29.74
Benzo(a)pyrene hydroxylase nmol OHBP/mg protein/hr	1.03 ±0.05	28.07^2 ±4.85	58.17^2 ±3.08	1.62^2 ±0.13
Cytochrome P-450 or P-448 nmol/mg protein	0.278 ±0.013	0.531^2 ±0.025	0.674^2 ±0.040	0.495^2 ±0.072
Ethyl isocyanide difference spectrum, ratio of 455:430 peaks	0.74 ±0.02	1.44^2 ±0.10	1.53^2 ±0.08	0.67^2 ±0.01

Rats were administered i.p. Aroclor 1254, 25 mg/kg/day for 6 days, or 3-MC, 25 mg/kg/day for 4 days, or PB, 75 mg/kg/day for 4 days and sacrificed 24 hours after the administration of the last dose.

[1] Each value represents mean ± S.E. for at least 5 rats.

[2] Value significantly different from control value.

benzo(a)pyrene hydroxylase and caused a marked change in the ratio of the 455:430 nm peaks of the ethyl isocyanide difference spectra (Table 3), these changes being similar to those observed when 3-MC is administered to male rats. However, as shown in Table 3, when Aroclor 1254 is administered at a dosage of 25 mg/kg/day - the dosage used in the male rat studies - PCB elicited only the 3-MC type of inductive effect i.e. benzo(a)pyrene hydroxylase activity was enhanced and the 455:430 nm ratio was increased. At this dosage, PCB did not significantly induce N-demethylase activity. Thus during gestation, the PB-type of inductive effect of PCB appears to be greatly diminished. As shown in Fig. 3, the pregnant rat requires Aroclor 1254, at a

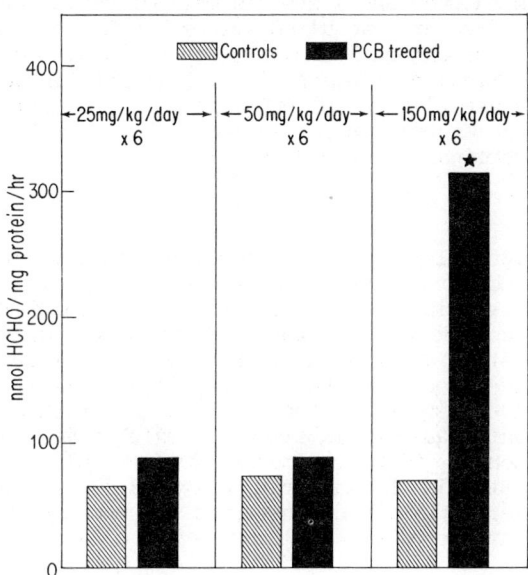

Fig. 3 Influence of various dosages of Aroclor 1254 on liver ethylmorphine N-demethylase activity in 19-day pregnant rats. Aroclor 1254 was administered for 6 days at dosages indicated in the figure and rats were sacrificed on the 7th day (19th day of pregnancy). Asterisk denotes value significantly different from control value ($p < 0.05$).

dosage of 150 mg/kg/day for 6 days to elicit the stimulation of N-demethylase activity. Thus the pregnant rat requires six times the male dosage to elicit the PB type of inductive effect. These data suggest the presence of a repressor substance during gestation which impairs the PB type of induction elicited by PCB. Nebert and coworkers (9) have postulated the existence of separate genetic loci which direct the synthesis of the P-450 and P-448 hemeproteins. The present studies suggest that repression of the P-450 hemeprotein synthesis by PCB during pregnancy involves only the P-450 locus but not the P-448 locus. It would be of importance to determine if similar dose related differences in inducibility of N-demethylase activity can be attributed to PB in the pregnant rat in comparison with the male or non-pregnant female rat.

The ability of environmental pollutants, such as PCB, to modify these P-450 mediated microsomal enzymes may have considerable significance, for example, in populations that are occupationally exposed to such agents as well as in the general population through inadvertent exposure to these long-lived pollutants. Since a number of endogenous steroid hormones also serve as substrates for the P-450 system, it would be of considerable importance to determine what effects on endogenous steroid metabolism in individuals exposed to PCB. Finally, recently acquired evidence seems to indicate that the induction of aryl hydrocarbon hydroxylase in human tissues may play a significant role in chemical carcinogenesis. The marked induction of benzo(a)pyrene hydroxylase elicited by PCB merits investigations on whether these chemicals are carcinogenic and/or increase the susceptibility of individuals to certain chemically induced cancer.

SUMMARY

The polychlorinated biphenyl mixture, Aroclor 1254 induces both cytochromes P-450 and P-448 and associated enzymic activities in the liver of the male rat. Besides being potent inducers of benzo(a)pyrene hydroxylase activity in the liver, PCB had a marked stimulatory effect on this enzymic activity in the lung, kidney, intestine and placenta. PCB, which exhibit both the PB and the 3-MC type inductive effects in the male rat liver, elicit primarily the latter type of inductive response in the liver in gestation at comparable doses. The pregnant rat requires six times the male dosage to elicit the PB type of inductive response. Of approximately 20 pure chlorinated isomers of biphenyl which were tested in the male rat, only 2,4,5,2',4',5'- and 2,3,4,2',4',5' hexachlorobiphenyls demonstrated the inducing properties of both the barbiturate and the polycyclic hydrocarbon classes of inducing substances.

ACKNOWLEDGEMENTS

Supported in part by USPHS Grant ES 01055. A. P. Alvares is a recipient of a Research Career Development Award 1 KO4 ES00010-02 from the National Institutes of Health.

REFERENCES

1. Price, H. A. and R. L. Welch. Occurrence of polychlorinated biphenyls in humans. Environ. Health Persp. 1, 73 (1972).

2. Alvares, A. P., D. R. Bickers and A. Kappas. Polychlorinated biphenyls: A new type of inducer of cytochrome P-448 in the liver. Proc. Natl. Acad. Sci. U.S.A. 70, 1321 (1973).

3. Bickers, D. R., J. Eiseman, A. Kappas and A. P. Alvares. Microscope immersion oils: Effects of skin application on cutaneous and hepatic drug-metabolizing enzymes. Biochem. Pharmacol. 24, 779 (1975).

4. Lake, B. G., R. Hopkins, J. Chakraborty, J. W. Bridges and D. V. W. Parke. The influence of some hepatic enzyme inducers and inhibitors on extrahepatic drug metabolism. Drug Metab. Disp. 1, 342 (1973).

5. Lu, A. Y. H. and W. Levin. Partial purification of cytochromes P-450 and P-448 from rat liver microsomes. Biochem. Biophys. Res. Commun. 46, 1334 (1972).

6. Ryan, D., A. Y. H. Lu, S. West and W. Levin. Multiple forms of cytochrome P-450 in phenobarbital and 3-methylcholanthrene-treated rats. J. Biol. Chem. 250, 2157 (1975).

7. Alvares, A. P. and P. Siekevitz. Gel electrophoresis of partially purified cytochromes P-450 from liver microsomes of variously-treated rats. Biochem. Biophys. Res. Commun. 54, 923 (1973).

8. Alvares, A. P. and A. Kappas. Induction of aryl hydrocarbon hydroxylase by polychlorinated biphenyls in the foeto-placental unit and neonatal livers during lactation. FEBS Letters 50, 172 (1975).

9. Goujon, F. M., D. W. Nebert and J. E. Gielen. Genetic expression of aryl hydrocarbon hydroxylase induction. IV. Interaction of various compounds with different forms of cytochrome P-450 and the effect on benzo-(a)pyrene metabolism in vitro. Molec. Pharmacol. 8, 667 (1972).

DEPRESSION OF HEPATIC CYTOCHROME P-450-DEPENDENT MONOOXYGENASE SYSTEMS WITH ADMINISTERED INTERFERON INDUCING AGENTS

Kenneth W. Renton and Gilbert J. Mannering

Department of Pharmacology, University of Minnesota, Minneapolis, Minnesota 55455

ABSTRACT

Cytochrome P-450-dependent monooxygenase activities and cytochrome P-450 levels were depressed in hepatic microsomes from rats treated with 12 interferon inducing agents of various types: small molecules (e.g. tilorone), an RNA virus (Mengo), a fungal mycophage (statolon), liver RNA, a synthetic double-stranded polynucleotide (poly rI · poly rC), a bacterial lipopolysaccharide (*E. coli* endotoxin) and an attenuated bacteria (*B. pertussis* vaccine). The results suggest that the depression of hepatic cytochrome P-450-dependent monooxygenase systems may be a general property of interferon inducing agents.

INTRODUCTION

The practice of administering agents such as phenobarbital and 3-methylcholanthrene to induce hepatic P-450 hemoprotein-dependent monooxygenase systems in laboratory animals has greatly facilitated studies of these systems during the past two decades (Reviewed by Conney, 1,2; Mannering, 3; and Remmer, 4). The chance observation made in our laboratory (5) that the administration of the interferon inducing agent, tilorone, caused a marked decrease in components of these monooxygenase systems in rats suggests a means of manipulating the activities of monooxygenase systems in hepatic microsomes in a direction opposite to that produced by inducing agents. These studies raised the question of whether the depressant effect of tilorone is peculiar to that compound or whether, as the studies to be presented strongly suggest, the effect is a general property of interferon inducing agents. Studies of the interaction of interferon inducing agents and inducers of hepatic monooxygenase systems may lead to both a better understanding of interferon induction and of the mechanisms regulating the maintainance of steady-state levels of the monooxygenases in the endoplasmic reticulum of the liver.

Effects of Tilorone on Hepatic Monooxygenase Systems

Tilorone (2,7-bis[2-(diethylamino)ethoxy]-fluorene-9-one HCl) in doses ranging from 10 to 100 mg/kg was administered orally to male Holtzman strain rats (180-200g) once daily for four days. The animals were killed 24 hr after the last administration and microsomal preparations were made from their livers. The effects of tilorone on microsomal protein, cytochrome

P-450, and monooxygenase activities are shown in Fig. 1. With the highest

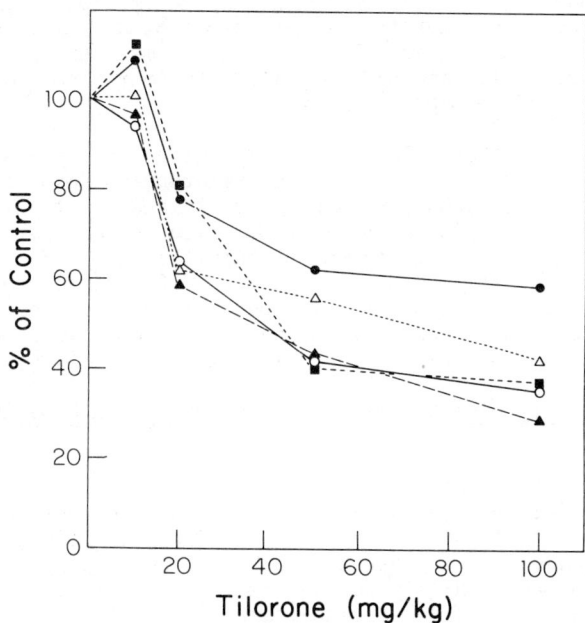

Fig. 1. Effects of tilorone on hepatic monooxygenase systems. Male Holtzman rats (180-200 g) were treated daily *per os* for 4 days with various doses of tilorone and killed 24 hr after the last dose. Hepatic microsomes were prepared and assays performed as described previously (5). Results are calculated on a per gram of liver basis and are expressed as the percent of corresponding control values obtained with untreated animals. Values are the means obtained from 4 rats considered individually. 100% of control (mean ± S.E.) = protein (●) 9.7 ± 0.9 mg; P-450 hemoprotein. (○), 8.7 ± 0.7 nmoles; ethylmorphine N-demethylase (▲), 4.80 ± 0.83 µmoles HCHO/hr; aniline p-hydroxylase (△), 0.46 ± 0.04 µmoles p-aminophenol/hr; benzo[a]pyrene hydroxylase (■), 0.22 ± 0.05 µmoles 8-OH benzo[a]pyrene/hr. All values obtained with rats which received 20 mg of tilorone or more are significantly different from control values (Student's t test, $P < .05$).

dose of tilorone (100 mg/kg), microsomal protein was lowered by 42%, P-450 hemoprotein by 65%, ethylmorphine N-demethylase activity by 70%, aniline p-hydroxylase activity by 58%, and benzo[a]pyrene hydroxylase activity by 62%. The 50 mg/kg dose was almost as effective as the 100 mg/kg dose. The animals appeared to be in good condition throughout the experimental period, even at the highest dose of tilorone administration. Body and liver weights were not affected by the drug and the livers showed no appreciable changes when examined by electron microscopy.

Tilorone did not lower the quantity of all components of the hepatic microsomes. The activity of NADPH-cytochrome c reductase was lowered about 20% in rats which had received the 50 mg/kg dose of tilorone, but neither NADH-cytochrome c reductase activity nor cytochrome b_5 was affected. Tilorone had no effect *in vitro*; preincubation of microsomes for up to 30 min with 0.1, 0.5, or 1.0 mM tilorone did not alter ethylmorphine N-demethylase activity or the P-450 hemoprotein content.

To further establish that tilorone was producing its depression of the monooxygenase system *in vivo*, the effects of tilorone on hexobarbital sleeping time and hexobarbital blood levels were determined. Rats given 20 mg of tilorone/kg daily for 4 days *per os* slept twice as long as control rats when challenged with 100 mg/kg of sodium hexobarbital (i.p.) 24 hr after the last dose of tilorone (control rats: 26.8 ± S.E. 2.8 min; tilorone-treated rats: 54.1 ± S.E. 4.3 min). Blood levels of hexobarbital 30 min after its administration were also elevated in tilorone-treated animals (control rats: 22.1 ± 3.5 µg/ml; tilorone-treated rats: 51.0 ± S.E. 5.7 µg/ml).

Effects of a Variety of Interferon Inducing Agents on Hepatic Monooxygenase Systems

Twelve known interferon inducing agents representing a wide variety of structures and molecular weights, were tested for their effects on hepatic monooxygenase systems of the rat: an RNA virus (Mengo), a fungal mycophage (statolon), hepatic RNA, a synthetic double stranded polynucleotide (poly rI · poly rC), a bacterial lipopolysaccharide (*E. coli* endotoxin), an attenuated bacteria (*B. pertussis* vaccine), and six small molecules (quinacrine, N-N-dioctadecyl-n'n'-bis(2-hydroxyethyl)propanediamine, tilorone, and three derivatives of tilorone). The interferon inducing agents, their doses, routes of administration, sources, and references to their interferon inducing activities are given in Table 1. Male, Holtzman strain rats (180-200 g) were used. The agents were given daily for 3 days and microsomes were collected 24 hr after the last dose except in the case of Mengo virus, which was administered once and microsomes collected 24 or 48 hr later.

Table 2 summarizes the effects of the 12 interferon inducing agents on ethylmorphine N-demethylase and aniline p-hydroxylase activities and microsomal protein, P-450 hemoprotein and cytochrome b_5 contents of hepatic microsomes from the rat. Ethylmorphine N-demethylase and aniline p-hydroxylase activities and the cytochrome P-450 content of microsomes were depressed whenever an interferon inducing agent was administered, although in a few cases, the loss was significant only at the $P < 0.1$ levels. Statistically significant losses of microsomal protein were seen only with tilorone and two of its analogues and with Mengo virus. Significant losses of cytochrome b_5 were observed with statolon, poly rI · poly rC, endotoxin and pertussis vaccine. Included in Table 2 are results obtained with the single stranded polynucleotides, poly rI and poly rC, neither of which is an interferon inducing agent (15). We consider it particularly pertinent that neither compound depressed the microsomal activities and components under investigation.

It is to be emphasized that with the exception of Mengo virus, which caused severe illness in the rats, none of the agents employed in this study caused a weight loss or other gross manifestation of toxicity.

TABLE 1 Administration regimen of interferon inducing agents

Agent	Dose (per kg body wt)		Source	Reference[a]
Mengo virus	5×10^{-7} pfu	(iv)	b	6
Statolon	50 mg	(ip)	Eli Lilly	7
Hepatic RNA (Type IV)	5 mg	(ip)	Sigma	8
Poly rI · poly rC	2.5 mg	(ip)	Sigma	9
E. coli endotoxin	5 mg	(ip)	Sigma	10
B. pertussis vaccine	2×10^{11} cells	(ip)	Eli Lilly	11
Tilorone	50 mg	(po)	Richardson-Merrill	12
RMI 11002[c]	50 mg	(ip)	Richardson-Merrill	d
RMI 11567[e]	50 mg	(ip)	Richardson-Merrill	d
RMI 11877[f]	50 mg	(ip)	Richardson-Merrill	d
CP 20,901[g]	50 mg	(ip)	Pfizer	13
Quinacrine	50 mg	(ip)	Sigma	14

[a] Selected reference to study showing interferon inducing activity of the agent.
[b] Supplied by Dr. P. G. W. Plagemann, Dept. of Microbiology, University of Minnesota.
[c] 3,6-bis(dimethylaminoacetyl)fluorene.
[d] Personal communication, Dr. G. J. Wright, Richardson-Merrill, Inc., Cincinnati, Ohio.
[e] 3,6-bis)dimethylaminoacetyl)dibenzofuran.
[f] 3,6-bis(dimethylaminoacetyl)dibenzothiophene.
[g] N-N-dioctadecyl-n'n'-bis(2-hydroxyethyl)propanediamine.

The relative potencies of the interferon inducing agents as depressors of the hepatic monooxygenase system cannot be determined from the data given in Table 2 because only a single dose of each was used and temporal aspects relative to administration and observance of effects were not considered. However, of the agents tested, E. coli endotoxin appeared to be the most potent, and a dose response study was performed using mice. The i.p. administration of 40, 400 and 4000 µg of endotoxin/kg of mouse caused losses of microsomal ethylmorphine N-demethylase activity of 32, 47, and 63%,

TABLE 2 Effects of interferon inducers on microsomal hepatic monooxygenase

INTERFERON INDUCER	MICROSOMAL PROTEIN (% of control)	ETHYLMORPHINE N-DEMETHYLASE (% of control)	ANILINE p-HYDROXYLASE (% of control)	P-450 HEMOPROTEIN (% of control)	CYTOCHROME b_5 (% of control)
Mengo virus 24 hr	79.1 ± 9.7	73.3 ± 16.4	83.3 ± 9.8	77.1 ± 4.5[b]	85.5 ± 6.1
Mengo virus 48 hr	46.0 ± 2.1[a]	58.3 ± 1.4[a]	82.6 ± 2.9[a]	80.3 ± 8.1[b]	107.1 ± 11.8
Statolon	84.5 ± 13.6	47.7 ± 6.9[a]	60.8 ± 1.8[a]	59.4 ± 1.8[a]	81.1 ± 4.8[a]
Heaptic RNA (Type IV)	91.0 ± 5.9	47.7 ± 6.4[a]	----	69.5 ± 2.7[a]	68.1 ± 5.6[a]
Poly rI · poly rC	112.3 ± 18.9	53.5 ± 1.6[a]	65.7 ± 5.7[a]	64.8 ± 5.1[a]	86.2 ± 2.0[a]
Poly rI	124.7 ± 13.4	104.4 ± 6.4	94.0 ± 11.6	93.2 ± 4.7	97.9 ± 4.1
Poly rC	113.4 ± 7.8	88.1 ± 11.5	107.5 ± 10.8	94.6 ± 4.5	102.5 ± 2.6
E. coli endotoxin	105.5 ± 4.6	27.1 ± 4.8[a]	69.6 ± 10.4[b]	54.9 ± 5.3[a]	79.2 ± 2.4[a]
B. pertussis vaccine	120.0 ± 16.0	58.6 ± 11.9[a]	68.5 ± 12.2[b]	77.0 ± 6.0[a]	79.8 ± 2.5[a]
Tilorone	63.2 ± 5.1[a]	50.5 ± 11.1[a]	77.4 ± 7.1[a]	64.5 ± 5.8[a]	91.4 ± 8.5
RMI 11002	90.1 ± 5.4	61.2 ± 9.7[a]	61.6 ± 10.4[a]	55.7 ± 5.9[a]	78.9 ± 4.1
RMI 11567	80.9 ± 5.4	72.1 ± 12.2	77.9 ± 10.9	72.0 ± 8.7[b]	85.2 ± 5.5
RMI 11877	66.1 ± 7.8[a]	27.4 ± 9.1[a]	47.1 ± 8.8[a]	49.4 ± 7.6[a]	80.6 ± 3.5
CP 20,901	68.3 ± 6.9[a]	42.0 ± 8.7[a]	33.6 ± 7.4[a]	44.8 ± 6.1[a]	----
Quinacrine	93.2 ± 8.8	55.4 ± 1.5[a]	53.9 ± 4.6[a]	62.0 ± 7.0[a]	89.7 ± 3.5

Values are the mean ± S.E. of the percentages of the individual control values for each agent; means ± S.E. of all of the controls were: microsomal protein, 11.28±0.68 mg/g of fresh liver; ethylmorphine N-demethylase, 306±22 nmole HCHO formed/mg of protein/hr; aniline p-hydroxylase, 37.44±2.10 nmole p-aminophenol formed/mg of protein/hr; P-450 hemoprotein, 0.756±.033 nmole/mg of protein; cytochrome b_5, 0.320±0.16 nmole/mg of protein.

[a] Significantly different from corresponding control ($P < 0.05$).
[b] Significantly different from corresponding control ($P < 0.1$).

respectively. Corresponding losses of P-450 hemoprotein were 23, 48 and 45%. The lowest dose is equivalent to about 1 µg of endotoxin per mouse.

DISCUSSION

We conclude from these studies that the depression of hepatic P-450 hemoprotein-dependent monooxygenase systems is a general property of interferon inducing agents. This conclusion is based on the knowledge that the great majority of xenobiotics either induce or have no effect on these monooxygenase systems, that few substances induce interferon synthesis, that the various kinds of interferon inducing agents are largely represented by the agents used in this study, and finally, because the twelve agents used in this study were selected solely for their known ability to induce interferon synthesis.

These studies raise fundamental questions regarding the mechanism of interferon induction, the mechanisms of maintainance of steady-state levels of hepatic monooxygenase systems, and the possibility that the two mechanisms may share components common to each. A current model of interferon synthesis has the inducing agent binding to a repressor protein in the exposed cell, which releases preexisting interferon mRNA for translation (16). Newly synthesized interferon mRNA can also be translated until halted by a repressor which is controlled by a repressor gene. The released interferon triggers the synthesis of antiviral protein which imparts antiviral properties to the cell. Interferon released from this cell initiates the process in other cells. This model provides several opportunities for speculation as to how interferon inducing agents might affect the synthetic or catabolic processes which govern the steady-state level of monooxygenase systems in the liver. The depression of the monooxygenase system might be caused by a) interferon itself, b) the transcriptional process whereby genes coding for the control of the monooxygenase system overlap those responsible for interferon synthesis, c) the repressor protein which blocks both interferon mRNA translation and an mRNA required for control of the monooxygenase system and d) the antiviral protein, which could act on the mechanism controlling the monooxygenase system.

Another highly speculative mechanism is offered which invokes membrane in an attempt to explain both induction and depression of the hepatic monooxygenase systems. A large percentage of the hepatic microsome is comprised of enzyme components of the monooxygenase system. The dense packing of these enzymes could render Michaelis-Menten kinetics inapplicable (17). Under these conditions, specific rate limitations for an overall reaction might be a matter of structural organization of the multienzyme complex in the membrane (18). In a recent review, Stier (18) has proposed that the physical state of the lipid components of the membrane may determine the arrangement of the densely packed multienzymes embedded in the lipid matrix, as well as the conformational states of the individual enzymes, and that a hormone or drug may induce a phase separation that could alter the nature of the monooxygenase system. Carrying this concept a step further, the organization in the membrane of those enzymes which are responsible for the synthesis or degradation of certain enzyme systems might similarly regulate the steady state levels of the monooxygenase systems and other systems. Accordingly, phenobarbital and other known inducers of the monooxygenase system and other systems (e.g., the unrelated system responsible for the synthesis of ascorbic acid) would produce their effects by inducing phase separation. Acting either directly or indirectly, interferon inducing agents, or interferon itself, would induce a kind of phase separation, which could produce an opposite effect on monooxygenase systems, possibly by inducing or activating degradative enzyme systems.

Interferon is not as harmless as was once believed. Interferon preparations inhibited cell division in adult mice (19), inhibited the multiplication of allogeneic spleen and syngeneic bone marrow cells in X-irradiated adult mice (20), and killed newborn mice by inflicting severe liver damage (21). Interferon treatment of uninfected cells has also been reported to inhibit the synthesis of macromolecules (22). The depression of hepatic monooxygenase systems by interferon inducing agents may be another specific example of the interferance of the synthesis of cellular macromolecules by interferon.

Our findings may have important clinical implications. Do viral infections cause a depression of drug metabolizing activity in man? In view of the potency of endotoxin in our experiments, the question is also raised as to whether rates of drug metabolism may be lowered in certain diseases of bacterial origin. Patients are currently being treated with potent interferon preparations and interferon inducing agents; it will be necessary to determine what effect these treatments may have on rates of drug metabolism in these patients.

We thank Mr. Dan Keyler and Ms. Viola Abbott for assistance. K. W. Renton is a recipient of a Medical Research Council of Canada Fellowship. This work was supported by a U.S. Public Health Service Grant 15477.

REFERENCES

1. A. H. Conney, Pharmacological implications of microsomal enzyme induction, Pharmacol. Rev. 19, 317 (1967).
2. A. H. Conney, in Fundamental of Drug Metabolism and Drug Disposition, (B. N. LaDu, H. G. Mandel, E. L. Way, eds.), Williams & Wilkins, Baltimore (1971).
3. G. J. Mannering, in Selected Pharmacological Testing Methods, (A. Burger, ed.), Marcel Dekker, New York (1968).
4. H. Remmer, Induction of Drug Metabolizing Enzyme System in the Liver. Europ. J. Clin. Pharmacol. 5, 116 (1972).
5. K. W. Renton and G. J. Mannering, Depresson of the hepatic cytochrome P-450 mono-oxygenase system by administered tilorone. Drug Metab. Disps. 4, 223 (1976).
6. J. B. Campbell, T. Grunberger, M. A. Kochman and S. L. White, A microplaque reduction assay for human and mouse interferon, Can. J. Microbiol. 21, 1247 (1975).
7. W. J. Kleinschmidt and E. B. Murphy, Interferon induction with statolon in the intact animal. Bact. Rev. 31, 132 (1967).
8. E. DeMaeyer, J. DeMaeyer-Guignard and L. Montagnier, Double stranded RNA from rat liver induces interferon in rat cells. Nature New Biol. 229, 109 (1971).
9. A. K. Field, A. A. Tytell, G. P. Lampson and M. R. Hilleman, Inducers of interferon and host resistance. II Multistranded synthetic polynucleotide complexes. Proc. Nat. Acad. Sci. 58, 1004 (1967).

10. M. Ho, Interferon-like viral inhibitor in rabbits after intravenous administration of endotoxin. Science 146, 1472 (1964).
11. C. Colby and M. J. Morgan, Interferon. Ann. Rev. Microbiol. 29, 131 (1975).
12. R. F. Krueger and G. D. Mayer, Tilorone hydrochloride, an orally acting antiviral agent. Science 169, 1213 (1970).
13. W. W. Hoffman, J. H. Korst, J. F. Niblack and J. F. Cronin, N,N-Dioctadecyl-N,N' Bis(2-hydroxyethyl)propanediamine: Antiviral activity and interferon stimulation in mice. Antimicrobiol Agents and Chemotherap. 3, 498 (1973).
14. E. T. Slaz, E. Szolsay, I. Stoger and M. Talas, Antiviral activity and induction of interferon-like substance by quinacrine and acranil. Antimicrob. Agents and Chemotherap. 3, 537 (1973).
15. D. C. Burke, in Interferon and Interferon inducers (Ed. N. B. Finter) North-Holland, Amsterdam (1973).
16. M. H. Ng and J. Vilcek, Interferons: Physicochemical properties and control of cellular synthesis. Adv. Prot. Chem. 26, 173 (1972).
17. S. Cha, Kinetic behavior at high enzyme concentrations. J. Biol. Chem. 245, 4814 (1970).
18. A. Stier, Lipid structure and drug metabolizing enzymes. Biochem. Pharmacol. 25, 109 (1976).
19. C. Frayssinet, I. Gresser, M. G. Tovey and P. Lindahl, Inhibitory effect of potent interferon preparations on the regeneration of mouse liver after partial hepatectomy. Nature 245, 146 (1973).
20. J. Cerrottini, K. T. Brunner, P. Lindahl and I. Gresser, Inhibitory effect of interferon preparations and inducers on the multiplication of transplanted allogeneic spleen cells and syngeneic bond marrow cells. Nature new Biol. 242, 152 (1973).
21. I. Gresser, M. G. Tovey, C. Mavry and I. Chourovlinkov, Lethality of interferon preparations for newborn mice. Nature 258, 76 (1975).
22. A. Sonnabend and R. M. Friedman, in Interferon and Interferon Inducers (Ed. N. B. Finter) North Holland, Amsterdam (1973).

ALTERATIONS OF HEPATIC MICROSOMAL MIXED FUNCTION OXYGENASE DEPENDENT H_2O_2 FORMATION DUE TO "UNCOUPLING" AND INDUCTION AS DEMONSTRATED *IN VIVO* BY ELIMINATION KINETICS OF ETHANOL IN GUINEA PIG*

A. G. Hildebrandt, L. Lehne, I. Roots and M. Tjoe

Institute for Clinical Pharmacology, Free University Berlin, D-1000 Berlin 45, Hindenburgdamm 30, Germany

ABSTRACT

By means of H_2O_2 determination it has been made possible to illustrate the complex regulation of H_2O_2 formation and degradation in microsomes and to characterize among others "uncoupling", which refers to a situation where microsomal cytochrome P-450 dependent mixed function reactions exhibit in presence of "uncouplers" more oxidase than oxygenase activities (1-3); e.g. in microsomes from livers of guinea pigs the addition of ethyl-morphine, in contrast to aminopyrine, significantly and acutely enhances rates of H_2O_2 formation.

Evidence for a possible physiological importance of "uncoupling" is demonstrated by <u>in vivo</u> experiments. The kinetics of elimination of ethanol from blood of guinea pigs are changed by application of ethylmorphine - but not by aminopyrine - in such a manner that a stimulation occurs if rate of H_2O_2 formation and ethanol concentrations are high enough (i.e. by phenobarbital treatment and by application of 2.5 g of ethanol per kg of bodyweight). However, demonstration of "uncoupling" <u>in vivo</u> by ethanol elimination kinetics is difficult at low concentrations of ethanol (1.0 g of ethanol per kg of bodyweight) and in naive animals where rates of H_2O_2 generation are low. Under such conditions ethylmorphine rather inhibits ethanol elimination.

The demonstrated effects by ethylmorphine <u>in vitro</u> on microsomal H_2O_2 generation as well as <u>in vivo</u> on ethanol elimination kinetics prove the participation of microsomal H_2O_2 formation in total liver H_2O_2 generation, a rate limiting role of H_2O_2 formation in ethanol elimination and points to the contribution of other enzymes in ethanol elimination in addition to alcohol dehydrogenase.

INTRODUCTION

The generation of H_2O_2 during the oxidation of NADPH by the liver microsomal electron transport system is now well established (4-7). Based upon investigations of microsomal H_2O_2 production and degradation, it was possible to demonstrate changes in net rates of microsomal H_2O_2 production and to explain such alterations as consequences of induction, species variation, and addi-

---
* This investigation was supported in part by Deutsche Forschungsgemeinschaft, Schwerpunktprogramm Biochemische Grundlagen der Arzneimittel- und Fremdstoffwirkungen.

tion to microsomes of substrates of microsomal mixed function oxidase cytochrome P-450 (1). The latter modification leads either to no variations of rates of H_2O_2 production (the situation which we relate to "coupled" monooxygenase activity if in presence of substrates of mixed function oxidase no changes in rates or extent of H_2O_2 production occur). The decrease in net rates or extent of H_2O_2 production observed in presence of substrates allows to assume a peroxidase activity of cytochrome P-450 which then catalyzes reactions with the oxidizing agent H_2O_2 (1,8,9).

This presentation, however, concerns the increase of H_2O_2 formation in microsomes which either is adaptive and derives from chronic treatment with e.g. Phenobarbital or spontaneous, due to the presence of "uncouplers". "Uncoupling" refers according to Hayaishi (10) to a situation where monooxygenase exhibits oxidase, rather than oxygenase activity under certain conditions. Okamato et al. (11) reported this for imidazolacetate monooxygenase if treated with mercurials. Salicylate monooxygenase was shown to catalyze the oxidation of NADH with formation of H_2O_2 when benzoate was substituted for salicylate (12). Similarily lysine monooxygenase acts under certain conditions as an aminooxidase (3). The reaction product of oxidase activity was in each case H_2O_2. However, until recently difficulties existed to prove increase in rates of H_2O_2 production during "uncoupling" in microsomes (1-3).

METHODS

In vitro analysis

Microsomes were prepared according to Hogeboom (14). H_2O_2 production in microsomes was determined as recently described (5). Formaldehyde was assayed according to Nash (15) and acetaldehyde formed during microsomal ethanol oxidation as described by Hildebrandt et al. (16).

In vivo analysis

Experiments were conducted with male albino guinea pigs, weighing 250-350 g. The food was a standard diet (Altromin[R]) and water was given ad libitum. Ethanol in blood was measured by a standard enzymic kit (Boehringer, Mannheim). The reaction was standardized by adding known amounts of ethanol to blood.

RESULTS AND DISCUSSION

Recently, we reported the occurrence of increased rates of H_2O_2 production from oxidation of NADPH in mammalian hepatic microsomes from male guinea pigs, especially when treated with pregnenolone-16α-carbonitrile, phenobarbital or when incubated in presence of ethylmorphine (Table I). The latter effect was explained by the situation described above as "uncoupling" which was established by several criteria (1,3).

The occurrence of "uncoupling" in microsomes as well as in other monooxygenase systems seems to be of more than academic interest, provided one is able to prove its significance <u>in vivo</u>. Since microsomal ethanol oxidation in presence of "uncouplers" largely parallels H_2O_2 formation (Table II) hypothetically ethanol elimination kinetics might become affected by "uncoupling" and thereby serving as a parameter to prove "uncoupling" <u>in vivo</u>.

TABLE 1   Alterations of Net Rates of H_2O_2 Production in Guinea Pig Liver Microsomes by Phenobarbital Treatment and/or Presence of Mixed Function Oxidase Substrates (mean ± SEM).

Treatment	n	no Substrate	Ethylmorphine	Aminopyrine
		nmoles mg^{-1} min^{-1}		
Controls	7	2.93 ± 0.43	5.59 P<0.005 ± 0.45	4.19 ±0.45 n.s.
Phenobarbital	4	8.0 ± 0.93	12.4 P<0.02 ± 0.84	9.52 ± 0.63 n.s.
P		<0.001	<0.001	<0.001

Microsomes from the livers of controls or guinea pigs treated with Phenobarbital (50 mg per kg per diem for 4 days) were diluted to 1.5 mg per ml in a buffer mixture containing 50 mM Tris-chloride buffer, pH 7.5, 150 mM KCl, 10 mM $MgCl_2$ and 0.5 mM sodium azide. Aminopyrine or ethylmorphine were 6.4 mM. The reaction was started by addition of 300 μM NADPH to a NADPH generating system, as previously described (5). Rates were determined from the slope and are expressed as nmoles · mg^{-1} · min^{-1}.

TABLE 2   Influence of Pretreatment of Guinea Pigs with Phenobarbital and of Addition of Ethylmorphine to Microsomes on Rates of Formaldehyde Formation from Methanol - Respectively Acetaldehyde from Ethanol Oxidation (nmoles · mg^{-1} · min^{-1}).

Substrate			no Substrate	Ethylmorphine
50 mM Methanol		Controls (n = 7)	5.17	8.57
		Phenobarbital (n = 4)	9.57	15.2
50 mM Ethanol		Controls (n = 3)	6.54	7.06
		Phenobarbital (n = 4)	9.8	15.0

Microsomes were prepared as described under Table I and incubated in a suspension of 1.5 mg of protein per ml of incubate in absence of sodium azide but in presence of 50 mM methanol or 50 mM ethanol and 2000 U/ml of catalase. Formaldehyde and acetaldehyde were determined as described under methods.

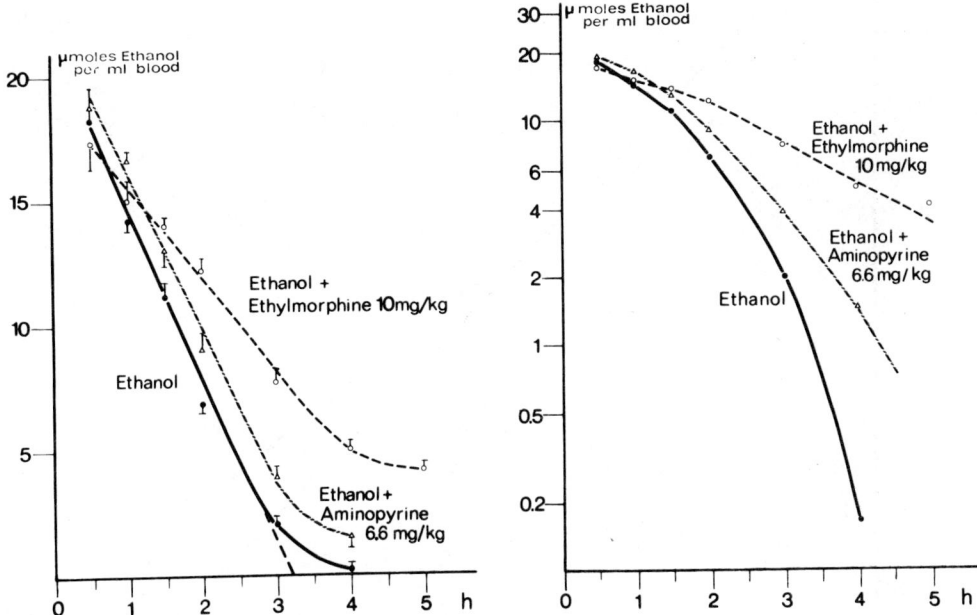

Fig. 1. Concentration of ethanol in blood of male guinea pigs at various times after intraperitoneal injection of 1 g/kg dose ethanol (solid circles). The simultaneous application of 10 mg/kg of ethylmorphine results in concentrations which are shown by open circles (o). Vertical lines indicate SEM (n=6). Lines were calculated for best fit of values and are demonstrated linearly (Fig. 1a) and semilogarithmically (Fig. 1b).

Consequently, if "uncoupling" does exist in vivo as well as in vitro, a spontaneous increase of H_2O_2 production by "uncoupling" conceivably could cause an increased metabolism of ethanol, provided that a) in vivo the production of H_2O_2 is rate limiting and b) that microsomal H_2O_2 formation could contribute to the overall rate of H_2O_2 formation in the liver, and c) that in spite of the predominant role of ADH, additional enzymes participate to a significant extent in the removal of ethanol from blood (17).

The concentration of ethanol in blood at various time intervals after the i.p. administration of 1 g of ethanol per kg of bodyweight is recorded in Fig. 1.

These data indicate that in naive guinea pigs peak concentration of ethanol disappears at a linear rate. This figure shows, in addition, a curve of mean blood ethanol concentrations in guinea pigs after an i.p. dose of 10 mg per kg of bodyweight of ethylmorphine given together with ethanol.

It is evident that administration of ethylmorphine concomitantly with ethanol affects the slope of ethanol disappearance leading to a significant difference between the two groups at various times of sampling. When regression lines were

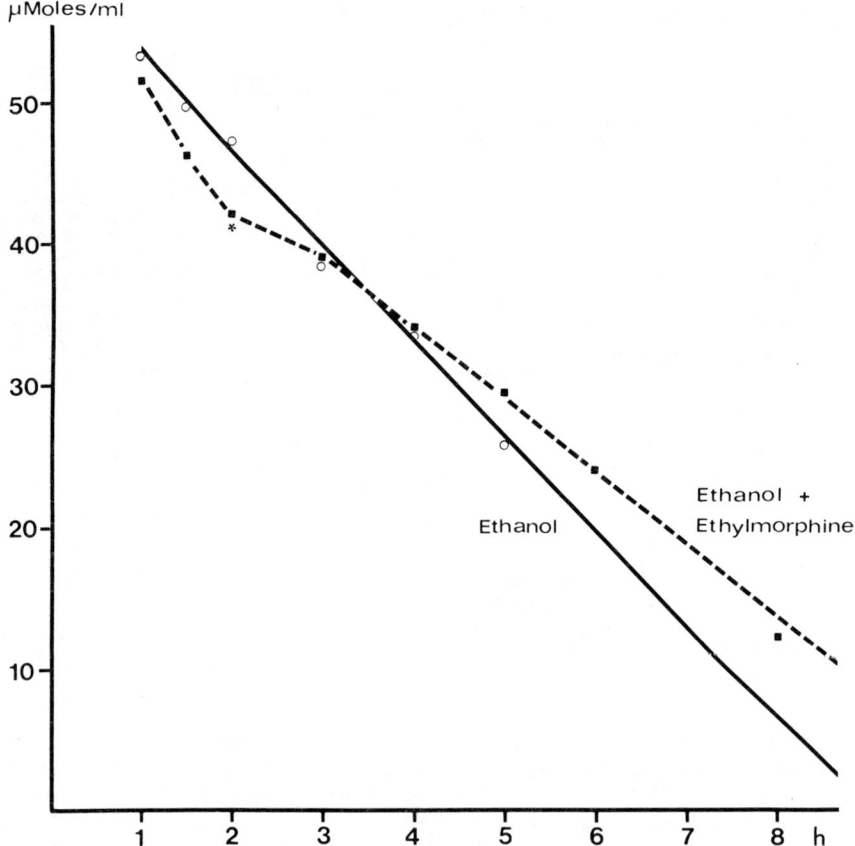

Fig. 2. Effect of Phenobarbital treatment on elimination of ethanol from male guinea pigs. For induction of microsomal enzymes and enhancement of microsomal H_2O_2 production, animals received 50 mg of phenobarbital (PB) per kg of bodyweight 72, 60, 48 and 24 hours prior to decapitation. Food and water were supplied ad libitum. 2.5 g of ethanol were given i.p. per kg of bodyweight. The influence of ethylmorphine (10 mg per kg) is shown by the dashed line curve. The asterisk indicates significant deviation from controls ($p<0.05$). Mean of 6 animals.

calculated it appeared that the regression curve from the controls showed a linear slope. In contrast to the effect of ethylmorphine, the elimination kinetics of ethanol were not significantly changed if aminopyrine was applied.

The results of these experiments indicate an inhibition rather than a stimulation of ethanol elimination in presence of ethylmorphine. The failure to observe an "uncoupling" in vivo might be overcome if higher ethanol concentrations were applied or under conditions where increased rates of microsomal H_2O_2 formation are present. As has been shown in Table I, treatment of guinea pigs with phenobarbital increases the rate of H_2O_2 formation in microsomes significantly; as ethylmorphine addition yields a further significant increase, stimulation of ethanol elimination might occur by such modifications.

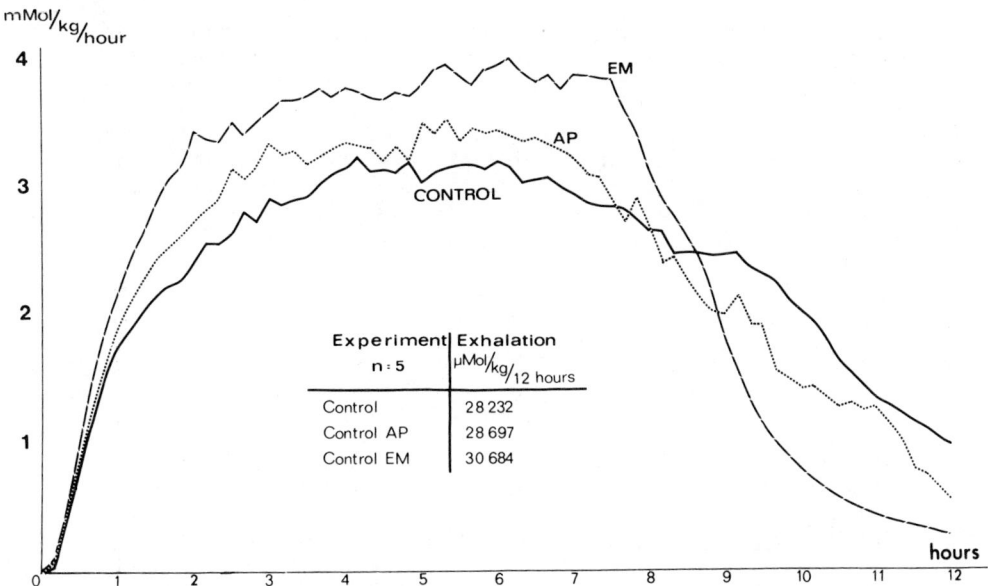

Fig. 3. $^{14}CO_2$ exhalation by guinea pigs after i.p. administration of 250 µCi of [1-^{14}C]-C_2H_5OH in 2.5 g of ethanol per kg bodyweight in absence (solid line curve) of substrate and after simultaneous application of equimolar amounts of either ethylmorphine (10 mg per kg, dashed line curve) or aminopyrine (6.6 mg per kg, dotted line curve).

Raising the ethanol concentration from 1 g to 2.5 g of ethanol and treatment of animals with PB leads to a situation with initially enhanced elimination kinetics (Fig. 2). As time passes, an inhibition becomes evident which leads to a crossover point if compared with the situation in absence of ethylmorphine.

These data suggest that demonstration of "uncoupling" in vivo can be considered as a situation which is made up by many controlling factors, e.g. ethanol concentration, rates of H_2O_2 formation and concentration of "uncouplers", i.e. ethylmorphine in this instance. The concentration of the "uncouplers" could become of crucial importance to prove "uncoupling" in vivo especially if one considers the rather short half life time of ethylmorphine in phenobarbital treated animals and the continuing decrease of ethanol concentration from blood. The demonstration of "uncoupling" by ethanol elimination kinetics further requires to obey the principles for the peroxidatic activity of catalase in liver as pointed out by Oshino et al.(18). Of additional concern are the effects of substrates of mixed function oxidase on ethanol oxidation in microsomes (16,19).

The importance of these findings is substantiated by the data given in Fig. 3. Rates of $^{14}CO_2$ exhalation from guinea pigs treated with ^{14}C-ethanol are increased in guinea pigs if 2.5 g of ethanol are given simultaneously with ethylmorphine. No similar increase is obtained when aminopyrine was applied instead of ethylmorphine in agreement with data shown in Table I.

Taken all together, changes in ethanol elimination due to addition of "uncouplers", e.g. ethylmorphine as reported in this study, and of perfluorinated hydrocarbons (unpublished) seem to reflect increased rates of H_2O_2 production in liver. These data support evidence for the assumption that for ethanol elimination under certain circumstances the production of H_2O_2 is rate limiting (17) and that microsomal H_2O_2 formation, as obtained during mixed function oxidation reactions, contributes to the overall rate of H_2O_2 formation in the liver. The drug induced effects on ethanol elimination kinetics imply the participation of additional enzymes aside from ADH in removal of ethanol from blood, at least at high ethanol concentration.

REFERENCES

(1) Hildebrandt, A.G., Tjoe, M. and Roots, I., Mono-oxygenase-linked hydrogen peroxide production and degradation in liver microsomal fractions, Biochem. Soc. Transactions 3, 807-811 (1975)

(2) Hildebrandt, A.G., Lehne, L., Tjoe, M. and Roots, I., "Uncoupling" of drug hydroxylating enzyme system as evidenced by ethanol elimination kinetics, Hoppe-Seyler's Z. Physiol. Chem., 357,1029 (1976)

(3) Ullrich, V. and Diehl, H.,Eur.J. Biochem. 20, 509-512 (1971)

(4) Hildebrandt, A.G., Speck, M. and Roots, I., Possible control of hydrogen peroxide formation and breakdown in rat liver microsomes during mixed function oxidation reactions, Biochem. Biophys. Res. Commun. 54, 968-975 (1973)

(5) Hildebrandt, A.G. and Roots, I., Reduced nicotinamide adenine dinucleotide phosphate (NADPH)-dependent formation and breakdown of hydrogen peroxide during mixed function oxidation reactions in liver microsomes, Arch. Biochem. Biophys. 171, 385-397 (1975)

(6) Thurman, R.G., Ley, H.G. and Scholz, R., Hepatic microsomal ethanol oxidation. Hydrogen peroxide formation and the role of catalase, Europ. J. Biochem. 25, 420-430 (1972)

(7) Werringloer, J., Hildebrandt, A.G. and Estabrook, R.W., Hydrogen peroxide formation and breakdown by the liver microsomal electron transport system, Abstr. 10th Int. Congr. Biochem. p. 292, Hamburg 1976

(8) Estabrook, R.W. and Werringloer, J., Active oxygen - fact or fancy, Microsomes and Drug Oxidations (V. Ullrich, A.G. Hildebrandt, R.W. Estabrook, A.H. Conney, eds.) Academic Press, New York, in press

(9) Hrycay, E.G. and O'Brien, P.J., Cytochrome P-450 as a microsomal peroxidase in steroid hydroperoxide reduction, Arch. Biochem. Biophys. 153, 480-494 (1972)

(10) Hayaishi, O., General properties and biological functions of oxygenases, Molecular mechanisms of oxygen activation (O. Hayaishi, ed.) 1-28, Academic Press, New York 1974

(11) Okamato, H., Nozaki, M. and Hayaishi, O., A role of sulfhydryl groups in imidazoleacetate monooxygenase,
Biochem. Biophys. Res. Commun. 32, 30-36 (1968)

(12) White-Stevens, R.H. and Kamin, H., Uncoupling of oxygen activation from hydroxylation in a bacterial salicylate hydoxylate,
Biochem. Biophys. Res. Commun. 38, 882-889 (1970)

(13) Nabazawa, T., Hori, K. and Hayaishi, O., Studies on monooxygenases manifestation of amino acid oxidase activity by L-lysine monooxygenase,
J. Biol. Chem. 247, 3439-3444 (1972)

(14) Hogeboom, G.H., Fractionation of cell components of animal tissue,
Methods in Enzymology I (S.P. Colowick and N.O. Kaplan, eds.)16-19, Academic Press, New York 1955

(15) Nash, T., The colorimetric estimation of formaldehyde by means of the Hantzsch reaction,
Biochem. J. 55, 416-421 (1953)

(16) Hildebrandt, A.G., Speck, M. and Roots, I., The effects of substrates of mixed function oxidase on ethanol oxidation in rat liver microsomes,
Naunyn-Schmiedeberg's Arch. Pharmacol. 281, 371-382 (1974)

(17) Videla, L., Bernstein, J. and Israel, Y., Metabolic alterations produced in the liver by chronic ethanol administration. Increased oxidative capacity. Biochem. J. 134, 507-514, 1973

(18) Oshino, N., Oshino, R. and Chance, B., The characteristics of the peroxidatic reaction of catalase in ethanol oxidation,
Biochem. J. 131, 555-567 (1973)

(19) Lieber, Ch.S. and de Carli, L.M., Hepatic microsomal ethanol-oxidising system. In vitro characteristics and adaptive properties in vivo,
J. Biol. Chem. 245, 2505-2512 (1970)

THE MICROSOMAL METABOLISM OF CARCINOGENIC AND/OR THERAPEUTIC HYDRAZINES

Russell A. Prough, Marguerite L. Coomes, and Danny L. Dunn

Department of Biochemistry, The University of Texas Health Science Center at Dallas, 5323 Harry Hines Boulevard, Dallas, Texas 75235

ABSTRACT

The metabolism of Procarbazine (N-isopropyl-p-[N'-methylhydrazino methyl]-benzamide) and other hydrazine derivatives has been studied using purified pig liver microsomal amine oxidase and rat liver microsomal fractions. The substrate specificity of the purified amine oxidase for hydrazines was studied; 1,1-dimethyl- and 1-methyl-1-phenylhydrazine were two of the best substrates. The metabolism of 1,1-dimethylhydrazine by liver microsomes has been shown to be dependent on the amine oxidase present in these fractions. Conversely, several other hydrazine derivatives, including Procarbazine, are metabolized exclusively by the cytochrome P-450-dependent monooxygenase of liver microsomal fractions. These results further support the observation that two major monooxygenases, the liver microsomal amine oxidase and the cytochrome P-450-dependent enzyme system, are involved in the metabolism of many nitrogenous compounds, including hydrazine derivatives.

INTRODUCTION

The disposition and metabolism of N-methylamines have been studied extensively (1) and the influence of N-oxidation on the pharmacology of nitrogenous compounds has been reviewed (2,3). Two microsomal monooxygenases have been described which can metabolize nitrogeneous compounds: the cytochrome P-450-dependent oxidase (1) and the flavoprotein, amine oxidase (4,5). These oxidases can be distinguished by their substrate specificities (2) and by the metabolites formed (5); i.e., the two oxidases catalyze the oxidation of sec. and tert. amines to distinctly different products. Further, a set of differential inhibitors have recently been used to assess the relative participation of these two oxidases in the N-demethylation and N-oxidation of N-methylamines (6). In this study, a specific antibody to NADPH-cytochrome c (P-450) reductase, an inhibitor of the cytochrome P-450-dependent hydroxylase, and methimazole (1-methyl-2-mercaptoimidazole), an alternate substrate inhibitor for the amine oxidase which does not yield formaldehyde as a product, were used to determine the relative contributions of the two monooxygenases.

Substituted hydrazine derivatives are highly reactive organic compounds used extensively in industry and medicine. Two recent reviews encapsulate much of the information relating to their metabolism (7,8). Prough, Wittkop, and Reed have shown that some N-methylhydrazines can be dealkylated to yield formaldehyde and/or methane in the presence of NADPH, O_2 and rodent liver microsomes (9). Formaldehyde formation appeared to be, in part, dependent on cytochrome P-450, but the methane production appeared to be independent of cytochrome P-450. Subsequently, purified liver microsomal amine oxidase was noted to N-oxidize several N-methylhydrazine derivatives (10). This report will provide evidence for the involvement of both mixed-function oxidases in the metabolism of this class of nitrogenous compounds.

MATERIALS

The hydrazine derivatives and methimazole were purchased from either the Aldrich Chemical Company or K & K Laboratories and Procarbazine was supplied by the Hoffman-LaRoche Company, Inc. All other chemicals were obtained from the Sigma Chemical Company and solvents were purchased from Burdick and Jackson, Inc. Male Sprague-Dawley CD (outbred) rats, 150-250 g, were supplied by the Charles River Breeding Laboratory, Inc. and maintained ad. libitum on water and laboratory chow. The purified pig liver microsomal amine oxidase was a generous gift from Dr. D.M. Ziegler, Clayton Foundation Biochemical Institute and Department of Chemistry, The University of Texas at Austin.

METHODS

Rats were injected in the peritoneal cavity with either corn oil, phenobarbital in saline (40 mg/kg), or 3-methylcholanthrene in corn oil (20 mg/kg) daily for 4 days and starved 18-24 hours prior to sacrifice. Microsomal fractions of liver were prepared as described by Remmer, et al. (11) and the protein concentration determined by the method of Lowry (12).

The N-oxidation of hydrazines by the purified oxidase was measured by observing the substrate-dependent oxidation of NADPH (10). Experiments utilizing liver microsomes as a source of enzyme were performed at 37° C with a reaction mixture consisting of 2 mg/ml microsomal protein, 3 mM DL-isocitrate, 0.8 I.U. isocitrate dehydrogenase, 0.5 mM $NADP^+$, 5 mM $MgSO_4$, and 0.05 M phosphate-0.05 M Tricine buffer at pH 7.9. The reaction mixture was initiated by addition of microsomes after a 3-minute preincubation period and aliquots removed every 1 - 1.5 minute during the linear portion of the reaction. The concentration of products was plotted as a function of time and specific activities (nmoles product formed per min per mg protein) were calculated from the slope of the plot. The production of formaldehyde from 1,1-dimethylhydrazine was measured by the method of Nash (13); 1.0 mM 1,1-dimethylhydrazine did not appreciably affect the recovery of formaldehyde and no increase in the rate of demethylation of 1,1-dimethylhydrazine was noted upon increasing the concentration. The formation of the azo derivative of Procarbazine (N-isopropyl-p-(N'-methylazo methyl]-benzamide, AZO) was determined by terminating the reaction with 1.0 N H_2SO_4. The AZO-metabolite was degraded by the acid to methylhydrazine (MH) and N-isopropyl-p-formylbenzamide (ALD). Under acid conditions, methylhydrazine reacts with Ehrlich's Reagent (p-dimethylaminobenzaldehyde) to yield a quinoid ion hydrazone which has an optical absorption maximum (14) at 465 nm ($\varepsilon=4.0 \times 10^3 M^{-1} cm^{-1}$). After addition of 1.0 N H_2SO_4 and Ehrlich's Reagent (0.05 M), the protein was precipitated by centrifugation at 12,000 x g and the concentrations of hydrazone measured at 465 nm. Alternatively, the aldehyde derivative was extracted from the reaction mixture with organic solvent and its concentration measured spectrophotometrically at 252 nm ($\varepsilon=2.6 \times 10^4 M^{-1} cm^{-1}$).

The hydrazines exhibited a single ninhydrin spot on silica gel G thin-layer chromatography using a solvent system consisting of 0.5 M oxalate in 1-butanol (1:4 v/v). The hydrazine concentration in a solution was quantitated by its ability to reduce potassium ferricyanide assuming an extinction coefficient for $K_3Fe(CN)_6$ at 420 nm of 1,030 $M^{-1} cm^{-1}$. A Waters Associates Series 200 liquid chromatograph equipped with a μbondapak-CN column was used to separate Procarbazine and its possible metabolites using an initial eluant of 20% methylene chloride in hexane. After eight minutes, the eluant concentration was changed to 65% methylene chloride in hexane. The absorption was monitored at 254 nm and the flow rate was 3 ml/min.

RESULTS AND DISCUSSION

N-Oxidation by the Purified Liver Microsomal Amine Oxidase

A number of hydrazines were tested as substrates for the amine oxidase and their Vmax and Km values can be seen in Table 1. Although monoalkyl-, monoaryl-, and arylalkylhydrazines can serve as substrates, only 1,1-dimethylhydrazine, 1-methyl-1-phenylhydrazine, and 1,2-dimethylphenylhydrazine are N-oxidized as effectively as N,N-dimethylaniline, one of the best N-methylamine substrates for the enzyme. In addition, the pH-maximal rate profile for 1,1-dimethyl- and 1-methyl-1-phenylhydrazine N-oxidation is identical to that of N,N-dimethylaniline N-oxidation. Studies aimed at defining the stoichiometry of the amine oxidase-catalyzed metabolic reaction have indicated that the N-oxidation is stoichiometric with regard to NADPH, oxygen, and either 1,1-dimethylhydrazine or 1-methyl-1-phenylhydrazine; that is, one molecule each of NADPH and oxygen are required to oxidize one molecule of hydrazine substrate. One product of 1,1-dimethylhydrazine N-oxidation can be shown to be formaldehyde and the rate of formation of formaldehyde equals the rate of NADPH or oxygen consumed in the reaction.

TABLE 1 Hydrazine Substrates for Purified Pig Liver Microsomal Amine Oxidase at pH 7.7, 25°C

Substrate	Km (mM)	Maximal Velocity (nmoles $NADP^+$ formed/min/mg)
3° - Amine		
N,N-Dimethylaniline	0.035	214
Alkylhydrazines		
Methylhydrazine	35.0	308
N-Butylhydrazine	7.0	334
1,1-Dimethylhydrazine	0.43	178
1,2-Dimethylhydrazine	6.0	80
Arylhydrazines		
Phenylhydrazine	1.8	228
Benzylhydrazine	7.0	200
1-Methyl-1-phenylhydrazine	0.08	254
1-Methyl-2-benzylhydrazine	2.0	210
1,2-Dimethylphenylhydrazine	0.38	226

The Metabolism of 1,1-Dimethylhydrazine by Liver Microsomal Fractions

In the presence of liver microsomes, 1,1-dimethylhydrazine is oxidatively metabolized to yield formaldehyde as a product. Using inhibitors of the two monooxygenases, either a specific, inhibitory antibody to NADPH-cytochrome c (P-450) reductase to inhibit cytochrome P-450-dependent reactions or 1-methyl-2-mercaptoimidazole to inhibit the amine oxidase-dependent reactions, the relative participation of the two enzyme systems of rat liver microsomes in the N-demethylation of 1,1-dimethylhydrazine can be seen in Table 2. The alternative substrate inhibitor of the amine oxidase nearly abolishes the N-demethylation reaction while the anti-reductase globulin has little or no effect on the reaction. These results show that the membrane-bound amine oxidase can N-oxidize hydrazines and that the amine oxidase is primarily responsible for the metabolism of 1,1-dimethylhydrazine by rodent liver microsomes.

TABLE 2   Effect of Inhibitors on 1,1-DMH Demethylation

Inhibitor	Rate (nmoles HCHO/min/mg)
	1.9
Methimazole	0.4
Anti-reductase	1.8

The Metabolism of Procarbazine by Liver Microsomal Fractions

Using high pressure liquid chromatography, the primary product of the microsomal metabolism of Procarbazine was isolated and identified based on its chromatographic mobility and mass spectrum. After 10 minutes of reaction, an incubation mixture containing 2 mM Procarbazine was quenched with an equal volume of chilled chloroform. Following centrifugation to remove protein, an aliquot of the chloroform phase was applied to a μbondapak-CN column and eluted with a methylene chloride: hexane solvent (see METHODS). As seen in Fig. 1, the principle metabolite had a retention time and mass spectrum ($M^+ = 219$) identical to the synthetic azo derivative of Procarbazine. However, if the reaction was terminated by addition of acid and extracted with chloroform, the principle metabolite was noted to be p-formyl-N-isopropyl-benzamide, ALD (based on chromatographic mobility and mass spectrum analysis, $M^+ = 191$). This result suggests that the stable product of the oxidative microsomal metabolism of Procarbazine is the azo derivative which hydrolyzes in acid to yield the aldehyde. This breakdown most probably is due to the acid catalyzed tautomerization of the azo derivative to a methyl hydrazone followed by hydrolysis to p-formyl-N-isopropyl-benzamide and methylhydrazine (Fig. 2). Using ^{14}C-ring labeled Procarbazine, more than 95% of the azo derivative of Procarbazine is converted to the aldehyde derivative in the presence of 1.0 N H_2SO_4.

Although the enzymatic conversion of Procarbazine to its azo derivative can be followed using liquid chromatography and ^{14}C-labeled substrate, the stoichiometric conversion by acid of the azo derivative to either the aldehyde derivative or methylhydrazine allows one to make a less complex and time consuming measurement of the primary oxidation of Procarbazine. The aldehyde derivative can be extracted from acidified reaction mixtures and its concentration measured at 252 nm. Alternatively, methylhydrazine can be simply measured as its quinoid ion hydrazone with Ehrlich's Reagent at 465 nm. Both assays yield nearly identical rates of formation of the azo derivative of Procarbazine.

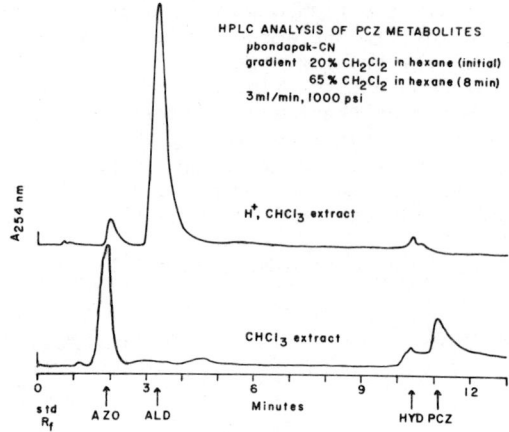

Fig. 1. The Separation of Procarbazine and Its Metabolites by HPLC. The arrows indicate the retention time for the authentic standards for AZO, N-isopropyl-p-(N'-methylazo methyl)-benzamide; ALD, N-isopropyl-p-formyl-benzamide; HYD, N-isopropyl-p-(N'-methylhydrazo methylene)-benzamide; and PCZ, N-isopropyl-p-(N'-methylhydrazine methyl)-benzamide (Procarbazine).

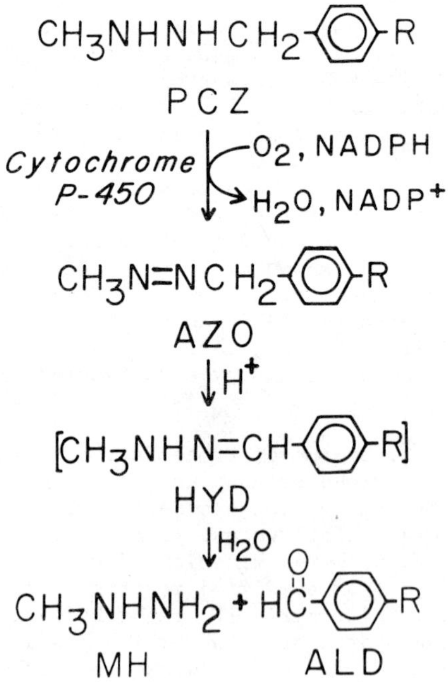

$$CH_3NHNHCH_2-\bigcirc-R$$
PCZ

Cytochrome P-450 $\quad \begin{matrix} \nearrow O_2, NADPH \\ \searrow H_2O, NADP^+ \end{matrix}$

$$CH_3N=NCH_2-\bigcirc-R$$
AZO
$\downarrow H^+$

$$[CH_3NHN=CH-\bigcirc-R]$$
HYD
$\downarrow H_2O$

$$CH_3NHNH_2 + H\overset{O}{C}-\bigcirc-R$$
MH ALD

Fig. 2. Scheme For the Oxidative Metabolism and Breakdown of Procarbazine.

Utilizing the assay for methylhydrazine, several characteristics of the oxidative reaction to form the azo-derivative of Procarbazine can be noted. The rate of metabolism of Procarbazine by liver microsomes from corn oil-treated rats was approximately 9.1 nmoles product formed per min per mg protein (Table 3); a rate which is 2-fold that of the N-demethylation of Benzphetamine (4.1 nmoles/min/mg protein), one of the best N-methylamine substrates of cytochrome P-450. Further, animal pretreatment with phenobarbital caused a 2.7 fold increase in the specific activity of Procarbazine oxidation. Administration of 3-methylcholanthrene did not increase this activity. The effect of several inhibitors of the two liver microsomal monooxygenases on Procarbazine oxidation can be seen in Table 4. As expected, oxygen and NADPH were required for activity. However, only inhibitors of cytochrome P-450-dependent reactions, such as carbon monoxide, metyrapone, or specific antibodies to liver microsomal NADPH-cytochrome c (P-450) reductase, were effective as inhibitors while the amine oxidase inhibitor, methimazole, was without effect.

Based on the inductive effect of phenobarbital and the inhibition patterns shown in Table 4, the primary oxidation of Procarbazine to its azo derivative appears to be a cytochrome P-450-mediated reaction. Two mechanisms of oxidation could account for this metabolic conversion: a dehydrogenation or an N-oxidation followed by dehydration. Several reports of cytochrome P-450-dependent N-hydroxylation reactions exist: 2-acetylaminofluorene (15), acetaminophen (16), and p-chloroacetanilide (17). Recently, Nelson et al. (18) have described the possible oxidation of N-acetylhydrazine and isopropylhydrazine and their resultant toxicity. These oxidations appear to be very similar to the oxidation of Procarbazine and as suggested by Nelson et al. may represent another N-oxidation reaction which cytochrome P-450 can support.

TABLE 3   Effect of Animal Pretreatment on Procarbazine Metabolism

Pretreatment	Rate (nmoles/min/mg)
Control	9.1
Phenobarbital	24.0
3-Methylcholanthrene	9.0

TABLE 4   Effect of Inhibitors on Procarbazine Metabolism

Addition or Deletion	Rate (nmoles/min/mg)
Control	9.4
$-O_2$	0.4
-NADPH	0.5
+Anti-reductase	2.3
+CO/O_2 (4:1)	4.0
+Metyrapone (1 mM)	3.4
+Methimazole (1 mM)	9.6

The liver microsomal monooxygenases involved in the metabolism of hydrazine derivatives appear to be determined by the structural features of the compound. Certain N,N-disubstituted hydrazines serve as excellent substrates for the amine oxidase, while Procarbazine and isopropylhydrazine are effectively N-oxidized by the cytochrome P-450-dependent mixed function oxidase. Chemical knowledge of the primary oxidation products of hydrazines resulting from their metabolism by these two enzyme systems may aid in understanding the carcinogenic, therapeutic, and/or toxic effects of this class of nitrogenous compounds.

ACKNOWLEDGEMENTS

This research was supported in part by grants from the American Cancer Society (BC-153) and the Robert A. Welch Foundation (I-616). M.L.C. is a National Science Foundation Pre-Doctoral Fellow and D.L.D. is a Robert A. Welch Foundation Post-Doctoral Fellow. The authors express their gratitude to Dr. D.M. Ziegler for his many helpful discussions and for samples of the purified pig liver microsomal amine oxidase.

References

1. J.W. Bridges, J.W. Gorrod, and D.V. Parke (Eds.), The Biological Oxidation of Nitrogen in Organic Molecules, Taylor and Francis, London, (1972).
2. J.W. Gorrod, Differentiation of Various Types of Biological Oxidation of Nitrogen in Organic Compounds, Chem.-Biol. Interactions 7, 289 (1973).
3. M.H. Bickel, The Pharmacology and Biochemistry of N-Oxides, Pharmacol. Rev. 21, 325 (1969).
4. D.M. Ziegler and F.H. Pettit, Microsomal Oxidases. I. The Isolation and Dialkylarylamine Oxygenase Activity of Pork liver Microsomes, Biochemistry 5, 2932 (1966).
5. D.M. Ziegler and C.H. Mitchell, Microsomal Oxidase IV: Properties of a Mixed-Function Amine Oxidase Isolated from Pig Liver Microsomes, Arch. Biochem. Biophys. 150, 116 (1972).
6. R.A. Prough and D.M. Ziegler, The Relative Participation of Liver Microsomal Amine Oxidase and Cytochrome P-450 in N-Demethylation Reactions, Arch. Biochem. Biophys. In Press (1976).
7. L.B. Colvin, Metabolic Fate of Hydrazines and Hydrazides, J. Pharmaceut. Sci. 58, 1433 (1969).
8. M.R. Juchau and A. Horita, Metabolism of Hydrazine Derivatives of Pharmacologic Interest, Drug Metab. Revs. 1, 71 (1972).
9. R.A. Prough, J.A. Wittkop, and D.J. Reed, Further Evidence on the Nature of Microsomal Metabolism of Procarbazine and Related Alkylhydrazines, Arch. Biochem. Biophys. 140, 450 (1970).
10. R.A. Prough, the N-Oxidation of Alkylhydrazines Catalyzed by the Microsomal Mixed-Function Amine Oxidase, Arch. Biochem. Biophys. 158, 442 (1973).
11. H. Remmer, H. Greim, J.B. Schenkman, and R.W. estabrook, Methods for the Elevation of Hepatic Microsomal Mixed-Function Oxidase Levels and Cytochrome P-450, Methods in Enzymology 10, 703 (1967).
12. O.H. Lowry, N.J. Rosebrough, A.L. Farr, and R.J. Randall, Protein Measurement with the Folin Phenol Reagent, J. Biol. Chem. 193, 265 (1951).
13. T. Nash, The Colorimetric Estimation of Formaldehyde by Means of the Hantzsch Reaction, Biochem. J. 55, 416 (1953).
14. L.F. Fieser and M. Fieser (Eds.), Reagents for Organic Synthesis, J. Wiley, N.Y., 1 (1967) p. 274.

15. P.H. Grantham, E.K. Weisburger, and J.H. Weisburger, Dehydroxylation and Deacetylation of N-hydroxy-N-2-fluorenylacetamide by Rat Liver and Brain Homogenates, Biochem. Biophys. Acta 107, 414 (1965).
16. W.Z. Potter, D.C. Davies, J.R. Mitchell, D.J. Jollow, J.R. Gillette and B.B. Brodie, Acetaminophen-Induced Hepatic Neurosis. III. Cytochrome P-450 mediated Covalent Binding in vitro, J. Pharm. Exptl. Ther. 187, 203 (1973).
17. J.A. Hinson, J.R. Mitchell, and D.J. Jollow, Microsomal N-Hydroxylation of p-Chloroacetanilide, Mol. Pharm. 11, 462 (1975).
18. S.D. Nelson, J.R. Mitchell, J.A. Timbress, W.R. Snodgrass and G.B. Corcoran III, Isoniazid and Iproniazid: Activation of Metabolites to Toxic Intermediates in Man and Rat, Science 193, 901 (1976).

ROLE OF HYDROXYLATED METABOLITES OF PHENYTOIN IN DOSE-DEPENDENCY

A. J. Glazko, T. Chang, E. Maschewske, A. Hayes and W. A. Dill

Research Laboratory, Parke-Davis & Co., Ann Arbor, Mich., U.S.A.

INTRODUCTION

Phenytoin (5,5-diphenylhydantoin, DPH, Dilantin®) exhibits some unusual pharmacokinetic behavior in laboratory animals and in man. These include a progressive decrease in plasma half-life as the plasma levels fall (1,2), a dependency of plasma half-life upon dosage (3,4), an abrupt non-linear increase in plasma levels of DPH with increasing dosage (5-7), a difference in the area-under-curve for oral and I.V. routes of administration (8,9), and a greatly delayed peak in the urinary excretion of the hydroxylated metabolite p-HPPH (5-(p-hydroxyphenyl)-5-phenylhydantoin) (1). DPH plasma levels exhibit a 10 to 20-fold variation in different subjects, due in part to genetic variations in hydroxylation capability (10), enzyme induction (4,10), and dose-dependent kinetics (3-7). The latter has been attributed to saturation of the enzyme system responsible for hydroxylation by the substrate DPH (3-5), and adherence to Michaelis-Menten kinetics has been offered as a simple explanation (11). However, we found that the addition of different concentrations of DPH to fortified 9000 xg rat liver supernates in vitro did not affect the rate of hydroxylation, whereas the addition of p-HPPH to this system inhibited hydroxylation in a competitive manner (12,13). In addition, the catechol and 3-O-methyl catechol metabolites of DPH were found to inhibit hydroxylation, and the hydroxylated metabolites of DPH, phenylbutazone and ripazepam exhibited cross-inhibition of the hydroxylation of these substrates in vitro (14). Recent observations by Stavchansky et al (15,16) indicate that p-HPPH inhibits the oxidation of hexobarbital and the N-dealkylation of ethylmorphine in vitro, and evidence has also been presented for the interaction of p-HPPH with cytochrome P-450 (17), indicating possible interference with the binding and metabolism of DPH as well as other drugs.

Shortly after our in vitro studies had been completed (12,13), Ashley and Levy reported elevated and greatly extended plasma levels of DPH in rats receiving loading doses of p-HPPH (18). Stavchansky et al reported that p-HPPH prolonged the hexobarbital sleeping time in mice (15,16) and also increased the duration of zoxazolamine paralysis in rats (15,19). Soda and Levy reported the inhibition of DPH metabolism by oxyphenylbutazone in rats (20), demonstrating cross-inhibition with this hydroxylated metabolite in vivo.

The present observations demonstrate a dependency of DPH plasma half-life in the Rhesus monkey upon dosage and presents the first quantitative data on the plasma levels of HPPH. After the administration of a loading dose of p-HPPH, followed by a 10 mg/kg dose of DPH, a sharp increase in the plasma half-life of DPH was observed, similar to that encountered with 50 mg/kg doses of DPH alone. These observations have been confirmed in numerous additional experiments with Rhesus monkeys which cannot be described here because of space limitations.

MATERIALS AND METHODS

^{14}C-Phenytoin: 5,5-[4-^{14}C]-diphenylhydantoin (^{14}C-DPH) was purchased from New England Nuclear Corp., Boston, Mass. This was dissolved in a 40% propylene glycol + 10% ethanol + 50% water for injection, adjusted to pH 12 with sodium hydroxide. The specific activity of the ^{14}CDPH was diluted with non-labelled DPH, also in solution, to provide a total ^{14}C dose of 50 µCi per monkey.

p-HPPH: 5-(p-hydroxyphenyl)-5-phenylhydantoin was synthesized in the Parke-Davis laboratories and dissolved in the 40% propylene glycol + 10% ethanol solvent for intravenous administration.

Radioactivity Measurement: Organic solvent extracts of plasma and urine (e.g., 3 ml ethyl acetate or 4 ml chloroform) were transferred to counting vials and evaporated to dryness. One ml of methanol and 12 ml of scintillator (PCS, Amersham/Searle Corp., Arlington Heights, Ill.) were added to dissolve the residue. Counting was carried out in a Packard Tri-Carb Liquid Scintillation Spectrometer, Model No. 3375, using external standardization technics.

Separation and Assay of DPH and HPPH: DPH and HPPH were separated by a modification of the solvent extraction technics described by Dill et al (21), and a colorimetric procedure was applied (21). The recovery of known amounts of DPH and HPPH added to normal plasma and urine specimens was determined, and appropriate corrections were applied to the analytical data. Distinction was made between conjugated and unconjugated HPPH by extracting the free HPPH from plasma with ethylene dichloride:2-ethylhexanol (2:1) (21) or from urine with ethyl acetate. The aqueous residue was adjusted to pH 5 and incubated overnight at 37° C with Glusulase (Endo Labs., Garden City, N. Y.). The HPPH liberated by enzymatic hydrolysis was then extracted and assayed, representing previously conjugated HPPH. Gas chromatography was also used to quantitate p-HPPH in the extracts before and after enzymatic hydrolysis and to establish the presence of m-HPPH (5-(m-hydroxyphenyl)-5-phenylhydantoin). Ethyl acetate extracts of urine (3 ml) were dried by the addition of anhydrous sodium sulfate, and the solvent was evaporated. An on-column methylation technic was employed for derivative formation, using trimethylphenyl ammonium hydroxide (50 µl) added to the residue. GLC was carried out on 3% OV-17 on GasChrom Q (Applied Sci. Lab., Inc., State College, Pa.) at 225° C. The internal standard was the p-tolyl derivative of DPH, developed earlier for DPH assays (22,23). All assay results are expressed in terms of DPH-equivalents.

Animal Experiments: Adult female Rhesus monkeys weighing 4 to 6 kg were used in these trials. DPH and HPPH were administered by slow intravenous infusion, extending over a 5-minute period. Animals were immobilized in restraining chairs during the course of the experiments. Venous blood samples were taken at the time periods indicated, heparinized, and the blood samples were centrifuged immediately to separate the plasma from the red cells. Complete urine collections were made by catheter during the first 24 hours after dosing and by chair collection thereafter.

RESULTS AND DISCUSSION

Effect of Dosage on Plasma Levels of DPH and HPPH: Two Rhesus monkeys were given I.V. doses of ^{14}C-DPH of 10 mg/kg and 50 mg/kg respectively. The plasma levels of DPH and total HPPH (free and conjugated) are shown in Fig. 1. At the 10 mg/kg dose level, the plasma level of DPH was 10 μg/ml 1 hr after dosing, and fell thereafter with typical first-order kinetics. The half-life was about 9 hr over the first 24 hr after dosing. The total HPPH levels reached a maximum of 2.5 μg/ml in 2-4 hr, and then fell slowly with a half-life of 16 hr. Approximately one-third of the HPPH was unconjugated in the monkey plasma, contrasting sharply with human plasma where most of the HPPH was conjugated (21). This monkey also showed approximately the same proportion of free HPPH after a 25 mg/kg I.V. dose of DPH, but another monkey showed a free/total HPPH ratio averaging about 0.27 for 10 and 25 mg/kg I.V. doses of DPH (data not shown). In the present study, the plasma half-life of free HPPH was about the same as that of the conjugated HPPH, although the levels were lower.

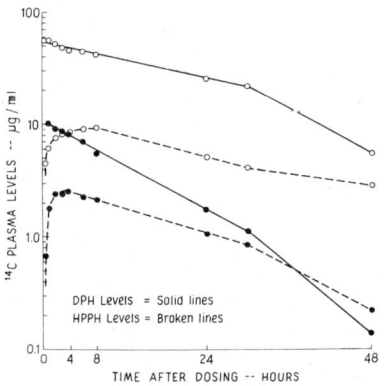

Fig. 1. Effect of DPH dosage on plasma levels and half-lives of DPH and HPPH. Solid lines = DPH levels; broken lines = HPPH levels. Open circles = 50 mg/kg I.V. dose; solid circles = 10 mg/kg I.V. dose.

The plasma levels for the monkey receiving a 50 mg/kg I.V. dose of ^{14}C-DPH are also shown in Fig. 1. The peak DPH levels were about 55 μg/ml, falling slowly over the first 24 hr period with a half-life of 22 hr, and then more rapidly as the plasma levels continued to fall with a half-life of about 9 hr. The total HPPH plasma levels reached a peak (9 μg/ml) about 4-8 hr after dosing, nearly double the time required at the lower dose level. The HPPH levels in this animal fell slowly with a plasma half-life of about 18 hr, close to that observed at the lower dose level. The extended half-life of HPPH was probably due to the slow formation and release of this metabolite from the tissues. The ratio of free/total HPPH in the plasma averaged 0.32, close to that observed at the lower dose level. This suggests that glucuronidation of HPPH has not been inhibited, consistent with our in vitro observations (13). The total recovery of ^{14}C in the 48 hr urine represented 80.4% of the dose for the 10 mg/kg animal, and 76.8% of the dose for the 50 mg/kg animal. The results of GLC and analysis of the urine are shown in Table 1.

The major excretion product was p-HPPH, with 80-90% appearing as conjugated metabolite. However, a significant amount of the meta-hydroxylated product (m-HPPH) also appeared in the urine of both monkeys, mainly as unconjugated metabolite. Although large amounts of m-HPPH have been identified in dog urine (24), no significant amounts were detected in human urine after single doses of the labelled drug (25,26). This is the first reported occurrence of m-HPPH as a minor metabolite in the urine of Rhesus monkeys.

Table 1

Urinary Excretion Rate for DPH Metabolites in Rhesus Monkeys Receiving Single Intravenous Doses of 10 mg/kg and 50 mg/kg DPH

I.V. Dose	Collection Period	DPH	m-HPPH		p-HPPH	
			Free	Conjugated	Free	Conjugated
(mg/kg)	(hr)	(mg/hr)	(mg/hr)	(mg/hr)	(mg/hr)	(mg/hr)
10	0 - 2	0.02	0.12	*	0.12	0.42
	2 - 4	0.04	0.22	*	0.24	0.86
	4 - 6	0.06	0.22	*	0.22	0.82
	6 - 8	0.02	0.20	*	0.20	0.78
	8 - 24	*	0.12	*	0.11	0.49
	24 - 48	*	0.03	*	0.05	0.18
50	0 - 2	0.19	0.22	0.14	0.26	1.51
	2 - 4	0.20	0.27	0.20	0.32	2.48
	4 - 6	0.12	0.30	0.20	0.32	2.67
	6 - 8	0.06	0.28	0.19	0.28	2.75
	8 - 24	0.04	0.24	0.16	0.26	2.22
	24 - 48	0.02	0.16	0.07	0.22	1.55

Peak excretion rates are underlined.
*Below limit of assay sensitivity.

Effect of p-HPPH on the Metabolic Disposition of ^{14}C-DPH: The preceding trial was repeated using the same animal that had received the single 10 mg/kg I.V. dose of labelled DPH, except that a 50 mg/kg I.V. dose of synthetic p-HPPH was administered 2 hr before a 10 mg/kg I.V. dose of ^{14}C-DPH. This was followed 2 hr later by a second 50 mg/kg I.V. dose of p-HPPH. Preliminary trials with p-HPPH in monkeys indicated (a) that absorption from the GI-tract was poor; (b) that I.V. doses produced good plasma levels of HPPH that fell rapidly over the first 8 hours with a half-life of about 2 hr, followed by a more extended rate of decline in later time periods; and (c) that urinary excretion accounted for 80-95% of the dose, mainly in the form of conjugated HPPH. The results of the present experiment are shown in Fig. 2.

In the absence of added HPPH the plasma DPH half-life was about 9 hr. The ^{14}C-HPPH levels rose to an early peak 2-4 hr after dosing and then declined with a half-life of 16 hr. In the presence of the added HPPH, the DPH plasma levels fell more slowly with a half-life of about 21 hr over the first 24-hr period, and then more rapidly (half-life 12 hr) in the 24-48 hr period after dosing. This closely resembled the rate of decline of plasma levels at the

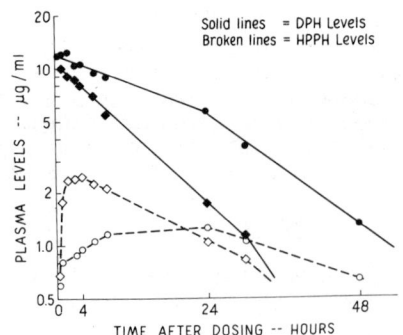

Fig. 2. Effect of loading doses of p-HPPH on plasma levels and half-lives of DPH and HPPH in a Rhesus monkey following a 10 mg/kg I.V. dose of ^{14}C-DPH. Solid lines = DPH levels; broken lines = HPPH levels. Squares = no HPPH was administered; circles = 50 mg/kg I.V. doses of p-HPPH given 2 hours before and 2 hours after the ^{14}C-DPH.

50 mg/kg dose level (Fig. 1). In addition, the plasma ^{14}C-HPPH levels in this trial rose very slowly, reached a peak in the 8-24 hr period after dosing, and then fell more slowly with an extended half-life of 25 hr.

The levels of total HPPH are not shown in Fig. 2 since the unlabelled HPPH used for the loading doses was not included in the assays. Comparative assay data for ^{14}C-labelled HPPH and total HPPH by colorimetric assay are shown in Table 2.

Table 2

Plasma Levels of DPH and HPPH in a Rhesus Monkey Receiving a 50 mg/kg I.V. Dose of p-HPPH 2 Hr Before and 2 Hr After a Single 10 mg/kg I.V. Dose of ^{14}C-DPH.

Time After DPH Dose	^{14}C-DPH	Conjugated HPPH		Free HPPH	
		^{14}C Assay	Colorimetric[a]	^{14}C Assay	Colorimetric[a]
(Hr)	(μg/ml)	(μg/ml)	(μg/ml)	(μg/ml)	(μg/ml)
0.5	11.8	0.30	26.6	0.30	6.4
1	12.0	0.46	25.8	0.35	4.4
2[b]	12.1	0.65	23.4	0.39	2.1
3	10.1	0.55	42.1	0.34	22.5
4	10.3	0.60	38.4	0.35	14.2
6	9.4	0.70	24.5	0.36	4.8
8	8.9	0.77	12.8	0.38	1.9
24	5.6	0.83	*	0.41	*
30	3.6	0.72	*	0.30	*
48	1.3	0.48	*	0.15	*

Peak plasma levels are underlined.
[a]Corrected for content of ^{14}C-HPPH.
[b]Second 50 mg/kg I.V. dose of HPPH was administered at this time.
*Below limit of assay sensitivity.

These observations indicate that the plasma levels of HPPH were several-fold higher than in the monkey receiving the 50 mg/kg I.V. dose of DPH alone (Fig. 1). However, the rate of fall of HPPH levels was greater because of diffusion into the tissue compartments. The extent of conjugation was about the same as observed previously, with approximately two-thirds of the plasma HPPH being conjugated with glucuronic acid.

The excretion data for total HPPH including unlabelled HPPH are shown in Table 3. The excretion pattern for the ^{14}C-labelled HPPH, produced as a metabolite of ^{14}C-DPH, was similar to that observed for the corresponding HPPH levels in blood plasma (Fig. 2). The peak ^{14}C excretion rate occurred in the 8-24 hr period with the loading doses of HPPH (Table 3), whereas the peak without loading doses occurred in the 2-4 hr period after dosing (Table 1). Also, the ^{14}C-HPPH peak excretion rate was about half that found in the animal without loading doses. However, the excretion of ^{14}C was about the same in 48 hr urine, with or without loading doses of HPPH, representing about 80 per cent of the dose. From the data presented, it is evident that the normal rate of hydroxylation of DPH is significantly reduced in the presence of excess HPPH, producing a marked increase in the plasma half-life of DPH.

Table 3

Urinary Excretion Rate for HPPH in a Rhesus Monkey Receiving a 50 mg/kg I.V. Dose of p-HPPH 2 Hr Before and 2 Hr After a Single 10 mg/kg I.V. Dose of ^{14}C-DPH

Collection Period	^{14}C-HPPH		p-HPPH	
	Free	Conjugated	Free	Conjugated
(Hr)	(mg/hr)	(mg/hr)	(mg/hr)	(mg/hr)
0-2	0.01	0.24	0.94	27.4
2-4	<0.01	0.30	_4.32_	_42.0_
4-6	0.05	0.33	1.37	29.7
6-8	0.11	0.45	1.12	19.0
8-24	_0.15_	_0.59_	0.41	5.0
24-48	0.10	0.47	0.11	1.1
48-72	0.03	0.20	0.03	0.28

GLC assays for m-HPPH and for DPH were below detection limits. Maximum excretion rates are underlined.

REFERENCES

1. Glazko, A. J., Chang, T., Baukema, J., Dill, W. A., Goulet, J. R. and Buchanan, R. A., Metabolic disposition of diphenylhydantoin in normal human subjects following intravenous administration. Clin. Pharmacol. Therap. 10, 498 (1969).

2. Arnold, K. and Gerber, N., The rate of decline of diphenylhydantoin in human plasma. Clin. Pharmacol. Therap. 11, 121 (1970).

3. Dayton, P. G., Cuccinell, S. A., Weiss, M. and Perel, J. M., Dose-dependence of drug plasma level decline in dogs. J. Pharmacol. Exper. Therap. 158, 305 (1967).

4. Gerber, N. and Arnold, K., Studies on the metabolism of diphenylhydantoin in mice. J. Pharmacol. Exper. Therap. 167, 77 (1969).

5. Remmer, H., Hirschmann, J. and Greiner, I., Die bedeutung von kumulation und elimination fur die dosierung von phenytoin (diphenylhydantoin). Deutsch Med. Wschr. 94, 1265 (1969).

6. Bochner, F., Hooper, W. D., Tyrer, J. H. and Eadie, M. J., Effect of dosage increment on blood phenytoin concentrations. J. Neurol. Neurosurg. Psychiatr. 35, 873 (1972).

7. Richens, A. and Dunlop, A., Serum phenytoin levels in management of epilepsy. The Lancet i, 247 (Aug. 9, 1975).

8. Lund, L., Pharmacokinetics of single and multiple doses of phenytoin in man. Europ. J. Clin. Pharmacol. 7, 81 (1974).

9. Jusko, W. J., Koup, J. R. and Alvan, G., Nonlinear assessment of phenytoin bioavailability. J. Pharmacokin. Biopharm. (in press) 1976.

10. Kutt, H., Wolk, M., Scherman, R. and McDowell, F., Insufficient para-hydroxylation as a cause of diphenylhydantoin toxicity. Neurol. (Minneap.) 14, 542 (1964).

11. Gerber, N. and Wagner, J. G., Explanation of dose-dependent decline of diphenylhydantoin plasma levels by fitting to the integrated form of the Michaelis-Menten equation. Res. Communic. Chem. Pathol. Pharmacol. 3, 455 (1972).

12. Glazko, A. J., Diphenylhydantoin. NAS/NRC/USP Conference on Bioavailability of Drugs, Washington, D. C., Nov. 22-23, 1971. Pharmacol. (Basel) 8, 163 (1972).

13. Borondy, P., Chang, T. and Glazko, A. J., Inhibition of diphenylhydantoin hydroxylation by 5-(p-hydroxyphenyl)-5-phenylhydantoin. Fed. Proc. 31, 582 (1972).

14. Glazko, A. J., Diphenylhydantoin metabolism, a prospective review. Drug Metab. Dispos. 1, 711 (1973).

15. Stavchansky, S. A., Cross-inhibition of drug metabolism by drug metabolites. Ph.D. Thesis, University of Kentucky (1974). University Microfilms, Ann Arbor.

16. Stavchansky, S. A., Lubawy, W. E. and Kostenbauder, H. B., Increase of hexobarbital sleeping time and inhibition of drug metabolism by the major metabolite of diphenylhydantoin. Life Sci. 14, 1535 (1974).

17. Stavchansky, S. A., Kostenbauder, H. B. and Lubawy, W. C., Kinetic and spectral studies of type I and type II compounds with rat hepatic microsomes in the presence of the major metabolite of diphenylhydantoin. Drug Metab. Dispos. 3, 557 (1975).

18. Ashley, J. J. and Levy, G., Inhibition of diphenylhydantoin elimination by its major metabolite. Res. Communic. Chem. Pathol. Pharmacol. 4, 297 (1972).

19. Lubawy, W. C., Kostenbauder, H. B. and Stavchansky, S. A., Cross inhibition of drug metabolism by drug metabolites. Increase of zoxazolamine paralysis time by the major metabolite of diphenylhydantoin. Res. Communic. Chem. Pathol. Pharmacol. 8, 75 (1974).

20. Soda, D. A. and Levy, G., Inhibition of drug metabolism by hydroxylated metabolites: cross-inhibition and specificity. J. Pharmaceut. Sci. 12, 1928 (1975).

21. Dill, W. A., Baukema, J., Chang, T. and Glazko, A. J., Colorimetric assay of 5,5-diphenylhydantoin (Dilantin) and 5-(p-hydroxyphenyl)-5-phenylhydantoin. Proc. Soc. Exptl. Biol. Med. 137, 674 (1971).

22. Chang, T. and Glazko, A. J., Quantitative assay of 5,5-diphenylhydantoin and 5-(p-hydroxyphenyl)-5-phenylhydantoin in plasma and urine of human subjects. Clin. Res. 16, 339 (1968).

23. Chang, T. and Glazko, A. J., Quantitative assay of 5,5-diphenylhydantoin (Dilantin) and 5-(p-hydroxyphenyl)-5-phenylhydantoin by gas-liquid chromatography. J. Lab. Clin. Med. 74, 145 (1970).

24. Atkinson, A. J., Jr., MacGee, J., Strong, J., Garteiz, D. and Gaffney, T. E., Identification of 5-meta-hydroxyphenyl-5-phenyl-hydantoin as a metabolite of diphenylhydantoin. Biochem. Pharmacol. 19, 2483 (1970).

25. Chang, T. and Glazko, A. J., Diphenylhydantoin biotransformation. In Antiepileptic Drugs (Eds., D. M. Woodbury, J. K. Penry and R. P. Schmidt) Raven Press, N. Y., pp. 149 (1972).

26. Chang, T., Young, R., Maschewske, E. Croskey, L., Smith, T. C., Buchanan, R. A. and Glazko, A. J., Metabolic studies with ^{13}C-^{14}C doubly-labelled phenytoin in human subjects. (Abstract) Amer. Epilep. Soc., Oct. 2, 1976 Dearborn, Mich. Epilepsia (in press, 1976).

SPECIES DIFFERENCES IN DRUG METABOLISM; *IN VIVO* PARAMETERS OF HEPATIC DRUG-METABOLISING ENZYME ACTIVITY IN THE BABOON

W. H. Down

Huntingdon Research Centre, Huntingdon, Cambridgeshire, England

ABSTRACT

The use of sub—human primates is widespread in toxicological studies, but in general there is little background metabolic data available in this species. In the present studies, the baboon and rat were compared, on the basis of their hepatic drug—metabolising enzyme activity, to evaluate the use of baboons as a laboratory model. Studies of hepatic drug—metabolising enzymes in man *in vivo* are difficult to correlate with those from animals.

Basal hepatic drug—metabolising enzyme activity *in vivo* has been determined in baboons and rats. These animals were then exposed to foreign compounds (p,p'—DDT or sodium phenobarbitone) to determine the toxicological and metabolic response of each species. The inducing agents were administered on an equivalent bodyweight basis for 14 days, and *in vivo* "indicators" of hepatic drug metabolism were measured at this time.

From the data obtained, it was apparent that the baboon may be considered as a model for laboratory investigations, having a similar profile in terms of hepatic drug—metabolising enzyme activity with that of the rat. Any similarity of baboon with man was not evident at this time.

INTRODUCTION

The rate of metabolism of certain drugs (for example antipyrine or phenylbutazone), as indicated by their plasma elimination half—lives, has been considered as an index of hepatic drug—metabolising enzyme activity (Ref 1,2). Similarly, the metabolism of steroids by animals and man can be regarded as an index, the urinary excretion of the metabolite of cortisol, $6-\beta$—hydroxycortisol, being a useful indicator of hepatic drug—metabolising enzyme activity (Ref 1,3). In man, the urinary excretion of D—glucaric acid, a product of glucuronic acid metabolism in the liver, has also been used to assess this activity (Ref 1,4); in rats a similar situation exists, but urinary ascorbic acid is used as the indicator of enzyme activity.

In the present study, these parameters were used to compare the basal hepatic drug—metabolising enzyme activity (and validity as laboratory models) of rats and baboon with that of other species, for example the rhesus monkey, dog (Ref 5) and man (Ref 6). Use of such *in vivo* parameters enables monitoring of enzyme induction in rats and baboons following the oral administration of known enzyme inducers of drug—metabolising enzyme activity, such as p,p'—DDT or sodium phenobarbitone (Ref 1,7)

METHODS

Male baboons of bodyweight range 3.0 — 5.5 kg were used in these studies; male Charles River rats of bodyweight approximately 250 g were used.

Prior to administration of either p,p'—DDT or sodium phenobarbitone, phenylbutazone (10 mg/kg) or antipyrine (100 mg/kg) were administered orally to rats and baboons for determination of the plasma elimination half—life of each drug. Blood samples were removed, and the plasma taken for the assay of antipyrine (Ref 2) or phenylbutazone (Ref 8). Daily urine samples were collected for the estimation of D—glucaric acid (Ref 9) and $6-\beta$—hydroxycortisol (Ref 10) excretion by baboons, and for ascorbic acid excretion by rats (Ref 11). The completion of urine collection from baboons was verified by measurement of urinary creatinine.

p,p'—DDT (dissolved in corn oil) or sodium phenobarbitone (dissolved in water) were administered daily to rats and baboons, each at a dose level of 15 mg/kg/day; control animals received the dose vehicle alone. After 14 days administration of each inducing agent, the plasma half—life of antipyrine and phenylbutazone was determined; at this time, daily urine samples were collected for measurement of D—glucaric acid and 6—β—hydroxycortisol excretion.

RESULTS

The rate of metabolism of antipyrine by baboons (plasma elimination half—life 2.6 ± 0.2 hr) and rats (2.4 ± 0.7 hr) was greater than that by man (10.9 — 15.0 hr), but was similar to that by rhesus monkeys or dogs (1.8 and 1.7 hr respectively) (Table 1). Phenylbutazone metabolism by the baboon and rat (plasma elimination half—life 4.5 ± 0.3 hr and 6.0 — 7.3 ± 1.1 hr respectively) was also more rapid than that by man (63.0 — 81.0 hr).

Urinary excretion of 6—β—hydroxycortisol by baboons (0.14 ± 0.02 μmole in 24 hr) was similar to that by man (0.25 — 0.71 μmole); daily urinary excretion of D—glucaric acid was, however, greater by man (12.0 — 34.3 μmole) than by the baboon (2.6 — 3.4 ± 0.2 μmole) (Table 1).

TABLE 1 *In vivo* indicators of hepatic drug—metabolising enzyme activity in various species

Parameter	Baboon*	Rat*	Rhesus monkey	Dog	Man
Antipyrine plasma half—life (hr)	2.6 ± 0.2	2.4 ± 0.7	1.8	1.7	12.0 10.9 — 15.0
Phenylbutazone plasma half—life (hr)	4.5 ± 0.3	7.3 ± 1.1 6.0	8.0	6.0	72.0 63.0 — 81.0
Urinary 6—β—hydroxycortisol excretion (μmole in 24 hr)	0.14 ± 0.02	—	—	—	0.25 — 0.71
Urinary D—glucaric acid excretion (μmole in 24 hr)	3.4 ± 0.2 2.6 ± 0.2	—	—	—	12.0 — 34.3

* Values are mean ± SEM

Oral administration of enzyme inducing agents (p,p'—DDT or sodium phenobarbitone at 15 mg/kg/day) to rats and baboons for 14 days caused an increase in the rate of metabolism of antipyrine and phenylbutazone in both species. In baboons, p,p'—DDT and sodium phenobarbitone increased the rate of antipyrine metabolism (plasma elimination half—life 2.0 ± 0.2 and 1.9 ± 0.2 hr respectively); the rate of phenylbutazone metabolism was increased similarly in this species (plasma half—life 3.8 ± 0.3 and 3.3 ± 0.7 hr respectively) (Table 2). Administration of p,p'—DDT or sodium phenobarbitone to rats caused similar increases in the rate of metabolism of phenylbutazone (plasma half—life 5.8 ± 0.4 and 4.3 ± 1.1 hr respectively).

Measurement of urinary metabolites in baboons (6—β—hydroxycortisol and D—glucaric acid) showed that oral administration of sodium phenobarbitone to baboons caused significant increases in the urinary excretion of 6—β—hydroxycortisol and D—glucaric acid; treatment of baboons with p,p'—DDT caused an increase in D—glucaric acid excretion, but had no effect on 6—β—hydroxycortisol excretion. The urinary excretion of ascorbic acid by rats was significantly increased by administration of either p,p'—DDT or sodium phenobarbitone (Table 3).

TABLE 2 Inducibility of the metabolism of drugs by baboons and rats

Species	Plasma half–life (hr)*	Control	p,p'–DDT treated ≠	Phenobarbitone treated ≠
Baboon	Antipyrine Phenylbutazone	2.6 ± 0.2 (23) 4.5 ± 0.3 (20)	2.0 ± 0.2 (12) 3.8 ± 0.3 (14)	1.9 ± 0.2 (11) 3.3 ± 0.7 (5)
Rat	Phenylbutazone	7.3 ± 1.1 (18)	5.8 ± 0.4 (6)	4.3 ± 1.1 (6)

* Values are mean ± SEM. (Figures in parentheses indicate number of estimations).
≠ Administered at 15 mg/kg/day for 14 days.

TABLE 3 Urinary excretion of endogenous metabolites by baboons and rats

Species	Urinary excretion (μmole/24 hr)*	Control	p,p'–DDT treated ≠	Phenobarbitone treated ≠
Baboon	6–β–hydroxycortisol D–glucaric acid	0.14 ± 0.02 (16) 2.6 ± 0.2 (21)	0.14 ± 0.03 (14) 4.7 ± 0.7 (14)	0.30 ± 0.04 (13) 5.6 ± 0.4 (10)
Rat	Ascorbic acid	19.5 ± 1.6 (36)	59.5 ± 3.7 (18)	52.7 ± 5.3 (18)

* Values are mean ± SEM. (Figures in parentheses indicate number of estimations).
≠ Administered at 15 mg/kg/day for 14 days.

CONCLUSION

From the results obtained, determinations of basal hepatic drug–metabolising enzyme activity and their inducibility have been made by means of *in vivo* parameters. Baboons appear to have a profile of drug–metabolising enzyme activity similar to that of the rat; similarities with man were not evident at this time, and comparison of sub–human primates with man (in terms of metabolic profile) is difficult. The availability of sub–human primates is becoming increasingly difficult; hence comparison of drug metabolism by alternative primates, for example, the rhesus monkey and marmoset (initially by means of *in vivo* parameters), may be of importance for future toxicological investigations.

REFERENCES

1. A.H. Conney, Pharmacological implications of microsomal enzyme induction, Pharmacol. Rev. 19, 317 (1967).

2. R.M. Welch, Y.E. Harrison and J.J. Burns, Implications of enzyme induction in drug toxicity studies, Toxicol. Appl. Pharmacol. 10, 340 (1967).

3. R. Kuntzman, M. Jacobson, W. Levin and A.H. Conney, Stimulatory effect of N–phenylbarbital (Phetharbital) on cortisol hydroxylation in man, Biochem. Pharmacol. 17, 565 (1968).

4. J. Hunter, J.D. Maxwell, M. Carrella, D.A. Stewart and R. Williams, Urinary D–glucaric acid excretion as a test for hepatic enzyme induction in man, The Lancet 1, 572 (1971).

5. Burns, J.J., Species differences in drug metabolism and toxicological implications, Proceedings of the European Society for the study of drug toxicity, Volume 11, 9, Excerpta Medica Foundation, Amsterdam, 1970.

6. J.O. Hunter and L.F. Chasseaud, Progress in Drug Metabolism (J.W. Bridges and L.F. Chasseaud eds.) Vol. 1, p.129, Wiley, New York (1976).

7. Conney, A.H., Fundamentals of Drug Metabolism and Drug Disposition (B.N. LaDu, H.G. Mandel and E.L. Way eds.), p. 253, Williams and Wilkins, Baltimore, 1971.

8. B. Herrmann, Über den Stoffwechsel des Butazolidin, Med. Expt. 1, 170 (1959).

9. C.A. Marsh, Metabolism of D—glucuronolactone in mammalian systems; identification of D—glucaric acid as a normal constituent of urine, Biochem. J. 86, 77 (1963).

10. J. Chamberlain, The determination of urinary 6—oxygenated cortisol in evaluating liver function, Clin. Chim. Acta. 34, 269 (1971).

11. J.H. Roe and C.A. Kuether, The determination of ascorbic acid in whole blood and urine through the 2,4—dinitro—phenylhydrazine derivative of dehydroascorbic acid, J.Biol. Chem. 147, 399 (1943).

EVIDENCE FOR BINDING OF LIDOCAINE TO TWO CATALYTICALLY DIFFERENT SITES OF LIVER MICROSOMAL CYTOCHROME P-450

Christer von Bahr*, Inger Hedlund, Bo Karlén[†] and Hans Grasdalen[‡]

*Depts. of Clin. Pharmacol. and Medicine, Huddinge University Hospital, Huddinge
† Div. Pharmacy, Dept. of Drugs, Nat. Board of Health and Welfare, Uppsala
‡ Dept. of Biophysics, Univ. of Stockholm

ABSTRACT

Addition of lidocaine to liver microsomes results in biphasic type I spectral titration curves. A high-affinity and a low-affinity binding phase exist. In this study we have found that microsomes from female rats have a marked high affinity phase, which can hardly be observed with microsomes from female guinea-pigs. Male rats were intermediate. On incubation of lidocaine at concentrations of 1 μM or less with female rat liver microsomes a larger fraction of the drug was aromatically hydroxylated than deethylated. The opposite was true for guinea-pig liver microsomes, and microsomes from male rats were intermediate. The ratio between the formation of deethylated and hydroxylated metabolite increases with the amount of added lidocaine in all microsomes. These data suggest that the two spectral phases represent different binding sites of cytochrome P-450 with a certain "catalytic specificity" - the "high affinity site" catalyzing aromatic hydroxylation and the "low affinity site" deethylation. Observed differential effects of pH and $MgCl_2$ concentration on aromatic hydroxylation and deethylation further strengthens the view that these reactions are catalyzed by different entities of cytochrome P-450.

INTRODUCTION

The type I spectral change is believed to reflect substrate interaction with cytochrome P-450 and to be related to metabolism. We recently found that the type I spectral titration curve of lidocaine was biphasic in rat liver microsomes (1). The affinity of the drug for the two phases was markedly different. On incubation of high concentrations (500 uM) of lidocaine with male rat liver microsomes the drug was mainly deethylated (2). In intact female rats, however, almost all excreted lidocaine is aromatically hydroxylated, while in female guinea-pigs less lidocaine is hydroxylated (3). To elucidate the mechanisms behind these differences we have carried out spectral and metabolic experiments with lidocaine in microsomes from male and female rats and female guinea-pigs.

MATERIALS AND METHODS

Male and female Sprague-Dawley rats (180-220 g) and albino female guinea-pigs (Axells strain) (400-600 g) were used. Radioactive lidocaine was obtained from New England Nuclear, USA and unlabelled lidocaine, ethylaminoaceto-2,6-xylidide (MEGX) and diethylaminoaceto-3-hydroxy-2,-xylidide (3-OH-Lid) from Astra Pharmaceutical Products, USA. Liver microsomes were isolated (4) and the concentrations of protein (5) and cytochrome P-450 (6) determined. The spectral changes produced by addition of lidocaine to suspension of microsomes in a 50 mM Tris-HCl buffer, pH 7.5, containing 0.15 M KCl were recorded. Radioactive lidocaine was incubated at 37°C in air for 2.5 min in a microsomal suspension containing 1 mg protein per ml, 50 mM Tris HCl buffer (pH 7.5), 5 mM $MgCl_2$, 5 µM $MnCl_2$ 5 mM D, L-isocitrate, 1 mM $NADP^+$ and 4 i.u. pig heart isocitric dehydrogenase per 2 ml incubation mixture. The concentrations of two metabolites primarily formed, MEGX and 3-OH-Lid (Fig. 1), were determined by liquid scintillation counting after extraction and separation by thin layer chromatography.

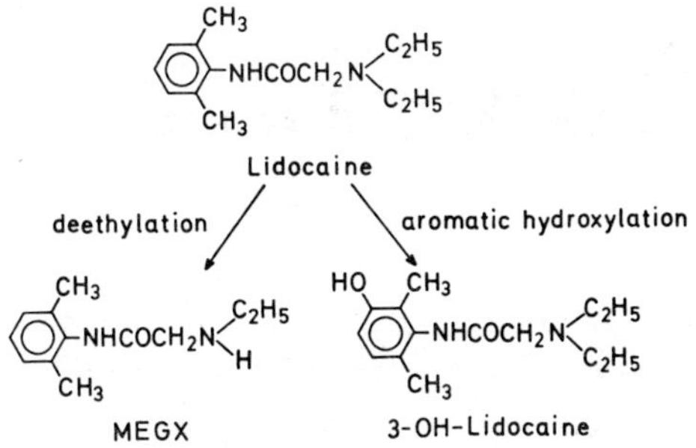

Fig. 1. Metabolism of lidocaine in liver microsomes

RESULTS

Figure 2 shows that the type I spectral change titration curve of microsomes from female rats had a marked high affinity phase, which could hardly be observed with microsomes from female guinea-pigs. Male rat liver microsomes were intermediate.

The effect of substrate concentration on these reactions is shown in Fig. 3. In female rat liver microsomes incubated with lidocaine at concentrations of 1 µM or less, a larger fraction of the drug was aromatically hydroxylated than deethylated. The opposite was true for female guinea-pig liver microsomes. Microsomes from male rats were intermediate in accordance with the spectral findings. At high concentration of lidocaine (10^{-4}M) deethylation was dominant in all microsomes.

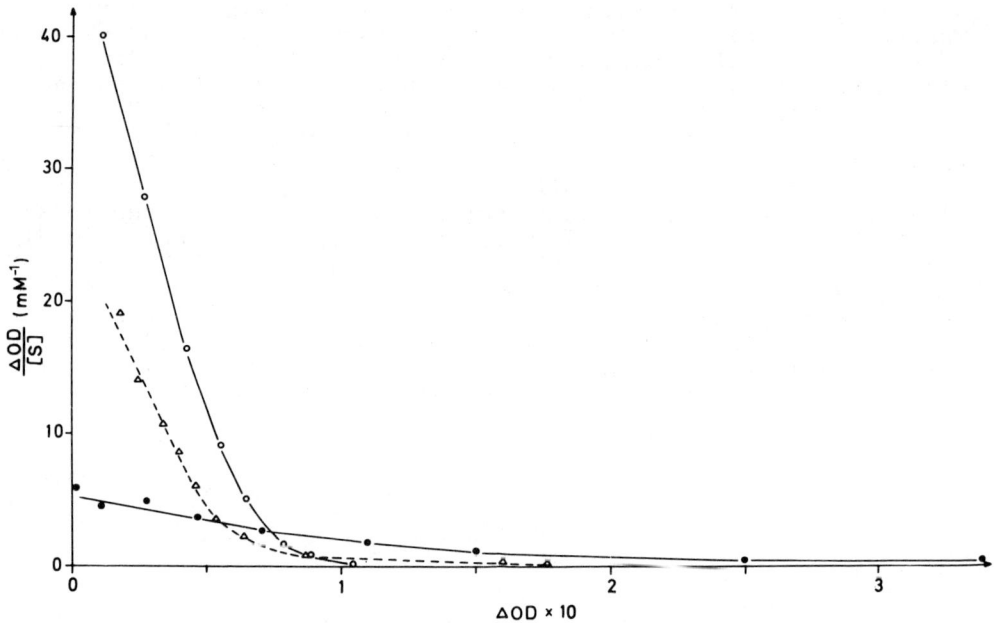

Fig. 2. Titration curves of the type I spectrum elicited on addition of lidocaine to different microsomes.

Δ OD refers to the absorbance change between 385 and 420 nm and [S] to concentration of lidocaine. The concentration of protein was 8 mg/ml and P-450 6.1 - 7.8 uM. O——O female rat; △ - △ male rat; ●——● female guinea-pig.

The effect of pH on the aromatic hydroxylation and deethylation in male rat liver microsomes is shown in Fig. 4. Deethylation was markedly influenced by pH and optimal at pH 7.9 whereas aromatic hydroxylation was much less affected by pH changes. Also changes in the concentration of $MgCl_2$ markedly influenced deethylation but not hydroxylation.

DISCUSSION

Many sex, strain, species and pretreatment differences in drug oxidation in the liver have been reported (7-13). Recently it has been more established that liver microsomes contain multiple fractions (or forms) of cytochrome P-450. Purification and reconstitution experiments with the microsomal enzyme system indicate that

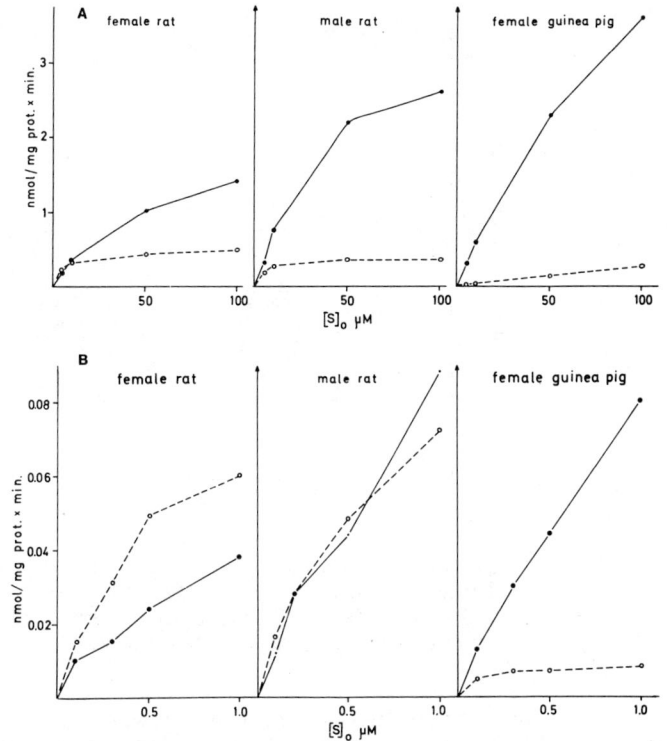

Fig. 3. Formation of MEGX (●——●) and 3-OH-lidocaine (O - - -O) from lidocaine, a) 5-100 µM, b) 0.1-1 µM, in suspension of different liver microsomes. $[S]_o$ = initial concentration of lidocaine.

different P-450 fractions have different substrate specificities (14, 15). The cytochrome moiety seems also to play a role in determining the site of oxidation reaction (15, 16). Knowing this it was tempting to hypothesize that the two recently found binding sites for lidocaine (1) represented two different cytochrome P-450 entities, each having a certain catalytic specificity. Differences in amounts of, and/or affinity for, different cytochromes P-450 could then explain observed sex and species differences in lidocaine oxidation.

In the present study we found in the microsomes from different

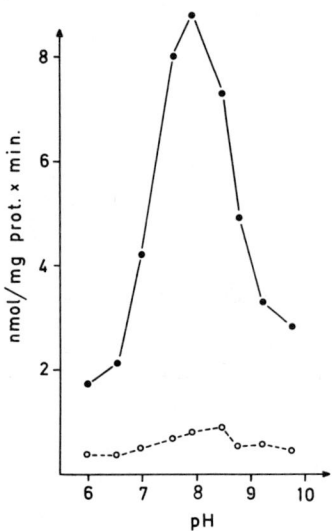

Fig. 4. Effect of pH on aromatic hydroxylation (O - - -O) and de-ethylation O———O of lidocaine in male rat liver microsomes.

species and sexes a positive relationship between the presence of a marked high affinity phase and aromatic hydroxylation at low substrate concentrations strongly indicating that the site corresponding to this phase catalyzes this reaction. In all microsomes the oxidation pattern was concentration dependent in such a way that the ratio between the formation of deethylated and hydroxylated metabolite increased with the amount of added lidocaine. This is also in accordance with the idea that the high affinity site catalyzes aromatic hydroxylation and the low affinity site deethylation - at low concentrations a relatively large fraction of lidocaine would interact with the high affinity site resulting in a large proportion of aromatically hydroxylated metabolite. Such a mechanism could explain the relative predominancy of aromatic hydroxylation <u>in vivo</u> (3) since the liver concentrations of lidocaine in that situation are low (3, 17). Finally, the finding that pH markedly affected the deethylation reaction but not the aromatic hydroxylation, also suggests that the two reactions are catalyzed by separate entities of cytochrome P-450. The marked influence of the $MgCl_2$ concentration on dealkylation but not the hydroxylation reaction strengthens this view.

ACKNOWLEDGEMENTS

This research was supported by grants from the Karolinska Institutet and the Wallenberg Foundation. One of us (H.G.) held a NATO Science Fellowship awarded by the Norwegian Research Council for Science and Technology.

REFERENCES

1. Grasdalen, H., Bäckström, D., Eriksson, L.E.G., Ehrenberg, A., Moldéus, P., von Bahr, C. and Orrenius, S., Heterogeneity of cytochrome P-450 in rat liver microsomes: selective interaction of metyrapone and SKF 525 A with different fractions of microsomal cytochrome P-450, FEBS Letters 60, 294 (1975).

2. Nyberg, G., Karlén, B., Hedlund, I., Grundin, R. and von Bahr, C., Extraction and metabolism of lidocaine in rat liver, Acta Pharmacol et Toxicol., in press (1976).

3. Keenaghan, J.B. and Boyes, R.N., The tissue distribution, metabolism and excretion of lidocaine in rats, guinea pigs, dogs and man, J. Pharmacol. & Exp. Ther. 180, 454 (1972).

4. Ernster, L., Siekevitz, P. and Palade, G., Enzyme structure relationship in the endoplasmic reticulum of rat liver. J. Cell Biol. 15, 541 (1962).

5. Lowry, O.H., Rosebrough, N.J., Farr, A.L. and Randal, R.J., Protein measurement with the folin phenol reagent, J. Biol. Chem. 193, 265 (1951).

6. Omura, T. and Sato, R., The carbon monoxide-binding pigment of liver microsomes. I. Evidence for its hemoprotein structure, J. Biol. Chem. 239, 2370 (1969).

7. Brodie, B.B. and Maickel, P., Metabolic factors controlling duration of drug action (Brodie & Erdes eds.), Comparative biochemistry of drug metabolism, Pergamon Press 6, 299, 1962.

8. Dingell, J.V., Sulser, F. and Gillette, J.R., Species differences in the metabolism of imipramine and desmethylimipramine, J. Pharmacol. 143, 14 (1964).

9. Conney, A.H., Pharmacological implications of microsomal enzyme induction, Pharmacol. Rev. 19, 317 (1967).

10. Schenkman, J.B., Frey, I., Remmer, H. and Estabrook, R.W., Sex differences in drug metabolism by rat liver microsomes, Mol. Pharmacol. 3, 516 (1967).

11. Davies, D.S., Gigon, P.L. and Gillette, J.R., Species and sex differences in electron transport systems in liver microsomes and their relationship to ethylmorphine demethylation, Life Sci. 8, 85 (1969).

12. Cohen, G.M. and Mannering, G.J., Sex-dependent differences in drug metabolism in the rat, Drug metabolism and disposition 2, 285 (1974).

13. Guarino, A.M., Gram, T.E., Gigon, P.L., Greene, F.E. and Gillette, J.R., Changes in Michaelis and spectral constants for aniline in hepatic microsomes from phenobarbital-treated rats.

13. Mol. Pharmacol. 5, 131 (1969).

14. Lu, A.Y.H., Kuntzman, R., West, S., Jacobson, M. and Conney, A.H., Reconstituted liver microsomal enzyme system that hydroxylates drugs, other foreign compounds, and endogenous substrates. II, J. Biol. Chem. 247, 1727 (1972).

15. Conney, A.H., Lu, A.Y.H., Levin, W., Somogyi, A., West, S., Jacobson, M., Ryan, E. and Kuntzman, R., Effect of enzyme inducers on substrate specificity of the cytochrome P-450´s, Drug metabolism and diposition 1, 199 (1973).

16. Lu, A.Y.H.,Levin,W.,West,S.B., Jacobson, M., Ryan, D., Kuntzman, R. and Conney, A.H., Reconstituted liver microsomal enzyme system that hydroxylates drugs, other foreign compounds, and endogenous substrates. VI, J. Biol. Chem. 248, 456 (1973).

17. Benowitz, N., Forsyth, R.P., Melmon, K.L. and Rowland, M., Lidocaine disposition kinetics in monkey and man I. Prediction by a perfusion model, Clin. Pharm. Ther. 16, 87 (1974).

DIFFERENTIAL INFLUENCE OF PHYSIOLOGICAL, PHARMACOLOGICAL AND PATHOLOGICAL ALTERATIONS ON HEPATIC AND EXTRAHEPATIC DRUG METABOLISM

Theodore E. Gram, Branimir I. Sikic, Charles L. Litterst and E. G. Mimnaugh

Laboratory of Toxicology, National Cancer Institute, National Institutes of Health, Bethesda, Maryland 20014, U.S.A.

SUMMARY

Factors known to influence hepatic drug metabolism were evaluated for their influence on drug metabolism in extrahepatic organs. The well-documented sex-related differences in drug metabolism in rat liver were not seen in either lung or kidney. Treatment of rats with phenobarbital elicited the expected increases in mixed function oxidase (MFO) parameters in liver without concomitant changes in these activities in lung and with only scattered and less marked changes in kidney. UDP-Glucuronyltransferase activity was undetectable toward two substrates in rabbit lung microsomes. Several activators which stimulated hepatic UDP-glucuronyltranferase several fold had no effect on the pulmonary enzyme. Treatment of rats with CCl_4 inhibited the hepatic and pulmonary MFO systems in a quantitatively and temporally similar manner but was without effect on the renal system. Increases in hepatic and renal but not pulmonary UDP-glucuronyltransferase were observed after CCl_4. Scurvy impaired most MFO activities in liver microsomes but evoked no change or much smaller changes in extrahepatic organs. It is concluded that some factors that influence hepatic drug metabolism produce similar effects in lung and kidney while the effects of other factors are organ-specific.

INTRODUCTION

Our laboratory has been interested for some years in the comparative aspects of drug metabolism by mammalian lung and liver. We published the first detailed and complete cell fractionation studies conducted with lung and a thorough evaluation of the subcellular distribution and general kinetic properties of the mixed function oxidase (MFO) system in lung and liver of rabbits (1-3). More recently, we have investigated conjugation reactions and species differences in the metabolism of xenobiotics by hepatic and pulmonary tissue (4,5). These studies have revealed many similarities and few differences between the two systems. However, in a continuing effort to arrive at a more complete understanding of the biological control mechanisms which regulate these activities in hepatic and extrahepatic tissues, we have studied the response of extrahepatic enzymes to some factors that have well-documented effects on drug metabolism by liver.

METHODS

Adult male and female Sprague-Dawley rats weighing between 150 and 250 g, male New Zealand rabbits weighing about 2.5 kg, and male Hartley guinea pigs weighing about 600 g, were used in these experiments. They were allowed free access to laboratory chow and tap water and were sacrificed by cervical dislocation. Lungs, livers, and kidneys were removed and cut into small pieces with scissors. Microsomal fractions were prepared in 0.15M KCl-50mM Tris (pH 7.4) and enzyme assays were conducted as described previously (5). Data were analyzed statistically by use of the Student t test (6).

RESULTS

In rat liver, there is a well-known sex difference in the metabolism of many drugs such as hexobarbital and this difference is reflected in the biological half-life and duration of action of the drug. Table 1 shows that this sex

TABLE 1   Lack of sex difference in extrahepatic microsomal drug metabolism in adult Sprague-Dawley rats[1]

	Liver	Lung	Kidney
Aminopyrine demethylase			
male	9.29 ± 1.16[2]	0.48 ± 0.17	0.71 ± 0.22
female	4.74 ± 1.37	0.47 ± 0.18	0.72 ± 0.06
Biphenyl-4-hydroxylase			
male	0.81 ± 0.10[2]	0.04 ± 0.01	0.03 ± 0.02
female	0.47 ± 0.06	0.03 ± 0.03	0.02 ± 0.01
Cytochrome P-450			
male	0.084 ± 0.005[2]	0.007 ± 0.003	0.012 ± 0.005
female	0.066 ± 0.007	0.006 ± 0.003	0.014 ± 0.007
NADPH cytochrome c reductase			
male	148 ± 11	42 ± 4	35 ± 6
female	153 ± 32	58 ± 15	32 ± 3

[1]Enzyme activities are expressed as nmoles product formed/min/mg protein (mean ± S.D.; n = 4).

[2]Male and female values significantly different at $P < .05$

difference in drug metabolism by rat liver microsomes was not accompanied by corresponding differences in rat lung or kidney microsomes. Similarly, no significant sex differences were noted in rabbit lung or kidney (unpublished data).

It has been repeatedly demonstrated that pretreatment of experimental animals with various enzyme inducers such as phenobarbital markedly enhances MFO activity in hepatic microsomes. However, the effect of such inducers on enzyme activities in extrahepatic tissues is less clear. In the present experiments, treatment of rabbits with phenobarbital (75 mg/kg once daily for 4 days, i.p.) produced the expected increases in hepatic microsomal biphenyl hydroxylase, aminopyrine demethylase, NADPH cytochrome c reductase and cytochrome P-450 levels but only scattered and minor increases in these activities in lung and kidney (Table 2). Pretreatment with 3-methylcholanthrene (25 mg/kg once daily for 3 days i.p.) significantly increased cytochrome P-450 levels in all three organs (Table 2).

TABLE 2 Effect of pretreatment of rabbits with phenobarbital (Pb) or 3-methylcholanthrene (3-MC) on microsomal enzyme activities[1]

	Liver	Lung	Kidney
Biphenyl-4-hydroxylase			
control	1.06 ± 0.25	1.08 ± 0.24	0.10 ± 0.02
Pb	2.97 ± 0.72[2]	1.37 ± 1.10	0.22 ± 0.02[2]
3-MC	0.90 ± 0.08	1.03 ± 0.34	0.12 ± 0.02
Aminopyrine demethylase			
control	4.05 ± 0.28	1.44 ± 0.92	0.69 ± 0.47
Pb	9.45 ± 0.44[2]	2.88 ± 0.20	1.43 ± 0.66
3-MC	------	------	------
NADPH cytochrome $\underline{c}$ reductase			
control	133 ± 17	85 ± 6	29 ± 1
Pb	194 ± 37[2]	97 ± 5	41 ± 11
3-MC	130 ± 17	110 ± 9[2]	31 ± 4
Cytochrome P-450			
control	0.127 ± 0.011	0.020 ± 0.001	0.014 ± 0.002
Pb	0.232 ± 0.028[2]	0.022 ± 0.006	0.030 ± 0.008[2]
3-MC	0.224 ± 0.034[2]	0.027 ± 0.004[2]	0.023 ± 0.005[2]

[1] Enzyme activities expressed as nmoles product formed/min/mg protein (mean ± S.D.; n = 4)

[2] Values significantly different from control at $P < .05$

During studies of the enzymatic conjugation of xenobiotics by rabbit lung it was observed that the "native" levels of UDP-glucuronyltransferase were below the limits of detection of our methods whereas the hepatic enzyme was readily demonstrable (7). Since it has been abundantly shown that this enzyme can be activated several fold in liver by detergents and other compounds (8), we thought it of interest to determine whether the enzyme was present in lung microsomes but in a latent form which required exogenous activators for expression. As can be seen in Table 3, a number of activators which are presumed to produce their effects through different mechanisms (allosteric effects, cationic and neutral detergents) stimulated rabbit liver UDP-glucuronyltransferase activity 3-5 fold but failed to activate the lung enzyme. These results (Table 3) were obtained with p-nitrophenol but similar results were obtained with o-aminophenol as substrate. Similarly, the use of additional activators such as Na deoxycholate, mersalyl acid, and diethylnitrosamine, alone or in combination were without effect on the pulmonary enzyme.

Administration of carbon tetrachloride (CCl_4) to rats produces rapid and extensive morphological and biochemical disruption of the endoplasmic reticulum of liver; this disruption is associated with a marked loss of microsomal MFO activity which is readily observable within 15 min. to 4 hours after administration and usually requires a week or more for complete recovery (9).

Male rats weighing 200-250 g were injected i.p. with 1.5 ml/kg CCl_4 dissolved in corn oil and sacrificed at various times. As can be seen in fig. 1, CCl_4 reduced hepatic cytochrome P-450 and biphenyl hydroxylase to less than 30% of control levels within 1-2 days. The effect in lung was similar in magnitude

TABLE 3 Effect of activators of microsomal UDP-glucuronyltransferase on the conjugation of p-nitrophenol by rabbit tissues[1,2]

Addition	Liver	Lung[4]
none ("native")	2.16 ± 0.67	N.D.
UDP-N-acetylglucosamine[3]	10.49 ± 2.01	N.D.
cetyltrimethylammonium Br	11.18 ± 1.58	N.D.
Triton X-100	8.15 ± 1.83	N.D.

[1] Enzyme activities are expressed as nmoles glucuronide formed/min/mg protein (mean ± S.D.; n = 4)

[2] All incubation mixtures contained saturating concentrations of UDPGA and Mg^{++}

[3] All activators were run at several concentrations and the data presented are those at which the greatest activation was observed. Measurable activities in lung microsomes were not observed at any of the concentrations of activators.

[4] Non-detectable activity; below limits of sensitivity of the method (<0.05 nmoles)

and time course (Fig. 1). It is noteworthy that despite these large changes in MFO activities in liver and lung, CCl_4 produced no corresponding alterations in kidney activities (Fig. 1).

In contrast to its effects on the microsomal MFO system, CCl_4 treatment elicited significant increases in UDP-glucuronyltransferase activity in liver (∼3 fold) and kidney (∼2 fold). In this instance, no changes in lung activity were observed (Fig. 1).

The effects of dietary ascorbic acid deficiency on hepatic drug metabolism have been well documented. Thus numerous investigators have demonstrated that 15-20 days of ascorbic acid deficiency reduces MFO activities of guinea pig liver to about 50% of control values, but no comparable studies have been performed with extrahepatic organs.

Adult male Hartley guinea pigs were fed an ascorbic acid deficient (<0.1mg/g) diet or the same diet supplemented with 2mg/g ascorbic acid (nutritionally adequate) for 0-25 days after which some of the scorbutic animals were transferred to a supplemented diet to study recovery. Table 4 shows that consumption of the deficient diet for as little as 14 days reduced tissue ascorbate to 10-30% of control levels. By 25 days, liver, lung, and adrenal ascorbate levels were less than 10% of control; renal levels appeared slightly more resistant to depletion. No deaths occurred during the 25 days of deficiency. Table 4 also demonstrates that 14-21 days of a scorbutigenic diet significantly decreased hepatic cytochrome P-450 and aminopyrine demethylase activities without significantly influencing these activities in lung. The effects in kidney were less consistent but cytochrome P-450 levels were

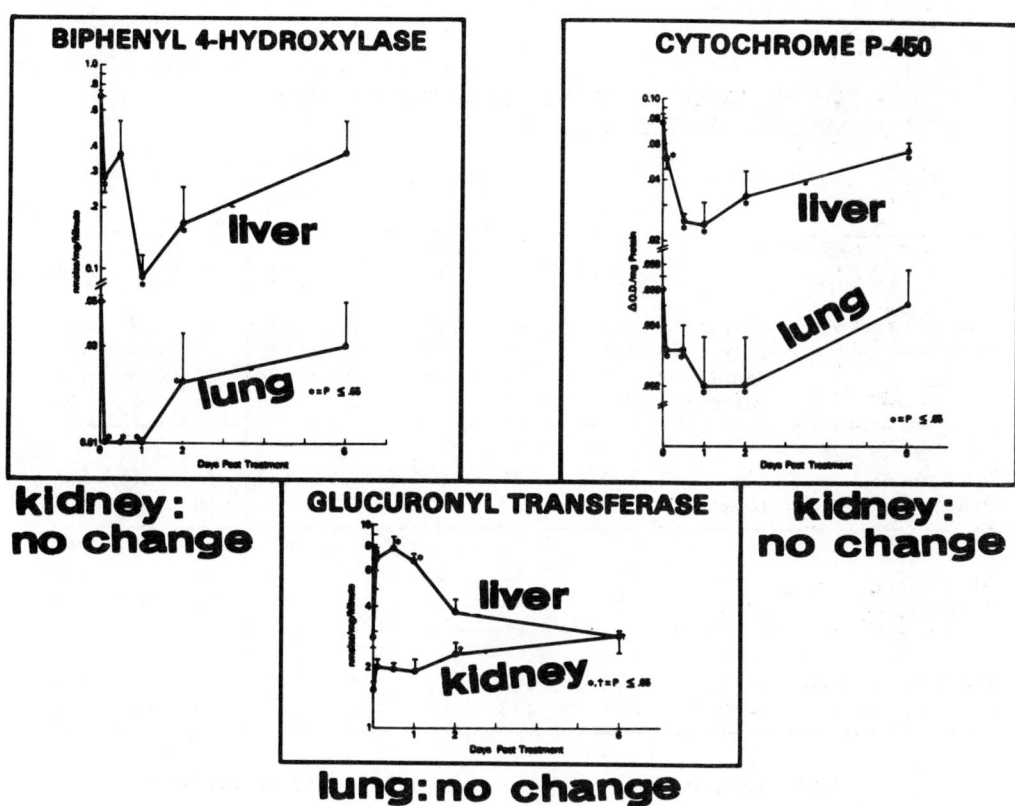

Fig. 1 The effect of treatment of rats with carbon tetrachloride (1.5 ml/kg i.p.) on microsomal enzyme activities in various organs. The substrate for glucuronyl transferase was p-nitrophenol.

significantly decreased at later time points (data not presented). The hepatic changes were readily reversible upon supplementation with dietary ascorbate, although direct addition of ascorbate to incubation mixtures was without effect. The effects of scurvy on MFO activity were not a general impairment of the entire microsomal electron transfer chain for the activity of NADPH cytochrome c reductase remained unchanged throughout the experiment (Table 4). Thus, the effects of scurvy were both organ- and enzyme-specific.

TABLE 4 Effect of scurvy and recovery on drug metabolism in guinea pig organs[1]

			Days Deficiency				Recovery		
		(control)	7	14	21	25	3	7	14
Tissue Ascorbic Acid Level ($\mu g/g$)	Liver	100	41	15	6	15	89	112	107
	Lung	100	20	14	6	12	77	103	114
	Kidney	100	50	31	19	33	83	115	151
	Adrenal	100	31	8	4	6	63	112	90
Cytochrome P-450	Liver	100	89	67^2	60^2	45^2	74^2	96	85
	Lung	100	88	92	75	77	86	92	93
Aminopyrine Demethylase	Liver	100	98	86	62^2	60^2	78	115	110
	Lung	100	86	90	86	84	121	90	92
NADPH Cytochrome c Reductase	Liver	100	102	100	105	120	102	93	103
	Lung	100	113	102	112	110	104	97	90

[1]Data are expressed as % of control analyzed on the same day (n = 4-6 at each point).

[2]Significantly different from control ($P < .05$).

DISCUSSION

The results of these studies may be summarized as follows:
1. The sex difference in drug metabolism by rat liver microsomes is not accompanied by corresponding differences in drug metabolism by lung or kidney microsomes.
2. The stimulation of hepatic drug metabolism produced by phenobarbital occurs without concomitant changes in lung and with only scattered changes which are of lesser magnitude in the kidney.

3. UDP-Glucuronyltransferase activity is not demonstrable in a "native" state in rabbit lung microsomes although it has been shown to be present in lungs of other species. This activity in rabbit lung is not revealed by the presence of activators (in vitro) which produced 3-5 fold activation of the hepatic enzyme (or by phenobarbital or 3-MC pretreatment).
4. The administration of CCl_4 to rats impaired MFO activities in liver and lung in a similar way but was without effect in kidney. On the other hand, CCl_4 stimulated or activated UDP-glucuronyltransferase activity in liver and to a lesser extent in kidney but was without effect in lung.
5. The production of scurvy in guinea pigs reduced several MFO activities of liver to less than 50% of control levels. Concomitantly, no changes or smaller changes were noted in renal activities and no changes were observed in pulmonary MFO activity.
6. It is concluded that some physiological, pharmacological, and pathological factors that influence hepatic drug metabolism produce similar effects in lung and kidney while other factors produce organ-specific effects. This lack of general similarity in responsiveness suggests that some aspects of drug metabolism in each organ may be under individual control and not interrelated.

REFERENCES

1. Hook, G. E., Bend, J. R., Hoel, D., Fouts, J. R. and Gram, T. E. J. Pharmacol. Expt. Ther. 182, 474-490, (1972).

2. Bend, J. R., Hook, G. E., Easterling, R. E., Gram, T. E. and Fouts, J. R. J. Pharmacol. Expt. Ther. 183, 206-217, (1972).

3. Gram, T. E. Drug Metab. Reviews 2, 1-32, (1973).

4. Gram, T. E., Litterst, C. L. and Mimnaugh, E. G. Drug Metab. Dispos. 2, 254-258, (1974).

5. Litterst, C. L., Mimnaugh, E. G., Reagan, R. L. and Gram, T. E. Drug Metab. Dispos. 3, 259-265, (1975).

6. Snedecor, G. W. Statistical Methods, The Iowa State University Press, Ames, Iowa, (1956).

7. Mimnaugh, E. G., Litterst, C. L. and Gram, T. E. Biochem. Pharmacol. 24, 1633-1635, (1975).

8. Dutton, G. J. Biochem. Pharmacol. 24, 1835-1841, (1975).

9. Recknagel, R. O. and Glende, E. A. Critical Rev. Toxicol. 2, 263-297, (1973).

METABOLISM OF HEXOBARBITAL ENANTIOMERS AND INTERACTION WITH CYTOCHROME P-450 IN MALE AND FEMALE MICE AND RATS

J. Noordhoek, A. P. Van Den Berg, E. M. Savenije-Chapel and E. Koopman-Kool

Department of Pharmacology, Medical Faculty, Erasmus University, Rotterdam, The Netherlands

ABSTRACT

The metabolism of hexobarbital enantiomers by 9000g liver supernatant was determined in mice and rats. Both male and female mice and rats metabolize hexobarbital stereospecifically. In mice the l-enantiomer is metabolized faster, whereas the opposite was observed in rats. The kinetics of metabolism and type I binding to cytochrome P-450 of the enantiomers was compared in female mice and male rats. In mice K_m equaled K_s for type I binding, but in rats K_m exceeded K_s. This suggests that in mice the type I binding site is the active site of the enzyme, whereas in rats this cannot be proved. In female rats the A_{max} for type I binding of the enantiomers differs, indicating a binding to different cytochrome P-450 subspecies. Comparison of the enhancement of NADPH-cyt. P-450 reductase activity by the enantiomers with their rate of metabolism at the same concentration revealed that in rats the reduction of the cytochrome P-450-substrate complex cannot be rate-limiting for the total reaction, whereas in mice this cannot be disproved. This would explain the apparent discrepancy between the kinetics of type I binding and metabolism in rats, which was not found in mice.

INTRODUCTION

Comparison of the kinetic parameters of metabolism and type I binding of substrates to cytochrome P-450 can be used to investigate the possibility that the type I binding site is the catalytic site of the enzyme. This method is valid only if the reduction of the substrate complex is rate limiting for the total reaction. Several authors have compared these parameters in different species and sexes (1,2,3). Although in some studies a similarity between K_m of metabolism and K_s of substrate binding was observed, in many others large differences were reported (4). Investigations conducted in this department, using a sex difference in mice for the metabolism of type I substrates (5,6), also revealed some discrepancies (7).
Instead of species and sex differences the use of different substrates may further clarify this problem. The ideal compounds for such studies are the enantiomers of chiral substrates, many of which are metabolized stereospecifically (8). Up till now investigations with benzphetamine (9) and other demethylated substrates (10) failed to give consistent parallels between the kinetic constants of metabolism and type I binding. This may be due to cyt. P-450-catalyzed metabolism of the substrates along pathways other than demethylation, which makes demethylation and substrate binding incomparable.

For our studies we chose hexobarbital as a substrate. In mice (11) and rats (12) this drug is converted almost exclusively into its 3-hydroxy metabolite, which is further dehydrogenated to 3-keto-hexobarbital. In unpublished experiments we found that in 9000g mouse-liver supernatant less than 2% of hexobarbital metabolism is due to demethylation. It appears, therefore, that only one reaction determines hexobarbital metabolism. Stereoselective metabolism of hexobarbital was demonstrated in rats (13,14,15,16), mice (15) and humans (17), but not in guinea-pigs (18). The kinetics of metabolism and type I binding of the enantiomers in untreated animals were compared only in guinea-pigs, in which no correlation was found (18).

In the present study the kinetics of metabolism and type I binding of hexobarbital enantiomers are compared in male and female rats and mice. In addition, the hexobarbital-induced enhancement of NADPH-cyt. P-450 reductase activity was investigated. The formation of a cyt. P-450 type I substrate complex is thought to be responsible for this enhancement, which is therefore suggested to be the expression of the reduction of this complex (19,20). If this reduction is rate-limiting in the hydroxylation reaction, differences in reduction enhancement by the enantiomers should be accompanied by similar differences in hydroxylation rate. Furthermore, a relationship between reductase stimulation and the magnitude of the type I spectral interaction is to be expected. This was also investigated.

MATERIAL AND METHODS

Animals
Male and female Swiss (CPB-SE) mice, at least 10 weeks old, and male and female Wistar (WAG) rats (TNO, Zeist) weighing 150-200 g were used.

Chemicals
d- and l-hexobarbital, $(\alpha)_D^{20} = +12.0°$; m.p.: $155°C$ and $(\alpha)_D^{20} = -12.5°$; m.p.: $154-155°C$, were kindly donated by Prof.Dr. J. Knabe, University of Saarland, Saarbrücken, Germany. Racemic hexobarbital was obtained from the O.P.G., Utrecht. All other chemicals and solvents used were of analytical grade purity.

Hexobarbital Metabolism
Hexobarbital metabolism was assayed in 9000g supernatant, which was prepared by centrifuging (20 min) a 1:2 (w/v) liver homogenate in 0.1 M Na-K phosphate buffer (pH 7.4) containing 0.1 mM EDTA. Enzyme incubations were done in the same buffer, containing 0.75 μmol NADP, 12.5 μmol glucose-6-P, 5 μmol nicotinamide, 12.5 μmol $MgCl_2$, hexobarbital as indicated and 1 ml (mouse) or 0.5 ml (rat) supernatant in a total volume of 3 ml. The incubates were shaken in 25 ml bottles at $37°C$ in air. The reaction was started by adding the supernatant to the prewarmed incubation medium and stopped after 5 min (rats) or 10 min (mice) by addition of 12 ml of heptane containing 1.5% n-amyl alcohol with shaking.

Hexobarbital was determined by means of GLC, as UV absorbing metabolites may interfere with spectrophotometric determination (21). Hexobarbital was extracted from the incubates with 12 ml heptane-amylalcohol containing 0.2 μmol methohexital as an internal standard, by shaking for 15 sec on a Vortex mixer as described by Breimer and Van Rossum (22). After centrifugation 10 ml of the organic layer were vigorously shaken for 5 minutes with 5 ml phosphate buffer pH 11 (23)   4 ml of which were acidified with 0.4 ml 12.5N HCl and snaken for 5 minutes with 20 ml $CHCl_3$. The chloroform layer was removed and evaporated to dryness. The residue was dissolved in 100 μl chloroform and 1 μl portions were injected into a Varian 2100 gaschromatograph, using a 1.8m x 4 mm column containing 3% OV-17 on 80-100 mesh Chromosorb W (HP). The following temperatures were used: injection port: $260°C$, oven: $210°C$ and FID detector $300°C$. In each series of determinations a standard curve was included, which proved to be linear and showed small day-to-day variations.

Difference Spectra and Cytochrome P-450 Reduction

Difference spectra and cyt. P-450 reduction were determined in microsomes. 9000g Supernatant was prepared from a 1:4 (w/v) liver homogenate and subsequently centrifuged at 75,000g for 90 min. The microsomal pellet was suspended in 0.1 M phosphate buffer (pH 7.4).
Spectra were recorded in an Aminco-Chance spectrophotometer in the split-beam mode. Microsomal suspensions contained 3 mg prot./ml. The difference between maximum and minimum absorbance was taken as the magnitude of spectral interaction.
NADPH-cyt. P-450 reductase activity was determined at $30^{\circ}C$ under anaerobic conditions (protein concentration 1 mg/ml), essentially as described by Gigon et al. (19). An Aminco DW-2 UV-VIS spectrophotometer was used in the dual wavelength mode. Reductase stimulation represents the difference between the initial rates of reduction with and without added substrate. Further details will be provided elsewhere (24).
Microsomal protein was estimated according to the method of Lowry (25), using bovine serum albumine as a standard.

Calculations

Kinetic constants for metabolism (K_m, V_{max}) and type I binding (K_s, A_{max}) were calculated using the method of Wilkinson (26). In some cases more than 10% of substrate was metabolized. Therefore, the substrate concentrations used in the calculations were corrected for substrate disappearance as described by Lee and Wilson (27).
Statistical significance of differences was tested by analysis of variance and Student's t-test.

RESULTS

Metabolism in Male and Female Mice

9000g Supernatant of livers from both male and female CPB-SE mice metabolizes l-hexobarbital faster than its optical antipode as can be seen in Table 1. Female mice metabolize both enantiomers faster than males, as was previously observed for racemic hexobarbital and other type I compounds (5,6). This result is in contrast to the work of McCarthy and Stitzel (15), who found not the l- but rather the d- enantiomer to be metabolized faster in their mice.

TABLE 1 Metabolism of hexobarbital enantiomers by 9000g liver supernatant (nmoles/min/incubate) of male and female mice (0.167 mM) and rats (0.333 mM) (means $\pm$ SE).

	Mice		Rats	
	male	female	male	female
d	7.8 $\pm$ 1.2 (7)	14.6 $\pm$ 0.8 (8)	73.7 $\pm$ 3.9 (3)	16.9 $\pm$ 0.8 (4)
l	13.4 $\pm$ 1.1 (7)	22.9 $\pm$ 0.8 (8)	25.8 $\pm$ 1.9 (4)	0.91 $\pm$ 0.40 (3)
	$p < 0.001$	$p < 0.001$	$p < 0.005$	$p < 0.001$

Kinetics of Metabolism and Type I Binding in Female Mice

Because the d-enantiomer of hexobarbital is metabolized very slowly in male mice (Table 1), our method of measuring hexobarbital metabolism by disappearance of substrate does not enable us to determine kinetic constants of this reaction in these animals. Similarly, the kinetics of type I binding could not be determined in male mice, due to the extremely small difference spectra produced by d-hexobarbital. Therefore, only values in females are given. Table 2 summarizes the results. The substrate concentrations used in metabolic studies were 3, 4.5, 6, 7.5 and 9 mM^{-1}. In binding studies 2, 4, 6, 8 and 10 mM^{-1} hexobarbital was used. Linear Lineweaver-Burk plots were obtained for metabolism as well as type I binding of the enantiomers. A significant difference ($P<0.02$) was found for the K_m values of the enantiomers, calculated with or without corrected substrate concentrations. No difference was found in V_{max}. For type I binding an analogous result was obtained. K_s values were different ($P<0.005$) whereas A_{max} values were not. The values for K_m, calculated using uncorrected substrate concentrations exceeded the analogous K_s, but calculation from corrected concentrations gave K_m values which did not exceed K_s.

TABLE 2 Kinetic constants of metabolism and type I binding of hexobarbital enantiomers in female mice and male rats (means ± SE).

	Mice		Rats	
	d	l	d	l
K_m (mM)	0.180 ± 0.029	0.092 ± 0.009	0.114 ± 0.014	0.225 ± 0.33
K_m (mM)*	0.112 ± 0.022	0.053 ± 0.005	0.071 ± 0.009	0.193 ± 0.026
K_s (mM)	0.114 ± 0.011	0.073 ± 0.004	0.014 ± 0.0004	0.036 ± 0.005
V_{max}	38.9 ± 2.3	37.3 ± 2.7	95.6 ± 5.4	44.3 ± 4.0
V_{max}*	35.4 ± 2.1	35.2 ± 2.2	90.6 ± 4.8	44.0 ± 3.3
A_{max}	18.0 ± 0.7	19.8 ± 0.6	24.5 ± 0.6	24.6 ± 2.0

*Corrected for substrate disappearance
n = 12 (K_m, V_{max}) or 5 (K_s, A_{max}) in rats and 8 in mice
V_{max} is expressed as nmol/min/incubate and A_{max} as ΔA/mg protein

Metabolism in Male and Female Rats

Table 1 gives the rate of metabolism of hexobarbital enantiomers in male and female rats. In this species the opposite result was obtained from that in mice. The d-enantiomer was metabolized faster than its antipode in males as well as in females, which is in contrast to the results of Furner et al. (13). Male animals metabolized each enantiomer faster than females.

Kinetics of Metabolism and Type I Binding in Rats

As can be seen from Table 1, female rats metabolized l-hexobarbital very slowly. Therefore, no kinetic studies on hexobarbital metabolism were carried out in these animals. Type I binding spectra, however, were of sufficient magnitude to study binding kinetics. Table 2 summarizes kinetic constants for metabolism and type I binding obtained in male rats. The concentrations used in metabolic studies were 1.5, 3, 4.5, 6, 7.5, 9 and 20.5 mM^{-1}, whereas 10, 50, 100, 150 and 200 mM^{-1} hexobarbital was used in binding studies. Linear

Lineweaver-Burk plots were obtained both for metabolism and type I binding. The values of K_m obtained in male rats using our method agree with those for racemic hexobarbital reported by Sitar and Mannering (28) and Kupfer and Rosenfeld (29), who used product formation instead of substrate disappearance for the determination of reaction rates. For the more slowly metabolized l-hexobarbital a higher K_m (P<0.001) and a lower V_{max} (P<0.001) were observed than for the d-enantiomer, as was also reported by McCarthy and Stitzel (15). For type I binding a significant parallel difference in K_s (P<0.01) was found whereas A_{max} values were similar. The K_m values were in this case higher than the K_s for the same enantiomer, even after correction for substrate disappearance. In female rats not only K_s (P<0.01) but also A_{max} (P<0.01) was different for the enantiomers, indicating that the nature of the type I binding sites are different for d- and l-hexobarbital. Hexobarbital concentrations of 2, 4, 6, 8 and 10 mM$_l$ were used in these determinations.

Enhancement of Cytochrome P-450 Reductase Activity

Due to the rather low enhancements of cyt. P-450 reduction produced by hexobarbital it was not possible to obtain reliable values in some cases in concentration regions where differences in the hydroxylation rates of the enantiomers are the most prominent. This relates in particular to the male mouse, of which no data are presented. In female mice at 1.0 mM no difference in reductase stimulations produced by the enantiomers could be detected (Table 3). However, at this concentration also no significant difference in hydroxylation rates was apparent. These findings cannot prove the involvement of cyt. P-450-complex reduction as a rate-limiting step, but the combined data on binding and hydroxylation kinetics and cyt. P-450 reduction leaves this possibility open.

TABLE 3 Enhancement of NADPH-cytochrome P-450 reductase activity, rate of metabolism, and magnitude of spectral interaction of d- and l-hexobarbital in female mice and male and female rats (means $\pm$ SE)

		reductase stimulation	rate of metabolism	type I spectral change
female mice	d	2.45 $\pm$ 0.11 (14)	31.6 $\pm$ 1.9 (8)	16.2 $\pm$ 0.7 (8)
1.0 mM	l	2.33 $\pm$ 0.11 (14)	33.4 $\pm$ 2.1 (9)	18.5 $\pm$ 0.6 (8)*
male rats	d	1.81 $\pm$ 0.14 (5)	53.5 $\pm$ 2.2 (12)	23.5 $\pm$ 0.9 (5)
0.1 mM	l	1.80 $\pm$ 0.13 (4)	15.5 $\pm$ 1.0 (12)	17.4 $\pm$ 0.6 (5)**
female rats	d	2.29 $\pm$ 0.16 (10)		8.0 $\pm$ 0.3 (8)
0.5 mM	l	1.83 $\pm$ 0.11 (10)*		5.5 $\pm$ 0.3 (8)**

Spectral changes and metabolic rates were derived from the kinetic studies presented in table 2 and are expressed in the units indicated there. Reductase stimulation is expressed as nmol cyt. P-450 reduced/min/mg prot.
* Significant difference between d and l: p<0.05
** Id. : p<0.001

In male rats d- and l-hexobarbital (0.1 mM) enhanced cyt. P-450 reduction to the same extent, although at this concentration the metabolic rates differred by a factor of 3.5. In female rats at 0.5 mM a small, but significant difference was found in reductase stimulation. In Table 1 it was shown that at a concentration of 0.33 mM d-hexobarbital was metabolized about 18 times faster than the l-enantiomer. These large discrepancies make it very unlikely that in rats the reduction of cyt. P-450-substrate complex is rate-limiting. Furthermore, Table 3 demonstrates statistically significant differences between the magnitudes of type I spectra of the enantiomers in all cases. This could mean either that reductase stimulation is not as dependent on type I binding as is currently believed, or that the initial rate of reduction of cyt. P-450-substrate complex is not only dependent on the concentration of this complex.

DISCUSSION

The present results indicate that, in the Swiss mouse strain we use, stereospecific metabolism of hexobarbital takes place. In contrast to the results of McCarthy and Stitzel (15) the l-enantiomer was metabolized faster. Apparently strain differences in stereoselectivity exist. The linear Lineweaver-Burk plots obtained for hexobarbital metabolism in female mice suggest a completely ordered system, as discussed by Gillette et al. (30). The difference found in K_m correlates with a similar difference in K_s for type I binding. The similarity of both V_{max} and A_{max} of the enantiomers indicate that they interact with the same cytochrome(s) P-450. The K_m values of the enantiomers, if corrected for substrate disappearance, did not exceed K_s, as was also observed for dl-hexobarbital (data not given). Furthermore, the rates of metabolism and the enhancements of NADPH-cytochrome P-450 reductase activity at the same concentration did not differ for the enantiomers. This suggests that in female mice the reduction of the cytochrome P-450-substrate complex is rate limiting and, moreover, because K_m and K_s values are similar, that the type I binding site is the active site of the enzyme.

In rats, the situation with regard to stereoselectivity and sex difference is the opposite of that in mice. In this species the d-enantiomer is metabolized faster and males metabolize both enantiomers faster than females. The stereoselectivity is seen in both males and females, whereas Furner et al. (13) found it only in males. Their in vivo results, however, clearly demonstrate that in females the d-enantiomer is eliminated faster. In male rats we found a difference in both K_m and V_{max} for the enantiomers. A similar difference was found for the K_s of type I binding, although, even after correction for substrate disappearance K_m exceeded K_s. This difference may be due to a fast product dissociation, which influences K_m but not K_s. Moreover K_s may be concentration range dependant (31) and the difference between K_m and K_s might be caused by the different concentration ranges used for the determination of these constants. The large difference in rates of metabolism of the enantiomers were not parallelled by similar differences in the enhancements of NADPH cytochrome P-450 reductase activity, measured at the same substrate concentration. This indicates that the reduction of the cytochrome P-450-substrate complex is not rate limiting for the total reaction, which would also explain why the kinetic constants of type I binding and metabolism are different. In female rats only type I binding could be investigated. Here again, the faster metabolized enantiomer had a lower K_s, but, in contrast to the results in males, also a higher A_{max}. The latter observation suggests that the enantiomers bind to different populations of type I binding sites in female rats. Recently Grasdalen et al. (32) demonstrated heterogeneity of type I binding sites in male rats, which may occur in females as well. Moreover Degkwitz et al. (1969) and Feller and Lubaway (1973) reported A_{max} differences for d- and l-hexobarbital in phenobarbital induced male rats, which may also be explained by binding to different type I site subspecies.

Comparing the kinetics of type I binding in female and male rats, a sex difference in the composition of cytochrome P-450 is apparent. This is concluded from the sex differences in the respective A_{max} values, which are much larger than the sex difference in cytochrome P-450 (data not given), as also observed for racemic hexobarbital by Schenkman et al. (1) and by the presence of an A_{max} difference for the enantiomers in females, which is absent in males. On the basis of other evidence El Defrawy et al. (33) came to a similar conclusion. In all experiments significant differences were found between the magnitudes of the spectral changes induced by the hexobarbital enantiomers at the concentrations used for the reductase stimulation assays. However, only in female rats a concomitant, but relatively small difference in reductase stimulation was observed. This apparent discrepancy between type I binding and reductase stimulation seems to be at variance with the concept that by type I binding an additional amount of cyt. P-450 is involved in the first phase of reduction, thereby causing an increase in initial reduction rate (19,20). In support with this we have observed simultaneous variations of type I binding and reductase stimulation in mice, using ethylmorphine as a substrate (24). The present results, however, are not necessarily in conflict with these findings. For we have also observed, that different substrates, which elicit type I spectra of comparable size, can produce diverging reductase stimulations:

		reductase stim. (nmol/min/mg prot.)	spectral change ($\Delta A \times 10^3$/mg prot.)
Hexobarbital (dl)	1.0 mM	2.46 $\pm$ 0.15	16.9 $\pm$ 0.4
Ethylmorphine	1.0 mM	6.52 $\pm$ 0.05	17.4 $\pm$ 0.2

(means $\pm$ SE of 4 determinations)

This suggests that the reductase stimulation is not simply the expression of an additional amount of cyt. P-450 being involved in the first fast phase of reduction. It could very well be that substrates activate similar amounts of cyt. P-450, but that the rate constants of cyt. P-450-complex reduction differ. For the hexobarbital enantiomers such differences in rate constants could be opposed to differences in amounts of complex formed, which occasionally might lead to reductase stimulations of similar size. Alternatively, or perhaps additionally, the binding sites for reductase stimulation and type I binding might be distinct (although tightly coupled) and have different substrate affinities, as suggested by Holtzman and Rumack (3). A significant difference in reductase stimulation was found only in a case where A_{max} values were different (female rats). This might indicate that type I and activation sites are present in equal amounts (situated on the same cyt. P-450 species), but that binding to the activation sites is less or not stereoselective.

ACKNOWLEDGEMENTS

The author wish to thank Prof.Dr. J. Knabe, University of Saarland, Saarbrücken, Germany, for a generous supply of hexobarbital enantiomers and the Netherlands' Foundation for Medical Research (FUNGO) for financial support.

REFERENCES

(1) J.B. Schenkman, I. Frey, H. Remmer and R.W. Estabrook, Sex differences in drug metabolism by rat liver microsomes, Mol.Pharmac. 3, 516 (1967).
(2) D.S. Davies, P.L. Gigon and J.R. Gilette, Species and sex differences in electron transport systems in liver microsomes and their relationship to ethylmorphine demethylation, Life Sci. 8, 85 (1969).
(3) J.L. Holtzman and B.H. Rumack, Kinetic studies on the control of ethylmorphine N-demethylase. The role of ethylmorphine activation of reduced nicotinamide adenine dinucleotide phosphate-cytochrome P-450 reductase, Biochemistry 12, 2309 (1973).
(4) G.J. Mannering, Microsomal enzyme systems which catalyse drug metabolism, In: "Fundamentals of drug metabolism and drug disposition", Williams and Wilkins, Baltimore, p.206 (1971).
(5) J. Noordhoek, The effect of castration and testosterone on some components of the microsomal drug metabolising enzyme system in mice, FEBS Letters 24, 255 (1972).
(6) J. Noordhoek, Sex difference in drug metabolism and substrate binding to cytochrome P-450 in mice, Proc.Europ.Soc. Drug Tox., Excerpta Medica, Amsterdam, p.99 (1975).
(7) J. Noordhoek, A.P. Van den Berg, E.M. Savenije-Chapel and E. Koopman-Kool, Sex differences in kinetics of ethylmorphine demethylation and type I binding to cytochrome P-450 in mice of different strains, Abstr. 916, Sixth Intern.Congr.Pharmac., Helsinki (1975).
(8) P. Jenner and B. Testa, The influence of stereochemical factors on drug disposition, Drug Metab.Rev. 2, 117 (1973).
(9) D.S. Hewick and J.R. Fouts, The metabolism in vitro and hepatic microsomal interaction of some enantiomeric drug substrates, Biochem.J. 117, 833 (1970).
(10) M.W. Anders, M.J. Cooper and A.E. Takemori, Kinetics of microsomal metabolism and binding of enantiomerically related substrates, Drug Metab. Disp. 1, 642 (1973).
(11) N. Gerber, R. Lynn, R. Holcomb, W.L. Weller, M.T. Bush, The metabolism of hexobarbital in mice and methodology for isolation and quantitation of its metabolites in vivo and in vitro, J.Pharmac.Exptl.Therap. 177, 234 (1971).
(12) R.R. Holcomb, N. Gerber and M.T. Bush, The metabolic fate of hexobarbital in the rat, J.Pharmac.Exptl.Therap. 188, 15 (1974).
(13) R.L. Furner, J.S. McCarthy, R.E. Stitzel and M.W. Anders, Stereoselective metabolism of the enantiomers of hexobarbital, J.Pharmac.Exptl.Therap. 169, 153 (1969).
(14) E. Degkwitz, V. Ullrich and H. Staudinger, Metabolism and cytochrome P-450 binding spectra of (+) and (-)-hexobarbital in rat liver microsomes, Hoppe-Seyler's Z.Physiol.Chem. 350, 547 (1969).
(15) J.S. McCarthy and R.E. Stitzel, Kinetic differences in the microsomal metabolism of the isomers of hexobarbital, J.Pharmac.Exptl.Therap. 176, 772 (1971).
(16) D.R. Feller and W.C. Lubaway, Interaction of hexobarbital enantiomers with rat liver microsomes, Pharmacology 9, 129 (1973).
(17) D.D. Breimer and J.M. Van Rossum, Pharmacokinetics of (+), (-) and (+) hexobarbitone in man after oral administration, J.Pharm.Pharmac. 25, 762 (1973).
(18) K. Gundermann, E. Degkwitz and H. Staudinger, Mischfunktionelle Oxygenierung von (+) und (-)-Hexobarbital und spektrale Änderungen des Cytochroms P-450 in der Leber ascorbinsäurefrei ernähter Meerschwein-

chen, Hoppe-Seyler's Z.Physiol.Chem. 354, 238 (1973).
(19) P.L. Gigon, T.E. Gram and J.R. Gilette, Studies on the rate of reduction of hepatic microsomal cytochrome P-450 by reduced nicotinamide adenine dinucleotide phosphate, Mol.Pharmac. 5, 109 (1969).
(20) H. Diehl, J. Schädelin and V. Ullrich, Studies on the kinetics of cytochrome P-450 reduction in rat liver microsomes, Hoppe-Seyler's Z. Physiol.Chem. 351, 1359 (1970).
(21) M.T. Bush and W.L. Weller, Metabolic fate of hexobarbital, Drug Metab. Rev. 1, 249 (1972).
(22) D.D. Breimer and J.M. Van Rossum, Rapid and sensitive gas chromatographic determination of hexobarbital in plasma of man using a nitrogen detector, J.Chromatogr. 88, 235 (1974).
(23) B.B. Brodie, J.J. Burns, L.C. Mark, P.A. Lief, E. Bernstein and E.M. Papper, The fate of pentobarbital in man and dog and a method for its estimation in biological material, J.Pharmac.Exptl.Therap. 109, 26 (1953).
(24) A.P. Van den Berg, J. Noordhoek, E.M. Savenije-Chapel and E. Koopman-Kool, Sex differences in hepatic microsomal drug metabolism in the mouse. The roles of type I binding and NADPH-cyt. P-450 reductase in ethylmorphine N-demethylation (submitted for publication).
(25) O.H. Lowry, N.J. Rosebrough, A.L. Farr and R.J. Randall, Protein measurement with the Folin phenol reagent, J.Biol.Chem. 193, 265 (1951).
(26) G.N. Wilkinson, Statistical estimations in enzyme kinetics, Biochem.J. 80, 324 (1961).
(27) H.-J. Lee and I.B. Wilson, Enzymic parameters: Measurement of V and K_m, Biochim.Biophys.Acta 242, 519 (1971).
(28) D.S. Sitar and G.J. Mannering, Determination of apparent kinetic constants of the microsomal hydroxylation of amobarbital, hexobarbital and pentobarbital, Drug Metab.Disp. 1, 663 (1973).
(29) D. Kupfer and J. Rosenfeld, A sensitive radioactive assay for hexobarbital hydroxylase in hepatic microsomes, Drug Metab.Disp. 1, 760 (1973).
(30) J.R. Gilette, H. Sasame and B. Stripp, Mechanism of inhibition of drug metabolic reactions, Drug Metab.Disp. 1, 164 (1973).
(31) J.B. Schenkman, Studies on the nature of the type I and type II spectral changes in liver microsomes, Biochemistry 9, 2081 (1970).
(32) H. Grasdalen, D. Bäckström, L.E.G. Eriksson, A. Ehrenberg, P. Moldéus, C. Von Bahr and S. Orrenius, Heterogeneity of cytochrome P-450 in rat liver microsomes: Selective interaction of metyrapone and SKF 525-A with different fractions of microsomal cytochrome P-450, FEBS Letters 60, 294 (1975).
(33) S. El Defrawy El Masry, G.M. Cohen and G.J. Mannering, Sex-dependent differences in drug metabolism in the rat. I. Temporal changes in the microsomal drug metabolising system of the liver during sexual maturation, Drug Metab.Disp. 2, 267 (1974).

THE MECHANISM OF DEGRADATION OF ENDOGENOUS HEME AND CYTOCHROME P-450 BY HEME OXYGENASE

Mahin D. Maines

The Rockefeller University, New York, N.Y. 10021, U.S.A.

ABSTRACT

1. Endogenous heme is metabolized both in vivo and in vitro by the microsomal heme oxygenase (MHO) system. Endogenous heme includes both the heme already incorporated into microsomal cytochromes as well as the "free heme" i.e., the heme not committed as a prosthetic group of cytochromes or enzymes.

2. The conversion of cytochrome P450 to P420 is a pre-requisite for the metabolism of its heme prosthetic group by the MHO.

3. It is suggested that metals may increase the rate of microsomal P450 degradation not only by inducing MHO but also by increasing the rate of conversion of cyt P450 to P420. The latter is accomplished by a) increasing the cellular content of a native "denaturant" of cyt P450; b) modifying the affinity of apo-P450 for heme; c) metal direct denaturing of the cytochrome.

INTRODUCTION

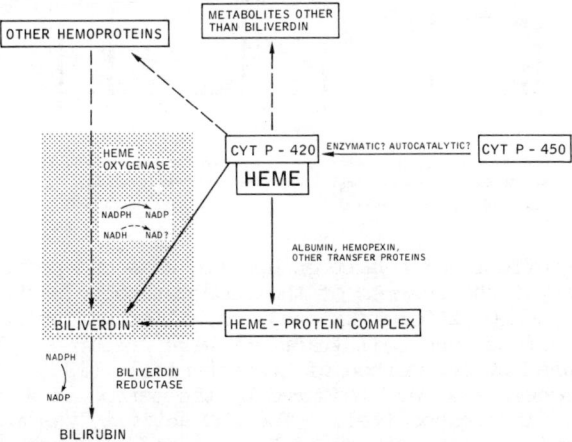

Fig. 1 Proposed mechanism of microsomal heme metabolism

Several years ago Maines and Anders (1,2) proposed that in order for the heme of cyt P450 to be degraded the cytochrome must be first converted to its nonreactive form, cyt P420, the heme of which would then be oxidatively degraded by the microsomal heme oxygenase (MHO) system to form biliverdin. The latter would, in turn, be reduced by the action of biliverdin reductase to bilirubin. The details of this scheme are shown in Fig. 1. The heme of cyt P420 may also be re-utilized by other hemoproteins or metabolized to products other than biliverdin.

In subsequent studies the observations of Maines and Kappas (3-11) as to the response of MHO and of microsomal heme and hemoprotein to a number of metals led to the proposal that MHO is involved in the endogenous breakdown of cellular heme (5,6). In these studies it was discovered that MHO responds to various metals of the first transition series and certain heavy metals by a rapid increase in its production. The induction of MHO activity was then shortly followed by a reduction of microsomal content of heme (3,4). The initial reduction in microsomal heme (0-4 hr) as shown in Fig. 2 was apparently mostly due to the inhibition of δ-aminolevulinate synthetase (ALAS), the rate limiting enzyme of heme biosynthesis (12). However, the reduction in the contents of heme and cyt P450 at later periods (4-24 hr) was most likely only due to the induced MHO activity, since cellular heme biosynthesis at this time was at normal or above normal levels.

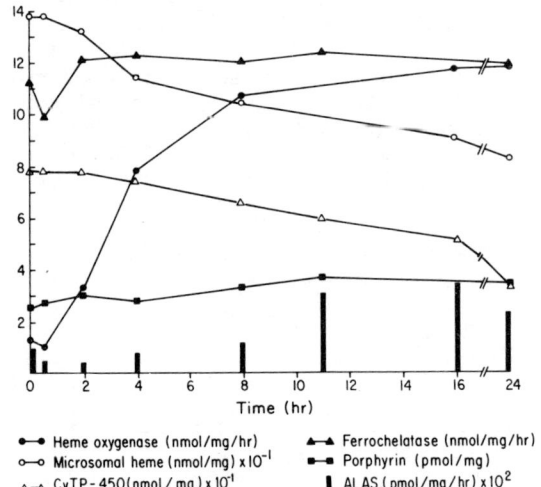

Fig. 2 Modifying effects of cobalt on hepatic heme and porphyrin contents and the activities of the enzymes of the heme metabolic pathways. Rats were treated with $CoCl_2 \cdot 6H_2O$ (250μmol/kg, sc) and killed at indicated intervals. The livers were perfused and cell fractions were prepared. The porphyrin content was measured by the method of Granick et al. (13). The activities of ALAS and ferrochelatase were measured by the method of Marver et al. (14) and Maines et al. (15) respectively. The MHO activity was measured as detailed previously (4). The methods of Paul et al. (16) and Omura and Sato (17) were used for heme and P450 measurements respectively.

In this report further evidence for degradation of cyt P450 by MHO is provided; also experimental data supporting the hypothesis that the conversion of P450 to P420 is necessary for its degradation is presented. In addition evidence for the degradation of "free heme" i.e., heme prior to its binding with various cytochromes by MHO is presented.

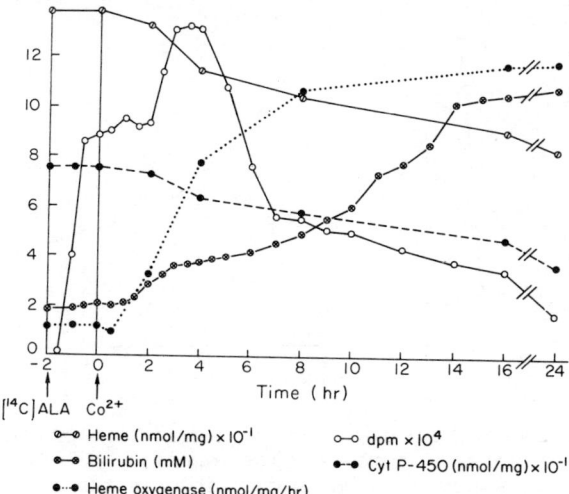

Fig. 3 The effect of induction of MHO on the microsomal contents of heme and hemoprotein and the biliary excretion of degradation products of microsomal heme. The experimental details are provided in the text.

In one series of experiments shown in Fig. 3 in vivo metabolism of microsomal heme and cyt P450 was studied. In these experiments hepatic microsomal heme of bile fistula rats was pulse labelled with 4-^{14}C ALA (8μci/100 g, iv) prior to the induction of MHO with Co^{2+} (250μmol/kg, sc). The bile was collected (in the dark, 0°) and analyzed for bilirubin content and ^{14}C-activity. These data were correlated with studies in which rats were treated with Co^{2+} and then killed at indicated intervals and the microsomal contents of cyt P450 and heme and the activity of MHO were measured. As shown, 2 hr after Co^{2+} injection MHO activity rapidly increased and approached its maximum after 8 hr. Concomitant with this increased enzyme activity were alterations in the microsomal contents of heme, cyt P450 and the constituents of bile samples described as follows: During the period of rapidly increasing MHO activity (2-8 hr) there was an abrupt transient increase in the biliary excretion of ^{14}C and bilirubin. The microsomal cyt P450 and heme contents significantly decreased. The decrease in microsomal P450 content and the increase in biliary bilirubin excretion continued for 24 hr. As expected for a pulse labelled study, the amount of radioactivity excreted into the bile progressively declined. These findings strongly suggest that the elevation of MHO activity had resulted in the degradation of microsomal heme and the heme moiety of cyt P450, leading to increased bilirubin excretion. It is reasonable to assume on the basis of these findings that the bulk of the ^{14}C excreted in the bile was derived from the breakdown of cyt P450 heme, since this cytochrome has the greatest requirement for heme synthesis.

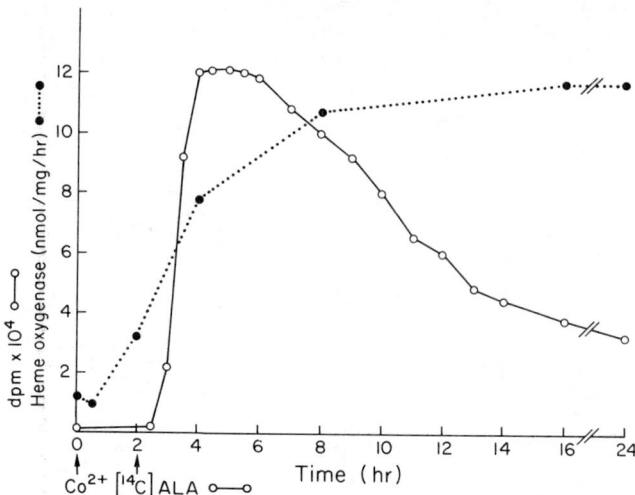

Fig. 4 Patterns of the biliary excretion of degradation product(s) of microsomal heme when the microsomal heme is pulse labelled following the induction of heme oxygenase. Rats were treated as described in the text for the experiment shown in Fig. 3 except that the sequence of injections of Co^{2+} and ^{14}C-ALA were reversed.

The nature of the association of the heme, which gave rise to the ^{14}C excreted in the bile, with the membranes of the endoplasmic reticulum was further investigated. Studies were designed to investigate whether the ^{14}C activity measured in the bile originated from the heme which was already incorporated into the microsomal membranes, i.e., heme formed prior to the induction of MHO activity; or originated from newly synthesized heme, i.e., the heme synthesized subsequent to induction of enzyme. In order to examine this question, as well as to unequivocally establish that the degradation of cyt P450 follows the induction of MHO, the biliary patterns of ^{14}C excretion from animals in which microsomal heme was labelled prior to the induction of MHO (Fig. 3), and that of animals in which the labelling of the microsomal heme was performed after the induction of MHO (2 hr after Co^{2+}) were compared (Fig. 4).

As shown, the patterns of biliary excretion of ^{14}C under the 2 experimental conditions were quite different. In the first experiment (Fig. 3) the rate of ^{14}C excretion was essentially unchanged for 2 hr following Co^{2+}. This lag period closely corresponded with the lenght of time which was required for MHO to attain high levels of activity. In contrast, when the activity of MHO had already been induced by the metal prior to the labelling of microsomal heme, this lag period was essentially absent (Fig. 4). Therefore it may be concluded that the decrease observed in heme content following MHO induction was due to the accelerated rate of degradation of preformed hemoprotein i.e. cty P450, as well as to the degradation of newly formed heme i.e. "free heme" before it was incorporated into the membranes of the endoplasmic reticulum.

The in vitro degradation of microsomal hemoprotein by MHO was studied (Fig. 5). In these experiments the microsomal contents of heme and hemoproteins of one group of rats were increased by phenobarbital treatment. 12 hr before sacrifice the microsomal heme was labelled with 10μci of ^{14}C-ALA. The MHO activity of a second group of rats was increased by treatment with Co^{2+}. The third group of animals served as the controls. The animals were sacrificed, the livers were perfused exhaustively in situ and the microsomal fractions were prepared from all three groups of animals. An aliquot of the microsomal fraction from the phenobarbital treated animals was treated with steapsin in order to convert cyt P450 to cyt P420. The details of this procedure for the preparation of "Co-binding pigment" (P420) are described elsewhere (18). The MHO activity was assayed utilizing the following combinations of preparations as the source of enzyme (E) and substrate (S): E=Co^{2+} induced microsomal preparation with (S) = microsomal fraction of phenobarbital treated (cyt P450) or (S) = cyt P420 prepared from the phenobarbital treated animals; E = control microsomal fraction with (S) = cyt P450 or (S) = cyt P420. Appropriate control assay mediums were utilized. The assay medium for MHO activity was the same as described before (4) except that it did not contain any exogenous heme. At the end of the incubation period (30 min, 37°) the bilirubin formed in the assay medium was extracted with $CHCl_3$. The absorption spectrum of the $CHCl_3$ layer was recorded and aliquots were removed for measurement of ^{14}C activity.

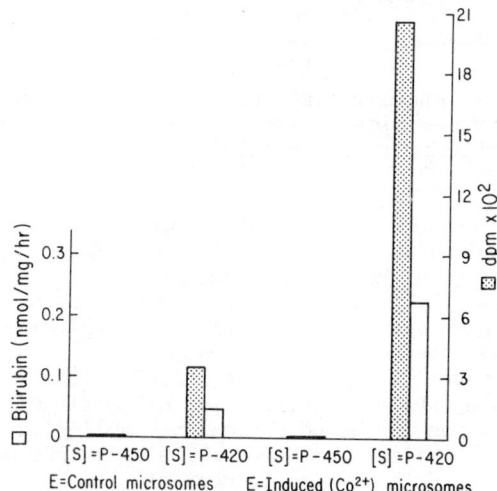

Fig. 5 In vitro degradation of microsomal hemoproteins by MHO activity. Experimental details are given in the text.

Figure 5 shows evidence for the in vitro degradation of the heme of P420 as measured by the amount of bilirubin formed by MHO activity and the amount of ^{14}C recovered in the bilirubin fraction. As shown cyt P420 served as substrate for both the induced and the control MHO. The amount of bilirubin formed by the induced preparation was 4-5 times greater than that formed by the control enzyme; also the ^{14}C activity of the $CHCl_3$ fraction was several fold (6-7x) greater when utilizing the induced enzyme. The observation that the relative amount of ^{14}C recovered in $CHCl_3$ was greater than that of the spectrally measured bilirubin could indicate the formation of some non-bilirubin heme degradation products. No ^{14}C activity or bilirubin were detected in the $CHCl_3$ extract of the MHO assay medium when using cyt P450 as the source of substrate. These findings indicate that the conversion of P450 to P420 is required for the metabolism of cyt P450.

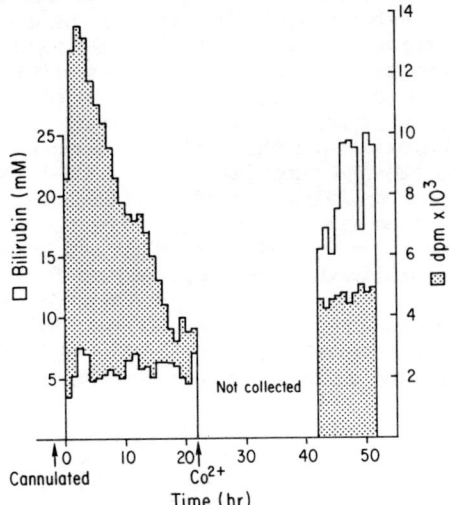

Fig. 6 Further studies demonstrating the degradation of microsomal hemoprotein by the heme oxygenase system. The heme moiety of hemoglobin was labelled with $2-^{14}C$-glycine and animals were treated as detailed in the text.

Further studies (Fig. 6) were carried out in order to unequivocally establish that the origin of the bilirubin excreted in the bile as the response to induction of MHO is microsomal heme. In these studies the heme moiety of hemoglobin was labelled and the patterns of biliary ^{14}C and bilirubin excretion were measured. The animals were treated with $2-^{14}C$-glycine (8μci/rat for 3 days, sc); after 5 days the bile duct was cannulated and hourly collection of bile samples were made. 24 hr after the cannulation of the bile duct, Co^{2+} was administered and bile collection was discontinued for 20 hr during which the MHO activity was increased maximally. Following this period, the hourly collection of the bile was resumed. The bile thus collected was analyzed for ^{14}C activity and the bilirubin content.

As Fig. 6 shows, the patterns of biliary excretion of bilirubin and of ^{14}C were totally dissimilar. As indicated prior to Co^{2+} injection, the bilirubin content of the samples was significantly less than that of the samples collected after MHO activity was induced. Conversely, the ^{14}C activity of the bile samples was significantly greater immediately following bile duct cannulation (before Co^{2+}). The distinct patterns of excretion of ^{14}C and formation of bilirubin indicate that these two components originate from different sources. It is apparent that the origin of the initial ^{14}C peak was the heme derived from the breakdown of R.B.C. hemoglobin. Accordingly, the secondary bilirubin peak most likely originated from the breakdown of microsomal hemoproteins. Following the surgical operation the blood released from the damaged vessels at the site of operation is the apparent source of the hemoglobin.

The data presented in Fig. 6 also provide some evidence for the increase in rate of degradation of non-microsomal heme following the induction of MHO. As shown, the ^{14}C activity of the bile samples following the initial peak, and prior to the Co^{2+} treatment, reached a plateau level which was maintained for several hours (20-24 hr after operation). However, the ^{14}C activity of bile samples after the induction of MHO (24-48 hr) was slightly but consistently elevated, and this elevation was maintained throughout the experiment. This finding implies that although the bilirubin derived from the accelerated rate of degradation of the cellular hemoproteins constitutes the bulk of the bilirubin peak observed following the induction of MHO, bilirubin originating from a somewhat accelerated rate of hemoglobin heme degradation also may have contributed to this peak. As the half-life of hemoglobin of red blood cells by far exceeds that of cytochromes (60 days vs. 7-48 hr) an overall increase in the rate of hemoglobin degradation could be expected to produce only a slight but steady increase in the rate of ^{14}C excretion. Of course since the induction of MHO by Co^{2+} takes place in many tissues (6) the rate of hemoglobin heme degradation most likely was elevated in a number of tissues and the source of ^{14}C activity recovered in the bile samples could be the hemoglobin degraded at various tissue sites.

DISCUSSION

The cytoplasmic hemoproteins such as tryptophane pyrrolase, cytochromes P450 cyt b_5 and catalase have half-lives which range from 2 to nearly 50 hr; that of cyt P450 is a biphasic one, ranging from approximately 7 to 16 hrs. Of the cytochromes mentioned, P450 is by far the most prevalent cytoplasmic hemoprotein with its cellular concentration being nearly twice that of b_5 and 150 times that of tryptophane pyrrolase. Furthermore, cyt P450 is converted to an inactive form, cyt P420; the latter form has been shown to be a reconstitutable cytochrome (18) probably consisting of a group of hemoproteins. Several years ago, on the basis of the relative affinities of cyt P450, P420 and the plasma heme binding proteins for heme (P450>hemopexin>albumin>P420) Maines and Anders suggested that the microsomal hemoprotein P450 is degraded by the MHO system; and it was further suggested that in order for this reaction to occur P450 must be first converted to P420, the apo-protein of which has a lower binding affinity for heme. Earlier evidence for the catabolism of endogenous heme by the MHO system has been provided (5,6) including the data presented here showing that in the presence of normal levels of heme synthesis the microsomal heme and cyt P450 contents are dramatically decreased. Additional direct evidence for these phenomena is provided here in studies in which the turnover of labelled microsomal heme and hemoglobin heme were monitored in bile fistula rats; and the oxidation of heme of cyt P420 to bilirubin but not that of cyt P450 by MHO were demonstrated. Moreover, these studies demonstrated that not only heme derived from P450 is degraded by the MHO system, but that cellular "free heme" i.e., heme prior to its incorporation into the apo-protein of various hemoproteins, is also a substrate for MHO. Since heme in biological systems is not present in unbound form, that is, the molecule is always associated with carrier proteins or peptides, in order for MHO to oxidize this heme, the enzyme must have a higher affinity for the heme molecule than the carrier moiety. Moreover, since as shown here, P420 heme is oxidized by the MHO system, but P450 heme is not, the affinity of the MHO for heme must be greater than that of the apo-P420, but less than that of apo-450. Therefore, in order for cyt P450 to be degraded in vivo by MHO, the cytochrome must be first converted to P420. The differential affinities of the proteins for heme discussed here necessitate the existence of an intermediary factor in the pathway of cyt P450 degradation.

This intermediary factor must be a P450 denaturing agent or a process which facilitates the conversion of P450 to P420. The increased rate of microsomal P450 degradation produced by metal ions and other inducers of MHO thus could be not only due to an increased rate of heme degradation, but must also involve an increased rate of conversion of P450 to P420.

The accelerated rate of conversion of P450 to P420 could be brought about through the following mechanisms: 1) by increasing the concentration of a cellular component which is the natural "denaturant" of P450. This increase could be brought about by increased synthesis of this "substance" or by the release of the pre-synthesized "denaturant" from a cellular compartment (e.g., lysosomes); 2) Co^{2+} directly interferes with proper binding of heme to the apo-P450 which could result in the formation of an aberrant form of the cytochrome, i.e., an apo-protein which binds heme less tightly than the unaltered apo-P450. This alteration could be brought about by the covalent binding of Co^{2+} at or near the receptor sites for the heme thus changing the protein configuration or the electronic potential of the heme binding site. 3) Lastly, the conversion of P450 to P420 could be accelerated through a direct effect of the metal on the hemoprotein causing its denaturation. Whichever the case may be, the data presented here unequivocally show that the heme of microsomal hemoproteins such as P450 is degraded by MHO and that for this to occur the heme of the hemoprotein must be labilized so that its affinity for MHO becomes greater than that for the protein to which it is bound.

REFERENCES

1. Maines, M. D. and Anders, M. W., in Microsomes and Drug Oxidations, The Williams & Wilkins Co., Baltimore, Md., 1973.
2. M. D. Maines, M. W. Anders and U. Muller-Eberhard, Mol. Pharm. 10, 204 (1974).
3. M. D. Maines and A. Kappas, Proc. Natl. Acad. Sci. USA 71, 4293 (1974).
4. M. D. Maines and A. Kappas, J. Biol. Chem. 250, 4171 (1975).
5. Maines, M. D. and Kappas, A., in Porphyrins in Human Disease, S. Karger Publ. Co., Basel, in press, 1976.
6. M. D. Maines and A. Kappas, Ann. Clin. Res. 8 (Suppl. 17), 39 (1976).
7. M. D. Maines and A. Kappas, Biochem. J. 154, 125 (1976).
8. M. D. Maines and A. Kappas, Proc. Natl. Acad. Sci. USA, in press (1976).
9. M. D. Maines and A. Kappas, J. Expt. Med. 141, 1400 (1975).
10. A. Kappas and M. D. Maines, Science 192, 60 (1976).
11. M. D. Maines, P. Sinclair and A. Kappas, J. Biol. Chem., in press (1976).
12. S. Granick and G. Urata, J. Biol. Chem. 238, 821 (1963).
13. S. Granick, P. Sinclair, S. Sassa and G. Grieninger, J. Biol. Chem. 250 9215 (1975).
14. H. S. Marver, D. P. Tschudy, M. G. Perlroth and A. Collins, J. Biol. Chem. 241, 2803 (1966).
15. M. D. Maines, V. Janoušek, J. M. Tomio and A. Kappas, Proc. Natl. Acad. Sci. USA 73, 1499 (1976).
16. K. G. Paul, H. Theorell and A. Akeson, Acta Chem. Scand. 7, 1284 (1953).
17. T. Omura and R. Sato, J. Biol. Chem. 239, 2370 (1964).
18. M. D. Maines and M. W. Anders, Mol. Pharm. 9, 219 (1973).

INDUCTION OF CYTOCHROME P-448 BY 3-METHYLCHOLANTHRENE IN THE RAT DURING INHIBITION OF PROTEIN SYNTHESIS *IN VIVO* *

Georg F. Kahl, Bernd Zimmer, Teresa Galinsky,
Hans G. Jonen and Regine Kahl

Department of Pharmacology, University of Mainz, Obere Zahlbacher Str. 67, D 6500 Mainz

ABSTRACT

Administration of cycloheximide in vivo during induction of rats with 3-methylcholanthrene prevents the increase in total cytochrome P-450 content usually seen under the influence of the inducer. The population of cytochromes P-450 in the livers of these animals is, however, similar to that in the completely induced animals. Microsomal aryl hydrocarbon hydroxylase activity and biphenyl-2-hydroxylation are enhanced severalfold and biphenyl-4-hydroxylation is enhanced twofold. Monooxygenase activity shows the same pattern of preferential inhibition as in microsomes from animals which had received the inducer only. The affinity of the reduced cytochromes for the ligand metyrapone is considerably decreased both after treatment with 3-methylcholanthrene and with 3-methylcholanthrene plus cycloheximide. A distinct blue shift of the absorbance maxima of the reduced P-450-CO complex and of the reduced P-450-metyrapone complex occurs after concomitant treatment with the inducer and the inhibitor of protein synthesis. The pattern of cytochrome P-450 bands after SDS polyacrylamide gel electrophoresis of microsomes from animals treated with both drugs is similar to that of MC-stimulated microsomes but different from that of control microsomes. These data show that during inhibition of protein synthesis in vivo 3-methylcholanthrene induces the same pattern of cytochromes P-450 as under normal conditions of induction while the increase in total cytochrome P-450 content is almost completely prevented.

INTRODUCTION

The changes in spectral and enzymic properties of the hepatic drug metabolizing enzyme system which occur after in vivo administration of polycyclic hydrocarbons have been ascribed to the formation of a specific hemoprotein, cytochrome P-448, by enzyme induction. First evidence for the dependence of the formation of cytochrome P-448 on de novo protein synthesis came from studies using inhibitors of protein synthesis in which the characteristic increase in total cytochrome P-450 content could no longer be found after 3-methylcholanthrene (MC) treatment (1). Later, the presence of induction-specific RNA was observed in cell cultures exposed to polycyclic hydrocarbons (2), and in the intact animal transcriptional events (3) as well as posttranscriptional processes (4) have been shown to follow MC administration.

---

* This study was supported by the Deutsche Forschungsgemeinschaft, Bonn-Bad Godesberg.

Partial purification of the cytochrome P-450 of liver microsomes from animals treated with a polycyclic hydrocarbon reveals the predominance of a hemoprotein with characteristic spectral properties and electrophoretic mobility (5) indicating that indeed induction-specific protein is formed under the influence of the inducer. This is further substantiated by immunochemical evidence obtained with an antibody against cytochrome P-448 which is now available (6).

On the other hand, it has been discussed if conformational changes of pre-existing cytochrome P-450 or alterations in the membrane environment of the hemoprotein may contribute to the changes seen in liver microsomes after MC treatment of the animals (7, 8).

This study is based on previous results in which differences in time course and sensitivity to the inhibitor of protein synthesis, cycloheximide (CH), suggested that certain changes in spectral properties of liver microsomes due to MC administration may not be connected with the synthesis of cytochrome P-448 (8). We now investigate the influence of cycloheximide on a number of parameters which are closely related to the formation of cytochrome P-448.

METHODS

Male Sprague-Dawley rats (150 g body weight) received 3 i.p. injections of 20 mg MC/kg dissolved in peanut oil in 12 h intervals. The last injection was given 36 h prior to sacrification. When the effect of inhibition of protein synthesis was to be studied, 1 mg CH/kg in saline was administered i.p. 3 h prior to each MC injection. Control animals received peanut oil plus saline or plus cycloheximide, respectively.

The preparation of microsomes (9), spectral measurements (9) and determination of benzpyrene hydroxylase (10), and biphenyl hydroxylase (11) were performed as described previously. SDS polyacrylamide gel electrophoresis was performed in a Desaga slab gel electrophoresis apparatus according to Laemmli (12) using a 10 % separating gel. The gels were fixed and stained overnight in a 10 % acetic acid/25 % isopropyl alcohol solution containing 0.015 % Coomassie Blue and were destained in a diffusion chamber using a 10 % acetic acid/10 % isopropyl alcohol solution.

RESULTS AND DISCUSSION

Table 1 demonstrates that the increase in total cytochrome P-450 content usually seen after pretreatment with MC can be prevented by the concomitant application of an inhibitor of protein synthesis during the induction period. An investigation of the properties of the cytochrome P-450 population in hepatic microsomes from animals in which MC induction has been inhibited by cycloheximide does, however, reveal that the characteristic features of MC induction are still present though an increase in total cytochrome P-450 content does not occur.

Table 1 shows that
a) a blue shift of the absorbance maximum of the reduced cytochrome P-450-CO complex by 1.5 nm is observed after combined MC + CH treatment as compared to a 2.5 nm blue shift after noninhibited MC induction,

b) a blue shift of the absorbance maximum of the reduced cytochrome P-450-metyrapone (MP) complex by 1.5 nm occurs after combined MC + CH treatment as compared to a 2 nm shift after noninhibited MC induction,

TABLE 1 Cytochrome P-450 content and spectral properties of rat liver microsomes after treatment with 3-methylcholanthrene and after combined treatment with 3-methylcholanthrene plus cycloheximide

Pretreatment	cytochrome P-450 content (nmoles/mg prot.)	λ_{max} of CO complex (nm)	K_s for metyrapone in reduced microsomes (μM)	λ_{max} of metyrapone complex (nm)
Oil	0.77 ± 0.04	449.5 ± 1.2	97*	445.1 ± 0.7
Oil + CH	0.46*	449.5*	122*	445.0*
MC	1.80 ± 0.11	447.0 ± 1.0	656 ± 132	442.9 ± 0.7
MC + CH	0.90 ± 0.08	448.0 ± 0.7	359 ± 86	443.6 ± 1.0

Values are means ± S.E.M. of 3 – 5 independent microsomal preparations. * n = 2.

c) the affinity loss of the reduced cytochromes for MP which is typical for MC-stimulated microsomes (9) can also but to a lesser extent be found after combined MC + CH treatment.

This is further substantiated by examination of the total amount and the differential inhibition of the enzymic activities in MC + CH microsomes. Figure 1 shows that after combined treatment with MC and CH aryl hydrocarbon hydroxylase (AHH) activity is induced to a marked extent (left part). Moreover, the specific inhibitor of cytochrome P-448-mediated drug metabolism, α-naphthoflavone (ANF) does also effectively inhibit benzpyrene hydroxylation in MC + CH microsomes. In contrast, constitutive AHH activity does not respond to the inhibitory action of ANF (Fig. 1, right part).

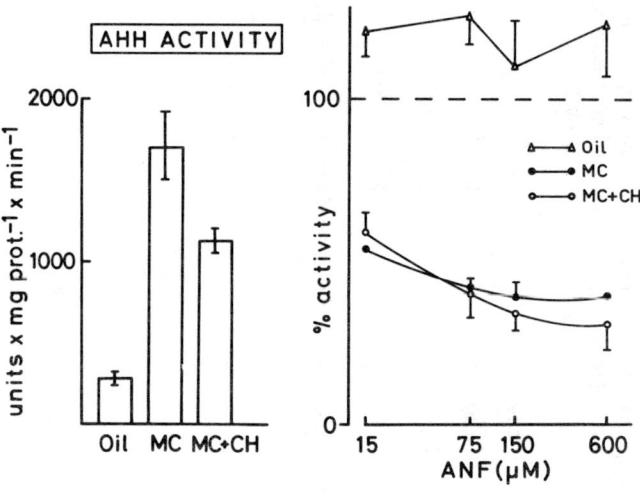

Fig. 1. Influence of cycloheximide treatment on induction of AHH activity by 3-methylcholanthrene and on inhibition of the induced activity by α-naphthoflavone. Benzo(a)pyrene 0.05 mM; 0.2 mg microsomal protein/ml.

Marked induction and preferential inhibition by ANF are also observed for biphenyl-2-hydroxylation and biphenyl-4-hydroxylation after combined treatment with MC + CH indicating that cytochrome P-448 must have been formed even during inhibition of protein synthesis in spite of the lack of increase in total cytochrome P-450 content (Fig. 2 and 3).

The different characteristics of enzymic activity in control and MC + CH microsomes are especially prominent in the case of biphenyl-2-hydroxylation (Fig. 2) which can be markedly activated in vitro by the addition of ANF in control microsomes. This phenomenon has been described previously for a number of carcinogens which are all able to enhance biphenyl-2-hydroxylation in vitro (13). In contrast, no activation but effective inhibition of this reaction by ANF is found after combined treatment with MC + CH.

Constitutive AHH activity has been shown to be sensitive to the inhibitor MP and a number of other agents which do not effect AHH activity in MC treated animals (14).

Induction of Cytochrome P-448

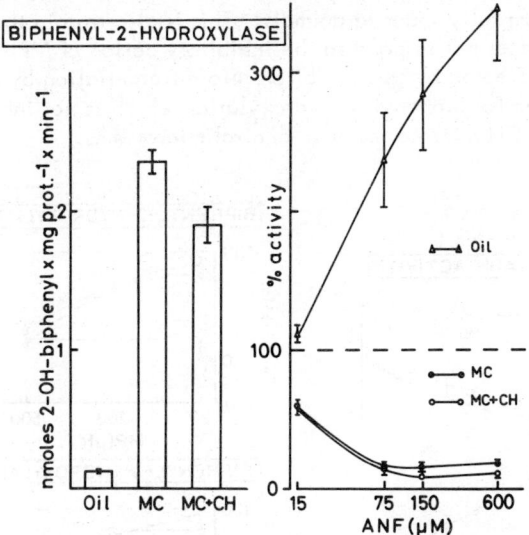

Fig. 2. Influence of cycloheximide treatment on induction of biphenyl-2-hydroxylation by 3-methylcholanthrene and on inhibition of the induced activity by α-naphthoflavone.
Biphenyl 2 mM; 1 mg microsomal protein/ml.

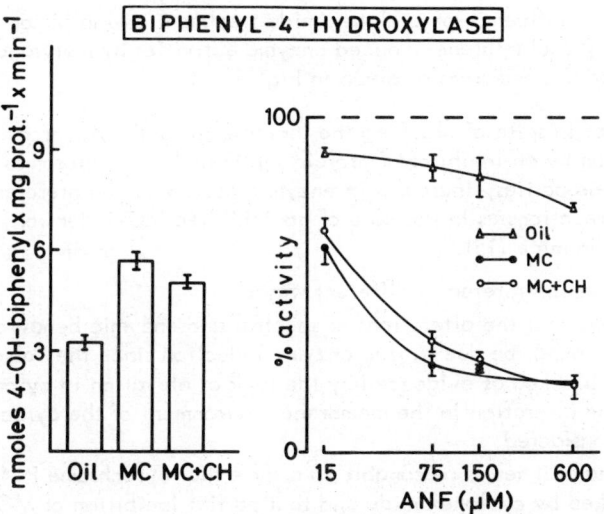

Fig. 3. Influence of cycloheximide treatment on induction of biphenyl-4-hydroxylation by 3-methylcholanthrene and on inhibition of the induced activity by α-naphthoflavone.
Biphenyl 2 mM; 1 mg microsomal protein/ml.

Figure 4 demonstrates that AHH activity in MC + CH microsomes is as insensitive to MP as it is in microsomes from fully induced animals. This is also true in the case of biphenyl-2-hydroxylation which does not respond to the inhibitory action of MP while the constitutive activity is markedly affected by this inhibitor. No differentiation by means of MP inhibition can be achieved for biphenyl-4-hydroxylation which is not inhibited either in MC microsomes and MC + CH microsomes or in control microsomes.

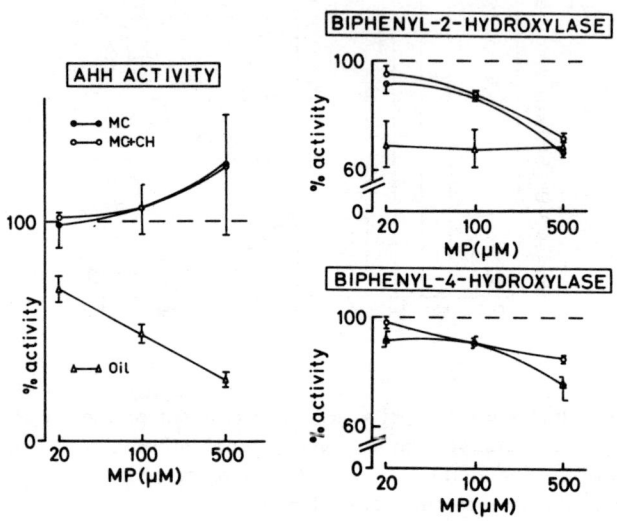

Fig. 4. Influence of cycloheximide treatment on inhibition of 3-methylcholanthrene-induced enzyme activities by metyrapone. Incubation conditions as given in Fig. 1 - 3.

These data show that in spite of blocking the increase in total cytochrome P-450 content after MC application by an inhibitor of protein synthesis liver microsomes show a similar pattern of spectral properties, increases in enzymic activities and preferential inhibition of monooxygenase reactions as in the case of noninhibited MC-induction. Similar results were also obtained in mice (15).

These findings can be interpreted in different ways:

1) One might argue that the alterations of spectral and enzymic behaviour seen after treatment with MC cannot be due to true enzyme induction since they also occur when protein synthesis is blocked as evidenced by the lack of elevation in cytochrome P-450 content. Instead, an alteration in the membrane environment of the cytochrome by the inducer might be considered.

2) On the other hand it seems reasonable to assume that cytochrome P-448 synthesis is only partially blocked by cycloheximide and that partial inhibition of MC-induced cytochrome P-448 synthesis and partial inhibition of noninduced synthesis of constitutive cytochrome P-450 sum up to give cytochrome P-450 levels near the control level while the relative proportions of the different cytochromes resemble that in the completely induced animal.

3) Finally it might be argued that formation of cytochrome P-448 does not occur by de novo synthesis but by the modification of preexisting hemoprotein in a posttranslational step not influenced by cycloheximide.

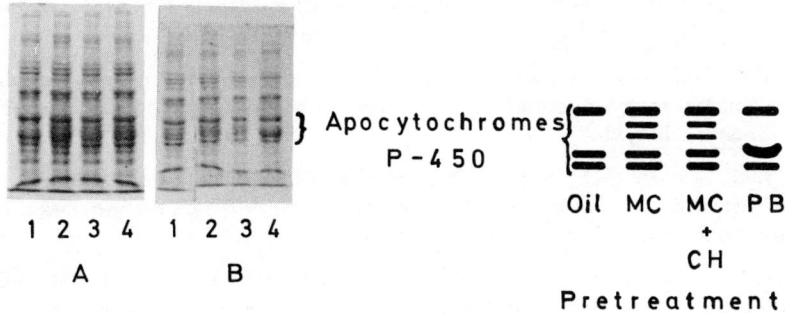

Fig. 5. SDS polyacrylamide gel electrophoretic pattern of microsomal apocytochromes P-450 after inhibition of protein synthesis by cycloheximide.
Gel A: 1, oil control; 2, MC; 3, 4, MC + CH
Gel B: 1, 2, 3, MC+CH; 4, phenobarbital (PB).
Samples of 32,5 resp. 13 (gel B, 3) μg protein were applied.

Assumption 1 can be excluded by experiments on the electrophoretic mobility of the microsomal proteins during SDS polyacrylamide gel electrophoresis. Figure 5 shows that the additional bands occurring after gel electrophoresis of MC microsomes can also be seen if the inhibitor of protein synthesis was administered concomitantly with the inducer. This could not easily be explained by changes in the phospholipid environment of the cytochrome. Also, relatively minor changes of a preexisting cytochrome P-450 should not lead to the appearance of a protein with different molecular weight. However, it cannot yet be concluded from our results which step in the synthesis of cytochrome P-450 is critical for the formation of cytochrome P-448 and if de novo synthesis or posttranslational events are of predominant importance for the appearance of this hemoprotein.

REFERENCES

1. A.P. Alvares, G. Schilling, W. Levin and R. Kuntzman, Alteration of the microsomal hemoprotein by 3-methylcholanthrene: effects of ethionine and actinomycin D, J. Pharmacol. exp. Ther. 163, 417 (1968)
2. D.W. Nebert and H.V. Gelboin, Substrate-inducible microsomal aryl hydroxylase in mammalian cell cultures, J. biol. Chem. 243, 6250 (1968)
3. E. Bresnick and H. Mossé, Activation of gene transcription in rat liver chromatin by 3-methylcholanthrene, Mol. Pharmacol. 5, 219 (1969)
4. K.D. Lanclos and E. Bresnick, The formation of poly-A-containing RNA in rat liver after administration of 3-methylcholanthrene, Chem.-biol. Interactions, 12, 341 (1976)

5. D.A. Haugen, T.A. van der Hoeven and M.J. Coon, Purified liver microsomal cytochrome P 450: Separation and characterization of multiple forms, J. biol. Chem. 250, 3567 (1975)

6. P.E. Thomas, A.Y.H. Lu, D. Ryan, S.B. West, J. Kawalek and W. Levin, Multiple forms of rat liver cytochrome P-450. Immunochemical evidence with antibody against cytochrome P-448, J. biol. Chem. 251, 1385 (1976)

7. Y. Imai and P. Siekevitz, A comparison of some properties of microsomal cytochrome P-450 from normal, methylcholanthrene-, and phenobarbital-treated rats, Arch. Biochem. Biophys. 144, 143 (1971)

8. R. Kahl, H.G. Jonen, G.F. Kahl, Effect of phenobarbital and 3-methylcholanthrene administration in vivo and isooctane extraction in vitro on metyrapone binding to reduced cytochrome P-450, Biochem. Pharmacol. 23, 2305 (1974)

9. H.G. Jonen, B. Huethwohl, R. Kahl and G.F. Kahl, Influence of pyridine and some pyridine derivatives on spectral properties of reduced microsomes and on microsomal drug metabolizing activity, Biochem. Pharmacol. 23, 1319 (1974)

10. J.E. Gielen, F.M. Goujon and D.W. Nebert, Genetic regulation of aryl hydrocarbon hydroxylase induction II. Simple mendelian expression in mouse tissue in vivo, J. biol. Chem. 247, 1125 (1972)

11. P.J. Creaven, D.V. Parke and R.T. Williams, A fluorimetric study of the hydroxylation of biphenyl in vitro by liver preparations of various species, Biochem. J. 96, 879 (1965)

12. U.K. Laemmli, Cleavage of structural proteins during the assembly of the head of bacteriophage T4, Nature 227, 680 (1970)

13. F.J. McPherson, J.W. Bridges and D.W. Parke, The effects of benzo(a)pyrene and safrole on biphenyl-2-hydroxylase and other drug-metabolizing enzymes, Biochem. J. 154, 773 (1976)

14. F.M. Goujon, D.W. Nebert and J.E. Gielen, Genetic expression of aryl hydrocarbon hydroxylase induction V. Interaction of various compounds with different forms of cytochrome P-450 and the effect on benzo(a)pyrene metabolism in vitro, Mol. Pharmacol. 8, 667 (1972)

15. G.F. Kahl, R. Kahl, K. Kumaki and D.W. Nebert, Association of the Ah locus with specific changes in metyrapone and ethyl isocyanide binding to mouse liver microsomes, J. biol. Chem., in press.

ACKNOWLEDGEMENTS

The able technical assistance of Mrs. Angelika Leonardi and Mrs. Gudrun Ritter is gratefully appreciated.

STIMULATION OF LIVER GROWTH AND MIXED-FUNCTION OXIDASE BY α-HEXACHLOROCYCLOHEXANE: SEPARATION OF INDUCTIVE PATHWAYS

Rolf Schulte-Hermann

Institut für Toxikologie und Pharmakologie, Philipps-Universität, 355 Marburg (West-Germany)

ABSTRACT

Xenobiotic compounds such as α-hexachlorocyclohexane (α-HCH) stimulate liver growth and hepatic mixed-function oxidase activity. The increase of DNA synthesis requires the consumption of dietary protein 5-8 hours before initiation of DNA replication and can be blocked by a low dose of actinomycin D (50 µg/kg) in the first few hours after α-HCH. In contrast, the increase of cytochrome P450 and of N-demethylation do not require dietary protein and are not sensitive to even higher doses of actinomycin D (500 µg/kg). The results suggest that the inductive pathways leading to cell proliferation and mixed-function oxidase multiplication diverge a few hours after initiation of the inductive process.

INTRODUCTION

Stimulation of hepatic mixed-function oxidase activity by xenobiotic compounds is frequently associated with liver growth. In fact, this association has been reported for more than 100 different compounds (see ref. 1 for review). Therefore the question arises: Are the increase of mixed-function oxidase and of liver mass necessarily coupled? Is liver growth a consequence of the accumulation of drug-metabolizing enzymes and of smooth endoplasmic membranes? Thus, an earlier, rather attractive hypothesis suggested that the growth response is triggered by the increased NADPH (and oxygen) turnover after multiplication of the drug metabolizing enzyme system (2, see also 1). Our results presented here suggest, however, that enzyme accumulation and growth are mediated in parallel - rather than sequentially.

RESULTS AND DISCUSSION

As a model compound we used the α-isomer of hexachlorocyclohexane (= benzene hexachloride, abbreviation: α-HCH), a potent inducer of liver growth and drug-metabolizing enzymes (3-5). As indicators of growth we studied DNA synthesis and mitotic activity. To increase the validity of measurements of DNA replication we used both ^{3}H-thymidine and ^{14}C-orotic acid, which are incorporated into DNA via different pathways. As shown in Fig. 1 the increase of DNA replication began 18 hours after α-HCH administration, was maximal for several hours, and then declined. Subsequently a steep increase in mitotic activity occured (Fig. 1 c). These temporal

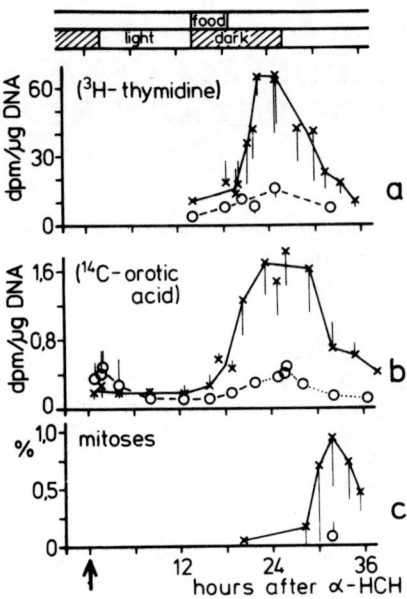

Fig. 1 Synchronization of α-HCH stimulated cell proliferation in rat liver.
Female rats (Wistar, SPF) were adapted for 3 weeks to a reversed light-dark rhythm (lights off from 9.oo to 21.oo o'clock) and to a feeding schedule which permitted access to food only from 9.oo to 14.oo o'clock (see top of the Fig.). The animals were used when they weighed 1oo g. 15o mg/kg α-HCH were dissolved in corn oil and administered orally by stomach tube at 19.oo o'clock (arrow). 1oo μCi/kg ^{3}H-thymidine (6.7 Ci/mmol) (Fig. 1a) or 5o μCi/kg ^{14}C-orotic acid (49 mCi/mmol) (Fig. 1 b) were injected i.v. 1 hour before sacrifice. The uptake of label into DNA was assayed according to standard procedures (6,7) as described in detail before (8-1o). In the experiment with ^{14}C-orotic acid RNA was removed according to (6,11). Mitotic activity (Fig. 1 c) was determined in histological sections by counting at least 2.ooo hepatocytes. o----o: untreated controls; x——x: α-HCH treated. Vertical bars indicate standard deviations. Each point is the mean of 2-4 (controls) or 4-6 (α-HCH treated) animals.

patterns reflect the known course and duration of the replicative cycle in adolescent rat liver: the S-phase lasts 8 hours, the G_2-phase 1 hour, and the M-phase 1-2 hours (1).

The steep slopes of increase and decrease of DNA synthesis and of mitotic activity suggest that the proliferating hepatocytes proceed through the replicative cycle almost in synchrony (Fig. 1). This high degree of synchrony was obtained by the feeding schedule used in the present experiment (see legend and top of Fig. 1). In rats fed ad libitum we observed much less synchrony (9). We have learned that the time of food consumption and the amount of the food consumed both determine the increase of DNA synthesis, and that feeding schedules as used here may indeed synchronize cell proliferation in the liver. The experimental approach which led to this result is published elsewhere (9,1o); here we show the crucial experiment suggesting that the protein component of the diet is the essential synchronizer of hepatocyte proliferation (Fig. 2). Rats received α-HCH and, beginning at the same time, a protein-free diet. In these animals the response to α-HCH was weak and delayed. Some animals received protein at the times indicated by the arrows. This was followed, at all times tested, by a steep additional increase of DNA synthesis which began 5-8 hours after protein (Fig. 2 a). In untreated rats protein feeding had only small effects (Fig. 2 b) indicating that protein was not mitogenic by itself. Rather it promoted the

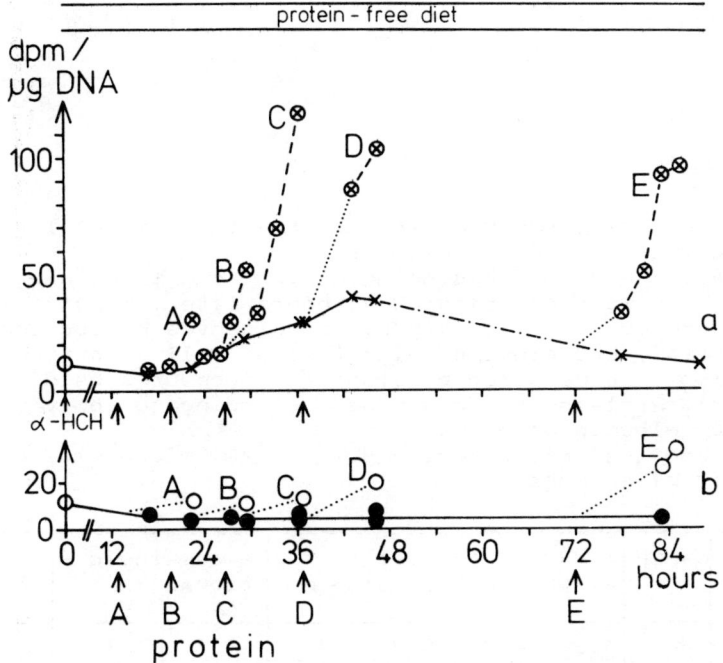

Fig. 2 Dietary protein as a determinant of hepatic DNA synthesis. Rats were adapted for 3 weeks to continuous lighting with food ad libitum. They received α-HCH at time "O" as indicated and, beginning at the same time, a protein-free diet (Fig. 2 a). Control rats received the protein-free diet, but no α-HCH (Fig. 2 b). No further treatment: x———x; ●———●. Groups A - E were treated, at the times indicated by arrows A - E, with protein hydrolysate (1o g/kg casein hydrolysate corresponding to 5o % of the protein content in the daily ration of the standard diet) and were then fed a complete diet: ⊗----⊗; O····O. ^{3}H-thymidine incorporation into DNA was measured. See legend to Fig. 1 for further details.

initiation of DNA synthesis only in cells which had previously been triggered for replication by α-HCH. Apparently the induction of hepatic DNA synthesis is controlled in 2 stages: α-HCH provides the first signal and protein the second. In the absence of protein cells triggered for replication are arrested at a critical point of their prereplicative phase and "wait" until protein provides release from the block (1o). Thus, consumption of dietary protein largely controls the increase of DNA synthesis in α-HCH treated rats.

Is there any evidence of a regulatory role of dietary protein also in the stimulation of hepatic mixed-function oxidase activity by α-HCH? We have measured, in the same experiment as shown in Fig. 2, the activity of 2 hepatic N-demethylases and the amount of cytochrome P45o (Table 1). Obviously, lack of dietary protein for 2 days did not decrease the magnitude of the response to α-HCH. Thus, the inductive process by which α-HCH increases the drug-metabolizing enzyme system apparently does not depend on dietary protein - in contrast with the induction of DNA synthesis.

Table 1 Effect of dietary protein on the increase of liver size and mixed-function oxidase.
Rats received α-HCH and, beginning at the same time, a protein-free diet. They were decapitated 48 hours later. Hepatic microsomes were prepared, and the rate of formaldehyde production from aminopyrine (AP) and ethylmorphine (EM) as well as cytochrome P45o concentration were measured by standard procedures as described before (5,8). Protein was determined according to Lowry (12) using bovine serum albumin as a standard.
*per 1oo g body weight. **nMol HCHOxmin^{-1}xmg microsomal protein^{-1}. (n) = number of animals.

	Food	α-HCH	liver weight*(g)	liver protein*(mg)	N-demethylation** AP	EM	P45o nMol / mg prot.
	normal	-	4.3±o.4(9)	618 ± 19	4.7	1.7	o.66
		+	5.3±o.6(9)	729 ± 56	9.3	4.2	o.94
Δ			+ 22 %	+ 18 %	+1oo%	+14o%	+ 43 %
	protein-free (2days)	-	3.8±o.3(8)	486 ± 8.6	4.6	1.4	o.47
		+	5.o±o.4(1o)	6o9 ± 35	1o.4	5.1	o.89
Δ			+ 3o %	+ 25 %	+125%	+275%	+ 9o %

Table 2 Effect of actinomycin D (50 µg/kg) on hepatic DNA synthesis and mitotic activity.
Rats adapted to the protocol described in Fig. 1 received α-HCH at 14.00 o'clock and at the same time an i.p. injection of 50 µg/kg actinomycin D. Incorporation of precursors into DNA was measured at 30 hours, and mitotic activity at 36 hours. *p < 0.02 vs. group 2.

	α-HCH	actinomycin D	^{3}H-thymidine dpm/µg DNA	^{14}C-orotic acid dpm/µg DNA	mitoses %
1	−	−	8.3 ± 5.4 (12)	0.33 ± 0.11 (2)	0.3 ± 0.2 (3)
2	+	−	58.6 ± 19.6 (26)	2.1 ± 0.6 (5)	1.4 ± 0.66 (5)
3	+	+	16.4 ± 8.1* (20)	0.66 ± 0.3* (5)	0.36 ± 0.24* (5)

A second measure to separate the inductive pathways is provided by use of actinomycin D. As shown in Table 2 the increase of DNA synthesis and of mitotic activity were prevented, if 50 µg/kg actinomycin D were administered simultaneously with α-HCH. That the induction of hepatic DNA synthesis is sensitive to such low doses of this inhibitor has been shown before by LIEBERMAN's group (13) and others (14). However, this high sensitivity exists specifically only during the begin of the inductive process (Fig. 3). The solid line shows again the time course of DNA synthesis as measured by thymidine uptake. 50 µg/kg of actinomycin D were administered at one of the time points indicated by arrows. A maximal inhibition was obtained only if actinomycin D was given simultaneously with α-HCH; already a few hours later the inhibitor produced no significant inhibition (Fig. 3, open circles). Apparently the critical, actinomycin − sensitive event is required only in the first few hours of the inductive pathway leading to DNA synthesis. If we assume inhibition of RNA synthesis as the mechanism of action of actinomycin D this finding suggests that an early step of the inductive pathway is the activation of certain genes, transcription of which is extremely sensitive to the inhibitor. It is worth adding that a "normal" sensitivity to actinomycin D persists throughout the prereplicative phase, since 1 mg/kg of the inhibitor prevent DNA synthesis until 0 − 2 hours before its initiation (15).

In contrast with the findings shown in Table 2 and Fig. 3, the stimulation of mixed-function oxidase activity by α-HCH was found completely insensitive to 50 µg/kg actinomycin D. Even 500 µg/kg did not depress the increase of cytochrome P450 and aminopyrine demethylase after α-HCH (Fig. 4). Likewise, the stimulation of demethylation of ethylmorphine and benzphetamine was not reduced by 500 µg/kg actinomycin D (to be published). The literature on the sensitivity to actinomycin D of mixed-function oxidase induction is controversial: most authors found inhibition (e.g. ref. 16, 17), but some observed no or weak inhibitory effects (18, 19). In our experiments, 2 mg/kg were required to block the increase, but this dose was not survived by most of our rats, and hence an interpretation of this finding is difficult.

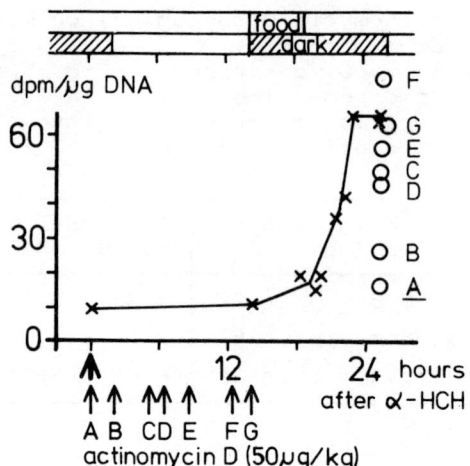

Fig. 3 Loss of sensitivity to actinomycin D in the early prereplicative phase.
Rats standardized as described in Fig. 1 were treated with α-HCH at 19.00 o'clock, and the time course of thymidine incorporation into DNA was measured: x———x. Several groups of α-HCH treated animals received, at one of the time points indicated by arrows A-G, 50 μg/kg actinomycin D i.p., and the effect on DNA synthesis was measured at 24 hours: O. Dash below letter marks significant inhibition.

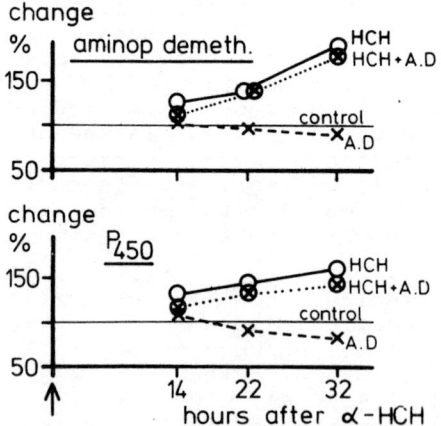

Fig. 4 Effect of actinomycin D (500 μg/kg) on the increase of mixed-function oxidase.
Rats standardized as described in Fig. 1 were treated, at "O" time (arrow), with α-HCH and/or 500 μg/kg actinomycin D (= A.D.) i.p. as indicated. Control values, set 100 %, were: 3,4; 3,9; 4,5 nMol HCHO min^{-1} x mg microsomal protein^{-1} and 0,48; 0,48; 0,65 nMol cytochrome P450 per mg microsomal protein.

The result of this study is summarized in Fig. 5: The inductive
pathways leading to DNA synthesis or mixed-function oxidase
multiplication differ by sensitivity to actinomycin D and by a
requirement for dietary protein. We cannot conclude on the initial
effects of α-HCH, and whether there are, initially, common regulatory steps. Our results do suggest, however, that the 2 pathways
can be separated at the latest a few hours after application of
the inducer. Therefore, the induction of cell proliferation by
xenobiotic compounds is not a <u>consequence</u> of the increase in
mixed-function oxidase activity and of the ensuing enhancement
of energy turnover. Rather, cell proliferation and enzyme multiplication appear to be mediated simultaneously via <u>parallel</u>
pathways.

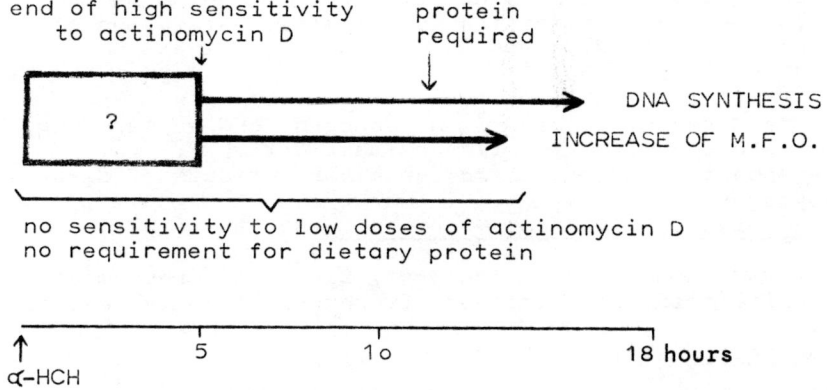

Fig. 5   Differences between inductive pathways leading to DNA
synthesis and increase of mixed-function oxidase (=M.F.O.)

REFERENCES

1. R. Schulte-Hermann , Induction of liver growth by xenobiotic
   compounds and other stimuli,
   <u>Crit. Rev. Toxicol.</u> 3, 97-158 (1974)

2. W. Kunz , G. Schaude, H. Schimassek, W. Schmid and M. Siess,
   Stimulation of liver growth by drugs. II. Biochemical analysis,
   proceedings european society for the study of drug toxicity
   excerpta medica foundation,
   <u>Proc. Eur. Soc. for the Study of Drug Toxicity VII,</u> 7, 138 (1966)

3. I. Schlicht , W. Koransky, S. Magour, and R. Schulte-Hermann,
   Größe und DNS-Synthese der Leber unter dem Einfluß körperfremder Stoffe,
   <u>Naunyn-Schmiedebergs Arch. Pharmakol. exp. Pathol.</u> 261,26 (1968)

4. R. Schulte-Hermann, , R. Thom, I. Schlicht, and W. Koransky,
   Zahl und Ploidiegrad der Zellkerne der Leber unter dem Einfluß

körperfremder Stoffe,
Naunyn-Schmiedebergs Arch.Pharmakol.exp.Pathol. 261,42 (1968)

5. W. Koransky, S. Magour, G. Noack, and R. Schulte Hermann, Über den Einfluß induzierender Substanzen auf Fremdstoff-Oxydasen und andere Redoxenzyme der Leber,
Naunyn-Schmiedebergs Arch.Pharmakol.exp.Pathol.263,281 (1969)

6. A. Fleck, and H.N. Munro, The precision of ultraviolet absorption measurements in the Schmidt-Thannhauser Procedure for nucleic acid estimation,
Biochim. Biophys. Acta, 55, 571 (1962)

7. K. Burton, A study of the conditions and mechanism of the diphenylamine reaction for the colorimetric estimation of deoxyribonucleic acid,
Bioch. J. 62, 315 (1956)

8. R. Schulte-Hermann, C. Leberl, H. Landgraf, and W. Koransky, Liver growth and mixed-function oxidase activity: dose-dependent stimulatory and inhibitory effects of α-hexachlorocyclohexane,
Naunyn-Schmiedeberg's Arch. Pharmacol. 285, 355 (1974)

9. R. Schulte-Hermann, and H. Landgraf, Circadiane rhythm of cell proliferation in rat liver: Synchronization by feeding habits,
Z. Naturforsch. 29c, 421 (1974)

1o. R. Schulte-Hermann, Two-stage control of cell proliferation induced in rat liver by α-hexachlorocyclohexane,
Cancer Res., in press (1977)

11. G. Schmidt, and S.J. Thannhauser, A method for the determination of desoxyribonucleic acid, ribonucleic acid, and phosphoproteins in animal tissues,
J. Biol. Chem. 161, 83 (1945)

12. O.H. Lowry, N.J. Rosebrough, A.L. Farr, and R.J. Randall, Protein measurement with the folin phenol reagent,
J. Biol. Chem. 193, 265 (1951)

13. M. Fujioka, M. Koga, and I. Lieberman, Metabolism of ribonucleic acid after partial hepatectomy,
J. Biol. Chem. 238, 34o1 (1963)

14. E. Oberdisse, C. Hochstrate, and H.J. Merker, Influence of drugs on induction of enzymes involved in DNA metabolism,
Proc. IV Internat. Congr. Pharmacol. 4, Schwabe, Basel,318(197o)

15. R. Schulte-Hermann, C. Leberl, and I. Ruberg, Stimulation of cell proliferation in rat liver by α-hexachlorocyclohexane or partial hepatectomy and end points during G_1 of the inhibitory action of ß-diethyl-aminoethylphenyldiallyl acetate · HCl (CFT 12o1), ß-diethylaminoethyldiphenylpropyl acetate · HCl (SKF

525-A) and actinomycin D,
Biochim. Biophys. Acta, in press (1976)
16. H.V. Gelboin, and N.R. Blackburn, The stimulatory effect of 3-methylcholanthrene on benzpyrene hydroxylase activity in several rat tissues: inhibition by actinomycin D and puromycin, Cancer Res. 24, 356 (1964)

17. S. Orrenius, J.I.E. Ericsson, and L. Ernster, Phenobarbital-induced synthesis of the microsomal drug-metabolizing enzyme system and its relationship to the proliferation of endoplasmic membranes,
J. Cell. Biol. 25, 627 (1965)

18. A.M. Cohen, and R.W. Ruddon, The role of gene activation in phenobarbital induction of hepatic drug-metabolizing enzymes, Pharmacologist 11, 313 (1969)

19. M. Bader, T. Thrum, and W. Klinger, Actinomycin D: Influence on induction of liver microsomal enzymes,
Acta biol. med. germ. 32, 99 (1974)

INDUCTION OF CYTOCHROME P-450 BY LONG-TERM INFUSION OF PHENOBARBITAL

Otto Rosenthal, Harry M. Vars, Heinz Schleyer,
David Y. Cooper, Sidney S. Levin and Joseph Touchstone

Harrison Department of Surgical Research and Department of Gynecology and Obstetrics, University of Pennsylvania, Philadelphia 19104, U.S.A.

ABSTRACT

A procedure is described for studying the induction of hepatic enzyme systems by long-term continuous intraperitoneal infusion of drugs at constant rates into rats. Its purpose is to prevent the fluctuations of the tissue levels of inducers associated with intermittent injections or their uptake with the drinking water. It was found that phenobarbital (PB), the inducer studied, can be infused at a rate of 80 mg/kg/day for at least three weeks without affecting activity, food consumption, and weight gain of the animals. Preliminary data are reported on the induction response to PB with this method in comparison with intermittent intraperitoneal administration. Up to a total dose of 320 mg PB/kg, induction of cytochrome P-450 in terms of specific microsomal concentration as well as total liver content was a linear function of the total dose irrespective of the method of administration used: daily injection of 80 mg/kg for four days, of 40 mg/kg every 48 hours for 16 days or continuous infusion at this dose level for six days--a novel and unexpected finding. The effect of higher accumulative doses of PB, up to 1680 mg/kg, was studied with the infusion technique. Maximal induction of PB was reached after a total dose of 560 mg/kg. The same held true for induction of demethylation activities toward codeine and ethylmorphine. The results indicate that the optimal or limiting dose for induction of PB, as observed by previous investigators, cannot be exceeded by continuous infusion of PB. Exploratory experiments have shown that urinary elimination of PB, which never exceeds 30% of the administered daily dose, drops to 9% when the limiting total dose is being reached, suggesting that conversion of the inducer to an inactive metabolite causes the limiting dose phenomenon. Experiments to establish such a relationship are in progress.

INTRODUCTION

Differences in the time course of induction of hepatic microsomal cytochrome P-450 and associated enzyme activities have been described by us (1) when either phenobarbital (PB) or 3-methylcholanthrene (MC) was administered intraperitoneally by intermittent injections. It was not clear, however, whether the prolonged induction characteristics of MC were caused solely by the prolonged time of exposure of the liver to this slowly released and eliminated agent or whether one was dealing with an induction mechanism specific for polycyclic aromatic hydrocarbons. We therefore devised a method for studying the induction effect of long-term exposure to constant dose levels of PB maintained by continuous intraperitoneal infusion. This procedure should avoid the daily fluctuations in plasma and tissue levels of the inducer that are associated with its intermittent intraperitoneal injection or its uptake with the drinking water.

In this paper we are describing in detail the procedure for long-term intraperitoneal infusion of drugs, presenting some preliminary data on the induction response to PB with this procedure, and comparing the data with those obtained with intermittent administration of PB.

METHODS

General Procedure

Male Sprague-Dawley rats were used having an initial weight of 200 to 250 grams. Care of the animals, dietary regimen, induction procedures, and enzyme assays have been previously described (1,2). Food was withdrawn 24 hours before sacrifice. Results have been reported in terms of specific P-450 concentration and specific enzyme activities per mg of microsomal protein as well as of total P-450 and protein content and total enzyme activity of the livers. Cytochrome P-450 and protein were assayed in the homogenate of the total liver. Total enzyme activities were computed by multiplying the microsomal turnover numbers (nanomoles product x min^{-1} x nmole $P-450^{-1}$) with the nmoles of P-450 in the total liver. Only the latter values are here reported.

Procedure for Continuous Intraperitoneal Infusion

The animal is lightly anesthetized with ether and a small paramedian incision is made in the abdominal skin. The exposed abdominal wall is pierced with a hypodermic needle of appropriate size through which the first three inches of a vinyl tubing (No. 24)* are introduced into the abdominal cavity. The needle is then withdrawn and removed over the outer end of the tubing anchored by a suture to the abdominal wall at the site where it emerges. From this point the outer end of the tubing, which is closed with a solid needle and guided by a hemostat, is passed subcutaneously around the animal's flank and through the dorsal skin above the spine at a point about one-third the distance from base of tail to base of head. An adhesive waist band is placed around the animal, covering the area of the abdominal incision but avoiding the dorsal exit point of the tubing. The anesthesia is then stopped. A lusteroid centrifuge tube (30 ml, 25 x 75 mm) is placed longitudinally on the back of the waist band and attached to it by means of two strips of adhesive tape slipped through a pair of slots in the side of the tube as shown in Fig. 1. Holes in the bottom of the tube and in the acetate films covering the open end of the lusteroid tube accommodate a height-adjustable iron rod which is mounted in the longitudinal axis of the cage. The infusion tubing is fixed with small adhesive strips to the outside of the acetate film, the rod inserted, the animal placed in the cage, and the height of the rod adjusted so that the animal can move freely to

---

* 3M Company, Irvington, #3002, #24 clear polyvinylchloride tubing

Fig. 1. Rat positioned in infusion cage.

and fro along the longitudinal axis of the cage and has unlimited access to food and water. Because the adhesive vest must not be too tight to avoid interference with the rat's respiration, blood circulation, and digestion, the animal still has some freedom of turn. At times this has been sufficient for him to sever the infusion tubing or to chew the harness. This is prevented by closing in the free space of the cage with baffles. In this way the extent to which the animal can turn is sufficiently reduced to preclude such damage. The infusion tube is then connected to the outlet tube of a syringe driven by a Harvard infusion pump set to deliver 5 ml infusion fluid over a 24-hour period. Infusion is stopped 10 minutes before sacrifice of the animal.

RESULTS

Induction by Different Techniques of PB Administration

To explore the influence of dosage and timing of intraperitoneal PB administration on the extent of induction of cytochrome P-450, one set of rats received 80 mg PB/kg once a day for four days by injection, our usual induction protocol; the second set 40 mg/kg every alternate day for 16 days; and the third set 40 mg/kg for six days as continuous infusion. As evident from Fig. 2a, up to an accumulative dosage of 320 mg/kg the rise in the specific concentration of microsomal P-450 was a linear function of the total amount of PB, regardless of the method of administration--a somewhat unexpected result. The same held true for the total amount of P-450 induced (Fig. 2b).

Induction by Prolonged Intraperitoneal Infusion

Using intermittent intraperitoneal administration, Remmer and Merker (3) in the dog and Ernster and Orrenius in the rat (4) observed that there is an optimal or limiting total dose of PB beyond which no further induction of hepatic endoplasmic reticulum and cytochrome P-450 occurs. The "limiting" total dose for the rat was 500 to 600 mg/kg. One of the goals of our studies was to explore whether this limiting dose could be exceeded if the inducer was administered by continuous intraperitoneal infusion at a constant daily rate of 80 mg/kg, i.e., under conditions we assumed the liver tissue would be

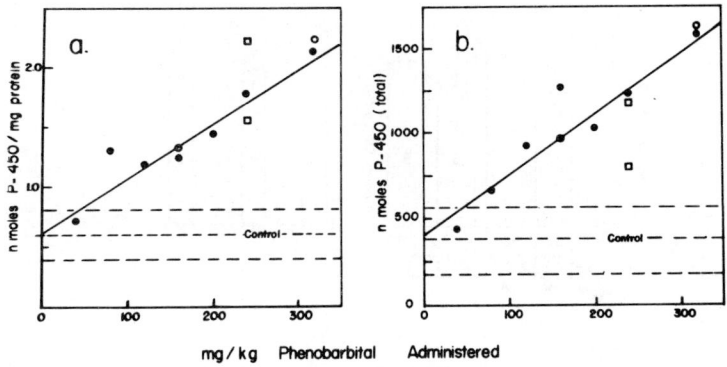

Fig. 2. Dose response of cytochrome P-450 induction to different types of phenobarbital (PB) administration.
Injections: ● 40 mg/kg every second day; ○ 80 mg/kg daily.
Infusion: □ 40 mg/kg/day for six days. Accumulative dose of PB on x-axis.
a) Specific concentration of microsomal P-450 (nmole/mg protein)
$y = 0.0048 x + 0.61$; $r = 0.945$.
b) Total amount (nmole) of P-450 in liver
$y = 3.79 x + 401.5$; $r = 0.93$.

exposed to constant plasma levels of the drug. Results of these experiments are summarized in Fig. 3. Amounts of PB infused as well as amounts of total protein and total P-450 and total enzyme activities of the liver are here computed for 100 grams of body weight. It is obvious from Fig. 3a that the protein and cytochrome P-450 contents after administration of 170 mg PB/100 g (Group C) were, on the average, not higher than after 60 mg/100 g. Similarly, as shown in Fig. 3b, demethylation activities of Group C with codeine or ethylmorphine as substrate were not significantly higher than those of Group A, although the scatter range of the values was large. Activity of benzphetamine demethylation of Group C **was** significantly higher than that of Group A. However, in view of the relatively low level of significance ($P < .02$) and the small number of experiments, the results cannot be considered as definitive and require further study. On the whole, our results corroborate those of the previous investigators.

The Fate of Phenobarbital

One of the explanations for the limiting dose of induction by PB has been accelerated elimination of the drug by urinary excretion as well as oxidative conversion to an inactive metabolite taking place in liver and/or extrahepatic tissues. We have started an exploratory study of this possibility. Three rats of 500 g body weight were used in order to have sufficient material available for several analytic procedures. The animals were kept in metabolic

572 O. ROSENTHAL et al.

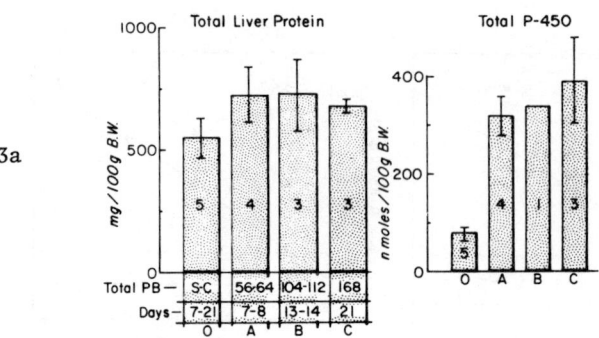

3a

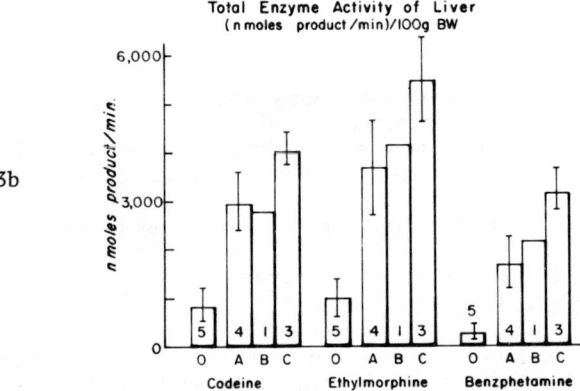

3b

Fig. 3. Dose response of total protein, total P-450 and total enzyme activity of liver to PB infusion. The accumulative dose as well as the mean liver values are calculated for 100 g body weight. In the bars, vertical lines indicate standard deviations, and numbers indicate the number of animals. S.C. designates saline controls.

cages equipped with funnels for collecting the urine in Erlenmeyer flasks and with wire screens above the flasks for separating the feces. Phenobarbital was quantitated gas-chromatographically by means of slight modifications of the methods of Baselt and Casarett (5) for urine and of Levy and Schwartz (6) for blood plasma. Phenobarbital was administered by single daily intraperitoneal injections of 76 to 80 mg/kg body weight. Fig. 4 shows the daily urinary PB excretion for animal No. 1 which received 13 injections (total dose 984 mg/kg). In addition, the plasma levels of PB at sacrifice are recorded for this animal as well as of the two others (Nos. 3 and 2) receiving respectively four injections (total dose 288 mg/kg) and 10 injections (total dose 800 mg/kg). The "limiting" dose for the induction process should have been reached after the seventh injection. Yet there was no accelerated urinary elimination but rather a marked decline. The smaller secondary

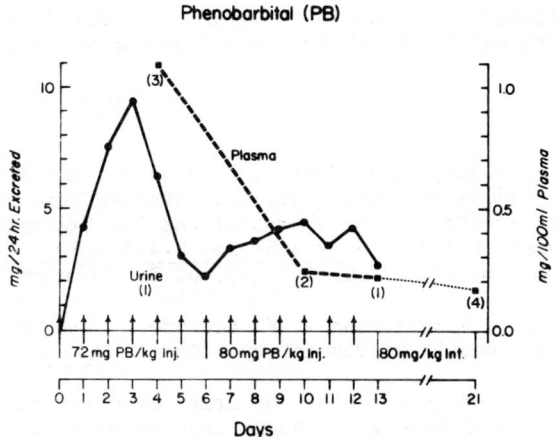

Fig. 4. Urinary excretion and plasma levels of PB.

increase shown in Fig. 4 was probably due to the increase of the dose. A similar decrease was found in rat No. 2 sacrificed at the 10th day.

Parallel to the decrease in urinary excretion, plasma levels of phenobarbital dropped from 1.1 mg/100 ml at day 4 (rat No. 3) to about 0.2 mg/100 ml at days 10 and 13 (rats No. 2 and 1). Interestingly, values of a similar order (0.17 mg/100 ml) were also found in an additional rat (No. 4) that had received a total of 1680 mg PB/kg by intraperitoneal infusion over a period of 21 days.

At sacrifice of animal No. 3, 20% of the total administered dose of 144 mg was recovered in the urine. Assays of liver, retroperitoneal fat, kidneys, and brain yielded the following values in terms of mg/100 g: 2.66, 2.36, 0.75, 0.48. Concentration of the agent above its plasma level did occur in liver and fat, but the mass of these tissues was not large enough to account for a significant portion of the missing 115 mg of PB. Unless there was substantial accumulation elsewhere in the body, the drug must have been converted to another product, probably by oxidative hydroxylation. For Butler (7) has shown that, in dog and man, the major conversion product is parahydroxyphenobarbital [5-ethyl-5-(p-hydroxyphenyl)barbituric acid] (p-OH-PB) excreted mainly as conjugate in an amount equal to or occasionally exceeding that of unchanged phenobarbital. Our exploratory experiments have shown that there is a compound in the urine of PB-treated rats that has the same R_F on T.L. chromatography and the same retention time on gas chromatography as an authentic sample of p-OH-PB kindly supplied by Dr. Butler.

The excretion of conjugated p-OH-PB has not been determined by us. Nevertheless, our preliminary experiments suggest that increased oxidative conversion of PB during prolonged administration might be responsible for the decline in urinary elimination of PB. Where such a conversion occurs, where the hydroxylation product is conjugated, and whether conversion to an inactive product is responsible for the limiting dose phenomenon are unsolved questions which are under investigation in our laboratory.

ACKNOWLEDGMENTS

This work was supported by Grants AM-04484 and CA 17618 from the National Institutes of Health and Grant BMS 74-01099 from the National Science Foundation.

REFERENCES

(1) D. Y. Cooper, H. Schleyer, J. Thomas, H. M. Vars, and O. Rosenthal. Comparison of the induction course, physical chemical interactions and photochemical action spectra of phenobarbital and 3-methylcholanthrene induced hepatic microsomal P-450. In Cytochromes P-450 and b_5 (Eds. D. Y. Cooper, O. Rosenthal, R. Snyder, and C. Witmer), p. 81. Plenum Press, New York, 1974.
(2) J. B. Schenkman, H. Remmer, and R. W. Estabrook, Spectral studies of drug interaction with hepatic microsomal cytochrome, Mol. Pharmacol. 3, 113-123 (1967).
(3) H. Remmer and H. J. Merker, Effect of drugs on the formation of smooth endoplasmic reticulum and drug metabolizing enzymes, Ann. N. Y. Acad. Sci. 125, 79-95 (1965).
(4) Ernster, L., and Orrhenius, S., Substrate-induced synthesis of the hydroxylating enzyme system of liver microsomes, Fed. Proc. 24(5), 1190-1199 (1965).
(5) Baselt, R. C., and Casarett, L. Y., Detection of drugs in urine for methadone treatment programs, J. Chromatogr. 57, 139-141 (1971).
(6) Levy, S. K., and Schwartz, T., New mixed bed column for rapid gas-liquid chromatographic determination of sedatives and dilantin in serum, Clin. Chem. Acta 54, 19-26 (1974).
(7) Butler, T. C., The metabolic hydroxylation of phenobarbital, J. Pharmacol. 113, 326-336 (1955).

METABOLISM AND MUTAGENICITY OF 2-ACYLAMINOFLUORENES AND RELATED COMPOUNDS

Snorri S. Thorgeirsson, Peter J. Wirth and
Wendel L. Nelson

Section on Molecular Toxicology, Developmental Pharmacology Branch, National Institute of Child Health and Human Development, National Institutes of Health, Bethesda, Maryland 20014 U.S.A.

ABSTRACT

There is a strong association between aromatic hydrocarbon responsiveness and N-hydroxylation of phenacetin, acetanilide, and p-chloroacetanilide in the livers from progeny and the appropriate backcrosses and intercross involving B6 and D2 mice. Mutagenic activity in the *Salmonella* test system of several 2-acylaminofluorenes were compared using the S-9 liver fractions from B6 and D2 mice. Pretreatment of B6 mice with 3-MC increased the mutagenic activity of all the analogs in the S-9 liver fraction. No increase in mutagenic activity is found in S-9 liver fraction from D2 mice after 3-MC pretreatment, indicating that mutagenic activity correlates with aromatic hydrocarbon responsiveness and N-hydroxylase activity. ANF, an inhibitor of N-hydroxylation of 2-AAF *in vitro*, inhibits the mutagenic activity of these analogs indicating that N-hydroxylation is the first step in the activation of N-acylaminofluorenes into mutagens.

INTRODUCTION

We have previously shown a strong genetic association between PCH[1] induced AHH and 2-AAF N-hydroxylase activities in the liver of responsive B6 mice compared with nonresponsive D2 mice (1). Furthermore, the mutagenicity of 2-AAF, assayed with liver fractions from B6 and D2 mice by the Ames system (2), showed the same genetic association as PCH-induced AHH activity (3). We have now examined the cytochrome P-450 dependent metabolism of other N-acetylarylamines in B6 and D2 mice, the structural features necessary for mutagenic activity among analogs of 2-AAF, and the association of AHH with the metabolic activation of these analogs to mutagens.

MATERIALS AND METHODS

The inbred B6 and D2 mice used in these studies were provided by the National Institutes of Health Animal Supply. Treatment of animals and isolation of liver fractions were done as previously described (1,3). The rate of N-

---

[1]Abbreviations: 2-AAF, 2-acetylaminofluorene; AHH, aryl hydrocarbon (benzo[a]pyrene) hydroxylase; 3-MC, 3-methylcholanthrene; ANF, α-naphthoflavone; B6, C57BL/6N inbred mouse; D2, DBA/2N inbred mouse; (B6D2)F$_1$ X D2, (C57BL/6N)(DBA/2N)F$_1$ X (DBA/2N) backcross; PCH, polycyclic hydrocarbon; DMSO, dimethylsulfoxide.

AAF was measured by high pressure liquid chromatography (4). N-hydroxylation of p-chloroacetanilide, acetanilide and phenacetin as well as p-hydroxylation and O-dealkylation of acetanilide and phenacetin respectively were measured by thin-layer chromatography [silica gel GF plates 250 μm, using the solvent systems: acetone/chloroform/ammonium hydroxide (50:50:1) and chloroform/methanol/ammonium hydroxide (92:7:0.5)] and authentic standards. The 2-acylaminofluorene derivatives were synthesized as previously described (5).

Mutagenesis assays were performed as recently described (2,3). Test compounds were routinely added to 2 ml of top agar in 0.1 ml DMSO, except at high concentrations when up to 0.5 ml of DMSO was used. Bacterial tester strain TA-1538 (2-3 X 10^9 bacterial/ml), 0.15 ml of cofactor solutions, 0.5 ml of the S-9 liver fraction containing 2 mg of protein, prepared from B6 or D2 mice (control or 3-MC pretreated), 8 μmoles of $MgCl_2$, 33 μmoles KCl, 5 μmoles of glucose-6-phosphate, 4 μmoles of NADP+ and 100 μmoles of sodium phosphate, per ml. In assays where ANF was used as an inhibitor, it was added (in 0.1 ml of DMSO) to bacteria, cofactors and top agar and mixed. The test compounds were added last. The plates were incubated at 37° for 48 hrs and then the colonies (histidine revertants) were counted. In control experiments using nutrient agar, no toxicity (reduction in bacterial growth) was observed for compounds at concentrations below 10 μg/plate, or in the range where mutagenic activity was obtained.

RESULTS AND DISCUSSION

The relationship between aromatic hydrocarbon responsiveness and N-hydroxylation of several N-acetylarylamines in a backcross involving B6 and D2 mice is shown in Table 1. The N-hydroxylation of acetanilide, phenacetin and p-

Table 1 *Metabolism of N-acetylarylamines in Inbred Strains of Mice. Monooxygenase Activities Genetically Linked with Polycyclic Aromatic Hydrocarbon Responsiveness*

Substrates	Responsive[a] (B6D2)F_1 X D2	Nonresponsive[a] (B6D2)F_1 X D2
	pmol $mg^{-1} min^{-1}$	pmol $mg^{-1} min^{-1}$
2-acetylaminofluorene N-hydroxylation	710 ± 42	102 ± 8
Acetanilide N-hydroxylation	163 ± 15	38 ± 2
p-hydroxylation	2200 ± 90	710 ± 52
Phenacetin N-hydroxylation	63 ± 4	17 ± 3
O-dealkylation	1060 ± 93	307 ± 37
p-chloroacetanilide N-hydroxylation	44 ± 3	12 ± 1

[a]The (B6D2)F_1 X D2 mice were phenotyped by the zoxazolamine test (9) three weeks prior to use. All the mice were given 3-MC (80 mg/kg^{-1} *i.p.*) 48 hrs before sacrifice. Procedures for isolating liver microsomes and analytical methods are given in *Materials and Methods*.

chloroacetanilide is highly correlated with the aromatic hydrocarbon responsiveness in the (B6D2)F$_1$ X D2 backcross and also in both (B6D2)F$_1$ and (B6D2)F$_1$ X B6 (data not shown). Similar association was found between aromatic hydrocarbon responsiveness and N-hydroxylation of 2-AAF in the appropriate backcrosses and intercross involving B6 and D2 mice (1). The p-hydroxylation of acetanilide and O-dealkylation of phenacetin are also highly associated with aromatic hydrocarbon responsiveness in the (B6D2)F$_1$ X D2 backcross (Table 1) which is in agreement with previous reports (6,7). The effect of 3-MC pretreatment on the relative mutagenic activity, expressed as revertants per nmole of test compound, for selective 2-acylaminofluorenes using S-9 liver fraction from B6 and D2 mice is illustrated in Table 2. None of the 2-acylaminofluorenes were mutagenic without the S-9 fraction and cofactors. Pretreatment of the mice with 3-MC increased the mutagenicity of all the compounds when the S-9 liver fraction from B6 mouse was used but no increase was observed in the D2 mouse. These results indicated that N-hydroxylation was required for the metabolic activation of 2-acylaminofluorenes to mutagens. To answer

Table 2 *Effect of 3-MC Pretreatment on Relative Mutagenicity (revertants per nanomole) of Selected 2-Acylaminofluorenes using S-9 Fractions from C57BL/6N (B-6) and DBA/2N (D-2) Mouse Livers. At Least Two Concentrations were Done.*

NHR	D-2 Control	3-MC	Ratio 3-MC/Control	B-6 Control	3-MC	Ratio 3-MC/Control
-C(=O)-CH$_3$ (2-AAF)	24	25	1.0	10	141	14
-C(=O)-CH$_2$CH$_3$	45	26	0.5	16	122	8
-C(=O)-C(CH$_3$)$_3$	0.5	0.4	0.8	~0	3.3	>10
-C(=O)-NH$_2$	24	17	0.7	6.2	176	26
-C(=O)-OCH$_3$	113	78	0.7	69	456	6
-C(=O)-CH$_2$Br	55	55	1.0	26	211	8
-C(=O)-CH$_2$OCH$_3$	44	27	0.6	10	104	10

the question of whether N-hydroxylation is required for mutagenesis of these compounds, ANF, a potent inhibitor of cytochrome P_1-450 dependent N-hydroxylation of 2-AAF *in vitro* (8), was added to the mutagenesis assays. Results with several analogs are shown in Table 3. Since ANF has been shown to block N-hydroxylation of 2-AAF and had no toxic effect on the bacteria, it is inferred that N-hydroxylation of the 2-acylaminofluorene is required for mutagenesis.

Table 3 *Effect of α-Naphthoflavone (ANF) on Mutagenicity of 2-Acylaminofluorenes. Liver Fractions (S-9) from 3-MC Treated B-6 Mice Were Used.*

NH–R	Colonies/μg.	Per cent Reduction .005 mM ANF	.05 mM ANF
$-\overset{O}{\underset{\|}{C}}-CH_3$ (2-AAF)	1250/2	40	95
$-\overset{O}{\underset{\|}{C}}-CH(CH_3)CH_3$	1500/10	70	>98
$-\overset{O}{\underset{\|}{C}}-O-CH_3$	1850/1	35	95
$-\overset{O}{\underset{\|}{C}}-CH_2Br$	1320/2	51	>98
$-\overset{O}{\underset{\|}{C}}-NHCH_3$	1350/5	84	>98
$-\overset{O}{\underset{O}{S}}-CH_3$	1400/2	86	>98

The differences in mutagenic activity among the 2-acylaminofluorene analogs could reflect differences in the rate of cytochrome P_1-450 dependent N-hydroxylation, but may also reflect the ability of the hydroxamic acids so formed to undergo subsequent activation steps (*e.g.* deacetylation) which have been shown to be important in 2-AAF mutagenicity (2). A proposed scheme for metabolic activation of 2-acylaminofluorenes is shown in Fig. 1.

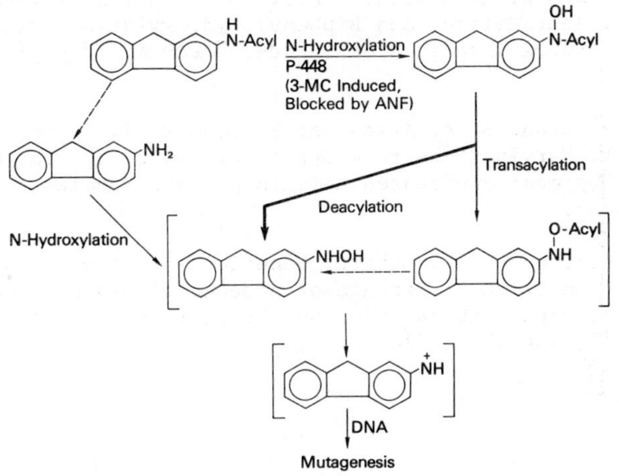

FIG. 1. A proposed scheme for metabolic activation of 2-acylaminofluorenes into mutagens.

REFERENCES

1. S. S. Thorgeirsson, J. S. Felton, and D. W. Nebert, Genetic Differences in the Aromatic Hydrocarbon-Inducible N-Hydroxylation of 2-Acetylaminofluorene and Acetaminophen-Produced Hepatotoxicity in Mice. *Mol. Pharmacol.* 11, 159-165 (1975).

2. B. N. Ames, L. D. Lee, and W. E. Durston, Improved Bacterial Test System for Detection and Classification of Mutagens and Carcinogens. *Proc. Nat. Acad. Sci. U.S.A.* 70, 782-786 (1973).

3. J. S. Felton, D. W. Nebert, and S. S. Thorgeirsson, Genetic Differences in 2-Acetylaminofluorene Mutagenicity *in vitro* Associated with Mouse Hepatic Aryl Hydrocarbon Hydroxylase Activity Induced by Polycyclic Aromatic Compounds. *Mol. Pharmacol.* 12, 225-233 (1976).

4. S. S. Thorgeirsson, and W. L. Nelson, Separation and Quantitative Determination of 2-Acetylaminofluorene and its Hydroxylated Metabolites by High Pressure Liquid Chromatography. *Anal. Biochem.*, in press (1976).

5. W. L. Nelson, and S. S. Thorgeirsson, Structural Requirements for Mutagenic Activity of 2-Acylaminofluorenes in the *Salmonella* Test System. *Biochem. Biophys. Res. Commun.* 71, 1201-1206 (1976).

6. P. J. Poppers, W. Levin, and A. H. Conney, Comparative Studies on Metabolism of Benzo[a]pyrene (BP) and Phenacetin (P) by Various Mouse Strains. *Pharmacologist* 16, 262 (1974).

7. S. A. Atlas, and D. W. Nebert, Genetic Association of Increases in Naphthalene, Acetanilide, and Biphenyl Hydroxylations with Inducible Aryl Hydrocarbon Hydroxylase in Mice. *Arch. Biochem. Biophys.* 175, 495-506 (1976).

8. S. S. Thorgeirsson, S. A. Atlas, A. R. Boobis, J. S. Felton, and D. W. Nebert, Varying Substrate Specificities of Cytochrome P_1-450 in Polycyclic Hydrocarbon-Treated Animals of Five Species. *Pharmacologist* 17, 217 (1975).

9. J. R. Robinson, and D. W. Nebert, Genetic Expression of Aryl Hydrocarbon Hydroxylase Induction. Presence or Absence of Association with Zoxazalamine, Diphenylhydantoin, and Hexobarbital Metabolism. *Mol. Pharmacol.* 10, 484-493 (1974).

IN VIVO PARAMETERS OF DRUG METABOLISM—DIFFERENCES IN SPECIFICITY TOWARDS INDUCING AGENTS

Ivar Roots, Beate Ley and Alfred G. Hildebrandt

Institute for Clinical Pharmacology, Free University Berlin, D-1000 Berlin 45, Hindenburgdamm 30, Germany

ABSTRACT

The occurrence or absence of enzyme induction in man can be sufficiently described by simultaneous application of four different in vivo parameters of drug metabolism (1,2). However, due to the existence of different types of inducers which provoke discernible states of induction, it can be presumed that the single parameters respond differently according to the type of inducer applied.

Six in vivo and two in vitro parameters have been examined systematically in the male guinea pig, applying phenobarbital, pregnenolone-16α-carbonitrile, 3-methyl-cholanthrene, rifampicine and spironolactone as inducers. Thus, a typical pattern of response or lack of response of the parameters is obtained which characterizes the inducers in vivo. Since various agents induce various types of cytochrome P-450, and likewise various inducers affect different in vivo parameters of drug metabolism, these investigations might lead to the cytochrome P-450 specificity of various in vivo parameters of drug metabolism.

The experiments further demonstrate that the lack of response of one parameter does not necessarily exclude the incidence of enzyme induction. In guinea pig, a combination of different parameters allows to establish an inducer-specific profile of in vivo parameters which perhaps can be transferred to situations existing in man.

INTRODUCTION

It has been shown that the presence or absence of induction of the drug hydroxylating enzyme system in man can be recognized by simultaneous application of several in vivo tests (1,2). These tests comprise the measurement of alterations of metabolism of endogenous substances (e.g. of cortisol and glucuronic acid by measuring urinary excretion rates of 6β-hydroxycortisol and D-glucaric acid, respectively), and the activity of serum enzymes (γ-glutamyltranspeptidase, β-glucuronidase). Besides these non-invasive parameters, the application of a test drug (e.g. aminopyrine, antipyrine) proved to be a valuable tool in evaluating enzyme induction. All these tests respond to phenobarbital treatment and can be regarded as in vivo parameters of enzyme induction in man. However, the existence of various types of inducing agents (e.g. phenobarbital, 3 methyl-cholanthrene) which stimulate different forms of cytochrome P-450 (3) requires further characterisation of induction. If the type of induction has to be determined in vivo, the palette of parameters used should contain tests which respond specifically to different kinds of induction or inducers. Two questions arise:
1) Can the effect of each type of inducer be characterized by a certain pattern of response of in vivo parameters?

2) If so, is it possible to correlate the inducing activity of unknown compounds with known types of inducers by a common pattern of in vivo parameters? To answer the first question and to obtain a relationship between inducers and parameters, the effect of known inducers on in vivo parameters has been studied in male guinea pigs (Table 1).

METHODS

Male guinea pigs (250-350 g) were subjected to different kinds of pretreatment: Phenobarbital (PB) 50 mg/kg i.p. daily for 7 days; 3-methylcholanthrene (MC) 26.5 mg/kg i.p. for 3 days; Rifampicine (Rif) 80 mg/kg p.o. for 5 days; Pregnenolone-16α-carbonitrile (PCN) 20 mg/kg p.o. for 2 days; Spironolactone (Spir) 100 mg/kg p.o. twice daily for 3 days.

Cytochrome P-450 and b_5 were measured in microsomal fractions according to Omura and Sato (4). For quantitation of D-glucaric acid, 4-aminoantipyrine, and antipyrine, similar methods were applied as described (1). A high pressure liquid chromatographic method (unpublished) has been developed for determination of 6β-hydroxycortisol (6β-OHF) in 24 hrs urine specimen; 17-hydroxycorticosteroids (17-OHCS) in urine were determined according to Sanghvi et al. (5). For collection of urine, two animals were kept in each metabolic cage. Hexobarbital sleeping time was determined upon administration of 50 mg/kg i.p. The dose for zoxazolamine paralysis time was 100 mg/kg i.p.

RESULTS AND DISCUSSION

To establish a relationship between induction in vivo and in vitro, cytochromes P-450 and b_5 were determined in vitro (Fig. 1). However, the determination of CO-binding spectra does not enable to differentiate the content of specifically induced forms of cytochrome P-450. An increase of cytochrome P-450 was obtained after PB, MC and rifampicine pretreatment. The lack of increase observed after treatment with PCN and spironolactone does therefore not necessarily exclude an increase of a specific form of cytochrome P-450. In contrast to cytochrome P-450, cytochrome b_5 remains unaffected by the different kinds of pretreatment.

Clear "yes" or "no" responses are desirable, i.e. an exceptional effect or the lack of any effect, to characterize inducers by parameters and vice versa.

TABLE 1   List of Parameters and Inducers Applied

In vivo Parameters	Inducers
1. 6β-Hydroxylation of Cortisol (urine)	1. Phenobarbital
2. Glucaric acid excretion (urine)	2. 3-Methylcholanthrene
3. Elimination of Aminopyrine (urine)	3. Rifampicine
4. Elimination of Antipyrine (serum)	4. Spironolactone
5. Hexobarbital sleeping time	5. PCN
6. Zoxazolamine paralysis time	

In vitro Parameters	Controls
1. Cytochrome P-450	1. Application of solvent
2. Cytochrome b_5	2. Food reduction

In Vivo Parameters of Drug Metabolism   583

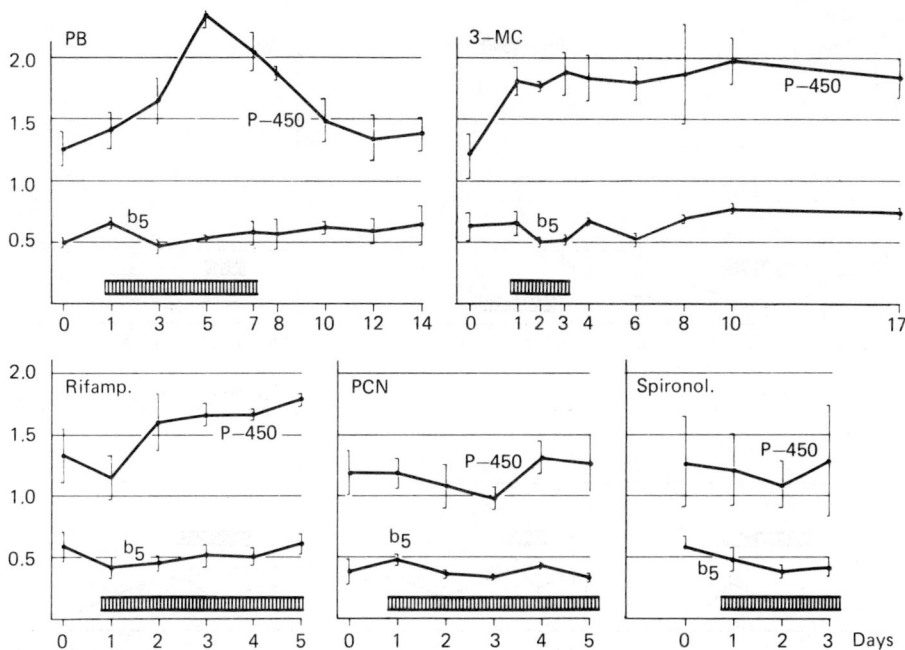

Fig. 1. Content of cytochrome P-450 and b_5 in liver microsomes of male guinea pigs after different kinds of pretreatment. Values are given in nmoles per mg of protein. Duration of pretreatment is indicated by the bars. Each point represents mean of 3 independent preparations of two livers.

Smaller, and in this respect uncharacteristic responses will be considered in the following only to a minor degree.

6ß-Hydroxycortisol

Cortisol is hydroxylated in 6ß-position by a hepatic mixed function oxidase. The extent of 6ß-hydroxycortisol production depends on the activity of the respective hydroxylase, and the degree of substrate (cortisol) supply. Urinary excretion rates of 6ß-hydroxycortisol have been applied to evaluate enzyme induction in vivo (6). In the present study this parameter responds by an about 3-5 fold increase only to phenobarbital and rifampicine pretreatment. This increase is partly unspecific and due to an increased supply of endogenous cortisol as reflected by increased excretion of urinary 17-OHCS. The ratio of 6ß-OHF/17-OHCS partly compensates for this effect (1,7). The significant increase of this ratio after phenobarbital and rifampicine enables to separate these inducers from MC, PCN and spironolactone.

D-Glucaric acid

Several parameters allow to further differentiate between phenobarbital and rifampicine effects. The clearest distinction, however, is obtained from measurements of urinary D-glucaric acid excretion rates (8), as this parameter only responds to PB pretreatment under conditions examined (Fig. 2).

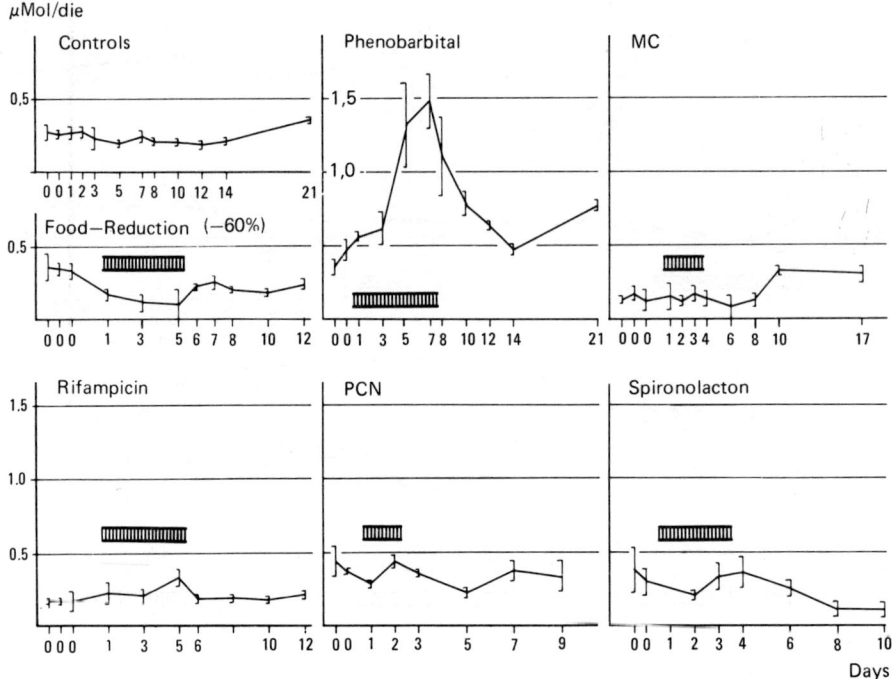

Fig. 2. Influence of various kinds of pretreatment on urinary D-glucaric acid excretion in male guinea pigs. Each point represents mean (+S.E.M.) obtained from six animals which were kept in pairs for urine collection.

Hexobarbital sleeping time

This classic test (9) shows a good response to phenobarbital pretreatment (shortening of sleeping time by about 50 %). No alteration is obtained after PCN, spironolactone or rifampicine treatment. Methylcholanthrene also shortens hexobarbital sleeping time in the guinea pig, but to a minor degree than PB (-25 %). A distinction between these two inducers by this parameter seems to be impracticable, though possible.

Zoxazolamine paralysis time

The underlying reaction of this parameter is an aromatic hydroxylation of zoxazolamine. This parameter seems to be inducer-specific for MC (-90 %). Phenobarbital produces a comparably little effect of -30 %; the other inducers do not alter the paralysis time significantly.

With the aid of the four in vivo parameters presented so far it is possible to describe a state of induction as obtained by the inducers PB, MC or rifampicine. None of these parameters is sensitive to PCN or spironolactone which both are effective enzyme inducers in rats (10,11). The following two tests, however, do respond to pretreatment with these substances.

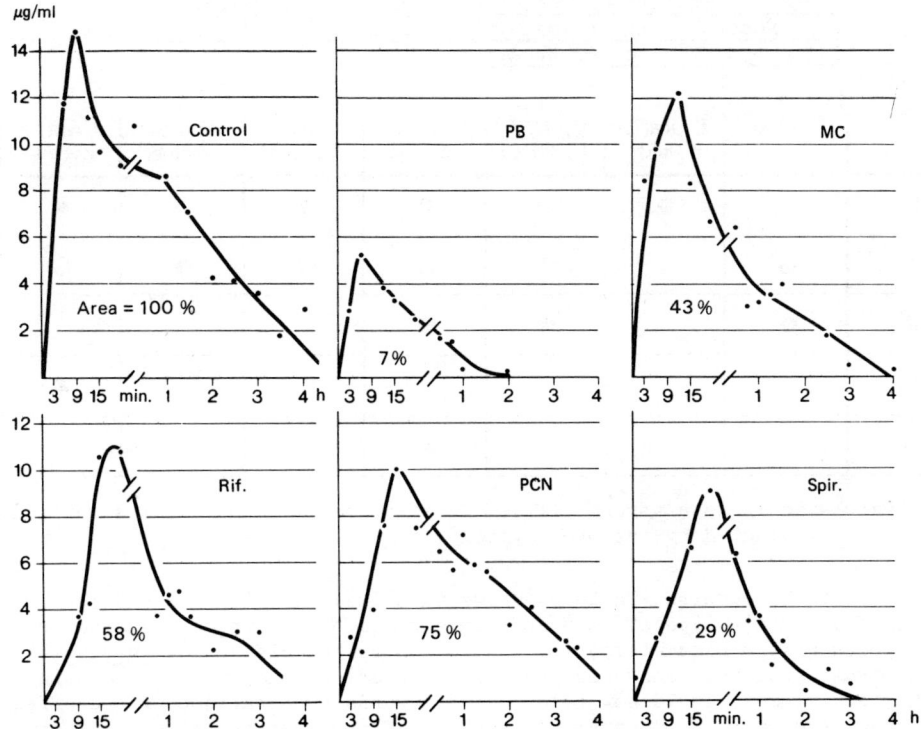

Fig. 3. Concentration-time curves of antipyrine in serum of male guinea pigs after different kinds of pretreatment. Each point represents one animal which was injected i.p. with 75 mg/kg of antipyrine at zero time and was killed at the times indicated. Area under the curves was measured in arbitrary concentration x time units and expressed in % of the value obtained for control animals.

Aminopyrine metabolism

After i.p. application of 70 mg of aminopyrine per kg of body weight, the urinary excretion within 24 hrs of one metabolite, 4-aminoantipyrine, has been determined. All inducers, with the exception of spironolactone, enhance urinary excretion of 4-aminoantipyrine: PB (+200 %), rifampicine (+120 %), PCN (50 %) and MC (+40 %). However, aminopyrine is almost completely excreted as 4-aminoantipyrine only after PB treatment.

Antipyrine metabolism

Fig. 3 shows that serum concentration-time curves of antipyrine following a dosage of 75 mg/kg i.p. are rather complex. Evaluating the area under the curves, however, provides a means by which the influence of the various inducers may be compared. Phenobarbital has the largest effect followed by spironolactone. Thus, aminopyrine and antipyrine prove to be suitable test substances to test for induction without distinguishing between different types of induction. The latter may be done by applying some of the other parameters mentioned above.

TABLE 2 Pattern of Response of In Vivo Parameters of Drug Metabolizing Enzyme System in the Guinea Pig after Different Kinds of Pretreatment

	P-450	Glucaric-acid	6β-OHF 17-OHCS	Amino-pyrine	Hexo-barbital	Zoxazol-amine	Anti-pyrine
PB	●	●	●	●	●	○	●
MC	●			○	○	●	○
Rif	●		●	●			○
Spir							●
PCN				●			

Dark symbols indicate a strong effect, relative smaller but still significant effects are indicated by the open symbols.

A summary of the results (Table 2) shows that it is possible to deduce the effect of a certain type of inducer from a certain pattern of response or lack of response of the parameters; i.e. induction can be classified in vivo by these parameters, thus answering the first question asked in the beginning. Further activities should now be directed towards a comparison of the effects of unknown agents to this parameter pattern as proposed above in the second point, in order to prove the applicability of this concept and to establish corresponding relationships of inducers and parameters in man.

An interesting basic aspect of this type of investigation concerns the demonstration of specific forms of cytochrome P-450 by in vivo parameters. The inducer specificity of various forms of cytochrome P-450 and the inducer specificity of in vivo parameters are necessary preconditions for this ultimate goal.

However, the parameters applied in this study partly lack this specificity. Ideally, a causal and quantitative relationship should exist between the response of a parameter and the activity of a certain species of cytochrome P-450. For instance, the drastic shortening of duration of the pharmacodynamic action of zoxazolamine by MC pretreatment directly reflects the known increase of cytochrome P-448 in vitro. In this respect, the 6β-hydroxycortisol, aminopyrine, hexobarbital and antipyrine tests are also considered to be "direct" tests, since the main underlying reactions are monooxygenase dependent.

On the other hand, parameters have been included which are only indirectly related to the monooxygenase system. Thus glucaric acid excretion primarily reflects changes in glucuronic acid metabolism. However, on an empirical basis, it proved to be a valuable tool to study enzyme induction in vivo (12, 13). Increases in glucaric acid excretion rates after application of many substances, considered to be enzyme inducers, correlate well with increases in other parameters (Table 2), e.g. pharmacokinetics of aminopyrine or cytochrome P-450 content (1,14).

Another aspect concerns both direct and indirect parameters: substances defined as enzyme inducers by their ability to raise the content of certain forms of cytochrome P-450 in microsomal fractions or to stimulate certain hydroxylation reactions in vitro often not only stimulate these parameters in vivo but exert further anatomical, physiological and biochemical effects. Thus, among other effects, phenobarbital stimulates liver growth, liver blood flow, and concentration of anion binding proteins in the hepatocyte. These unspecific effects may strongly influence the response of several in vivo parameters applied in this study. Vice versa, tests being specific for, e.g. liver blood flow, may prove efficient though indirect parameters for enzyme induction as obtained after phenobarbital treatment.

From these considerations it follows that the choice of suitable parameters depends on the question to be answered:

1) To recognize in vivo the application of a certain enzyme inducer (PB, PCN etc.) only a few tests are necessary as shown in Table 2. These parameters may either be direct ones and reflect the activity of the cytochrome P-450 species induced, or may be indirect and unspecific ones utilizing indicative effects.

2) If a compound, the inductive properties of which are not known, is to be examined by comparison with known types of inducers, a complete pattern of responses must be determined which allows to recognize a certain type of inducer and to exclude other types of inducers. Thus, compounds having a parameter profile different from that of one standard inducer, can also be classified.

3) To detect specific forms of cytochrome P-450 and evaluate their activity in vivo, all indirect parameters seem to be useless. Parameters must be applied which measure the activity of a reaction being specific for one certain form of cytochrome P-450 as directly as possible. The performance of the test must be such that all unspecific effects (changes in absorption or in the volume of distribution, bile flow, pharmacodynamic action of inducer etc.) are either minor or can be taken into consideration. Suitable test reactions may be learned from in vitro studies with purified cytochrome P-450 fractions obtained from the animal species to be studied.

As it will be difficult to estimate by direct or indirect methods under in vivo conditions all or many types of hydroxylation reactions, the characterization of one cytochrome P-450 form by one selected parameter should be aimed at. This means that from in vivo studies cytochrome P-450 will not be defined by its substrate specificity but by its parameter specificity.

4) In the clinical situation, a more descriptive approach to enzyme induction is reasonable, since induction in its strict sense of stimulation of cytochrome P-450 content and/or activity is often secondary to other biochemical and physiological changes introduced by enzyme inducers.

This study provides further evidence that in vivo enzyme induction cannot be sufficiently characterized by one single test, but only by a well reflected combination of several parameters.

REFERENCES

(1) A.G. Hildebrandt, I. Roots, M. Speck, K. Saalfrank and H. Kewitz, Evaluation of in vivo parameters of drug metabolizing enzyme activity in man after administration of clemastine, phenobarbital or placebo, Europ. J. clin. Pharmacol. 8, 327-336 (1975)

(2) I. Roots, K. Saalfrank and A.G. Hildebrandt, Comparison of methods to study enzyme induction in man, in: Cytochromes P-450 and b_5 (D.Y. Cooper, O. Rosenthal, R. Snyder and C. Witmer, eds.), 485-502, Plenum Press, New York, 1975

(3) A.H. Conney, A.Y.H. Lu, W. Levin, A. Somogyi, S. West, M. Jacobson, D.Ryan and R. Kuntzman, Effect of enzyme inducers on substrate specificity of the cytochrome P-450's, Drug Metab. and Dispos. 1, 199-209 (1973)

(4) T. Omura, R. Sato, The carbon monoxide-binding pigment of liver microsomes. II. Solubilization, purification and properties. J. biol. Chem. 239, 2379-2385 (1964)

(5) A. Sanghvi, C. Wight, B. Parikh and H. Desai, Urinary 17-hydroxycorticosteroid determination with p-hydrazinobenzenesulfonic acid-phosphoric acid, Am. J. Clin. Pathol. 60, 684-690 (1973)

(6) R. Kuntzman, M. Jacobson, K. Schneideman and A.H. Conney, Similarities between oxidative drug metabolising enzymes and steroid hydroxylases in liver microsomes, J. Pharmac. exp. Ther. 146, 280-285 (1964)

(7) A.H. Conney, Pharmacological implications of microsomal enzyme induction, Pharmacol. Rev. 19, 317-366 (1967)

(8) C.A. Marsh, L.M. Reid, Changes in D-glucaric acid excretion induced by stimulators of ascorbic acid biosynthesis, Biochim. biophys. Acta (Amst.) 78, 726-728 (1963)

(9) H. Remmer, Der beschleunigte Abbau von Pharmaka in den Lebermikrosomen unter dem Einfluss von Luminal, Naunyn-Schmiedebergs Arch. exp. Path. Pharmak. 235, 279-290 (1959)

(10) H. Selye, Hormones and resistance, Springer, Berlin, 1971

(11) R. Stripp, R.H. Menard, N.G. Zampaglione, M.E. Hamrick and J.R. Gillette, Effect of steroids on drug metabolism in male and female rats, Drug Metab. and Dispos. 1, 216-221 (1973)

(12) E.M. Aarts, Evidence for the function of D-glucaric acid as an indicator for drug induced enhanced metabolism through the glucuronic acid pathway in man, Biochem. Pharmacol. 14, 359-363 (1965)

(13) J. Hunter, J.D. Maxwell, M. Carrella, A.D. Stewart and R. Williams, Urinary D-glucaric acid excretion as a test for hepatic enzyme induction in man, Lancet 1, 572-575 (1971)

(14) J. Hunter, J.D. Maxwell, D.A. Stewart and R. Williams, Urinary D-glucaric acid excretion and total liver content of cytochrome P-450 in guinea pigs: Relationship during enzyme induction and following inhibition of protein synthesis, Biochem. Pharmacol. 22, 743-747 (1973)

STUDIES ON THE MECHANISM BY WHICH DISULFIRAM AND DIETHYLDITHIOCARBAMATE AFFECT DRUG METABOLISM

M. Marselos, P. Alakuijala, M. Lang and R. Törrönen

Department of Physiology and Chemistry, University of Kuopio, SF-70101 Kuopio 10, Finland

ABSTRACT

Treatment of adult male rats with disulfiram or diethyldithiocarbamate (300 mg/kg, *per os*, x4) severely depressed the microsomal content of cytochrome P-450 and enhanced the activity of UDPglucuronosyltransferase. Also the hepatic contents of copper, zinc and magnesium were markedly changed. Disulfiram increased the levels of copper and zinc, while diethyldithiocarbamate decreased to the same extent all the metals determined. When phenobarbital was given concurrently, the effects of disulfiram on cytochrome P-450 and the metal contents were significantly depressed, and those of diethyldithiocarbamate were almost completely abolished.

The observed changes of the drug-metabolizing enzymes do not correlate with those of the metal contents. The accumulation of copper in the liver after treatment with disulfiram, however, strongly indicates that chemical intrahepatic cholestasis may be the underlying mechanism for the action of disulfiram on microsomal drug metabolism.

INTRODUCTION

Several studies have shown that disulfiram severely inhibits microsomal drug metabolism (1-3), and prolongs the half-life of drugs concurrently administered (4,5). Despite the overall impairment of the mixed-function oxidase, however, the activity of NADPH-cytochrome *c* reductase is increased by disulfiram treatment (3). On the other hand, enzymes of the D-glucuronic acid pathway are enhanced, indicating an increased glucuronidation capacity (6). These paradoxal effects of disulfiram suggest that this compound cannot be considered exclusively either as an inducer nor as an inhibitor of drug metabolism.

Disulfiram is reduced *in vivo* to diethyldithiocarbamate (7), which is a potent chelating agent with a high affinity for copper. Administration, therefore, of either compound could be expected to alter the content and function of metals in liver and other tissues. In fact, disulfiram and its reduced derivative inhibit copper-containing enzymes (8,9).

It has been reported that copper is essential for the synthesis of heme in the liver mitochondria (10), and it is plausible that changes in the content of this metal may affect the level of the microsomal hemoprotein, cytochrome P-450. Chelation of copper in the hepatic cell could be the underlying mechanism for the reported depression of cytochrome P-450 by disulfiram and diethyldithiocarbamate (1,3,11). In addition to copper, several other trace elements are able to influence the activity of enzymes involved in the metabolism of drugs (12).

In this study, the effects of disulfiram and diethyldithiocarbamate on the hepatic content of copper, zinc and magnesium, were compared to those on cytochrome P-450 and UDPglucuronosyltransferase. These parameters were also examined after treatment with phenobarbital solely and in combination with disulfiram or diethyldithiocarbamate.

MATERIALS AND METHODS

Treatment of the animals

Thirty five adult male albino rats (Wistar/Af/Han/Mol/Han 67) were used. The animals were housed in groups of five, in plastic cages with birch-chips as bedding. Tap water and standard rat chow (Hankkija Ltd., Finland) were provided *ad libitum*, and the diurnal lighting was 14 hr light and 10 hr dark. Disulfiram (tetraethylthiuram disulfide, Antabus, Dumex A/S, København, Denmark) was prepared as a fine suspension in distilled, deionized water (6%,w/v), and thoroughly mixed before every administration. Sodium diethyldithiocarbamate (trihydrate) (Merck AG, Darmstadt, W.Germany) was dissolved in distilled water (6%, w/v). Both compounds were given by gastric intubation (300 mg/kg daily, for 4 days). Control animals received intragastrically a respective volume of distilled water. Phenobarbital (Merck) was dissolved (2%) in 0.5 N NaOH, and the pH was adjusted to 8.6 by 1 N HCl. The drug was given subcutaneously at a dose of 80 mg/kg daily, for 4 days. Control animals were injected with a respective volume of saline.

Tissue preparation

The rats were killed with a blow on the head and bled by cutting the cervical vessels. The livers were homogenized in four volumes of 0.25 M sucrose with a Teflon-pestle Potter-Elvehjem tissue grinder. Part of this crude homogenate was solubilized by sodium deoxycholate (Merck) (final conc 2.5%), so that 10 ml corresponded to 1 g fresh wt. The rest was centrifuged at 10 000 g for 15 min and the postmitochondrial supernatant was recentrifuged at 105 000 g, in order to obtain the microsomal fraction. Microsomes were resuspended in KCl (0.15 M), and part of them solubilized by sodium deoxycholate (final conc 2.5%).

Microsomal drug-metabolizing enzymes

The level of cytochrome P-450 and the activity of UDPglucuronosyltransferase were measured in the microsomes as previously described (13,14).

Trace element analyses

All reagents used were analytical grade, and were prepared in distilled, deionized water. During the treatment of the samples, acid-washed glassware and plastic automatic micropipettes were used for preventing any possible contamination. Copper (Cu), zinc (Zn) and magnesium (Mg) were measured by an atomic absorption spectrophotometer (Perkin Elmer, model 306), using the method of flame atomization. The tissue preparations used were solubilized to homogeneity by sodium deoxycholate. In all cases, the unknown sample was added to external standards. For the elimination of possible quenching, series of standards were run, prepared with KCl and sodium deoxycholate in the same concentration as in the tissue preparations. All measurements were in triplicate.

Cu was determined by adding 150 μl of the unknown sample to 350 μl of standard solutions (0.0, 0.5, 1.5 and 2.0 μg/ml). The standards were obtained from a stock solution (1 mg Cu/ml), prepared by dissolving 1 g metallic Cu in a small volume of 50% HNO_3 and by diluting up to 1 l. with 1% HNO_3.

Determinations of Zn were done after addition of 150 μl of the unknown sample to 1 ml of external standard solutions (0.0, 0.5, 1.5 and 2.0 μg/ml). The stock standard solution (0.5 mg Zn/ml) was prepared by dissolving 0.5 g metallic Zn in a small volume of 50% HCl, and diluting to 1 l. with 1% HCl.

Mg was determined from 50 μl of unknown sample and 2 ml of external standard solutions (0.0, 0.25, 0.5, 0.75 and 1.0 μg/ml). The stock standard solution (1 mg Mg/ml) was prepared by dissolving 1 g Mg in HCl, as described for Zn.

Protein determination was carried out by the biuret method (15), using bovine serum albumin (Sigma, St. Louis, U.S.A.) as standard.

Statistical analysis of the results was made by Student's t-test. Values of P less than 0.05 were considered significant.

RESULTS

Treatment with disulfiram and diethyldithiocarbamate

Treatment with disulfiram increased the microsomal protein and phospholipid contents by 15% (P<0.05). A similar, but not statistically significant trend, was noticed after treatment with diethyldithiocarbamate. The level of cytochrome P-450 was uniformly decreased by both compounds, while the activity of UDPglucuronosyltransferase was greatly enhanced (Fig. 1).

Administration of disulfiram or diethyldithiocarbamate affected all the tested trace metals, measured either in the whole homogenate or in the microsomes. However, the changes were significantly different between the two compounds (Table 1,A). Disulfiram produced a 3-fold increase in the amount of Cu in the liver parenchyma and also in the isolated microsomes. An increase was noticed also in the amount of Zn when measured from the crude homogenate. Mg was less affected than the other metals. Its content in the hepatic homogenate was decreased, while it showed a slight increase in the microsomal fraction.

As expected, all metals tested were diminished by diethyldithiocarbamate due to the chelating properties of this compound. The effect of diethyldithiocarbamate was more pronounced in the whole hepatic tissue than in the microsomes. In the crude homogenate, Cu and Zn were decreased by 20%, while Mg showed only a slight decline (10%). In the microsomes, although the concentration of all metals tended to decrease, only Cu was statistically different compared to the controls.

Treatment with phenobarbital alone, and in combination with disulfiram or diethyldithiocarbamate

Administration of phenobarbital alone increased both the level of cytochrome P-450 and the activity of UDPglucuronosyltransferase. Concurrent administration of phenobarbital with disulfiram, however, resulted in a suppression of the induction of cytochrome P-450 produced by phenobarbital alone. Diethyldithiocarbamate brought about the same effect, but to a lesser degree than disulfiram (Fig. 1). In the case of UDPglucuronosyltransferase, combination of the tested compounds led to an additive enhancement of the activity of this enzyme (Fig. 1). Animals treated with phenobarbital alone had in general depressed hepatic contents of Cu, Zn and Mg, both in the crude homogenate and in the microsomal

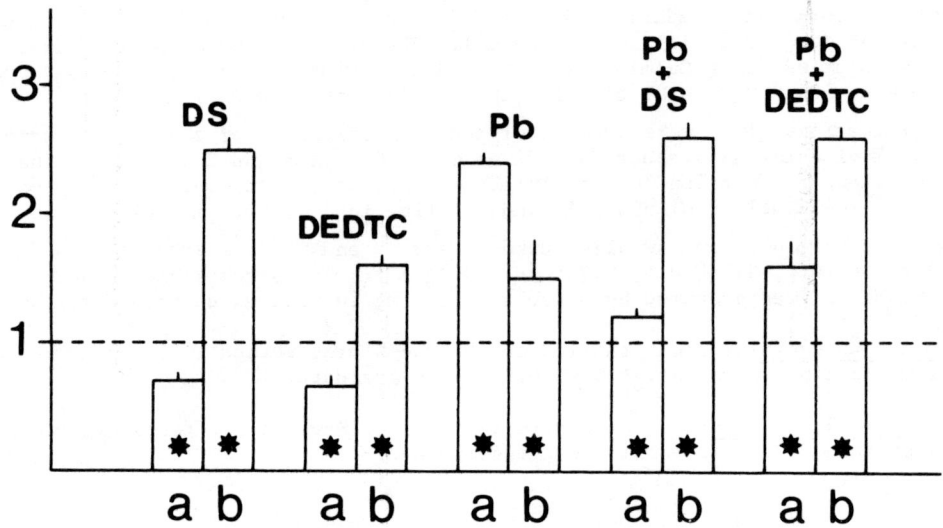

Fig. 1. Microsomal drug-metabolizing enzymes.
Relative content of cytochrome P-450 (a) and activity of UDPglucuronosyltransferase (b) in rats treated with disulfiram (DS), diethyldithiocarbamate (DEDTC) and phenobarbital (Pb) solely, or in combination (Pb+DS, Pb+DEDTC). The mean values of five experimental animals ($\pm$ SE) are compared to the untreated controls (5), considered to be unity. Control values were for cytochrome P-450 0.5$\pm$0.01 nmol/mg protein, and for UDPglucuronosyltransferase 0.44$\pm$0.01 nmol/min/mg protein. The asterisk denotes P<0.05.

fraction. The amount of Zn was markedly decreased in both preparations tested, but that of Mg showed a significant change only in the crude homogenate (Table 1,B). Concurrent administration of phenobarbital and disulfiram diminished to some extent the effects of disulfiram, as can be seen in the case of Cu and Zn (Table 1,A,B). On the contrary, the disulfiram-produced decrease in the hepatic Mg content (10%) was potentiated by concomitant administration of phenobarbital. Animals treated with phenobarbital and diethyldithiocarbamate in combination had lower hepatic levels of Cu, Zn and Mg. It was, however, only Mg which differed statistically from the controls. Thus the slight decrease observed after treatment with diethyldithiocarbamate alone was completely abolished by phenobarbital.

DISCUSSION

Trace elements are necessary for the normal function of the cell, since they are cofactors in many important enzymic systems of intermediary metabolism (16,17). The enzymes of drug metabolism are also influenced by the presence of metals, as has been shown by *in vivo* and *in vitro* studies. A common finding by many investigators is the inhibitory effect of Zn on the activity of UDPglucuronosyltransferase, while Mg and Cu activate or have no effect on this enzyme (18-22). Our results show that administration of disulfiram and diethyldithiocarbamate produces a large increase in the activity of UDPglucuronosyltransferase, which cannot be interpreted as due to changes in the micro-

TABLE 1 HEPATIC CONTENTS OF COPPER, ZINC AND MAGNESIUM

PANEL A Changes produced by disulfiram (DS) or diethyldithiocarbamate (DEDTC). The data quoted are the means of five animals (±SD), and are expressed as ng/mg protein. * Denotes $P<0.05$.

	Control	DS	DEDTC
Crude homogenate			
Cu	24.1 ± 3.0	71.3 ± 12.1 *	19.0 ± 2.2 *
Zn	166.1 ± 8.6	200.0 ± 12.2 *	127.0 ± 12.3 *
Mg	836.3 ± 45.2	767.4 ± 43.0 *	739.2 ± 30.4 *
Microsomes			
Cu	9.5 ± 1.0	23.3 ± 4.2 *	8.2 ± 0.6 *
Zn	129.4 ± 15.8	129.3 ± 9.9	120.3 ± 2.5
Mg	1102.0 ± 144.0	1106.0 ± 111.2	968.4 ± 65.3

PANEL B The same parameters as above, after treatment with phenobarbital (Pb) alone, and in combination with disulfiram or diethyldithiocarbamate (Pb+DS, Pb+DEDTC).

	Control	Pb	Pb+DS	Pb+DEDTC
Crude homogenate				
Cu	18.0 ± 2.0	15.1 ± 1.0	48.3 ± 7.0 *	17.4 ± 1.2
Zn	115.1 ± 24.3	80.2 ± 7.2 *	129.1 ± 13.2 *	98.4 ± 4.4
Mg	1017.0 ± 152.0	708.0 ± 152.0 *	726.0 ± 54.0 *	719.0 ± 30.0 *
Microsomes				
Cu	10.4 ± 1.4	8.3 ± 1.5	20.4 ± 2.5 *	11.1 ± 0.2
Zn	155.7 ± 14.0	118.2 ± 17.4 *	163.1 ± 20.2	147.5 ± 9.4
Mg	1076.0 ± 91.0	962.0 ± 142.0	1062.0 ± 70.0	1154.0 ± 70.0

somal trace element contents. Similarly, the depression of cytochrome P-450 by disulfiram and diethyldithiocarbamate is not due to the depletion of Cu from the hepatic cell and the consequent inhibition of heme synthesis, as has been suggested (10). Although both compounds decreased the amount of cytochrome P-450, they affected in an opposite way the level of Cu.

The accumulation of Cu in the liver after disulfiram treatment may indicate mobilization of the metal from other tissues, increased Cu-binding capacity of the hepatic proteins, or decreased Cu excretion. The last alternative seems most probable although no further investigation was pursued on these lines. The bile is the major pathway for Cu excretion (23), and a disturbance in the biliary production and excretion could readily affect the rate of Cu elimination from the liver. In fact treatment of rats with disulfiram (100 mg/kg, x3) produced a 6-fold increase in the plasma concentration of total bilirubin, which indicates a severe cholestasis (24). Furthermore, metabolism of disulfiram and diethyldithiocarbamate yields carbon disulfide, which has also been found to produce bile retention (25). It is thus plausible that disturbed bile secretion is responsible for the accumulation of Cu in the liver. In the case of diethyldithiocarbamate, chelation with this water-soluble compound mediates perhaps a compensatory excretion of Cu in the urine. The contribution of the bile in the excretion of Zn is less compared to that of Cu (17), and this may explain the moderate accumulation of Zn in the liver after treatment with disulfiram. On the other hand, the lack of any change in the hepatic Mg content can be attributed to the fact that this metal is readily excretable via the urine (26).

Chronic administration of phenobarbital mobilizes Cu from the muscles and increases its level in the liver (27). Treatment with phenobarbital for only three days has been found effective in elevating the amount of Cu and Zn in the microsomal fraction but not in the whole liver (28,29). In our experiments, phenobarbital did not significantly affect the amount of hepatic Cu, while it produced even a slight decrease in the Zn content of the whole liver, as well as of the microsomes. These discrepancies cannot be given a satisfactory explanation, since a similar dose of phenobarbital was used in our studies. The modified atomic absorption method used by us for the determination of metals is simple, accurate and reproducible. Moreover, the control values obtained in the present study are very well compatible with those previously reported for the rat liver (28-30).

Treatment with phenobarbital could somewhat diminish the increase of Cu and Zn produced by disulfiram. This gives further support to the hypothesis that these metals are retained in the liver due to impaired bile excretion, since phenobarbital itself acts in an opposite way enhancing the formation and excretion of bile (31).

Our results indicate that the changes in drug-metabolizing enzymes produced by disulfiram, diethyldithiocarbamate or phenobarbital are not mediated by alterations in the hepatic microsomal metal contents. The decrease in the detectable amount of cytochrome P-450 may be due to binding with disulfiram, which behaves as a type I substrate (3). On the other hand, both disulfiram and diethyldithiocarbamate react with -SH groups (32), and this could trigger inactivation of cytochrome P-450 (33). In addition to these direct actions, an indirect impairment of the microsomal mono-oxygenase system could be expected due to the generation of carbon disulfide (34).

The possibility exists, however, that biliary retention produced by disulfiram, diethyldithiocarbamate, or carbon disulfide, could be partly responsible for the changes found in the microsomal membranes and the activity of the membrane-bound cytochrome P-450 and UDPglucuronosyltransferase. This view is supported by the observations that also cholestasis brings about an increase of the microsomal protein and phospholipids, it inactivates the hydroxylation enzymes, and it enhances the activity of UDPglucuronosyltransferase (35,36).

REFERENCES

1. B. Stripp, F.E. Greene and J.R. Gillette, Disulfiram impairment of drug metabolism by rat liver microsomes, J. Pharmacol. exp. Ther. 170, 347 (1969).
2. T. Honjo and K.J. Netter, Inhibition of drug demethylation by disulfiram *in vivo* and *in vitro*, Biochem. Pharmacol. 18, 2681 (1969).
3. M.A. Zeimatis and F.E. Greene, Impairment of hepatic microsomal drug metabolism in the rat during daily disulfiram administration, Biochem. Pharmacol. 25, 1355 (1976).
4. O.V. Olesen, Disulfiram (Antabuse) as inhibitor of phenytoin metabolism, Acta pharmacol. et toxicol. 24, 317 (1966).
5. N.J. Giarman, F.H. Flick and J.M. White, Prolongation of thiopental anesthesia in the mouse by premedication with tetraethylthiuram disulfide ("antabuse"), Science 114, 35 (1951).
6. W.R.F. Notten and P.Th. Henderson, Effect of disulfiram on the urinary D-glucaric acid excretion and activity of some enzymes involved in drug metabolism in guinea-pig, Arch. int. Pharmacodyn. 205, 199 (1973).
7. J.H. Strömme, Effects of diethyldithiocarbamate and disulfiram on glucose metabolism and glutathione content of human erythrocytes, Biochem. Pharmacol. 12, 705 (1963).
8. R.E. Heikkila, F. Gabbat and G. Cohen, *In vivo* inhibition of superoxide dismutase in mice by diethyldithiocarbamate, J. Biol. Chem. 251, 2182 (1976).
9. M. Goldstein, B. Anagnoste, E. Lauber and M.R. McKeregham, Inhibition of dopamine-β-hydroxylase by disulfiram, Life Sci. 3, 763 (1964).
10. G.S. Wagner and T.R. Tephly, A possible role of copper in the regulation of heme biosynthesis through ferrochelatase, Adv. Exp. Med. Biol. 58, 343 (1975).
11. A.L. Hunter and R.A. Neal, Inhibition of hepatic mixed-function oxidase activity *in vitro* and *in vivo* by various thionosulfur-containing compounds, Biochem. Pharmacol. 24, 2199 (1975).
12. T.C. Campbell and J.R. Hayes, Role of nutrition in the drug-metabolizing enzyme system, Pharmacol. Rev. 26, 171 (1974).
13. T. Omura and R. Sato, The carbon monoxide-binding pigment of liver microsomes. II. Solubilization, purification and properties, J. Biol. Chem. 239, 2379 (1964).
14. O. Hänninen and R. Puukka, Effect of digitonin on UDPglucuronosyltransferase in microsomal membranes, Finn. Chem. J. B 43, 451 (1970).
15. A.G. Gornall, C.J. Bardawill and M.M. David, Determination of serum proteins by means of the biuret reaction, J. Biol. Chem. 177, 751 (1949).
16. J.G. Reinhold, Trace elements - A selective survey, Clin. Chem. 21, 476 (1975).
17. R.E. Burch, K.J. Hahn and J.F. Sullivan, Newer aspects of the roles of zinc, manganese, and copper in human nutrition, Clin. Chem. 21, 501 (1975).
18. I. Schällhammer, D.S. Poll and M.H. Bickel, Liver microsomal β-glucuronidase and UDPglucuronosyltransferase, Enzyme 20, 269 (1975).

19. D. Zakim, J. Goldenberg and D.A. Vessey, Effects of metals on the properties of hepatic microsomal uridine diphosphate glucuronosyltransferase, Biochemistry 12, 4068 (1973).
20. G.W. Lucier, O.S. McDaniel and H.B. Matthews, Microsomal rat liver UDP-glucuronosyltransferase: Effects of piperonyl butoxide and other factors on the enzyme activity, Arch. Biochem. Biophys. 145, 520 (1971).
21. J. Frei, H. Birchmeier and E. Schmid, Multiplicity and specificity of UDPglucuronosyltransferase, Enzym. biol. clin. 11, 385 (1970).
22. T.R. Devereux and J.R. Fouts, N-Oxidation and demethylation of N,N-dimethylaniline by rabbit liver and lung microsomes. Effects of age and metals, Chem.-Biol. Interactions 8, 91 (1974).
23. G.W. Evans, Copper homeostasis in the mammalian system, Physiol. Rev. 53, 535 (1973).
24. P.G. Tonkes, The glucaric acid pathway of glucuronic acid metabolism, Ph.D. Thesis, Univ. New South Wales, Australia (1973).
25. J.D. Gibson and R.J. Roberts, Effect of carbon disulfide on liver function *in vivo* and in the isolated perfused liver, J. Pharmacol. exp. Ther. 181, 176 (1972).
26. H.A. Harper, Review of Physiological Chemistry, Lange Medical Publications, Los Altos, California (1973).
27. D. Hilderbrand, M.S. Fahim, D.G. Hall and E. Pickett, in: Trace Substances in Environmental Health, (ed. D.D. Hemphill), 5, 235, University of Missouri, Columbia (1972).
28. A.E. Moffit, Jr., J.R. Dixon, F.C. Phipps and H.E. Stokinger, The effect of benzpyrene, phenobarbital, and carbon tetrachloride on subcellular metal distribution and microsomal enzyme activity, Cancer Res. 32, 1148 (1972).
29. A.E. Moffit, Jr. and S.D. Murphy, Effect of excess and deficient copper intake on rat liver microsomal enzyme activity, Biochem. Pharmacol. 22, 1453 (1973).
30. V. Albergoni, A. Cassini, N. Favero and G.P. Rocco, Effect of penicillamine on some metals and metalloproteins in the rat, Biochem. Pharmacol. 24, 1131 (1975).
31. R.J. Roberts and G.L. Plaa, Effect of phenobarbital on the excretion of an exogenous bilirubin load, Biochem. Pharmacol. 16, 827 (1967).
32. J.H. Strömme, Interactions of disulfiram and diethyldithiocarbamate with serum proteins studied by means of a gel filtration technique, Biochem. Pharmacol. 14, 381 (1965).
33. H.S. Mason, J.C. North and M. Vanneste, Microsomal mixed-function oxidations: the metabolism of xenobiotics, Fed. Proc. 24, 1172 (1965).
34. F. De Matteis, Covalent binding of sulphur to microsomes and loss of cytochrome P-450 during the oxidative desulphuration of several chemicals, Mol. Pharmacol. 10, 849 (1974).
35. F. Hutterer, P.G. Bacchin, I.H. Raisfeld, J.B. Schenkman, F. Schaffner and H. Popper, Alteration of microsomal biotransformation in the liver in cholestasis, Proc. Soc. Exp. Biol. Med. 133, 702 (1970).
36. E. Hietanen, A. Auranen and Ch. Savvakis, The rate of biotransformation reactions and cellular viability in the livers of patients with biliary diseases, Acta Hepato-Gastroenterol. 22, 170 (1975).

ALTERATIONS IN THE ACTIVITIES OF DRUG METABOLIZING ENZYMES IN RAT LIVER MICROSOMES BY DIETARY AND *IN VITRO* INCORPORATED CHOLESTEROL

Matti Lang and Osmo Hänninen

Department of Physiology, University of Kuopio, SF-70101 Kuopio 10, Finland

ABSTRACT

Feeding the rats with a 4% cholesterol diet was able to increase the microsomal cholesterol to protein ratio. Dietary cholesterol elevated also the microsomal phospholipid to protein ratio, the microsomal content of cytochrome P-450 and the specific activities of microsomal NADPH cytochrome c reductase and p-nitroanisole-O-demethylase. Microsomal UDPglucuronosyltransferase activity was also elevated, although the increase could be demonstrated only after trypsin digestion of microsomes.
On the contrary, when cholesterol was incorporated into microsomes in vitro, a decrease in the activities of p-nitroanisole-O-demethylase and UDPglucuronosyltransferase took place. NADPH cytochrome c reductase activity and the amount of cytochrome P-450 was not affected by the cholesterol incorporation. When the binding of the fluorescent probe 1,8-ANS in microsomes was studied, it was found that dietary cholesterol and the in vitro incorporation of cholesterol had an opposite effect on the binding properties of probe molecules in microsomal membranes.

INTRODUCTION

Biological membranes contain variable amounts of cholesterol as an essential component of their structure. There exists much evidence that cholesterol affects membranes by restricting the mobility of fatty acyl side chains of phospholipid molecules (1). This leads to an increased solidity of the membrane and to alterations in the phospholipid-protein interactions (1,2). It has been shown, that a decrease in the activities of some membrane bound enzymes, including the microsomal cytochrome P-450 system, takes place if the amount of membrane cholesterol is increased in vitro. The decrease is probably because of the increased solidity of the membrane structure (2,3). However, we have found that if rats are fed with a high cholesterol diet, an increase, instead of a decrease, in the activities of microsomal drug metabolizing enzymes takes place (4,5).
In this study we have tried to find an explanation for the different effect of dietary- and in vitro incorporated cholesterol on the activities of microsomal drug metabolizing enzymes. The quantities of structural components and the enzyme activities have been compared with those of control microsomes. In addition to this, the changes in the binding properties of 1,8-ANS mole-

cules in microsomes have been analyzed. 1,8-ANS binds readily in microsomes, mainly near the polar end of phospholipid molecules (6) and its spectral properties are sensitive to the environment of the binding site. Accordingly, possible changes in the physico-chemical properties of microsomes due to cholesterol may be revealed as changes in the emission spectra.

MATERIALS AND METHODS

Animals
Male wistar rats of the strain Af/Han/Mol/(Han 67), weighing 240+15 g were used. The rats were fed a 4% cholesterol diet ad lib. for four weeks. The diet was prepared by mixing cholesterol (Merck AG, Darmstadt, Germany) with standard rat food (Hankkija Ltd. Finland). Control animals received the same food without cholesterol.

Microsomes
Rats were killed by a blow on the head and bled by cutting the cervical vessels. The livers were removed and homogenized in 4 vol. of 0.25 M ice cold sucrose with a Potter-Elvehjem type homogenizer. After spinning the homogenate at 10000 g/15 min, the microsomal fraction was sedimented at 105000 g/60 min. The microsomal pellet was then suspended in 0.15 M KCl and recentrifuged. The washed microsomes were suspended in 0.1 M Tris-HCl buffer solution pH 7.4 (7). The in vitro incorporation of cholesterol in microsomes was carried out according to Duppel and Ullrich (3). Microsomes were incubated in the cholesterol suspension for 30 min. at $30^{\circ}C$ in 0.1 M Tris-HCl pH 7.4. In trypsin digestion of microsomes 1 mg of trypsin (Type III, Sigma Chem. Co., USA) was added to 1 ml of microsomes and incubated for 30 min. at $38^{\circ}C$. The reaction was stopped with 1.2 mg trypsin inhibitor (Type II-O, Sigma). The microsomal protein content was determined by using the method of Gornall et al. (8), as modified by Lang et al. (4). The phospholipid content was determined by measuring the released inorganic phosphate after sulphuric acid hydrolysis of microsomal phospholipids (9). Results are expressed as equivalents to lecithin (Sigma, Type II-E, Mw 734). The amount of cholesterol in microsomes was determined as described by Abell et al. (10) and modified by Anderson and Keys (11).
The activity of microsomal NADPH cytochrome c reductase was measured according to Phillips and Langdon (12), p-nitroanisole-O-demethylase according to Netter (13) and UDPglucuronosyltransferase according to Isselbacher (14) and Hänninen (15). Arylhydrocarbon hydroxylase activity was determined according to Wattenberg (16) and Nebert and Gelboin (17) by measuring the amount of hydroxylated 3,4-benzpyrene in a Perkin-Elmer MPF-3A spectrofluorometer. The amount of cytochrome P-450 was determined according to Omura and Sato (18).

Fluorescence Measurements
1-Anilinonaphtalene-8-sulphonic acid (1,8-ANS, Mg-salt from Serva, Heidelberg) was used as fluorescence probe. Microsomal suspensions and 1,8-ANS solutions were prepared in 0.1 M Tris-HCl pH 7.4. Measurements were carried out at $25^{\circ}C$ using a Perkin-Elmer MPF-3A spectrofluorometer and 3 ml quartz cells. The wavelength for excitation was 382 nm and for emission 466 nm, band widths 8 nm. To reveal the binding chracteristics of 1,8-ANS molecules in microsomes, Scatchard plots were constructed as has previously been described in detail by Dallner and Azzi (19) and DiAugustine et al. (20).

RESULTS

Alterations in Membrane Composition and Enzyme Activities

Dietary cholesterol elevated both the phospholipid to protein and the cholesterol to protein ratio in hepatic microsomes (Table 1). The increased ratios were also seen in trypsin treated microsomes, where about 33% of microsomal protein has been released (Table 1). The cholesterol to protein ratio in microsomes could also be elevated by incubating microsomes with cholesterol. Extra cholesterol remained in the membranes also after trypsin digestion (Table 1).

TABLE 1 Alterations in Microsomal Membrane Composition

	Phospholipid to protein ratio		Cholesterol to protein ratio	
	Native microsomes	Trypsin trtd. microsomes	Native microsomes	Trypsin trtd. microsomes
Control	0.26	0.29	0.015	0.019
Cholesterol diet.	0.33	0.39	0.033	0.042
Cholesterol in vitro			0.022	0.024

Cholesterol and phospholipid to protein ratios in control microsomes and in microsomes from rats on a high cholesterol diet, and cholesterol to protein ratio in microsomes, where cholesterol was incorporated in vitro. Protein content in microsomal suspensions was about 23 mg/ml.

After cholesterol feeding of rats, an enhancement in the amount of microsomal cytochrome P-450 and in the activities of NADPH cytochrome c reductase and p-nitroanisole-O-demethylase took place. The activity of arylhydrocarbon hydroxylase remained unchanged (Fig. 1). Microsomal UDPglucuronosyltransferase activity was elevated after the cholesterol feeding of rats but the elevation could be measured only if microsomes were first treated with trypsin (Fig. 1). When cholesterol was incorporated into microsomes in vitro, a decrease in the activity of p-nitroanisole-O-demethylase and UDPglucuronosyltransferase took place. The decreased activity of UDPglucuronosyltransferase was found both in native and in trypsin treated microsomes (Fig. 1). The measurable amount of microsomal cytochrome P-450 and the activities of NADPH cytochrome c reductase and arylhydrocarbon hydroxylase were not affected by the in vitro incorporation of cholesterol into microsomes (Fig. 1).

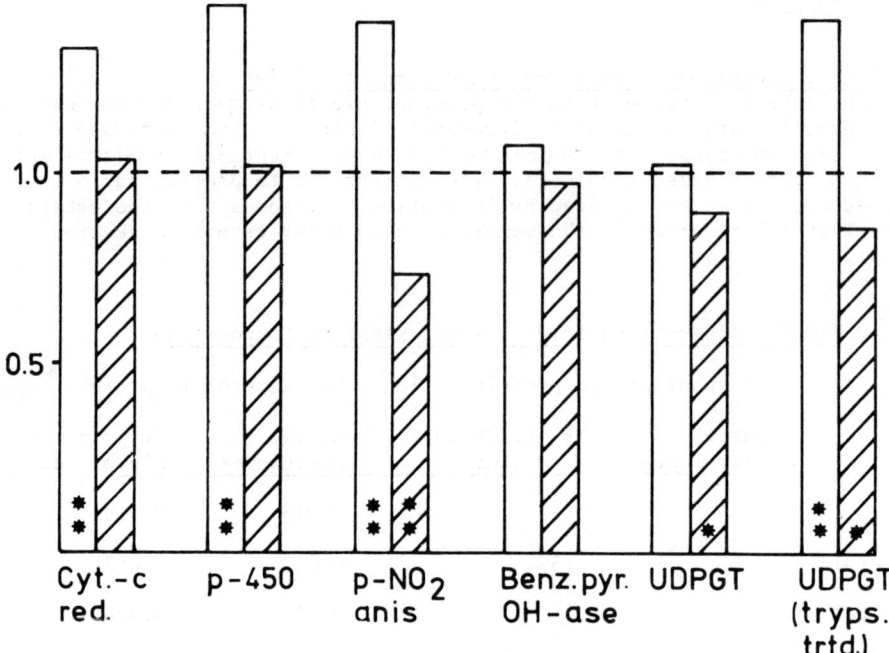

Fig. 1. Relative content of microsomal cytochrome P-450 and activities of NADPH cytochrome c reductase, p-nitroanisole-O-demethylase, arylhydrocarbon hydroxylase and UDPglucuronosyltransferase (from native and trypsin digested microsomes) after feeding rats with a high cholesterol diet (open columns) and after the in vitro incorporation of cholesterol in microsomes (shaded columns). Control values: Cytochrome P-450 0.31 nmol/mg prot. and NADPH cytochrome c reductase 105 nmol, p-nitroanisole-O-demethylase 0.59 nmol, benzpyrene hydroxylase 0.088 nmol, UDPglucuronosyltransferase (p-nitrophenol) 0.50 nmol and after trypsin digestion 3.6 nmol per min. and mg protein respectively.

Changes in the Binding of 1,8-ANS in Microsomes

If 1,8-ANS was added to a microsomal suspension an enhancement in the fluorescence intensity of the 1,8-ANS molecules took place due to the incorporation of the molecules in the microsomal membrane. It was found, however, that the fluorescence was enhanced more in the case of microsomes from rats on a high cholesterol diet than in the case of control microsomes. On the other hand, when 1,8-ANS was added in microsomes, where cholesterol was incorporated in vitro, the increase in the fluorescence was somewhat less than in control microsomes. Similar effects could also be demonstrated when trypsin digested microsomes were used. Fig. 2 presents the Scatchard plots demonstrating the binding of 1,8-ANS in microsomes. The slope of the plots reveals the affinity of the probe molecules in microsomes and it can be seen that there are high and low affinity binding sites. The cholesterol feeding appeared to increase the affinity of 1,8-ANS molecules to the high affinity binding site in microsomes. The increased affinity due to cholesterol feeding is still present after trypsin treatment of microsomes (Fig. 2). The incorporation of choleste-

Alterations in the Activities of Drug Metabolizing

rol in microsomes in vitro did not enhance the affinity of 1,8-ANS molecules in microsomes, but on the contrary slightly depressed it (Fig. 2).

Dietary cholesterol

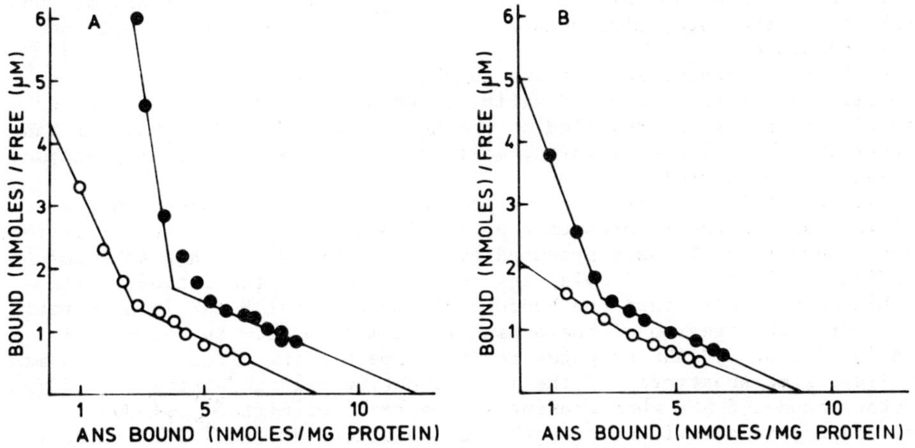

Cholesterol incorporated in vitro

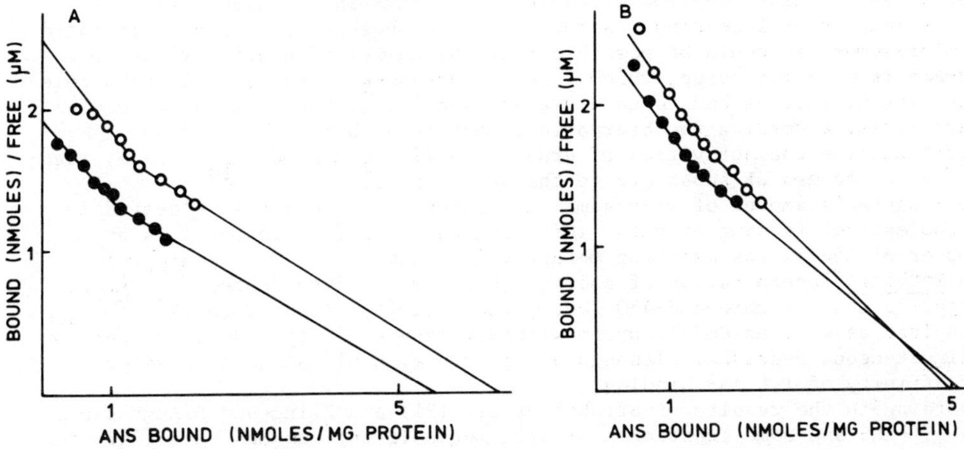

Fig. 2. The effect of dietary cholesterol and cholesterol incorporated in vitro on Scatchard plots demonstrating the binding of 1,8-ANS molecules in liver microsomes. Spots: high cholesterol microsomes, circles: control microsomes. A: Native microsomes, B: Trypsin digested microsomes.

DISCUSSION

The elevated cholesterol to protein and phospholipid to protein ratios could still be detected in microsomes although they had been treated with trypsin. This indicates that the additional cholesterol or phospholipid molecules in microsomes, due to cholesterol feeding or the in vitro incorporation, are not loosely bound to the outer microsomal surface but have penetrated into the microsomal membrane.

Dietary cholesterol caused an increase in the binding of 1,8-ANS in microsomes (partly due to the increased affinity and partly due to the increased capacity of microsomes) as revealed by the Scatchard plots. On the other hand, there occured also an increase in the activities of drug metabolizing enzymes due to dietary cholesterol.

An essential point is whether or not a common factor in microsomes caused both the increase in the fluorescence of 1,8-ANS and in the enzyme activities. Eling and DiAugustine (6) have produced evidence, suggesting that an important binding site for 1,8-ANS molecules in microsomes is the cationic site of lecithin molecules so that the hydrophobic moiety of 1,8-ANS is penetrating into the hydrophobic region of the phospholipid molecules. Furthermore it has been found that 1,8-ANS competes for the same binding sites in microsomes with at least some substrates of the mixed function oxidase system (6,7,20). On the other hand 1,8-ANS when present in the reaction mixture is able to decrease the enzyme activities of both hydroxylation (20) and glucuronidation (our preliminary observations). 1,8-ANS has however not been found to be metabolized via mixed function oxidase system although there is some evidence that a type II like spectrum is caused by 1,8-ANS in microsomes (20).

Because cholesterol feeding has increased the phospholipid to protein ratio in microsomes, it could be possible that the number of binding sites for 1,8-ANS is also increased, which leads to enhanced affinity of 1,8-ANS molecules for microsomes and to an elevated capacity to bind 1,8-ANS molecules. These sites, according to observations mentioned above, may also be of importance for the activities of drug metabolizing enzymes, perhaps by regulating the access of substrate to the active site.

The measurable amount of microsomal cytochrome P-450 was also elevated due to cholesterol feeding of rats, which indicates that cholesterol is an inducer of the microsomal drug metabolizing enzymes.

The in vitro incorporation of cholesterol in microsomes depressed, without affecting the cytochrome P-450 level, the activity of p-nitroanisole-O-demethylase as well as UDPglucuronosyltransferase activity. On the other hand a simultaneous decrease, although a slight one, could also be detected in the affinity of 1,8-ANS binding.

According to the results of Strobel et al. (21) and Eling and DiAugustine, (6) 1,8-ANS and some substrates of drug metabolizing enzymes have a greater affinity for unsaturated fatty acids than for saturated ones. On the other hand it is known that cholesterol is able to increase the membrane solidity, an effect which can also be caused by lowering the temperature or by increasing the saturation degree of fatty acyl side chains in microsomal phospholipids (2). It may thus be possible that both the slight decrease in the affinity of 1,8-ANS molecules in microsomes and the decrease in the enzyme activities caused by the in vitro incorporation of cholesterol is due to the hampered penetration of molecules into the microsomal membrane.

It can be noticed that the effect of dietary or in vitro incorporated cholesterol on the activity of arylhydrocarbon hydroxylase is much weaker than on the activity of p-nitroanisole-O-demethylase, which indicate that cholesterol induced changes in the enzyme activities, depend on the substrate.

The results indicate that cholesterol may affect the activities of microsomal drug metabolizing enzymes partly by modifying the membrane structure and partly by inducing the enzymes.

REFERENCES

1. W. Kleeman and H.M. McConnel, Interactions of proteins and cholesterol with lipids in bilayer membranes, Biochim. Biophys. Acta 419, 206 (1976).
2. D. Papahadjopoulos, N. Cowden and H. Kimelberg, Role of cholesterol in membranes, effect on phospholipid protein interactions, membrane permeability and enzyme activity, Biochim. Biophys. Acta 330, 8 (1973).
3. W. Duppel and V. Ullrich, Membrane effects on drug mono-oxygenation activity in hepatic microsomes, Biochim. Biophys. Acta 426, 399 (1976).
4. M. Lang, M. Laitinen, E. Hietanen and H. Vainio, Modification of microsomal membrane components and induction of hepatic drug biotransformation in rats on a high cholesterol diet, Acta Pharmacol. et toxicol. 39, 273 (1976).
5. M. Laitinen, Enhancement of hepatic drug metabolism with dietary cholesterol in rat, Acta Pharmacol. et toxicol. 39, 241 (1976).
6. T. E. Eling and R.P. DiAugustine, The role for phospholipids in the binding and metabolism of drugs by hepatic microsomes, Biochem. J. 123, 539 (1971).
7. D. J. Birkett, Interaction of some drugs, metal ions and alcohols with rat liver microsomes as studied with fluorescent probe, Clin. Exp. Pharmacol. Physiol. 1, 415 (1974).
8. A. G. Gornall, C. J. Bardawill and M. M. David, Determinations of serum proteins by means of the biuret reaction, J. Biol. Chem. 177, 751 (1949).
9. G. R. Bartlett, Phosphorous assay in column chromatography, J. Biol. Chem. 234, 466 (1959).
10. L.L. Abell, B. B. Levy, B. B. Brodie and F. E. Kendall, A simplified method for the estimation of total cholesterol in serum and demonstration of its specificy, Biol. Chem. 195, 357 (1952).
11. J. T. Anderson and A. Keys, Cholesterol in serum and lipoprotein fraction Clin. Chem. 2, 145 (1956).
12. A. H. Phillips and R. G. Langdon, Hepatic triphosphopyridine nucleotide cytochrome c reductase: Isolation, characterization and kinetic studies, J. Biol. Chem. 237, 2652 (1962).
13. K. J. Netter, Eine Methode zur direkten Messung der O-Demethylierung in Lebermikrosomen und ihre Anvendung auf die Mikrosomenhemmwirkung von SKF 525-A, Naunyn-Scmiedeberg's Arch. Pharmakol. 238, 292 (1960).
14. K. J. Isselbacher, Enzymatic mechanisms of hormone metabolism. II. Mechanisms of hormonal glucuronide formation, Rec. Progr. Hormonal Res. Commun. 12, 134 (1956).
15. O. Hänninen, On the metabolic regulation in the glucuronic acid pathway in the rat tissues, Ann. Acad. Sci. Fenn. A2, 142, 1 (1968)
16. W. L. Wattenberg, J. L. Leong and P. J. Strand, Benzpyrene hydroxylase activity in the gastrointestinal tract, Cancer Res. 22, 1120 (1962)
17. D. W. Nebert and H. V. Gelboin, Substrate inducible microsomal arylhydrocarbon hydroxylase. II. Cellular responses during enzyme induction, J. Biol. Chem. 244, 6242 (1968).
18. T. Omura and R. Sato, The carbon monoxide binding pigment of liver microsomes, J. Biol. Chem. 239, 2370 (1964).
19. G. Dallner and A. Azzi, Structural properties of rough and smooth microsomal membranes; a study with fluorescent probes, Biochim. Biophys.

Acta 255, 589 (1972).
20. R. P. DiAugustine, T. E. Eling and J. R. Fouts, The interaction of a fluorescent probe with rat hepatic microsomes, Chem.-Biol. Interactions 2, 17 (1970).
21. H. W. Strobel, A. Y. H. Lu, J. G. Heidema and M. J. Coon, Phosphatidylcholine requirement in the enzymatic reduction of hemoprotein P-450 and in fatty acid, hydrocarbon and drug hydroxylation, J. Biol. Chem. 245, 4851 (1970).

STEROID-16 α-HYDROXYLASE IN RAT LIVER: BIOCHEMICAL AND BIOLOGICAL PROPERTIES

Pierre Kremers, Ari Azhir-Amirsoleymanie, Jean De Graeve and Jacques E. Gielen

Laboratoire de Chimie Médicale, Institut de Pathologie, Université de Liège, B-4000 Sart-Tilman par Liège 1, Belgium

INTRODUCTION

The cytochrome P 450 linked oxygenases are able to hydroxylate naturally ocurring steroids on specific carbons of the ring skeleton, the 6α, 6β, 7α and 16α positions being the most frequently hydroxylated in the rat liver (1).

To overcome the methodological difficulties in measuring steroid hydroxylase activities accurately, we have recently developed new and precise assays for progesterone (2), pregnenolone (2) and testosterone (3) 16α-hydroxylase. These assays are based on the release of a 16α-positioned atom of tritium into the incubation medium during the enzymatic hydroxylation. This same principle was already applied to the assay of cholesterol-7α-hydroxylase (4) and of progesterone and pregnenolone-17α-hydroxylase (5).

On a quantitative basis, the activity of the steroid-16α-hydroxylase seems to vary extensively, not only with the steroid used as a substrate, but also as a function of different physiological parameters (age, sex, strain...) and pharmacological treatments (1).

Here we will describe some of the fundamental biochemical properties of the rat liver steroid-16α-hydroxylase, using progesterone, pregnenolone and testosterone as substrates of the enzymic reaction. The influence of the sex of the animals and of the administration of different chemical inducers will also be studied.

MATERIALS AND METHODS

The sources of the chemicals have been reported previously (4,5). $[16^3H]$-progesterone (26 Ci/mmol) and $[16^3H]$-pregnenolone (26 Ci/mmol) were obtained from I.R.E. (Fleurus, Belgium). $[16^3H]$-testosterone was prepared from $[16^3H]$-progesterone (3). 16α-cyanopregnenolone was kindly donated by Upjohn, S.A.

150 g Sprague-Dawley rats (Centre des Oncins, Lyon, France) were sacrificed by cervical dislocation and carefully bled. The livers were removed and rapidly cooled in isotonic KCl. Isolated microsomes were prepared as previously described (4).

The incubation mixture for the 16α-hydroxylase assay was made of phosphate buffer (0.15 M; pH 7.4), NADP (1 mM), glucose-6-phosphate (5 mM), glucose-6-phosphate dehydrogenase (1 I.U./ml), magnesium chloride (4 mM) and 0.1 ml of microsomal suspension in a final volume of one ml. The substrates (0.5 μCi per tube), solubilized in the incubation medium with the help of tween 80

(2,4,5) were added at a final concentration of 200 μM for progesterone and 50 μM for pregnenolone and testosterone.

The samples were incubated at 37°C for 20 min, the enzymatic reaction stopped by addition of 1.5 ml of trichloracetic acid (20%) and the water distilled under reduced pressure as described in detail elsewhere (4). The radioactivity present in the distilled water was finally counted and converted into specific enzymatic activity (nmol of hydroxylated substrate per min and per mg of protein).

RESULTS

Steroid-16α-Hydroxylase Affinity for the Different Substrates

The water solubility of the steroids being very limited, the introduction of a solubilizing agent (tween 80) in the incubation medium was necessary in order to reach a saturating concentration of substrates. Under these conditions, it was possible to determine apparent K_m for the different substrates using the Lineweaver Burk graphical method. The measured values were respectively 7.5, 8.0 and 100 μM for testosterone, pregnenolone and progesterone.

Cofactor Requirement

Regardless of the steroid used as the enzyme substrate, the 16α-hydroxylase, like the other monooxygenases, displays an absolute requirement for atmospheric oxygen and reduced NADPH. NADH alone is unable to support the enzymatic hydroxylation, but has a synergistic effect when added together with low (suboptimal) concentrations of NADPH (Fig. 1).

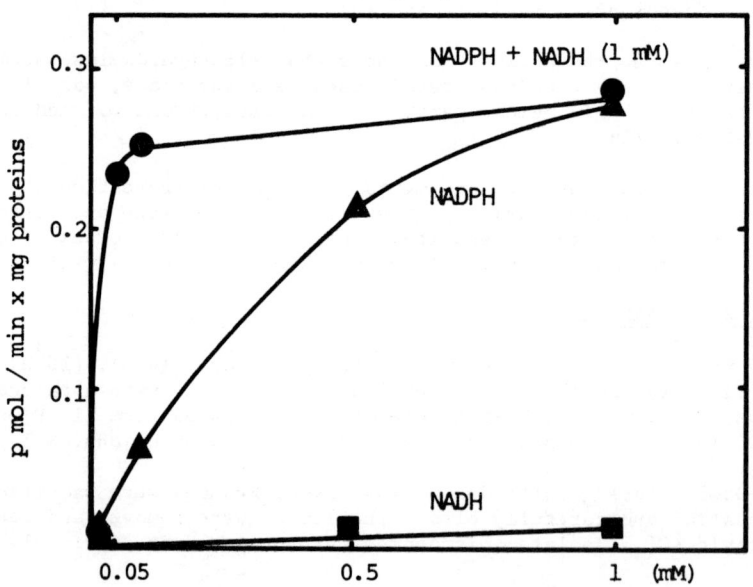

Fig. 1. Synergistic Effect of NADPH plus NADH on pregnenolone-16α-hydroxylase activity *in vitro*.

Tritiated pregnenolone was incubated under the optimal conditions (See methods section) except for the co-factors which were added at the indicated concentrations.

This phenomenon which can be compared to the partial uncoupling effect described by Ullrich et al. (6) for n-hexane, was described with each of the three substrates (Results not shown for progesterone and testosterone).

Competition Between the Substrates
Figure 2 shows the *in vitro* action of the indicated steroids on the progesterone (Fig. 2A) or pregnenolone (Fig. 2B) 16α-hydroxylase activity. Although the latter was inhibited very well by both pregnenolone and testosterone, the former was only inhibited by testosterone. Pregnenolone did not modify progesterone-16α-hydroxylase activity even at a concentration which was five times higher than the one of the substrate, progesterone.

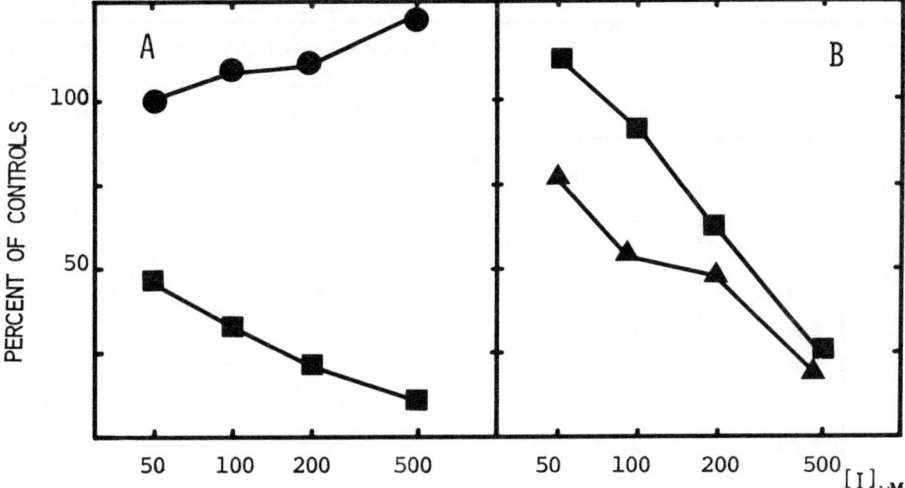

Fig. 2. *In vitro* influence of testosterone (■), pregnenolone (●) or progesterone (▲) on the progesterone or pregnenolone 16α-hydroxylase activity.
Tritiated progesterone (A) or pregnenolone (B) were incubated under the optimal conditions described in the methods section. The competing steroids were mixed at the indicated concentrations with the substrate and the tween 80 prior to the addition of the incubation medium. The results are expressed as the percentage of the activity found in the presence of the substrate alone.

Sex Dependency
As reported earlier by others (7,8) an important sex difference in the 16α-hydroxylase activity was observed for the three substrates. The highest male/female ratio being observed with testosterone. On a quantitative basis, testosterone-16α-hydroxylase activity was also highest when compared to the other substrates in both male and female animals (Table 1).

Effects of Chemical Inducers
The administration of phenobarbital, 3-methylcholanthrene or cyanopregnenolone affected the hydroxylase activities in different ways. Phenobarbital very significantly induced progesterone and pregnenolone-16α-hydroxylase in both male and female animals, but to a much lesser extent than the testosterone-16α-hydroxylase activity. 24 hours after a methylcholanthrene injection, the

16α-hydroxylases are inhibited to roughly the same extent (40-60%) in male and female rats. Cyanopregnenolone administration determined quite a different pattern of answers: the hydroxylase activities were all induced in female animals (very significantly for progesterone-16α-hydroxylase: 6 fold increase), but surprisingly, if the progesterone and pregnenolone-16α-hydroxylase activities were slightly enhanced in male rats, testosterone-16α-hydroxylase activity was not induced at all. None of those inducers modified the enzymatic activities when added *in vitro* into the incubation medium up to concentrations corresponding to 2 times the substrate concentration.

TABLE 1   Effects of Chemical Inducers on Steroid-16α-Hydroxylase Activities

TREATMENT	SEX	STEROID-16α-HYDROXYLASE		
		Pregnenolone	Progesterone	Testosterone
NaCl	male	208 ± 46	72 ± 20	539 ± 156
	female	67 ± 20	34 ± 05	97 ± 20
PB	male	863 ± 71*	483 ± 76*	778 ± 135*
	female	240 ± 53*	140 ± 23*	143 ± 30*
MC	male	166 ± 52	50 ± 17	200 ± 61*
	female	31 ± 15*	21 ± 11	57 ± 07*
CPN	male	469 ± 62*	173 ± 23*	476 ± 39
	female	365 ± 93*	113 ± 25*	187 ± 39*

Pregnenolone, progesterone and testosterone-16α-hydroxylase activities as a function of the sex of the animals and of the i.p. administration of phenobarbital (PB; 50 mg/kg daily for 4 days), 3-methylcholanthrene (MC; 80 mg/kg 24 hours before sacrifice) or 16α-cyanopregnenolone (CPN; 50 mg/kg daily for 3 days). * indicates a $P < 0.05$ when the male or female treated groups were compared to the corresponding NaCl treated animals.

DISCUSSION

The experiments described in this report clearly demonstrated the decisive advantages of the tritium exchange method for assaying the microsomal steroid hydroxylases. Providing that a specifically tritiated substrate is available and that its tritium is not released by non-enzymatic reactions during the various manipulations, such methods are technically very easy to realize, allow the performance of many assays at the same time and are applicable not only on purified subcellular fractions, but also on crude tissular preparations, on tissue cultures and on biopsy samples.

The *in vitro* sensitivity of the steroid-16α-hydroxylase to the action of various inhibitors in addition to its *in vivo* response to the administration of chemical inducers varies as a function of the steroid used as the enzymatic substrate. No clear explanation can presently account for those differences except that it is very likely that different steroid-16α-hydroxylases exist in the rat liver. This phenomenon would have to be compared to the 7α-hydroxylase situation. In that case, cholesterol-7α-hydroxylase behaves differently in many ways from the 7α-hydroxylases active on pregnenolone, testosterone or taurodeoxycholic acid (9,10). To solve this important problem of steroid multiplicity, the obvious way will be to solubilize, purify and reconstitute the microsomal enzymatic complex responsible for those reactions.

REFERENCES

1 Hamberg, M., Samuelsson, B., Bjorkhem, I., and Danielsson, H., Molecular Mechanisms of Oxygen Activation, edited by Hayaishi, D., Academic Press, Inc. p. 30-76, 1974.

2 Kremers, P., De Graeve, J., and Gielen, J., Measure of Steroid Hydroxylase Activities by Tritium Exchange Methods. III. Progesterone and Pregnenolone-16α-Hydroxylase. Submitted for publication.

3 De Graeve, J., Azhir-Amirsoleymanie, A., Kremers, P., and Gielen, J., Measure of Steroid Hydroxylase Activities by Tritium Exchange Methods. IV. Testosterone and Dehydroepiandrosterone-16α-hydroxylase. In preparation.

4 Van Cantfort, J., Renson, J., and Gielen, J. Rat Liver Cholesterol-7α-Hydroxylase. Development of a New Assay Based on the Enzymic Exchange of the Tritium Located on the 7α Position of the Substrate, Eur. J. Biochem. 55, 23-31, 1975.

5 Kremers, P., New Assay for Pregnenolone and Progesterone-17α-Hydroxylase Based on the Specific Substitution of a Tritium Situated on Carbon 17, Eur. J. Biochem. 61, 481-486, 1976.

6 Staudt, H., Lichtenberger, F., and Ullrich, V., The Role of NADH in Uncoupled Microsomal Monooxygenations, Eur. J. Biochem. 46, 99-106, 1974.

7 Heinrichs, N.L., Feder, H.N., and Colas, A., The Steroid 16α-Hydroxylase System in Mammalian Liver, Steroids 7, 91-98, 1966.

8 Einarsson, K., Gustafsson, J.A., and Stenberg, A., Neonatal Imprinting of Liver Microsomal Hydroxylation and Reduction of Steroids, J. Biol. Chem. 248, 4987-4997, 1973.

9 Einarsson, K., and Johansson, G., Effect of Carbon Monoxyde and Phenobarbital on Hydroxylation of Bile Acids by Rat Liver Microsomes, FEBS letter, 4, 177-180, 1969.

10 Johansson, G., Oxidation of Cholesterol, 3β-hydroxy-5-pregnen-20-one and 3β-hydroxy-5-androsten-17-one by Rat Liver Microsomes, Eur. J. Biochem. 21, 68-79, 1971.

CORRELATION BETWEEN DRUG, LIPID AND CARBOHYDRATE METABOLISM IN THE RAT AFTER CHRONIC BETA-ADRENORECEPTOR ANTAGONIST TREATMENT

Erik Fellenius, Karl Olof Borg, Berit Eklund,
Kurt-Jürgen Hoffmann, Britt-Marie Magnusson,
Inger Skånberg, Marita Wallborg and Boel Wallin

AB Hässle, Analytical Chemistry and Biochemistry, Fack, S-431 20 Mölndal, Sweden

ABSTRACT

Rats were treated for fourteen months with the two beta-receptor blockers propranolol* (non-selective) and metoprolol* (cardio selective). Activities of hepatic and cardiac enzymes, the level of blood-borne substrates representative for different metabolic steps and different parameters representative for drug metabolism were correlated to each other in all fifteen animals. The chosen enzymes represent aerobic and anaerobic breakdown of glucose, anabolism and catabolism of glycogen, the pentosphosphate shunt, β-oxidation of fatty acids, citric acid cycle and the electron transport chain. The following reactions within drug metabolism were studied: epoxide metabolism, O- and N-demethylation, aromatic and aliphatic hydroxylation, kinetic parameters (K_s, K_m, V_{max}, ΔOD_{max}) in the metabolism of metoprolol and cyt-P-450 content. Several correlations were found within lipid and carbohydrate metabolism and the parameters within drug metabolism were also correlated to each other. There were very few correlations between drug metabolism and lipid-carbohydrate metabolism. However, glycogen phosphorylase activity in the heart was negatively correlated to several enzymes in drug metabolism, probably through a common hormonal influence. Glutathione-S-epoxide transferase was positively correlated to the liver activity of glycogen phosphorylase, hexokinase and malic enzyme. This result demonstrates the possible role of glycolysis and malic enzyme for the supply of NADPH in epoxide metabolism. To summarize, lipid - carbohydrate metabolism and drug metabolism seems to be functionally interrelated to a very limited extent.

INTRODUCTION

There are several possibilities for interaction between lipid - carbohydrate metabolism and drug metabolism. NADPH, oxygen (1,2,3,4) the energy charge (1,5) and c-AMP (6) are all well-known modifiers of lipid - carbohydrate metabolism and drug metabolism.

The aim of the present study was to correlate several parameters within lipid - carbohydrate and drug metabolism in order to get further information on possible functional relationships in vivo. A correlation analysis was therefore made after chronic treatment of rats with the non-selective beta-receptor antagonist propranolol and the cardioselective beta$_1$-receptor antagonist meto-

---

* propranolol = Inderal®
  metoprolol = Seloken®

prolol (7). The effect of each drug on lipid - carbohydrate and drug metabolism will be presented elsewhere (8). The present report only contains an evaluation of the correlation analysis made on all the material since it was assumed that a functional relationship should be independent of the pretreatment of the animal.

MATERIAL AND METHODS

Fifteen rats were divided into three groups: control, propranolol and metoprolol. The propranolol and metoprolol rats received 1.7 and 2.9 mM, respectively of the drugs in the drinking water, corresponding to a daily drug intake of about 50 and 250 µmol/kg, respectively. The animals were killed after fourteen months and analyzed for content of blood-borne substrate and enzyme activities as shown in figure 1. The methods for determination of the different parameters will be presented elsewhere (8).

A correlation analysis was made with a computer on all the rats regardless of pretreatment. All correlations with a $r > 0.5$ and a probability value of $p < 0.05$ according to Student´s t-test were considered statistically significant.

RESULTS AND DISCUSSION

Correlation within lipid - and carbohydrate metabolism

Several correlations between enzyme activities within the liver were found (Fig. 1). Thus, HK* activity was positively correlated to GS, I-form and G-6--PDH and negatively correlated to PFK activity. The activity of the latter enzyme was negatively correlated to GP-activity. CS-activity was positively correlated to PFK-activity. All these correlations demonstrate a close functional relationship within liver lipid - and carbohydrate metabolism and confirm results presented previously by several other authors (8,9,10). Measurement of certain key enzyme activities is thus a useful tool to demonstrate functional relationships.

In contrast to several correlations between enzyme activities within the liver there was no single correlation within the heart tissue. Instead, several correlations between enzyme activities in the heart were correlated to the level of blood-borne substrates. Thus, the glucose level in the blood was negatively correlated to LDH activity in the heart and positively correlated to HK activity. The blood lactate level was positively correlated to GS total

---

*
Abbreviations: GP = glycogen phosphorylase (EC 2.4.1.1.), LDH = lactate dehydrogenase (EC 1.1.1.27), HK = hexokinase (EC 2.7.1.1.), GS = glycogen synthase (EC 2.4.1.11), PFK = phosphofructokinase (EC 2.7.1.11), ACDH = 3-hydroxyacyl-CoA-dehydrogenase (EC 1.1.35), CS = citrate synthetase (EC 4.1.3.7), ME = malic enzyme (EC 1.1.1.40), G-6-PDH = glucose-6-phosphate dehydrogenase (EC 1.1.1.49), cyt.ox = cytochrome-c-oxidase, epoxide hydratase (EC 4.2.1.63), glutathione-S-epoxide transferase (EC 2.5.1.18)

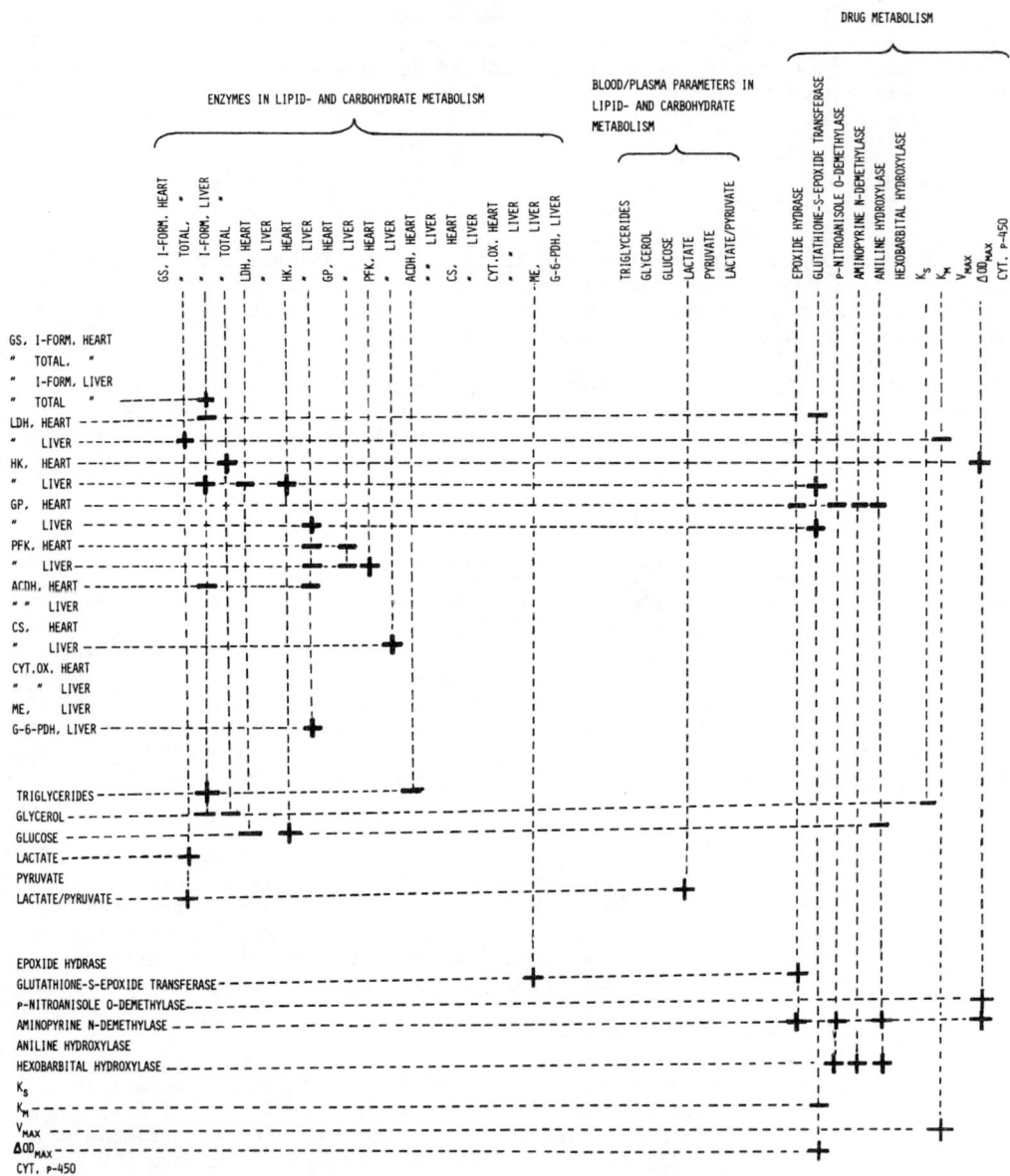

FIGURE 1. CORRELATION BETWEEN DRUG- LIPID AND CARBOHYDRATE METABOLISM AFTER FOURTEEN MONTHS TREATMENT OF RATS WITH THE BETA-RECEPTOR ANTAGONISTS, PROPRANOLOL AND METOPROLOL. THE K_s, K_m, V_{max} AND ΔOD_{max} REFER TO DETERMINATION WITH METOPROLOL AS SUBSTRATE. THE TOTAL NUMBER OF OBSERVATIONS ARE FIFTEEN. CORRELATIONS WITH A $r > 0.5$ AND A VALUE OF PROBABILITY OF $p < 0.05$ ACCORDING TO STUDENT´S TEST HAVE BEEN INCLUDED. A POSITIVE CORRELATION IS SHOWN BY A + AND A NEGATIVE BY A -.

activity in the heart. The triglyceride level was negatively correlated to cardiac tissue ACDH activity. These results demonstrate that the metabolic fluxes within the heart are to a great extent dependent on the plasma levels of substrates rather than the activity of the enzyme. The blood-borne substrate levels are influenced by metabolic fluxes within the liver which might explain the correlation between enzyme activities within the heart and the liver.

Correlation within drug metabolism

Very many parameters within drug metabolism were positively correlated to each other (Figure 1). Thus, the activity of glutathione-S-epoxide transferase was positively correlated to epoxide hydratase activity. The activity of this enzyme was also positively correlated to amino N-demethylase which in turn was positively correlated to p-nitroanisole-O-demethylase and aniline hydroxylase. Hexobarbital hydroxylase activity was positively correlated to the activity of p-nitroanisole-O-demethylase, aminopyrine N-demethylase and aniline hydroxylase. The K_m and V_{max} values for metoprolol were positively correlated to each other. The cyt-P-450 content was not correlated to any of the parameters within drug metabolism. Our results demonstrate a very close functional association of parameters in drug metabolism.

Correlation between lipid - carbohydrate metabolism and drug metabolism

Generally, very few correlations were found between the liver and heart enzyme activities and blood-borne parameters on the one hand and parameters within drug metabolism on the other hand. Glutathione-S-epoxide transferase was positively correlated to the liver activity of GP, HK and ME. Since epoxide metabolism is a part of the mixed oxidation function of drugs, and thus NADPH-utilization, this can explain why the NADPH-producing enzyme, malic enzyme, will be correlated to glutathione-S-epoxide transferase. Breakdown of glycogen by GP and activation of glucose by HK will supply the malic enzyme with substrate.

GP-activity in the heart was negatively correlated to four enzyme activities within liver drug metabolism, e.g. epoxide hydrase, p-nitroanisole-O-demethylase, aminopyrine-N-demethylase and aniline hydroxylase. A possible explanation is that liver drug metabolism and GP-activity in the heart is influenced by a common blood-borne factor, e.g. a hormone. C-AMP may very well mediate such a hormonal action, since it is known that c-AMP decreases drug metabolism (6,12) and increases heart GP-activity within the heart (13).

REFERENCES

1. R.G.Thurman & R.Scholz, Mixed Function Oxidation in Perfused Liver. The Effect of Aminopyrine on Oxygen Uptake, Europ.J.Biochem. 10, 459 (1969)

2. H.Sies & K.H. Summer, Hydroperoxide-Metabolizing Systems in Rat Liver, Europ.J.Biochem. 57, 503 (1975)

3. O.Junge & K.Brand, Mixed Function Oxidation of Hexobarbital and Generation of NADPH by the Hexose Monophosphate Shunt in Isolated Rat Liver Cells, Arch.Biochem.Biophys. 171, 398 (1975)

4. R.Grundin, P.Moldéus, H.Vadi, S.Orrenius, C.von Bahr, D.Bäckström & A. Ehrenberg, Drug Metabolism in Isolated Rat Liver Cells, Adv. Exp. Biol. 5, 251 (1975)

5. E. Quagliariello, S.Papa, A.J. Meijer & J.M. Tager, In Mitochondria, Structure & Function, Ed. by L. Ernster & Z. Drakota, Academic Press, New York (1968) p.335

6. M. Weiner, C.G. Buteshaugh, D.A. Blake, Inhibition of hepatic drug metabolism by cyclic $3^-,5^-$-adenosine monophosphate, Res. Comm.Chem. Path. Pharmacol. 3, 249 (1972)

7. B.Åblad, K.O.Borg, E.Carlsson, G.Johnsson, T.Malmfors & C-G. Regårdh, Animal and human pharmacological studies on metoprolol - a new selective adrenergic β_1-receptor antagonist, Acta Pharmacol. Toxicol. 35, 1 suppl.V (1975)

8. E. Fellenius, K.O. Borg, K-J. Hoffmann, B. Eklund, B-M. Magnusson, I.Skånberg, B.Wallin & M. Wallborg, Effect of chronic administration of propranolol and metoprolol on lipid,- carbohydrate - drug metabolism and pharmacokinetic of the drugs in the rat. To be published

9. A. Bass, D. Brdicka, P.Eyer, S.Hofer, D.Pette, Metabolic Differentiation of Distinct Muscle Types at the Level of Enzymatic Organization, Europ. J. Biochem. 10, 198 (1969)

10. A-Ch. Bylund, T. Bjure, G. Cederblad, J. Holm, K. Lundholm, K.A. Änqvist, M. Sjöström & T. Scherstén, Physical Training in Man. Skeletal Muscle Metabolism in Relation to Muscle Morphology and Running Ability. Submitted for publication.

11. E.A. Newsholme & C. Start, Regulation in Metabolism, John Wiley & Sons, New York, (1973)

12. W.E. Ross & W.W. Oppelt, Effect of dibutyryl cyclic AMP on the hepatic mixed function oxidase system, Pharmacologist 12, 200 (1970)

13. G.A. Robison, R.W. Butcher & E.W. Sutherland, Cyclic AMP, Academic Press, New York, (1971)

CONVERSION OF TRICHLOROETHYLENE TO CARBON MONOXIDE BY MICROSOMAL CYTOCHROME P450

P. S. Traylor, W. Nastainczyk and V. Ullrich

Department of Physiological Chemistry, University of the Saarland, Homburg-Saar, GFR

ABSTRACT

Microsomes prepared from livers of rats pretreated either with phenobarbital or 3,4-benzo(a)pyrene have been found to convert trichloroethylene to carbon monoxide in the presence of NADPH and dioxygen. Evidence is presented for the involvement of cytochrome P450 in this conversion.

INTRODUCTION

Trichloroethylene is known to yield trichloroacetic acid and trichloroethanol as urinary excretion products in all species of experimental animals studied (1). Research involving its metabolism in isolated perfused rat livers (2) and in rat liver microsomes (3) have also identified chloral and/or chloral hydrate as a significant product of oxidation. The recent finding that addition of 2,2,3-trichlorooxirane (trichloroethylene oxide) to rat liver perfusates (2,4) results in the formation of chloral, trichloroethanol and trichloroacetic acid indicates the following pathway for the metabolism of trichloroethylene.

This reaction scheme accords with the previous observation (5) that oral administration of ^{36}Cl-labelled trichloroethylene to rats yields urinary trichloroacetic acid and trichloroethanol of unchanged specific activities.

RESULTS

We here report the production of carbon monoxide from trichloroethylene by microsomes prepared from livers of rats pretreated either with phenobarbital or 3,4-benzo(a)pyrene. Incubation of trichloroethylene with microsomes in the presence of NADPH, followed by reduction with sodium dithionite results in a difference spectrum with a peak at 451 nm. Virtually an identical difference spectrum is obtained upon addition of carbon monoxide to a mixture of microsomes and NADPH, followed by reduction with sodium dithionite (Fig. 1).

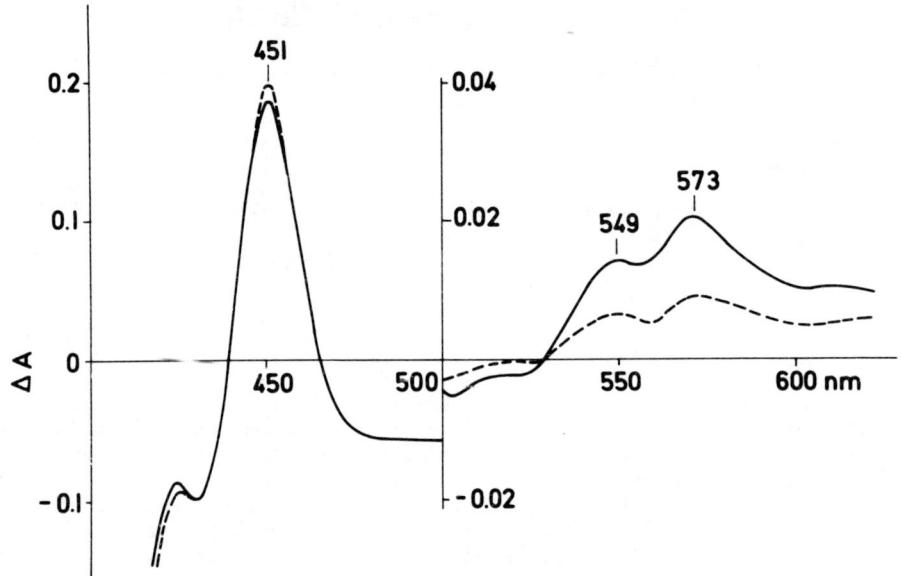

Fig. 1. Difference spectra of reduced microsomes in the presence of CO and after incubation with trichloroethylene.

Each cuvette contained 6 mg of microsomal protein (1.7 nmol cytochrome P450/mg) from phenobarbital pretreated rats and 10^{-3}M NADPH in 3 ml of 0.1 M Tris-HCl buffer, pH 7.6. The solid line was obtained after 10 min. incubation at room temperature with 10 µmol trichloroethylene prior to addition of 2 mg sodium dithionite. The dashed line was obtained after saturation with carbon monoxide prior to addition of 2 mg sodium dithionite.

Stepwise addition of hemoglobin resulted in a progressive decrease in absorbance at 451 nm, accompanied by an increase in absorbance at 419 nm. This observation illustrated in Fig. 2, can be sensibly interpreted as the transfer of carbon monoxide from the reduced cytochrome P450 complex to hemoglobin, forming carboxyhemoglobin.

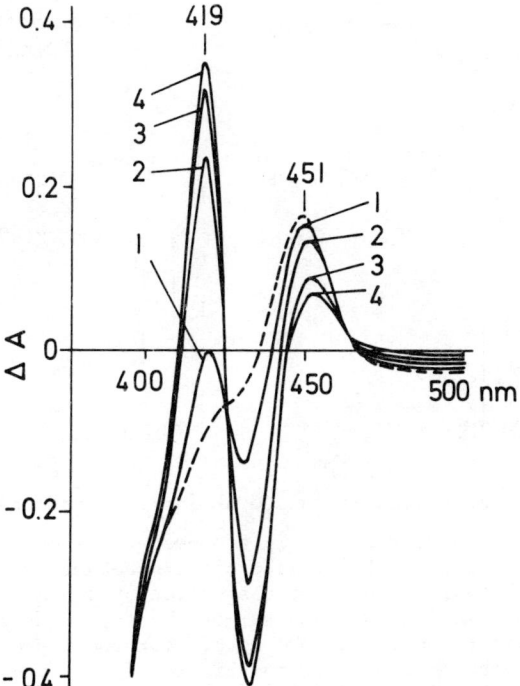

Fig. 2. Effect of increasing concentrations of rat hemoglobin on the 451 nm peak.

Each cuvette contained 3.3 mg of microsomal protein (2.4 nmol cytochrome P450/mg) and 10^{-3}M NADPH in 3 ml of 0.1 M Tris-HCl buffer, pH 7.6. The dashed line represents difference spectrum after 4 min. incubation with 10 μmol trichloroethylene at room temperature followed by reduction with sodium dithionite. Solid lines 1 to 4 represent difference spectra after cumulative additions of 3, 9, 15 and 21 μl of approximately 10^{-5}M hemoglobin solution.

Unambiguous identification of the product as carbon monoxide was achieved through infrared spectroscopy. Rat liver microsomes were incubated with NADPH in the presence or absence of trichloroethylene, and each suspension flushed with N_2. Released gases were collected in an evacuated 10-meter gas cell and the region 2200 cm^{-1} to 2000 cm^{-1} was scanned. Carbon monoxide is readily identified by characteristic absorptions centering at 2170 cm^{-1} and at 2110 cm^{-1} (7). In Fig. 3 are presented infrared spectra of the gas phase showing no detectable CO production by microsomes and NADPH alone, but a clear CO spectrum in the presence of trichloroethylene.

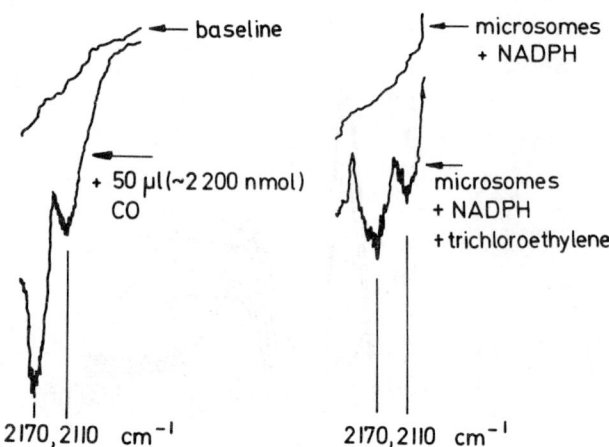

Fig. 3. Identification of CO by infrared spectroscopy.

All spectra were recorded in a 10-meter gas cell using the Perkin-Elmer 325 infrared spectrophotometer. Spectra on the left side represent absorption by 3 atm. N_2, with and without added CO. Spectrum on the top right was obtained after incubating 300 mg of microsomal protein (2.7 nmol cytochrome P450/mg) from livers of rats pretreated with phenobarbital, with 42 mg NADPH in 85 ml of Tris-HCl buffer, pH 7.6, at 30°C for 30 mins. The microsomal suspension was then flushed with N_2; the resulting vapors were passed through 8 M aqueous NaOH, anhydrous $CaCl_2$, 2 traps cooled with liquid N_2 and thence into an evacuated gas cell. Pressure in the cell was adjusted to 3 atm. with N_2 before scanning. Spectrum on the bottom right was recorded after identical treatment except that 48 mg trichloroethylene was included in the incubation mixture.

Having identified the product as carbon monoxide, we then studied the kinetics of its formation. This could conveniently be measured in microsomal suspensions containing varying amounts of trichloroethylene by adding excess hemoglobin and monitoring increases in the peak at 421 nm (maximum[*]) relative to the absorption at 431 nm (isosbestic point).

There was observed an orderly increase in CO production rates with increasing concentrations of trichloroethylene (Fig. 4), and from a Lineweaver-Burk plot of the rate data, a K_m value of 5.9×10^{-4} M was obtained.

---

[*]The absorption maximum at 419 nm for the hemoglobin-CO complex in the presence of $Na_2S_2O_4$ (Fig. 2) is shifted to 421 nm in difference spectra containing NADPH.

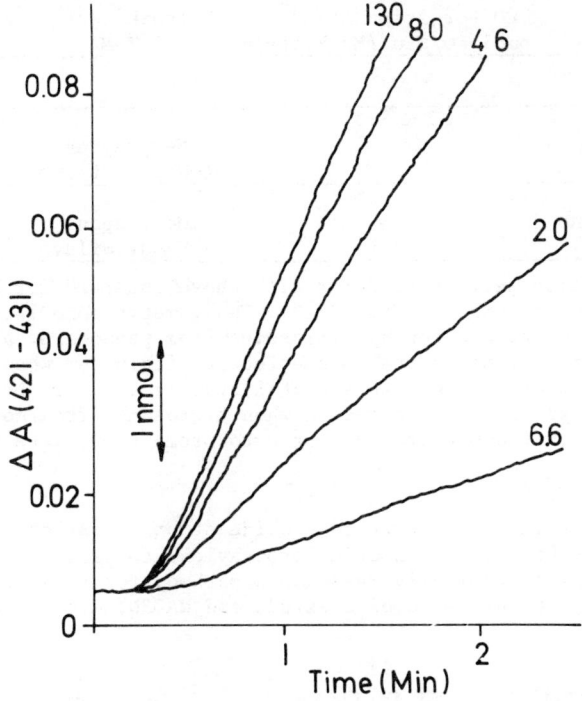

Fig. 4. CO production with increasing concentrations of trichloroethylene

Each cuvette contained 1.9 mg of microsomal protein (2.5 nmol cytochrome P450/mg) from phenobarbital pretreated rats in 1.5 ml of 0.1 M Tris-HCl buffer, pH 7.6, and 20 ul of approximately 10^{-5}M rat hemoglobin solution. Additionally, NADPH and an NADPH-regenerating system consisting of isocitrate, Mg^{2+}, Mn^{2+} and isocitrate dehydrogenase were present. Trichloroethylene was introduced as a 0.1 M methanolic solution; the concentrations on the graph are shown in mM units. These experiments were carried out at 37°C.

The difference spectrum of rat liver microsomes with trichloroethylene, measured at room temperature, showed a peak at 387 nm and a trough at 423 nm. From a Lineweaver-Burk plot of the difference in absorption (387 nm - 423 nm) measured at trichloroethylene concentrations ranging from 3.3×10^{-5} to 7.3×10^{-4}M, a substrate dissociation constant, K_s, of 5.9×10^{-4}M was obtained. Thus, there appears to be a quantitative relationship between the affinity of trichloroethylene for cytochrome P450 and the rate of CO production.

Further evidence that cytochrome P450 was involved in this conversion was provided by inhibition studies summarized in Table 1.

TABLE 1 Effect of P450 Inhibitors on CO Production

Pretreatment	CO Formation at 37°C nmol CO/Min./Mg Protein	Inhibitors 10^{-4} M	Inhibition
Controls	~ 0.5	-	-
Phenobarbital	2.5 - 3.3	-	(0)
		Metyrapone	84 %
		∢-Naphthoflavone	20 %
3,4-Benzo(a)pyrene	1.3 - 1.6	-	(0)
		Metyrapone	20 %
		∢-Naphthoflavone	50 %

The observed inhibition pattern conforms with those observed for hydroxylation reactions mediated by cytochrome P450 (8,9). Thus, metyrapone is observed to effectively inhibit CO production by microsomes from phenobarbital-pretreated rats, whereas naphthoflavone is the more effective inhibitor when microsomes from 3,4-benzo(a)pyrene-induced rats are utilized. It can also be seen that CO production was significantly depressed when microsomes from control rat livers, containing lower concentrations of cytochrome P450, were utilized.

DISCUSSION

The spectral and kinetic data presented provide strong evidence for a cytochrome P450-mediated conversion of trichloroethylene to carbon monoxide by rat liver microsomes. A chemically reasonable pathway for this conversion, which takes place in the presence of dioxygen and NADPH, can be formulated as follows:

$$\underset{H}{\overset{Cl}{C}}=\underset{Cl}{\overset{Cl}{C}} \xrightarrow[P450]{[O]} \underset{H}{\overset{Cl}{C}}\underset{Cl}{\overset{O}{\underset{\diagdown}{C}}}\underset{}{\overset{Cl}{C}} \xrightarrow{H_2O} H-\underset{O}{\overset{Cl}{C}}-\underset{Cl}{\overset{OH}{\underset{|}{C}}}-Cl$$

$$\downarrow$$

$$CO + HCl \longleftarrow \underset{O}{\overset{H}{\underset{\parallel}{C}}}\overset{Cl}{\diagdown}$$

$$\underset{}{\overset{O}{\underset{\parallel}{C}}}\overset{}{\diagup}$$

$$\underset{H}{\overset{}{C}}\underset{Cl}{\diagdown} + HCl \longleftarrow H-\underset{Cl}{\overset{OH}{\underset{|}{C}}}-Cl$$

The first two steps, oxygenation at a carbon-carbon π-bond by cytochrome P450-dependent monooxygenase to form an epoxide which can subsequently hydrolyze to the corresponding diol, are already well documented for other systems (10). Regardless of the exact mechanism by which the carbon-carbon σ bond is broken, the third step is formally analogous to the bond-cleaving step in the classical iodoform reaction. And finally, all products resulting from the cleavage reaction are thermodynamically unstable and kinetically reactive towards decomposition to hydrochloric acid and carbon monoxide.

In humans, excretion as trichloroethanol and trichloroacetic acid accounts for the major portion of trichloroethylene retained after inhalation (11). Therefore, its major metabolic pathway appears to proceed through chloral hydrate. It seems probable that conversion to carbon monoxide would not play a significant role in the reported toxicity of trichloroethylene to humans.

ACKNOWLEDGMENT

This work was supported by the Deutsche Forschungsgemeinschaft, Sonderforschungsbereich 38.

REFERENCES

(1) R.J. Defalque, Pharmacology and toxicology of trichloroethylene,
Clin. Pharmacol. Therap. 2, 665 (1961).

(2) G. Bonse, T. Urban, D. Reichert and D. Henschler, Chemical reactivity, metabolic oxirane formation and biological reactivity of chlorinated ethylenes in the isolated perfused rat liver preparation,
Biochem. Pharmacol. 24, 1829 (1975).

(3) K.H. Byington and K.C. Leibman, Metabolism of trichloroethylene in liver microsomes II. Identification of the reaction product as chloral hydrate,
Mol. Pharmacol. 1, 247 (1965).

(4) G. Bonse, and D. Henschler, Molecular aspects to the oxirane-formation of trichloroethylene and other chlorinated ethylenes,
Naunyn-Schmiedeberg's Arch. Pharmacol. 293 (suppl.), 253 (1976).

(5) J.W. Daniel, The metabolism of ^{36}Cl-labelled trichloroethylene and tetrachloroethylene in the rat,
Biochem. Pharmacol. 12, 795 (1963).

(6) U. Frommer, V. Ullrich and Hj. Staudinger, Hydroxylation of aliphatic compounds by liver microsomes,
Hoppe-Seyler's Z. Physiol. Chem. 351, 903 (1970).

(7) Herzberg, G., Molecular Spectra and Molecular Structure, Vol I,
Van Nostrand, New York, 1950.

(8) V. Ullrich, U. Frommer and P. Weber, Differences in the O-dealkylation of 7-ethoxycoumarin after pretreatment with phenobarbital and 3-methylcholanthrene,
Hoppe-Seyler's Z. Physiol. Chem. 354, 514 (1973).

(9) V. Ullrich, P. Weber and P. Wollenberg, Tetrahydrofurane - an inhibitor for ethanol-induced liver microsomal cytochrome P450,
Biochem. Biophys. Res. Commun. 64, 808 (1975).

(10) V. Ullrich, Enzymatic hydroxylations with molecular oxygen,
Angew. Chem. internat. edit. 11, 701 (1972).

(11) V. Bartoníček, Metabolism and excretion of trichloroethylene after inhalation by human subjects,
Brit. J. Industr. Med. 19, 134 (1962).

STEREOSELECTIVE *IN VITRO* AROMATIC OXYGENATION OF CHIRAL 1,4-BENZODIAZEPIN-2-ONES

Slobodan Rendić, Vitomir Šunjić, Franjo Kajfež and Nikola Blažević

Department of Biomedical and Biochemical Research, CRC, 33048 S. Giovanni al Natisone (UD) Italy.
Faculty of Pharmacy and Biochemistry, University of Zagreb, Yugoslavia

Biotransformation pathways and pharmacology of drugs belonging to the group of 1,4-benzodiazepin-2-ones are being extensively investigated by many authors (1, 2, 3). In these investigations structure-activity relationships are of special interest. However, in spite of assiduous work, no final conclusions can be reached as to the correlations of stereochemical properties and pharmacological activity. The interpretive difficulties encountered are due to the complex dynamic stereochemistry of the compounds in question, involving different conformations of the seven-membered ring, and hindered rotation of the 5-phenyl group. Some biological effects might, therefore be explained by the preference of assymetric biological systems to bind particular conformations of various prochiral and chiral 1,4-benzodiazepine derivatives. Additionally, in biotransformation of chiral derivatives, their absolute configurations may be essential.

The most frequent in vivo biotransformations of 1,4-benzodiazepine derivatives are: N(1)-dealkylation, hydroxylation of the heterocyclic and aromatic rings, hydrolysis of the lactam and azomethyne bonds, N(4)-oxidation and formation of O-glucuronides. Actual metabolism depends on the substituents. Stereoselectivity in oxidative metabolism, presumably residing in cytochrome P-450, has been repeatedly reported for various drugs (4, 5). In the field of prochiral 1,4-benzodiazepines there is only one report on stereoselective C(3)-hydroxylation (6). Differences in biotransformation pathways of chiral, enantiomeric 1,4-benzodiazepines could further reveal stereoselectivity of cytochrome P-450 in oxidative processes, such as N(1)-dealkylation and oxygenation of heterocyclic and aromatic rings, therefore several chiral 1,4-benzodiazepin-2-ones were used in the biotransformation experiments described subsequently. Furthermore we considered it interesting to find out whether aromatic oxygenation of chiral 1,4-benzodiazepin-2-ones were catalysed by cytochrome P-450 species different from those catalysing C(3)-hydroxylations.

Accordingly, in continuation of our work on chiral 1,4-benzodiazepin-2-ones (7, 8), we have made a systematic study of stereoselectivity in biotransformation of enantiomeric 7-chloro-1,3-dihydro-3(S- and R-)-methyl-5-phenyl-2H-1,4-benzodiazepin-2-ones (compound 2, S- and R-form) and their N(1)-methyl derivatives (compound 1, S- and R-form). These experiments were carried out _in vitro_, using postmitochondrial and microsomal fractions of rat liver as enzyme sources. In a recent preliminary report (9) we have presented our first results showing stereoselectivity in aromatic oxygenation of some chiral 1,4-benzodiazepin-2-ones.

The actual isolated product of C(3)-hydroxylation of compounds S-1 and R-1 was 12. This compound originated from nonenzymatic rearrangement of 4, an intermediate metabolic product. N(1)-Demethylated metabolites S-2 and R-2 have also been isolated from both enantiomers of 1. These results show that enzymatic hydroxylation in C(3) position of 1 and 2, as well as N(1)-demethylation of 1, are nonstereoselective. Aromatic hydroxylation gave two isomeric products, as concluded from their mass-, u.v. and c.d. spectra. Extensive mass spectrometric investigation (10), however, failed to give any definite clue about the site of hydroxylation. Since in previous _in vitro_ studies no aromatic hydroxylation was observed, only extrahepatic formation of the products hydroxylated in the aromatic rings was suggested (11). Hepatic monooxygenases convert aromatic compounds into areneoxides, and products of hydroxylation in the aromatic rings of 1,4-benzodiazepines might be proposed (Fig. 1.).

To unambiguously determine the position of _in vitro_ aromatic hydroxylation in two isomeric pairs of the isolated metabolites 1 (S-form) and 2 (S-form), an extensive synthetic program has been elaborated in order to prepare possible isomeric hydroxyaryl metabolites (16). Comparison of the mass-, u.v. and c.d. spectra of synthetised hydroxyaryl isomers, as well as those of corresponding benzophenones, with spectra of the isolated metabolites and their products of hydrolysis, allowed unambiguous assignation of structures to the metabolites as 3- and 4-hydroxy-1,4-benzodiazepin-2-ones, (compounds 9 and 11, and 7 and 10, respectively). According to these results Figure 2 is proposed as a possible picture of biotransformations of chiral 1,4-benzodiazepin-2-ones.

In summary, results presented in this paper suggest that oxyge-

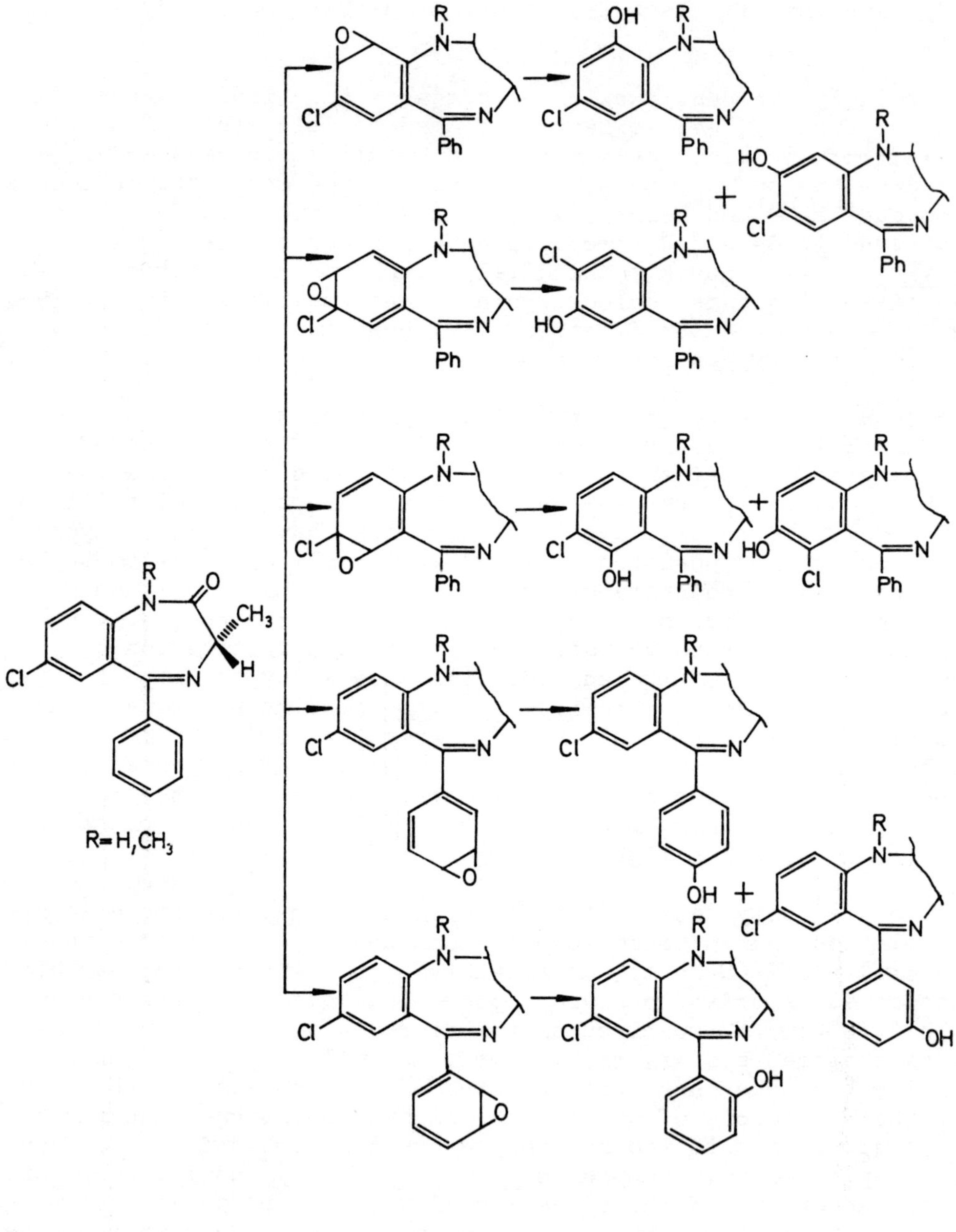

Fig. 1. Arene-oxides and products of hydroxylation proposed according to NIH shift.

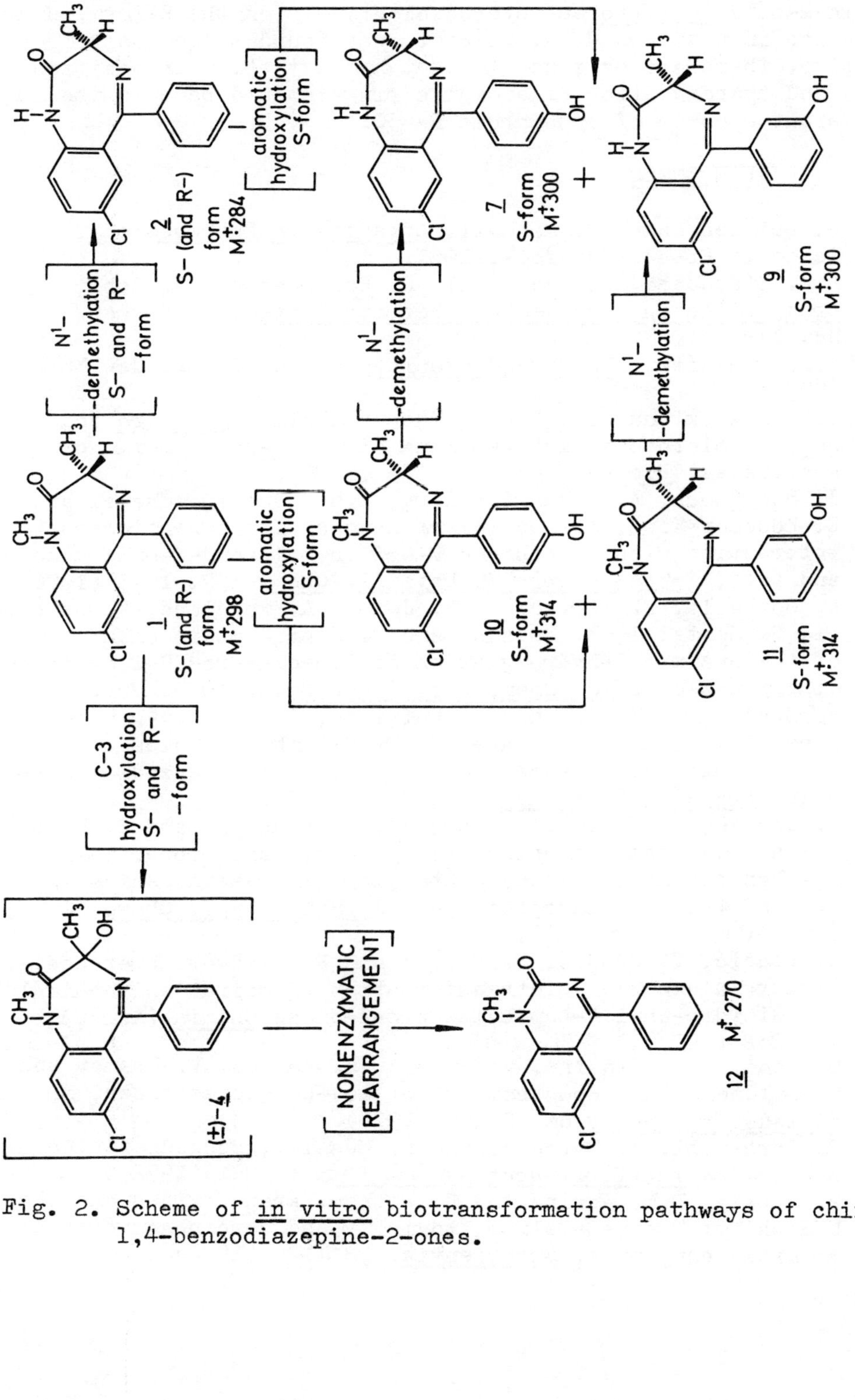

Aromatic Oxygenation of Chiral 1,4-Benzodiazepin-2-Ones

Fig. 2. Scheme of *in vitro* biotransformation pathways of chiral 1,4-benzodiazepine-2-ones.

nases, involved in aromatic oxygenation of chiral 1,4-benzodiazepin-2-ones in vitro act stereoselectively on the S-form. Hydroxylation in position C(3), however, was found to be nonstereoselective. Therefore we conclude that hydroxylation in position C(3) and hydroxylation of aromatic rings should be catalysed by different species of cytochrome P-450.

REFERENCES

(1) G. Zbinden and L.O. Randall, Advances in Pharmacology, Academic Press, New York, 1967.
(2) L.H. Sternbach, L.O. Randall, R. Banziger and H. Lehr, Drugs Affecting the Central Nervous System, M. Dekker, New York, 1968.
(3) M.A. Schwartz, The Benzodiazepines, Raven Press, New York, 1973.
(4) D.S. Hewick and J.R. Fouts, The metabolism in vitro and hepatic microsomal interactions of some enantiomeric drug substrates, Biochem. J. 117, 833 (1970).
(5) J. Noordhoek, A.P. Van den Berg, E.M. Savenije-Chapel and E. Koopman-Kool, Metabolism of hexobarbital enantiomers and interaction with cytochrome P-450 in male and female mice and rats, Hoppe-Seler's Z. Physiol. Chem. 357, 1045 (1976).
(6) A. Corbella, P. Gariboldi, G. Jommi, A. Forgione, F. Marcucci, P. Martelli, E. Mussini and F. Mauri, Stereochemistry of the enzymic 3-hydroxylation of 1,3-dihydro-2H-1,4-benzodiazepin-2-ones, J. Chem. Soc. Chem. Comm. 1973, 721.
(7) V. Šunjić, F. Kajfež and J. Kuftinec, Chiral 1,4-benzodiazepine.VII. Cyclisation rates of 2-(N-alpha-ammoniumacyl)-amino-5-chloro-benzophenones in the chiral 1,4-benzodiazepin-2-ones, Arzneim.-Forschung, 25, 340 (1975).
(8) V. Šunjić, F. Kajfež, I. Štromar, N. Blažević and D. Kolbah, Chiral 1,4-benzodiazepines.V. Synthesis and properties of 1,4-benzodiazepin-2-ones containing alpha-amino acids as a part of the 1,4-diazepine ring, J. Heterocycl. Chem. 10, 591 (1973).
(9) S. Rendić, V. Šunjić, F. Kajfež and P. Mildner, Stereoselectivity in enzymic biotransformation of chiral and achiral 1,3-dihydro-2H-1,4-benzodiazepin-2-ones, Chimia (Aarau), 28, 232 (1974),
(10) S. Rendić, L. Klasinc, V. Šunjić, F. Kajfež, V. Kramer and P. Mildner, Mass Spectrometry of 1,4-benzodiazepines, Biomed. Mass Spectrom. 2, 97 (1975).
(11) S. Garattini, F. Marcucci and E. Mussini, Benzodiazepine metabolism in vitro, Drug Metab. Rev. 1, 291 (1972).
(12) W.J. Daly, D.M. Jerina and B. Witkop, Arene oxides and the NIH shift: The metabolism, toxicity and carcinogenicity of aromatic compounds, Experientia, 28 1129 (1972).

(13) G.J. Kasparek, P.Y. Bruice, T.C. Bruice, H. Yagi and D.M. Jerina, Multiple pathways for aromatization of 8,9-indan oxide, J. Am. Chem. Soc. 95, 6041 (1973).
(14) A.M. Jeffery and D.M. Jerina, Novel rearrangements during dehydration of nucleophile adducts of arene oxides, J. Am. Chem. Soc. 97, 4427 (1975).
(15) H.G. Selander, D.M. Jerina and J.W. Daly, Metabolism of chlorobenzene with hepatic microsomes and solubilized cytochrome P-450 systems, Arch. Biochem.Biophys. 168, 309 (1975).
(16) D. Kolbah, N. Blažević, M. Hannoun, F. Kajfež, T. Kovač, S. Rendić and V. Šunjić, Stereoselective in vitro oxygenations of chiral 1,4-benzodiazepin-2-ones, submitted for publication.

EFFECTS OF DISEASE STATES ON DRUG DISPOSITION IN MAN

Elliot S. Vesell

Department of Pharmacology, The Pennsylvania State University College of Medicine, Hershey, Pennsylvania 17033

INTRODUCTION

Although drugs remain our strongest and most frequently used weapon in combating human disease, this powerful weapon can be a double-edged sword. Many recent discussions of drug abuse, drug interactions and adverse drug reactions have overemphasized these negative aspects, thereby creating the false impression that drugs are above all harmful and should be entirely avoided. It is true that development of new, exceedingly potent compounds and exposure of large populations to them require greater care and attention to the details of drug administration to reduce the high incidence of adverse drug reactions and to maximize therapeutic effects. For example, for many potent drugs with low therapeutic indices adjustment of drug dosage to individual patient needs can be achieved more satisfactorily by measuring drug concentrations in biological fluids. Furthermore, certain adverse reactions could be either eliminated entirely or reduced in number and severity if the dose of potent drugs with low therapeutic indices were selected more frequently on the basis of the pharmacokinetic properties of these drugs.

Unfortunately, until very recently the profound effects exerted by some disease states on drug disposition have been inadequately investigated. Furthermore, the effects of many factors, in addition to disease, on drug disposition have been insufficiently investigated in man, although their significant influence on drug disposition in experimental animals has been established (Table 1). Some biologically determining factors which have been identified as contributing to large interindividual variations in human drug disposition, such as genetically controlled differences in drug metabolism, are generally inadequately considered when physicians prescribe drugs; too frequently all patients receive the same dose of a drug, regardless of individual requirements. Such uniformity of dose produced fewer untoward results in previous decades when drugs of low potency and high therapeutic index were more commonly used; but now with the advent of potent compounds with low therapeutic indices a small change in dose can have profound clinical consequences.

Factors Affecting Drug Disposition in Normal Subjects

To place into perspective the effects of some disease states on drug disposition, I shall describe certain characteristics of drug disposition in normal individuals. Without controlling such environmental conditions as diet or exposure to chemicals encountered at work or at home, studies on normal nonmedicated adult volunteers living in a basal state demonstrated that the very large interindividual variations in elimination rates of such drugs

TABLE 1 A Partial List of Variables Affecting Drug Disposition in Experimental Animals

Variables in the external environment	Variables in the internal environment	Pharmacologic variables

Air exchange and composition
Barometric pressure
Cage design-materials (crowding, exercise)
Cedar and other softwood bedding
Cleanliness
Coprophagia
Diet (food and water)
Gravity
Hepatic microsomal enzyme induction or inhibition by insecticides, piperonyl butoxide, heavy metals, detergents, organic solvents, ammonia, vinyl chloride, aerosols containing eucalyptol, etc.
Handling
Humidity
Light cycle
Noise level
Temperature
	Adjuvant arthritis
	Age
	Alloxan diabetes
	Cardiovascular function
	Castration and hormone replacement
	Circadian and seasonal variations
	Dehydration
	Disease
	  hepatic, renal, malignant, endocrine (thyroid, adrenal)
	Estrous cycle
	Fever
	Gastrointestinal function, patency and flora
	Genetic constitution (strain and species differences)
	Hepatic blood flow
	Infection
	Malnutrition, starvation
	Pregnancy
	Sex
	Shock (hemorrhagic or endotoxic)
	Stress
		Drugs
		  acute vs. chronic administration, bioavailability, dose, withdrawal, presence of other drugs or food, routes of administration, volume of material injected, tolerance, vehicle, etc.

as antipyrine, bishydroxycoumarin, ethanol, halothane, nortriptyline, phenylbutazone, phenytoin and salicylates are under predominantly genetic control. Twin studies disclosed that most of the interindividual variations, which ranged from 4- to 40-fold depending on the drug, disappeared within identical twinships but were retained to some extent within fraternal twinships (1-4) (Fig. 1). These results, as well as family studies on bishydroxycoumarin (5), phenylbutazone (6) and nortriptyline (7) metabolism, suggested that large interindividual variations in the disposition of many commonly used drugs were maintained mainly by genetic, rather than environmental, factors.

In most of these studies the only limitations imposed on the environment of these normal volunteers was that they were not taking drugs for one month prior to the study and this did not constitute a major change in their habits, since they did not customarily receive drugs. Nevertheless, these studies did not systematically investigate or attempt to quantitate the influence on drug disposition of such variables as age, sex, cigarette smoking or consumption of coffee, tea or alcohol. It has subsequently become clear that age plays a small role in controlling the rate of elimination of certain drugs; antipyrine half-life was 16.5% longer in older than in younger subjects (8); a poor correlation existed between age and antipyrine half-life

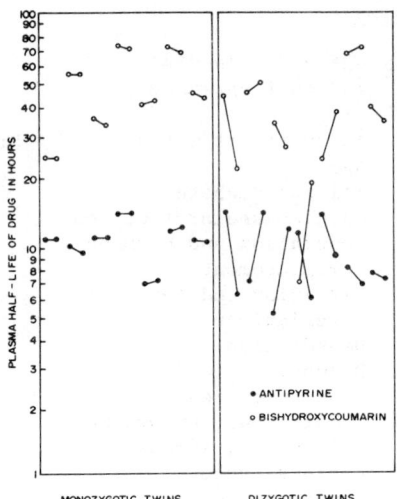

Fig. 1. Plasma half-lives of bishydroxycoumarin and antipyrine were measured separately at an interval of more than 6 months in healthy monozygotic (identical) and dizygotic (fraternal) twins. A solid line joins the values for each set of twins for each drug. Note that intratwin differences in the plasma half-life of both bishydroxycoumarin and antipyrine are smaller in monozygotic than in dizygotic twins. However, some DZ twins resemble MZ twins in having very small intratwin differences.

($r = -0.25$) (8). Although changes in rates of drug metabolism with age are small, significant alterations in drug disposition do occur over very much shorter periods of time. We observed that in some, but not all, normal male volunteers the half-lives of both phenacetin and acetaminophen are approximately 15% shorter at 2:00 p.m. than at 6:00 a.m. (Fig. 2) (9). Our own studies (10), as well as those of others (11), revealed that chronic ethanol ingestion increased the metabolism of drugs in some, but not all, subjects. Moreover, cigarette smoking altered phenacetin metabolism (12,13) and accelerated antipyrine metabolism (14). On an isocaloric diet, alterations of the carbohydrate to protein ratio changed antipyrine and theophylline metabolism (15). Low protein content was associated with retarded metabolism of these drugs (15). Nevertheless, in obese subjects starvation for 7 to 10 days failed to alter antipyrine or tolbutamide metabolism (16). Occupational exposure to certain environmental chemicals, such as DDT, accelerated phenylbutazone and cortisol (17), as well as antipyrine (18), metabolism. Thus, some of the environmental factors listed in Table 1 for experimental animals also affect drug disposition in man.

If environmental agents were mainly responsible for large variations in drug disposition among normal subjects, it would be anticipated that with increasing age there would be increasing rates of drug metabolism. This expectation

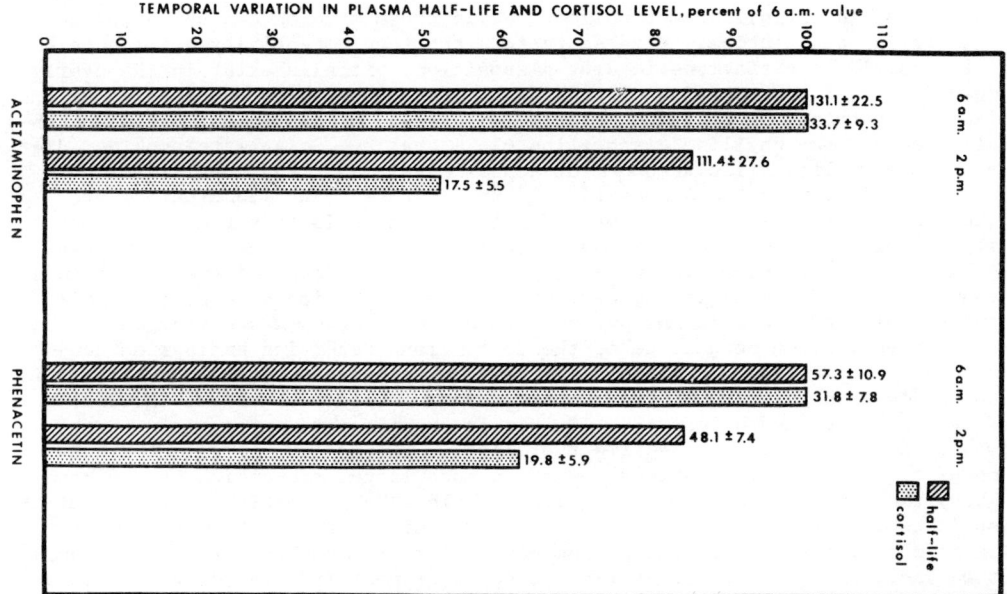

Fig. 2. Temporal variations in mean plasma half-life (min) and mean plasma free 11-hydroxycorticoids (μg/100 ml) in normal males after a single oral dose of either acetaminophen or phenacetin. Data are graphed as percent of control, using the mean 6:00 a.m. value as control. Mean ± SD are indicated above bars (9).

is based on the fact that most environmentally encountered chemicals such as DDT, PCB (polychlorinated biphenyls) and polycyclic hydrocarbons are potent inducing agents; thus, increasing exposure to these compounds with time should be accompanied by accelerated drug metabolism with age. However, interindividual variations in drug disposition cannot be explained simply on the basis of increasing exposure with time to potent inducing agents. This conclusion is based on the fact that drug metabolism is most active in the pediatric age (19), less so in adulthood (8) and least in old age (8). This pattern is opposite to that expected were environmental chemicals alone responsible for large interindividual variations in drug disposition. Clearly other factors must be involved. One commonly encountered environmental chemical, the birth control pill, inhibits hepatic drug metabolism (20,21), but this potential environmental effect on the observed decrease in rates of drug metabolism with age can be excluded for several reasons. First of all, the main age effects are observed in the pediatric and geriatric groups where birth control pills are not taken and second, decreased drug elimination occurs with age in groups composed entirely of males.

Because of the introduction of plasma antipyrine half-lives as a convenient, simple test for assessing rates of hepatic antipyrine metabolism (2) and the further refinement of this test to permit each individual to serve as his own control in investigating effects of drugs, environmental chemicals or other conditions on hepatic antipyrine metabolism (3), studies have been undertaken to measure the relative contributions of a number of different factors, such as age, diet, cigarette smoking, etc., to large interindividual variations in antipyrine metabolism. While these studies can be useful in identifying qualitatively diverse factors that can potentially contribute to interindividual variations in drug disposition, pitfalls exist in the overly quantitative interpretation of such data. For example, recent studies on factors contributing to large interindividual variations in antipyrine metabolism and in theophylline disposition claim that age, cigarette smoking, diet and antifertility pills account for approximately half the total interindividual variation in the disposition of these drugs. Such conclusions are questionable for several reasons. The first reason is that the small subpopulation examined in each study is unrepresentative of the entire population of the particular area with respect to the incidence of the factor under examination. For example, to determine the contribution of diet to interindividual variations in antipyrine metabolism, a study was performed in which a relatively large percentage of the total group receiving antipyrine consisted of vegetarians. However, in the total population of London, from which the study group was drawn, the number of individuals who are vegetarians is exceedingly low, much less than 1%. It is erroneous to extrapolate to the whole London population the effect of diet in this specially selected subpopulation because vegetarians compose a much larger proportion of the study group than in the complete population. While in a population where 50% of the subjects are vegetarians diet may contribute significantly to large interindividual variations in antipyrine metabolism, in another very large population where there are no vegetarians absence of meat from the diet obviously cannot be considered as contributory to the interindividual variations observed. Thus, the principle must be firmly established that quantitative extrapolation of contributions of a particular factor from one population to another is valid only if the proportion of individuals exposed to the particular factor is comparable in the two populations.

The second problem in studies of this type concerns the confounding of different factors under a single rubric. For example, smoking and diet are often examined as independent variables, representative exclusively of "environmental" factors. However, smoking and diet are not isolated traits, but rather are inextricably connected with genetic and other environmental characteristics. Smoking habits have been correlated with intellectual, behavioral and genetic characteristics of an individual; and smoking in high school students is highly correlated to parental smoking habits (22). Thus, rather than a simple single factor that consists of exposing an individual to potent chemicals that can induce drug metabolizing enzymes, smoking should be considered as a more complex trait that is closely associated with other environmental conditions as well as with genetic factors. For these reasons, studies attributing large variations in drug metabolism to different smoking habits of a population confound smoking with other environmental as well as with genetic factors.

Table 2 provides a means of comparing the magnitude of interindividual and intraindividual variations in drug distribution. The half-lives of both antipyrine and bishydroxycoumarin are highly reproducible in a given individual; intraindividual variation is approximately 10 to 20% for each drug. By contrast, the range of interindividual variations for bishydroxycoumarin extends over 1,000% and for antipyrine approximately 400%. Intraindividual variations may be attributed to environmental factors that fluctuate in their temporal impact on an individual; such fluctuating environmental factors are of small consequence compared to large, genetically controlled, interindividual variations in drug disposition (Table 2).

TABLE 2 Reproducibility of Plasma Half-lives of Bishydroxycoumarin and Antipyrine in Normal Volunteers

	BISHYDROXYCOUMARIN				ANTIPYRINE		
	PLASMA HALF-LIFE (HRS)				*PLASMA HALF-LIFE (HRS)*		
Volunteer	*Initial*	*Repeat*	*% Change*	*Volunteer*	*Initial*	*Repeat*	*% Change*
D.H.	46.0	45.0	2.2	T.Ch.	10.6	11.3	6.6
D.W.	44.0	42.5	3.4	T.C.	12.5	11.4	8.8
Ge.L.	72.0	66.8	7.2	G.Z.	7.7	7.5	2.6
Gu.L.	69.0	70.8	2.6	T.L.	12.7	13.2	3.9
Ja.T.	74.0	70.8	4.3	R.F.	22.4	20.5	8.5
Jo.T.	72.0	73.6	2.2	M.R.	13.1	13.3	1.5
Ja.H.	7.0	7.5	7.1	C.H.	9.5	10.0	5.3
Je.H.	19.0	18.4	3.2	P.M.	5.8	6.0	3.5

Although a disease process may alter the pharmacokinetic parameters of a given drug in an individual by only 20%, this change, while negligible compared to the much larger interindividual variations that exist among normal subjects, can be clinically significant for a potent drug with a low therapeutic index. Such small changes in rates of drug elimination may require compensation by appropriate alterations in dose.

Effects of Disease on Drug Metabolism

Only recently was abnormal drug metabolism demonstrated convincingly to accompany various disease states. To explain why this association remained unproven so long, it is useful to contrast the problems of investigating disease states in laboratory animals and human beings. In experimental animals the effect of disease states on drug disposition can be carefully measured because the environmental and genetic characteristics of the laboratory animal can be determined and controlled rigorously; accordingly, Table 1 lists several disease states in which appropriate studies on animals showed that drug disposition was altered. Rates of drug elimination during these disease states were compared to those that prevailed under normal basal conditions in the same animal. Development of a highly reproducible model in the animal permitted precise quantitation of the effects on drug disposition produced by a specific pathological lesion of known intensity and severity.

By contrast, in human disease each patient represents an almost unique constellation of genetic and environmental variables. Thus, variation in drug disposition due to these factors is immense. Furthermore, a very broad spectrum exists in the severity, intensity and duration of most human diseases; for a single patient large changes in a disease process and its effects occur commonly with time, even when only a single organ is affected. Generally, however, several organs are simultaneously involved. The functional status of the cardiovascular, endocrine, hepatic, hormonal and nervous systems fluctuate during many disease processes, thereby producing fluctuations with time in drug disposition. In addition, such variables as age, sex, genetic constitution, diet, weight, body size, exposure to other drugs, environmental chemicals, etc., differ among patients. In addition to these complications, the normal rate of drug elimination for the patient prior to disease is rarely known. Thus, it is often impossible to determine whether the rate of drug elimination measured in a given patient during a disease constitutes a change from the basal, normal rate.

Many different disorders and pathological states can alter the normal rates and pathways for drug absorption, distribution, biotransformation, excretion, interaction with receptor sites or various combinations of these. Since each drug has a distinct profile for these 5 processes, the extent to which a disease that affects these processes will alter the distribution of a drug depends on the particular drug. Therefore, it is hazardous to extrapolate how pathological processes will affect other drugs on the basis of pathological effects that alter the distribution of one drug. Furthermore, the clinical consequences of a change in drug distribution produced by disease will be determined not so much by the absolute amount of that change but rather by the therapeutic index of the drug as well as by certain genetic and environmental characteristics of the patient. For example, a change of 200% in the plasma half-life of antipyrine, salicylates or penicillin will probably have little or no clinical consequence, whereas a change of 20% in the plasma half-life of digoxin, procainamide or lidocaine may prove critical. By the same token, a change of 1% in the albumin binding of warfarin, which normally is 99% bound, may have profound toxicological results, whereas a change of 1% in the albumin binding of probenecid, which normally is 75% bound, has negligible clinical consequences. Here the crucial factor is the percent change in the free, but not the bound, portion of the drug.

Effects of disease states on the rate of absorption of the drug in the body depend on many factors including the site of drug administration. If the drug is administered orally, the influence of disease on rates of drug absorption from the gastrointestinal tract will depend on the nature of the disease process, whether it affects the areas in the gut where the drug is normally absorbed, how the disease alters the normal physiological volume, pH, temperature, viscosity, surface tension, and composition of the gastrointestinal secretions and contents (23). Rates of drug absorption may also be influenced by whether or not food is present, the nature and quantity of bile salts and bacterial flora, the rate of splanchnic blood flow, prior diet and food intake as well as gastrointestinal motility (23). Until recently little was known about how disease states altered these factors in man; however, large interindividual differences in rates of absorption of many orally administered drugs occur in hospital patients (24-26) as well as in normal volunteers (27-29). For example, a 7-fold range in the amount of tetracycline absorbed was reported in 6 fasting, healthy subjects (30). Variations in gastric emptying may contribute significantly to large interindividual differences in rates of drug absorption because numerous physiological conditions, such as posture and autonomic activity as well as the temperature, volume, viscosity and tonicity of gastric contents, can change gastric emptying time, thereby altering rates of drug absorption. Grossly impaired absorption of paracetamol occurs in patients with delayed gastric emptying and pyloric stenosis. In patients with slow gastric emptying, L-dopa may be ineffective (31). Therapeutic failure of orally administered drugs usually accompanies gastric stasis (32-34). In patients with achlorhydria, aspirin was absorbed significantly faster and plasma salicylate concentrations were higher than in controls (Fig. 3) (32). Prescott (35) reported the interesting observation that acetaminophen plasma concentrations were significantly higher after oral administration in 12 convalescent hospital patients in bed than in 7 age and sex matched healthy ambulant volunteers.

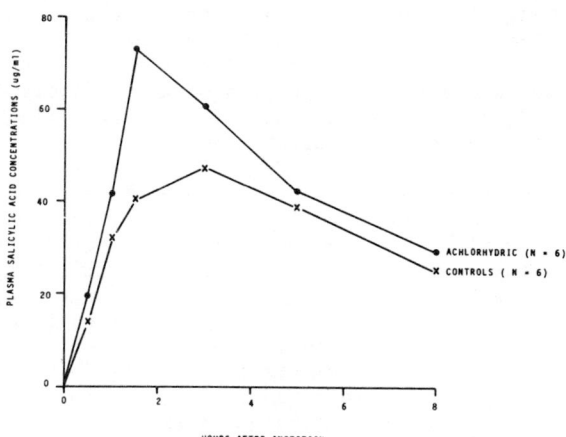

Fig. 3. Mean plasma salicylate concentrations in achlorhydric and control patients following an oral dose of 900 mg of acetylsalicylic acid. Reproduced from Prescott (32).

Since rates of absorption of orally administered pills and capsules are dependent on rates of their dissolution and dispersion, some of the factors enumerated above can alter such rates, thereby contributing to variations in drug absorption. It is interesting and somewhat surprising that the absorption of p-aminosalicylic acid and isoniazid was unchanged by gastrectomy performed for peptic ulcer, although in some patients complete failure of ethionamide absorption occurred (36). Furthermore, gastrectomy failed to alter the absorption of sulfisoxazole, quinidine or ethambutol unless vagotomy had been performed, thereby slowing gastric emptying (37).

In jejunal disease folic acid absorption is diminished (38). In ileal disease the transport of bile acids may be impaired, as well as the enterohepatic transport of many lipid soluble drugs, since bile acids promote the gastrointestinal absorption of fat and certain fat soluble compounds including many drugs and vitamins A, D, K and E. Ileal disease may be associated with impaired B_{12} absorption, since B_{12} is absorbed in the ileum after it forms a complex with intrinsic factor, a substance produced by the gastric parietal cell. Defective function of the ileum secondary to surgical removal or disease through interference with B_{12} absorption can lead to pernicious anemia. B_{12} absorption can also be impaired in gastric diseases where parietal cell function is abnormal, thereby producing intrinsic factor deficiency and pernicious anemia. In some patients with pernicious anemia precipitating or blocking antibodies to intrinsic factor have been identified (39,40). In addition, regional enteritis as well as such other disorders of the gastrointestinal mucosa as tropical sprue, celiac disease and Whipple's disease can produce B_{12} malabsorption. In steatorrhea fat soluble drugs and vitamins may be lost in the feces, thereby producing deficiency of the fat soluble vitamins. Of the fat soluble vitamins, deficiency of vitamin D is by far the most significant chronic problem in gastrointestinal disease.

Some active drugs are produced after metabolism of the inactive form by gut bacteria. The best example of this is cleavage of salicylazosulfapyridine, the drug of choice in the treatment of chronic ulcerative colitis. Diseases or drugs that change the nature of the gastrointestinal flora can affect the disposition of other drugs metabolized by gut bacteria. Possibly the effect of large doses of charcoal on the gastrointestinal absorption of phenacetin is mediated by gut bacteria and induction of AHH activity in these bacteria by charcoal (41). Another recent example is the reduction of the absorption of digoxin by neomycin administration (42).

The binding of many drugs to albumin is altered in several disease states. For example, the phenytoin binding was decreased in plasma from 15 uremic patients (43). The size of the unbound fraction correlated well with blood urea nitrogen, serum creatinine and clinical state of the patient. In addition to phenytoin, other organic acids including congo red, sulfonamides, thyroxine, diphenylhydantoin, tryptophan, clofibrate, fluorescein and methyl red exhibited decreased protein binding in uremia. By contrast, most organic bases bind normally to plasma from uremic patients. A clinically significant neutral compound, digitoxin, exhibited decreased plasma binding in uremic patients (44). Disease states including cirrhosis and nephritis are associated with hypoproteinemia and hypoalbuminemia. In hypoproteinemia and hypoalbuminemia the unbound fraction of most drugs is elevated compared to the unbound fraction that exists with normal protein and albumin concentrations. These situations illustrate the danger of selecting drug dose solely on the basis of total drug concentrations in plasma, rather than on the basis of the free concentration, since only the free form is pharmacologically active. In cirrhosis, the plasma binding of the organic bases quinidine, diazepam and triamterene, as well as of the organic acid fluorescein are all decreased.

Since poor renal function is associated with decreased excretion of several drugs, it is important to modify normal drug dosage in uremic patients. Methods for this have been presented by Dettli and others (45-48). Dettli (46) described a linear relationship between the overall elimination rate constant (k_e) and the endogenous creatinine clearance ($\dot{V}'_{cr}$): $k_e = k_{nr} + \delta \cdot \dot{V}'_{cr}$ where k_{nr} is the mean extrarenal elimination rate constant in anuric patients and δ is a constant relating $\dot{V}'_{cr}$ to the renal elimination rate constant (k_r) of the drug. This equation can be used for about 40 drugs; simple nomograms have been devised which allow the estimation of the rate of drug elimination in the individual patient with kidney disease from the value of $\dot{V}'_{cr}$ (46,49). However, there are a large number of drugs whose apparent elimination rates are unaltered by uremia, including tolbutamide, phenobarbital, histamine, phenacetin, diphenylhydantoin, antipyrine, quinidine, propranolol and vitamin D (43).

Much less study has been devoted to effects of disease states on receptor sites and kinetics of formation and dissociation of drug receptor complexes. Comparison of the basal state of receptors under normal and disease conditions should facilitate analysis of drug action. Advances in receptor isolation by techniques such as affinity labeling now encourage careful in vitro studies to assess how diseases affect receptors. For example, such techniques have the potential of revealing whether cholinergic receptors from myasthenic skeletal muscle function normally in response to transmitter

Effects of Disease States

and in the kinetics of drug binding and dissociation. Diseases can also alter the local environment of receptors, thereby changing their conformation. A classical example of how environmental conditions alter the binding of a drug to its receptor is that of digoxin with receptors in cardiac muscle, where the efficacy of the drug in producing inotropic and chronotropic effects can be appreciably altered by changes in potassium ion concentration or pH near the receptor.

Effect of Fever on Drug Disposition

In the course of many diseases fever develops either acutely or chronically. Thus, fever is a concomitant of many diseases; yet, until recently, the effects of fever on drug distribution were unexamined. In 1975, it was shown that etiocholanolone-induced fever in normal volunteers was associated with a prolongation of plasma antipyrine half-life in 11 of 14 volunteers who developed fever (50) (Figs. 4 and 5). Recently these results have been confirmed by Trenholme et al (51), who demonstrated retardation of quinine elimination both in etiocholanolone-induced fever and in fever associated with experimentally induced malaria. The effects of fever on drug distribution need to be studied both with more drugs and in other febrile states before generalizations concerning the role of fever in drug disposition and appropriate dosage compensation can be made.

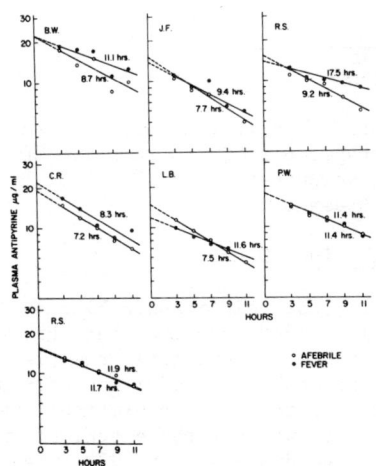
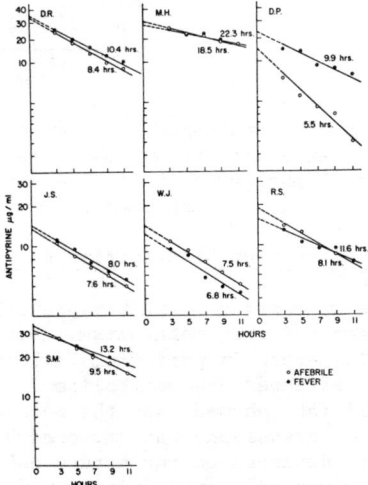

Figs. 4 and 5. Plasma antipyrine half-life after an oral dose of 10 mg/kg antipyrine in aqueous solution given to normal volunteers in an afebrile state (o) and again during production of fever by etiocholanolone administration (●).
Note prolongation of plasma antipyrine half-life by fever in most, but not all, volunteers. Data from Elin et al (50).

Effect of Thyroid Disease on Drug Disposition

An extensive literature based on experimental animals described various effects produced by altered thyroid function on drug disposition. In 1973 patients with hyperthyroidism were shown to exhibit accelerated antipyrine metabolism, whereas patients with hypothyroidism displayed prolongation of plasma antipyrine half-life (52) (Fig. 6). These observations were confirmed in 1974 (53) and extended in 1975 (54) (Fig. 7) to other drugs. Thus, for drugs of low therapeutic index that are metabolized by the hepatic mixed function oxidases, the appropriate dose can be significantly altered by the functional status of the thyroid.

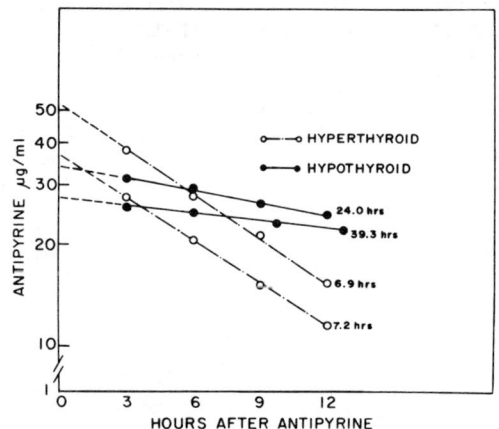

Fig. 6. Prolonged antipyrine half-lives in 2 hypothyroid patients and shortened antipyrine half-lives in 2 hyperthyroid patients relative to the normal antipyrine half-lives in 36 healthy volunteers of 10.9 ± 0.5 hrs. Reproduced from Vesell and Passananti (52).

Studies on Drug Distribution in Patients with Liver Disease

For many years uncertainty prevailed concerning the relationship between liver disease and drug metabolism. Levi et al (55) in 1968 revealed that this ambiguity arose in part from failure to consider how concomitantly administered drugs affected the metabolism of the drug under study. In 1971 Cookesly and Powell (56) showed that the metabolic pathway for the hepatic biotransformation of prednisone was affected by liver disease and that the extent of this effect depended on the nature and severity of the disease. In 1972 Faigle (57) reported that in patients with severe bilharziasis, metabolism of the schistosomicide Niridazole (nitrothiamidazol) was impaired, as revealed by higher blood levels of the drug and a higher incidence of adverse effects than in normal subjects. In 1973 Branch et al (58) reported that serum antipyrine half-life was prolonged in patients with liver disease, an observation confirmed in 1974 by Andreasen et al (59) and by Andreasen and Vesell (60) (Fig. 8). Also in 1974, Hepner and Vesell (61) reported that aminopyrine metabolism was deranged in a large percentage of patients with liver disease and proposed an aminopyrine breath test (ABT) in which $^{14}CO_2$ was measured in breath after the administration of 4-dimethyl-^{14}C-aminoantipyrine

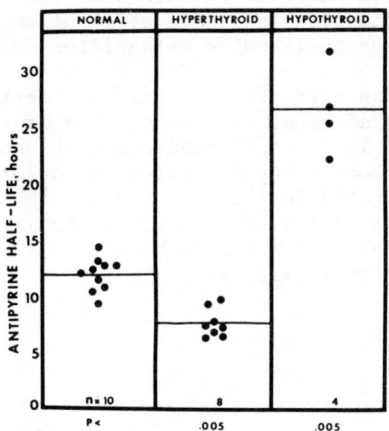

Fig. 7. Values for plasma antipyrine half-life in normal subjects and in patients with hyperthyroidism and hypothyroidism. Each closed circle represents the value for a single individual.
Reproduced from Vesell and Passananti (52).

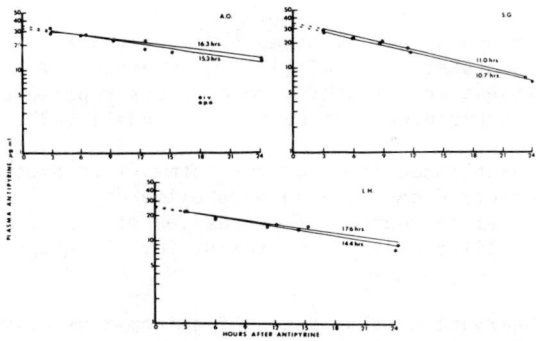

Fig. 8. The decay of antipyrine after oral and intravenous administration to 3 patients with cirrhosis of the liver. The oral and intravenous doses were separated by an interval of 8 days.
Reproduced from Andreasen and Vesell (60).

(^{14}C-aminopyrine), as a sensitive, clinically convenient test of hepatic function and drug metabolism. The virtue of this method was that it measured production of a metabolite, rather than disappearance of the parent drug. It is now abundantly clear from numerous studies that the rate of disappearance of cold aminopyrine or antipyrine in plasma of a patient correlates highly with the rate of production of their major metabolites (62).

Further studies utilizing patients with various hepatic disorders disclosed that the ABT discriminated among different types of hepatic dysfunction (63) (Fig. 9). For example, in 16 of 21 subjects with cholestasis not caused by malignant disease, the mean $^{14}CO_2$ excretion was normal. The mean percentage of administered ^{14}C excreted in $^{14}CO_2$ in 2 hours in control patients was 7.1 ± 1.3 (SD)% and significantly less ($p < 0.01$) in patients with portal cirrhosis (2.6 ± 1.2%), fatty liver (4.7 ± 1.1%), hepatitis (2.6 ± 1.4%) and hepatic malignancy (3.5 ± 1.8%) (63) (Fig. 9).

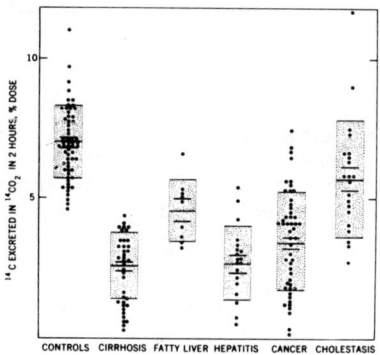

Fig. 9. The percentage of administered ^{14}C excreted as $^{14}CO_2$ in breath 2 hrs after oral administration of [^{14}C]aminopyrine. Transverse lines represent mean ± SEM; hatched areas represent SD.
Reproduced from Hepner and Vesell (63).

It is now firmly established that various forms of hepatocellular disease are associated with decreased capacity to metabolize drugs. However, it is presently unclear whether the extent of depression of hepatic microsomal function in individuals with liver disease is similar for all drugs metabolized by the liver.

In light of our observations that rates of aminopyrine elimination are reduced in almost all patients with parenchymal liver disease and the observations by Branch (58) and by Andreasen (59,60) that liver disease is also accompanied by reductions in rates of hepatic antipyrine metabolism, we can place into context the observations of Williams et al (64) on the lack of change in warfarin disposition during acute viral hepatitis and of Shull et al (65) that oxazepam disposition is normal during acute viral hepatitis and cirrhosis. Between these extremes, a group of drugs apparently exists where intermediate or moderate effects in disposition occur with liver disease.

Clindamycin is an example of this group (66). Thus, liver disease affects drug disposition in different ways depending on the particular drug. A range of effects, from no change whatever in disposition to significant retardation, has been reported. For drugs with high hepatic extraction ratios, greater than 0.8, such as propranolol and lidocaine, alterations in blood flow accompanying liver disease can produce large changes in hepatic clearance of the compound. For drugs with very low hepatic extraction ratios, less than 0.2, such as antipyrine and aminopyrine, large variations exist in the extent to which hepatocellular disease alters their rates of metabolism; these variations cannot be due to abnormal liver blood flow. They may be attributable in part to the existence of multiple molecular forms of hepatic cytochrome P-450 and the differential effects that a particular hepatic disorder might exert on these forms.

Heterogeneous effects of liver disease on drug disposition are in accord with previous studies in normal individuals that established lack of correlation in rates of hepatic biotransformation of different drugs. In normal individuals rates of hepatic biotransformation of antipyrine did not correlate with rates of hepatic biotransformation of such other drugs as phenylbutazone and bishydroxycoumarin (3), an observation confirmed by Kadar et al (67). No inherent incompatibility exists between failure to correlate rates of drug metabolism in the basal state and high correlation of these rates when the hepatic drug-metabolizing system is induced or inhibited. One explanation for this observation focuses on the heterogeneity, both qualitative and quantitative, of cytochrome P-450, the multiple forms of which probably each act on different drug substrates (68). When an inducing or inhibiting compound is administered, one or a few of these forms are selectively altered. Consequently, a greater uniformity of response occurs in different subjects as identified by administration of a test drug if that test drug is the substrate for the particular cytochrome P-450 that has been altered. Although large interindividual variations in the extent of induction as measured by percent change in plasma antipyrine half-life from the basal level have been reported (69), an inducing agent such as phenobarbital reduced interindividual variations in rates of antipyrine elimination measured in unmedicated normal volunteers. A more uniform rate of antipyrine metabolism was produced by phenobarbital for all subjects in the study, even though some subjects changed more than others to attain this given rate (69). These observations are compatible with and help explain lack of significant correlations among rates of elimination of some drugs in the basal, unmedicated state, but significant correlation in the induced state of drug metabolism.

Finally, a discussion of the biological significance of correlations among certain pharmacokinetic values of drugs seems appropriate. The principal aim of such correlations among rates of elimination of different drugs is to attempt to predict the behavior of one drug from that of another. The predictive value is the square of the correlation coefficient. In clinical practice the usefulness of predictive values of less than 90% is limited; to achieve a predictive value of 90%, the correlation coefficient must be 0.95. With lower correlation coefficients, the difference between the predictive value and the correlation coefficient is larger. For example, an r value of 0.8 has a predictive value of 0.64, and an r of 0.6 has a predictive value of 0.36, indicating that in only 1 of 3 patients would such a prediction be expected to be accurate. It should be stressed that a low p value should not be misconstrued to indicate that the correlation is biologically significant. The p value does not provide information on the likelihood of a biologically significant relationship, but rather on the probability of reproducing the best line that can be drawn through as many of the data points as possible. Thus, a correlation coefficient of 0.6 or 0.3 may have a p value of 0.001, but the correlation itself may be without biological significance. Furthermore, studies on small numbers of subjects may yield a very high correlation between the rates of elimination of several drugs; these correlations should be rechecked with larger groups, since in the past some high correlations obtained with small numbers of volunteers have not been reproducible, even in the same laboratory (70). Present data suggest that the extent to which hepatic dysfunction alters rates of elimination of exogenously administered compounds removed by the liver varies according to the substrate. Therefore, when patients with hepatic dysfunction receive drugs eliminated through hepatic mechanisms, the rate of elimination of one drug should not be extrapolated from pharmacokinetic knowledge of another; future work may disclose drugs whose correlations in liver disease are so high that they can be used for such predictions. However, the results in patients with liver disease agree in principle with most observations in normal, uninduced volunteers (71).

REFERENCES

1. E.S. Vesell and J.G. Page, Genetic control of drug levels in man: phenylbutazone, Science 159, 1479 (1968).
2. E.S. Vesell and J.G. Page, Genetic control of drug levels in man: antipyrine, Science 161, 72 (1968).
3. E.S. Vesell and J.G. Page, Genetic control of dicumarol levels in man, J. Clin. Invest. 47, 2657 (1968).
4. E.S. Vesell, Advances in pharmacogenetics, Prog. Med. Genet. 9, 291 (1973).
5. A. Motulsky, Pharmacogenetics, Prog. Med. Genet. 3, 49 (1964).
6. J.A. Whittaker and D.A. Price Evans, Genetic control of phenylbutazone metabolism in man, Brit. Med. J. 4, 323 (1970).
7. M. Åsberg, D.A. Price Evans and F. Sjöqvist, Genetic control of nortriptyline kinetics in man: a study of relatives of propositi with high plasma concentrations, J. Med. Genet. 8, 129 (1971).
8. R.E. Vestal, A.H. Norris, J.D. Tobin, B.H. Cohen, N.W. Shock and R. Andres, Antipyrine metabolism in man: influence of age, alcohol, caffeine, and smoking, Clin. Pharmacol. Ther. 18, 425 (1975).
9. C.A. Shively and E.S. Vesell, Temporal variations in acetaminophen and phenacetin half-life in man, Clin. Pharmacol. Ther. 18, 413 (1975).
10. E.S. Vesell, J.G. Page and G.T. Passananti, Genetic and environmental factors affecting ethanol metabolism in man, Clin. Pharmacol. Ther. 12, 192 (1971).
11. R.M.H. Kater, G. Roggin, F. Tobon, P. Zieve and F.L. Iber, Increased rate of clearance of drugs from the circulation of alcoholics, Amer. J. Med. Sci. 258, 35 (1969).
12. E.J. Pantuck, K.-C. Hsiao, A. Maggio, K. Nakamura, R. Kuntzman and A.H. Conney, Effect of cigarette smoking on phenacetin metabolism, Clin. Pharmacol. Ther. 15, 9 (1974).
13. A.H. Beckett and E.J. Triggs, Enzyme induction in man caused by smoking, Nature 216, 587 (1967).
14. P. Hart, G.C. Farrell, W.G.E. Cooksley and L.W. Powell, Enhanced drug metabolism in cigarette smokers, Brit. Med. J. 2, 147 (1976).
15. A.P. Alvares, K.E. Anderson, A.H. Conney and A. Kappas, Interactions between nutritional factors and drug biotransformation in man, Proc. Natl. Acad. Sci. 73, 2501 (1976).
16. M.M. Reidenberg and E.S. Vesell, Unaltered metabolism of antipyrine and tolbutamide in fasting man, Clin. Pharmacol. Ther. 17, 650 (1975).

17. A. Poland, D. Smith, R. Kuntzman, M. Jacobson and A.H. Conney, Effect of intensive occupational exposure to DDT on phenylbutazone and cortisol metabolism in human subjects, Clin. Pharmacol. Ther. 11, 724 (1970).
18. B. Kolmodin, D.L. Azarnoff and F. Sjöqvist, Effect of environmental factors on drug metabolism: decreased plasma half-life of antipyrine in workers exposed to chlorinated hydrocarbon insecticides, Clin. Pharmacol. Ther. 10, 638 (1969).
19. A.P. Alvares, S. Kapelner, S. Sassa and A. Kappas, Drug metabolism in normal children, lead-poisoned children, and normal adults, Clin. Pharmacol. Ther. 17, 179 (1975).
20. K. O'Malley, I.H. Stevenson and J. Crooks, Impairment of human drug metabolism by oral contraceptive steroids, Clin. Pharmacol. Ther. 13, 552 (1972).
21. D.E. Carter, J.M. Goldman, R. Bressler, R.J. Huxtable, C.D. Christian and M.W. Heine, Effect of oral contraceptives on drug metabolism, Clin. Pharmacol. Ther. 15, 22 (1974).
22. B.L. Borland and R.P. Rudolph, Relative effect of low socio-economic status, parental smoking and poor scholastic performance on smoking in high school students, Soc. Sci. Med. 9, 27 (1975).
23. R.R. Levine, Factors affecting gastrointestinal absorption of drugs, Digest. Dis. 15, 171 (1970).
24. B.K. Armstrong, A.W. Ukich and P.M. Goatcher, Plasma salicylate levels in rheumatoid arthritis produced by four different salicylate preparations, Med. J. Aust. 2, 181 (1970).
25. J. Koch-Weser, Pharmacokinetics of procainamide in man, Ann. N.Y. Acad. Sci. 179, 370 (1971).
26. J.D. Nelson, S. Shelton, H.T. Kusmiesz and K.C. Haltalin, Absorption of ampicillin and children with acute shigellosis, Clin. Pharmacol. Ther. 13, 879 (1972).
27. L.F. Prescott, R.F. Steel and W.R. Ferrier, The effects of particle size on the absorption of phenacetin in man, Clin. Pharmacol. Ther. 11, 496 (1970).
28. H. Schröder and D.E.S. Campbell, Absorption, metabolism and excretion of salicylazosulfapyridine in man, Clin. Pharmacol. Ther. 13, 539 (1972).
29. B. Beermann, K. Hellström and A. Rosen, On the metabolism of propantheline in man, Clin. Pharmacol. Ther. 13, 212 (1972).
30. L.F. Prescott and J. Nimmo, Generic inequivalence - clinical observations, Acta pharmacol. (Kbh.) 29, Suppl. 1, 288 (1971).
31. J.R. Bianchine, L. Rivera-Calimlim, J.P. Morgan, C.A. Sujuvne and L. Lasagna, Metabolism and absorption of L-3,4-dihydroxyphenylalanine in patients with Parkinson's disease, Ann. N.Y. Acad. Sci. 179, 126 (1971).
32. L.F. Prescott, Gastrointestinal absorption of drugs, Med. Clin. N. Am. 58, 907 (1974).
33. R.C. Heading, J. Nimmo, L.F. Prescott and P. Tothill, The dependence of paracetamol absorption on the rate of gastric emptying, Brit. J. Pharmacol. 47, 415 (1973).
34. J. Nimmo, R.C. Heading, P. Tothill and L.F. Prescott, Pharmacological modification of gastric emptying: effects of propantheline and metoclopramide on paracetamol absorption, Brit. Med. J. 1, 587 (1973).
35. L.F. Prescott, Pathological and physiological factors affecting drug absorption, distribution, elimination, and response in man, in Concepts in Biochemical Pharmacology, Part 3, Springer-Verlag, Berlin-Heidelberg-New York, pg. 241 (1975).
36. M.J. Mattila, A. Friman, T.K.I. Larmi and R. Koskinen, Absorption of ethionamid, isoniazid and aminosalicylic acid from the post-resection gastrointestinal tract, Ann. Med. exp. Fenn. 47, 209 (1969).

37. V.M.K. Venho, J. Jussila and S. Aukee, Drug absorption in man after gastric surgery, Fifth International Congress on Pharmacology, Abstract #1445, 241 (1972).
38. G.W. Hepner, C.C. Booth, J. Cowan, A.V. Hoffbrand and D.L. Mollin, Absorption of crystalline folic acid in man, Lancet 2, 302 (1968).
39. J.M. Fisher, C. Rees and K.B. Taylor, Intrinsic factor antibodies in gastric juice of pernicious-anaemia patients, Lancet 2, 88 (1966).
40. S.G. Schade, M. Muckerheide, P. Feick and R.F. Schilling, Occurrence in gastric juice of antibody to a complex of intrinsic factor and vitamin B_{12}, N. Engl. J. Med. 275, 528 (1966).
41. E.J. Pantuck, K.-C. Hsiao, R. Kuntzman and A.H. Conney, Intestinal metabolism of phenacetin in the rat: effect of charcoal-broiled beef and rat chow, Science 187, 744 (1975).
42. J. Lindenbaum, R.M. Maulitz and V.P. Butler, Inhibition of digoxin absorption by neomycin, Gastroenterology 71, 399 (1976).
43. M.M. Reidenberg, Kidney disease and drug metabolism, Med. Clin. N. Am. 58, 1059 (1974).
44. D.W. Shoeman and D.L. Azarnoff, The alteration of plasma proteins in uremia as reflected in their ability to bind digitoxin and diphenylhydantoin, Pharmacology 7, 169 (1972).
45. L. Dettli, Individualization of drug dosage in patients with renal disease, Med. Clin. N. Am. 58, 977 (1974).
46. L. Dettli, Translation of pharmacokinetics to clinical medicine, in Pharmacology and Pharmacokinetics, Plenum Press, New York, pg. 69 (1974).
47. W.M. Bennett, I. Singer and C.H. Coggins, Guide to drug usage in adult patients with impaired renal function, J. Am. Med. Assoc. 223, 991 (1973).
48. M.M. Reidenberg, Renal Function and Drug Action, Saunders, Philadelphia (1971).
49. J.G. Wagner, Biopharmaceutics and Relevant Pharmacokinetics, Hamilton Press, Hamilton (1971).
50. R.J. Elin, E.S. Vesell and S.M. Wolff, Effects of etiocholanolone-induced fever on plasma antipyrine half-lives and metabolic clearance, Clin. Pharmacol. Ther. 17, 447 (1975).
51. G.M. Trenholme, R.L. Williams, K.H. Rieckmann, H. Frischer and P.E. Carson, Quinine disposition during malaria and during induced fever, Clin. Pharmacol. Ther. 19, 459 (1976).
52. E.S. Vesell and G.T. Passananti, Inhibition of drug metabolism in man, Drug Metab. Disp. 1, 402 (1973).
53. M. Eichelbaum, G. Bodem, R. Gugler, C. Schneider-Deters and H.J. Dengler, Influence of thyroid status on plasma half-life of antipyrine in man, N. Engl. J. Med. 290, 1040 (1974).
54. E.S. Vesell, J.R. Shapiro, G.T. Passananti, H. Jorgensen and C.A. Shively, Altered plasma half-lives of antipyrine, propylthiouracil and methimazole in thyroid dysfunction, Clin. Pharmacol. Ther. 17, 48 (1975).
55. A.J. Levi, S. Sherlock and D. Walker, Phenylbutazone and isoniazid metabolism in patients with liver disease in relation to previous drug therapy, Lancet 1, 1275 (1968).
56. W.G.E. Cookesley and L.W. Powell, Drug metabolism and interaction with particular reference to the liver, Drugs 2, 177 (1971).
57. J.W. Faigle, Blood levels of a schistosomicide in relation to liver function and side effects, Acta Pharmacol. Toxicol. [Suppl] (Kbh.) 3, 233 (1971).
58. R.A. Branch, C.M. Herbert and A.E. Read, Determinants of serum antipyrine half-lives in patients with liver disease, Gut 14, 569 (1973).

59. P.B. Andreasen, L. Ranek, B.E. Statland and N. Tygstrup, Clearance of antipyrine - dependence of quantitative liver function, Eur. J. Clin. Invest. 4, 129 (1974).
60. P.B. Andreasen and E.S. Vesell, Comparison of plasma levels of antipyrine, tolbutamide, and warfarin after oral and intravenous administration, Clin. Pharmacol. Ther. 16, 1059 (1974).
61. G.W. Hepner and E.S. Vesell, Assessment of aminopyrine metabolism in man by breath analysis after oral administration of ^{14}C-aminopyrine, N. Engl. J. Med. 291, 1384 (1974).
62. E.S. Vesell, G.T. Passananti, P.A. Glenwright and B.H. Dvorchik, Studies on the disposition of antipyrine, aminopyrine, and phenacetin using plasma, saliva, and urine, Clin. Pharmacol. Ther. 18, 259 (1975).
63. G.W. Hepner and E.S. Vesell, Quantitative assessment of hepatic function by breath analysis after oral administration of [^{14}C]aminopyrine, Ann. intern. Med. 83, 632 (1975).
64. R.L. Williams, W.L. Schary, T.F. Blaschke, P.J. Meffin, K.L. Melmon and M. Rowland, Influence of acute viral hepatitis on disposition and pharmacologic effect of warfarin, Clin. Pharmacol. Ther. 20, 90 (1976).
65. H.J. Schull, G.R. Wilkinson, R. Johnson and S. Schenker, Normal disposition of oxazepam in acute viral hepatitis and cirrhosis, Ann. intern. Med. 84, 420 (1976).
66. G.R. Avant, S. Schenker and R.H. Alford, The effect of cirrhosis on the disposition and elimination of clindamycin, Am. J. Digest. Dis. 20, 223 (1975).
67. D. Kadar, T. Inaba, L. Endrenyi, G.E. Johnson and W. Kalow, Comparative drug elimination capacity in man - glutethimide, amobarbital, antipyrine and sulfinpyrazone, Clin. Pharmacol. Ther. 14, 552 (1973).
68. P.E. Thomas, A.Y.H. Lu, D. Ryan, S.B. West, J. Kawalek and W. Levin, Multiple forms of rat liver cytochrome P-450: immunochemical evidence with antibody against cytochrome P-448, J. Biol. Chem. 251, 1385 (1976).
69. E.S. Vesell and J.G. Page, Genetic control of the phenobarbital-induced shortening of plasma antipyrine half-lives in man, J. Clin. Invest. 48, 2202 (1969).
70. E.S. Vesell, G.T. Passananti and P.A. Glenwright, Anomalous results of studies on drug interaction in man. III. Disulfiram and antipyrine, Pharmacology 13, 481 (1975).
71. S.E. Smith and M.D. Rawlins, Prediction of drug oxidation rates in man: lack of correlation with serum gamma-glutamyl transpeptidase and urinary excretion of D-glucaric and 6β-hydroxycortisol, Eur. J. Clin. Pharmacol. 7, 71 (1974).

ACKNOWLEDGMENT

This work was supported in part by Grant CA-16536 from the National Institutes of Health.

ISOLATION OF RAT LIVER EPOXIDE HYDRATASE: PROPERTIES AND SUBSTRATE SPECIFICITY OF THE PURE ENZYME

Philip Bentley and Franz Oesch

Section of Biochemical Pharmacology, Institute of Pharmacology, University of Mainz, Obere Zahlbacher Strasse 67, D-6500 Mainz, F.R.G.

ABSTRACT

The microsomal enzyme epoxide hydratase has been purified to homogeneity as judged by electrophoretical, ultracentrifugal and immunological criteria and by C- and N-terminal analysis. The preparation procedure consisted of solubilisation using the non-ionic detergent cutscum, $(NH_4)_2SO_4$ precipitation, ion-exchange chromatography on DEAE-cellulose and cellulose phosphate and hydrophobic chromatography on butyl-sepharose. The product was detergent-free, had a relatively high content of hydrophobic amino acids and tended to aggregate in aqueous solutions. The protein had a minimum molecular weight of 49,000 ± 500 with a sedimentation coefficient of $S_{20w} \simeq 3$. Antibodies raised against the homogeneous preparation precipitated the entire hydratase activity in solubilised microsomes towards benzo(a)pyrene 4,5-(K-region-)oxide and styrene oxide as substrates. Moreover, the pure enzyme had a very broad substrate specificity, hydrating both arene and alkene oxides. The same general relationship between the hydration velocities of 6 K-region epoxides of polycyclic hydrocarbons was obtained with rat liver microsomal fractions and pure epoxide hydratase. These findings suggest that the enzyme isolated is the only one present in the rat liver microsomal fractions which is responsible for the hydration of such substrates.

INTRODUCTION

Epoxide hydratase [EC 4.2.1.63] is located in the endoplasmic reticulum of mammalian liver cells and catalyses the conversion of epoxides to the corresponding trans-dihydrodiols (1). Epoxides are produced during the microsomal metabolism of many compounds including polycyclic hydrocarbons (2), clinically used drugs (3) and some naturally occurring steroids (4). Many metabolically produced monofunctional epoxides derived from polycyclic hydrocarbons are very mutagenic (5-7) and dihydrodiol epoxides of benzo(a)pyrene and benz(a)anthracene are thought to be responsible for the major part of the in vivo binding of the parent hydrocarbons to DNA (8). These findings would suggest that epoxide hydratase may play a dual role in the activation of polycyclic hydrocarbons to mutagens and carcinogens, being responsible for removing the harmful epoxides but also providing the dihydrodiol precursors of the dihydrodiol epoxides.

The results of previous studies using guinea pig liver preparations (9) suggested that more than one form of epoxide hydratase existed. This may be very important since different forms of an enzyme may preferentially attack different portions of a large lipophilic molecule, as is the case with the microsomal monooxygenases (10,11).

To clarify the role of epoxide hydratase in the metabolism of polycyclic hydrocarbons and to gain insight into the possible existence of multiple forms of epoxide hydratase we decided to purify the enzyme(s). We report here the purification of epoxide hydratase from rat liver microsomes to homogeneity and some of the properties and substrate specificity of the pure enzyme.

MATERIALS AND METHODS

Butyl-sepharose was prepared by a modification of the method of Er-el and Shalteil (12). [7-^{3}H]Styrene oxide (13) and [^{3}H]benzo(a)pyrene 4,5-oxide (14) were synthesised as described. The tritiated polycyclic hydrocarbon epoxides, phenanthrene 9,10-oxide; benz(a)anthracene 5,6-oxide; dibenz(a,h)anthracene 5,6-oxide; 7-methylbenz(a)anthracene 5,6-oxide and 3-methylcholanthrene 11,12-oxide were a generous gift from Dr. P. Sims, Institute of Cancer Research, London.

Enzyme Assays

Styrene oxide hydration was measured as described (13). Hydration of epoxides derived from polycyclic hydrocarbons was measured as described (15). All buffers contained 0.3 mM EDTA unless otherwise stated.

Purification of Epoxide Hydratase

50 male Sprague-Dawley rats (200-250 g) were killed by a blow on the head. The livers were removed, washed in ice cold 10 mM sodium phosphate buffer, pH 7.0, containing 0.25 M sucrose and homogenised in the same buffer (3 ml/g liver) using a Braun MX3 blender at full speed. Homogenisation and all subsequent steps were performed at 0-4° C. The homogenate was centrifuged at 10,000 g for 15 min and the supernatent (fraction 1) was further centrifuged for 1 hr at 100,000 g. The resultant crude microsomal pellet was resuspended in 10 mM sodium phosphate buffer, pH 7.0, containing 1 % cutscum to give a final volume equal to that of fraction 1 and stirred for 20 min (fraction 2). $(NH_4)_2SO_4$ (140 g/l) was added, the solution was stirred for 20 min and centrifuged for 20 min at 10,000 g yielding 3 phases, a floating viscous brown layer, a red intermediate phase and a small pink sediment. The intermediate phase was removed by aspiration and discraded. The remaining two phases were combined, dissolved in a small volume of 5 mM sodium phosphate buffer, pH 7.0, and dialysed against 5 x 4 l of the same buffer over a 36 hr period. The dialysate was cleared by centrifugation (fraction 3) and applied to a DEAE-cellulose column (800 ml vol) which had previously been equilibrated with 5 mM sodium phosphate buffer, pH 7.0. Epoxide hydratase was not bound to the column and the column was washed with the equilibration buffer until no further activity was eluted. The active fractions were combined (fraction 4) and applied directly to a column of cellulose phosphate (2.5 x 36 cm) which had previously been equilibrated with 5 mM sodium

phosphate, pH 7.0. A fraction of the enzyme and the major part of the detergent flowed through this column and was discarded. The column was washed with 5 mM and 50 mM sodium phosphate buffer, pH 7.0, until no further protein was eluted as judged by the absorbance at 280 nm measured in a LKB Uvicord II flow cell. The epoxide hydratase was then eluted by 50 mM sodium phosphate buffer, pH 7.0, containing 0.5 M NaCl. Active fractions were combined and concentrated by overnight dialysis against 50 mM sodium phosphate buffer, pH 7.0, containing 1 mM EDTA and sufficient $(NH_4)_2SO_4$ to bring the total volume (both inside and outside the dialysis bag) to a concentration of 430 g/l. The hydratase, which precipitated under these conditions was collected by centrifugation, dissolved in a small volume of 50 mM sodium phosphate buffer, pH 7.0, and dialysed against 2 x 1 l of the same buffer. The dialysate (fraction 5) was then applied to a column of butyl-sepharose (0.9 x 15 cm) which had previously been equilibrated with 5 mM sodium phosphate buffer, pH 7.0. The column was washed with the equilibration buffer until no more protein was eluted. The epoxide hydratase was then eluted using 5 mM sodium phosphate buffer, pH 7.0, containing 0.05 % cutscum. The active fractions were combined (fraction 6) and rechromatographed on a very small column of cellulose phosphate (0.9 x 3 cm) to give the final detergent free preparation (fraction 7).

RESULTS AND DISCUSSION

Purification of Epoxide Hydratase

The results of a typical enzyme purification are shown in Table 1.

The epoxide hydratase was solubilized with the non-ionic detergent cutscum which had proved satisfactory for solubilisation of the enzyme from guinea pig liver micro-

TABLE 1 Purification of Rat Liver Epoxide Hydratase

Fraction	Volume (ml)	Protein (mg)	Units*	Specific activity**	Purification	Yield %
(1) 10,000 g supernatant	1660	49800	82557	1.67	1	100
(2) Solubilised microsomes	1660	11454	102107	8.71	5.3	124
(3) $(NH_4)_2SO_4$ precipitate	360	5184	99270	19.15	11.5	120
(4) DEAE cellulose effluent	500	740	69300	93.8	56	84
(5) Cellulose phosphate effluent	6	80	24000	300	179	29
(6) Butyl-sepharose effluent	50	21	14500	690	415	17.7
(7) Final preparation	4.8	15.7	8100	516	310	9.8

* 1 unit is defined as that amount of enzyme catalysing the formation of 1 nmole styrene glycol per min.
** nmoles styrene glycol per min per mg protein.

somes and human liver microsomes (1). After solubilisation and $(NH_4)_2SO_4$ precipitation the preparation was applied to a column of DEAE-cellulose. The effluent from this column contained epoxide hydratase of a high purity in relatively high yields. However, large amounts of activity were lost during chromatography on cellulose phosphate. This step was essential for removal of the detergent, a prerequisite of hydrophobic chromatography. Hydrophobic arms of different length and geometry were coupled to sepharose. Selective adsorption of epoxide hydratase followed by elution with low concentrations of detergent proved most satisfactory when n-butyl residues were used as hydrophobic arms. The detergent was then removed using a second very small column of cellulose phosphate. When enzyme activity was measured with styrene oxide as substrate the final preparation was purified about 300 fold compared to 10,000 g supernatant and 70 fold compared to liver microsomal fraction with yield of 10 %.

Criteria of Purity

The enzyme preparation gave a single protein band when analysed by SDS-gel electrophoresis in gels of a 10 % acrylamide concentration and sedimented as a single symmetrical peak when examined in an analytical ultracentrifuge in the presence of SDS ($S_{20w} \simeq 3$). These findings suggest that the enzyme preparation contained a single polypeptide chain. Furthermore, Ouchterlony double diffusion analysis of the pure preparation using antibodies raised against the pure enzyme in New Zealand white rabbits showed a single precipitation line. Thus the preparation was homogeneous as judged by electrophoretical, ultracentrifugal and immunological criteria. The results of C-terminal and N-terminal amino acid analysis (see below) also indicated the presence of a single protein species.

Effect of Detergent on Enzyme Activity

The specific activity of the final preparation measured using styrene oxide as substrate was between 520 and 690 units per mg protein when measured in the presence of 0.05 % cutscum and 450 to 520 units per mg protein when measured after the removal of cutscum. The lost activity could be restored by addition of cutscum to the reaction medium. Detergent to protein ratios between 0.3 and 5 mg cutscum per mg protein gave an apparent activation of 30 to 40 % when compared to the activity of the enzyme in the absence of detergent. This loss of activity upon removal of detergent is probably caused by aggregation of the protein.

Properties of the Pure Enzyme

In the absence of SDS the enzyme did not migrate into the separating gel during polyacrylamide gel electrophoresis (5 % gel concentration) and sedimented in the ultracentrifuge as a polydisperse system with a major band at $S_{20w} \simeq 14.5$. In the presence of 0.2 % SDS the protein showed a single band with a sedimentation coefficient. $S_{20w} \simeq 3$. The minimum molecular weight estimated from SDS-gel electrophoresis in the presence of proteins of known molecular weight was 49,000 ± 500. This is similar to that suggested by the sedimentation coefficient measured in the presence of 0.2 % SDS. Thus in the absence of detergents the enzyme does exist as a high molecular weight aggregate. This aggregation of epoxide hydratase can in part be explained by the amino acid analysis. The enzyme has a relatively high content of aromatic and hydrophobic amino acid residues. In fact the hydrophilic amino acids lysine, histidine, arginine, aspartic acid, asparagine, threonine, serine, glutamic acid and glutamine account for only 44 % of the sum of amino acids in the protein. The high content of hydrophobic amino acids could explain the aggregation of the enzyme in aqueous solutions. The content of tryptophan and tyrosine was determined both chemically and spectrophotometrically. The two methods gave very similar results. This confirmed that the enzyme was free of detergent, since the cutscum has a very strong absorption at 277 nm and would, if present, interfere with the spectrophotometric determinations.

No free N-terminal amino acid could be detected as either $[^{14}C]$-dinitrofluorobenzene or phenylhydantoin-derivatives. C-terminal analysis using a variety of methods indicated that the C-terminal amino acid was either glutamine or asparagine.

Substrate Specificity of the Pure Enzyme

The pure epoxide hydratase was active with a wide variety of substrates including both arene and alkene oxides. The specific activity of the pure enzyme towards six K-region epoxides of polycyclic hydrocarbons is compared with that of rat liver microsomal fractions in Table 2.

The pure epoxide hydratase was active with all six K-region epoxides. Moreover, the same general relationship between the hydration velocities of the different substrates was obtained with pure enzyme and rat liver microsomal fractions. In both cases the rate of hydration of phenanthrene 9,10-oxide > 7-methylbenz(a)anthracene 5,6-oxide ≃ benz(a)anthracene 5,6-oxide ≃ benzo(a)pyrene 4,5-oxide > 3-methylcholanthrene 11,12-oxide > dibenz(a,h)anthracene 5,6-oxide. However, quantitatively there were considerable differences in the ratio of activities of the two preparations towards the individual epoxides (the specific activity of the pure epoxide hydratase was between 18 and 66 fold higher than that of the microsomal fraction).

The hydration of all the K-region epoxides except phenanthrene 9,10-oxide was non-linear with time when catalysed by the pure enzyme. This is shown for benzo(a)pyrene 4,5-oxide in Fig. 1. Production of dihydrodiols from all the K-region oxides was linear with time for at least 10 min with liver microsomal fractions at the protein concentrations used. The reason for the non-linearity of diol production with time when pure epoxide hydratase was used is not known. It is not caused by instability of the enzyme since the pure enzyme is exceptionally stable even in dilute solutions. The epoxides themselves did not irreversibly inactivate the pure enzyme and no inhibition by K-region trans-dihydrodiols or phenols could be detected. The difficulty in determining the initial rate of hydration with such substrates using pure epoxide hydratase does, however, introduce a considerable error into the specific activities as shown in Table 2. In view of this error the variation in the "relative purification" measured with these substrates is not large enough to suggest that another enzyme species with a different substrate specificity is lost during purification. The only discarded fraction from the enzyme purification which showed substantial epoxide hydratase activity was the flow through from the cellulose phosphate column. This fraction,

TABLE 2 The Activity of Microsome bound and Pure Epoxide Hydratase with K-Region Epoxides of Polycyclic Hydrocarbons

Substrate	Activity (nmoles diol·min^{-1}·mg protein^{-1})		Activity compared liver microsomal fraction A/B
	Pure epoxide hydratase (A)	Liver microsomal fraction (B)	
Phenanthrene 9,10-oxide	1077 ± 152	27.9 ± 3.60	39
7-Methylbenz(a)anthracene 5,6-oxide	165 ± 18	9.38 ± 1.29	18
Benz(a)anthracene 5,6-oxide	201 ± 49	6.76 ± 0.84	30
Benzo(a)pyrene 4,5-oxide	240 ± 63	7.41 ± 0.81	32
3-Methylcholanthrene 11,12-oxide	36.6 ± 6.8	1.43 ± 0.18	26
Dibenz(a,h)anthracene 5,6-oxide	23.2 ± 4.8	0.35 ± 0.16	66

All assays were performed at 37° C with incubation times of 10 min with liver microsomal fractions and 5 min with pure epoxide hydratase. In each case two protein concentrations were used, one being twice as high as the other, the average higher protein concentrations (μg protein/0.5 ml incubation volume) with pure epoxide hydratase and liver microsomal fractions were respectively: phenanthrene 9,10-oxide, 3.0, 123; 7-methylbenz(a)anthracene 5,6-oxide, 6.1, 123; benz(a)anthracene 5,6-oxide, 6.1, 370; benzo(a)pyrene 4,5-oxide, 6.1, 183; 3-methylcholanthrene 11,12-oxide, 24.4, 185 and dibenz(a,h)anthracene 5,6-oxide, 24.4, 760. Results represent means ± SD (n=8).

which usually contained less than 15 % of the total enzyme activity, had the same relative activity towards several substrates as the cellulose phosphate column fraction which possessed the major portion of the epoxide hydratase activity (fraction 5).

With four of the six substrates the specific activity of the pure enzyme was about 30 fold higher than that of liver microsomal fractions (32 ± 5.4). This is much less than measured with styrene oxide as substrate (approximately 70 fold). However, immunoprecipitation studies (16) showed that the activities towards benzo(a)pyrene 4,5-oxide and styrene oxide were simultaneously precipitated from solubilised microsomes by antibodies against the pure epoxide hydratase. This finding together with the results discussed above suggests that rat liver microsomal fractions contain a

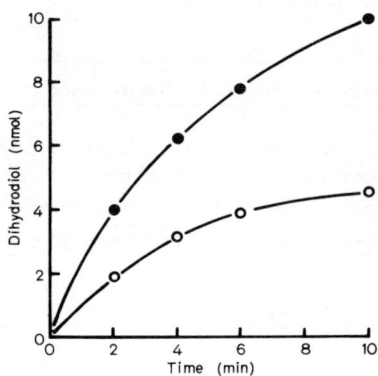

Fig. 1 Hydration of benzo(a)pyrene 4,5-oxide by pure epoxide hydratase. Incubations contained 4 (O) and 8 (●) µg epoxide hydratase.

single epoxide hydratase which accepts substrates with structures as different as benzo(a)pyrene 4,5-oxide, 3-methylcholanthrene 11,12-oxide and styrene oxide. It does not of course exclude the existence of separate epoxide hydratases for other classes of epoxides which have not been investigated.

Acknowledgements

We thank Mrs. P. Dent for technical assistance and Mr. A.J. Sparrow for preparation of the [^{3}H] benzo(a)pyrene 4,5-oxide. This work was supported by the Deutsche Forschungsgemeinschaft.

REFERENCES

(1) Oesch, F., Mammalian epoxide hydrases: Inducible enzymes catalyzing the inactivation of carcinogenic and cytotoxic metabolites derived from aromatic and olefinic compounds, Xenobiotica 3, 305 (1973).
(2) Jerina, D.M. and Daly, J.W., Arene oxides: A new aspect of drug metabolism, Science 185, 573 (1974).
(3) Horning, M.G., Butler, C.M., Nowlin, J. and Hill, R.M., Drug metabolism in the human neonate, Life Sci. 16, 651 (1975).
(4) Breuer, H. and Knuppen, R., The formation and hydrolysis of 16α, 17α-epoxoestratriene-3-ol by rat liver tissue, Biochim. Biophys. Acta 49, 620 (1961).

(5) Glatt, H.R., Oesch, F., Frigerio, A. and Garattini, S., Epoxides metabolically produced from some known carcinogens and from some clinically used drugs. I. Differences in mutagenicity. Int. J. Cancer 16, 783 (1975).
(6) Wood, A.W., Goode, R.L., Chang, R.L., Levin, W., Conney, A.H., Yagi, H., Dansette, P.M. and Jerina, D.M., Mutagenic and cytotoxic activity of benzo(a)pyrene 4,5-, 7,8- and 9,10-oxides and the six corresponding phenols. Proc. Natl. Acad. Sci. (USA) 72, 3176 (1975).
(7) Ames, B.N., Sims, P. and Grover, P.L., Epoxides of carcinogenic polycyclic hydrocarbons are frameshift mutagens, Science 176, 47 (1972).
(8) Sims, P., Grover, P.L., Swaisland, A., Pal, K. and Hewer, A., Metabolic activation of benzo(a)pyrene proceeds by a diol epoxide, Nature 252, 326 (1974).
(9) Oesch, F., Jerina, D.M. and Daly, J.W., Substrate specificity of hepatic epoxide hydratase in microsomes and a purified preparation: Evidence of homologous enzymes, Arch. Biochem. Biophys. 144, 253 (1972).
(10) Holder, G., Yagi, H., Dansette, P., Jerina, D.M., Levin, W., Lu, A.Y.H. and Conney, A.H., Effects of inducers and epoxide hydrase on the metabolism of benzo(a)pyrene by liver microsomes and a reconstituted system: Analysis by high pressure liquid chromatography, Proc. Natl. Acad. Sci. (USA) 71, 4356 (1974).
(11) Wiebel, F.J., Selkirk, J.K., Gelboin, H.V., Haugen, D.A., Van der Hoeven, T.A. and Coon, M.J., Position-specific oxygenation of benzo(a)pyrene by different forms of purified cytochrome P 450 from rabbit liver, Proc. Natl. Acad. Sci. (USA) 72, 3917 (1975).
(12) Er-el, Z. and Shalteil, S., Hydrophobic chromatography in the resolution of interconvertible forms of glycogen phosphorylase, FEBS Letts. 40, 142 (1974).
(13) Oesch, F., Jerina, D.M. and Daly, J.W., A radiometric assay for hepatic epoxide hydrase activity with 7-^{3}H-styrene oxide, Biochim. Biophys. Acta 227, 685 (1971).
(14) Dansette, P. and Jerina, D.M., A facile synthesis of arene oxides at the K-region of polycyclic hydrocarbons. J. Amer. Chem. Soc. 96, 1224 (1974).
(15) Bentley, P., Schmassmann, H.U., Sims, P. and Oesch, F., Epoxides derived from various polycyclic hydrocarbons as substrates of homogeneous and microsome bound epoxide hydratase, Eur. J. Biochem. in the press.
(16) Oesch, F. and Bentley, P., Antibodies against homogeneous epoxide hydratase provide evidence for a single enzyme hydrating styrene oxide and benzo(a)-pyrene 4,5-oxide, Nature 259, 53 (1976).

REDUCTION OF A TERTIARY AMINE N-OXIDE AND AN ARENE OXIDE IN RELATION TO THE REDUCTION RATE OF CYTOCHROME P-450 IN RAT LIVER MICROSOMES

Ryuichi Kato, Kazuhide Iwasaki and Hideyo Noguchi

Research Laboratories, Fujisawa Pharmaceutical Co., Ltd., 1-6, 2-Chome Kashima, Yodogawa-ku, Osaka 532, Japan

ABSTRACT

The reductions of a tertiary amine N-oxide and an arene oxide are catalyzed by the reduced form of cytochrome P-450 of liver microsomes. FMN, FAD, riboflavin, methyl viologen and benzyl viologen stimulated the anaerobic reduction of a tertiary amine N-oxide and an arene oxide by liver microsomes, and these stimulatory effects were completely inhibited by carbon monoxide. Spectral studies indicated that FMN or methyl viologen is reduced by NADPH-cytochrome c reductase and that reduced FMN or methyl viologen is reoxidized by cytochrome P-450 in the presence of a tertiary amine N-oxide or arene oxide.

In the presence of FMN or methyl viologen, xanthine oxidase-hypoxanthine system rapidly reduced tertiary amine N-oxide through the reduction of cytochrome P-450: the maximum reduction rate of tiaramide N-oxide was about 3600 nmoles/mg protein/min. These results indicated that the maximum turnover rate of cytochrome P-450 was about 4000 times per min and that the reduction of the tertiary amine N-oxide by liver microsomes is the rate-limiting step.

INTRODUCTION

In previous papers, we have reported that the reduced form of cytochrome P-450 in liver microsomes catalyzes both the reduction of tertiary amine N-oxides to corresponding tertiary amines and of benzo(a)pyrene 4,5-oxide to benzo(a)pyrene (1-3). These reductions are dependent on NADPH and are inhibited by both carbon monoxide and oxygen (2,3).

It is likely that the reductions of tertiary amine N-oxides and benzo(a)pyrene 4,5-oxide are catalyzed by the same mechanism. The reduction of tertiary amine N-oxides was inhibited by benzo(a)-pyrene 4,5-oxide, and the reduction of benzo(a)pyrene 4,5-oxide was inhibited by tertiary amine N-oxides. Both reductions were inhibited by cumene hydroperoxide and octylamine, but they were not inhibited by SKF-525A (2,3).

It was found that the addition of FMN markedly stimulated the rate of reduction of NADPH-dependent tertiary amine N-oxide and benzo(a)pyrene 4,5-oxide by liver microsomes under anaerobic conditions, and these stimulated activities were almost completely inhibited by carbon monoxide (2,3).

In the present communication, the stimulatory effect of FMN on tertiary amine N-oxide and benzo(a)pyrene 4,5-oxide reduction was investigated by comparison with the effect of methyl viologen on the microsomal electron transport system, especially the relationship between the rate of cytochrome P-450 reduction and the rate of tertiary amine N-oxide reduction, by using an artificial electron transport system such as the xanthine oxidase-hypoxanthine system.

METHODS

Male rats of the Sprague-Dawley strain were 7 weeks old when used. Liver microsomes were prepared as described in a previous paper (2). The microsomes were resuspended in 1.15% KCl and recentrifuged to remove traces of contaminating hemoglobin. The reduction rate of tertiary amine N-oxides was determined by measuring formation of the corresponding tertiary amines utilizing gas chromatography as described in the previous paper (2). The reduction rate of benzo(a)pyrene 4,5-oxide was determined by measuring the formation of benzo(a)pyrene by spectrofluorometry after separation of the metabolite and substrate by thin-layer chromatography as described in a previous paper (3).

The standard incubation mixtures (2.5 ml) consisted of various amounts of the microsomal fraction, substrates (2.5 μmoles for tertiary amine N-oxides and 0.25 μmoles of benzo(a)pyrene 4,5-oxide), NADPH (5 μmoles), $MgCl_2$ (25 μmoles) and pH 7.4 phosphate buffer (150 μmoles), unless otherwise specified. All incubations were carried out under an atmosphere of nitrogen at 37 °C. NADPH, glucose-6-phosphate dehydrogenase, xanthine oxidase, hypoxanthine, FMN, FAD, riboflavin, methyl viologen and benzyl viologen were purchased from Sigma Chemical Co., St. Louis. Benzo(a)pyrene 4,5-oxide was kindly donated by Prof. T. Watabe (Tokyo College of Pharmacy).

RESULTS AND DISCUSSION

The addition of FMN, FAD or riboflavin markedly stimulated NADPH-dependent reductions of tiaramide N-oxide, imipramine N-oxide and N,N-dimethylaniline N-oxide to corresponding tertiary amines. The maximum stimulation was obtained with riboflavin at a concentration of 1 mM, whereas 100 μM of FMN produced the maximum effect. FAD showed the least stimulatory effect. The addition of methyl viologen or benzyl viologen also markedly stimulated

NADPH-dependent reduction of tertiary amine N-oxide. Benzyl viologen was more effective than methyl viologen at a concentration of 100 μM - 1 mM.

The reduction of benzo(a)pyrene 4,5-oxide by liver microsomes was stimulated by the addition of FMN, FAD or riboflavin, as well as methyl viologen or benzyl viologen. The stimulated reduction activities were almost completely inhibited under an atmosphere of carbon monoxide.

Since the tertiary amine N-oxides and benzo(a)pyrene 4,5-oxide were hardly reduced chemically by the reduced form of FMN or methyl viologen formed through addition of sodium dithionite under an anaerobic condition, the involvement of cytochrome P-450 for the stimulatory mechanism of flavins and viologens was suggested. Moreover, these results suggested that added flavins or viologens are reduced through microsomal NADPH-cytochrome c reductase, and the reduced forms of flavins or viologens then reduce cytochrome P-450, and the reduced form of cytochrome P-450 in its turn reduces the tertiary amine N-oxides or arene oxides. The reductions of FAD and methyl viologen by microsomal NADPH-

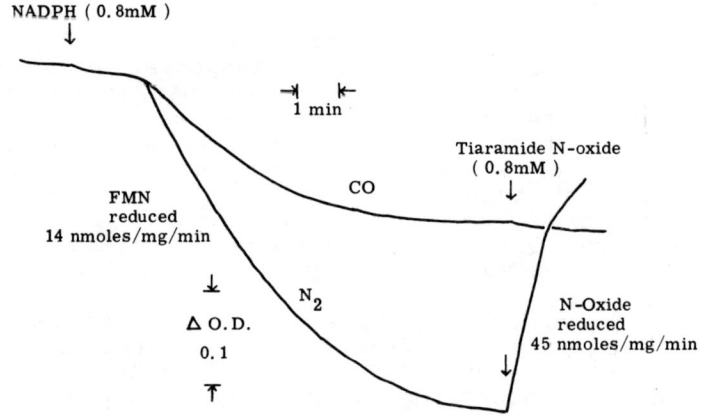

Fig. 1. NADPH-dependent reduction of FMN and its reoxidation by tiaramide N-oxide and liver microsomes.

The reduction of FMN was measured by determination of the decrease in absorbance at 450 nm in an atmosphere of nitrogen or carbon monoxide. The concentration of microsomal protein was 0.5 mg/ml, and that of FMN was 50 μM. The volume of the reaction mixture was 2.5 ml. The arrows indicate the addition of 0.05 ml of NADPH and tiaramide N-oxide solutions. One minute after the addition of tiaramide N-oxide, the formation of tiaramide was determined by gas chromatography.

cytochrome c reductase have been demonstrated (4,5). To confirm this hypothesis, spectrophotometric studies on the reduction and oxidation of FMN or methyl viologen by liver microsomes were carried out. As shown in Fig. 1, FMN was rapidly reduced by addition of NADPH.

The maximum rate of FMN reduction was about 14 nmoles/mg protein/min and the initial rate of $FMNH_2$ oxidation induced by tiaramide N-oxide was about 33 nmoles/mg protein/min by the spectrophotometric measurement.

In the presence of carbon monoxide, the optical change of FMN was masked by formation of the cytochrome P-450-CO complex, but the reoxidation of $FMNH_2$ induced by tiaramide N-oxide was clearly inhibited by carbon monoxide as shown in Fig. 1. The formation of tiaramide was 45.3 and 0.75 nmoles/mg protein/min, respectively, under an atmosphere of nitrogen and carbon monoxide.

Similarly, spectrophotometric studies indicated that methyl viologen is rapidly reduced by liver microsomes on addition of NADPH and reoxidized on addition of tiaramide N-oxide. Carbon monoxide completely inhibited the oxidation of reduced methyl viologen.

The addition of xanthine oxidase and hypoxanthine to liver microsomes, instead of NADPH, rapidly reduced FMN, and then $FMNH_2$ was reoxidized rapidly on addition of tiaramide N-oxide. However, in the absence of microsomes, the xanthine oxidase-hypoxanthine system with or without FMN did not appreciably reduce tiaramide N-oxide.

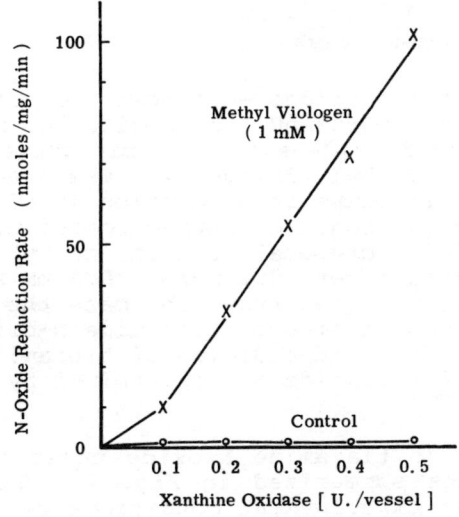

Fig. 2. Xanthine oxidase-dependent reduction of tiaramide N-oxide by liver microsomes.
The concentrations of hypoxanthine and methyl viologen were 4 mM and 1 mM, respectively. The concentration of microsomal protein was 1 mg/ml, and the samples were incubated anaerobically for 10 min.

The reduction rate of tiaramide N-oxide in the presence of methyl viologen was increased in parallel with the amount of added xanthine oxidase (Fig. 2).

As shown in Fig. 3, methyl viologen was rapidly reduced by the xanthine oxidase-hypoxanthine system under an atmosphere of nitrogen, and on addition of tiaramide N-oxide the reduced form of methyl viologen was instantly oxidized. The formation of tiaramide was 1.8 µmoles/mg protein/30 sec. The reoxidation of reduced methyl viologen was completely inhibited under an atmosphere of carbon monoxide.

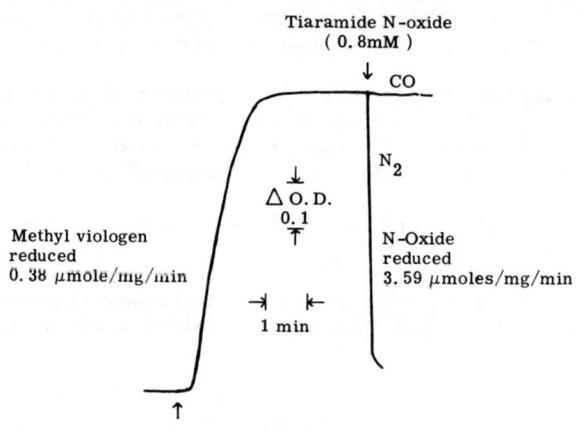

Fig. 3. Xanthine oxidase-dependent reduction of methyl viologen and its reoxidation by tiaramide N-oxide and liver microsomes.

The reduction of methyl viologen was measured by determining the increase in absorption at 600 nm in an atmosphere of nitrogen or carbon monoxide. The concentrations of microsomal protein, methyl viologen and xanthine oxidase were 0.2 mg/ml, 0.2 mM and 0.2 U/ml, respectively. The arrows indicate the addition of 0.05 ml of hypoxanthine or tiaramide N-oxide. Thirty seconds after the addition of tiaramide N-oxide the formation of tiaramide was determined by gas chromatography.

The reduction rates of tiaramide N-oxide under various experimental conditions are summarized in Fig. 4. The formation of tiaramide under these experimental conditions was completely inhibited by carbon monoxide. These results indicate that the reduction rate of tertiary amine N-oxide depends on the presence of cytochrome P-450 and its reduction system.

In the absence of oxygen, the reduction-oxidation cycle of cytochrome P-450 is stopped in the reduced form of cytochrome P-450. On addition of the tertiary amine N-oxide, the reduced form of cytochrome P-450 is oxidized; thus a reduction-oxidation cycle of cytochrome P-450 is started. The maximum reduction rate of tiaramide N-oxide in the presence of the xanthine oxidase system and methyl viologen was about 3600 nmoles/mg protein/min.

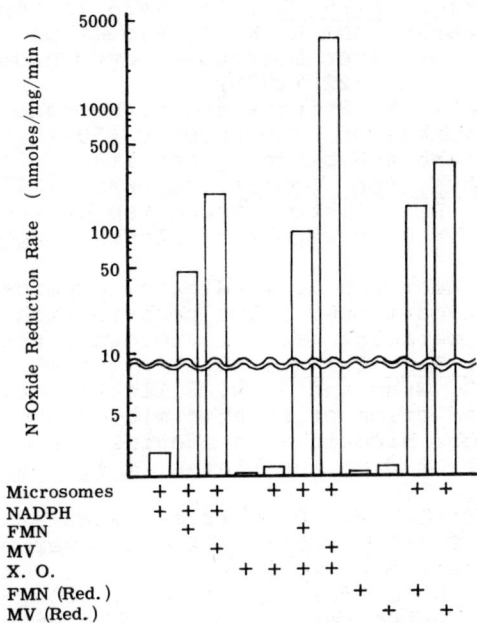

Fig. 4. Stimulatory effect of FMN and methyl viologen (MV) on reduction of tertiary amine N-oxide by liver microsomes.
FMN (Red) and MV (Red) indicate 0.4 mM of reduced FMN and 4 mM of reduced methyl viologen reduced chemically by sodium dithionite. X.O. indicates 4 mM hypoxanthine and 1 U xanthine oxidase.

This indicated that the turnover rate of the reduction-oxidation cycle of cytochrome P-450 is about 4,000 times per min. It has been reported that the maximum turnover rate of the reduction-oxidation cycle of cytochrome P-450 in rat liver microsomes under an atmosphere of air is about 12-15 times per min, and in the presence of a substrate it is about 15-25 times per min (6,7).

These results therefore suggested that the rate-limiting step for tertiary amine N-oxide reduction exists in the reduction of cytochrome P-450 and that the substrate-binding and metabolite-releasing steps of cytochrome P-450 appear to be relatively rapid. These results give valuable information for further kinetic studies and interpretations on the relationship between the microsomal electron transport system and the reduction-oxidation cycle of cytochrome P-450 during drug oxidation (8,9).

REFERENCES

(1) M. Sugiura, K. Iwasaki, H. Noguchi and R. Kato, Evidence for the involvement of cytochrome P-450 in tiaramide N-oxide reduction, Life Sci. 15, 1433 (1974).
(2) M. Sugiura, K. Iwasaki and R. Kato, Reduction of tertiary amine N-oxides by liver microsomal cytochrome P-450, Mol. Pharmacol. 12, 322 (1976).
(3) R. Kato, K. Iwasaki, T. Shiraga and H. Noguchi, Evidence for the involvement of cytochrome P-450 in reduction of benzo(a)pyrene 4,5-oxide by rat liver microsomes, Biochem. Biophys. Res. Commun. 70, 681, (1976).
(4) J. J. Kamm and J. R. Gillette, Mechanism of stimulation of mammalian nitro reductase by flavins, Life Sci. 2, 254 (1963).
(5) J. S. Bus, S. D. Aust and J. E. Gibson, Superoxide- and singlet oxygen-catalyzed lipid peroxidation as a possible mechanism for paraquat (methyl viologen) toxicity, Biochem. Biophys. Res. Commun. 58, 749 (1974).
(6) P. L. Gigon, T. E. Gram and J. R. Gillette, Studies on the rate of reduction of hepatic microsomal cytochrome P-450 by reduced nicotinamide adenine dinucleotide phosphate: effect of drug substrates, Mol. Pharmacol. 5, 109 (1969).
(7) H. Diehl, J. Schädelin and V. Ullrich, Studies on the kinetics of cytochrome P-450 reduction in rat liver microsomes, Z. Physiol. Chem. 351, 1359 (1970).
(8) F. P. Guengerich, D. P. Ballou and M. J. Coon, Spectral intermediates in the reaction of oxygen with purified liver microsomal cytochrome P-450, Biochem. Biophys. Res. Commun. 70, 951 (1976).
(9) T. Matsubara, J. Baron, L. L. Peterson and J. A. Peterson, NADPH-cytochrome P-450 reductase, Arch. Biochem. Biophys. 172, 463 (1976).

MICROSOMAL OXIDATION OF ISOPROTERENOL AND IRREVERSIBLE PROTEIN BINDING OF METABOLITES

M. Scheulen, H. Kappus, and H. M. Bolt

Institute of Toxicology, University of Tübingen, Wilhelmstraße 56, D-7400 Tübingen, W. Germany

ABSTRACT

During incubation of ^{3}H-isoproterenol with rat liver microsomes and NADPH-regenerating system, oxidation products of isoproterenol are irreversibly bound to protein. Isolated microsomal NADPH-cytochrome c reductase also catalyzes this reaction. Irreversible binding is inhibited by glutathione in both systems. This confirms the previous theory that microsomal oxidation and irreversible protein binding of catechols are effected by NADPH-cytochrome c reductase. However, inhibition experiments with carbon monoxide and with (4-phenyl)-phenyl-4(5)-imidazole indicate that also the hemoprotein, cytochrome P-450, should partly be involved in microsomal transformation of the catechol isoproterenol to reactive intermediates.

INTRODUCTION

Previous experiments (1) have shown that rat liver microsomes oxidize DOPA (L-3,4-dihydroxyphenylalanine) and DOPAmine (3-hydroxytyramine) to metabolites which irreversibly bind to protein. It was suggested that this reaction may be attributed to the action of microsomal NADPH-cytochrome c reductase. Because the closely similar NADPH-dependent oxidation of isoproterenol could be inhibited by DOPA and DOPAmine (2), also oxidation of isoproterenol was studied in detail. The experiments suggest that, although participation of the hemoprotein, cytochrome P-450, is not an essential requirement for oxidation and irreversible protein binding of isoproterenol, cytochrome P-450 may be partly involved in the transformation of this catechol to reactive intermediates.

METHODS

dl-[7-^{3}H]Isoproterenol was obtained from the Radiochemical Centre, Amersham, U. K..

Liver microsomes from male Wistar rats (200 - 260 g) were prepared as described by Remmer et al. (3). Microsomal NADPH-cytochrome c reductase (EC 1.6.2.4) was solubilized and purified by the method of Omura and Takesue (4). Microsomal incubations comprised of rat liver microsomes, bovine serum albumin (BSA), NADPH-regenerating system, and the substrate ^{3}H-isoproterenol were carried out as described elsewhere (5, 6). After incubation, microsomes and supernatant were separated by ultracentrifugation (5). Microsomal pellets were washed twice with Tris-buffer and dialyzed against aq. dist. as described (1). This treatment removed all free and reversibly bound radioactivity from the microsomal pellet. That removal was complete was proved by additional solvent extractions and measurement of radioactivity of the remaining supernatants and determination of radioactivity irreversibly bound to protein as it has been done in the case of imipramine (7).

The supernatants of ultracentrifugation, which contained the BSA added to the microsomal incubation mixtures, were treated with charcoal suspension and dialyzed in order to remove reversibly bound radioactive material as already described (1). The supernatants thus obtained only contained radioactive metabolites of isoproterenol which were irreversibly attached to the protein. This was proved as shown in Table 1. Gel filtration on Sephadex G-100 (1) gave the albumin peak which was of the same specific radioactivity as the starting material. When this albumin eluate was lyophilized and reincubated with an excess of cold isoproterenol, no radioactivity was removed from the protein. This demonstrated that binding of the isoproterenol metabolites generated during microsomal incubation to the protein was irreversible.

Incubations of ^{3}H-isoproterenol with tyrosinase (EC 1.10.3.1) and BSA were performed as previously described using estrogens as substrates (6). Incubations with xanthine oxidase / hypoxanthine and incubations with isolated NADPH-cytochrome c reductase plus NADPH-regenerating system were carried out as described elsewhere, using DOPA and DOPAmine as substrates. Radioactivity was determined in Hyamine hydroxide containing Bray's scintillator (8). Protein was measured by Lowry's method (9).

RESULTS

Rat liver microsomes catalyze irreversible binding of isoproterenol to microsomal protein as well as to BSA, if the latter is added to the incubation (Table 2). Binding depends on NADPH and on enzymatically active microsomes, and is inhibited by glutathione. Glutathione is thought to trap the reactive oxidation products involved in the binding reaction. This concept has been well documented for the catechol estrogens (5, 6, 10 - 12). However, carbon monoxide (95 %) inhibits isoproterenol binding by only 20 %.

Previous experiments demonstrated an irreversible binding of DOPA and DOPAmine to proteins, catalyzed by isolated liver microsomal NADPH-cytochrome c reductase (1). Table 3 gives the characteristics of irreversible binding of isoproterenol to proteins, as catalyzed by microsomes and by the isolated reductase. Compared to the amount of reductase present in our microsomal preparation (0.18 U/mg microsomal protein), the isolated enzyme catalyzes 76 % of the binding which occurs in microsomes. Both enzymic systems show similar behaviour towards inhibition by glutathione and by p-hydroxymercuribenzoate. They are not influenced by addition of catalase. However, a newly synthesized inhibitor of some cytochrome P-450-dependent oxidations, (4-phenyl)-phenyl-4(5)-imidazole (1 mM), inhibits microsomal irreversible binding of isoproterenol by 27 %, whereas it does not influence binding in the reductase system.

It has been shown that tyrosinase and a superoxide (O_2^-) generating system comprised of xanthine oxidase and hypoxanthine also catalyze irreversible binding of DOPA to proteins (1). The appropriate data for isoproterenol are included in Table 3. Irreversible binding of isoproterenol also in these two enzymic model systems depends on active enzyme, is depressed by glutathione, but not by catalase or by the inhibitor (4-phenyl)-phenyl-4(5)-imidazole.

Table 4 shows the amount of irreversible binding of isoproterenol by human liver microsomes. Human liver microsomes were prepared from biopsy material which was obtained during laparotomy by courtesy of Dr. V. Hempel, Dept. of Anesthesiology, University of Tübingen. Binding is about half of that observed with rat liver microsomes, but shows identical behaviour with respect to omission of NADPH, addition of glutathione, and boiling of the enzymic source.

TABLE 1 Examination of Irreversibility of Binding of Isoproterenol Metabolites to BSA after Microsomal Incubation ($\bar{x} \pm SD$).

		nmol isoproterenol metabolites bound by 1 mg microsomal protein to 100 mg BSA
a)	supernatant after charcoal treatment and dialysis	13.5 ± 0.93 (n = 8)
b)	after gel filtration	12.8
c)	gel filtration peak after lyophilization	13.9
d)	c) after re-incubation with cold isoproterenol and dialysis	13.1 ± 0.72 (n = 4)

TABLE 2 Irreversible Binding of Isoproterenol Metabolites in vitro to Microsomal Protein and to BSA. One Milliliter of Incubation Contained 250 nmol Isoproterenol, 100 nmol BSA, 1 mg Microsomal Protein, and NADPH-Regenerating System. Incubation Time was 45 Minutes at 37° C ($\bar{x} \pm SD$).

	nmol isoproterenol metabolites per mg microsomal protein x 45 min		
	bound to:		
	a) 1 mg microsomal protein	b) 100 nmol BSA	total a) plus b)
complete microsomal system (n = 8)	11.90 ± 2.50	13.50 ± 0.93	25.4 ± 3.4
without NADPH-regenerating system (n = 8)	5.41 ± 1.40	3.21 ± 0.31	8.6 ± 1.7*
boiled microsomes (n = 4)	2.15 ± 0.35	2.66 ± 0.06	4.8 ± 0.4*
under 95 % CO/5 % O_2 (n = 4)	9.32 ± 1.60	10.90 ± 0.48	20.2 ± 2.1
with 1 mM glutathione (n = 4)	5.32 ± 0.32	3.30 ± 0.09	8.6 ± 0.4*

* $p < 0.001$, compared to complete system

TABLE 3 Irreversible Binding of Isoproterenol Metabolites to BSA by Different Enzymic Systems. Incubation Time was 45 min at 37° C (n = 4). The Term "Reductase" Refers to NADPH-Cytochrome c Reductase, Isolated from Rat Liver Microsomes. XOD = Xanthine Oxidase. PPI = (4-Phenyl)-phenyl-4(5)-imidazole ($\bar{x} \pm$ SD).

	Binding (nmol) by:			
	1 mg microsomal protein	180 mU reductase	40 mU XOD	26 U tyrosinase
complete system	13.50 ± 0.93	10.30 ± 0.40	4.98 ± 0.52	8.90 ± 0.75
without NADPH or hypoxanthine	3.21 ± 0.31	2.72 ± 0.25	1.21 ± 0.09	-
boiled enzyme	2.66 ± 0.06	2.30 ± 0.11	0.87 ± 0.09	0.96 ± 0.11
with 1 mM glutathione	3.30 ± 0.09	2.07 ± 0.23	1.01 ± 0.14	0.63 ± 0.02
with catalase (50,000 U/ml)	13.60 ± 2.00	10.20 ± 1.50	4.72 ± 0.83	9.05 ± 1.10
with 5 % ethanol	11.70 ± 1.10	9.45 ± 0.58	4.92 ± 0.44	8.32 ± 0.53
with 5 % ethanol plus 1 mM PPI	9.65 ± 0.39	9.87 ± 0.84	-	-

TABLE 4 Irreversible Binding of Isoproterenol Metabolites to Microsomal Protein and BSA by <u>Human</u> Liver Microsomes. Conditions were as Stated in Tab. 2, except that 200 nmol BSA/ml was Used. Figures Give nmol Isoproterenol Metabolites Bound.

	subj. 1 (40 y)	subj. 2 (59 y)	subj. 3 (65 y)	$\bar{x} \pm$ SD
complete system:				
bound to 1 mg microsomal protein	6.92	6.13	4.35	5.80 ± 1.30
to 200 nmol BSA	18.20	21.60	15.80	18.60 ± 2.90
without NADPH:				
1 mg micros. protein	1.99	1.05	1.12	1.39 ± 0.52
200 nmol BSA	4.08	4.71	4.30	4.36 ± 0.32
boiled microsomes:				
1 mg micros. protein	-	0.73	-	
200 nmol BSA	-	3.77	-	
with 1 mM glutathione:				
1 mg micros. protein	-	-	0.61	
200 nmol BSA	-	-	8.10	

DISCUSSION

The experiments indicate that, similar to DOPA, isoproterenol is converted by rat liver microsomal enzymes to reactive metabolites which irreversibly bind to protein. A formally related reaction is catalyzed by tyrosinase which is known to form o-quinones from catechol substrates (6), or by the superoxide generating system comprised of xanthine oxidase and hypoxanthine. That hydrogen peroxide is not involved in all of these oxidations can be seen from the lacking effect of catalase (Table 3). The implication of superoxide in binding of catechol substrates by isolated NADPH-cytochrome c reductase has been discussed previously (1). This consideration is based on the finding of Aust et al. (13) that rat liver microsomal NADPH-cytochrome c reductase may generate superoxide.

Our present findings on irreversible protein binding of isoproterenol by isolated NADPH-cytochrome c reductase stress the view that this particular enzyme, at least partly, also may be responsible for irreversible protein binding of isoproterenol, when catalyzed by liver microsomes. However, it ought to be considered that, if related to the content of reductase in microsomes, the isolated enzyme only catalyzes 76 % of that binding that microsomes do. Conversely, 27 % of the microsomal binding is inhibited by large amounts (1 mM) of the inhibitor (4-phenyl)-phenyl-4(5)-imidazole, and 20 % is inhibited by carbon monoxide. Substituted imidazoles are thought to occupy the type-II binding site at the heme of cytochrome P-450, thereby inhibiting cytochrome P-450-dependent oxidations (14). Therefore, it is reasonable to assume that, under in vitro conditions, microsomal oxidation of isoproterenol to metabolites which are capable of irreversibly attaching to proteins, should mainly be due to action of the microsomal NADPH-cytochrome c reductase. However, also cytochrome P-450 seems to be involved, accounting for some 25 - 30 % of transformation of isoproterenol to the reactive metabolites.

Further experiments (unpublished) demonstrated that irreversible binding of isoproterenol to proteins also occurs under in vivo conditions. Our experiments with human liver microsomes indicate that the present findings in rats should also be relevant for the human species.

REFERENCES

(1) M. Scheulen, P. Wollenberg, H. M. Bolt, H. Kappus, and H. Remmer, Irreversible binding of DOPA and DOPAmine metabolites to protein by rat liver microsomes, Biochem. Biophys. Res. Commun. 66, 1396 (1975).
(2) M. Scheulen, P. Wollenberg, H. Kappus, and H. M. Bolt, Microsomal oxidation of catechols and irreversible protein binding of metabolites, Arch. Pharmacol. Suppl. 293, R 50 (1976).
(3) H. Remmer, H. Greim, J. B. Schenkman, and R. W. Estabrook, Methods for the elevation of hepatic microsomal mixed function oxidase levels and cytochrome P-450, Methods in Enzymology 10, 703 (1967).
(4) T. Omura and S. Takesue, A new method for simultaneous purification of cytochrome b-5 and NADPH-cytochrome c reductase from rat liver microsomes, J. Biochem. 67, 249 (1970).
(5) H. Kappus, H. M. Bolt, and H. Remmer, Irreversible protein binding of metabolites of ethinylestradiol in vivo and in vitro, Steroids 22, 203 (1973).

(6) H. M. Bolt and H. Kappus, Irreversible binding of ethinylestradiol metabolites to protein and nucleic acids as catalyzed by rat liver microsomes and mushroom tyrosinase, J. Steroid Biochem. 5, 179 (1974).

(7) H. Kappus and H. Remmer, Irreversible protein binding of ^{14}C-imipramine with rat and human liver microsomes, Biochem. Pharmacol. 24, 1079 (1975).

(8) G. A. Bray, A simple efficient liquid scintillator for counting aqueous solutions in a liquid scintillation counter, Anal. Biochem. 1, 279 (1960).

(9) O. H. Lowry, N. J. Rosebrough, A. L. Farr, and R. J. Randall, Protein measurement with the folin phenol reagent, J. Biol. Chem. 193, 265 (1951).

(10) F. Marks and E. Hecker, Metabolism and mechanism of action of estrogens. XII. Structure and mechanism of formation of water soluble and protein-bound metabolites of estrone in rat liver microsomes in vitro and in vivo, Biochim. Biophys. Acta 187, 250 (1969).

(11) E. Kuss, Mikrosomale Oxidation des Oestradiols 17ß. 2-Hydroxylierung und 1- bzw. 4-Thioätherbildung mit und ohne 17-Hydroxyl-Dehydrogenierung, Hoppe-Seyler's Z. Physiol. Chem. 352, 817 (1971).

(12) S. D. Nelson, J. R. Mitchell, E. Dybing, and H. A. Sasame, Cytochrome P-450-mediated oxidation of 2-hydroxy-estrogens to reactive intermediates, Biochem. Biophys. Res. Commun. 70, 1157 (1976).

(13) S. D. Aust, D. L. Roerig, and T. C. Pederson, Evidence for superoxide generation by NADPH-cytochrome c reductase of rat liver microsomes, Biochem. Biophys. Res. Commun. 47, 1133 (1972).

(14) C. F. Wilkinson, K. Hetnarski, and L. J. Hicks, Substituted imidazoles as inhibitors of microsomal oxidation and insecticide synergists, Pesticide Biochem. Physiol. 4, 299 (1974).

BENZO (α) PYRENE METABOLISM BY MICROSOMES AND ISOLATED EPITHELIAL CELLS FROM RAT SMALL INTESTINE

Roland Grafström, Sidney J. Stohs, M. Danny Burke, Peter Moldéus and Sten Orrenius

Department of Forensic Medicine, Karolinska Institutet, S-104 01 Stockholm, Sweden

The intestinal mucosa exhibits drug monooxygenase activity which has been compared to the activities found in the liver and other extrahepatic tissues (1-4). The most extensively investigated intestinal monooxygenase has been the benzo(α)pyrene (BP) monooxygenase which is enhanced 4-30-fold by pretreatment of the animals with 3-methylcholanthrene (MC) (1,3-5). Diets have also been found to markedly alter intestinal drug metabolism (5-9) and dietary induction has been suggested to cause the routine occurrence of cytochrome P-450 and BP monooxygenase in the small intestine (5,7).

However, one of the major problems in studying intestinal drug metabolism has been the isolation of microsomes with stable cytochrome P-450 (2,4,8,9), and as a result many investigations have relied on the use of crude mucosal homogenates or the 10,000 g supernatant fraction (1,3,5-7,9,10). Furthermore, xenobiotic metabolism in intact, isolated intestinal cells has not been previously reported.

We have recently developed a method for isolation of microsomes from the small intestinal mucosa of the rat, which possess stable cytochrome P-450 and monooxygenase activity (11). In addition, we have isolated rat intestinal epithelial cells (12), which offer a convenient drug metabolizing system that is more physiological, than the use of subcellular fractions. This communication describes some properties of the intestinal BP monooxygenase studied with these two experimental models.

METHODS

Male Sprague-Dawley rats (175-200 g) were used. The animals were allowed free access to food (Anticimex Avelsfoder R3, supplied by Anticimex AB, Södertälje, Sweden) and water, unless otherwise specified. All enzyme inducers were suspended or dissolved in corn oil and given orally. MC (20 mg/kg) was given as a single dose 24 h prior to sacrifice. Rats receiving phenobarbital (Pb, 80 mg/kg) or pregnenolone-16α-carbonitrile (PCN, 20 mg/kg) were treated daily for three days. Control animals received an equivalent volume of corn oil. Animals which were fasted for 48 h were given either corn oil or MC in corn oil 24 h prior to sacrifice.

All animals were sacrificed by decapitation between 8-9 a.m. The upper villous layer of the mucosa from the first 25 cm segment of the duodenal end of the small intestine was used for all preparations. Intestinal microsomes were isolated as previously described, including trypsin inhibitor (5 mg/g wet weight small intestine), 20% (v/v) glycerol and 3 I.U. heparin/ml in the mucosal homogenization medium (11). Intact epithelial cells were isolated by differentially scraping the mucosa, in combination with a collagenase plus hyaluronidase digestion to obtain the free cells according to a method we have recently reported (12). Intestinal microsomes were kept in 0.15 M KCl + 0.05 M Tris-Cl, pH 7.8 (11). Krebs medium, pH 7.8, supplemented with 0.0625 M Tris-Cl, 2% albumin, 10% glycerol, 5 mM glucose, 25 I.U. heparin/ml and 500 I.U. benzyl penicillin/ml was used for storing the cells (12). Liver microsomes were isolated from rats given MC intraperitoneally for three days, according to the method of Ernster et al (13).

BP monooxygenase was measured according to the fluorometric procedure of Dehnen et al (14), as we have previously described (11), using 150 µM BP as final substrate concentration. The BP monooxygenase assay selects primarily for 3-hydroxy and 9-hydroxy-BP (15) and one unit of activity is defined as the formation of one pmole product/min using 3-hydroxy-BP as the reference standard. BP monooxygenase activity in the intestinal cells was measured within 1 h after cell preparation in the presence of 1.5 mM salicylamide, which was added in acetone to the preincubation medium. α-Naphthoflavone (α-NF) and rotenone were added in acetone prior to preincubation. Acetone was present in a concentration not exceeding 1% and was found to have no effect on BP metabolism.

The metabolites of BP produced by intestinal microsomes from MC-treated rats were determined quantitatively with high pressure liquid chromatography (HPLC) using a modification of the methods of Selkirk et al (16,17) and Holder et al (18) as previously described (11). The identification of BP metabolites is based on their retention times in the HPLC as compared to authentic BP-derivatives (graciously provided by Dr H.V. Gelboin, NIH). The microsomes (0.1 mg microsomal protein) were incubated for 5 min with 150 µM 7,10-^{14}C-BP (10 µcuries/pmole). The microsomal epoxide hydrase activity was determined by the method of Oesch et al (19) as modified by Seidegaard et al (20) using styrene oxide as the substrate. Microsomal protein was determined according to Lowry et al (21) using bovine serum albumin as the standard. Cells were always examined for viability based on trypan blue exclusion in Krebs medium containing 0.16% trypan blue, and all cell concentrations were determined by counting the cells in a Buerker chamber. A trypan blue exclusion of 90-95% was observed immediately after cell preparation. After 2,5-3 h storage of the cells on ice, approximately 20-25% of the cells lost the ability to exclude trypan blue.

RESULTS AND COMMENTS

Various BP monooxygenase activities of intestinal microsomes from fasted and fed rats receiving corn oil, MC, Pb or PCN are given in Table 1. The intestinal monooxygenase responded rapidly and extensively to oral pretreatment of the rats with MC. As compared to fed control animals, the activity increased by 29-fold with a maximum observed within 24 h after a single dose of MC. The BP monooxygenase activity of intestinal microsomes from 24 h MC-treated rats was, on a per mg protein basis, as high as the maximally induced level of this enzyme in liver microsomes occurring after three days daily i.p. administration of MC (22). When the vehicle, corn oil, was given to fed rats 24 h prior to sacrifice no differences in BP monooxygenase activities were observed as compared to the fed rats. Fasting of the rats for 48 h markedly decreased the activities to approximately 1/6th the level observed in animals having free access to food. However, fasting of the rats for another 24 h with a total of 72 h starvation, did not further impair the BP monooxygenase. In this case, the lack of any exogenous inducers for three days did not abolish the BP monooxygenase activity. These results provide arguments which should be considered in the discussion whether the routine occurrence of cytochrome P-450-mediated metabolism in the intestine is due to dietary induction (5,7).

TABLE 1 Influence of various pretreatments on benzo(α)pyrene monooxygenase activity of rat intestinal microsomes

Pretreatment	pmoles product formed/ mg protein/minute
Fed ad.lib.	82 ± 6
Corn oil, 1 day	83 ± 6
MC, 1 day	2425 ± 90
PB, 3 days	100 ± 9
PCN, 3 days	126 ± 10
Fasted, 2 days	14 ± 2
Fasted, 3 days	18 ± 3
Fasted, 2 days + corn oil, 1 day	26 ± 3
Fasted, 2 days + MC, 1 day	1009 ± 38

Administration of corn oil to the 48 h fasted rats, resulted in a 2-fold increase in BP monooxygenase when compared to the rats that were only fasted. The most striking response of BP monooxygenase to MC induction was obtained for the MC-treated rats which had been fasted. A 72-fold increase in activity occurred in the intestinal microsomes of MC-treated rats as compared to rats which had only been fasted and a 39-fold increase was observed when compared to microsomes from corn oil-treated and fasted rats. However, the activity seen in the fasted, MC-treated rats was less than one half of that obtained with fed MC-treated rats. Pb and PCN were administered three days to give maximal induction of the intestinal cytochrome P-450. Both inducers poorly enhanced the BP monooxygenase activity, PCN produced a 1.5-fold increase and an insignificant increase was observed after PB treatment.

TABLE 2 Comparison of benzo(a)pyrene metabolite patterns in liver and intestinal microsomes from MC-treated rats

Benzo(a)pyrene metabolites	Liver	Small intestine
	% of total metabolites	
More polar than 9,10-dihydrodiol	1.5	7.0
9,10-dihydrodiol	18.0	0.5
4,5-dihydrodiol	11.5	1.5
7,8-dihydrodiol	10.5	3.0
1,6-quinone	10.0	11.0
3,6-quinone	12.0	12.5
6,12-quinone	3.5	7.5
4,5-oxide	-	14.5
9-hydroxy	6.5	14.0
3-hydroxy	14.0	17.0
Other	12.5	11.5
	100 %	100 %

The pattern of BP metabolites produced by intestinal microsomes of 24 h MC-treated rats is compared to the pattern in liver microsomes, isolated from rats receiving MC i.p. for three days, in Table 2. The identification of BP metabolites, which is based on the retention times in the HPLC, is not absolute since other BP derivatives than those listed in Table 2, are known to have identical retention times (17,18). When examining the per cent of each metabolite formed the major product was 3-hydroxy-BP in both the intestinal and hepatic microsomes with large amounts formed of the 1,6- and 3,6-dione derivatives of BP. However, the distribution of BP metabolites for the intestinal microsomes also revealed large amounts of 9-hydroxy-BP and a product which was probably the 4,5-oxide. The identification of this metabolite was based on two points: co-chromatography with authentic 4,5-oxide and incubation of the intestinal ^{14}C-BP-4,5-oxide metabolite with liver microsomes which yielded only two peaks of radioactivity upon re-chromatography, one peak corresponding to authentic 4,5-oxide and another co-chromatographing with authentic BP-4,5-dihydrodiol, this metabolite presumably being a product of the liver microsomal epoxide hydrase. Consistently, the liver microsomal distribution of BP metabolites revealed the presence of large amounts of dihydrodiols, in sharp contrast to the results for intestinal microsomes. In fact, the BP metabolite pattern in intestinal microsomes was similar to the pattern observed for BP being metabolized by hepatic microsomes in the presence of 1,2-epoxy-3,3,3-trichloropropane (TCPO), an inhibitor of epoxide hydrase (17). These findings indicated that intestinal microsomes may have very low epoxide hydrase activity.

The intestinal microsomal epoxide hydrase activities - in control rats and after pretreatment of the rats with MC, Pb and PCN - are presented in Table 3. Neither of the inducers produced any significant change in the level of this enzyme. As expected, the intestinal epoxide hydrase was far less active than the hepatic epoxide hydrase, as determined by the use of styrene oxide as substrate. In fact, the activity was less than 10% of the activity observed in control hepatic microsomes (20), in agreement with the recent observations of James et al (23).

TABLE 3 Epoxide hydrase activity of rat intestinal microsomes

Treatment	pmoles styrene glycol/ mg protein/minute
Fed ad.lib.	667 ± 63
MC, 1 day	636 ± 30
PB, 3 days	725 ± 32
PCN, 3 days	702 ± 14

The influence of α-NF and rotenone on the BP monooxygenase activity in intestinal microsomes and isolated epithelial cells is shown in Table 4 and Fig. 1, respectively. Rats received either MC or the corn oil vehicle 24 h prior to sacrifice. A 7 μM concentration of α-NF produced a 96% decrease in BP monooxygenase activity of intestinal microsomes from MC-treated rats, and at 70 μM α-NF complete inhibition occurred. However, in microsomes from control rats, 7 μM α-NF resulted in a 4.5-fold stimulation of BP monooxygenase, while at the 10 times higher concentration partial inhibition occurred. Analogously, α-NF stimulated BP metabolism in control intestinal cells but inhibited BP metabolism in the cells isolated from MC-treated rats (Fig. 1). At 180 μM, α-NF caused a 3.7-fold increase in the monooxygenation of BP in control intestinal cells, although it was highly inhibitory in MC-induced cells. At higher concentrations, α-NF was also inhibitory in control cells. Our results with α-NF thus suggest the presence in the small intestine of two distinct types of BP monooxygenase, in agreement with what has been reported for liver by Wiebel and Gelboin (24).

TABLE 4 Influence of α-naphthoflavone and rotenone on benzo(α)-pyrene monooxygenase in intestinal microsomes from control and 3-methylcholanthrene-treated rats

Inhibitor	Control microsomes	MC microsomes
	pmoles product formed/mg protein/minute	
α-Naphthoflavone, μM		
0	65± 3	2116±122
7.0	296±22	82± 25
70.0	50± 3	0
Rotenone, μM		
0	83± 6	2407± 77
125	87± 8	2303±130
312.5	123± 3	2334±100
625	123±11	2157±194
1250	118± 3	1857± 49

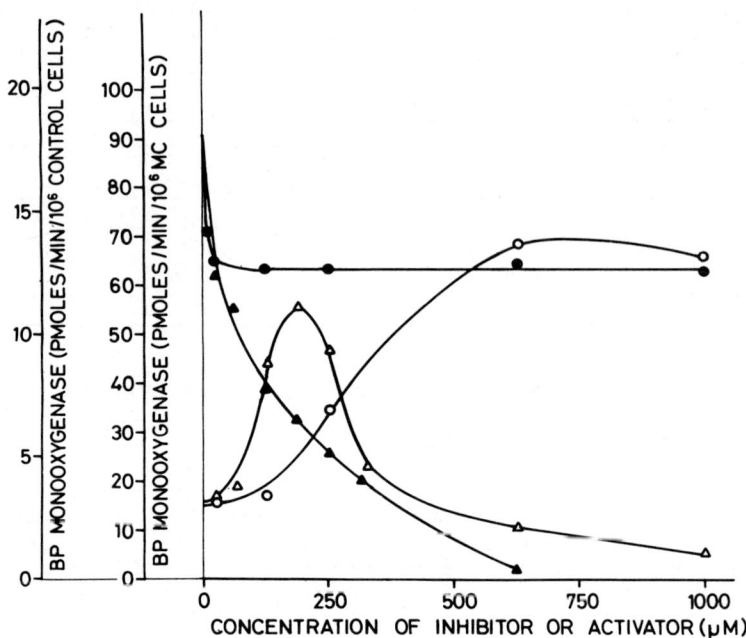

Fig. 1. BP metabolism by intestinal cells isolated from control and MC-treated rats. Cell concentrations of 5×10^5 cells/ml were used for MC induced cells and 2×10^6 cells/ml for control cells. Incubations were performed at 37° for 5 min in the supplemented Krebs medium as described in Methods. Each value is the mean of 4 individual experiments with different cell preparations. △——△, control cells in the presence of various concentrations of α-NF; ο——ο, control cells in the presence of various concentrations of rotenone; ▲——▲, MC-induced cells in the presence of various concentrations of α-NF; ●——●, MC-induced cells in the presence of various concentrations of rotenone.

When testing the effects of various respiratory inhibitors on the isolated epithelial cells, we made the unexpected observation that rotenone had a marked stimulatory effect on BP monooxygenase activity in control cells. Upon further investigation the effect of rotenone was found to be similar to that of α-NF, i.e. stimulation of BP monooxygenase in control intestinal microsomes and cells but inhibition of the MC-induced activity (Table 4 and Fig. 1). These findings provide further evidence for the formation of a distinct hemoprotein upon MC treatment and suggest that rotenone, like α-NF, may be of use to differentiate between various species of cytochrome P-450.

In summary, low levels of cytochrome P-450 and BP monooxygenase seem to be present in the rat small intestine epithelial cells and the microsomal fraction derived therefrom. The hemoprotein concentration and BP monooxygenase activity varies considerably under the influcence of dietary factors. Of the known enzyme inducers, MC has a marked stimulatory effect on both the cytochrome P-450 level and BP monooxygenase activity, whereas PB and PCN are virtually without effect. The observed differences in BP metabolite patterns between MC-induced liver and intestinal microsomes are probably due to the much lower epoxide hydrase activity in the latter tissue. Finally, α-NF and rotenone exert different effects - stimulation or inhibition - on the MC-induced and noninduced intestinal BP monooxygenase, indicating the formation of a distinct form of cytochrome P-450 upon MC treatment. It is suggested that rotenone, like α-NF, may be of use in differentiating between different cytochrome P-450 species.

ACKNOWLEDGEMENT

This study was supported by NIH contract 1 CP 33363.

REFERENCES

(1) B. G. Lake, R. Hopkins, J. Chakraborty, J. W. Bridges and D. V. W. Parke, The influence on some hepatic enzyme inducers and inhibitors on extrahepatic drug metabolism, Drug. Met. Disp. 1, 342 (1973).
(2) R. S. Chaabra, R. J. Pohl and J. R. Fouts, A comparative study of xenobiotic-metabolizing enzymes in liver and intestine of various animal species, Drug. Met. Disp. 2, 443 (1974).
(3) H. V. Gelboin and N. R. Blackburn, The stimulatory effect of 3-methylcholanthrene on benzpyrene hydroxylase activity in several rat tissues: inhibition by actinomycin D and puromycin, Cancer Res. 24, 356 (1964).
(4) N. G. Zampaglione and G. J. Mannering, Properties of benzpyrene hydroxylase in the liver, intestinal mucosa and adrenal of untreated and 3-methylcholanthrene treated rats, J. Pharmacol. Exp. Therap. 185, 676 (1973).
(5) L. W. Wattenberg, J. L. Leong and P. J. Strand, Benzpyrene hydroxylase activity in the gastrointestinal tract, Cancer Res. 22, 1120 (1962).
(6) L. W. Wattenberg, Studies of polycyclic hydrocarbon hydroxylases of the intestine possibly related to cancer, Cancer Res. 28, 99 (1971).
(7) E. Hietanen and O. Hänninen, Variable activities of drug-metabolizing enzymes in the gut of the rat after feeding with different pelleted diets, Comp. Gen. Pharmacol. 5, 255 (1974).
(8) H. Hoensch, C. H. Woo and R. Schmid, Cytochrome P-450 and drug metabolism in intestinal villous and cryptcells of rats: effects of dietary iron, Biochem. Biophys. Res. Commun. 65, 399 (1975).
(9) M. Marselos and M. Laitinen, Starvation and phenobarbital treatment effects on drug hydroxylation and glucuronidation in the rat liver and small intestinal mucosa, Biochem. Pharmacol. 24, 1529 (1975).
(10) C. Lehrmann, V. Ullrich and W. Rummel, Phenobarbital inducible drug monooxygenase activity in the small intestine of mice, Naunyn-Schmiedeberg's Arch. Pharmacol. 276, 89 (1973).

(11) S. J. Stohs, R. C. Grafström, M. D. Burke, P. W. Moldéus and S. G. Orrenius, The isolation of rat intestinal microsomes with stable cytochrome P-450 and their metabolism of benzo-(α)pyrene, Arch. Biochem. Biophys. (in press) (1976).
(12) S. J. Stohs, R. C. Grafström, M. D. Burke and S. G. Orrenius, Benzo(α)pyrene metabolism by isolated rat intestinal epithelial cells, Arch. Biochem. Biophys. (accepted for publication) (1976).
(13) L. Ernster, P. Siekevitz and G. E. Palade, Enzyme-structure relationships in the endoplasmic reticulum of rat liver. A morphological and biochemical study, J. Cell Biol. 15, 541 (1962).
(14) W. Dehnen, R. Tomingas and J. Roos, A modified method for the assay of benzo(α)pyrene hydroxylase, Analyt. Biochem. 53, 373 (1973).
(15) G. Holder, H. Yagi, W. Levin, A. Y. H. Lu and D. M. Jerina, Metabolism of benzo(α)pyrene. An evaluation of the fluorescence assay, Biochem. Biophys. Res. Commun. 65, 1363 (1975).
(16) J. K. Selkirk, R. G. Croy and H. V. Gelboin, Benzo(α)pyrene metabolites: Efficient and rapid separation by high-pressure liquid chromatography, Science 184, 169 (1974).
(17) J. K. Selkirk, R. G. Croy and H. V. Gelboin, Isolation by high pressure liquid chromatography and characterization of benzo(α)pyrene-4,5-epoxide as a metabolite of benzo(α)-pyrene. Arch. Biochem. Biophys. 168, 322 (1975).
(18) G. Holder, H. Yagi, P. Dansette, D. M. Jerina, W. Levin, A. Y. H. Lu and A. H. Conney, Effects of inducers and epoxide hydrase on the metabolism of benzo(α)pyrene by liver microsomes and reconstituted system: Analysis by high pressure liquid chromatography, Proc. Nat. Acad. Sci. USA 71, 4356 (1974).
(19) F. Oesch, D. M. Jerina and J. Daly, A radiometric assay for hepatic epoxide hydrase activity with [7-^{3}H] styrene oxide, Biochim. Biophys. Acta 227, 685 (1971).
(20) J. E. Seidegaard, J. W. DePierre, M. S. Moron, K. A. M. Johannesen and L. Ernster, Epoxide hydratase in rat lung, Proceedings of "Microsomes and Drug Oxidation" by Pergamon Press (1976).
(21) O. H. Lowry, N. J. Rosebrough, A. L. Farr and R. I. Randall, Protein measurement with the folin phenol reagent, J. Biol. Chem. 193, 265 (1951).
(22) H. Vadi, P. Moldéus, J. Capdevila and S. Orrenius, The metabolism of benzo(α)pyrene in isolated rat liver cells, Cancer Res. 35, 2083 (1975).
(23) M. O. James, J. R. Fouts and J. R. Bend, Hepatic and extrahepatic metabolism in vitro of an epoxide (8-^{14}C-styrene oxide) in the rabbit, Biochem. Pharmacol. 25, 187 (1976).
(24) F. J. Wiebel, J. C. Leutz, L. Diamond and H. V. Gelboin, Aryl hydrocarbon (benzo(α)pyrene) hydroxylase in microsomes from rat tissues: Differential inhibition and stimulation by benzoflavone and organic solvents. Arch. Biochem. Biophys. 144, 78 (1971).

CHARACTERIZATION OF THE DRUG MONOOXYGENASE SYSTEM IN MOUSE SMALL INTESTINE

Peter Wollenberg and Volker Ullrich

Department of Physiological Chemistry, University of the Saarland, Homburg-Saar, GFR

ABSTRACT

The microsomal fraction of small intestinal mucosa from phenobarbital-pretreated mice has been prepared by a special method. The monooxygenase activity of this preparation has been measured with the 7-ethoxycoumarin test and has been compared with the activity from mice and rat liver microsomes. The monooxygenase system turned out to be different with respect to inhibitors as well as the ligand spectra of cytochrome P450.

INTRODUCTION

According to our previous studies on monooxygenase activities in mouse small intestine (1,2) we noticed some peculiar properties of this enzyme system. First, the activity in normal mice was rather high compared to rat and rabbit, although considerable variations were observed over a longer period of time. Also the activity increased several-fold after pretreatment with phenobarbital which was not seen in rats. The induction by phenobarbital could also be verified in an *in-vitro* system of the isolated small intestine (3). Since this organ shows a continous proliferation with a half-life of about one day, it seemed worthwhile to use the small intestine for studies on the inducing effect of phenobarbital. The present study attempts to characterize the nature of the induced cytochrome P450 involved, because of the well-known heterogeneity of this main component of the drug monooxygenase system. Earlier investigations on the microsomal fraction of the mouse small intestine mucosa were hampered by the fact that rapid inactivation of the monooxygenase system occurred. This could be prevented by addition of trypsin inhibitor but then aggregation and precipitation of microsomal membranes was observed. Therefore, a new preparation of microsomal membranes was developed first.

MATERIALS AND METHODS

A home bred strain of white male mice was used, which was not inducible by ß-naphthoflavone. The animals (about 25 g body weight) were pretreated by two daily intraperitoneal injections of 80 mg/kg of phenobarbital in isotonic NaCl. They obtained a standard diet (Altromin ®) and water ad libitum. After killing by cervical dislocation the body was opened and the small intestine perfused with ice-cold isotonic NaCl from the vena portae. 10 cm of the small intestine were excised starting 3 cm behind the pylorus. The segment was rinsed with ice-cold isotonic KCl solution, opened longitudinally and spread on an ice cooled glass plate. The mucosa was scraped off and immediately frozen in liquid nitrogen. This preparation could be stored for about two weeks at -60°C without loss of activity. For the preparation of one batch of microsomes the pooled tissue beads from ten segments were homogenized in 10 ml 10 mM Tris-HCl (pH 7.6) containing 250 mM sucrose in a Potter-Elvejhem glass-teflon homogenizer (10 strokes, 450 rpm). The 20 000 xg supernatant was passed over a Sepharose 2 B column which was equilibrated with the homogenization medium. The microsomal fraction was collected and diluted to 5 mg of protein per ml. Rat and mouse liver microsomes were prepared conventionally (4). Protein was determined according to Gornall (5) and the ethoxycoumarin-O-dealkylation as described previously (6). Spectral studies were performed on an Aminco DW-2 spectrophotometer. Microliter additions of ligands dissolved in methanol were made to 0.5 ml of microsomal suspension with 0.01 M Tris buffer pH 7.6 containing 1.15 % KCl. Pure solvent was added to the reference cell. An extinction coefficient for the cytochrome P450-CO complex of 91 cm^2 mMol (7) was used. All chemicals used were commercially available.

RESULTS AND DISCUSSION

Microsomes of the small intestinal mucosa prepared according to the method described contained 0.1 nMol of cytochrome P450 per mg of protein. Surprisingly the maximum of the reduced CO-difference spectrum was consistently found at 452 nm. Liver microsomes from phenobarbital pretreated mice or rats, however, yielded the peak at 450 nm. This finding prompted us to further characterize the nature of this cytochrome P450 by using other ligands to the heme iron, like phosphines, mercaptides, sulfides or nitrogen-containing compounds.

Table 1 shows that some qualitative and quantitative differences between the three preparations can be detected. With metyrapone the oxidized preparations all showed maxima at 425 nm but the trough was at a lower wavelength with small intestinal microsomes. In the reduced state two peaks at 424 and 446 nm were seen, whereas in mouse liver only the 425 peak and in rat liver only the 446 nm absorption was present. NADPH in the presence of metyrapone could reduce only cytochrome P450 in rat liver microsomes. The other ligands also showed small differences. In the reduced state pentamethylene sulfide caused the formation of a rather large ligand spectrum in mouse small intestine compared to mouse and rat liver, however the affinity of the binding process was two orders of magnitude lower in small intestinal microsomes. In view of our previous findings on the ligand binding in microsomes from differently induced rats (8, 9,10) these results can probably be explained by the heterogeneity of microsomal cytochrome P450.

A second approach to prove this heterogeneity consists in the use of specific inhibitors for the different species. In rat liver 7,8-benzoflavone has been shown to block rather selectively the O-dealkylation of 7-ethoxycoumarin catalyzed by the cytochrome induced by polycyclic hydrocarbons, whereas metyrapone inhibits the same reaction in microsomes from phenobarbital-pretreated rats (11). Tetrahydrofuran was an inhibitor for microsomes from ethanol pretreated rats (12). The effect of these three inhibitors upon the O-dealkylation catalyzed by the various microsomal preparations are listed in Table 2.

7,8-benzoflavone did not inhibit, but metyrapone was effective, although significantly lower in hepatic mouse liver microsomes. The best differentiation was obtained with ethanol which rather selectively blocked the O-dealkylation in mouse liver microsomes.

These results show that after phenobarbital induction the cytochrome P450 fraction is not only different in the same organs of different species, but also

TABLE 1 Spectral Absorptions of Microsomal Cytochrome P450 Ligand Complexes from Mouse Small Intestine, Mouse Liver and Rat Liver. The Maxima and Minima Correspond to the Soret Absorption Peaks in the Difference Spectra. Reduced Complexes were obtained by Addition of Sodium Dithionite to Sample and Reference Cuvettes

Ligand		Mouse small intestine				Mouse liver			Rat liver			
		min.	max.	K_s	ΔA/nMol P450	min.	max.	ΔA/nMol P450	min.	max.	K_s	ΔA/nMol P450
Carbon monoxide	red.		452				450			450		
Metyrapone	ox.	390	425	5×10^{-7}		394	425		395	446	1×10^{-6}	
	red.	424,	446				425			446		
Diethylphenyl phosphine	red.		460	3×10^{-6}	.066		442, 461	.063		459	1×10^{-5}	.086
Pentamethylene	ox.	395	433	1×10^{-4}	.04	398	434	.04	400	435	2.5×10^{-6}*	.05
sulfide	red.		450	1×10^{-4}	.062		450	.03		449	2.8×10^{-6}	.03
-Toluene-thiol	ox.	412	377, 466	7×10^{-5}	0.066	416	378, 469	.067	421	379, 471	3×10^{-5}	0.053

*biphasic, high affinity constant

TABLE 2 Cytochrome P450 Content, O-Dealkylation and Inhibition Pattern of Microsomes from Mouse Small Intestine, Mouse Liver and Rat Liver. The Values Represent the Average of Five Determinations

		Mouse small intestine	Mouse liver	Rat liver
Cytochrome P450 (nMol/mg prot.)		.1	1.8	1.8
7-Ethoxycoumarin O-dealkylation (nMol umbelliferone/nMol P450 · min)		7.8	2.2	1.2
% Inhibition by				
Metyrapone	(2×10^{-5} M)	71 ± 8	48 ± 10	72 ± 7
Ethanol	(1×10^{-2} M)	0	40 ± 5	0
7,8 Benzoflavone	(1×10^{-6} M)	0	5	3

in the different organs of one species.

The O-dealkylation activity for 7-ethoxycoumarin turned out to be surprisingly high in small intestinal microsomes when calculated on the basis of the cytochrome P450 content. The reason for this is not yet understood. It could be either an increased turnover of this cytochrome or a more effective binding of the substrate or a tighter coupling of the system. Irrespective of the reason for this high molecular activity it supports the suggested significance of the drug monooxygenase system in small intestine for the metabolism of orally administered drugs and foreign compounds (3).

ACKNOWLEDGEMENT

This work was supported by the Deutsche Forschungsgemeinschaft, Sonderforschungsbereich 38, Teilprojekt L.

REFERENCES

1) Ch. Lehrmann, V. Ullrich and W. Rummel, Phenobarbital inducible drug monooxygenase activity in the small intestine of mice,
   Naunyn-Schmiedeberg's Arch. Pharmacol. 276, 89 (1973).

2) V. Ullrich and P. Weber, A direct test for monooxygenase activity of intact small intestine using surface reflectance fluorimetry,
   Biochem. Pharmacol. 23, 3309 (1974).

3) R. Scharf and V. Ullrich, In vitro induction by phenobarbital of drug monooxygenase activity in mouse isolated small intestine,
   Biochem. Pharmacol. 23, 2127 (1974).

4) H. Remmer, H. Greim, J.B. Schenkman and R.W. Estabrook, Methods for the elevation of hepatic microsomal mixed function oxidase levels and cytochrome P450,
   Methods in Enzymology, Acad. Press, New York (1967).

5) A.G. Gornall, C.J. Bardawill and M.M. David, Determination of serum proteins by means of the biuret reaction,
   J. Biol. Chem. 177, 751 (1949).

6) V. Ullrich and P. Weber, The O-dealkylation of 7-ethoxycoumarin by liver microsomes,
   Hoppe-Seyler's Z. Physiol. Chem. 353, 1171 (1972).

7) T. Omura and R. Sato, The carbon monoxide binding pigment of liver microsomes,
   J. Biol. Chem. 239, 2370 (1964).

8) D. Mansuy, W. Duppel, H.H. Ruf and V. Ullrich, Phosphines as ligands to microsomal cytochrome P450,
   Hoppe Seyler's Z. Physiol. Chem. 355, 1341 (1974).

9) W. Nastainczyk, H.H. Ruf and V. Ullrich, Ligand binding of organic sulfides to microsomal cytochrome P450,
   Eur. J. Biochem. 60, 615 (1975).

10) W. Nastainczyk, H.H. Ruf and V. Ullrich, Binding of thiols to microsomal cytochrome P450,
    Chem. biol. Interactions 14, 251 (1976).

11) V. Ullrich, P. Weber and P. Wollenberg, Tetrahydrofurane – an inhibitor for ethanol-induced liver microsomal cytochrome P450,
    Biochem. Biophys. Res. Commun. 64, 808 (1975).

FORMATION OF ALKYLATION AND CARBAMYLATION INTERMEDIATES AND CYTOCHROME P-450 CATALYZED MONOOXYGENATION OF THE 2-CHLOROETHYLNITROSOUREAS CCNU AND METHYL CCNU

Donald J. Reed and Hubert E. May

Biochemistry and Biophysics Dept., Oregon State University, Corvallis, Oregon 97331, USA

ABSTRACT

Carcinostatic 1-(2-chloroethyl)-3-alkyl-1-nitrosoureas degrade to alkylating and carbamylating intermediates. Pharmacokinetic studies with these drugs have shown a complex pattern of metabolites that form within two min after drug administration to rats. CCNU[+] half life is estimated to be less than five min in phenobarbital-induced rats. Monooxygenation of the cyclohexyl ring of CCNU occurred within two min and monohydroxylated CCNU metabolites were observed in all tissues examined. Covalent labeling was determined. Additional metabolites were observed that may result from nitroso group reduction. Spectral evidence for the participation of cytochrome P-450 in nitroso group reduction during the incubation of CCNU with microsomes and NADPH is presented. The consequence of metabolism of CCNU and Methyl CCNU prior to formation of alkylating and carbamylating intermediates is discussed.

INTRODUCTION

The 2-chloroethylnitrosoureas BCNU, CCNU and Methyl CCNU are carcinostatic and effective against a variety of animal tumors and certain human neoplasms (1). Nitrosoureas are known to be both alkylating and carbamylating agents (2,3,4,5). However, lack of knowledge concerning both the pharmacokinetics of these agents and their chemical nature has hindered the understanding of their mechanism(s) of action. Evidence is presented which demonstrates that CCNU undergoes rapid monooxygenation in vivo to yield cis-3-, trans-3-, cis-4-, and trans-4-hydroxy derivatives. These metabolites could be isolated along with a varying amount of unchanged CCNU from every organ examined of rats administered ^{14}C CCNU. Metabolites more polar than the hydroxy CCNU derivatives were observed indicating that in vivo reduction of the nitroso group of CCNU may occur. Spectral studies with microsomes in the absence of O_2 support this tentative conclusion.

[+]The abbreviations used are: BCNU, 1,3-bis(2-chloroethyl)-1-nitrosourea (NSC 409962); CCNU, 1-(2-chloroethyl)-3-cyclohexyl-1-nitrosourea (NSC 79037); Methyl CCNU, 1-(2-chloroethyl)-3-(trans-4-methylcyclohexyl)-1-nitrosourea (NSC 95441).

METHODS AND MATERIALS

Chemicals

CCNU, synthesized by Parke Davis, [cyclohexyl-1-^{14}C] CCNU synthesized by Monsanto Chemical Co., were provided by the National Cancer Institute. All other chemical were reagent grade and commercially available.

Animals

All experiments were conducted with Sprague Dawley male rats or liver microsomes prepared after phenobarbital induction (100 mg/kg body wt for 4 to 5 days). At least one week prior to each in vivo experiment, a surgically implanted cannula was placed in the vena cava of each rat.

Metabolite isolation

All tissues and organs were damp blotted, weighed and homogenized in 1.15% KCl. An aliquot of each homogenate was combusted in a Packard model 306 Sample Oxidizer for ^{14}C determination. The remaining portion of each homogenate was extracted with hexane (3 x 10 ml), ether (3 x 10 ml), and finally ether-methanol (70:30 v/v) (1 x 40 ml). All solvent extracts were dried with anhydrous sodium sulfate and evaporated in vacuo at room temperature. The residues were dissolved in methylene chloride, centrifuged, decanted, the decantate evaporated and the residue dissolved in methanol. Acetonitrile was added to the methanol solutions and evaporated to dryness with a stream of dry N_2. The final residue was dissolved in the HPLC solvent.

Covalent binding of ^{14}C

The procedure used was that described by Maling et al (6) with the modification that samples were oxidized with a Packard model 306 Sample Oxidizer prior to assaying ^{14}C content by liquid scintillation counting.

High-performance liquid chromatography

HPLC of ether extracts of organs from rats administered [cyclohexyl-1-^{14}C] CCNU in vivo 2 min prior to sacrifice was with a LiChrosorb (5 µ) silica column (3.2 x 250 mm). The solvent system was isooctane-methylene chloride-2-propanol 930:63:24 (v/v) at a flow rate of 1.2 ml/min and a pressure of 2,000 psi.

Difference spectra

Difference spectra with phenobarbital-induced rat liver microsomes were performed with an Aminco DW-2 spectrophotometer as previously described (7).

RESULTS

Organ Distribution of CCNU and Metabolites

Administration of [cyclohexyl-1-^{14}C] CCNU (30 mg/1g body wt) by rapid i.v. infusion (Emulphor vehicle) via a vena cava cannula was followed by sacrifice of the rats after 2 min. The highest organ level of ^{14}C was in the liver; 16.6% of the total ^{14}C administered (Table 1). Blood (as 7% of body wt) ^{14}C was 13.1% while the small intestine had a higher level of ^{14}C than either the kidneys or lungs because of enterohepatic circulation of ^{14}C from administered CCNU as determined in other experiments with bile-duct cannulated rats.

TABLE 1 Organ ^{14}C Distribution Two Minutes After Administration of [Cyclohexyl-1-^{14}C] CCNU to Rats

Organ	Percent of Total ^{14}C Administered	Percent of Organ ^{14}C Extracted by		
		Hexane	Ether	Methanol-Ether
Blood	13.1	23	34	31
Liver	16.6	6	22	53
Kidney	4.2	33	21	44
Sm. Intest.	7.1	51	20	27
Brain	2.2	65	27	8
Lung	1.4	26	37	35

	Percent as Residue	Covalently Bound ^{14}C Expressed as Percent of Organ ^{14}C
Blood	14	10
Liver	22	3
Kidney	3	<1
Sm. Intest.	2	<1
Brain	1	<1
Lung	2	2

Isolation of Hydroxy CCNU Metabolites

Prior experiments (8) had shown that in vitro microsomal metabolism of CCNU occurred very rapidly (up to 19 nmoles/min/mg protein). In vivo metabolism of CCNU was found to occur so rapidly also that the distribution of metabolites was examined at 2, 10 and 45 min intervals after administration of [cyclohexyl-1-^{14}C] CCNU. Such extensive metabolism occurred within two min that a more detailed study was made of the metabolic pattern in the various organs. Hexane extraction of the organ homogenates removed only unchanged ^{14}C CCNU as shown by HPLC of the extracts. The low amount of CCNU found in the liver suggested a very extensive metabolism of CCNU in this organ. Ether extraction removed almost exclusively (75-80%) the monohydroxylated CCNU isomers as shown by HPLC analysis of the ether extracts. The presence of these isomers in the liver is illustrated in Figure 1. Very similar ratios of isomers were observed in the blood, small intestine and brain.

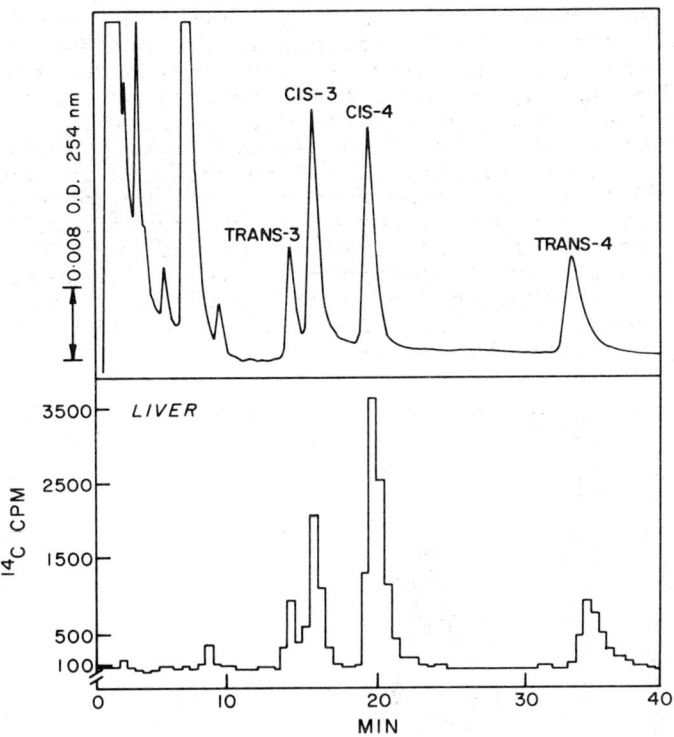

Fig. 1. Liquid chromatogram of hydroxy CCNU metabolites. The metabolites were isolated from liver by ether extraction following sacrifice of the rats two min after administration of [cyclohexyl-1-^{14}C] CCNU. Metabolite identification was as previously described (8).

Methanol-Ether Extraction of Polar Metabolites

Methanol-ether extraction removed the remaining (20-25%) hydroxy CCNU derivatives from the ether-extracted homogenates. It can be noted that the brain and small intestine homogenates contained essentially only unchanged CCNU and the monohydroxylated isomers (Table 1). In contrast, more than 40% of the ^{14}C present in the liver was metabolites other than the monohydroxy CCNU isomers. Efforts thus far to characterize these metabolites have failed except to demonstrate that about 50% of the ^{14}C present can be converted to 2,4-dinitrophenyl monohydroxycyclohexylamine derivatives by mild alkaline hydrolysis in the presence of 1-fluoro-2,4-dinitrobenzene. The apparent absence of typical glucuronide or sulfate conjugates led to the conclusion that the polarity of the metabolites could be due to the formation of N-hydroxy and/or N-amino derivatives via the reduction of the nitroso moiety. Such derivatives would be expected to be unstable and undergo chemical degradation at a much faster rate that CCNU or the hydroxy CCNU derivatives.

In previous studies, the half life of CCNU in buffer at physiological pH was found to be 48 min (9). Since the rate of nonenzymic degradation did not increase in the presence of serum, liver microsomes or other tissues, very little degradation of CCNU should occur during the two min metabolism period. Yet, extensive formation of metabolites more polar than the hydroxy CCNU metabolites was observed in certain organs (Table 1). Also, substantial covalent binding or ^{14}C to proteins in both the liver and blood occurred. These data support the speculation that unstable metabolites may be formed by the enzymic reduction of the nitroso moiety of either CCNU or the hydroxy CCNU metabolites. Efforts were therefore directed towards determining if a characteristic absorption band could be formed by the binding of CCNU to reduced cytochrome P-450 and detected by difference spectra measurements (Figure 2).

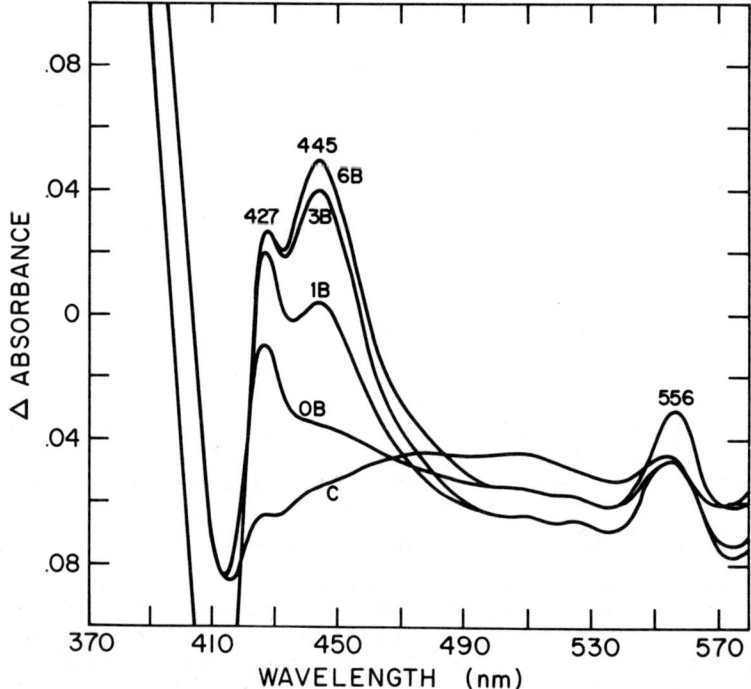

Fig. 2. Difference spectra of liver microsomes and CCNU in the presence of NADPH. Microsomes (1.33 mg protein/ml) were placed in both reference and sample cuvettes. CCNU (final conc. 0.53 mM) in 10 μl of acetone was added to the sample and 10 μl acetone in the reference cuvette. A type I binding spectrum was observed (7). Both cuvettes were then gassed for 2 min with N_2 and NADPH (0.48 mM) was added to the sample cuvette. Development of the 445-446 nm absorption band was recorded at 0, 1, 3 and 6 min (B). Both cuvettes were flushed with O_2 for 1 min and the difference spectrum recorded (C).

An Anaerobic CCNU Dependent Absorption Band With Reduced Microsomes

Anaerobic interaction of CCNU with microsomes from phenobarbital induced rats in the presence of NADPH, NADPH and NADPH generating system or dithionite gave difference spectra characterized by an absorption band at 445-446 nm. The band was not observed in the presence of O_2 and once formed it could be rapidly eliminated by admission of O_2 into the cuvette (figure 2). Dithionite as a reductant could also cause the formation of the CCNU complex in the absence of O_2. The complex could be rapidly replaced by a 450 nm peak by the replacement of N_2 with CO.

DISCUSSION

CCNU undergoes rapid monooxygenation *in vivo* to form cis-3-, trans-3-, cis-4-, and trans-4-hydroxy CCNU. The formation of these metabolites *in vivo* is in agreement with the rapid hydroxylation of CCNU observed with liver microsome preparations (8). From the distribution of the hydroxy CCNU metabolites in the various organs and the low amount of unchanged CCNU in the liver, it would appear that the liver is the major site of the monooxygenation reaction. Enterohepatic circulation results in the small intestine being exposed to a relatively high concentration of administered CCNU and its metabolites (10). Brain tissue was found to contain both CCNU and the hydroxy CCNU metabolites but very little of other ^{14}C labeled metabolites after two min following ^{14}C CCNU administration. CCNU and its 4-hydroxy metabolites have been compared in regards to antitumor activity (11). These data support the conclusion that the antitumor activity of CCNU is due primarily to its metabolites. Mouse lung tissue has been shown to possess CCNU hydroxylation activity (12). The relatively high proportion of metabolites in rat lung tissue after two min (Table 1) would suggest that rat lung is capable of metabolizing CCNU to some extent, even to form the more polar metabolites.

The extensive *in vivo* formation of unstable metabolites that are more polar than the hydroxy CCNU derivatives is of major interest. It is proposed that the formation of these metabolites could be related to the anaerobic interaction of CCNU with microsomal cytochrome P-450 after reduction with NADPH or dithionite. Other workers have reported the conversion of the nitroso group of nitrosamines to N-hydroxy and to N-amino substituents. The conversion of N-nitrosomorpholine to N-amino morpholine by guinea pig liver homogenates in the presence of NADPH has been described (13). Incubation of dimethylnitrosamine in the presence of rat liver microsomes and pH 5 enzymes resulted in the formation of several metabolites including N,N-dimethylhydrazine (14). This evidence is supported by the original observation of Imai and Sato (15) that certain nitrogenous ligands exhibited spectra with maxima around 425 amd 445 nm in the presence of dithionite reduced microsomes. Finally it is worth noting that a carbon nitroso intermediate has been proposed as being responsible for the 455 nm absorbing complex formed during the cytochrome P-450 dependent oxidative metabolism of amphetamine (16).

Much remains to be understood about the pharmacokinetics of the 2-chloroethylnitrosoureas. The formation of stable covalent bonds with cellular constituents including macromolecules has been demonstrated (2,3,17,18). The major reactive intermediates that cause covalent interaction appear to be a 2-chloroethyl carbonium ion (4,5,9) that causes alkylation and isocyanates that undergo carbamylation reactions (2,3,4,17). The major urinary metabolites reflect these covalent bond interactions. They include thiodiacetic acid and S-carboxymethyl-N-acetylcysteine (4), hydroxycyclohexylamines as free bases and metabolites that can be decomposed by acid hydrolysis to yield hydroxycyclohexylamines (Reed, D.J., unpublished results, 1976). Investigations have been reported that describe specific interactions that could represent major cytotoxic effects of the nitrosoureas (19,20). The rapid metabolism of the nitrosoureas to form multiple metabolites prior to the formation of alkylating and carbamylating intermediates underscores the complexity of these cellular effects.

REFERENCES

(1) Carter, S.K., Schabel, F.M., Jr., Broder, L.E. and Johnston, T.P., 1,3-Bis (2-chloroethyl)-1-nitrosourea (BCNU) and Other Nitrosoureas in Cancer Treatment: A Review. Advan. Cancer Res., 16, 273-332, 1972.
(2) Wheeler, G.P., Bowdon, B.J. and Struck, R.F., Carbamoylation of Amino Acids, Peptides, and Proteins by Nitrosoureas, Cancer Res., 35, 2974-2984, 1975.
(3) Schmall, B., Cheng, C.J., Fujimura, S., Gersten, N., Grunberger, D. and Weinstein, I.B., Modification of Proteins by 1-(2-Chloroethyl)-3-cyclohexyl-1-nitrosourea (NSC 79037) In Vitro, Cancer Res., 33, 1921-1924, 1973.
(4) Reed, D.J. and May, H.E., Alkylation and Carbamoylation Intermediates from the Carcinostatic 1-(2-Chloroethyl)-3-cyclohexyl-1-nitrosourea (CCNU), Life Sciences, 16, 1263-1270, 1975.
(5) Colvin, M., Cowens, J.W., Brundett, R.B., Kramer, B.S. and Ludlum, D.B., Decomposition of BCNU (1,3-Bis(2-chloroethyl)-1-nitrosourea) in Aqueous Solution, Biochem. Biophys. Res. Commun., 60, 515-520, 1974.
(6) Maling. H.M., Eichelbaum, F.M., Saul, W., Spies, I.G., Brown, E.A.B. and Gillette, J.R., Protection Against Carbon Tetrachloride-induced Hepatotoxicity Produced Pretreatment with Dibenamine [N-(2-chloroethyl)dibenzylamine], Biochem. Pharmacol., 23, 1479-1491, 1974.
(7) May, H.E., Boose, R. and Reed, D.J., Hydroxylation of the Carcinostatic 1-(2-Chloroethyl)-3-cyclohexyl-1-nitrosourea (CCNU) by Rat Liver Microsomes, Biochem. Biophys. Res. Commun., 57, 426-433, 1974.
(8) May, H.E., Boose, R. and Reed, D.J., Microsomal Monooxygenation of the Carcinostatic 1-(2-Chloroethyl)-3-cyclohexyl-1-mitrosourea. Syntehsis and Identification of Cis and Trans Monohydroxylated Products, Biochem., 14, 4723-4730, 1975.
(9) Reed, D.J., May, H.E., Boose, R.B., Gregory, K.M. and Beilstein, M.A., 2-Chloroethanol Formation as Evidence for a 2-Chloroethyl Alkylating Intermediate During Chemical Degradation of 1-(2-Chloroethyl)-3-cyclohexyl-1-nitrosourea and 1-(2-Chloroethyl)-3-(trans-4-methylcyclohexyl)-1-nitrosourea, Cancer Res., 35, 568-576, 1975.

(10) Oliverio, V.T., Vietzke, W.M., Williams, M.K. and Adamson, R.H., The Absorption, Distribution, Excretion and Biotransformation of the Carcinostatic 1-(2-Chloroethyl)-3-cyclohexyl-1-nitrosourea in Animals, Cancer Res., 30, 1330-1337, 1970.

(11) Johnston, T.P., McCaleb, G.S. and Montgomery, J.A., Synthesis and Biologic Evaluation of Major Metabolites of N-(2-Chloroethyl)-N'-cyclohexyl-N-nitrosourea, J. Med. Chem., 18, 634-637, 1975.

(12) Hill, D.L., Kirk, M.C. and Struck, R.F., Microsomal Metabolism of Nitrosoureas, Cancer Res., 35, 296-301, 1975.

(13) Süss, R., Zur Wirkungsweise der Nitrosamine, Z. Naturforsch., 20b, 714, 1965.

(14) Grilli, S. and Prodi, G., Identification of Dimethylnitrosamine Metabolites In Vitro, Gann, 66, 473-480, 1975.

(15) Imai, Y. and Sato, R., Anomalous Spectral Interaction of Reduced P-450 with Ethyl Isocyanide and Some Other Lipophilic Ligands, J. Biochem., Tokyo, 62, 464-473, 1967.

(16) Mansuy, D., Beaune, P., Chottard, J.C., Bartoli, J.F., and Gans, P., The Nature of the "455 nm Absorbing Complex" Formed During the Cytochrome P-450 Dependent Oxidative Metabolism of Amphetamine, Biochem. Pharmacol., 25, 609-612, 1976.

(17) Cheng, C.J., Fujimura, S., Grunberger, D. and Weinstein, I.B., Interaction of 1-(2-Chloroethyl)-3-cyclohexyl-1-nitrosourea (NSC 79037) with Nuclei Acids and Proteins In Vivo and In Vitro, Cancer Res., 32, 22-27, 1972.

(18) Connors. T.A. and Hare, J.R., The Binding of ^{14}C Labelled 1-(2-Chloroethyl)-3-cyclohexyl-1-nitrosourea (CCNU) to Macromolecules of Sensitive and Resistant Tumors, Br. J. Cancer, 30, 477-480, 1974.

(19) Ludlum, D.B., Kramer, B.S., Wang, J. and Fenselau, C., Reaction of 1,3-Bis(2-chloroethyl)-1-nitrosourea with synthetic polynucleotides, Biochem., 14, 5480-5485, 1975.

(20) Woolley, P.V., III, Dion, R.L., Kohn, K.W. and Bono, Jr., V.H., Binding of 1-(2-Chloroethyl)-3-cyclohexyl-1-nitrosourea to L1210 Cell Nuclear Proteins, Cancer Res., 36, 1470-1474, 1976.

PROPOSED MECHANISM FOR THE REDUCTIVE GLYCOSIDIC CLEAVAGE OF DAUNOMYCIN (NSC 82151) AND ADRIAMYCIN (NSC 123127) *

David W. Yesair, Suzanne McNitt and Linda Bittman

Arthur D. Little, Inc., Cambridge, Mass., U.S.A.

INTRODUCTION

The anthracycline drugs, adriamycin (NSC 123127) and daunomycin (NSC 82151) are clinically useful cancer chemotherapeutic agents (1). Their therapy results in dose-related hematopoietic and cardiac toxicity; the latter being irreversible. The pharmacokinetics of both drugs are generally similar among several animal species (2-7) and man (8-10). The metabolism of the anthracycline drugs that occurs *in vivo* is shown in Fig. 1 (11-22). Enzymes effecting the reduction of the 13-carbonyl are located in the 100,000 x g supernatant of liver and other tissues,

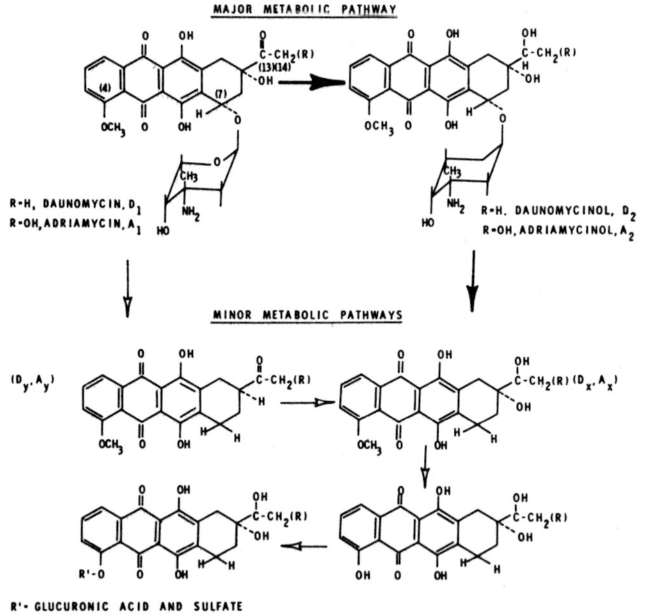

Fig. 1. Metabolism of anthracycline drugs.

---

* This work was supported by Contract NO1-CM-53849 from DCT, NCI, NIH, DHEW.

require NADPH and oxygen, and are similar to the general class of reductases found in the cytosol fraction (23,24). The reductive glycosidic cleavage of the anthracycline drugs to yield a reduced aglycone and daunosamine, requires microsomes and NADPH, and this metabolic pathway appears to be unique and unprecedented.

We have proposed a mechanism (Fig. 2) to effect the reductive cleavage of the amino sugar, daunosamine, from the anthracycline drugs. The lability of the protons in the dihydroquinone system of the anthracycline drugs appears essential for this proposed reaction mechanism.

Calendi *et al.* have characterized the binding of several cations to daunomycin in terms of visible spectral changes and the release of the anthracycline from a DNA-daunomycin complex by magnesium ions (25). Others (26) have further detailed the effects of cuprous and magnesium ions on a DNA-daunomycin complex, especially with regard to their pH dependence. Recently, we have evaluated the pharmacodynamic significance of metal chelates of the anthracycline drugs (NSC 82151 and NSC 123127)(27). The order of the stability constants of metal chelates of dihydroxyanthraquinone is generally as follows: Pb>Cu>Ni>Pd>Ca>Zn>Mg and dihydroxyanthraquinone forms chelate compounds at a molar ratio of 1:1 in neutral and alkaline media, which are probably polymeric (28,29).

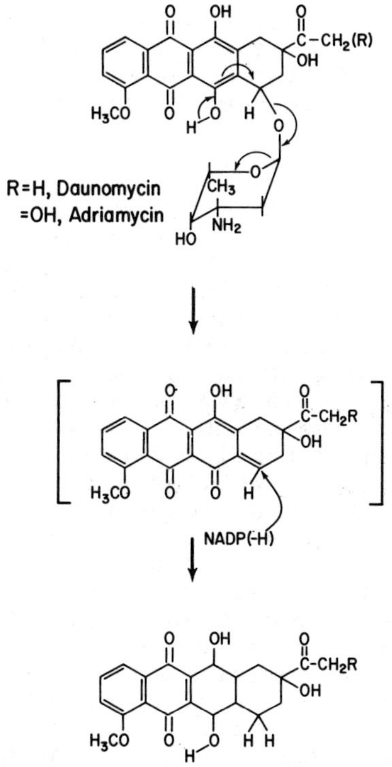

Fig. 2. Proposed mechanism for reductive cleavage of anthracycline drugs.

In these studies, we have evaluated the effect of metal cations on the reductive glycosidic cleavage and the reduction of the 13-carbonyl of both daunomycin and adriamycin. If the lability of the protons of the dihydroquinone system of the anthracycline drugs were important for the reductive glycosidic cleavage, we would anticipate a differential effect of using metal chelates of the anthracycline drugs in these two enzymatic pathways; i.e., preferential inhibition of reductive glycosidic cleavage.

METHODS

Daunomycin and adriamycin were supplied by Farmitalia, Societa Farmaceutica Italia 20146, Milano, Italy.

Chelates of the anthracycline drugs were obtained by mixing the anthracycline drugs in ethanol with varying concentrations of cations in ethanol or by mixing aqueous buffered solutions of both. The ultraviolet and visible absorption spectra were recorded on a Beckman DK-2 recording spectrophotometer. The chloride salts of the metal ions were used in all studies.

The technique for handling livers and kidneys of rats, the preparation of the enzyme systems, and the reaction conditions have been described by Asbell *et al.* (11,13). The metabolites and parent drugs were separated by thin-layer chromatography on silica gel, using two solvent systems composed of $CHCl_3$:methanol: $HOAc:H_2O$ by volume, I (100:50:14:6) and II (179:4:5:12). The fluorescent drugs and their metabolites were eluted from silica gel with 0.3N HCl in 100% ethanol and quantitated with an Hitachi Perkin-Elmer MPF 2A fluorescence spectrophotometer (3,11).

RESULTS

In order to evaluate whether the protons of the dihydroquinone system of the anthracycline drugs are exchangeable, we showed that the absorption spectra of the anthracycline drugs varied as a function of increased pH showing two new absorption maxima at 445 and 585 nm and decreasing the absorption region at 475-500 nm (Fig. 3). Chelates of metals with the anthracycline drugs also showed two new absorption maxima at about 530 and 570 nm (Fig. 4) and the absorption in the 475-500 nm region was only slightly decreased by metal ions. The maximum yield of absorbency at 567 nm appeared to be obtained at a molar ratio of cation to drug of 1:1 for cuprous ions (Fig. 5) as well as for other ions such as magnesium, manganese, nickel and cobalt.

In subsequent studies, we have evaluated the effect of both magnesium and copper ions on the metabolism of daunomycin and adriamycin by 1,000 x g homogenates of kidney and liver under aerobic (air) and "anaerobic" (nitrogen) conditions. Daunomycin is metabolized by homogenates of kidney of rats via a $D_1 \rightarrow D_2 \rightarrow D_x$ pathway, whereas by homogenates of liver via a $D_1 \rightarrow D_y \rightarrow D_x$ pathway. Adriamycin, in contrast, is metabolized by homogenates of both kidney and liver of rats via a $A_1 \rightarrow A_y \rightarrow A_x$ pathway.

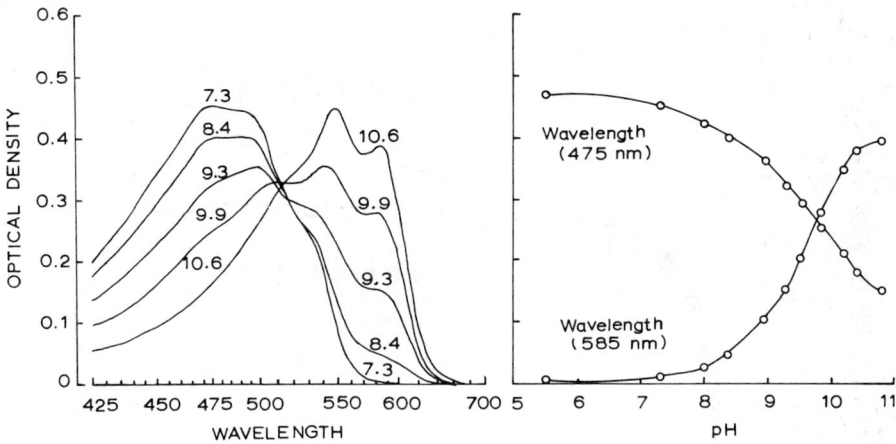

Fig. 3. Effect of pH on the absorption spectrum of daunomycin.

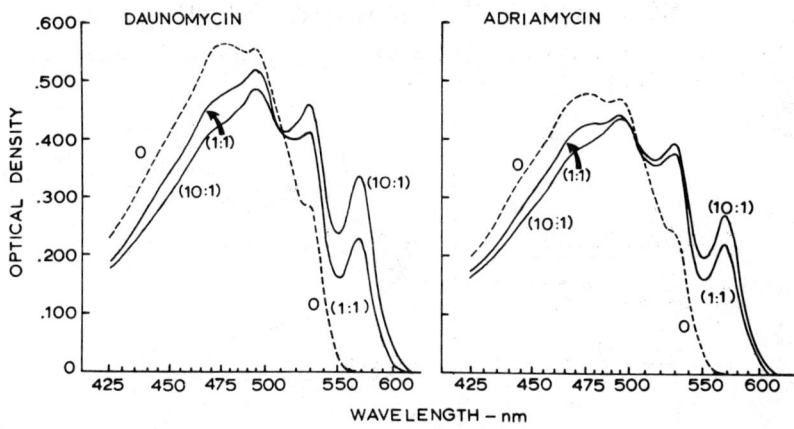

Fig. 4. Absorption spectra of copper chelates with anthracycline drugs.

It is seen in Fig. 6 that Mg^{++} at low molar ratios with respect to daunomycin (<10:1) has no effect on the metabolism of daunomycin (D_1) to daunorubicinol (D_2) by kidney preparations nor on the more complex pattern ($D_1 \rightarrow D_y \rightarrow D_x$) of metabolism of daunomycin by liver. At high molar ratios of Mg^{++}, metabolism of daunomycin was prevented probably via a mechanism of chelation of the anthracycline drug and inhibition of the enzyme *per se*.

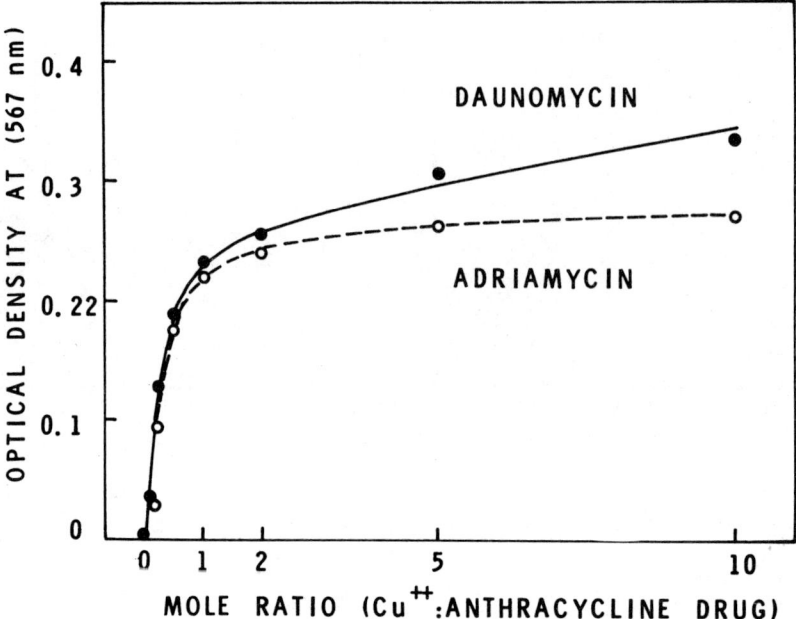

Fig. 5. Estimate of chelate formation between Cu^{++} and anthracycline drugs in ethanol.

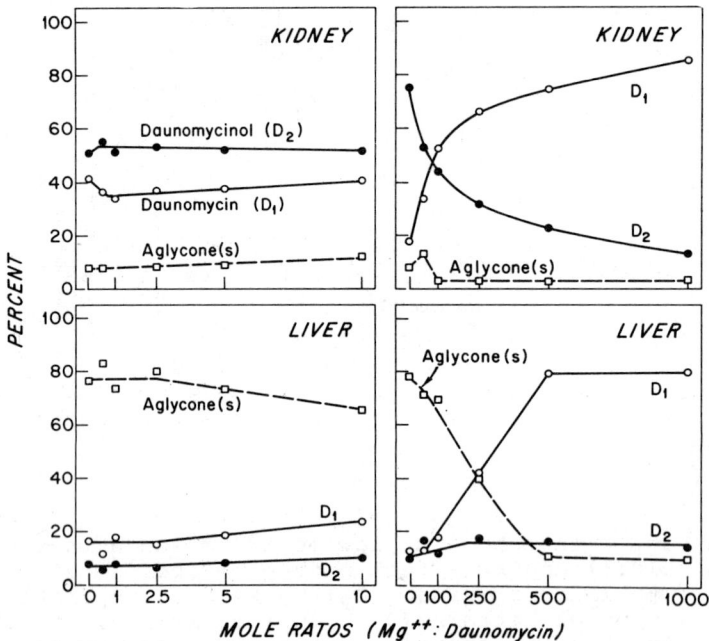

Fig. 6. Effect of Mg^{++} on the aerobic metabolism of daunomycin by 1,000 x g homogenates of both kidney and liver.

Reductive Glycosidic Cleavage of Daunomycin

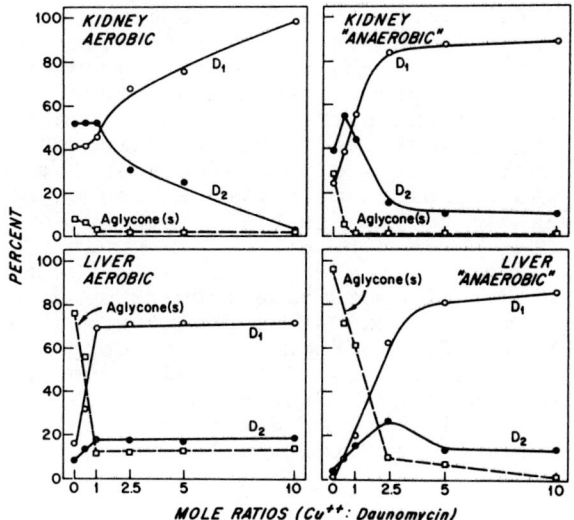

Fig. 7. Effect of Cu^{++} on the aerobic and "anaerobic" metabolism of daunomycin by 1,000 x g homogenate.

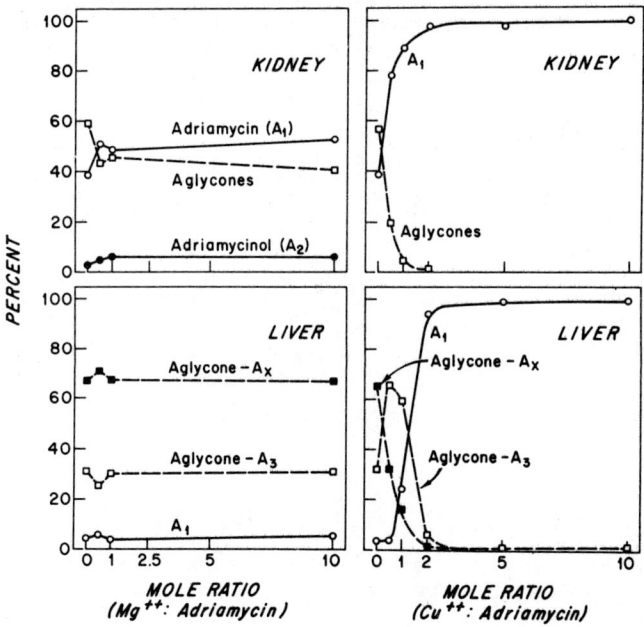

Fig. 8. Effect of Mg^{++} and Cu^{++} on the "anaerobic" metabolism of adriamycin by 1,000 x g homogenate.

Cuprous chelates of the anthracycline drugs, whose stability is much greater than magnesium chelates in aqueous solution, show a marked inhibition of the reductive glycosidic cleavage at low molar ratios of cuprous ions and daunomycin (Fig. 7). At molar ratios of 1:1, in which the reductive glycosidic cleavage is markedly or totally inhibited, the reduction of the 13-carbonyl was either stimulated or was not affected. This reduction of the 13-carbonyl was inhibited only with excess cuprous ions which probably was due to the inhibition of the enzyme *per se* by cuprous ions.

Similar studies with adriamycin gave results comparable to those with daunomycin (Fig. 8). Magnesium chelates, which are unstable in aqueous solution, had no effect, whereas copper chelates at 1:1 molar ratio markedly inhibited the reductive glycosidic cleavage of adriamycin. The reduction of the 13-carbonyl and the normal glycosidic cleavage of daunosamine to yield the aglycone (A_3) was stimulated at low molar ratios of 1:1.

DISCUSSION

These findings *in toto* (Fig. 9) indicate that the dihydroquinone system of the anthracycline drugs is involved in the unprecedented reductive cleavage of the daunosamine from the anthracycline drugs. In addition, it has been shown that the reductive cleavage requires NADPH. This reaction will occur in the presence of nitrogen or carbon monoxide, suggesting that a typical microsomal mono-oxygenase system is not an obligatory component.

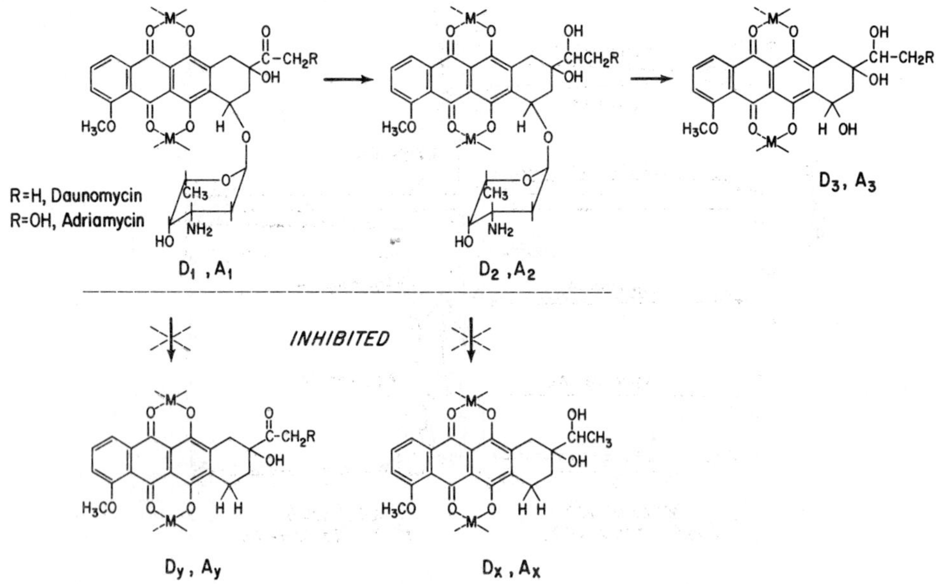

Fig. 9. Pathway for metabolism of metal (M) chelates of anthracyclines.

The essential requirement for NADPH in the reductive glycosidic cleavage pathway would suggest that the hydride ion of NADPH attacks the nucleophilic carbon at position 7 of the postulated intermediate (Fig. 2). This possibility is intriguing, since other nucleophiles may interact; such as, glutathione, an adenine or guanine of nucleic acid, and methionine and cysteine of protein. The possibility, therefore, of covalently binding the aglycone of the anthracycline drugs to biological constituents might relate to the incidence of irreversible cardiotoxicity with the anthracycline drugs -- total dose is important, whereas the schedule of administering the adriamycin is unimportant (1). At present, we are engaged in evaluating the possibility of substituting nucleophiles for NADPH in this novel reductive glycosidic pathway for the anthracycline drugs.

REFERENCES

(1) R.H. Blum and S.K. Carter, Adriamycin. A new anticancer drug with significant clinical activity, Ann. Int. Med. 80, 249 (1974).

(2) D.W. Yesair, M.A. Asbell, R. Bruni, F.J. Bullock and E. Schwartzbach, Pharmacokinetics and metabolism of adriamycin and daunomycin, International Symposium on Adriamycin, Springer-Verlag, Berlin, 1972.

(3) D.W. Yesair, E. Schwartzbach, D. Shuck, E.P. Denine and M.A. Asbell, Comparative pharmacokinetics of daunomycin and adriamycin in several animal species. Cancer Res. 32, 1177 (1972).

(4) N.R. Bachur and J.C. Cradock, Daunomycin metabolism in rat tissue slices, J. Pharmacol. Exptl. Therap. 175, 331 (1970).

(5) J.M. Finkel, K.T. Knapp and L.T. Mulligan, Fluorometric determination of serum levels and urinary excretion of daunomycin (NSC 82151) in mice and rats, Cancer Chemother. Rept. 53(1), 159 (1969).

(6) A. Rusconi, G. DiFronzo and A. DiMarco, Distribution of tritiated daunomycin (NSC 82151) in normal rats. Cancer Chemother. Rept. 52(1), 331 (1968).

(7) J.C. Cradock, M.J. Egorin and N.R. Bachur, Daunorubicin biliary excretion and metabolism in the rat, Arch. Int. Pharmacodyn. Ther. 202, 48 (1973).

(8) D.S. Alberts, N.R. Bachur and J.L. Holtzman, The pharmacokinetics of daunomycin in man, Clin. Pharmacol. Ther. 12, 96 (1971).

(9) R.S. Benjamin, C.E. Riggs, Jr. and N.R. Bachur, Pharmacokinetics and metabolism of adriamycin in man, Clin. Pharmacol. Ther. 14, 592 (1973).

(10) R.S. Benjamin, P.H. Wiernik and N.R. Bachur, Adriamycin chemotherapy -- Efficacy, safety and pharmacologic basis of an intermittent single high-dosage schedule, Cancer 33, 19 (1974).

(11) M.A. Asbell, E. Schwartzbach, F.J. Bullock and D.W. Yesair, Daunomycin and adriamycin metabolism via reductive glycosidic cleavage, J. Pharmacol. Exptl. Therap. 182, 63 (1972).

(12) F.J. Bullock, R.J. Bruni and M.A. Asbell, Identification of new metabolites of daunomycin and adriamycin, J. Pharmacol. Exptl. Therap. 182, 70 (1972).

(13) M.A. Asbell, E. Schwartzbach, I. Wodinsky and D.W. Yesair, Metabolism of daunomycin (NSC 82151) *in vitro* and the chemotherapeutic activity of isolated metabolites *in vivo*, Cancer Chemother. Rept. 56, 315 (1972).

(14) N.R. Bachur, Daunorubicinol, a major metabolite of daunorubicin: Isolation from human urine and enzymatic reactions, J. Pharmacol. Exptl. Therap. 177, 573 (1971).

(15) N.R. Bachur and J.C. Cradock, Daunomycin metabolism in rat tissue slices, J. Pharmacol. Exptl. Therap. 175, 331 (1970).

(16) N.R. Bachur and M. Gee, Daunorubicin metabolism by rat tissue preparations, J. Pharmacol. Exptl. Therap. 177, 567 (1971).

(17) N.R. Bachur and D.H. Huffman, Daunorubicin metabolism: estimation of daunorubicin reductase, Brit. J. Pharm. 43, 828 (1971)

(18) D.H. Huffman and N.R. Bachur, Daunorubicin metabolism in acute myelocytic leukemia, Blood 39, 637 (1972).

(19) D.H. Huffman and N.R. Bachur, Daunorubicin metabolism by human hematological components, Cancer Res. 32, 600 (1972).

(20) D.H. Huffman, N.R. Bachur and M. Gee, Hematological metabolism of daunomycin in man, Clin. Res. 18, 472 (1970).

(21) D.H. Huffman, R.S. Benjamin and N.R. Bachur, Daunorubicin (Dl) metabolism in acute myelogenous leukemia, Clin. Res. 19, 493 (1971).

(22) D.H. Huffman, R.S. Benjamin and N.R. Bachur, Daunorubicin metabolism in acute nonlymphocytic leukemia, Clin. Pharmacol. Ther. 13, 895 (1972).

(23) K. Lieberman, Reduction of ketones in liver cytosol, Xenobiotica 1, 97 (1971).

(24) N.R. Bachur and R.L. Felsted, Oxisuran reduction by rabbit tissue preparations, Drug Metab. Dispos. 4, 239 (1976).

(25) E. Calendi, A. DiMarco, M. Reggiani, B. Scarpinato and L. Valentini, On physico-chemical interactions between daunomycin and nucleic acids, Biochim. Biophys. Acta 103, 25 (1965).

(26) M.M. Fishman and I. Schwartz, Effect of divalent cations on the daunomycin-deoxyribonucleic acid complex, Biochem. Pharmacol. 23, 2147 (1974).

(27) D.W. Yesair, L. Bittman and E. Schwartzbach, Pharmacodynamic significance of metal chelates of the anthracycline drugs (NSC 82151 and NSC 123127), Proc. Amer. Assoc. Cancer Res. 15, 72 (1974).

(28) R. Suemitsu, Studies on metal chelate compounds of anthraquinone derivatives. Part I. Magnesium chelate compounds of 1-hydroxy and 1,4-dihydroxyanthraquinone, Agric. Biol. Chem. 27, 1 (1963).

(29) R. Suemitsu, A spectrophotometric study on α-hydroxyanthraquinone bivalent metal chelate compounds, Science and Engineering Rev. Doshisha University, 5, 1 (1965).

COVALENT BINDING OF STYRENE OXIDE TO RAT LIVER MACROMOLECULES *IN VIVO* AND *IN VITRO*

Jukka Marniemi, Else-Maj Suolinna, Niilo Kaartinen and Harri Vainio

Department of Physiology, University of Turku, Turku, and Department of Industrial Hygiene and Toxicology, Institute of Occupational Health, Helsinki, Finland

ABSTRACT

Styrene oxide was found to bind covalently to homogenate, microsomes and protein and nucleic fractions of the rat liver after intraperitoneal administration in vivo. Styrene oxide also bound covalently in vitro when incubated with different liver fractions, and reduced glutathione decreased this binding. The free glutathione content of the liver was decreased by styrene oxide in vivo.

INTRODUCTION

Styrene (phenyl ethylene, vinyl benzene) is widely used in the manufacturing of plastics and synthetic resins. Styrene is metabolized by liver microsomes to its epoxide, styrene oxide (1). Epoxides have been implicated in hepatotoxicity, mutagenicity and carcinogenicity (2). Styrene oxide has been shown to be a base substitution mutagen (3, 4) and a weak carcinogen when painted on mouse skin (5). Since covalent binding to tissue macromolecules appears to be a necessary link in many chemically induced toxic manifestations (6), we wanted to study whether the first metabolic derivative of styrene, styrene oxide, is bound irreversibly to liver proteins and nucleic acids.

EXPERIMENTAL

Adult male Wistar rats fed ad libitum on commercial pellets (Hankkija Oy, Finland) were used. The animals were given $[7-^3H]$ styrene oxide (sp. act. 15.5 mCi/mmole, NEN Chemicals GmbH, Dreieichenhain, West Germany) intraperitoneally together with unlabeled styrene oxide (50 and 200 mg/kg of body weight, Kochlight Laboratories Ltd., Colnbrook, England) both dissolved in corn oil. After the indicated time periods the livers were removed and cooled in 0°C 0.25 mol/l sucrose and homogenized in this solution to give a 20% (w/v) homogenate. The microsomal fraction was isolated by 60 min centrifugation at 105 000 x g after a presedimentation of cell debris, nuclei and mitochondria. Protein and nucleic acid fractions were isolated essentially according to Diamond et al. (7). The liver was homogenized in 1% sodium triisopropylnaphthalene sulfonate-6% sodium 4-aminosalicylate (15 ml per g of liver) and extracted with phenol-m-cresol-8-hydroxyquinoline-water (500:70:0.5:55 by weight). Total nucleic acids were precipitated from the aqueous phase with two volumes of ethanol and washed 3 times with ethanol and ether. (No attempt was made to check contamination with protein or glycogen in this fraction.) The proteins were precipitated from the phenol phase with methanol and washed with ethanol and ether (3 times with each).

The covalent binding was measured using a modification of the washing procedure of Jollow et al. (8): homogenate and microsomal fraction (0.5 ml) were precipitated with two volumes of 0.9 mol/l trichloroacetic acid (TCA). After centrifugation (1000 x g, 15 min) the precipitate was resuspended into 3 ml of 0.6 mol/l TCA, shaken vigorously for 15 min and recentrifuged. This procedure was repeated twice. In the next step the precipitate was extracted three times with 3 ml of 80% aqueous methanol followed by one pure methanol and three diethyl ether washes. The tightness of the binding was confirmed by extracting the final precipitate with several organic solvents (diethyl ether, ethanol, chloroform, heptane and acetone) for six hours and also by solubilizing the precipitate with 1 N NaOH and repeating the precipitation and washing. After the last extraction the precipitate was collected by filtration on a filter paper and it was oxidized in a sample oxidizer (model 306, Packard, Instrument Co. Downers Grove, Ill, U.S.A., modified by Niilo Kaartinen, to be published). The final radioactivity was measured with liquid scintillation counting.

Reduced glutathione was determined according to Ellman (9). Protein was determined with the biuret method (10).

RESULTS AND DISCUSSION

Covalently bound radioactivity was found in the total homogenate and microsomal fraction of rat liver after intraperitoneal administration of tritium labeled styrene oxide (Table 1).

The binding was highest one hour after administration. With a styrene oxide dose of 50 mg/kg only 2-3% of the radioactivity found in the liver was covalently bound, whereas with the dose of 200 mg/kg this proportion was up to 30% (Table 1). This dose treshold apparently reflects the capacity of the liver to metabolize styrene oxide. Covalently bound styrene oxide appeared both in crude protein and nucleic acid fractions (Table 2). The relative binding to protein fraction increased when the styrene oxide dose was increased from 50 mg/kg to 200 mg/kg.

TABLE 1 Covalent binding of styrene oxide to rat liver in vivo after intraperitoneal administration (nmoles/g of liver)

	HOMOGENATE			MICROSOMES		
50 mg/kg	Total radio-activity	Covalently bound	% of total	Total radio-activity	Covalently bound	% of total
1 hr	340	8.2	2.4	12	3.1	26
4 hr	140	3.9	2.8	5.4	2.4	44
12 hr	95	2.5	2.6	4.3	0.5	12
200 mg/kg						
1 hr	1400	170	12	67	18	27
4 hr	450	150	33	30	13	43
12 hr	510	71	14	29	11	38

Results are means from 3-6 animals.

TABLE 2 Covalently bound styrene oxide (nmoles/g of liver) in rat liver protein and nucleic fractions after intraperitoneal administration in vivo

50 mg/kg	Protein fraction	Nucleic acid fraction
1 hr	5.2	0.80
4 hr	3.6	0.37
12 hr	2.4	0.43
200 mg/kg		
1 hr	78	2.6
4 hr	65	3.5
12 hr	55	4.1

For other explanations see "Experimental" and Table 1.

Styrene oxide was also found to bind covalently to liver macromolecules in vitro when incubated in the presence of homogenate or microsomal fraction (Fig. 1). Reduced glutathione decreased this binding. Ten millimolar glutathione decreased the binding to the homogenate by 80-90%. When protein or nucleic acid fraction (20 mg) isolated from rat liver was incubated (20 min) with styrene oxide (4 μmoles), about 3.7 and 0.3%, respectively, became covalently bound.

Styrene oxide also appeared to decrease the glutathione content of the liver in vivo. (Fig. 2.) One hour after the injection of the dose of 200 mg/kg the glutathione content was only about 23% of the control. However, after 12 hours it had already returned to almost normal values (Fig. 2.).

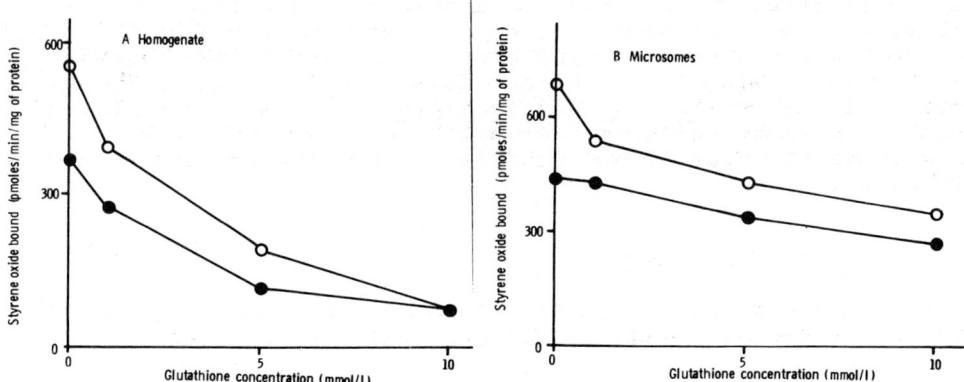

Fig. 1. Covalent binding of tritium labeled styrene oxide (4 umoles) in vitro to rat liver homogenate (15 mg prot.) and microsomes (7.2 mg prot.) when incubated in the presence of different concentrations of reduced glutathione. Open circles: 10 min incubation; black spots: 20 min incubation.

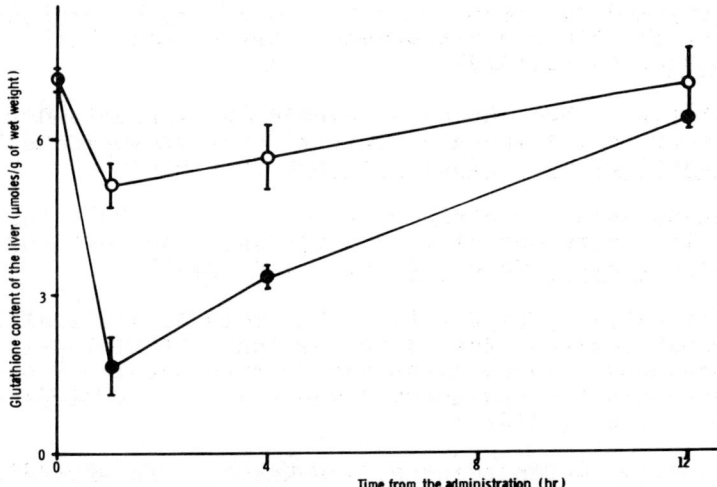

Fig. 2. The effect of styrene oxide administration in vivo on reduced glutathione concentration of the liver. Means ± their standard errors obtained from 4-6 animals are indicated. Open circles: 50 mg/kg of styrene oxide; black spots: 200 mg/kg of styrene oxide.

The present results show that styrene oxide can be bound covalently to liver and that this binding is dependent on the glutathione content of the liver. Hepatotoxicity of styrene has not yet been reported (11), but especially at the conditions of reduced glutathione content or low epoxide hydratase (E.C. 4.2.1.63) activity, it appears a distinct possibility. Fortunately, on the other hand, styrene enhances more the activity of epoxide hydratase than that of epoxide forming mono-oxygenase (12). On the basis of the current data studies of the covalent binding of styrene in vivo and in vitro with concomitant histological and pathological investigation of the liver seem a worthwhile project. These studies are now in progress in our laboratory.

ACKNOWLEDGEMENTS

This study has been supported by grants from U.S. Public Health Service (AM-06018) and Juho Vainio Foundation, Finland.

REFERENCES

1. K.C. Leibman and E. Ortiz, Oxidation of styrene in liver microsomes, Biochem. Pharmacol. 18, 552 (1969).

2. D.M. Jerina and J.W. Daly, Arene oxides: A new aspect of drug metabolism, Science (Wash. D.C.) 185, 573 (1974).

3. P. Milvy and A.J. Garro, Mutagenic activity of styrene oxide (1,2-epoxyethylbenzene), a presumed styrene metabolite, Mutation Res. 40, 15 (1976).

4. H. Vainio, R. Pääkkönen, K. Rönnholm, V. Raunio and O. Pelkonen, A Study on the mutagenic activity of styrene and styrene oxide, Scand. J. Work. Environm. Health, in press.

5. B.L. Van Duuren, S. Nelson, L. Orris, E.D. Palmes and F.L. Schmitt, Carcinogenicity of epoxides, lactones, and peroxy compounds, J. Natl. Cancer Inst. 31, 41 (1963).

6. J.R. Gillette, A perspective on the role of chemically reactive metabolites of foreign compounds in toxicity-I. Correlation of changes in covalent binding of reactivity metabolites with changes in the incidence and severity of toxicity, Biochem. Pharmacol 23, 2785 (1974).

7. L. Diamond, V. Defendi and P. Brookes, The interaction of 7,12-dimethylbenz(a)anthracene with cells sensitive and resistant to toxicity induced by this carcinogen, Cancer Res. 27, 890 (1967).

8. D.J. Jollow, J.R. Mitchell, W.Z. Potter, D.C. Davis, J.R. Gillette and B.B. Brodie, Acetaminophen-induced hepatic necrosis. II. Role of covalent binding in vivo, J. Pharmacol. Exp. Therap. 187, 195 (1973).

III OUTLOOK

THE REGULATION OF HUMAN DRUG METABOLISM BY NUTRITIONAL FACTORS

A. Kappas,[1] A. P. Alvares,[1] K. E. Anderson,[1]
W. A. Garland,[2] E. J. Pantuck,[3] and A. H. Conney[2]

The Rockefeller University,[1] New York, N.Y. 10021, Department of Anaesthesiology, Columbia University College of Physicians and Surgeons,[3] New York, N.Y. 10032, and the Department of Biochemistry and Drug Metabolism, Hoffmann-La Roche, Inc.,[2] Nutley, N.J. 07110

The present report summarizes the results of some early studies which we have been carrying out on the regulation of drug oxidation rates by nutritional factors in man. It is widely recognized that the activity of the cytochrome P-450 dependent "mixed function oxidase" system is greatly altered in man and in animals by chemical exposures of diverse kinds (1,2). In animals there are also considerable data which attest to the ability of variations in nutritional composition or status to influence drug metabolism (3). There is, however, very little information on this general subject in man. Many reasons account for the lack of such studies in humans - and perhaps an important one is the instinctive, and understandable conviction that foods and drugs somehow represent fundamentally "different" kinds of chemical substances. Yet this is of course not the case at all. Humans receive the bulk of their exposure to exogenous chemicals through their diet; and if the essential elements of the foodstuffs themselves - that is the proteins, carbohydrates and fats (or specific components of these macronutrients) which constitute the diet - could be shown to have significant influences on drug metabolism rates in man, the potential clinical implications of such findings would be quite considerable.

We have initiated a series of studies (4,5,6,7) to examine the possibility that nutritional factors can significantly influence drug oxidation in man and the results of two such investigations are briefly described here. These studies focussed on a) the influence of variations in the carbohydrate and protein composition of the diet on antipyrine and theophylline plasma elimination rates in normal subjects; and b) the effects, in similar individuals, of the manner of food preparation - specifically the charcoal broiling of meat - on the metabolic disposition of phenacetin.

The first study involved 6 normal volunteer subjects studied over an 8-week period, divided as follows. Period 1 (Days 1-14) in which the subjects ate their ordinary home diet; Period 2 (Days 15-28) in which they ate a low carbohydrate (CHO)-high protein (PRO) diet in which all 3 meals were prepared in the Metabolic Diet Kitchen of the Rockefeller University Hospital; Period 3 (Days 29-42) in which they similarly ate Metabolic Diet Kitchen meals prepared to have a high carbohydrate (CHO)-low protein (PRO) composition; and Period 4 (Days 43-56) in which they returned to the ordinary diet they ate at home.

Plasma half-lives of antipyrine and theophylline were studied after single oral doses of each drug (antipyrine dose 18 mg/kg; theophylline dose 5 mg/kg) in each individual during each test period. The low CHO-high PRO diet (Period 2, Days 15-28) was comprised of 35% carbohydrate; 44% protein and 21% fat. The high protein content of this diet was achieved by inclusion of protein-rich foods, such as meat, fish, egg whites etc., as well as use of a liquid dietary supplement Sustacal®. The composition of the high CHO-low PRO diet in Period 3 (Days 29-42) was 70% carbohydrate; 10% protein and 20% fat. The caloric intake during each of the test Periods 2 and 3 ranged from 2400-2500 calories per day per subject. For comparative purposes the composition of the average American home diet is 50% carbohydrate, 15% protein and 35% fat.

The results of the half-life studies of antipyrine and theophylline determined in each of the 4 test periods are shown in Table 1. For each subject there was a significant and sometimes marked decrease of antipyrine half-life in the shift from a home diet (Period 1) to a low CHO-high PRO diet (Period 2); in all but one there was a reversal of this phenomenon in the shift back to a high CHO-low PRO diet (compare Period 3 with Period 2); and there were not significant differences between the two home diet periods (Periods 1 and 4) and Period 3. Similar observations were recorded in the study of theophylline plasma elimination rates during the same diet periods (Table 1), confirming that the shift from an average home diet composition to one comprised of low carbohydrate and high protein, results in a major acceleration of oxidation rates of these two prototype drugs and thus of a substantial decrease of their plasma half-lives. In contrast, a high carbohydrate-low protein diet reverses this phenomenon and restores drug metabolism rates to the levels characteristic of those produced by the average American home diet. Supplementations of the diet, in other studies conducted with these individuals, indicated that protein additions, or carbohydrate additions, alone shifted drug metabolism rates in the directions indicated from the data in Table 1, although single macronutrient substitutions had lesser effects than the combined isocaloric substitutions carried out in Periods 2 and 3.

Antipyrine metabolism is inducible by the barbiturate class of inducing agents in man (8); in contrast theophylline metabolism is not significantly altered by phenobarbital in humans (9) but its metabolism is accelerated in chronic cigarette smokers (10). Although the proximate mechanism by which alterations in the macronutrient composition of the diet affected antipyrine and theophylline metabolism is not known with certainty, it is not unreasonable to suppose that these dietary changes affected in some manner the specific mono-oxygenases catalyzing the oxidative biotransformation of these two test drugs.

The second study which I would like to summarize concerns the effect of the charcoal broiling of meat on the metabolic disposition of phenacetin in man. Phenacetin is a prototype drug whose oxidative biotransformation may be partially dependent upon cytochrome P-448, the microsomal heme protein which is inducible by polycyclic hydrocarbon carcinogens such as 3-methylcholanthrene. In any event, cigarette smokers have markedly lower levels of phenacetin in their blood after a single oral dose of the drug than do normals although the plasma half-life of phenacetin or the levels of its N-dealkylated metabolite (N-acetyl-p-aminophenol, or APAP) are not altered (11); thus the metabolite/unaltered drug ratio in plasma in smokers is

TABLE 1 Antipyrine and Theophylline Half-lives in Subjects Maintained on Various Diets

Subject	Period 1 Home diet-1 Days 1-14	Period 2 Low CHO High PRO Days 15-28	Period 3 High CHO Low PRO Days 29-42	Period 4 Home diet-2 Days 43-56
		Antipyrine, half-life in hrs		
1	12.5	8.3	17.5	13.0
2	15.0	10.2	18.0	15.8
3	17.4	9.0	17.0	14.5
4	14.4	9.0	11.0	11.9
5	14.5	9.8	9.7	15.0
6	23.5	11.0	20.5	15.0
Mean ± S.E.	16.22±1.59	9.55±0.40	15.62±1.74	14.20±0.60
Subject		Theophylline, half-life in hrs		
1	8.0	5.7	8.0	6.2
2	6.8	5.6	9.6	6.5
3	13.0	6.3	9.2	9.2
4	7.2	3.6	5.4	7.8
5	7.0	4.6	7.1	7.8
6	6.7	5.6	6.4	7.7
Mean ± S.E.	8.12±0.99	5.23±0.39	7.62±0.66	7.53±0.66

Each subject was maintained on each test diet for 14 days. Antipyrine half-lives were determined on day 10 and theophylline half-lives were determined on day 14 of each test diet period. Drug studies were also carried out immediately prior to the start of the test diet study (home diet 1) as well as on the day 10th and 14th (home diet 2) after the test diets.

greatly increased suggesting that smoking enhances phenacetin metabolism in the gastrointestinal tract and/or during its first pass through the liver. Since charcoal broiling of meat is a widespread habit in this and other countries; and since charcoal broiled beef markedly stimulates the activity of an intestinal enzyme system in animals that oxidizes phenacetin to APAP (12), we investigated the possibility that feeding humans charcoal broiled meat would enhance their oxidative metabolism of the drug.

Nine normal volunteers were studied during 4 test periods as follows: Period 1 (home diet); Period 2 (control hospital diet with meat cooked over burning charcoal but protected from the charcoal by being in aluminum foil); Period 3 (an identical hospital diet, except that the meat was broiled over burning charcoal in absence of aluminum foil); Period 4 (returned to the control hospital diet as in Period 2). The amount of meat provided during the meals ingested in Periods 2 and 3 was a total of 14 ounces/day (1 small steak and

1 small hamburger); Periods 2 and 4 lasted 7 days each. During Period 3 the control hospital diet was fed for the first 3 days and the charcoal broiled diet for the subsequent 4 days. Phenacetin was administered in a dose of 900 mg orally to each of the 9 subjects at the end of each study period and plasma as well as urine levels of the drug and its main metabolite APAP were quantitated by a newly developed gas chromatographic-mass spectrometric method. The details of these studies will be published elsewhere (5) and only a brief summary of the findings are reported here. In Table 2, it is shown that the mean peak plasma level of phenacetin was markedly diminished following the period (Period 3) of short-term ingestion of charcoal broiled meat as compared with the period on the customary home diet (Period 1) or the periods during which the ingested meat was protected by aluminum foil (Periods 2 and 4) during its broiling. The half-life of phenacetin did not change significantly in any of the 4 test periods, nor did the plasma concentrations of the unconjugated phenacetin metabolite APAP. Thus, the concentration of APAP to phenacetin in plasma markedly increased during the charcoal broiled beef diet (Period 3) as compared with Periods 1, 2 and 4 and the increase in metabolite/unaltered drug was observed in the earlier time periods studied during the plasma half-life determination of phenacetin. Table 3 depicts the substance of these findings in terms of a summation of the total areas under the plasma concentration of phenacetin - time curve (during the period 0-7 hours after administration of the drug). It is evident from this table that there was a very marked enhancement of phenacetin metabolism in the 9 normal subjects after only 4 days ingestion of charcoal broiled meat; and that in the absence of an alteration in the plasma half-life of the drug this dietary effect on the phenacetin oxidation rate most likely reflected its greatly enhanced biotransformation either in the intestine and/or during its first pass through the liver.

TABLE 2 Peak Plasma Concentration of Phenacetin during the Test Period

Test Period	Phenacetin (ng/ml)
Period 1 - Home diet	2492±709
Period 2 - Control hospital diet	1628±424
Period 3 - Charcoal broiled beef diet	352±80
Period 4 - Control hospital diet	1885±642

The values shown are mean values ± S.E. for the group of 9 subjects. Peak plasma concentrations of the drug occurred at either 1 or 2 hours. The values for Periods 1, 2 and 4 are not significantly different from each other. Period 3 differes significantly from Period 1 ($P<0.01$); from Period 2 ($P<0.01$) and from Period 4 ($P<0.05$).

These two studies make it clear that the composition of the foods which humans eat as well as the manner of food preparation may profoundly influence the rates at which drugs and other exogenously derived chemicals are metabolized. The test drugs employed in these studies are biotransformed principally by the microsomal "mixed function oxidase" system and

TABLE 3 Area Under the Plasma Concentration of Phenacetin - Time Curve (0-7 Hours) during the Test Periods

Test Period	Area ($\mu g/ml \cdot min$)
Period 1 - Home diet	260±70
Period 2 - Control hospital diet	170±40
Period 3 - Charcoal broiled beef diet	37±8
Period 4 - Control hospital diet	174±53

The values shown are mean values ± S.E. for the group of 9 subjects. The values for Periods 1, 2 and 4 are not significantly different from each other. Period 3 differs significantly from Period 1 ($P<0.01$); from Period 2 ($P<0.01$); and from Period 4 ($P<0.025$).

thus the nutritional influences on drug metabolism demonstrated here presumably reflect alterations in the activities of enzymes coupled to this system. However, it is evident that nutrition dependent changes in the rates of drug metabolism need not be reflected solely in altered rates of plasma elimination of drugs due to enhanced activity of drug metabolizing enzymes in liver since the marked increase in phenacetin metabolism which was demonstrated was unaccompanied by an alteration of its plasma half-life.

The proximate chemical mechanisms by which the nutrition-evoked changes in drug metabolism reported here occur are not known; they could reflect the action of specific macronutrients themselves (i.e. carbohydrate, protein) and/or their ratios; the effects of chemicals associated, naturally or by artifact, with these macronutrients; or in the charcoal-broiled beef studies, the actions of combustion products of the meat itself or the volatilized components of which become absorbed to the meat. In any case a very wide range of possible mechanisms by which nutritional components can influence human drug metabolism - and thus alter the biological or toxic actions of chemicals - must be taken into account in evaluating these phenomena. It is of great potential medical importance to analyze to the extent possible such mechanisms in clinical studies; and indeed, this appears to us to be a potentially very fruitful area of clinical pharmacological investigation. For example, such nutritional-pharmacological interactions may occur in those individuals who undertake weight-reducing regimens of unusual type; in malnourished subjects; in post-operative patients who receive glucose intravenously as a sole form of nourishment; in individuals with special dietary restrictions, such as vegetarians; and in large numbers of patients whose disease processes (i.e. diabetes, obesity etc.) require, or are associated with significant nutritional restrictions. The studies described here raise the possibility that these various groups might show substantial deviations from the normal in the rates at which they metabolize, or respond therapeutically to, drugs.

It should be noted finally that the microsomal monooxygenases which metabolize drugs and other foreign chemicals are also involved in the biotransformation of endogenous steroid hormones. It would be of great interest to determine if steroid metabolism in man can be altered by changes in the macronutrient composition of the diet. Nutritional influences on hormone metabolism would inevitably be reflected in changes in the biological actions of hormones and would therefore be expected to have considerable physiological impact for man.

ACKNOWLEDGEMENTS

Supported in part by the National Institutes of Health Grants GM-09069 and ES-01055. A. P. Alvares is a recipient of a Research Career Development Award 1 KO4 ES-00010-02 from the National Institutes of Health.

REFERENCES

1. A. H. Conney, Pharmacological implications of microsomal enzyme induction, Pharmacol. Rev. 19, 317 (1967).

2. A. Kappas and A. P. Alvares, How the liver metabolizes drugs, Scien. Am. 232, 22 (1975).

3. T. C. Campbell and J. R. Hayes, Role of nutrition in the drug-metabolizing enzyme system, Pharmacol. Rev. 26, 171 (1974).

4. A. P. Alvares, K. E. Anderson, A. H. Conney and A. Kappas, Interactions between nutritional factors and drug biotransformation in man, Proc. Natl. Acad. Sci. U.S.A. 73, 2501 (1976).

5. A. H. Conney, E. J. Pantuck, K.-C. Hsiao, W. A. Garland, K. E. Anderson, A. P. Alvares and A. Kappas, Enhanced phenacetin metabolism in humans fed charcoal-broiled beef, Clin. Pharmacol. Ther. 19, December issue, (1976).

6. E. J. Pantuck, K.-C. Hsiao, A. H. Conney, W. A. Garland, A. Kappas, K. E. Anderson and A. P. Alvares, Effect of charcoal-broiled beef on phenacetin metabolism in man, Science 188, in press (1976).

7. A. Kappas, K. E. Anderson, A. H. Conney and A. P. Alvares, The influence of dietary protein and carbohydrate on the metabolism of antipyrine and theophylline, Clin. Pharmacol. Ther. 19, December issue (1976).

8. E. S. Vesell and J. G. Page, Genetic control of the phenobarbital induced shortening of plasma antipyrine half-lives in man, J. Clin. Invest. 48, 2202 (1969).

9. K. D. Piafsky, D. S. Sitar and R. P. Ogilvie, Effect of phenobarbital on theophylline kinetics, Clin. Res. 23, 610A (1975).

10. J. Jenne, H. Nagasawa, R. McHugh, F. MacDonald and E. Wipe, Decreased theophylline half-life in cigarette smokers, Life Sciences 17, 195 (1975).

11. E. J. Pantuck, R. Kuntzman and A. H. Conney, Decreased concentration of phenacetin in plasma in cigarette smokers, Science 175, 1248 (1972).

12. E. J. Pantuck, K.-C. Hsiao, R. Kuntzman and A. H. Conney, Intestinal metabolism of phenacetin in the rat: Effect of charcoal broiled beef and rat chow, Science 187, 744 (1975).

THE BAY-REGION THEORY: A QUANTUM MECHANICAL APPROACH TO AROMATIC HYDROCARBON-INDUCED CARCINOGENICITY

Donald M. Jerina and Roland E. Lehr

Laboratory of Chemistry, National Institutes of Health, National Institute of Arthritis, Metabolism and Digestive Diseases, Bethesda, Maryland 20014

Although many alternant polycyclic aromatic hydrocarbons have long been recognized as chemical carcinogens, the intimate details of how these molecules induce neoplasia remain unknown. Presently, the most widely accepted concept regarding the action of these and other carcinogens is that chemical modification of some critical cellular molecule must occur (1). DNA offers an attractive biological target for such damage since stored genetic information would be directly affected. Other molecules, such as RNA or protein cannot, however, be excluded. If the carcinogen is chemically reactive, it can act directly. If, as is the case for the hydrocarbons, the carcinogen is relatively inert, it must first be converted into a reactive metabolite.

The past several years have seen dramatic advances in our understanding of the nature of the reactive metabolite(s) which is generated from the ubiquitous environmental carcinogen benzo[a]-pyrene (BP). Studies of the metabolism induced binding of BP and BP metabolites to DNA in vitro and in cultured cells (2,3) have provided evidence that a BP 7,8-diol-9,10-epoxide accounts for most of the binding of BP to DNA which occurs in vivo (4). The diastereomeric forms of the 7,8-diol-9,10-epoxide, in which the benzylic 7-hydroxyl group is either cis or trans to the epoxide oxygen, have been synthesized and their structures have been assigned (5,6). As predicted (7), the isomer in which these two groups are cis is chemically more reactive due to anchimeric

assistance by an intramolecular hydrogen bond (5). Both isomers
are highly mutagenic toward bacterial and mammalian cells (8-12).
Studies of the carcinogenicity of BP 7,8-oxide and BP 7,8-
dihydrodiol (13,14), which are metabolic precursors of the 7,8-
diol-9,10-epoxides, have established that both are potent carcinogens and provide critical evidence that such diol epoxides may
be ultimate carcinogenic forms of BP. Reviews of the evidence
for this pathway are available (15,16).

The rather compelling case for BP 7,8-diol-9,10-epoxides as ultimate carcinogens has prompted our consideration of the possible
general importance of diol epoxides to the carcinogenicity of
alternant polycyclic aromatic hydrocarbons. The effects on carcinogenicity of methyl and fluoro substituents at different ring
positions is of interest in that substitution on angular benzo-
rings was found to cause significant decreases in activity (17,18).
These substituent effects suggested that, for a given hydrocarbon,
an epoxide which forms part of a "bay region" on a saturated,
angular benzo-ring should be the most mutagenic and carcinogenic.
The most simple example of such a "bay region" epoxide is 1,2,3,
4-tetrahyrophenanthrene-3,4-epoxide.

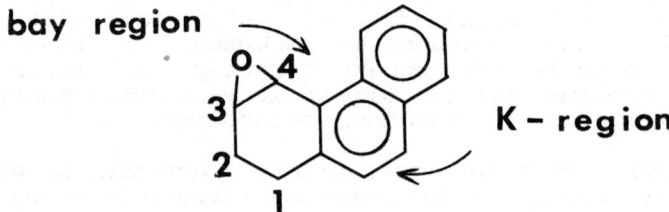

The hydroxyl groups in the dihydrodiol precursors in effect alter
the substrate to a dihydroaromatic system and thereby provide a
convenient metabolic pathway to the "bay region" epoxide on the
saturated ring. The hydroxyl groups may also have an effect on
biological activity, as has been discussed elsewhere (12). Of
particular importance to the present discussion is the unusually
high reactivity of "bay region" epoxides of saturated rings compared to their "non-bay-region" counterparts (Yagi, unpublished)
and the selectivity of reagents such as lead tetraacetate toward
the 10-position of 7,8,9,10-tetrahydro-BP (19). This unusual
reactivity at benzylic "bay region" positions on saturated,
angular benzo-rings prompted quantum mechanical calculations of
the ease of carbonium ion formation at such positions (15).
Perturbational molecular orbital calculations (20) provide a convenient means of assessing the π-electron energy change which

occurs upon conversion of an aromatic hydrocarbon into an aryl benzylic carbonium ion:

$$\xrightarrow{\Delta E_{deloc}}$$

This transformation serves as the appropriate model for assessing the relative ease of carbonium ion formation such as would occur on heterolytic opening of the benzylic oxirane C-O bond of a diol epoxide:

$$\xrightarrow{\Delta E_{deloc}}$$

The results of these calculations (15) clearly predicted that benzylic carbonium ions on saturated benzo-rings are most easily formed (larger values of ΔE_{deloc}) when the carbonium ion forms part of a "bay region" of the hydrocarbon. This point is illustrated in Fig. 1 for benzo[a]anthracene.

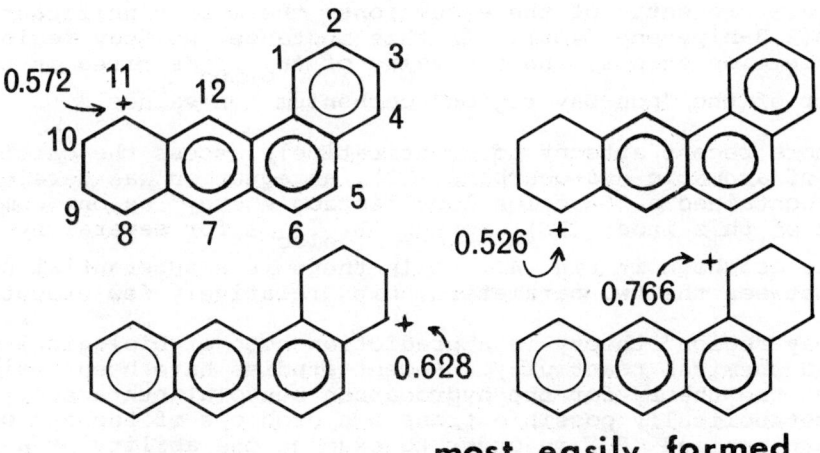

Fig. 1. Values of $\Delta E_{deloc}/\beta$ for carbonium ion formation at various benzylic positions of tetrahydrobenzo[a]anthracene derivatives.

In addition, comparison of the values of $\Delta E_{deloc}/\beta$ for the "bay region" carbonium ions of a series of carcinogenic and noncarcinogenic hydrocarbons showed a marked correlation between carcinogenicity and calculated ease of carbonium ion formation, although certain exceptions were noted (15). Further, the calculations revealed examples of easily formed benzylic carbonium ions derived from tetrahydro benzo-rings at "non-bay regions" and some of these have been cited (15). Values of $\Delta E_{deloc}/\beta$ for "bay region" and other benzylic carbonium ions for a number of hydrocarbons are presented in Table 1 and Fig. 2 along with the relative carcinogenic activities of the parent hydrocarbons. The degree of carcinogenic activity is dependent upon factors such as species, strain, age, sex and pharmacological state of animals as well as upon the method of administration of the carcinogen; unfortunately, tabulations of relative carcinogenicity for a large series of hydrocarbons in which factors such as these have been carefully controlled are unavailable. These factors present serious difficulties in any attempted correlation between relative carcinogenic activity and calculated theoretical parameters.

Among the numerous attempts to correlate structure with carcinogenic activity, the quantum mechanical "K-region" theory of the Pullmans (21) has received the widest attention. In this theory, the hydrocarbon must possess an active "K-region" to be carcinogenic. If the molecule also possesses a "L-region," this region must be rather inactive in order for the hydrocarbon to remain carcinogenic. Since virtually all molecules which have "bay regions" also have "K-regions," we have sought to determine whether the Pullmans' combined "K-region" index shows any correlation with our calculations of $\Delta E_{deloc}/\beta$ at "bay regions". The covariance of these parameters is shown in Fig. 3. In general, there is a substantial correlation, although some marked exceptions are present. Of the exceptions, the most significant is naptho[2,3-b]pyrene (#10). In this instance, no "bay region" carbonium ion exists, and the value of $\Delta E_{deloc}/\beta$ cited is the highest of the "non-bay region" carbonium ion values.

In a more recent attempt to quantitatively assess the carcinogenicity of aromatic hydrocarbons (22), an equation was developed which contained a "K-region localization energy" as one component. A plot of this index $(LE)_K$ versus $\Delta E_{deloc}/\beta$ for several hydrocarbons is shown in Fig. 4. Again there is a substantial correlation between the two parameters, with relatively few exceptions.

The "bay region" theory is a prediction both of biological activity and chemical reactivity. Recent studies have been designed to test the theory for the hydrocarbon benzo[a]anthracene. The five metabolically possible trans dihydrodiols of the hydrocarbon were synthesized (23) in order to examine the ability of a highly purified and reconstituted cytochrome P-448 system to metabolize them to compounds (presumably diol epoxides) mutagenic toward bacteria. As a substrate for the formation of mutagens, the 3,4-dihydrodiol was found to be ten times more active than the other

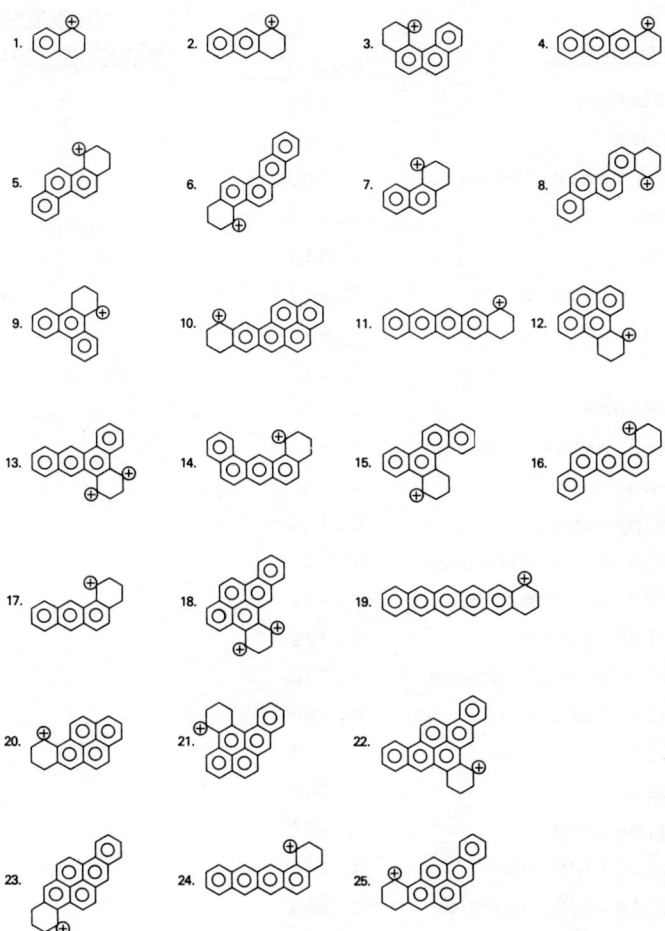

Fig. 2. Structures of benzylic carbonium ions from tetrahydro benzo-ring derivatives of the parent aromatic hydrocarbons cited in Table 1 and for whose formation the cited values of $\Delta E_{deloc}/\beta$ correspond.

TABLE 1 Approximate relative carcinogenicity of various hydrocarbons and the highest values of $\Delta E_{deloc}/\beta$ for the formation of benzylic carbonium ions from their tetrahydro benzo-ring derivatives

Parent hydrocarbon	$\Delta E_{deloc}/\beta$ [a]	Approximate relative carcinogenicity[b]
1. naphthalene	0.488	−
2. anthracene	0.544	−
3. benzo[c]phenanthrene	0.600	+[c]
4. tetracene	0.628	−
5. chrysene	0.640	+[c]
6. benzo[b]chrysene	0.647	−
7. phenanthrene	0.658	−
8. picene	0.662	−
9. triphenylene	0.664	−
10. naphtho[2,3-b]pyrene	0.690	++
11. pentacene	0.710	−
12. benzo[c]pyrene	0.714	+[c]
13. dibenzo[a,c]anthracene	0.722	+[c]
14. dibenzo[a,j]anthracene	0.722	+[c]
15. benzo[g]chrysene	0.728	++
16. dibenzo[a,h]anthracene	0.738	++
17. benzo[a]anthracene	0.766	+
18. dibenzo[a,e]pyrene	0.778	+++
19. hexacene	0.782	?[d]
20. benzo[a]pyrene	0.794	++++
21. dibenzo[a,l]pyrene	0.808	++++
22. tribenzo[a,e,i]pyrene	0.818	++
23. dibenzo[a,h]pyrene	0.845	++++
24. benzo[a]tetracene	0.846	−[e]
25. dibenzo[a,i]pyrene	0.870	++++

a) This value corresponds to the highest value calculated for the formation of a benzylic carbonium ion on a tetrahydro benzo-ring of the parent hydrocarbon. The structure of the carbonium ion is shown in Fig. 2. b) These estimates were based upon carcinogenicity indexes cited in Arcos and Argus (34) and Herndon (22). c) These compounds are very weak carcinogens, which are cited to have carcinogenicity index values <5 (34). d) To our knowledge, the carcinogenicity of hexacene has not been determined. e) It may be pertinent that this compound is calculated to have a very reactive "L-region" (21).

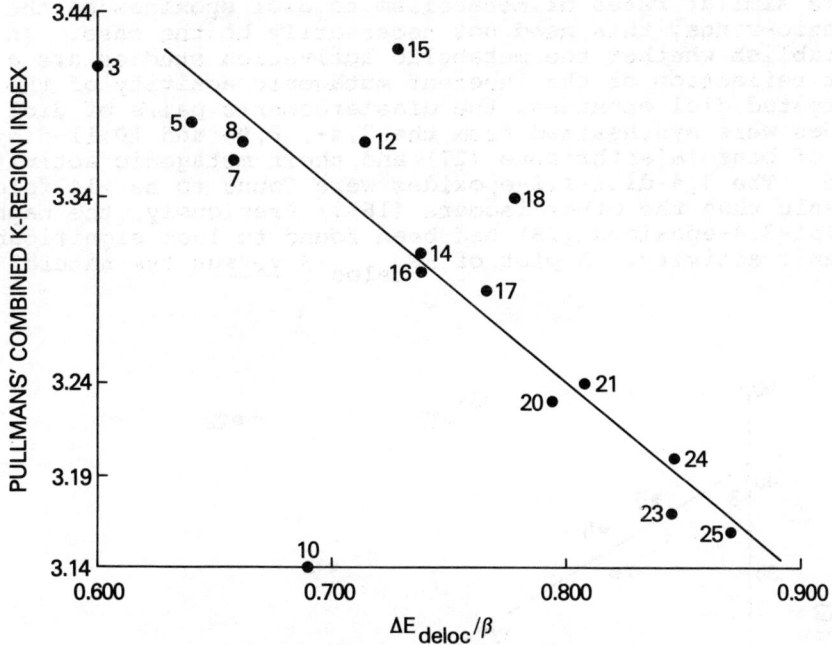

FIG. 3. PLOT OF HIGHEST $\Delta E_{deloc}/\beta$ VALUE (table 1) vs. THE PULLMANS' COMBINED K-REGION INDEX

dihydrodiols or the parent hydrocarbons (24). Since the 3,4-dihydrodiol can form "bay region" epoxides at the 1,2-position, which were predicted to be the most reactive (Fig. 1), the quantum

BA 3,4-diol-1,2-epoxides

mechanical predictions are borne out. Similar support for the "bay region" concept has been found through metabolic activation of dihydrodiols from BP (25,26). As anticipated (Table 1), dihydrodiols from smaller hydrocarbons such as naphthalene, anthracene or phenanthrene could not be activated to produce significant numbers of mutations (16). Although these arguments presume similar rates of metabolism to diol epoxides on the terminal benzo-rings, this need not necessarily be the case. In order to establish whether the metabolic activation studies are a direct reflection of the inherent mutagenic activity of the anticipated diol epoxides, the diastereomeric pairs of diol epoxides were synthesized from the 3,4-, 8,9- and 10,11-dihydrodiols of benzo[a]anthracene (27) and their mutagenic activity was tested. The 3,4-diol-1,2-epoxides were found to be >14-fold more mutagenic than the other isomers (16). Previously, the naphthalene 1,2-diol-3,4-epoxides (28) had been found to lack significant mutagenic activity. A plot of $\Delta E_{deloc}/\beta$ <u>versus</u> the natural

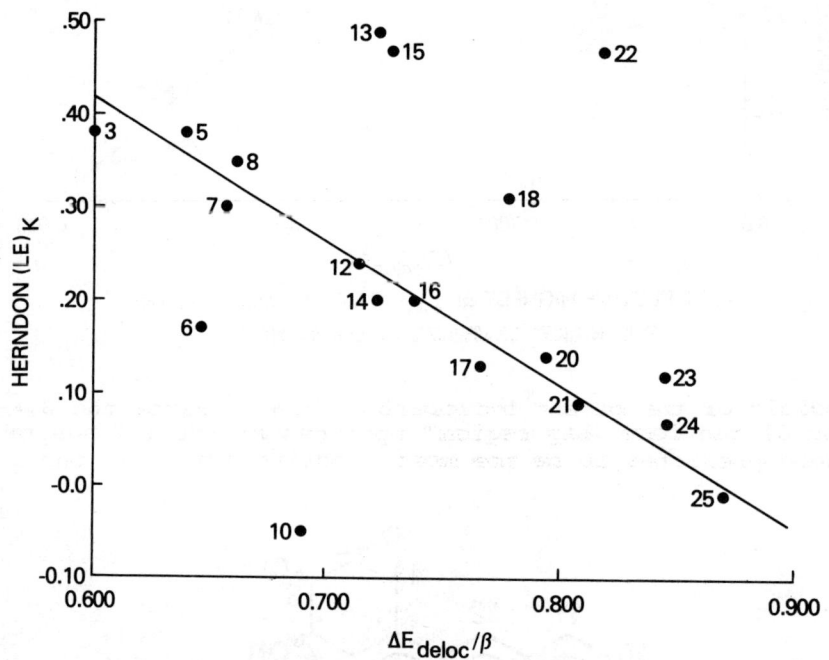

FIG. 4. PLOT OF HIGHEST $\Delta E_{deloc}/\beta$ VALUE (table 1) vs. HERNDON'S K-REGION LOCALIZATION ENERGY

logarithm of the second order rate constant (ln k) for the reaction of a series of diol epoxides with p-nitrothiophenolate in t-butyl alcohol (Fig. 5) established a fairly linear relationship between the calculated parameters and ln k, which indicates that the perturbational molecular orbital calculations do provide a reasonable prediction of chemical reactivity both for diol epoxides from a given hydrocarbon and for diol epoxides from different hydrocarbons. Taken together, the biological and chemical data provide significant support for the "bay region" theory.

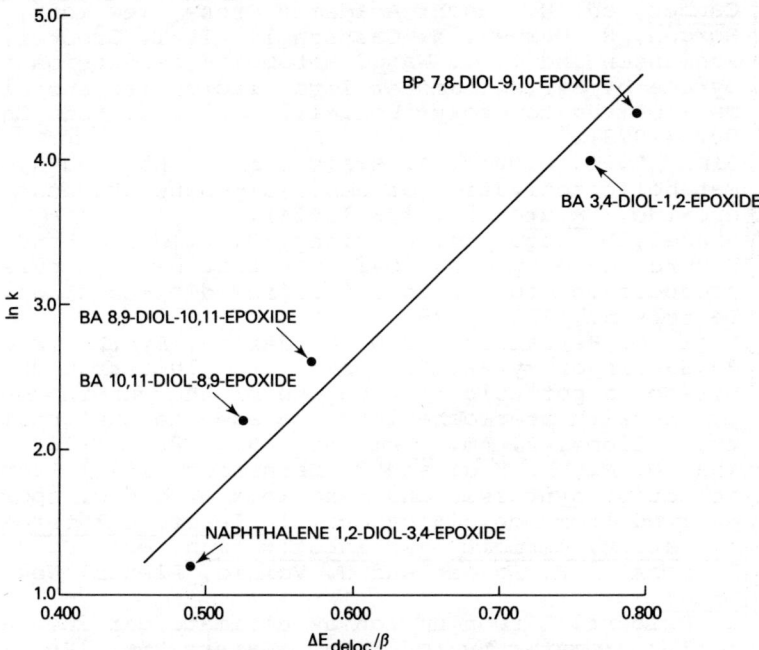

FIG. 5. PLOT OF $\Delta E_{deloc}/\beta$ vs. ln k FOR THE REACTION OF A SERIES OF DIASTEREOMERIC DIOL EPOXIDES (oxirane oxygen cis to the benzylic hydroxyl group) WITH SODIUM P-NITROTHIOPHENOLATE IN t-BuOH

The substantial correlation between the calculations of the "K-region" theory and those of the "bay region" theory is curious since the former is based on the parent hydrocarbon and the latter is based on a specific class of metabolite which requires three enzymatic steps in its formation. Since the rates and specificities of the numerous cytochrome P-450 enzymes (29,30) and of epoxide hydrase (31,32) are unknown for these transformations, the above calculations of quantum mechanical parameters and their correlation with biological activity may be as good as can be expected. Some of the pitfalls of computational approaches to carcinogenicity have been reviewed (33).

ACKNOWLEDGEMENT

We are deeply indebted to our colleagues both from the Laboratory of Chemistry at the NIH and from the Department of Biochemistry and Drug Metabolism at Hoffmann La Roche, whose efforts laid the foundation for the present calculations.

REFERENCES

(1) Miller, E. C. and J. A. Miller (1974) Biochemical mechanisms of chemical carcinogenesis, in The Molecular Biology of Cancer, ed. H. Busch, Academic Press, New York, 377.

(2) A. Borgen, H. Darvey, N. Castagnoli, T. T. Crocker, R. E. Rasmussen and I. Y. Wang, Metabolic conversion of benzo[a]pyrene by Syrian hamster liver microsomes and binding of metabolites to deoxyribonucleic acid, J. Med. Chem. 16, 502 (1973).

(3) P. Sims, P. L. Grover, A. Swaisland, K. Pal and A. Hewer, Metabolic activation of benzo[a]pyrene proceeds by a diol epoxide, Nature 252, 326 (1974).

(4) P. Daudel, M. Duquesne, P. Vigny, P. L. Grover and P. Sims, Fluorescence spectral evidence that benzo[a]pyrene-DNA products in mouse skin arise from diol-epoxides, FEBS Letters 57, 250 (1975).

(5) H. Yagi, O. Hernandez and D. M. Jerina, Synthesis of (±)-$7\beta,8\alpha$-dihydroxy-$9\beta,10\beta$-epoxy-7,8,9,10-tetrahydrobenzo[a]pyrene, a potential metabolite of the carcinogen benzo[a]pyrene with stereochemistry related to the antileukemic triptolides, J. Am. Chem. Soc. 97, 6881 (1975).

(6) Jerina, D. M., H. Yagi and O. Hernandez (1976) Stereoselective synthesis and reactions of a diol epoxide derived from benzo[a]pyrene, in Reactive Intermediates: formation, Toxicity and Inactivation, eds. D. Jollow, J. Kocsis, R. Snyder and H. Vainio, Plenum, New York, in press.

(7) P. B. Hulbert, Carbonium ion as ultimate carcinogen of polycyclic aromatic hydrocarbons, Nature 256, 146 (1975).

(8) Conney, A. H., A. W. Wood, W. Levin, A. Y. H. Lu, R. L. Chang, P. G. Wislocki, R. L. Goode, G. M. Holder, P. M. Dansette, H. Yagi and D. M. Jerina (1976) Metabolism and biological activity of benzo[a]pyrene and its metabolic products, in Reactive Intermediates: Formation, Toxicity, and Inactivation, eds. D. Jollow, J. Kocsis, R. Snyder and H. Vainio, Plenum, New York, in press.

(9) P. G. Wislocki, A. W. Wood, R. L. Chang, W. Levin, H. Yagi, O. Hernandez, D. M. Jerina and A. H. Conney, High mutagenicity and toxicity of a diol epoxide derived from benzo[a]pyrene, Biochem. Biophys. Res. Commun. 68, 1006 (1976).

(10) E. Huberman, L. Sachs, S. K. Yang and H. V. Gelboin, Identification of mutagenic metabolites of benzo[a]pyrene in mammalian cells, Proc. Natl. Acad. Sci. USA 73, 607 (1976).

(11) R. F. Newbold and P. Brookes, Exceptional mutagenicity of a benzo[a]pyrene diol epoxide in cultured mammalian cells, Nature 261, 52 (1976).

(12) A. W. Wood, P. G. Wislocki, R. L. Chang, W. Levin, A. Y. H. Lu, H. Yagi, O. Hernandez, D. M. Jerina and A. H. Conney, Mutagenicity and cytotoxicity of benzo[a]pyrene benzo-ring epoxides, Cancer Res. 36, 3358 (1976).

(13) W. Levin, A. W. Wood, H. Yagi, P. M. Dansette, D. M. Jerina and A. H. Conney, Carcinogenicity of benzo[a]pyrene 4,5-, 7,8- and 9,10-oxides on mouse skin, Proc. Natl. Acad. Sci. USA 73, 243 (1976).

(14) W. Levin, A. W. Wood, H. Yagi, D. M. Jerina and A. H. Conney, (±)-trans-7,8-dihydroxy-7,8-dihydrobenzo[a]pyrene: A potent skin carcinogen when applied topically to mice, Proc. Natl. Acad. Sci. USA 73, in press (1976).

(15) Jerina, D. M., R. E. Lehr, H. Yagi, O. Hernandez, P. M. Dansette, P. G. Wislocki, A. W. Wood, R. L. Chang, W. Levin and A. H. Conney (1976) Mutagenicity of benzo[a]pyrene derivatives and the description of a quantum mechanical model which predicts the ease of carbonium ion formation from diol epoxides, in In Vitro Metabolic Activation in Mutagenesis Testing, eds. F. J. de Serres, J. R. Bend and R. M. Philpot, Elsevier, Amsterdam, in press.

(16) Jerina, D. M., R. E. Lehr, M. Schaefer-Ridder, H. Yagi, J. M. Karle, D. R. Thakker, A. W. Wood, A. Y. H. Lu, D. Ryan, S. West, W. Levin and A. H. Conney (1977) Bay region epoxides of dihydrodiols: A concept which explains the mutagenic and carcinogenic activity of benzo[a]pyrene and benzo[a]anthracene, in Origins of Human Cancer, eds. H. Hiatt, J. D. Watson and I. Winsten, Cold Spring Harbor Laboratory, Cold Spring Harbor, New York, in press.

(17) Jerina, D. M. and J. W. Daly (1976) Oxidation at carbon, in Drug Metabolism, eds. D. V. Parke and R. L. Smith, Taylor and Francis Ltd., London, in press.

(18) Jerina, D. M., H. Yagi, W. Levin and A. H. Conney (1977) Carcinogenicity of benzo[a]pyrene, in Alfred Benzon Symposium X. Drug Design and Adverse Reactions, eds. H. Bundgaard, P. Juul and H. Kofod, Munksgaard, Copenhagen, in press.

(19) G. A. R. Kon and E. M. F. Roe, The direct hydroxylation of 1':2':3':4'-tetrahydro-3:4-benzpyrene, J. Chem. Soc. 143 (1945).

(20) Dewar, M. J. S. (1969) The Molecular Orbital Theory of Organic Chemistry, McGraw Hill, New York, 214-217, 304-306.

(21) A. Pullman and B. Pullman, Electronic structure and carcinogenic activity of aromatic molecules, Adv. Cancer Res. 3, 117 (1955).

(22) W. C. Herndon, Theory of carcinogenic activity of aromatic hydrocarbons, Trans. N.Y. Acad. Sci. 36, 200 (1974).

(23) R. E. Lehr, M. Schaefer-Ridder and D. M. Jerina, Synthesis and properties of the vicinal trans dihydrodiols of anthracene, phenanthrene, and benzo[a]anthracene, J. Org. Chem., in press (1977).

(24) A. W. Wood, W. Levin, A. Y. H. Lu, D. Ryan, S. B. West, R. E. Lehr, M. Schaefer-Ridder, D. M. Jerina and A. H. Conney, Mutagenicity of metabolically activated benzo[a]anthracene 3,4-dihydrodiol: evidence for bay region activation of carcinogenic polycyclic hydrocarbons, Biochem. Biophys. Res. Commun. 72, 680 (1976).

(25) A. W. Wood, W. Levin, A. Y. H. Lu, O. Hernandez, D. M. Jerina and A. H. Conney, Metabolism of benzo[a]pyrene and benzo[a]pyrene derivatives to mutagenic products by highly purified hepatic microsomal enzymes, J. Biol. Chem. 251, 4882 (1976).

(26) C. Malaveille, H. Bartsch, P. L. Grover and P. Sims, Mutagenicity of non-K-region diols and diol-epoxides of benz[a]anthracene and benzo[a]pyrene in S. Typhimurium TA 100, Biochem. Biophys. Res. Commun. 66, 693 (1975).

(27) R. E. Lehr, M. Schaefer-Ridder and D. M. Jerina, Synthesis and reactivity of diol epoxides derived from non-K-region trans-dihydrodiols of benzo[a]anthracene, submitted.

(28) Jerina, D. M., H. Yagi, O. Hernandez, P. M. Dansette, A. W. Wood, W. Levin, R. L. Chang, P. G. Wislocki and A. H. Conney (1976) Synthesis and biologic activity of potential benzo[a]pyrene metabolites, in Polynuclear Aromatic Hydrocarbons: Chemistry, Metabolism and Carcinogenesis, eds. R. Freudenthal and P. W. Jones, Raven Press, New York, 91.

(29) D. A. Haugen, T. A van der Hoeven and M. J. Coon, Purified liver microsomal cytochrome P-450. Separation and characterization of multiple forms, J. Biol. Chem. 250, 3567 (1975).

(30) D. Ryan, A. Y. H. Lu, J. Kawalek, S. B. West and W. Levin, Highly purified cytochrome P-448 and P-450 from rat liver microsomes, Biochem. Biophys. Res. Commun. 64, 1134 (1975).

(31) A. Y. H. Lu, D. Ryan, D. M. Jerina, J. W. Daly and W. Levin, Liver microsomal epoxide hydrase. Solubilization, purification and characterization, J. Biol. Chem. 250, 8283 (1975).

(32) P. Bentley and F. Oesch, Purification of rat liver epoxide hydratase to apparent homogeneity, FEBS Letters 59, 291 (1975).

(33) J. D. Scribner, Molecular orbital theory in carcinogenesis research, J. Natl. Cancer Inst. 55, 1035 (1975).

(34) J. C. Arcos and M. R. Argus, Molecular geometry and carcinogenic activity of aromatic compounds. New perspectives, Adv. Cancer Res. 11, 305 (1968).

THE ACTIVATION AND INDUCTION OF BIPHENYL HYDROXYLATION AND CHEMICAL CARCINOGENESIS

Dennis V. Parke

Department of Biochemistry, University of Surrey, Guildford, U.K.

The phenomenon of drugs stimulating their own metabolism by induction of the microsomal enzymes is now well known and the consequences of long-term administration of drugs are being recognised (1, 2). The microsomal enzymes are also concerned in the activation of many chemical carcinogens and induction of these enzymes may diminish or potentiate the mutagenic and carcinogenic effects of these toxic chemicals, dependent on the relative periods of administration (3,4,5). With the hepatic mixed function oxygenases, three broad classes of enzyme inducing agent have been recognised, namely, drugs - which increase the activities of cytochrome P-450 and its associated reductases, steroids - which increase the activities of the reductases but not cytochrome P-450, and carcinogenic polycyclic hydrocarbons - which increase cytochrome P-448, but not cytochrome P-450, and do not significantly affect the reductases. This induction of the hepatic microsomal mixed function oxidases involves increased synthesis of enzyme proteins and appears to result from a deficiency of enzyme activity consequent upon prolonged occupation of the active site of the enzyme by a surfeit of substrate. Inducers of these enzymes are therefore generally xenobiotics which are slowly metabolised substrates of the mixed function oxidases (6). In the case of carcinogenic inducers, the synthesis of cytochrome P-448 instead of cytochrome P-450 has been attributed to a change in the microsomal environment which results in the conversion of cytochrome P-450 to P-448, these cytochromes being different physical forms of a single molecular species (7).

Biphenyl is a model substrate for the microsomal mixed-function oxidases which can be hydroxylated in two different modes, yielding 2- and 4-hydroxybiphenyl, fluorescent-products which may be determined simultaneously (8). Pretreatment of animals with various foreign compounds produces different effects on the stimulation of these two modes of hydroxylation of biphenyl (9). Phenobarbitone and other non-carcinogenic microsomal enzyme inducing compounds administered to the rat selectively stimulate biphenyl 4-hydroxylation whereas carcinogens such as 3-methylcholanthrene, benzo(a)pyrene, or safrole stimulate the 2-hydroxylation of biphenyl as well as biphenyl 4-hydroxylation. It was later shown that the increase in 2-hydroxylation of biphenyl produced by carcinogens, such as safrole, administered orally to rats comprised two phases, the first occurring between 2-12 hours, and the second occurring subsequent to this (10). The first phase, activation, was independent of de novo protein synthesis, but the second phase was true enzyme induction involving protein synthesis, inhibited by actinomycin or aminotriazole. It has also been shown that the pre-incubation of normal hepatic microsomal preparations with various carcinogens in the presence of an NADPH-regenerating system similarly produces enhancement of biphenyl 2-hydroxylation but not biphenyl 4-hydroxylation. Preincubation of the microsomes with non-carcinogens, however, produced no significant increase in the hydroxylation in either the 2- or 4-positions (11).

To determine this time sequence of increased activity of biphenyl 2-hydroxylase more precisely, the chemical carcinogen, safrole, was administered intraperitoneally, instead of orally as previously. After the intraperitoneal administration an increase in the 2-hydroxylation of biphenyl (activation) was apparent after two hours, whereas biphenyl 4-hydroxylation was even inhibited. The activation of biphenyl 2-hydroxylase reached a maximum 4-6 hours after administration of safrole and, in contrast to the previous oral experiments, fell again to basal values after 10 hours. The second phase of increased biphenyl 2-hydroxylase activity (enzyme induction) occurred at about 24 hours after the intravenous safrole at which time increases in cytochrome P-450 and other parameters of drug metabolism were also seen (12). Intravenous administration of 3-methylcholanthrene or benzo(a)pyrene produced similar effects to safrole. These effects of carcinogens contrast markedly with the effects of the mixed-function oxidases produced by intravenous administration of phenobarbitone, which produced no increase in biphenyl 2-hydroxylase or any other mixed-function oxidase during the first 8 hours after administration. The inhibition of the increased biphenyl 2-hydroxylase activity by actinomycin D in the second phase of stimulation, but not the initial phase, again supports the conclusion that only the second phase is true enzyme induction involving de novo protein synthesis.

The extent of enhancement of biphenyl 2-hydroxylase in the early phase of stimulation by safrole and polycyclic carcinogens administered in vivo shows good correlation with that found when these compounds are added to hepatic microsomal preparations containing an NADPH-regenerating system. In contrast, addition of phenobarbitone to microsomal preparations in vitro produced no significant changes (12).

Activation of the microsomal mixed-function oxidases in vitro is not unique, and has been observed in the 4-hydroxylation of aniline by ethyl isocyanide (13), acetone and pentan-2-one (14), 2,2-bipyridyl (15), volatile anaesthetics (16), and metyrapone (17). However, the activation of biphenyl 2-hydroxylation by addition of carcinogens to microsomal preparations in vitro, does appear to be unique, and no other manifestation of enzymic activity of the microsomal mixed-function oxidases has yet been found which exhibits this activation in vitro exclusively by carcinogens. Possibly, this is due to some special physico-chemical characteristic of the substrate biphenyl.

It is particularly noteworthy that cytochrome P-450, NADPH-cytochrome P-450 reductase and NADPH-cytochrome c reductase are not significantly affected by preincubation of microsomes with benzo(a)pyrene or phenobarbitone in vitro. Moreover, there was no enhancement of the hydroxylation of substrates such as aniline or ethoxyresorufin, which are induced by pretreatment with benzopyrene in vivo (12).

Induction of microsomal enzymes by phenobarbitone is usually characterised by an increase in the apparent Vmax without any change in the apparent Km whereas induction by 3-methylcholanthrene or benzo(a)pyrene usually involves changes in both apparent Vmax and Km (18). In the metabolism of biphenyl by hamster liver microsomes, induction by phenobarbitone caused a five-fold increase in the apparent Vmax of biphenyl 4-hydroxylation with a decrease in Km, and a three-fold increase in the apparent Vmax offset by an eight-fold increase in the apparent Km for the 2-hydroxylation of biphenyl (19). After induction with 3-methylcholanthrene the Vmax for biphenyl 4-hydroxylation was increased 50% with no significant change in the Km; for the 2-hydroxylation no significant change in the apparent Vmax occurred but there was a 75% decrease in the apparent Km. The results of these experiments with hamster liver microsomes were interpreted by Burke & Bridges (19) as follows. In the normal liver microsomes of hamster two sites for the hydroxylation of biphenyl are present, one exclusively for the 4-hydroxylation of biphenyl and the other permitting both 2- and 4-hydroxylation with Km and Vmax values tending to favour 2-hydroxylation. With this model, the selective induction of 4-hydroxylation by phenobarbitone is explained by de novo protein synthesis of microsomal enzymes possessing the former type of hydroxylation sites, probably cytochrome P-450, while the effects of 3-methylcholanthrene and other carcinogens could be explained by assuming the formation of a haemoprotein with the latter hydroxylation site, probably cytochrome P-448.

A study of the kinetic parameters of biphenyl 2-hydroxylation and its stimulation by benzo(a)pyrene, both when administered to Wistar albino rats intraperitoneally and when added to rat liver microsomes in vitro, show that the activation of this hydroxylase is accompanied by both an increase in Vmax and a decrease in Km (12). Moreover, there is good correlation in the extents of increased enzyme activity and kinetic parameters, between activation of biphenyl 2-hydroxylase produced by the in vivo and the in vitro administration of benzo(a)pyrene (see Table 1). At 8 hours after intravenous administration of benzo(a)pyrene the biphenyl 2-hydroxylase activity, together with Km and Vmax, have returned almost to normal. At 24 hours, when true enzyme induction is becoming manifest, the Km is again reduced, with no significant change in Vmax. When a second dose of benzo(a)pyrene was administered to the rats intraperitoneally 6 hours after the first dose, or benzo(a)pyrene was added to liver microsomes obtained 8 hours after intravenous administration of the polycyclic hydrocarbon no further activation occurred, indicating that the phenomenon of activation of biphenyl 2-hydroxylase is finite and not repeatable within the life of the membrane.

These observations suggest that activation of biphenyl hydroxylase is a highly specific change in the physico-chemical state of the endoplasmic reticulum, produced by interaction with carcinogenic chemicals or their metabolites and resulting in a change in the nature of the cytochrome P-450 which, although apparently reversible, cannot be truly so since it is not repeatable.

Evidence has been obtained that carcinogens require to undergo metabolism before they are able to evoke activation of biphenyl 2-hydroxylase activity (20). Although biphenyl 2-hydroxylase is not entirely dependent on NADPH and is able to utilise NADH as coenzyme to a considerable extent, substitution of NADH for NADPH in the in vitro activation by safrole does not result in the usual increase in biphenyl 2-hydroxylase activity. This has been attributed to the absence of safrole metabolism, which has a specific requirement for NADPH, and the lack of activation of the biphenyl 2-hydroxylase is presumably due to the failure to form the active safrole metabolite. Similarly, addition of glutathione diminished the in vitro activation of biphenyl 2-hydroxylase, presumably because of the known interaction of glutathione with the reactive metabolites of safrole (20). Simultaneous addition of oestradiol and benzo(a)pyrene to liver microsomes also diminished the in vitro activation of biphenyl 2-hydroxylase by the polycyclic carcinogen (20). Although this could result from competitive inhibition of the oxidative

TABLE 1 Effects of benzo(a)pyrene in vivo and in vitro on the activation of rat hepatic microsomal biphenyl 2-hydroxylase

Time after intraperitoneal administration (h)	Biphenyl 2-hydroxylase activity (nmol/min per mg microsomal protein)		K_m (10^{-4}M)	V_{max} (nmol/min per mg microsomal protein)
	single dose	repeat dose		
0	0.23 ± 0.02		5.1 ± 2.1	2.2 ± 1.4
4	0.80 ± 0.04		0.2 ± 0.06	7.6 ± 2.1
8	0.28 ± 0.04	0.29 ± 0.01	3.9 ± 1.2	3.2 ± 2.3
24	0.45 ± 0.03	0.49 ± 0.02	1.4 ± 0.9	0.8 ± 0.7
48	0.65 ± 0.02			
Benzo(a)pyrene added in vitro	0.81		0.14 ± 0.2	6.7 ± 0.6

The intraperitoneal dose of benzo(a)pyrene was 75 mg/kg body weight, and 1 mM benzo(a)pyrene was added to rat liver microsomes.

(Data from McPherson et al, 1976a (12).

activation of the benzo(a)pyrene by oestradiol it could also be attributed to a stabilisation of the ribosomes on the endoplasmic reticulum.

Carcinogens in the presence of NADPH are known to cause dissociation of the ribosomes from the endoplasmic reticulum (degranulation), whereas oestradiol favours their association (21). It is thus tempting to equate the activation of biphenyl 2-hydroxylase with degranulation of the endoplasmic reticulum, but certain dissimilarities such as the relative instability and rapid appearance of biphenyl 2-hydroxylase activation compared with degranulation, suggest that these two phenomena although closely related are not strictly identical (20).

Subfraction of liver microsomes of rats and mice has shown that the smooth endoplasmic reticulum (SER) contains a cytochrome with a CO-reduced spectral maximum at 448.8 nm whereas the rough reticulum contains a cytochrome with a maximum at 450.1 nm. However, it was considered that the SER cytochrome is not identical with the cytochrome P-448 produced by pretreatment of animals with carcinogenic hydrocarbons (22). Biphenyl 2-hydroxylase seems to be unique among the microsomal oxygenases in that both the 'basal' and the 'induced' forms of this enzyme are associated solely with cytochrome P-448 (23). For although cytochrome P-448 is able to catalyse both the 2- and 4-hydroxylation of biphenyl, only the 4-hydroxylation is catalysed by cytochrome P-450.

The in vitro activation of biphenyl 2-hydroxylase is currently being developed into a short-term test for chemical carcinogens (11). However, since the 'basal' form of this enzyme, quite unlike the basal forms of benzo(a)pyrene hydroxylase or any other known microsomal oxygenase, is associated solely with cytochrome P-448, activation of biphenyl 2-hydroxylase may also be of value in determining the genetic disposition to chemical carcinogenesis. For the activation of this enzyme would not only measure the potential for increased aryl hydrocarbon hydroxylation, and thus the potential for activation of chemical carcinogens, but may also indicate the genetic disposition to degranulation of the endoplasmic reticulum, a pathological manifestation intimately associated with malignant cell transformation.

At least 80% of human cancer is thought to be of chemical origin and much should be preventable by recognition of these chemicals and elimination of them from the environment (24). Long-term animal studies for the screening of potential chemical carcinogens are extremely time consuming and costly, and furthermore the results obtained are often equivocal and indecisive. Short-term tests of high predictive potential would therefore be of inestimable value in any programme aimed at prophylactically reducing the incidence of human cancer. There is a well-established correlation between carcinogenicity and the alkylation of DNA by environmental chemicals, which may be expressed as gene mutation, changes in gene expression, chromosomal damage, teratology, etc., and the Ames test and other short-term tests based on the detection of DNA damage in bacterial or mammalian cells have proved to have high predictive potential (24).

However, the close correlation between the carcinogenic potential of chemicals and the activation of biphenyl 2-hydroxylase and degranulation of the endoplasmic reticulum suggests that there may be other phenomena of molecular pathology which are intrinsic to malignant cell transformation. In addition to metabolically activating carcinogens the endoplasmic reticulum is also responsible for the synthesis of glycoproteins, which are vital components of

TABLE 2 Effect of pretreatment of rats with carbenoxolone on activation and induction of biphenyl 2-hydroxylase by the carcinogen benzo(a)pyrene.

Pretreatment	Incubation with benzopyrene	Liver biphenyl 2-hydroxylase (nmol/min per mg protein)	percentage increase over controls
None	0	0.32 ± 0.02	
None	1mM	1.14 ± 0.11	+260
Carbenoxolone, 80 mg/kg/day	1mM	0.84 ± 0.08	+160
Carbenoxolone, 160 mg/kg/day	1mM	0.72 ± 0.04	+120
Benzopyrene, 50 mg/kg i.p.	0	1.07 ± 0.06	+235
Benzopyrene, 50 mg/kg i.p. plus carbenoxolone, 80 mg/kg/day	0	0.31 ± 0.01	0

Male Wistar rats were pretreated with carbenoxolone orally for 7 days; benzopyrene was administered once only by i.p. injection, when given, and the animals killed 24 h after the last dose. Liver microsomes were prepared and biphenyl 2-hydroxylase determined in vitro according to McPherson et al, 1976a (12).

the cell surface glycocalyx that determine the mitotic potential of the cell and its immune characteristics. Glycoproteins are also involved in the mechanisms of cellular immuno-surveillance. As a result of degranulation of the endoplasmic reticulum it is likely that the synthesis of glycoproteins will be impaired or their characteristics altered. Moreover, the polyribosomes of the cytosol will be disposed to the synthesis of endogenous proteins which will result in growth of the cell, increased DNA synthesis, and an increased rate of cell division. Degranulation of the endoplasmic reticulum may therefore be associated with a switchover of the basal metabolism of the cell from glycoprotein synthesis and immune competence to a state of rapid growth (hypertrophy) and rapid cell division (hyperplasia) with loss of immune competence, which would potentiate the damage to the DNA and facilitate the establishment of a clone of malignantly-transformed cells (25, 26). It is likely that the activation of biphenyl 2-hydroxylase may be a measure of this process.

Glycoproteins are also found as components of mucus, and carbenoxolone, a drug which is known to stimulate the synthesis of gastric glycoproteins and the production of gastric mucus is also known to reduce the rate of DNA synthesis and cell division of the gastric epithelium (26). Carbenoxolone also inhibits the rate of DNA synthesis of various malignant tumour cell lines in culture (N. Bishun, private communication). It therefore seemed relevant to study the effects of carbenoxolone on the activation by carcinogens of biphenyl 2-hydroxylation. As might have been predicted, microsomal preparations from animals pretreated with carbenoxolone did not show the same extent of carcinogen-induced activation of biphenyl 2-hydroxylase as did microsomes from untreated animals (see Table 2).

It would therefore appear that the activation of biphenyl 2-hydroxylase following the administration of carcinogens in vivo or the incubation of microsomal preparations with carcinogens plus NADPH in vitro leads to the metabolic activation of the carcinogen with the formation of a highly reactive metabolite which, in addition to alkylating the nuclear DNA, also damages the endoplasmic reticulum - the site of formation of the reactive metabolite. As a consequence of this, the metabolism of the cell is switched to the foetal state of rapid growth and rapid cell division which potentiates the damage to the DNA and thus contributes to the malignant transformation (25). If the activation of biphenyl 2-hydroxylase monitors this toxic process this enzyme assay would provide a short-term test for monitoring another aspect of chemical carcinogenesis, that has long been suspected but not identified, and should prove to be a valuable complementary addition to the Ames test and other short-term procedures for detecting potential chemical carcinogens.

REFERENCES

(1) Parke, D.V. Enzyme Induction, Plenum Press, London, 1975.

(2) Breckenridge, A. Enzyme Induction, Plenum Press, London, 1975.

(3) P.H. Grantham, J.H. Weisburger & E.K. Weisburger, Effect of the antioxidant butylated hydroxytoluene (BHT) on the metabolism of the carcinogens N-2-fluorenylacetamide and N-hydroxy-N-2-fluorenylacetamide. Fd.Cosmet.Toxicol., 11, 209 (1973).

(4) C. Peraino, R.J.M. Fry & E. Staffeldt, Reduction and enhancement by phenobarbital of hepatocarcinogenesis induced in the rat by 2-acetylamino fluorene. Cancer Res., 31, 1506 (1971).

(5) R.S. Yamamoto, J.H. Weisburger & E.K. Weisburger, Controlling factors in urethane carcinogenesis in mice: effect of enzyme inducers and metabolic inhibitors. Cancer Res., 31, 483 (1971).

(6) C. Ioannides & D.V. Parke, Mechanism of induction of hepatic microsomal drug metabolizing enzymes by a series of barbiturates. J.Pharm. Pharmac., 27, 739 (1975).

(7) Y. Imai & P. Siekevitz, A comparison of some properties of microsomal cytochrome P-450 from normal, methylcholanthrene and phenobarbital treated rats. Arch.Biochem.Biophys., 144, 143 (1971).

(8) P.J. Creaven, D.V. Parke & R.T. Williams, A fluorimetric study of the hydroxylation of biphenyl in vitro by liver preparations of various species. Biochem.J., 96, 879 (1965).

(9) P.J. Creaven & D.V. Parke, The stimulation of hydroxylation by carcinogenic and non-carcinogenic compounds. Biochem. Pharmac., 15, 7 (1966).

(10) D.V. Parke & H. Rahman, The induction of hepatic microsomal enzymes by safrole. Biochem.J., 119, 53P (1970).

(11) F.J. McPherson, J.W. Bridges & D.V. Parke, In vitro enhancement of hepatic microsomal biphenyl 2-hydroxylation by carcinogens. Nature (Lond), 252, 488 (1974).

(12) F.J. McPherson, J.W. Bridges & D.V. Parke. The effects of benzopyrene and safrole on biphenyl 2-hydroxylation and other drug-metabolizing enzymes. Biochem. J., 154, 773 (1976a).

(13) Y. Imai & R. Sato, Activation and inhibition of microsomal hydroxylation by ethyl isocyanide. Biochem.Biophys.Res.Commun., 25, 80 (1966).

(14) M.W. Anders, Acetone enhancement of microsomal aniline para-hydroxylase activity. Arch.Biochem.Biophys., 126, 269 (1968).

(15) D.R. Buhler & M.E. Rasmusson, The oxidation of drugs by fishes. Comp.Biochem.Physiol., 25, 223 (1968).

(16) K. Korten & R.A. Van Dyke, Acute interaction of drugs - 1. The effect of volatile anesthetics on the kinetics of aniline hydroxylase and amino pyrine demethylase in rat hepatic microsomes. Biochem.Pharmacol., 22, 2105 (1973).

(17) K.J. Netter, G.F. Kahl & K. Magnusson, Kinetic experiments on the binding of metyrapone to liver microsomes. Naunyn-Schmiedeberg's Arch. Pharmakol.Exp.Pathol., 265, 205 (1969).

(18) A.P. Alvares, G.R. Schilling & R. Kuntzman, Differences in the kinetics of benzpyrene hydroxylation by hepatic drug-metabolizing enzymes from phenobarbital and 3-methylcholanthrene-treated rats. Biochem.Biophys. Res.Commun., 30, 588 (1968).

(19) M.D. Burke & J.W. Bridges, Biphenyl hydroxylations and spectrally apparent interactions with liver microsomes from hamsters pre-treated with phenobarbitone and 3-methylcholanthrene. Xenobiotica, 5, 357 (1975).

(20) F.J. McPherson, J.W. Bridges & D.V. Parke, Studies on the nature of the in vitro enhancement of biphenyl 2-hydroxylation provoked by some chemical carcinogens. Biochem.Pharmacol., 25, 1345 (1976b).

(21) D.J. Williams & B.R. Rabin, Disruption by carcinogens of the hormone dependent association of membranes with polysomes. Nature (Lond), 232, 102 (1971).

(22) R.B. Mailman, L.G. Tate, K.E. Muse, L.B. Coons & E. Hodgson, The occurrence of multiple forms of cytochrome P-450 in hepatic microsomes from untreated rats and mice. Chem.Biol.Interactions, 10, 215 (1975).

(23) S.A. Atlas & D.W. Nebert, Genetic association of increases in napthalene, acetanilide and biphenyl hydroxylations with inducible aryl hydrocarbon hydroxylase in mice. Arch.Biochem.Biophys., In publication (1976).

(24) B.A. Bridges, Short term screening tests for carcinogens. Nature (Lond), 261, 195 (1976).

(25) Parke, D.V. A Treatise of Surgical Oncology Plenum Press, London, 1977.

(26) Parke, D.V. & Symons, A. Proceedings of the International Symposium on Mucus Plenum Press, London, 1977.

INTEGRATED ELECTRON TRANSFER: IRON STATES AND REGULATION

I. C. Gunsalus

Biochemistry Department, University of Illinois, Urbana, IL 61801

Integrated electron transfer and substrate hydroxylation is catalyzed by multicomponent monoxygenase systems (1). The heme proteins of these systems, termed P450, exhibit broad ranges of specificity in substrate binding and in the regulation of the energy input required for dioxygen cleavage. All P450's have been found to be a single heme containing polypeptide chain with a molecular weight of approximately 50,000. Multiple forms of the P450 heme protein are inducible in hepatic microsomes, and can be classed by their affinities for various substrates and the ensuing mixed function oxidation velocities (2).

Further descriptive classification of these multiple P450's is made possible due to slight differences in the optical absorption spectra on substrate association and carbon monoxide binding to the ferrous state. Apparently, all accept redox energy directly from pyridine nucleotide driven flavoprotein dehydrogenases with, at least in one case, two electrons being transferred in a single step (3). Although this reaction has received extensive investigation, no identity of a redox acceptor in addition to the obvious heme group has been postulated.

In contrast, the synthetic monoxygenase systems from mammalian adrenal mitochondria (4) and microbial sources (5) are more substrate selective and also exhibit a high specificity for a single electron redox carrier intermediary between the flavoprotein and P450 active center (6). This small molecular weight protein contains an active site of two inorganic sulfur atoms bound to two antiferromagnetically coupled iron atoms which are tied to the polypeptide backbone via cysteine linkages (7). Redox transfer has been found to occur in two one equivalent steps with precise regulation accomplished through modulation of potential via protein and substrate ligation (8,9).

The necessity of multiprotein complexes in both electron transfer and oxygen cleavage reactions has been clearly demonstrated (8,10). As reported by Pederson et al. in this volume, these two required redox transfers show saturation kinetics with maximal rates of ~40 sec^{-1} for the ferric-ferrous heme transition and ~17 sec^{-1} for reduction of the ferrous-oxy intermediate. This latter process, which leads to oxygen cleavage and product formation, contains as scientific puzzles the chemistry of the mixed function oxidation processes and requires further resolution and dissection.

Owing to the common usage of optical spectroscopy to characterize the P450 monoxygenase proteins and for future reference, we present in Table 1 the summary of extinction coefficients for the purified and recrystallized bacterial P450 system (11).

TABLE 1 Extinction Coefficients of Cytochrome $m^{(a)}$ and Putidaredoxin$^{(b)}$

Form	Wavelength (nm)	E_{mM} (mM^{-1} cm^{-1})
m^o	280	68.3
	360	36.7
	417	115.0
	535	11.6
	569	11.9
m^{os} (c)	280	63.3
	391	102.0
	510	13.0
	540	11.2
	646	5.4
m^r (d)	408	76.7
	540	15.1
m^{rs} (c,d)	408	86.5
	542	16.0
m^r_{CO} (d)	447	121.0
	550	14.3
m^{rs}_{CO}	446	120.0
	550	14.0
$m^{rs}_{CO} - m^{rs}$	446-490	92.8
	446	88.0
Pd^o	275	23.6
	280	22.8
	325	15.6
	415	11.1
	455	10.4

ABBREVIATIONS: m^o = oxidized P450; m^{os} = oxidized substrate bound P450; m^r = reduced P450; m^{rs} = reduced substrate bound P450; m^{rs}_{CO} = carbon monoxide substrate adduct of ferrous P450

(a) Determined in 50 mM potassium phosphate pH 7.0 at 25°

(b) Determined in 50 mM Tris-Cl pH 7.4, 10 mM 2-mercaptoethanol at 25°

(c) D-(+) camphor added to a concentration of 100 μM

(d) Reduction with $Na_2S_2O_4$

In all aspects compared to date, the adrenal cortex steroid 11β hydroxylase system as isolated by Katagiri (4), Kimura (12), and others conforms to reaction processes of the microbial P450 system. The cholesterol side chain cleavage system, although representing a multimeric association of hydroxylation on activities, also appears to have similar features.

Two important advances relating to the overall reaction scheme of hepatic P450 have emerged recently, and are discussed in this volume. One relates to the stimulatory effect of cytochrome b_5 in the generation of hydroxylated substrate and the second to a specific requirement for choline phosphoglyceride in both product formation, electron transfer, and hydrogen peroxide generation. This latter reaction, occurring concomitant with substrate hydroxylation, may be viewed as an uncoupling or suboptimal product of pyridine nucleotide oxidation. As such, it represents a partition between the reversible dissociation of oxygen from the ferrous enzyme and autoxidation via superoxide or H_2O_2 production with a ferric heme end product.

Perhaps one of the most exciting discoveries during the past two years has been the demonstration that the microsomal P450 heme proteins, in the ferric state, as well as the microsomes from which they are isolated, are capable of utilizing peroxides or peracids to form oxygenated product (13). Similar reactions are also observed with the microbial camphor metabolizing system.

Sligar et al., reporting in this volume, describe a spectral intermediate in this reaction and document a plausible scheme for the hydroxylation that utilizes an acylation of bound dioxygen by an available carboxylic acid group. Two pathways of substrate hydroxylation are then possible. One is via an oxene species, $[FeO]^{+3}$, which is analogous to the compound I intermediate observed in peroxidases; the other involves an enzyme bound peracid. The former has been discussed at length by Estabrook (13), Coon (14), and others (15,16), and the latter by Hamilton (17). Further studies of the possible compound I type intermediates in both the mammalian and microbial P450's are certainly required and should represent an exciting area of exploration in the coming years.

Studies on the microsomal and bacterial P450 systems have also been presented at this meeting by Douzou, Debey, and their colleagues. Using low temperature and mixed solvent techniques, they have been able to stabilize the ferrous-oxygenated forms of the enzyme both with and without substrate. Using the microbial P450, kinetic studies of the breakdown of this intermediate on further electron donation to form hydroxylated substrate and water have been carefully documented. Detailed analysis has again demonstrated the essentiality of the P450-iron-sulfur protein complex in the redox and product forming couples.

In summary, the detailed mechanistic and thermodynamic studies of the P450 heme proteins, the flavoprotein reductases, and, in the mitochondrial and microbial systems, the $Fe_2S_2^*Cys_4$ redoxins, reflect the essential features of the two electron monooxygenase reactions and the role of multienzyme complexes in the various processes. These systems provide an ideal model for the chemistry of the centers active in oxygen cleavage and electron transfer. We look forward to further extensions in the understanding of the biological systems represented by this important enzymatic process which is so generally distributed and essential to cellular processes.

REFERENCES

(1) Gunsalus, I. C., Pederson, T. C., and Sligar, S. G., Oxygenase-Catalyzed biological hydroxylations, Annual Review of Biochemistry 44, 377 (1975).

(2) Haugen, D. A., van der Hoeven, T. A., and Coon, M. J., Purified liver microsomal cytochrome P450 -- Separation and characterization of multiple forms, J. Biol. Chem. 250, 3567 (1975).

(3) Guengerich, F. P., Ballou, D. P., and Coon, M. J., Purified liver microsomal cytochrome P-450 -- Electron accepting properties and oxidation reduction potential, J. Biol. Chem. 250, 7405 (1975).

(4) Katagiri, M., Takemori, S., Tagaki, E., and Suhara, K, Methods in Enzymology, Biological Oxidations Part C (In press).

(5) Hedegaard, J., and Gunsalus, I. C., Mixed function oxidation: IV. An induced methylene hydroxylase in camphor oxidation, J. Biol. Chem. 240, 4038 (1965).

(6) Tyson, C. A., Lipscomb, J. D., and Gunsalus, I. C., The roles of putidaredoxin and P450 complexes in methylene hydroxylation, J. Biol. Chem. 247, 5777 (1972).

(7) Gunsalus, I. C., Meeks, J. R., Lipscomb, J. D., Debrunner, P. G., and Münck, E, Bacterial monoxygenases -- The P450 cytochrome system, In Molecular Mechanisms of Oxygen Activation (Osamu Hayaishi, ed.), Academic Press, New York, 1974.

(8) Sligar, S. G., and Gunsalus, I. C., A thermodynamic model of regulation: Modulation of redox equilibria in camphor monoxygenase, Proc. Nat. Acad. Sci. USA 73, 1078 (1976).

(9) Sligar, S. G., Coupling of spin, substrate and redox equilibria in cytochrome P450, Biochem. 15, 5399 (1976).

(10) Lipscomb, J. D., Sligar, S. G., Namtvedt, M. J., and Gunsalus, I. C., Autoxidation and hydroxylation reactions of oxygenated cytochrome $P450_{cam}$, J. Biol. Chem. 251, 1116 (1976).

(11) Garg, G., Gunsalus, I. C., Toscano, W. A., and Wagner, G., J. Biol. Chem. (In press) (1976).

(12) Kimura, T., and Suzuki K., J. Biol. Chem. 242, 485 (1967).

(13) Rahimtula, A. D., O'Brien, P. J., Hrycay, E. G., Peterson, J. A., and Estabrook, R. W., Possible higher valence states of cytochrome P-450 during oxidative reactions, Biochem. Biophys. Res. Comm. 60, 695 (1974).

(14) Nordblom, G., White, R., and Coon, M. J., Arch. Biochem. Biophys. 175, 524 (1976).

(15) Gustafson, J., Hrycay, E., and Ernster, L., Arch. Biochem. Biophys. 174, 440 (1976).

(16) Rahimtula, A. D., and O'Brien, P. J., Hydroperoxide dependent O-Dealkylation reactions catalyzed by liver microsomal cytochrome P450, Biochem. Biophys. Res. Comm. 62, 268 (1975).

(17) Hamilton, G., In Molecular Mechanisms of Oxygen Activation (O. Hayaishi, ed.), Academic Press, New York, 1974, P. 405.

PURIFICATION OF LIVER MICROSOMAL CYTOCHROME P-450: HOPES AND PROMISES

W. Levin

Department of Biochemistry and Drug Metabolism, Hoffmann-LaRoche Inc., Nutley, New Jersey 07110

During the past 20 years, many investigators have studied the metabolism of endogenous substrates, drugs and chemical carcinogens by a group of nonspecific enzymes localized in the endoplasmic reticulum of liver and other tissues. The versatility of these enzymes is unique in biochemistry, since they catalyze the metabolism of an almost limitless number of compounds through widely diverse types of reactions, such as the oxidation of alkanes, the epoxidation of aromatic compounds, the dealkylation of secondary or tertiary amines, deaminations, dehalogenations and the oxidative cleavage of ethers and organic thiophosphate esters. The metabolism of foreign compounds by mammalian liver was first described by Mueller and Miller (1,2), Axelrod (3,4) and Brodie et al. (5). These investigators established that the enzyme system is localized in liver microsomes and that the reaction requires both NADPH and molecular oxygen. Studies by Conney et al. (6,7) and Remmer and coworkers (8,9) established that the metabolism of drugs and foreign compounds by liver microsomes can be enhanced by the in vivo treatment of animals with a variety of compounds. Cooper, Estabrook and coworkers (10,11) demonstrated the role of cytochrome P-450 in these enzymatic reactions. The importance of this enzyme system for the metabolism of steroids, drugs and other xenobiotics has become increasingly apparent during the last 10 years. Today we live in a society that has become increasingly aware of the potential dangers of drugs, pollutants, food additives and carcinogenic compounds. Research aimed at a better understanding of the metabolic effects evoked by such compounds has grown at a rapid pace since the First International Symposium on Microsomes and Drug Oxidations was held eight years ago (12).

Perhaps the greatest challenge we as pharmacologists, toxicologists and biochemists face in the future is to define and better understand man's ability to metabolize such substances and to better understand how to control and regulate their toxic effects at the cellular and molecular level. Although much has been learned about the inducibility, specificity and other properties of this enzyme system through in vitro studies with liver microsomes, it seems likely that some of the more puzzling questions will not be answered without the purification and characterization of the individual components of this enzyme system. Perhaps one of the more complex aspects of this work has centered around the multiplicity of cytochrome P-450's in the endoplasmic reticulum of liver and other tissues. Of the three components of the microsomal enzyme system (cytochrome P-450, NADPH-cytochrome c reductase and lipid) involved in drug metabolism, cytochrome P-450 is undoubtedly the most important because of its vital role in oxygen activation, substrate binding and in determining the overall substrate specificity of the enzyme system. The broad substrate specificity of this enzyme system and the effects of age, sex, drug pretreatment and the species and strain of animal used for metabolism studies indicate that a substantial number of cytochromes exist which are under different regulatory control. As this Symposium draws to a close, I would like to discuss some advances that have been made in characterizing multiple forms of cytochrome P-450 and to speculate about some of the challenges that we face in future years to more fully understand the complex role of this rather unique enzyme system, which plays a dual metabolic role--the pharmacological and toxicological inactivation of a variety of xenobiotic compounds and the metabolic activation of other foreign compounds resulting in the formation of highly toxic products. It has been estimated that 60-80% of all human cancers are caused by environmental factors (13-15), and many chemicals in our environment are metabolically activated to ultimate carcinogens by the mixed-function oxidase system.

Purification of Cytochrome P-450

In 1968, Lu and Coon and co-workers (16,17) first solubilized the liver microsomal hydroxylation system and chromatographically resolved the system into three components which were identified as cytochrome P-450, NADPH-cytochrome c reductase and phosphatidylcholine. All three components have since been shown to be required for the metabolism of a variety of substrates, including drugs, chemical carcinogens, steroids and fatty acids (18). Since the procedure used for the initial solubilization and resolution of the components of the hydroxylation system was not necessarily an optimal first step for the further purification of each individual component, alternate methods have been developed for the solubilization and purification of these enzymes (19-27). These procedures generally involve the solubilization of liver microsomal protein with sodium cholate, eventually followed by column chromatography in the presence of a non-ionic detergent such as Emulgen (19,22, 24-27) or Renex (20,21,23). Liver microsomal cytochrome P-450's from several different species have now been highly purified (Table 1). Although several of these studies have also resulted in the isolation of different forms of liver microsomal cytochrome P-450 from the same animal, the problem of isolating and identifying all of the forms of cytochrome P-450 that exist in an animal species has not yet been resolved. Perhaps the four most important criteria that have been used in identifying multiple forms of cytochrome P-450 are spectral properties, catalytic activity, molecular weight using SDS-polyacrylamide

Table 1. Highly Purified Forms of Liver Microsomal Cytochrome P-450[a]

Species	Preatment	Cytochrome	Absorption Maximum of Reduced CO-Complex	Specific Content (nmoles/mg p)	References
Rabbit	Phenobarbital	P-450	450	17	Imai & Sato (19)
	Phenobarbital	P-450 LM$_2$	451	18	van der Hoeven et al (20)
	Phenobarbital	P-450 LM$_4$	448	14	Haugen et al. (23)
	ß-Naphthoflavone	P-450 LM$_4$	447.5	12	Haugen et al. (23)
	3-Methylcholanthrene	P-448	447	18	Kawalek et al. (24)
	Control	P-450	448	14	Philpot & Arinc (25)
Rat	Phenobarbital	P-450	450	17	Ryan et al. (22)
	3-Methylcholanthrene	P-448	447	18	Ryan et al. (22)
	Aroclor 1254	P-450 & P-448	448	17	Ryan et al. (27)
Mouse	Phenobarbital	P-450 A$_2$	451	16	Huang et al. (26)
	Phenobarbital	P-450 C$_2$	450	17	Huang et al. (26)

[a]Several other forms of liver microsomal cytochrome P-450 have been partially purified and are reported in the above references.

gel electrophoresis and antigenic properties. Ideally, at least several of these parameters should be used to establish the multiplicity of cytochrome P-450.

A. *Spectral properties.* Of the different criteria available to establish the multiplicity of cytochrome P-450, differences based only on spectral properties are difficult to interpret and may be more subject to error than other criteria. Although spectral parameters such as ethyl isocyanide binding and the absorption maximum of the reduced-CO complex of cytochrome P-450 have been useful in establishing different forms of purified cytochrome P-450, the effect of lipid, detergents and other contaminating materials may cause problems in the interpretation of other spectral parameters. For example, highly purified cytochrome P-448 from 3-methylcholanthrene-treated rats and rabbits have identical CO-difference spectra (Fig. 1A), but the purified rat hemeprotein is almost exclusively a low-spin ferric hemeprotein (28,29), whereas the hemeprotein purified from rabbits exists as a mixture of high- and low-spin forms (24,30) that can be interconverted by the use of different detergents (Fig. 1B and 1C). These same detergents have no effect on the absolute oxidized spectrum of rat cytochrome P-448.

Haugen and Coon (31) have purified a cytochrome P-448 (P-450 LM_4) from β-naphthoflavone-treated rabbits, using a different purification procedure, and have obtained similar results. Is this spectral difference in the hemeprotein purified from two different species a property of the protein <u>per se</u> or due to some small molecule which is tightly bound to the rabbit enzyme which is not removed during purification? This question remains unanswered at the present time, but it appears that the high-spin nature of rabbit cytochrome P-448 (P-450 LM_4) is unrelated to pretreatment with 3-methylcholanthrene or β-naphthoflavone, since cytochrome P-450 LM_4 purified from untreated or phenobarbital-treated rabbits has similar spectral properties (31). Fortunately, in the case of rabbit cytochrome P-448 and rat cytochrome P-448, other criteria (molecular weight on SDS gels, catalytic activity and antigenic properties) have been used to establish that these two hemeproteins have different apoproteins (24).

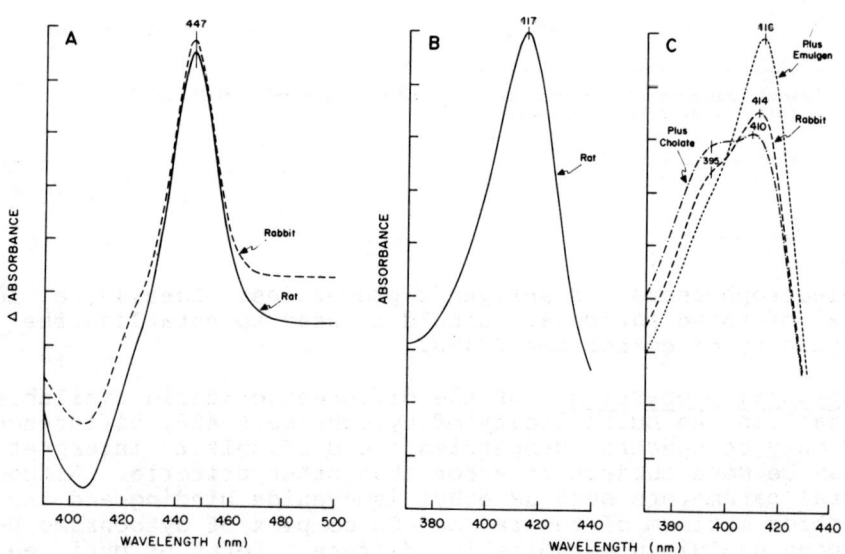

Fig. 1   CO-difference spectra (A) and absolute oxidized spectra of highly purified rat (B) and rabbit (C) cytochrome P-448 (22,24). In Fig. 1C, 0.5 mg of sodium cholate or Emulgen 911 was added per nmole cytochrome P-448.

Spectral properties have, nevertheless, been used as strong evidence for the presence of multiple forms of cytochrome P-450 in the same animal (23,32,33). Comai and Gaylor (32) separated three forms of cytochrome P-450 from liver microsomes as judged by their ability to bind cyanide. The microsomes were treated with protease, solubilized with deoxycholate and fractionated on a DEAE-cellulose column. Based on the affinity of cytochrome P-450 to bind cyanide, three forms of the hemeprotein having binding constants of 0.5, 1.5 and 5.0 mM were eluted from the column. The three forms of cytochrome P-450 were induced to different extents by prior treatment of the animals with 3-methylcholanthrene, phenobarbital or ethanol. Ryan et al. (33) separated at least two forms of cytochrome P-450 from both phenobarbital and 3-methylcholanthrene-treated rats after solubilization of liver microsomes with sodium cholate and column chromatography in the presence of Emulgen 911. These forms of cytochrome P-450 had different CO-maxima and different ethyl isocyanide binding properties. Similar differences in the CO-absorption maxima of multiple forms of cytochrome P-450 isolated from the same animal have been obtained by Haugen et al. (23). Philpot and Arinc (25) and Huang et al. (26).

B. Catalytic activity. Although one of the most useful criteria for characterizing multiple forms of cytochrome P-450 is the differential ability of various forms to catalyze the metabolism of a variety of substrates, problems can also arise using this method of identification. One major problem is that the different forms of cytochrome P-450 that have been isolated to date have different, but overlapping, substrate specificity (Table 2). For example, highly purified cytochrome P-450 and P-448 from phenobarbital- or 3-methylcholanthrene-treated rats (22) metabolize a variety of substrates at different rates, but each form has at least some activity for most substrates that have been assayed to date. This can make interpretation of results with respect to catalytic activity difficult if one of the separated forms has poor catalytic activity towards many substrates when compared to another form separated from the same animal. In the studies of Ryan et al. (33), Haugen et al. (23) and Philpot and Arinc (25), one of the separated forms of cytochrome P-450 (or P-448) had a lower catalytic activity for all substrates studied than the other form. This could be the result of an alteration in the hemeprotein during isolation, the presence of high amounts of detergent or other factors which are inhibitory, or that the catalytic activity of the hemeprotein was not studied with a sufficient number of substrates. It is also possible that the particular form isolated has an inherently poor activity for all substrates, but this seems less likely. Haugen et al. (23) and Huang et al. (26) have succeeded in isolating different forms of cytochrome P-450 from phenobarbital-treated rabbits and mice which show true differential substrate specificity towards different substrates (Table 2). That is, at least one of the isolated forms of the hemeprotein has better catalytic activity towards some substrates compared to another form and vice versa. This is

Table 2. Substrate Specificity of Various Forms of Liver Microsomal Cytochrome P-450

Substrate	Rat[a]		Rabbit[b]			Mouse[c]			
	P-450	P-448	LM_2	LM_4	$LM_{1,7}$	A_1	A_2	C_1	C_2
Benzphetamine	30-50	2-3	66	3	7.5	6	5	45	8
Benzo[a]pyrene	0.1-0.2	2.5-5.0	0.04	trace	0.5	0.3	0.3	0.1	0.1
Ethylmorphine	5-7	N.D.[d]	6.1	3.0	3.0	1	5	3	12
Testosterone									
7α-OH	0.7	1.0	0.02	0.02	0.02	0.06	0.73	0.08	0.15
16α-OH	1.5	0.2	0.43	0.02	0.07	<0.01	0.03	1.64	0.06
6α-OH	0.1	0.2	0.02	0.02	0.32	0.09	0.01	0.93	0.88

[a]Taken from the data of Ryan et al. (22) and Thomas et al. (35).
[b]Taken from the data of Haugen et al. (23).
[c]Taken from the data of Huang et al. (26).
[d]N.D. = not determined.

also clearly the case when comparing the catalytic activity of cytochrome P-450 and P-448 purified from rats treated with either phenobarbital or 3-methylcholanthrene (Table 2).

Another major problem in determining the catalytic activity of a purified cytochrome P-450 arises when the purified preparation has poor catalytic activity for a substrate compared to microsomes, as is the case for ethylmorphine N-demethylation and the 6β-hydroxylation of testosterone by purified cytochrome P-450 from phenobarbital-treated rats. The simplest explanation for this observation is that a form(s) of cytochrome P-450 has been lost during the purification procedure.

Alternatively, a particular form may have been inactivated during purification but may still be present in the purified preparation. Another possibility is that reconstitution of the system with phosphatidylcholine and NADPH-cytochrome c reductase may be quite different for some substrates when compared to other substrates. Interestingly, antibody prepared against purified cytochrome P-450 was equally effective in inhibiting the microsomal metabolism of both benzphetamine and ethylmorphine N-demethylation. Since we have demonstrated the specificity of the antibody preparation using both the reconstituted system (34,35) and liver microsomes (36), it may be that the form of cytochrome P-450 responsible for most of the ethylmorphine activity is present in the purified preparation, or two different cytochromes with similar antigenicity catalyze the demethylation of both ethylmorphine and benzphatamine.

C. *SDS-polyacrylamide gel electrophoresis*. SDS-gel electrophoresis has been one of the most useful methods available to determine the purity of a membrane-bound protein. Since membrane-bound proteins such as cytochrome P-450 usually contain some detergent and tend to aggregate upon purification, other determinants of molecular weight, such as Sephadex columns and ultracentrifugation, are usually of no value in determining purity. In addition, SDS gel electrophoresis offers the potential of separating proteins that differ in molecular weight by as little as 1,000 daltons.

Alvares and Siekevitz (37) first described differences in the minimum molecular weight of the major protein component of partially purified preparations of cytochrome P-450 and P-448 obtained from phenobarbital- and 3-methylcholanthrene-treated rats, respectively. The studies of Ryan et al. (22) using highly purified rat liver cytochrome P-450 and P-448 have established that these proteins differ in molecular weight by approximately 5,000. Similar studies by Haugen et al. (23) have shown that rabbit cytochrome P-450 and P-448 also differ in molecular weight by approximately 4,000. With the increased use of SDS gel electrophoresis to establish different forms of cytochrome P-450, it became increasingly apparent that different gel systems varied considerably in their ability to separate proteins in the 40,000-60,000 molecular weight range (the minimum molecular weight range of all highly purified preparations of cytochrome P-450 described to date). As gel systems with better resolution were used, the problem of multiple forms of cytochrome P-450 became even more complex. It is now apparent that several preparations of apparently pure cytochrome P-450, as determined by a single protein band in some gel systems, contain several proteins differing in molecular weight by as little as 1,000 (Fig. 2). For example, highly purified cytochrome P-450 obtained from phenobarbital-treated rats and cytochrome P-448 obtained from rats (22) or rabbits (24) treated with 3-methylcholanthrene showed a single band of 50,000 molecular weight, using the system of Weber and Osborn (38). This gel system does not detect any difference in the molecular weight of cytochrome P-450 and cytochrome P-448 (39). However, the purified cytochrome P-450 and P-448 obtained from rats can be separated (22) using the gel system of Neville (40). In addition, a qualitatively minor protein band is now apparent in each preparation that differs in molecular weight from the major component by 4,000-5,000 (22). Using this gel system, purified cytochrome P-448 from rats and rabbits have apparently identical molecular weights of 53,000. Even if longer gels were used in the system of Neville (40), no separation of these two proteins could be observed. Perhaps the SDS-gel system which results in the grestest resolution of proteins in this molecular weight range is the system of Laemmli (41), which is capable of clearly separating rat and rabbit cytochrome P-448 (24). These hemeproteins differ in molecular weight by 2,000 in this gel system (24). This gel system has resolved proteins in the highly purified cytochrome P-450 preparation from phenobarbital-treated rats into four proteins having molecular weights of 43,000-47,000, all of which are probably hemeproteins (see below). This gel system has been used by Haugen et al. (23) to establish the purification and separation of different molecular

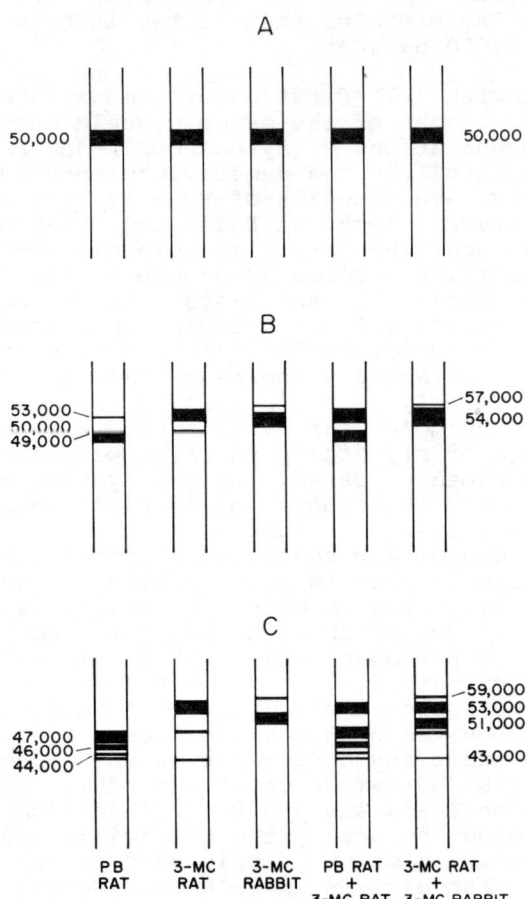

Fig. 2 SDS-gel electrophoresis of highly purified cytochrome P-450 and P-448 from phenobarbital- (PB) and 3-methylcholanthrene-treated (3-MC) rats and rabbits (22,24). A, B and C represent the gel profiles obtained using the SDS-gel electrophoresis methods of Weber and Osborn (38), Neville (40) and Laemmli (41), respectively.

weight forms of cytochrome P-450 and P-448 obtained from rabbits.

In addition to the different resolving properties of the various SDS-gel systems, each give slightly different molecular weights for the same protein when using the same protein standards. These results clearly indicate that molecular weights obtained by different investigators using a variety of gel systems cannot be directly compared. Perhaps more important is the resolving power of the gel system used to determine the purity of the enzyme preparation, and whether multiple forms of cytochrome P-450 are present in the same preparation. If a highly purified cytochrome P-450 preparation such as that reported in Fig. 2 shows two to three minor protein components, how does one know if these are contaminating proteins or other forms of cytochrome P-450? One of the most helpful methods in detecting cytochrome P-450 on SDS gels is the benzidine-H_2O_2 staining procedure originally described by Welton and Aust (42). This method has permitted the detection of small amounts of cytochrome P-450 (as cytochrome P-420) on SDS-gels, based on the peroxidase nature of this hemeprotein (43). The benzidine-H_2O_2 method has been very useful but has some shortcomings, such as the poor photographic quality of the gels and the very rapid loss of color from the stained bands. We have recently developed an improved procedure (44) using tetramethylbenzidine and have been able to modify the method so that the gels possess a clear background, the stained bands are stable for several weeks, and the same gel can be destained and restained for protein. Using this method, we have established that the four protein bands observed in the cytochrome P-450 preparation from phenobarbital-treated rats are hemeproteins, based on their peroxidase activity.

As with the other parameters that have been used to establish the multiplicity of cytochrome P-450, SDS-gel electrophoresis also has some problems that have not yet been reconciled. It is well known that some proteins bind SDS differentially compared to other proteins, which can influence the mobility of the protein (45,46). Obviously, the buffer system used, the pH of the buffers, the degree of cross-linking of the gels, the use of sulfhydryl reagents and heating the sample prior to application to the gels are influencing factors in the mobility of proteins (45,47-50). Given the number of forms of cytochrome P-450 that have thus far been described, it is reasonable to assume that not all forms of cytochrome P-450 will differ sufficiently in molecular weight to be separated by SDS-gel electrophoresis alone. Thus, other more sensitive criteria for establishing whether a particular purified preparation of cytochrome P-450 contains a single protein species will have to be developed.

D. <u>Immunological</u> <u>properties</u> <u>of</u> <u>cytochrome</u> P-450. One of the most encouraging recent developments for characterizing multiple forms of cytochrome P-450 has been the production of antibodies against highly purified preparations of cytochrome P-450 and

P-448 from phenobarbital (PB)- and 3-methylcholanthrene (MC)-treated rats (34,35). These antibodies have been shown to be quite specific in their reactions with cytochrome P-450 obtained from different sources (Table 3). Anti-PB-P-450 and anti-MC-P-448 react well with their homologous antigens but show poor cross-reaction with their heterologous antigens. Rabbit cytochrome P-448 cross-reacts poorly with the antibody prepared against rat cytochrome P-448, suggesting that these hemeproteins have different antigenic sites (24,34). Thus, rabbit cytochrome P-448 and rat cytochrome P-448 have been shown to be different hemeproteins as judged not only by their different catalytic and physical properties, but also by their different immunological properties (24,34). Similarly, rabbit cytochrome P-450 LM_2 is immunochemically related to, but not identical with, rat cytochrome P-450 (35). In addition, rabbit cytochrome P-450 has been shown to be immunologically different from rabbit cytochrome P-448 (51). Both rat antibodies show poor, or no cross-reaction with purified cytochrome P-450 obtained from $B6D2F_1$ mice, beef adrenal mitochondrial cytochrome P-450 and Pseudomonas putida cytochrome P-450 (Table 3). These results demonstrate the high degree of specificity of the antibodies prepared against rat cytochrome P-450 and P-448.

Table 3. Ouchterlony Double-Diffusion Analysis of Various Forms of Cytochrome P-450 with Antibodies Prepared Against Rat Cytochrome P-450 and P-448

Species	Pretreatment	Cytochrome	Relative Extent of Precipitin Band Formation	
			Anti-P-450	Anti-P-448
Rat	Phenobarbital	P-450	+++	+
	3-Methylcholanthrene	P-448	+	+++
	Aroclor 1254	P-450 & P-448	++	++
Rabbit	Phenobarbital	P-450 LM_2[a]	+	-
	3-Methylcholanthrene	P-448	-	+
Mouse	Phenobarbital	P-450 A_2	±	±
	Phenobarbital	P-450 C_2	±	±
Beef Adrenal Mitochondria		P-450[b]	-	-
Pseudomonas putida		P-450[c]	-	-

Antibodies were prepared against highly purified rat liver cytochrome P-450 and P-448 from phenobarbital and 3-methylcholanthrene-treated rats, respectively (34,35). The extent of precipitin band formation is recorded on a 3-point scale with "+++" indicating strong reaction and "-" indicating no visible reaction.

[a] Kindly supplied by Dr. M. J. Coon.
[b] Kindly supplied by Dr. D. Y. Cooper.
[c] Kindly supplied by Dr. J. Peterson.

Perhaps the most intriguing result obtained using these two
antibodies has been the suggestion that the purified cytochrome
P-450 preparation from phenobarbital-treated rats has at least
four forms of cytochrome P-450 and the purified cytochrome P-448 preparation from 3-methylcholanthrene-treated rats contains
at least two forms of cytochrome P-448 or P-450 (34,35). It
is interesting to note that the SDS-gel system of Laemmli (41)
also confirms this observation (see Fig. 2). The use of antibodies prepared against different forms of cytochrome P-450
holds great promise in the identification and characterization
of multiple forms of cytochrome P-450 in liver and extrahepatic
tissues of different species.

Concluding Remarks

The results of studies performed in many laboratories during
the last decade have established the multiplicity of liver microsomal cytochrome P-450 in the metabolism of endogenous and foreign compounds. I have tried to summarize some of this data
with specific reference to some of the problems involved in
the purification of multiple forms of cytochrome P-450. Although much has been learned through the use of these procedures,
a great deal of work is still required to identify and characterize the many forms of this unique hemeprotein. To my knowledge,
no one has yet separated and identified all of the forms of
liver microsomal cytochrome P-450 present in a single animal
species. It is quite apparent that improved purification procedures which result in a higher yield of purified hemeprotein
are needed to accomplish these goals. Development of such procedures will help our understanding of the species, age and
sex differences that exist in the metabolism of xenobiotics.
Indeed, even more difficult will be the purification and characterization of cytochrome P-450 from extrahepatic tissues.
The recent results of Arinc and Philpot (52) on the partial
purification of a cytochrome P-450 from rabbit lung is the first
successful attempt in this regard.

Hopefully, a better understanding of these enzymes at the molecular level will provide insight into problems involved in extrapolating to man the pharmacological and toxicological effects
of compounds in animal studies. For example, how accurately
do the results of studies with a chemical that is metabolically
activated to a carcinogen in an animal predict the toxicity
of the chemical in man? This is perhaps one of the most important questions we face in the future in light of the vast
number of environmental contaminants in man's environment.
Although our present knowledge of the cytochrome P-450 system
does not permit us to answer such questions, it is likely that
as our understanding of this unique enzyme system expands, we
shall be able to extrapolate the results of animal data to man
with greater certainty than we can at the present time. Needless to say, the eventual separation, purification and characterization of multiple forms of cytochrome P-450 from human
tissues will be of great value in determining the metabolic
factors involved in the activation of chemical carcinogens in
man.

Acknowledgment

I am grateful to my colleagues Anthony Lu, Dene Ryan, Susan West, Joseph Kawalek, Mou-Tuan Huang and Paul Thomas for their contributions to many of the studies described in this paper. I wish to thank Mrs. Candace Caso and Mrs. MaryAnn Augustin for their assistance in the preparation of this manuscript.

References

1. Mueller, G. C., and Miller, J. A., J. Biol. Chem. 180, 1125-1136 (1949).
2. Mueller, G. C., and Miller, J. A., J. Biol. Chem. 185, 145-154 (1950).
3. Axelrod, J., J. Biol. Chem. 214, 753-763 (1955).
4. Axelrod, J., J. Pharmacol. Exp. Therap. 117, 322-330 (1956).
5. Brodie, B. B., Axelrod, J., Cooper, J. R., Gaudette, L., LaDu, B. N., Mitoma, C., and Udenfriend, S., Science. 121, 603-604 (1955).
6. Conney, A. H., Miller, E. C., and Miller J. A., Cancer Res. 16, 450-459 (1956).
7. Conney, A. H., Miller, E. C., and Miller J. A., J. Biol. Chem. 228, 753-766 (1957).
8. Remmer, H., Naturwiss. 45, 189-190 (1958).
9. Remmer, H., and Alsleben, B., Klin. Wschr. 36, 332-333 (1958).
10. Cooper, D. Y., Levine, S., Narasimhulu, S., Rosenthal, O., and Estabrook, R. W., Science 147, 400-402 (1965).
11. Omura, T., Sato, R., Cooper, D. Y., Rosenthal, O., and Estabrook, R. W., Fed. Proc. 24, 1181-1189 (1965).
12. Symposium on Microsomes and Drug Oxidations, Ed. Gillette, J. R., Conney, A. H., Cosmides, G. J., Estabrook, R. W., Fouts, J. R., and Mannering. G. J., pp. 1-547, Academic Press, New York (1969).
13. World Health Organization, Prevention of Cancer, Tech. Rept. Serial 276, Geneva (1964).
14. Boyland, E., Prog. Exp. Tumor Res. 11, 222-234 (1969).
15. Epstein, S. S., Cancer Res. 34, 2425-2435 (1974).
16. Lu, A. Y. H., and Coon, M. J., J. Biol. Chem. 243, 1331-1332 (1968).
17. Strobel, H. W., Lu, A. Y. H., Heidema, J., and Coon, M. J., J. Biol. Chem. 245, 4851-4854 (1970).
18. Lu, A. Y. H., and Levin, W., Biochem. et Biophys. Acta 344, 205-240 (1974).
19. Imai, Y., and Sato, R., Biochem. Biophys. Res. Commun. 60, 8-14 (1974).
20. van der Hoeven, T. A., Haugen, D. A., and Coon, M. J., Biochem. Biophys. Res. Commun. 60, 569-575 (1974).
21. Digman, J. D., and Strobel, H. W., Biochem. Biophys. Res. Commun. 63, 845-852 (1975).
22. Ryan, D., Lu, A. Y. H., Kawalek, J., West, S. B., and Levin, W., Biochem. Biophys. Res. Commun. 64, 1134-1141 (1975).
23. Haugen, D. A., van der Hoeven, T. A., and Coon, M. J., J. Biol. Chem. 250, 3567-3570 (1975).
24. Kawalek, J. C., Levin, W., Ryan, D., Thomas, P. E., and Lu, A. Y. H., Mol. Pharmacol. 11, 874-878 (1975).
25. Philpot, R. M., and Arinc, E., Mol. Pharmacol. 12, 483-494 (1976).

26. Huang, M.-T., West, S. B., and Lu, A. Y. H., J. Biol. Chem. 251, 4659-4665 (1976).
27. Ryan, D., Thomas, P. E., Lu, A. Y. H., West, S., and Levin, W., The Pharmacologist 18, 241 (1976).
28. Levin, W., Lu, A. Y. H., Ryan, D., West, S., Kuntzman, R., and Conney, A. H., In Cytochromes P-450 and b_5: Structure, Function and Interaction (Ed. Cooper, D. Y., Rosenthal, O., Snyder, R., and Witmer, C.), Vol. 58, pp. 1-24, Plenum Press, N.Y. (1975).
29. Stern, J. O., Peisach, E., Peisach, J., Blumberg, W. E., Lu, A. Y. H., West, S., Ryan D., and Levin, W., In Cytochromes P-450 and b_5: Structure, Function and Interaction (Ed. Cooper, D. Y., Rosenthal, O., Snyder, R., and Witmer, C.), Vol. 58, pp. 189-202, Plenum Press, N.Y. (1975).
30. Kawalek, J. C., and Lu, A. Y. H., Mol. Pharmacol. 11, 201-210 (1975).
31. Haugen, D. A., and Coon, M. J., submitted for pub. (1976).
32. Comai, K., and Gaylor, J. L., J. Biol. Chem. 248, 4947-4955 (1973).
33. Ryan, D., Lu, A. Y. H., West, S., and Levin, W., J. Biol. Chem. 250, 2157-2163 (1975).
34. Thomas, P. E., Lu, A. Y. H., Ryan, D., West, S. B., Kawalek, J., and Levin, W., J. Biol. Chem. 251, 1385-1391 (1976).
35. Thomas, P. E., Lu, A. Y. H., Ryan, D., West, S. B., Kawalek, J., and Levin, W., Mol. Pharmacol. 12, 746-758 (1976).
36. Thomas, P. E., Lu, A. Y. H., West, S. B., Ryan, D., Miwa, G., and Levin W., submitted for publication (1976).
37. Alvares, A. P., and Siekevitz, P., Biochem. Biophys. Res. Commun. 54, 923-929 (1973).
38. Weber, K., and Osborn, M., J. Biol. Chem. 244, 4406-4412 (1969).
39. Levin, W., Lu, A. Y. H., Ryan, D., West, S., Kuntzman, R., and Conney, A. H., Archiv. Biochem. Biophys. 153, 543-553 (1972).
40. Neville, D. M., J. Biol. Chem. 246, 6328-6334 (1971).
41. Laemmli, U. K., Nature 227, 680-685 (1970).
42. Welton, A. F., and Aust, S. D., Biochem. Biophys. Res. Commun. 56, 898-906 (1974).
43. Hrycay, E. G., and O'Brien, P. J., Archiv. Biochem. Biophys. 147, 14-27 (1971).
44. Thomas, P. E., Ryan, D., and Levin, W., Anal. Biochem. 75, 168-175 (1976).
45. Tung, J. -S., and Knight, C. A., Anal. Biochem. 48, 153-163 (1972).
46. Robinson, N. C., and Tanford, C., Biochemistry 14, 369-377 (1975).
47. Chrambach, A., and Rodbard, D., Science, 172, 440-451 (1971).
48. Stoklosa, J. T., and Latz, H. W., Anal. Biochem. 68, 358-370 (1975).
49. Camacho, A., Carascosa, J. L., Vinuela, E., and Salas, M., Anal. Biochem. 69, 395-400 (1975).
50. Deutsch, D. G., Anal. Biochem. 71, 300-303 (1976).
51. Dean, W. L., and Coon, M. J., Fed Proc. 35, 1535 (1976).
52. Arinc, E., and Philpot, R. M., J. Biol. Chem. 251, 3213-3220 (1976).

ACTIVE OXYGEN—FACT OR FANCY*

Ronald W. Estabrook and Jurgen Werringloer

Department of Biochemistry, University of Texas Health Science Center, Dallas, Texas 75235

INTRODUCTION

The central role of oxygen in the function of complex metabolic reactions associated with cellular homeostasis has fascinated scientists since the discovery of oxygen by Priestly slightly over 2 centuries ago followed closely by the classic studies of Lavoisier. Much has been written about the fitness of oxygen for the development and maintenance of the unique environment and life-form that now exists on earth - although our knowledge remains rudimentary. Only a very small segment of this overall process is of direct concern to those of us experimentally examining the manner in which oxygen participates in mixed-function oxidation reactions - albeit, reactions recognized to be of great significance to the organism as evidenced by the broad spectrum of interests encompassed by reports at this Symposium related to the pharmacological effect of various chemicals, the initiation of chemical carcinogenesis, and the metabolism of normal cellular constituents such as steroids and fatty acids.

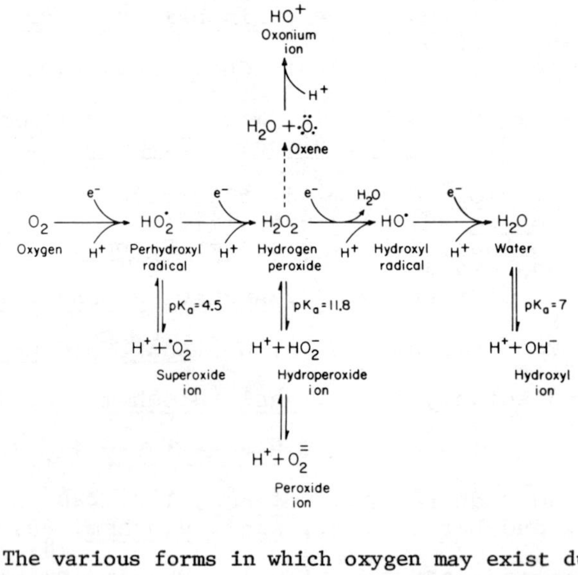

Fig. 1. The various forms in which oxygen may exist during its stepwise reduction to water.

*Supported in part by a grant (NIGMS-16488) from the USPHS.

The number of hypotheses that have been proposed to identify "active oxygen" are limited only by the number of forms in which oxygen may exist, some of which are shown in Fig. 1. In this context it is truly humbling to return and reread the classic review of H. S. Mason published nearly 20 years ago (1); many of the hypotheses now in vogue are merely a recasting of concepts developed over two decades ago and one must frankly admit that firm experimental evidence on which to select one alternative over another is still very meager. One must ask: will the next 20 years show the same limited degree of progress? Most important is a consideration of what evidence will be required and what new methodologies must be developed to <u>convincingly</u> establish the criteria for defining "active oxygen"? Frankly, no simple answer is currently available to respond to these questions. It is proposed to briefly discuss here two facets of studies currently under examination in a number of laboratories (2-7) which may bear on the question "active oxygen".

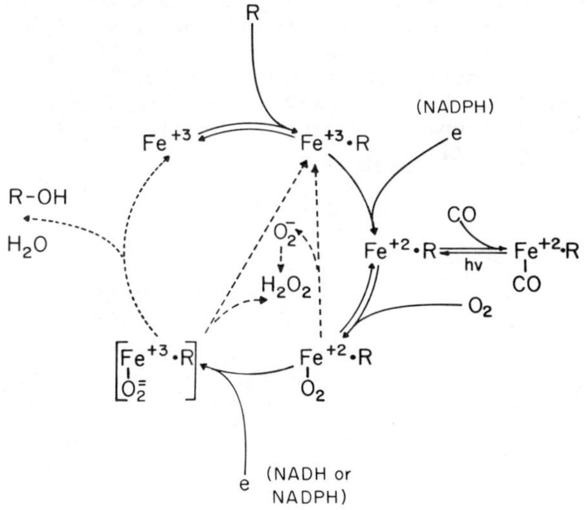

Fig. 2. Schematic representation of cytochrome P-450 function illustrating two possible sites for hydrogen peroxide generation.

The Source of Hydrogen Peroxide During Cytochrome P-450 Function

The generation of hydrogen peroxide during NADPH oxidation by liver microsomes was demonstrated in 1957 by Gillette, Brodie, and LaDu (8) when they observed the associated catalatic oxidation of methanol to formaldehyde. About that time, Mitoma, et al. (9) suggested the potential role of a peroxide-enzyme complex as the active hydroxylating species of oxygen for some mixed-function oxidation reactions. These observations were essentially neglected during the enthusiasm to understand the role of cytochrome P-450 as it interacted with various substrates and underwent enzymatic reduction although the analogy to peroxidase catalyzed reactions was suggested early (10). The controversial proposal (11,12) for a unique enzyme system of microsomes (possibly involving directly cytochrome P-450) functional in the metabolism of alcohols, rekindled an interest in understanding the mechanism for the generation of hydrogen peroxide during NADPH oxidation. Recent studies by Thurman et al. (13), Hildebrandt and Roots (14), and Boveris, et al. (15) have firmly established the fact that hydrogen peroxide is generated concomitant with NADPH oxidation by liver microsomes. Based on the premise that an understanding of hydrogen peroxide formation might provide a much needed clue for the understanding of "active oxygen" formed during cytochrome P-450 function, studies have been carried out in our laboratory to determine the source of hydrogen peroxide.

It was first considered that hydrogen peroxide might arise from the protonation of a peroxide ion complex of the iron of ferric cytochrome P-450 (Fig. 2) resulting from the two electron reduction and interaction of oxygen with cytochrome P-450. An alternate would be the generation of hydrogen peroxide by the dismutation of superoxide anions dissociating from the one electron reduced complex (16) of cytochrome P-450 and oxygen (the so-called oxy-cytochrome P-450 species). The distinction between these two alternatives hinges on the validity of the proposed scheme illustrating the functioning of two separate and identifiable sites of one electron transfer reactions during the catalytic cycle of substrate hydroxylation involving cytochrome P-450. However, this hypothesis may be in error if mammalian cytochrome P-450 accepts simultaneously two electrons as described by Coon et al. (17).

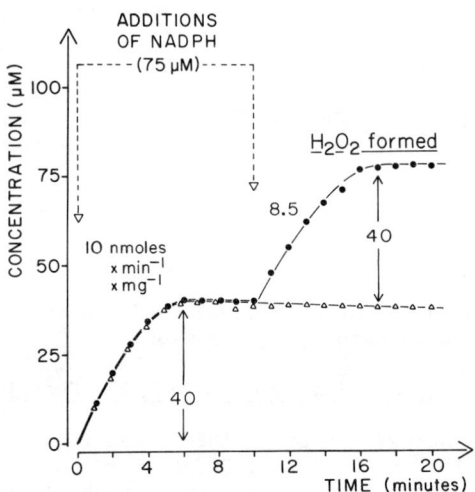

Fig. 3. The formation of hydrogen peroxide during NADPH oxidation by rat liver microsomes. Microsomes from the livers of phenobarbital treated rats were diluted to 1 mg protein per ml in a buffer mixture containing 50 mM Tris-chloride buffer, pH 7.5, 150 mM KCl, 5 mM $MgCl_2$, 2 mM 5'-AMP, and 1 mM sodium azide. The reaction was initiated by the addition of NADPH. Hydrogen peroxide concentration was determined colorometrically using ferrous ammonium sulfate and potassium thiocyanate (14).

Studies of the stoichiometry of hydrogen peroxide production relative to oxygen utilization and NADPH oxidation have been described in detail by Werringloer et al. (18). As shown in Fig. 3 repeated additions of NADPH (in the absence of an NADPH generating system) results in the incremental formation of hydrogen peroxide unincumbered by the presence of hydrogen peroxide, i.e. the amount of hydrogen peroxide formed with the second addition of NADPH is the same as that observed with the first addition of NADPH. When suitable precautions are taken to inhibit adventitious catalase by the addition of sodium azide or the destruction of NADPH by the microsomal pyrophosphatase, a significant amount of the oxygen utilized or NADPH oxidized can be accounted for by the production of hydrogen peroxide, although the presence of endogenous substrates which undergo simultaneous metabolism does complicate the exact definition of stoichiometry. When such corrections are applied, a stoichiometry of 1 oxygen utilized per 1 NADPH oxidized per 1 hydrogen peroxide formed can be calculated - a result which is compatible with either of the two alternatives described above for the formation of hydrogen peroxide.

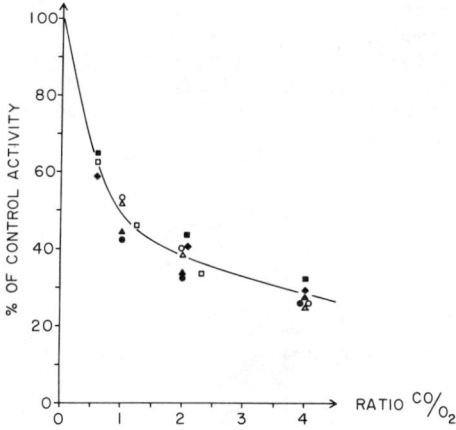

Fig. 4. The inhibition by carbon monoxide of hydrogen peroxide formation and the N-demethylation of ethylmorphine during NADPH oxidation by liver microsomes. Rat liver microsomes from phenobarbital treated animals were suspended in a buffer mixture similar to that described in Fig. 3. The microsomes were incubated in a special reaction vessel designed for rapid stirring while gassing the solution with various mixtures of $CO:O_2:N_2$. The oxygen concentration was maintained at 20 percent. The reaction was initiated by addition of 200 μM NADPH. In some experiments (open symbols) 5 mM ethylmorphine was added and the rate of formaldehyde formation determined. NADH was added together with NADPH in experiments designated by a triangle.

Prior to proceeding further, it was deemed necessary to establish that the generation of hydrogen peroxide, indeed, does involve cytochrome P-450. Therefore, a series of studies were carried out comparing the effects of carbon monoxide inhibition, the influence of varying steady state concentrations of NADPH (to determine a "Km" for reduced pyridine nucleotide) and the energy of activation at various temperatures for hydrogen peroxide formation and the N-demethylation of ethylmorphine. In each instance identical results were obtained. As illustrated in Fig. 4, the pattern of inhibition by carbon monoxide for these two reactions is superimposable. Although not conclusive, the weight of evidence is strongly in favor of a role for cytochrome P-450 in the generation of hydrogen peroxide.

Extending our consideration of the scheme describing the two possible alternatives for the generation of hydrogen peroxide, it is noted (Fig. 2) that a main difference concerns the proposed role of a second electron reduction step for the conversion of oxy-cytochrome P-450 to a state equivalent to the peroxide anion complex of the ferric hemeprotein. It has been proposed (19, 20) that this reaction is the basis for an NADH synergism observed during the metabolism of many drug substrates by cytochrome P-450. NADH support of cytochrome P-450 catalyzed reactions carried out in the presence of NADPH results in nearly 2-fold increase in both the rate and extent of the reaction when substrates such as ethylmorphine are employed. The generation of hydrogen peroxide, however, is <u>not</u> subject to a synergistic effect by NADH. Therefore, we conclude that the favored reaction for the generation of hydrogen peroxide involves the generation of the superoxide anion by dissociation of the oxy-cytochrome P-450 complex. We are aware of the frailities of the hypotheses necessary for this interpretation, but until better evidence is produced this difference in the synergistic effect of NADH remains as a pivotal argument for the interpretation of the source of hydrogen peroxide.

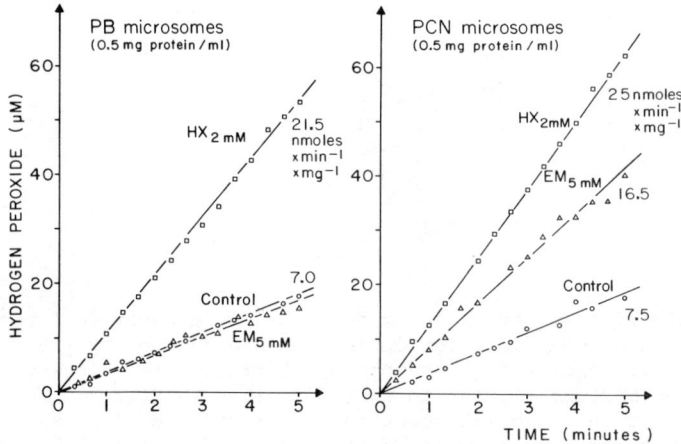

Fig. 5. The effect of substrates for mixed function oxidation reactions on the rate of hydrogen peroxide formation during NADPH oxidation by liver microsomes. Liver microsomes prepared from male rats pretreated with phenobarbital (PB) or pregnenelone-16α-carbonitrile (PCN) were diluted to 0.5 mg per ml in a reaction mixture containing 50 mM Tris-chloride buffer, pH 7.5, 150 mM KCl, 10 mM $MgCl_2$, 2 mM 5'AMP, 50 mM methanol, and substrates as indicated, i.e. 2 mM hexobarbital or 5 mM ethylmorphine. The rate of hydrogen peroxide generated was determined indirectly (14) by the measurement of formaldehyde in the presence of 2,200 I.U. of catalase. The reaction was initiated by the addition of 200 μM NADPH.

A point of primary interest is the question of the constancy of the rate of hydrogen peroxide production, i.e. what factors perturb the rate of generation of hydrogen peroxide? As yet, we have no unifying principle that permits the prediction of changes that may occur in the rate of hydrogen peroxide production, but as shown in Fig. 5, the presence of hexobarbital triples the rate of hydrogen peroxide formation when using microsomes from phenobarbital or pregnenolone 16α-carbonitrile (PCN) treated rats. In contrast, ethylmorphine has no influence on the rate of hydrogen peroxide formation when using liver microsomes from phenobarbital treated animals but does double the rate of hydrogen peroxide formation when using microsomes from PCN treated animals. Further, codeine, for example, suppresses the rate of hydrogen peroxide formation. From the studies it is apparent that the oft-used "endogenous rate" of NADPH oxidation or oxygen utilization, which is largely associated with the generation of hydrogen peroxide, is not a constant and cannot be merely substrated as a fixed value when evaluating the stoichiometric balance of drug metabolism (21).

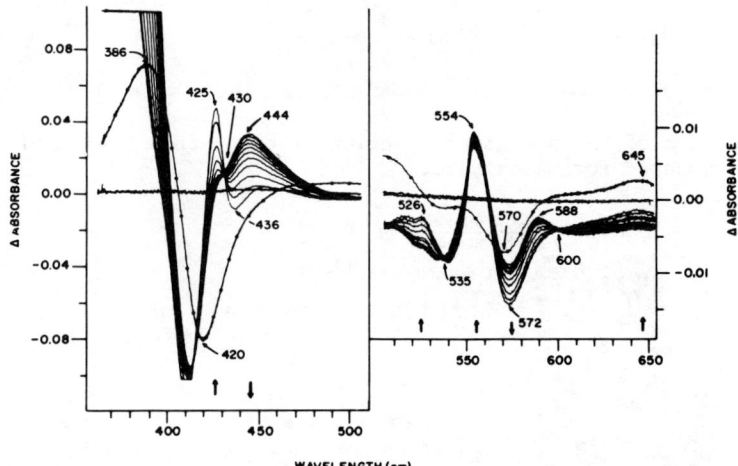

Fig. 6. Spectral changes associated with oxycytochrome P-450 during the aerobic steady state of hexobarbital oxidation. A suspension of rat liver microsomes (2 mg protein per ml) was divided equally into two cuvettes and the difference spectra were recorded with an Aminco DW-2 spectrophotometer in the split-beam mode. An aliquot of 2 mM hexobarbital was added initially to the contents of the sample cuvette and substrate binding determined by the spectral change shown by the trace with solid circles. An equal aliquot of hexobarbital was then added to the contents of the reference cuvette. NADH (200 µM) was added to both cuvettes and NADPH (200 µM) to the contents of the sample cuvette and spectra recorded every 30 seconds. The transient appearance of oxy-cytochrome P-450 is observed at 444 nm.

One possible clue to the understanding of hydrogen peroxide production is the apparent relationship between the magnitude of the spectral change (16) seen during the aerobic steady state of drug metabolism, attributable to oxy-cytochrome P-450 (Fig. 6) and the rate of hydrogen production. As might

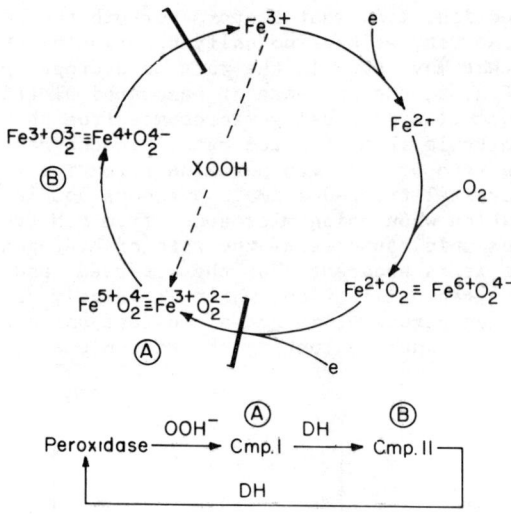

Fig. 7. Comparison of the peroxidatic action of cytochrome P-450 and the intermediates proposed for peroxidase.

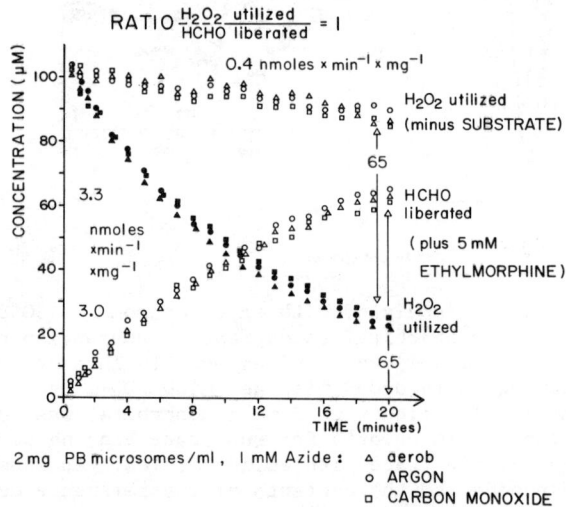

Fig. 8. The hydrogen peroxide dependent N-demethylation of ethylmorphine as catalyzed by liver microsomes. Rat liver microsomes from phenobarbital treated animals were diluted to 2 mg protein per ml in a reaction mixture similar to that described in Fig. 3. Where indicated 5 mM ethylmorphine was added. The reaction was initiated by the addition of 100 μM hydrogen peroxide. Special reaction vessels were used to permit equilibration with various gas mixtures and to permit sampling the reaction at the times indicated. The changes in the concentrations of hydrogen peroxide utilized or formaldehyde formed were determined colorometrically.

be predicted from the scheme presented in Fig. 2, an increase in the steady state concentration of oxy-cytochrome P-450 would provide a greater opportunity for its dissociation to form the superoxide anion and hence, a greater rate of hydrogen peroxide formation. As yet, our results are preliminary, but such a relationship does seem consistent with the studies of various substrates measured to date.

The Utilization of Hydrogen Peroxide

The observation that cytochrome P-450 can function as a peroxidase was initially proposed by Hrycay and O'Brien (3) and this finding has been expanded by O'Brien and his colleagues and in our own laboratory (22). Initially, organic peroxides, such as cumene hydroperoxide, have been employed to support hydroxylation reactions catalyzed by cytochrome P-450. Indeed, we drew attention (22) to the similarities between cytochrome P-450 and peroxidase by demonstrating the presence of a free radical specie formed on interaction of cumene hydroperoxide with cytochrome P-450 comparable to that observed for complex 1 of peroxidase or the higher valence state of iron for hemoglobin and myoglobin. From these studies a scheme was proposed (Fig. 7) where further reactions of cytochrome P-450 with oxygen would be equivalent to the transitions observed with peroxidase. In our earlier studies (22) we failed to observe a reaction with hydrogen peroxide - this failure was due to an oversight because of our neglect to consider the role of adventitious catalase in destroying the added hydrogen peroxide. More recent studies (Fig. 8) have shown the ability to utilize hydrogen peroxide to support the cytochrome P-450 dependent reaction for the N-demethylation of ethylmorphine with a stoichiometry of 1 hydrogen peroxide utilized per formaldehyde formed. This reaction occurs as rapidly anaerobically as aerobically. Further, it was insensitive to inhibition by carbon monoxide. However, the reaction is influenced by NADPH under anaerobic conditions where it gains the characteristics of carbon monoxide inhibition.

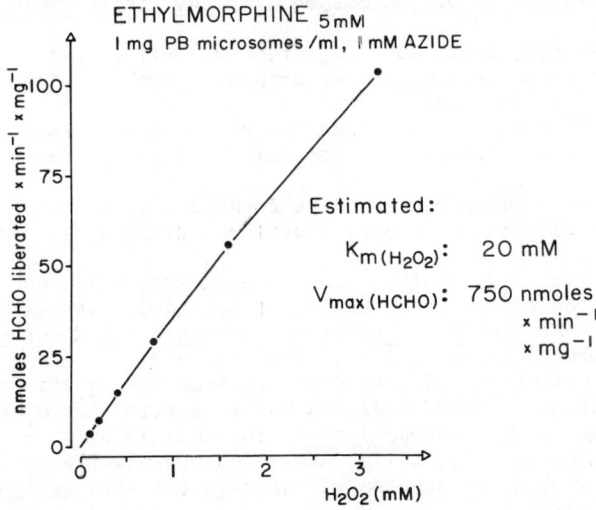

Fig. 9. The influence of varying concentrations of hydrogen peroxide on the initial rate of N-demethylation of ethylmorphine by rat liver microsomes. A series of experiments similar to those described in Fig. 8 were carried out using varying initial concentrations of hydrogen peroxide.

The affinity of the reaction for hydrogen peroxide is rather poor as shown in Fig. 9 with an estimated Km of about 20 mM. It should be noted, however, that the rates obtained are in great excess of those obtained when NADPH supports the aerobic N-demethylation of ethylmorphine. This result is consistent with the interpretation that the interaction of the second electron required for the reduction of oxy-cytochrome P-450 may dictate the overall rate of NADPH support of cytochrome P-450 function; when hydrogen peroxide is used as the donor this step is by-passed. One must seriously question, however, the validity of hydrogen peroxide driven reactions as representative of NADPH-supported cytochrome P-450 dependent hydroxylation reactions (24).

SUMMARY

Critical to the further understanding of cytochrome P-450 function is the definition of the reaction of oxygen with the substrate molecule to be hydroxylated. The preponderance of the evidence to date suggests that cytochrome P-450 plays a central role in the activation of oxygen although we cannot exclude a concomitant role for cytochrome P-450 in the activation of the substrate molecule. Present interest in hydrogen peroxide formation and utilization, may provide the needed scapel for the incisive dissection of the mysterious corpus embodied by the ternary complex of an organic substrate molecule, a poorly understood hemeprotein, and an undefined state of oxygen. Is "active oxygen" a fact or fancy? Only the development of new methodologies and the ingenuity of the laboratory scientist can firmly established a creditable answer to that question. Progress has been accomplished but much more remains to be learned.

References

1. Mason, H.S., Mechanisms of Oxygen Metabolism, Adv. in Enzymology, 19, 79-233, 1957.
2. Kadlubar, F.F., Morton, K.C., and Ziegler, D.M., Microsomal-catalyzed hydroperoxide-dependent C-oxidation of amines. Biochem. Biophys. Res. Comm., 54, 1255-1261, 1973.
3. Hrycay, E.G. and O'Brien, P.J., Cytochrome P-450 as a microsomal peroxidase in steroid hydroperoxide reduction. Arch. Biochem. Biophys., 153, 480-494, 1972.
4. Werringloer, J. and Estabrook, R.W., The formation of hydrogen peroxide during microsomal electron transport reactions, Z. Physiol. Chem., 357, 1063, 1976.
5. Coon, M.J., Nordblom, G.D., White, R.E., Guengerich, F.P., and Ballou, D.P., Studies on the mechanism of action of purified liver microsomal cytochrome P-450. Abstracts of The Tenth International Congress of Biochemistry, Hamburg, Germany, July, 1976, pg. 348.
6. Sasame, H.A. and Gillette, J.R., Evidence against synergistic effect of NADH participation in NADPH mediated P-450 reactions as a scavenger of peroxide species., The Pharmacologist, 18, 242, 1976.
7. Sligar, S.G., Namtvedt, M.J., Ellis, R.V., and Gunsalus, I.C., The chemistry and enzymology of oxygenated cytochrome P-450. Z. physiol. chem., 357, 1056, 1976.
8. Gillette, J.R., Brodie, B.B., and LaDu, B.N., The oxidation of drugs by liver microsomes: on the role of TPNH and oxygen. J. Pharm. Exp. Therap., 119, 532-540, 1957.

9. Mitoma, C., Posner, H.S., Reitz, H.C., and Udenfriend, S. Enzymatic hydroxylation of aromatic compounds, Arch. Biochem. Biophys., 61, 431-441, 1956.
10. Estabrook, R.W., Hildebrandt, A., Remmer, H., Schenkman, J.R., Rosenthal, O., and Cooper, D.Y., The role of cytochrome P-450 in microsomal mixed function oxidation reactions. In Biochemie des Sauerstoffs (edited by B. Hess and Hj. Staudinger), 19. Colloquium der Gesellschaft fur Biologische Chemie, Springer-Verlag, Berlin, 1968, pg. 142-177.
11. Lieber, C.S. and DeCarli, L.M., Hepatic microsomal ethanol-oxidizing system. In vitro characteristics and adaptive properties in vivo. J. Biol. Chem., 245, 2505-2512, 1970.
12. Alcohol and Aldehyde Metabolizing Systems, edited by R. G. Thurman, T. Yonetani, J. R. Williams, and B. Chance, Academic Press, New York, 1974.
13. Thurman, R.G., Ley, H.G., and Scholz, R., Hepatic microsomal ethanol oxidation. Hydrogen peroxide formation and the role of catalase. Eur. J. Biochem., 25, 420-430, 1972.
14. Hildebrandt, A.G. and Roots, I. Reduced nicotinamide adenine dinucleotide phosphate (NADPH)-dependent formation and breakdown of hydrogen peroxide during mixed function oxidation reactions in liver microsomes. Arch. Biochem. Biophys., 171, 385-397, 1975.
15. Boveris, A., Oshino, N., and Chance, B., The cellular production of hydrogen peroxide. Biochem. J., 128, 617-630, 1972.
16. Estabrook, R.W., Hildebrandt, A.G., Baron, J., Netter, K., and Leibman, K., A new spectral intermediate associated with cytochrome P-450 function in liver microsomes, Biochem. Biophys. Res. Comm., 42, 132-139, 1971.
17. Guengerich, F.P., Ballou, D.P., and Coon, M.J., Purified liver microsomal cytochrome P-450. Electron-accepting properties and oxidation-reduction potential. J. Biol. Chem., 250, 7405-7414, 1975.
18. Werringloer, J., Hildebrandt, A., and Estabrook, R.W., Hydrogen peroxide formation and breakdown by the liver microsomal electron transport system. Abstracts of the Tenth International Congress of Biochemistry, Hamburg, Germany, July, 1976, pg. 292.
19. Cohen, B. and Estabrook, R.W., Microsomal electron transport reactions. III. Cooperative interactions between reduced diphosphopyridine nucleotide and reduced triphosphopyridine nucleotide linked reactions. Arch. Biochem. Biophys., 143, 54-65, 1971.
20. Correia, M.A. and Mannering, G.J., Reduced diphosphopyridine nucleotide synergism of the reduced triphosphopyridine nucleotide-dependent mixed function oxidase system of hepatic microsomes. I. Effects of activation and inhibition of the fatty acyl coenzyme A desaturation system. Mol. Pharm., 9, 455-469, 1973.
21. Jeffrey, E. and Mannering, G.J., Discrepancy in the measurement of reduced triphosphopyridine nucleotide oxidized during ethylmorphine N-demethylation due to the presence of a nucleotide pyrophosphatase. Mol. Pharm., 10, 1004-1008, 1974.
22. Rahimtula, A.D., O'Brien, P.J., Hrycay, E.G., Peterson, J.A., and Estabrook, R.W., Possible higher valence states of cytochrome P-450 during oxidative reactions. Biochem. Biophys. Res. Comm., 60, 695-702, 1974.
23. Yonetani, T. and Schleyer, H., Studies on cytochrome c peroxidase. IX. The reaction of ferrimyoglobin with hydroperoxides and a comparison of the peroxide-induced compounds of ferrimyoglobin and cytochrome c - peroxidase. J. Biol. Chem., 242, 1974-1979, 1967.
24. Griffin, B.W., Mechanism of hemeprotein catalysis of N-demethylation reactions supported by cumene hydroperoxide. Abstracts of The International Congress of Biochemistry, Hamburg, Germany, July, 1976, pg. 348.

INDEX

Cytochrome P450 will be listed under "P450"

Acetanilide	576
ADH	104
Adrenocortical microsomes	119
Adrenoreceptor, antagonist	610
Adriamycin	688
Affinity chromatography	111, 170
Alprenolol	297
6-Aminonicotinamide	317
Aminopyrine	64, 128, 309, 492, 582, 612
Aminopyrine, elimination	640
Amphetamines	284
Androstenedion	67, 213
Aniline	88, 162, 236
Anilinonaphthalene-8-sulphonic acid	597
Anthracyclines, metabolism	694
Antimycin A	319
Antioxidants	214
Antipyrine	387, 517, 582
Antipyrine, plasma half-life	630, 704
Aroclor 1254	477
Aryl Hydrocarbon hydroxylase	426, 427
2-Arylaminofluorenes	575
Ascorbic acid	518, 532
Azide	262
Baboon	516
Bacillus megaterium	377
Benzo(a)pyrene	11, 88, 128, 155, 162, 170, 188, 213, 251, 285, 362, 395, 477, 600, 667, 724, 740
Benzo(a)pyrene, metabolites	400, 403, 411, 435, 443, 455
Benzo(a)pyrene, protein binding	438
1,4-Benzodiazepin-2-ones	622

Benzphetamine	12, 88, 128, 155, 188, 238, 527, 740
Bilirubin	543
Biphenyl	88, 178, 251, 556, 721
Biphenyl, polychlorinated	416
Bishydroxycoumarin	630
Boundary lipids	1
Bromobenzene	302
t-Butyl hydroperoxide	205, 309
Caboxolane	727
Caffeine	128
Camphor-monooxygenase	59, 275
Carbon disulfide	469
Carbon monoxide	242, 615
Carbon tetrachloride	241, 530
Carcinogenicity	409, 459, 467, 500, 709
Carcinogenicity, test	721
Catalase	105
Catinomycin D	563
Cell culture, human fetal	411
Cell culture, human lymphozytes	418
Cell culture, hybrid	426
Cell culture, primate	415
Cetavlon	364
Charcoal broiled meat	706
Chirality	622
4-Chloranilides	39
4-Chloroacetanilide	576
Chloroethylnitrourease	680
3-Chloroperbenzoic acid	219, 282
Chloroperoxidase	205
Cholestasis	589
Cholesterol	597
Cigarette smoke	457
Cobaltous chloride	340, 349, 546
Codein	572
Conjugation	297, 448
Coumarin	188

Cumene hydroperoxide	89, 205, 210, 219, 269
Cyclohexane	59, 91, 227
Cycloheximide	365, 551
Cysteamine	354
Cytochrome b_5	31, 90, 167, 212, 309, 325, 582, 732
Daunomycin	688
Deoxycholate	31, 153
11-Deoxycortisol	172
Desulfuration	269
Detoxication, extracorporal	391
Deuterium, isotope effect	271
Diaphorase, DT	362
Dicoumarol	362
Diethyldithiocarbamate	589
Digitonin	307
N,N-Dimethylaniline	346
5,5-Diphenylhydantoin	508
Disulfiram	589
DNA, metabolite binding	395
DNA-Synthesis	559
Drug binding	296
Drug elimination, age effect	387
Drug elimination, diet	713
Drug elimination, disease effect	628
Drug elimination, phenytoin	511
Drug response, individual differences	385
Drug toxicity	391
Drug-drug interaction	442
Electron transfer	730
Epoxide hydratase	152, 170, 364, 403, 438, 442. 459, 612, 671
Epoxide hydratase, assay	460
Epoxide hydratase, inhibitors	463
Epoxide hydratase, purification	646
Epoxide hydratase, substrate specificity	650
Epoxide, reduction	654

Epoxides	698
Ethanedithiol	358
Ethanol	103, 319
Ethanol, elemination	491
7-Ethoxycoumarin	12, 34, 59, 155, 162, 178, 188, 219, 250, 294, 678
Ethyl isocyanide	187
Ethylmorphin	14, 128, 263, 285, 326, 331, 492, 540, 572, 740, 754
Fatty acids	320
Filorone	484
Fluorescence, polarization	124
Fluroxene	76
Glucaric acid	517, 582
Glutathione	302, 447, 700
Halomethanes	240
Halothane	254
Heme degradation	543
Heme oxygenase	543
Hemin	340
Hemoglobin	242
Hemoglobin, carboxy	616
Hemoproteins	232
Hepatocytes, nuclear membrane	396
γ-Hexachlorocyclohexane	559
Hexobarbital	19, 297, 612
Hexobarbital, enantiomers	534
Housefly	144
Hydrazine derivates	500
Hydrocortisone	386
Hydrogen peroxide	82, 205, 261, 492, 749
Imidazol	236
Imipramine	351
Interferon inducing agents	484
Intestinal microsomes	668, 675
Iodosobenzene	219, 271
3-Iodosobenzoic acid	219

Iron-sulfur protein	377
Isoelectric focusing	380
Isoproterenol	661
Isosafrole	247
K-Region	646
K-Region epoxides	711
Lidocaine	520
Linoleic acid hydroperoxide	323
Lipase	346
Lipid peroxidation	336
Lipids, 32 P-NMR	6
Lipids, fluidity	1, 119
Lipids, protein interaction	2, 67
Lipids, lateral organization	1
Lipids, phase transitions	3, 119
Liposomes	31
Little skate	160
Liver microsomes, anaerobic incubation	240, 256, 654
Liver microsomes, cholesterol content	599
Liver microsomes, electron transport	31, 261
Liver microsomes, membrane	1, 7, 31
Liver microsomes, metabolite binding	661, 699
Liver microsomes, metal content	584
Liver microsomes, NMR	2
Liver microsomes, organization	7
Liver microsomes, spin label	5
Liver necrosis	467
Low temperature	23, 220
Luminal	24
Lung microsomes	460
Lung, perfused	448, 453
Menadion	349
MEOS	103
Mersalyl	10
Metabolites, covalent binding	455
Metoprolol	610

Metyrapone	178, 213, 244, 257, 354, 556, 620, 678
Mitochondria-RER complex	339
Mitotic activity	563
Monooxygenase, clinical impact	385
Monooxygenase, depression	489
Monooxygenase, diet	669
Monooxygenase, diet-protein	562
Monooxygenase, diet-lipids	612
Monooxygenase, effect of cholesterol	597
Monooxygenase, gene location	426
Monooxygenase, genetic effect	423
Monooxygenase, individual differences	423
Monooxygenase, inducers	581
Monooxygenase, induction	160, 421, 430, 453, 529, 559
Monooxygenase, reconstitution	152, 160
Monooxygenase, reducing equivalents	307, 315
Monooxygenase, seasonal change	420
Monooxygenase, sex differences	170, 523, 536, 607
Monooxygenase, specie difference	517
Monooxygenase, stoichiometry	331
Monooxygenase, uncoupled	492
Monooxygenase, diet-carbohydrates	612
Mutagenicity	409, 435, 575, 710
Mutagenicity, test	439, 442, 576
NADH, synergistic effect	267, 372
NADH-cytochrome c reductase	325
$NADH/NAD^+$-ratio	301, 307
NADPH-cytochrome b_5-reductase	31
NADPH-cytochrome c-reductase	354
NADPH-generation	300, 307
NADPH-cytochrome P450-reductase	9, 17, 31, 59, 68, 86, 104, 111, 152, 160, 354, 362, 377, 538, 599, 664
$NADPH/NADP^+$-ratio	317
α-Naphthoflavone	179, 213, 296, 397, 554, 578, 620, 671, 678
Nicotinamide nucleotide	307
NIH-shift	624

Nitroacetophenone, 2-bromo-4'-	346
4-Nitroanisole	14, 88, 249, 287, 316, 354, 599, 612
NMR, relaxation	232
Nortriptyline	388
N-oxidation	346, 502, 576
N-oxide, reduction	654
Octylamine	227
Organ spectrophotometry	309
Oxene donor	219
Oxygen, labeled	269
P 448	12, 41, 142, 186, 476, 551, 740
P 448, antibody	190
P 450	9, 31
P 450 cam	26, 202, 275
P 450 cam, spin state	47
P 450 125-J-labeled	99
P 450, composition	96
P 450, active oxygen	748
P 450, amine complexes	193
P 450, antibody	190, 743
P 450, apoprotein	339, 557
P 450, autoxidation	203
P 450, biosynthesis	339
P 450, carbene complex	240, 254
P 450, cyanide complex	105
P 450, delipidated	67
P 450, destruction	76, 323, 543
P 450, electrophoretic mobility	82, 154, 186, 742
P 450, inducers	581, 591, 620
P 450, induction	39, 59, 79, 127, 136, 179, 188, 211, 224, 288, 324, 350, 355, 366, 370, 416, 505, 551, 568, 669, 744
P 450, inhibitors	505, 665
P 450, isoelectric focusing	97, 149
P 450, ligands	193, 226, 275, 358, 676
P 450, metabolite binding	682
P 450, metabolite complex	247, 281, 284

P 450, molecular activity	90
P 450, multiple forms	39, 82, 95, 104, 127, 136, 144, 177, 185, 324, 352, 385, 520, 557, 676, 737
P 450, optical spectra	86, 127, 166, 211, 220, 224, 242, 249, 257, 328, 328, 478, 616, 684, 738, 753
P 450, oxenoid complex	204, 218, 270
P 450, oxycomplex	24, 90, 202, 278, 730
P 450, peptides	98
P 450, photoaffinity labels	100
P 450, potentials	310
P 450, purification	82, 104, 144, 152, 160, 185, 377, 735
P 450, reaction mechanism	88, 192, 207, 210, 266, 275, 754
P 450, redox state	308
P 450, reduction	359
P 450, sex difference	177, 327
P 450, spin state	47, 86, 193, 224
P 450, stereospecificity	383, 408, 534, 622
P 450, substrate complex	194, 232, 248, 520, 537
P 450, substrate specificity	740
P 450, thioether complex	193
P 450, thiol complexes	193
Paracetamol	303
Parathion	271, 468
Pentose phosphate shunt	318
Peracetic acid	271
Perfusion, liver	308, 316
Phenacetin	576, 706
Phenobarbital, infusion	568
Phenobarbital, plasma concentration	573
Phenylbutazone	517
Phenylimidazole, N-	170
Phenytoin	508
Phosphatidylcholin	377
6-Phosphatogluconate	318
Photodissociation	255
Photoreduction	276
Piperonylbutoxide	287

Plasma half-life, genetic effects	630
Plasma half-life, temporal variation	631
Plasma, drug concentration	389, 510, 516
Polycyclic aromatic hydrocarbons	709
Procarbazine	500
Progesterone	382, 607
Propanolol	297, 610
Prostaglandin A 1	370
Putidaredoxin	203, 275
Radicals	215
Rhesus monkey	509
RNA-synthesis	563
Rotenone	671
Safrole	722
Saliva, drug concentrations	390
Salizylamide	296
Singlet oxygen	215
SKF 525 A	213, 297
Sodium chlorite	205
Sodium peracetate	205
Sodium periodate	205, 219, 383
Steroid monooxygenase	377
Steroid-16α-monooxygenase	605
Stopped flow method	53, 276
Styrene oxide	698
Substrate complex	224
Sulfur, labeled	470
Sulfur, protein binding	470
Sulfur, thiono-	467
Superoxide anion	202, 267, 336, 664, 750
Superoxide dismutase	335
Testosterone	188, 607, 740
Theophylline, plasma half-life	704
Toxic metabolites	301
Trichloropropane, 1,2-epoxy-3,	364
Trimethylamine	351
Triton N-101	172

Triton X-100	71
Trout liver	435
UDP-glucoronyl transferase	174, 530, 589, 597
Xanthine oxidase	656
Xylidine	232
Zinc	592
Zoxazolamine	188